Book of Abstracts of the 62nd Annual Meeting of the European Federation of Animal Science

EAAP - European Federation of Animal Science

The European Federation of Animal Science wishes to express its appreciation to the
Ministero delle Politiche Agricole e Forestali (Italy) and the
Associazione Italiana Allevatori (Italy)
for their valuable support of its activities.

Book of Abstracts of the 62nd Annual Meeting of the European Federation of Animal Science

Stavanger, Norway, 29th August - 2nd September 2011

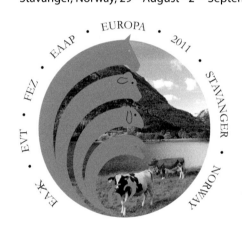

ISBN 978-90-8686-177-4
e-ISBN: 978-90-8686-731-8
DOI: 10.3920/978-90-8686-731-8

ISSN 1382-6077

First published, 2011

© Wageningen Academic Publishers
The Netherlands, 2011

Welcome to EAAP 2011 in Stavanger – Norway

On behalf of the Norwegian Organising Committee I am pleased to invite you to attend the 62nd Annual Meeting of the EAAP which takes place in Stavanger from the 29th of August to the 2nd of September. This will be the third time that Norway has hosted this meeting, with the last time being in Lillehammer in 1996.

The main theme of the meeting will be **'The importance of animal production for food supply, food quality and environment'**, which are hot topics in view of the current worldwide demands for both the livestock industry and human society. The program will cover all aspects of scientific achievements within animal production, including genetics, physiology, nutrition, management and health. We are also planning topics related to fish farming and the relationship between the green and blue sectors.

We will have the opportunity to attend a select number of presentations and study posters from a great number of scientists from the whole of Europe, as well as take part in workshops and discussions of the latest and most relevant research in the field of Animal Science and Aquaculture. This is a unique occasion for updating knowledge and acquiring new ideas, and we especially encourage young scientists and students to attend.

We are sure that all of you will have a productive meeting from a scientific point of view and that you will also enjoy the social events, landscape and hospitality in Norway.

Torbjørn Auran
President of the Norwegian Organising Committee

National organisers of the 62nd eaap annual meeting

Local Organising Committee

President
- **Dr. Torbjørn Auran** (torbjorn.auran@fkf.no)
  Felleskjøpet Fôrutvikling, Trondheim

Vice-president
- **Prof. Odd Vangen** (odd.vangen@umb.no)
  Dept. of Animal and Aquacultural Sciences, Norwegian University of Life Sciences, Ås

Executive Secretary
- **Dr. Ina Andersen-Ranberg** (ina.ranberg@norsvin.no)
  Norsvin, Norwegian Pig Breeders' Association, Dept. of Animal and Aquacultural Sciences,
  Norwegian University of Life Sciences, Ås

Members
- **Dr. Lars Bævre** (lars.bavre@tine.no)
  Tine Rådgiving, Ås
- **Dr. Bente Fredriksen** (bente.fredriksen@animalia.no)
  Animalia, Meat and Poultry Research Centre, Oslo
- **Dr. (Prorector) Halvor Hektoen** (Halvor.hektoen@veths.no)
  Dept. of Production Animal Clinical Sciences, Norwegian School of Veterinary Science, Oslo
- **Ms. Siv Kristin Holt** (Siv.Kristin.holt@geno.no)
  Geno, Breeding and A. I. Association, Særheim
- **Mr. Sjur Erik Kvåle** (sjur-erik.kvåle@lmd.dep.no)
  Ministry of Agriculture and Food, oslo
- **Dr. (Head of department) Torstein Steine** (torstein.steine@umb.no)
  Norwegian University of Life Sciences, Ås
- **Mr. Asbjørn Schjerve** (Asbjorn.schjerve@norsvin.no)
  Norsvin, Norwegian Pig Breeders' Association, Hamar

Local Scientific Committee

President
- **Prof. Odd Vangen** (odd.vangen@umb.no)
  Dept. of Animal and Aquacultural Sciences, Norwegian University of Life Sciences, Ås

Secretary
- **Dr. Ina Andersen-Ranberg (**ina.ranberg@norsvin.no**)**
  Norsvin, Norwegian Pig Breeders' Association, Dept. of Animal and Aquacultural Sciences, Norwegian University of Life Sciences, Ås

Animal Genetics
- **Prof. Theo Meuwissen** (theo.meuwissen@umb.no)
  Dept. of Animal and Aquacultural Sciences, Norwegian University of Life Sciences, Ås

Animal Nutrition
- **Prof. Odd Magne Harstad** (odd.harstad@umb.no)
  Dept. of Animal and Aquacultural Sciences, Norwegian University of Life Sciences, Ås
- **Prof. Harald Volden** (harald.volden@tine.no)
  Tine Rådgiving, Ås/ Dept. of Animal and Aquacultural Sciences, Norwegian University of Life Sciences, Ås

Animal Management and Health
- **Prof. Olav Østerås** (olav.osteras@veths.no)
  Dept. of Production Animal Clinical Sciences, Norwegian School of Veterinary Science, Oslo

Animal Physiology
- **Dr. Arne Ola Refsdal** (arne.ola.refsdal@geno.no)
  Geno Breeding and A. I. Association, Hamar

Cattle Production
- **Dr. Bjørg Heringstad** (bjorg.heringstad@umb.no)
  Dept. of Animal and Aquacultural Sciences/Geno, Norwegian University of Life Sciences, Ås.

Sheep and Goat Production
- **Ass. Prof. Tormod Ådnøy** (tormod.adnoy@umb.no)
  Dept. of Animal and Aquacultural Sciences, Norwegian University of Life Sciences, Ås

Pig Production
- **Ass. Prof. Nils Petter Kjos** (nils.kjos@umb.no)
  Dept. of Animal and Aquacultural Sciences, Norwegian University of Life Sciences, Ås

Horse Production
- **Ass. Prof. Dag Austbø** (dag.austbo@umb.no)
  Dept. of Animal and Aquacultural Sciences, Norwegian University of Life Sciences, Ås

Livestock Farming Systems
- **Ass. Prof. Tore Framstad** (Tore.Framstad@nvh.no)
  Dept. of Production Animal Clinical Sciences, Norwegian School of Veterinary Science, Oslo

EAAP Program Foundation

Aims
EAAP aims to bring to our annual meetings, speakers who can present the latest findings and views on developments in the various fields of science relevant to animal production and its allied industries. In order to sustain the quality of the scientific program that will continue to entice the broad interest in EAAP meetings we have created the "EAAP Program Foundation". This Foundation aims to support:
- Invited speakers with a high international profile by funding part or all of registration and travel costs.
- Delegates from less favoured areas by offering scholarships to attend EAAP meetings.
- Young scientists by providing prizes for best presentations.

The "**EAAP Program Foundation**" is an initiative of the Scientific committee (SC) of EAAP. The Foundation aims to stimulate the quality of the scientific program of the EAAP meetings and to ensure that the science meets societal needs. The Foundation Board of Trustees oversees these aims and seeks to recruit sponsors to support its activities.

Sponsorships
1. Meeting sponsor – from 6000 euro
- acknowledgements in the final booklet with contact address and logo
- one page allowance in the final booklet
- advertising/information material inserted in the bags of delegates
- advertising/information material on a stand display
- acknowledgement in the EAAP Newsletter with possibility of a page of publicity
- possibility to add session and speaker support (at additional cost to be negotiated)

2. Session sponsor – from 3000 to 5000 euro
- acknowledgements in the final booklet with contact address and logo
- one page allowance in the final booklet
- advertising/ information material in the delegate bag
- ppt at beginning of session to acknowledge support and recognition by session chair
- acknowledgement in the EAAP Newsletter.

3. Speaker sponsor – from 2500 euro (cost will be defined according to speakers country of origin)
- half page allowance in the final booklet
- advertising/ information material in the delegate bag
- recognition by speaker of the support at session
- acknowledgement in the EAAP Newsletter

4. Registration Sponsor – equivalent to a full registration fee of the Annual Meeting
- acknowledgements in the booklet with contact address and logo
- advertising/information material in the delegate bag

The Association
EAAP (The European Federation of Animal Science) organises every year an international meeting which attracts between 900 and 1500 people. The main aims of EAAP are to promote, by means of active co-operation between its members and other relevant international and national organisations, the advancement of scientific research, sustainable development and systems of production; experimentation, application and extension; to improve the technical and economic conditions of the livestock sector; to promote the welfare of farm animals and the conservation of the rural environment; to control and optimise the use of natural resources in general and animal genetic resources in particular; to encourage the involvement of young scientists and technicians. More information on the organisation and its activities can be found at www.eaap.org.

Contact and further information
If you are interested to become a sponsor of the 'EAAP Program Foundation' or want to have further information, please contact the EAAP Secretariat (eaap@eaap.org, Phone +39 06 44202639).

Acknowledgements

Felleskjøpet
- more than 100 years of feed innovation!

European Federation of Animal Science (EAAP)

President: Kris Sejrsen
Secretary General: Andrea Rosati
Address: Via G.Tomassetti 3, A/I
I-00161 Rome, Italy
Phone: +39 06 4420 2639
Fax: +39 06 4426 6798
E-mail: eaap@eaap.org
Web: www.eaap.org

63rd EAAP Annual meeting of the European Federation of Animal Science

27 - 31 August 2012, Bratislava, Slovakia

Local Organisers: Animal Production Research Centre Nitra (APRC) with the support of the Ministry of Agriculture and Rural development of the Slovak Republic

President: Zsolt Simon, Minister of Agriculture of the Slovak Republic

Vice-President: Dana Peškovičová, Director of Animal Production Research Centre Nitra

Secretary: Peter Polák, Animal Production Research Centre Nitra

Emails: eaap2012@cvzv.sk / polak@cvzv.sk

ICAR 2012 Conference and INTERBULL Meeting

28 May - 1 June 2012, Cork, Ireland

www.icar2012.ie
Icar Secretariat: icar@eaap.org

Scientific Programme EAAP 2011

Monday 29 August 8.30 – 12.30	Monday 29 August 14.00 – 18.00	Tuesday 30 August 8.30 – 12.30	Tuesday 30 August 14.00 – 18.00
Session 1 Methodological developments in genomic selection (Joint Session of Interbull symposium and Animal Genetic Resources symposium) Chair: V. Ducrocq	**Session 10** Environmental value of animal genetic resources (Animal Genetic Resources symposium) Chair: G. Gandini (EAAP WG AnGR, ERFP, FAO)	**Plenary session** Co-Evolution of research, production and marketing of animal products	**Session 19** Architecture of quantitative traits Chair: L. Varona
Session 2 FABRE Technology platform: Research updates and White Paper on Food Security and Climate Change Chair: A.-M. Neeteson	**Session 11** Genomic selection - application, ownership, and economics in dairy cattle (Interbull symposium) Chairs: R. Reents/M. Coffey		**Session 20** The best cow for environmental efficiency Chair: M. Klopcic
Session 3 Welfare, ethics and behaviour in pig production Chair: A. Velarde	**Session 12** Link between rumen genome, nutrition and milk protein synthesis (REDNEX session) Chair: A. van Vuuren		**Session 21** Breeding programmes and genetic evaluation of horses Chair: T. Arnason
Session 4 Applied cattle breeding and genomics Chair: G. Thaller	**Session 13** Animal Task Force: Food security policy perspectives for industry and research Chair: P. Vriesekoop.		**Session 22a** Mastitis in non-dairy animals Chair: L. Bodin **Session 22b** Alternatives to drugs for parasite control in sheep and goats Chair: M. Gauly
Session 5 Free communications in horse production Chair: M. Saastamoinen	**Session 14** *Breeding for resource and environmental efficiency* Chair: P. Løvendahl		Consumer studies of the consequences of stopping piglet castration **Session 23** Chair: M.-A. Oliver
Session 6 Free communications in sheep and goats production Chair: C. Papachristoforou	**Session 15** Feeding practices and social constraints in horse housing Chair: E. Sondergaard		Interactions between nutrition, genetics and health **Session 24** Chair: G. van Duinkerken
Session 7 *Life-cycle assessment of livestock production* (SOLID/ANIMALCHANGE Symposium Chair: P. Gerber	**Session 16** Challenges of rangeland farming systems (economics, grazing, reproduction, health and welfare) Chair: T. Ådnøy		**Session 25** *Ecological intensification and ecosystem services of livestock farming systems* Chairs: M. Tichit/J. Hermansen
Session 8 Liver metabolism in dairy cows during the transition period Chair: A. v. Dorland	**Session 17** Use of new technologies for breeding and management of pigs and cattle Chair: P. Knap		**Session 26** Animal disease models for health and robustness Chair: J.L. Sartin
Session 9 Impact of farm technology on nutrition and feeding strategies Chair: K. de Koning	**Session 18** Free communications: Parameter estimation and molecular genetics Chair: G. Pollott		**Session 27** Interaction of diseases and production systems; risk evaluation and management Chair: M. Pearce
		Poster Session (viewing (18.00 – 19.00))	

Themed sessions shown in **_bold italics_**:
The importance of animal production for food supply, food quality and environment

Wednesday 31 August 8.30 – 12.30	Wednesday 31 August 14.00 – 18.00	Thursday 1 September 8.30 – 12.30	Thursday 1st September 14.00 – 18.00
Session 28 Building intellectual capital in animal science (WAAP session) Chair: N. Casey ——— **Session 29** New and advanced technologies applied in the dairy herd (Industry session) Chair: A. Rosati ——— **Session 30** Free communications: animal breeding methodology Chair: B. Gredler ——— **Session 31** **_Sustainable use of antibiotics_** Chair: A. Kuipers ——— **Session 32** Management of replacements of females Chair: E. Ugarte ——— **Session 33** Nutrition and welfare Chair: J.-E. Lindberg ——— **Session 34** Improving neonatal survival Chair: H. Spoolder	Business meeting, future programme and elections (14.00 – 15.00) followed by free communications on: (15.00 – 18.00) ——— **Session 35** Selection and genetic gain Chair: H. Simianer ——— **Session 36** Cattle production Chair: C. Lazzaroni ——— **Session 37** Advances in the Horse Commission Working Groups: from tradition to new challenges Chair: N. Miraglia ——— **Session 38** Sheep and Goats production Chair: F. Ringdorfer ——— **Session 39** Pig production Chair: P. Knap ——— **Session 40** Animal Nutrition Chair: J.-E. Lindberg ——— **Session 41** Livestock Farming Systems Chair: S. Ingrand ——— **Session 42** Animal Physiology Chair: M. Vestergaard ——— **Session 43** Animal Management and Health Chair: G. Das	Food Quality symposium **Session 44** **_Milk and meat product quality_** (am) Chair: J. Hocquette **Session 53** **_Gastronomic quality of animal products_** (pm) Chair: J. Hocquette/L. Aass ——— Aquaculture Symposium **Session 45** Breeding applications in industry (am) Chair: K. Kolstad **Session 54** Sustainable aquaculture for the future (pm) Chair: M. Øverland ——— Sustainability symposium **Session 46** **_Sustainable cattle farming_** (am) Chair: I. Stokovic/M. Klopcic **Session 55** **_Social pillar of sustainability_** (pm) Chair: K Eilers ——— Horse network workshop **Session 47** Equine education 1 (am) Equine education 2 (pm) Chair: A. Ellis	
		Session 48 Genetics of health, behaviour and functional traits in dogs (Companion Animal Working Group) Chair: P. Carnier ——— **Session 49** Breeding value estimation in sheep Chair: M. Schneeberger ——— **Session 50** Free communications: pig nutrition Chair: D. Torrallardona ——— **Session 51** Comparative lactation biology Chair: C. Knight ——— **Session 52** Causes and consequences of mortality and premature culling of breeding animals Chair: L. Boyle	**Session 56** Increasing the nutritive value of raw materials and compound feed by feed technology (Industry session) Chair: M. Helander ——— **Session 57** Animal Fibre Science (Animal Fibre Working Group) Chair: C. Renieri ——— **Session 58** Free communications: Genomic selection and genome-wide association studies Chair: J. Szyda
Poster Session (viewing (18.00 – 19.00)			

Commission on Animal Genetics

Prof. Dr Simianer	President	University of Goettingen
	Germany	hsimian@gwdg.de
Dr Baumung	Vice-President	FAO Rome
	Italy	Roswitha.Baumung@fao.org
Dr Meuwissen	Vice-President	Norwegian University of Life Sciences
	Norway	theo.meuwissen@umb.no
Dr Szyda	Vice-President	Agricultural University of Wroclaw
	Poland	szyda@karnet.ar.wroc.pl
Dr Ibañez	Secretary	IRTA
	Spain	noelia.ibanez@irta.es
Dr De Vries	Industry rep.	CRV

Commission on Animal Nutrition

Dr Lindberg	President	Swedish University of Agriculture
	Sweden	jan-eric.lindberg@huv.slu.se
Dr Bailoni	Vice-President	University of Padova
	Italy	Lucia.bailoni@unipd.it
Dr Connolly	Vice-President/	
	Industry rep.	Alltech
	Ireland	aconnolly@alltech.com
Mrs Tsiplakou	Secretary	Agricultural University of Athens
	Greece	eltisplakou@aua.gr
Mr Van Duinkerken	Secretary	Wageningen University
	Netherlands	gert.vanduinkerken@wur.nl

Commission on Animal Management & Health

Dr Fourichon	President	Oniris INRA
	France	Christine.fourichon@oniris-nantes.fr
Dr Spoolder	Vice-President	ASG-WUR
	Netherlands	Hans.spoolder@wur.nl
Prof. Dr Krieter	Vice-President	University Kiel
	Germany	jkrieter@tierzucht.uni-kiel.de
Mr Pearce	Vice-President/	
	Industry rep.	Pfizer
	United Kingdom	Michael.C.Pearce@pfizer.com
Dr Edwards	Secretary	University of Newcastle Upon Tyne
	United Kingdom	Sandra.edwards@ncl.ac.uk
Mr Das	Secretary	University of Goettingen
	Germany	gdas@gwdg.de

Commission on Animal Physiology

Dr Vestergaard	President	Aarhus University
	Denmark	Mogens.Vestergaard@agrsci.dk
Dr Kuran	Vice-President	Gaziosmanpasa University
	Turkey	mkuran@gop.edu.tr
Dr Driancourt	Vice president/	
	Industry rep.	Intervet
	France	Marc-antoine.driancourt@sp.intervet.com
Dr Bruckmaier	Vice-President	University of Bern
	Switzereland	Rupert.bruckmaier@physio.unibe.ch
Dr Quesnel	Secretary	INRA Saint Gilles
	France	Helene.quesnel@rennes.inra.fr
Dr Scollan	Secretary	Institute of Biological, Environmental and rural sciences
	UK	ngs@aber.ac.uk

Commission on Livestock Farming Systems

Dr Bernués Jal	President	CITA
	Spain	abernues@aragon.es
Dr Hermansen	Vice-President	DIAS
	Denmark	john.hermansen@agrsci.dk
Dr Leroyer	Vice president/	
	industry rep.	ITAB
	France	Joannie.leroyer@itab.asso.fr
Dr Matlova	Vice-President	Res. Institute for Animal Production
	Czech Republic	matlova.vera@vuzv.cz
Dr Eilers	Secretary	Wageningen University
	Netherlands	karen.eilers@wur.nl
Dr Ingrand	Secretary	INRA/SAD
	France	ingrand@clermond.inra.fr

Commission on Cattle Production

Dr Kuipers	President	Wageningen UR
	Netherlands	abele.kuipers@wur.nl
Dr Thaller	Vice-President	Animal Breeding and Husbandry
	Germany	Georg.Thaller@tierzucht.uni-kiel.de
Dr Lazzaroni	Vice-president	University of Torino
	Italy	carla.lazzaroni@unito.it
Dr Coffey	Vice president/	
	Industry rep.	SAC, Scotland
	UK	mike.coffey@sac.ac.uk>
Dr Hocquette	Secretary	INRA
	France	hocquet@clermont.inra.fr
Dr Klopcic	Secretary/	
	Industry rep.	University of Ljublijana
	Slovenia	marija.klopcic@bf.uni-lj.si

Commission on Sheep and Goat Production

Dr Schneeberger	President	ETH Zurich
	Switzerland	markus.schneeberger@inw.agrl.ethz.ch
Dr Ringdorfer	Vice-President	LFZ Raumberg-Gumpenstein
	Austria	
	ferdinand.ringdorfer@raumberg-gumpenstein.at	
Dr Bodin	Vice President	INRA-SAGA
	France	Loys.bodin@toulouse.inra.fr
Dr Papachristoforou	Vice President	Agricultural Research Institute
	Cyprus	Chr.Papachristoforou@arinet.ari.gov.cy
Dr Milerski	Secretary/	
	Industry rep.	Research Institute of Animal Science
	Czech Republic	m.milerski@seznam.cz
Dr Ugarte	Secretary/	
	Industry rep.	NEIKER-Tecnalia
		eugarte@neiker.net

Commission on Pig Production

Dr Knap	President/	
	Industry rep.	PIC International Group
	Germany	pieter.knap@pic.com
Dr Lauridsen	Vice-President	Aarhus University
	Denmark	charlotte.lauridsen@agrsci.dk
Dr Torrallardona	Vice-President	IRTA
	Spain	David.Torrallardona@irta.es
Dr Pescovicova	Secretary	Research Institute of Animal Production
	Slovak Republic	peskovic@vuzv.sk
Dr Velarde	Secretary	IRTA
	Spain	antonio.velarde@irta.es

Commission on Horse Production

Dr Miraglia	President	Molise University
	Italy	miraglia@unimol.it
Dr Burger	Vice president	Clinic Swiss National Stud
	Switzerland	
	Dominique.burger@mbox.haras.admin.ch	
Dr Janssen	Vice president	BIOSYST
	Belgium	Steven.janssens@biw.kuleuven.be
Dr Lewczuk	Vice president	IGABPAS
	Poland	d.lewczuk@ighz.pl
Dr Saastamoinen	Vice president	MTT Agrifood Research Finland
	Finland	markku.saastamoinen@mtt.fi
Dr Palmer	Vice president/	
	Industry rep.	CRYOZOOTECH
	France	ericpalmer@cryozootech.com
Dr Holgersson	Secretary	Swedish University of Agriculture
	Sweden	Anna-Lena.holgersson@hipp.slu.se
Dr Hausberger	Secretary	CNRS University
	France	Martine.hausberger@univ-rennes1.fr

Session 01. Methodological developments in genomic selection (joint Interbull and Animal Genetic Resources working group symposium)

Date: 29 August 2011; 08:30 - 12:30 hours
Chairperson: Ducrocq

Session 02. Fabre Technology platform: research updates and White Paper on Food Security and Climate Change

Date: 29 August 2011; 08:30 - 12:30 hours
Chairperson: Neeteson-Van Nieuwenhoven

Session 03. Welfare, ethics and behaviour in pig production

Date: 29 August 2011; 08:30 - 12:30 hours
Chairperson: Velarde

Session 04. Applied cattle breeding and genomics

Date: 29 August 2011; 08:30 - 12:30 hours
Chairperson: Thaller

Session 05. Free communications in horse production

Date: 29 August 2011; 08:30 - 12:30 hours
Chairperson: Saastamoinen

Session 06. Free communications in sheep and goat production

Date: 29 August 2011; 08:30 - 12:30 hours
Chairperson: Papachristoforou

Session 07. Life-cycle assessment of livestock production (SOLID/ANIMALCHANGE Symposium)

Date: 29 August 2011; 08:30 - 12:30 hours
Chairperson: Gerber

Session 08. Liver metabolism in dairy cows during the transition period

Date: 29 August 2011; 08:30   12:30 hours
Chairperson: Van Dorland

Session 09. Impact of farm technology on nutrition and feeding strategies

Date: 29 August 2011; 08:30 - 12:30 hours
Chairperson: De Koning

Session 10. Environmental value of animals genetic resources (Animal Genetic Resources symposium session)

Date: 29 August 2011; 14:00 - 18:00 hours
Chairperson: Gandini

Session 11. Genomic selection - application, ownership, and economics in dairy cattle (Interbull symposium)

Date: 29 August 2011; 14:00 - 18:00 hours
Chairperson: Reents and Coffey

Session 12. Link between rumen genome, nutrition and milk protein synthesis

Date: 29 August 2011; 14:00 - 18:00 hours
Chairperson: Van Vuuren

Session 13. Animal Task Force: Food security policy perspectives for industry and research

Date: 29 August 2011; 14:00 - 18:00 hours
Chairperson: Vriesekoop

Session 14. Breeding for resource and environmental efficiency

Date: 29 August 2011; 14:00 - 18:00 hours
Chairperson: Løvendahl

Session 15. Feeding practices and social constraints in horse housing

Date: 29 August 2011; 14:00 - 18:00 hours
Chairperson: Søndergaard

Session 16. Challenges of rangeland farming systems (economics, grazing, reproduction, health and welfare)

Date: 29 August 2011; 14:00 - 18:00 hours
Chairperson: Ådnøy

Poster **Session 16 no. Page**

Session 17. Use of new technologies for breeding and management of pigs and cattle

Date: 29 August 2011; 14:00 - 18:00 hours
Chairperson: Knap

Theatre **Session 17 no. Page**

Session 18. Free communications: parameter estimation and molecular genetics

Date: 29 August 2011; 14:00 - 18:00 hours
Chairperson: Pollott

Session 19. Architecture of quantitative traits

Date: 30 August 2011; 14:00 - 18:00 hours
Chairperson: Varona

Theatre   **Session 19 no.**   **Page**

Cytoplasmic line effects for birth weight and preweaning growth traits in the Asturiana de los Valles beef cattle breed 18 123
Pun, A., Goyache, F., Cervantes, I. and Gutiérrez, J.P.

Session 20. The best cow for environmental efficiency

Date: 30 August 2011; 14:00 - 18:00 hours
Chairperson: Klopcic

Theatre	Session 20 no.	Page
invited Improving environmental sustainability of the dairy cow Capper, J.L.	1	123
invited Interactions between dairy cow nutrition and environmental burdens Sebek, L.B.J.	2	124
The best dairy cow system for future in terms of environment, economics and social criteria - worldwide Hemme, T.	3	124
High milk production and good reproduction: is it possible? Zeron, Y., Galon, N. and Ezra, E.	4	125
Impact of breeding strategies using genomic information on fitness and health Egger-Danner, C., Willam, A., Fuerst, C., Schwarzenbacher, H. and Fuerst-Waltl, B.	5	125
Genetic analysis of calf vitality in Dutch dairy cows Van Pelt, M.L. and De Jong, G.	6	126
Future breeding goals in beef cattle: effect of changed production conditions on economic values Åby, B.A., Vangen, O., Aass, L. and Sehested, E.	7	126
Economic selection indexes for various breeds under different farming and production systems in Slovenia Klopcic, M., Veerkamp, R.F., Zgur, S., Kuipers, A., Dillon, P. and De Haas, Y.	8	127
The relationship between diet characteristics, milk urea, nitrogen excretion and ammonia emissions in dairy cows Bracher, A., Münger, A., Stoll, W., Schlegel, P. and Menzi, H.	9	127
Genetic relationships between energy balance, fat protein ratio, body condition score and disease traits in Holstein Friesians Buttchereit, N., Stamer, E., Junge, W. and Thaller, G.	10	128
Fat-to-protein-ratio in early lactation as an indicator of herdlife for first lactation dairy cows Bergk, N. and Swalve, H.H.	11	128
Genetic associations between feed efficiency and fertility in beef cattle Crowley, J.J., Evans, R.D. and Berry, D.P.	12	129

Poster	Session 20 no.	Page
Efficiency of milk production in farms with a high share of grasslands Litwinczuk, Z., Teter, W., Chabuz, W. and Stanek, P.	13	129

Session 21. Breeding programmes and genetic evaluation of horses

Date: 30 August 2011; 14:00 - 18:00 hours
Chairperson: Árnason

Theatre **Session 21 no. Page**

Poster **Session 21 no. Page**

Session 22a. Mastitis in non-dairy animals

Date: 30 August 2011; 14:00 - 15:45 hours
Chairperson: Bodin

Session 22b. Alternatives to drugs for parasite control in sheep and goats

Date: 30 August 2011; 16:15 - 18:00 hours
Chairperson: Gauly

Light treated bucks induce a well synchronized estrus and LH peak during anestrous season
by male effect in north Moroccan goats 5 151
Chentouf, M. and Bister, J.L.

Session 23. Consumer studies of the consequences of stopping piglet castration

Date: 30 August 2011; 14:00 - 18:00 hours
Chairperson: Oliver

Session 24. Interactions between nutrition, genetics and health

Date: 30 August 2011; 14:00 - 18:00 hours
Chairperson: Van Duinkerken

Session 25. Ecological intensification and ecosystem services of livestock farming systems

Date: 30 August 2011; 14:00 - 18:00 hours
Chairperson: Hermansen and Tichit

Theatre	Session 25 no.	Page
invited Livestock farming with care *Scholten, M.C.T., Gremmen, H.G.J. and Vriesekoop, P.W.J.*	1	161
Towards a framework based on contributions from various disciplines to study adaptive capacities of livestock farming systems *Douhard, F., Friggens, N.C. and Tichit, M.*	2	162
Crops and fodder for sustainable organic low input dairy systems *Marie, M. and Bacchin, M.*	3	162
Effect of uncertainty on GHG emissions and economic performance for increasing milk yields in dairy farming *Zehetmeier, M. and Gandorfer, M.*	4	163
invited Livestock farming systems and ecosystem services: from trade-offs to synergies *Magda, D., Tichit, M., Durant, D., Lauvie, A., Lécrivain, E., Martel, G., Roche, B., Sabatier, R., De Sainte Marie, C. and Teillard, F.*	5	163
invited It's all about livestock ecology! *Meerburg, B.G. and Vriesekoop, P.W.J.*	6	164
Ecological intensification of livestock farming systems: landscape heterogeneity matters *Tichit, M., Sabatier, R. and Doyen, L.*	7	164

Session 26. Animal disease models for health and robustness

Date: 30 August 2011; 14:00 - 18:00 hours
Chairperson: Sartin

Theatre	Session 26 no.	Page
invited Endotoxemia as a model for evaluating naturally occurring and nutritionally-induced variations in the stress and innate immune responses of cattle and swine *Carroll, J.A., Burdick, N.C., Randel, R.D., Welsh, Jr., T.H., Chase, Jr., C.C., Coleman, S.W. and Sartin, J.L.*	1	165
invited Heat stress in farrowing sows under piglet-friendly thermal environments *Malmkvist, J. and Pedersen, L.J.*	2	165
invited Underlying infections enhance the physiological responses of cattle to endotoxemia *Sartin, J.L., Walz, P., Givens, D. and Elsasser, T.H.*	3	166
invited Impacts and mechanisms underlying stress-associated post-translational nitration of protein structure and function in signaling pathways *Elsasser, T., Ischiropoulos, H., Collier, R. and Sartin, J.*	4	166

Changes in expression of TGF-β, CTGF, collagen and elastin associated with seasonal time of sampling in bovine claw laminar tissue
La Manna, V. and Galbraith, H.

Poster   **Session 26 no. Page**

Session 27. Interaction of diseases and production systems; risk evaluation and management

Date: 30 August 2011; 14:00 - 18:00 hours
Chairperson: Pearce

Theatre   **Session 27 no. Page**

Poster   **Session 27 no. Page**

Session 30. Free communications: animal breeding methodology

Date: 31 August 2011; 08:30 - 12:30 hours
Chairperson: Gredler

Poster   **Session 30 no. Page**

Session 31. Sustainable use of antibiotics

Date: 31 August 2011; 08:30 - 12:30 hours
Chairperson: Kuipers

Session 32. Management of replacements of females

Date: 31 August 2011; 08:30 - 12:30 hours
Chairperson: Ugarte Sagastizabal

Session 33. Nutrition and welfare

Date: 31 August 2011; 08:30 - 12:30 hours
Chairperson: Lindberg

Session 34. Improving neonatal survival

Date: 31 August 2011; 08:30 - 12:30 hours
Chairperson: Spoolder

Poster   **Session 34 no. Page**

Session 35. Free communications: selection and genetic gain

Date: 31 August 2011; 14:00 - 18:00 hours
Chairperson: Simianer

Theatre   **Session 35 no. Page**

Poster **Session 35 no. Page**

Session 36. Free communications in cattle production

Date: 31 August 2011; 14:00 - 18:00 hours
Chairperson: Lazzaroni

Theatre **Session 36 no. Page**

Session 37. Advances of the Horse Commission Working Groups: from tradition to new challenges

Date: 31 August 2011; 14:00 - 18:00 hours
Chairperson: Miraglia

Session 38. Free communications in sheep and goat production

Date: 31 August 2011; 14:00 - 18:00 hours
Chairperson: Ringdorfer

Poster Session 38 no. Page

Session 39. Free communications in pig production

Date: 31 August 2011; 14:00 - 18:00 hours
Chairperson: Knap

Poster   **Session 39 no. Page**

Session 40. Free communications in animal nutrition

Date: 31 August 2011; 14:00 - 18:00 hours
Chairperson: Lindberg

Poster   Session 40 no. Page

Session 41. Free communications in livestock farming systems

Date: 31 August 2011; 14:00 - 18:00 hours
Chairperson: Ingrand

Session 42. Free communications in animal physiology

Date: 31 August 2011; 14:00 - 18:00 hours
Chairperson: Quesnel

Session 43. Free communications in animal management and health

Date: 31 August 2011; 14:00 - 18:00 hours
Chairperson: Das

Session 44. Milk and meat product quality (Food Quality symposium)

Date: 01 September 2011; 08:30 - 12:30 hours
Chairperson: Hocquette

Poster

Session 45. Breeding applications in industry (Aquaculture symposium)

Date: 01 September 2011; 08:30 - 12:30 hours
Chairperson: Kolstad

Poster **Session 45 no.** **Page**

Session 46. Sustainable cattle farming (Sustainability symposium)

Date: 01 September 2011; 08:30 - 12:30 hours
Chairperson: Štokovic and Klopcic

Theatre **Session 46 no.** **Page**

Poster

Session 47. Equine education (Horse network workshop)

Date: 01 September 2011; 08:30 - 18:00 hours
Chairperson: Ellis

Theatre

Session 48. Genetics of health, behaviour and functional traits in dogs (Companion Animal working group)

Date: 01 September 2011; 08:30 - 12:30 hours
Chairperson: Carnier

Session 49. Breeding value estimation in sheep

Date: 01 September 2011; 08:30 - 12:30 hours
Chairperson: Schneeberger

Session 50. Free communications: pig nutrition

Date: 01 September 2011; 08:30 - 12:30 hours
Chairperson: Torrallardona

Poster   **Session 50 no. Page**

Session 51. Comparative lactation biology

Date: 01 September 2011; 08:30 - 12:30 hours
Chairperson: Knight

Session 52. Causes and consequences of mortality and premature culling of breeding animals

Date: 01 September 2011; 08:30 - 12:30 hours
Chairperson: Boyle

Theatre	Session 52 no.	Page
invited Premature removal and mortality of commercial sows *Engblom, L., Stalder, K. and Lundeheim, N.*	1	364
On-farm mortality of cows in Swedish dairy herds *Alvåsen, K., Jansson Mörk, M., Hallén Sandgren, C., Thomsen, P.T. and Emanuelson, U.*	2	364
Association of herd demographics and biosecurity with cattle longevity and udder health in dairy herds *Brouwer, H., Bartels, C.J.M. and Van Schaik, G.*	3	365
Claw and foot healt: Early diagnostics and prevention of foot lesions in dairy cattle *Tamminen, P., Häggman, J., Pastell, M., Tiusanen, J. and Juga, J.*	4	365

Poster	Session 52 no.	Page
Effect of inbreeding and estimation of genetic parameters for heifer mortality in Austrian Brown Swiss cattle *Fuerst-Waltl, B. and Fuerst, C.*	5	366
Association between herd characteristics and on-farm cow mortality in Swedish dairy herds *Jansson Mörk, M., Alvåsen, K. and Hallén Sandgren, C.*	6	366

Session 53. Gastronomic quality of animal products (Food Quality symposium)

Date: 01 September 2011; 14:00 - 18:00 hours
Chairperson: Hocquette and Aass

Theatre	Session 53 no.	Page
invited Can eating quality genetics be incorporated into the Meat Standards Australia lamb grading system? *Pannier, L., Pethick, D., Ball, A., Jacob, R., Mortimer, S. and Pearce, K.*	1	367
Prediction of beef eating quality in France using the Meat Standards Australia (MSA) system *Legrand, I., Hocquette, J.F., Polkinghorne, R.J. and Pethick, D.W.*	2	367
High energy supplement post-weaning does not enhance marbling in beef cattle *Greenwood, P.L., Siddell, J.P., Mc Phee, M.J., Walmsley, B.J. and Pethick, D.W.*	3	368
Contribution of Protected Denomination of Origin (PDO) beef for sustainable agriculture and meat quality in Portugal *Costa, P., Alfaia, C.M., Pestana, J.M., Bessa, R.J.B. and Prates, J.A.M.*	4	368
Traditional Corsican meat and dairy products move upmarket: could local consumers be excluded? *Casabianca, F. and Linck, T.*	5	369

Consumer gastronomic evaluation of wild and farmed brown trout (Salmo trutta)   6   369
Pohar, J.

Breed effects and heritability for concentration of fatty acids in milk fat determined by
Fourier transform infrared spectroscopy   7   370
Lopez-Villalobos, N., Davis, S.R., Lehnert, K., Mellis, J., Berry, S., Spelman, R.J. and Snell, R.G.

Poster   **Session 53 no. Page**

Analysis of the red meat price changes over the last 25 years and effects of import decisions
in Turkey   8   370
Aydin, E., Aral, Y., Can, M.F., Cevger, Y., Sakarya, E. and Isbilir, S.

Session 54. Sustainable aquaculture for the future (Aquaculture symposium)

Date: 01 September 2011; 14:00 - 18:00 hours
Chairperson: Øverland

Theatre   **Session 54 no. Page**

Industrial response to increased sustainability demand in aquaculture   1   371
Wathne, E. and El-Mowafi, A.

Resource use in Norwegian salmon production   2   371
Ytrestøyl, T., Ås, T.S., Berge, G.M., Sørensen, M., Thomassen, M. and Åsgård, T.

Microbes: a sustainable aquafeed resource for the future   3   372
Mydland, L.T., Romarheim, O.H., Landsverk, T., Skrede, A. and Øverland, M.

Optimal feeding rate for genetically improved Nile tilapia   4   372
Storebakken, T., Kumar Chowdhury, D. and Gjøen, H.M.

In vitro assays are useful tools in the value chain of aquaculture products   5   373
Moyano, F.J., Morales, G. and Márquez, L.

Interaction between physical quality, feed intake and nutrient utilization in modern fish feeds 6   373
Sørensen, M., Åsgård, T. and Øverland, M.

Sodium diformate and extrusion temperature affects nutrient digestibility and
physical quality of diets with fishmeal or barley protein concentrate for rainbow trout
(Oncorhynchus mykiss)   7   374
Morken, T., Kraugerud, O.F., Barrows, F.T., Sørensen, M., Storebakken, T. and Øverland, M.

Emerging trends and research needs in sustainable aquaculture   8   374
Fiore, G., Natale, F. and Hofherr, J.

Session 55. Social pillar of sustainability (Sustainability symposium)

Date: 01 September 2011; 14:00 - 18:00 hours
Chairperson: Eilers

Theatre   **Session 55 no. Page**

invited Local breed farmers providing ecosystem services and sustainable development   1   375
Soini, K.

Session 56. Increasing nutritive value of raw materials and compound feed by feed technology (industry session)

Date: 01 September 2011; 14:00 - 18:00 hours
Chairperson: Helander

Session 57. Animal fibre science (Animal Fibre working group)

Date: 01 September 2011; 14:00 - 18:00 hours
Chairperson: Renieri

Theatre   Session 57 no. Page

Poster   Session 57 no. Page

Session 58. Free communications: genomic selection and genomewide association studies

Date: 01 September 2011; 14:00 - 18:00 hours
Chairperson: Szyda

Poster   Session 58 no.   Page

The single step: genomic evaluation for all

Legarra, A., Misztal, I. and Aguilar, I., INRA, UR 631, 31326 Castanet-Tolosan, France;
andres.legarra@toulouse.inra.fr

Due to complexity of real populations, there is a long tradition in animal breeding to consider all data simultaneously. This has lead to optimal theoretical solutions: the mixed model equations, the (inverse of) the numerator relationship matrix, multiple trait evaluations. In genomic evaluations, not all animals in a population are genotyped. This leads to use of pseudo-data (DYDs), which are uncertain, correlated, and sometimes hard to define. It has been also shown that after implantation of genomic selection, pedigree-based genetic evaluations (and DYD's) will be biased, unless genomic information is included in the estimation. However, this is essentially a missing data problem: genotypes are lost. Missing data problems can be solved by 'imputation' (data augmentation). Use of linkage (disequilibrium) analysis to impute is still very challenging in a general pedigree beyond one or two generations. However, under mild assumptions (no major genes), relationships serve to predict genotypes, and thus a linear model can be implemented by using a modified relationship matrix, H, whose inverse is $H^{-1} = A^{-1} + [0\ 0;\ 0\ G^{-1}-A_{22}^{-1}]$, where G is a genomic relationship matrix. BLUP equations can be used at a low computational cost (number of SNPs x number of individuals). Beyond unexpected – but statistically and genetically sound – features (i.e. negative relationships), main problems of the method are: (1) what is the effect of selection? (2) how to deal with genes of large effect? The effect of selection is akin to drift, increasing overall relationships, changing the genetic base and creating bias. A correction based on Wright's F_{st} removes this bias for pedigree-based selection. As for major genes, more credit can be given in G to SNPs of 'large' effect; or, a hierarchical model can be written to consider all unknowns and a non-linear method for SNPs effects. These remain open questions.

Comparison of different methods to calculate genomic predictions – results from SNP-BLUP, G-BLUP and one-step H-BLUP

Koivula, M., Strandén, I. and Mäntysaari, E.A., MTT Agrifood Research Finland, Biotechnology and Food Research, Biometrical Genetics, FI-31600 Jokioinen, Finland; minna.koivula@mtt.fi

Genomic selection has become an attractive approach to estimate breeding values because it allows breeders to select animals early in life. Several strategies to use genomic data in genomic evaluations have been proposed. The aim of this study was to compare different approaches to predict genomic breeding values. After edits the data consisted of 6145 genotyped Nordic Red breed bulls with 37996 SNP markers. Two different phenotypic data for production and mastitis traits were used in the analyses. First, deregressed proofs (DRP) and variance components were calculated for the milk, protein, fat and mastitis using the estimated breeding values (EBV) from 2005. Genomic breeding values were calculated using different BLUP models: marker SNP-BLUP, genomic breeding value G-BLUP and one-step approach H-BLUP. EBV data from 2010 was used for the validations. Of genotyped bulls, the reference bulls had EBVs in the 2005 and in the 2010 data, and the candidate bulls had index only in the 2010 data. Interbull protocol was used in the comparison of DGV's and model validation. For the candidate bulls, SNP BLUP and G-BLUP gave the same DGV's (e.g. correlation of DGV's between SNP-BLUP and G-BLUP for protein was 0.999) but there was a difference to DGV's from H-BLUP (correlations of SNP-BLUP to G-BLUP and H-BLUP were about 0.96). For all the traits, SNP-BLUP and G-BLUP got the same validation reliabilities R^2, but for H-BLUP they were slightly better. For example for protein, R^2 was 13% higher in H-BLUP than in SNP-BLUP or G-BLUP. Therefore, there seems to be slight advantage of using H-BLUP instead of other genomic evaluation models.

LDLA: a unified approach to the use of genomic and pedigree information in genomic evaluations

Meuwissen, T.[1], Luan, T.[1] and Woolliams, J.[2], [1]Norwegian University of Life Sciences, Institute Animal Science and Health, Søråsveien 5B, 1432 Ås, Norway, [2]University of Edinburgh, The Roslin Institute and Royal (Dick) School of Veterinary Studies, Easter Bush, Roslin, Midlothian EH25 9RG, United Kingdom; theo.meuwissen@umb.no

In practical genomic evaluations, often many animals are phenotyped and pedigree recorded but not genotyped, which makes it hard to include these in the genomic evaluation training data set. In the literature methods have been suggested to combine pedigree and marker information into one overall relationship matrix, H, which encompasses genotyped and non-genotyped animals. These methods fall into two categories: (1) methods that rely on linkage disequilibrium (LD) to extract the marker genotype data; and (2) methods that use linkage analysis (LA) to combine genotype and pedigree data. Here we combined both methods into a combined approach (LDLA) to make best use of both information sources. Two more corrections were needed to render the LDLA approach virtually unbiased: (1) when combining the LD-based relationship matrix (G) with the LA-based relationship matrix, relationship and inbreeding coefficients needed to be corrected to the same base population, where both these coefficients are assumed to equal 0; and (2) the LD-based relationship matrix needed to be regressed in order to account for the uncertainty in the marker based estimates of relationship. In a simulation study, the genomic evaluations of the LDLA approach proved more accurate and considerably less biased than previous methods for combining pedigree and genomic data.

Estimation of genomic breeding values for traits with high and low heritability in Brown Swiss bulls

Biscarini, F.[1], Kramer, M.[1], Bapst, B.[2] and Simianer, H.[1], [1]University of Göttingen, Animal Breeding and Genetics, Albrecht-Thaer weg, 3, 37075 Göttingen, Germany, [2]Swiss Brown Cattle Breeders Federation, Chamerstrasse, 56, 6300 Zug, Sweden; mkramer@gwdg.de

In this study we assess some methodological aspects of the estimation of genomic breeding values (GEBVs) in Brown Swiss bulls. Milk yield ($h^2=0.33$) and non-return rate (NRR, $h^2=0.09$) were the considered traits. A dataset of 1,138 Brown Swiss bulls with official EBVs for the mentioned traits was available. After filtering, 38,779 informative SNPs positioned on the 29 autosomes and the sex chromosomes were used. For the estimation of GEBVs, the methods GBLUP and BayesC were compared. For GBLUP, a genomic relationship matrix (G-matrix) derived according to Astle & Balding was used: this way of constructing the relationship matrix from markers has not been applied in animal genetics before, and gave a higher likelihood in the estimation of variance components compared to other tested methods. The use of EBVs or de-regressed proofs (DRPs) as dependent variables in the model was compared. Also the influence of including or not SNP markers placed on the sex chromosome was investigated. All comparisons were based on the accuracy of estimates obtained through cross-validation. Preliminary results show that EBVs gave better results than DRPs in the estimation of GEBVs with GBLUP: the accuracy of GEBVs with EBVs vs DRPs was 0.743 (s.d. 0.028) and 0.727 (0.031) for milk yield, and 0.726 (0.031) and 0.650 (0.038) for NRR. Accuracy for milk yield from EBVs was 0.754 (0.027) with BayesC. No appreciable difference was observed when SNPs on the sex chromosomes were excluded: the accuracy was 0.740 (0.028) for milk yield and 0.726 (0.031) for NRR. In general, accuracies of GEBVs for milk yield were higher than those for NRR; however, the relative increase in accuracy compared to standard BLUP is likely to be higher for traits with low heritability, for which greater benefits of genomic selection are expected.

Accuracy of genomic predictions using subsets of SNP markers in dual purpose Fleckvieh cattle
Gredler, B.[1], Anche, M.T.[1], Schwarzenbacher, H.[2], Egger-Danner, C.[2] and Sölkner, J.[1], [1]University of Natural Resources and Life Sciences Vienna, Gregor Mendel Str. 33, 1180 Vienna, Austria, [2]ZuchtData EDV-Dienstleistungen GmbH, Dresdner Str. 89/19, 1200 Vienna, Austria; birgit.gredler@boku.ac.at

SNP-selection strategies to generate subsets of SNP were evaluated for estimation of direct genomic breeding values (DGV) in dual purpose Fleckvieh cattle in Austria and Germany. 5,556 Fleckvieh bulls genotyped with the Illumina Bovine SNP50TM Beadchip were used in the analysis. The reference set consists of bulls born before 2003 and the validation set of bulls born 2003 to 2006. The full set of SNP contained 41,008 SNPs. DGV were estimated using SNP-BLUP and BayesB methodology. De-regressed breeding values of the traits total merit index, milk yield, fat percentage, protein yield, length of productive life, female fertility, male calving ease and somatic cell count were used as dependant variables in genomic predictions. The 100, 300, 500, 1,000, and 3,000 most influential SNP were selected according to their absolute size of the estimated regression coefficient using the full set of 41,008 SNP estimated with BayesB. In addition, genomic predictions were performed using the SNP included on the Illumina GoldenGate Bovine3K Genotyping Beadchip. Furthermore, subsets of 5K, 10K, 20K, and 30K consisting of the Illumina 3K chip plus 2K, 7K, 17K and 27K randomly selected SNP, respectively, were used. Accuracy of direct genomic breeding values was calculated as the correlation between DGV and the estimated breeding values from routine genetic evaluations in Austria and Germany of bulls in the validation set. Results show that BayesB slightly out performed SNP-BLUP for almost all traits independent of subset of SNP. Using the Illumina 3K chip accuracies were in the range of 67 to 94% of the accuracies achieved using the full set of SNP for all traits applying BayesB. Highest accuracies were always achieved with the full set of SNP. However, accuracies of some combinations of traits and subsets of SNP were close to the full set results.

Description of the French genomic evaluation approach
Croiseau, P.[1], Hozé, C.[1,2], Fritz, S.[2], Guillaume, F.[3], Colombani, C.[4], Legarra, A.[4], Baur, A.[2], Robert-Granié, C.[4], Boichard, D.[1] and Ducrocq, V.[1], [1]INRA, UMR1313, GABI, Domaine de vilvert, 78352 Jouy en Josas, France, [2]UNCEIA, 149 rue de Bercy, 75595, Paris, France, [3]Institut de l'élevage, 149 rue de Bercy, 75595, Paris, France, [4]INRA, UMR631, SAGA, Chemin de Borde Rouge, Auzeville, 31320 Castanet-Tolosan, France; pascal.croiseau@jouy.inra.fr

Worldwide, a large panel of genomic evaluation methodologies has been proposed and compared. When reference populations of adequate size are used, these methods appear to lead to similar accuracies of prediction of future bull's performances. However, computing requirements and time may be different, to such an extent that some of the better methods cannot be applied routinely. In France, a strong experience on Marker Assisted evaluation and Selection (MAS) has been accumulated over the last decade. A major advantage of MA-BLUP used in MAS is that the size of the mixed model equations system is manageable and computing requirements are limited. To build on this experience, a genomic evaluation methodology was developed to combine SNP corresponding to large QTL detected via Linkage Disequilibrium and Linkage analysis (LDLA) or by other genomic selection approaches such as Elastic-Net (EN) or Sparse PLS. Different scenarios were compared for a set of 1172 progeny tested bulls from the Montbéliarde breed. The 222 youngest ones formed the validation population. For six different traits – milk and protein yield, rear udder width, height at sacrum, conception rate and somatic cell counts – the correlations between the predicted and observed phenotypes (daughter yield deviations) were calculated. A MA-BLUP including 400-500 QTL detected by LDLA or by EN and followed by haplotypes of 3-5 SNP provided the best correlations. This leads to computation times of genomic evaluations which are less sensitive to future increases in number of animals or SNP chip densities. This approach has been routinely applied for genomic evaluations in the French Holstein, Montbéliarde and Normande breeds since June 2010.

Recent trends in genomic selection in dairy cattle
De Roos, A.P.W., CRV, P.O. Box 454, 6800 AL Arnhem, Netherlands; sander.de.roos@crv4all.com

Genomic selection is causing a revolution in dairy cattle breeding. As a result of the introduction of affordable 50K SNP panels and the genotyping of thousands of progeny tested bulls, genomic breeding values for many traits can now be obtained with 50-75% reliability, directly after birth, and at very little costs. This enables large scale screening of bull dams and young bulls for progeny testing. Secondly, superior young bulls are being marketed directly and used sires of sons. It is expected that genomic selection, in combination with a reduction in the generation interval, can double the rate of genetic improvement in dairy cattle. The largest opportunities and challenges for the future are the application of genomic selection to novel traits that are difficult to improve in traditional schemes and the extension to genomic selection across multiple populations. The availability of 750K SNP data and genome sequence data gives opportunities to detect and use markers very close to QTL, which could result in genomic predictions that are persistent across populations and generations. The decreasing cost of genotyping makes large scale genotyping of commercial animals possible. This has several advantages, including their use in reference populations for novel traits and the use of genomic information in herd management processes. With the progressively increasing size of data sets, the computational demands for genotype imputation and genomic evaluations will increase proportionally, and may become a limiting factor. In the future complete genomic information will be available for large numbers of commercial animals, leading to new applications for not only genetic improvement, but also optimisation of herd management and livestock production chains. Organisations need to strengthen their relationships with research and industry partners to remain competitive and to add value for their customers and the dairy production chain.

The contribution of linkage and linkage disequilibrium information to the accuracy of genomic selection
Luan, T.[1], Woolliams, J.A.[2], Ødegård, J.[3] and Meuwissen, T.H.E.[1], [1]Norwegian University of Life Sciences, Animal and Aquacultural Sciences, Arboretveien 6, N-1432, Ås, Norway, [2]The Roslin Institute and Royal (Dick) School of Veterinary Studies, University of Edinburgh, Midlothian EH25, 9RG, United Kingdom, [3]Nofima Marin, Arboretveien 2, N-1432, Ås, Norway; tu.luan@umb.no

The prediction of genome-wide breeding values (GW-EBV) in genomic selection (GS) is generally thought to be achieved by using linkage disequilibrium (LD) between the markers and QTL. In 2007 Habier *et al.* showed that GS implicitly also uses linkage analysis information. Here we studied how much of the reliability of the prediction of GW-EBV is due to linkage analysis information and how much due to LD that already existed in the founders of the pedigree. The study was performed on milk yield, fat yield and protein yield. GW-EBV was predicted using genotype data (45,888 SNPs) and phenotypic data of 255 British Holstein bulls. The method of best linear unbiased prediction was used to predict GW-EBV with LD information. Breeding value prediction using only linkage analysis (LA) information was performed by applying the Fernando and Grossman approach to marker assisted selection across all loci, using 5 generations of pedigree data. To investigate the effect of the number of generations of pedigree used on the accuracy of the GW-EBV prediction, LA based G matrix was set up using 1, 2, 3, 4 and 5 generations of pedigree, respectively. In addition, we performed GW-EBV prediction using both LA and LD information. The study showed that the reliability of the prediction of GW-EBV using LA information achieved similar accuracy to that of the prediction using LD information. About all the reliability of GS could be explained by LA information. The reliability of GW-EBV prediction didn't get improved when using both LA and LD information. The results of using different number of generations of pedigree showed that the reliability was reduced if less than 3 generations of linkage analysis information was used.

Genome-wide prediction of complex traits under population stratification and hidden relatedness

Gonzalez-Recio, O.[1], Forni, S.[2], Gianola, D.[3,4], Weigel, K.A.[4], Rosa, G.J.M.[3,4] and Wu, X.-L.[4], [1]INIA, Mejora Genética Animal, Ctra. La Coruña km 7,5, 28040 Madrid, Spain, [2]Genus Plc, 100 Bluegrass Commons Blvd. Ste 2200., Hendersonville, TN, USA, [3]University of Wisconsin, Animal Science, 1675 Observatory Dr., WI 53076, USA, [4]University of Wisconsin, Dairy Science, 1675 Observatory Dr., WI 53076, USA; gonzalez.oscar@inia.es

Genome-enabled evaluation and prediction of complex traits (e.g. yield or disease resistance) are difficult when individuals from genetically distant populations are included in the training/learning samples. Multi-breed/line, multi-environment or across-country genomic evaluations of animals of the same breed provide examples. It seems reasonable to expect that an increase in training sample size due to joining different populations would increase accuracy of prediction. However, this has not been confirmed, which may be due to heterogeneous additive, epistatic or genotype x environment effects, or even different LD levels or LD phases across populations. Non-parametric approaches have shown some advantages when predicting target variables using whole genome information. Here, a non-parametric approach is proposed to deal with population stratification while accounting for cryptic genetic interactions. Data from two purebred swine lines (A and B) and from a crossbred line (C) from PIC North America, Genus PIC, were used (923, 919 and 700 individuals, respectively), with genotypes for 5,302 SNPs. First, a random forest approach was used to detect population stratification, and to create a between-individuals similarity matrix. Then, a multi-trait reproducing Kernel Hilbert spaces regression was used to predict scrotal hernia incidence in the adjoined population, supplemented with the first three eigen-vectors of the similarity matrix used as covariates. This approach was compared to the Bayesian Lasso (BL) with regards to predictive ability. BL showed better predictive ability in the purebred lines, whereas RKHS was slightly better in the crossbreds.

The use of two types of mating to profit from dominance in genomic selection

Varona, L.[1] and Toro, M.A.[2], [1]Universidad de Zaragoza, c/ Miguel Servet 177, 50013 Zaragoza, Spain, [2]Universidad Politecnica de Madrid, Ciudad Universitaria, 50040 Madrid, Spain; lvarona@unizar.es

With the recent availability of very dense panels of SNPs and the advent of Genomic Selection it is seems natural that methods to use dominance variation should be revisited. In a previous work we considered the use of mate allocation. Here we start recognizing that, as it is well known, the most appealing way of using dominance effects is to carry out two types of matings: a) matings from wich the population will be propagated; b) matings to obtain commercial animals. Although the straightforward method to apply this methodology is to implement a crossing system between breeds that profit from heterosis it can also we carried out in a single population. Here we explore this possibility. In each generation commercial population is produced by planned matings that maximize the overall (additive + dominanc) genetic merit of the offspring. But this commercial population is a transient one not longer used to propagate the scheme. On the contrary the animals to propagate the population are produced implementing maximum coancestry matings where coancestry refers to the molecular coancestry based on SNPs involved. We show that this strategy could increase the expected response about (2-16%) depending on the proportion of dominance variance.

Estimating inbreeding coefficients from dense panels of biallelic markers and pedigree information
Bouquet, A., Sillanpää, M.J. and Juga, J., University of Helsinki, Department of Agricultural Sciences, P.O.Box 27, 00014 Helsinki, Finland; alban.bouquet@helsinki.fi

The aim of this simulation study was to compare performances of inbreeding (F) estimators using dense panels of biallelic markers for different types of population structures. Simulated populations were made up of 1000 individuals per generation and differed by the level of selection applied: no selection or a strong phenotypic selection for 10 discrete generations on a trait of heritability 0.5. Individual genotypes for 5400 markers (18 per cM) were available for animals in the last 5 generations whereas pedigrees were recorded on the last 8 generations. Different distributions of marker allele frequencies (AF) were simulated by setting the population-scaled mutation rate to 1 and 10. Ten replicates were carried out per scenario. Analysed estimators were the correlation and regression estimators proposed by VanRaden to build the genomic relationship matrix. Other estimators included the original Ritland's estimator and a modified Ritland's estimator accounting for pedigree information. F estimates were correlated and regressed to the true simulated values to assess the precision and bias of analysed estimators, respectively. Main results showed that the performance of estimators depends on both population structure and distribution of marker AF. The estimate accuracy increased in situation of selection. Across scenarios, Ritland's estimators were slightly more correlated with true values than the correlation estimator and were less sensitive to marker AF distribution. Accounting for pedigree information into Ritland's estimator only slightly increased estimation accuracy in the tested cases. Adapting those estimators to assess coancestry coefficients led to similar conclusions. Thus, results about Ritland's estimator, which can be also used with multiallelic markers, are promising both to build the genomic relationship matrix for genomic evaluations and to better assess genetic diversity in selected populations.

Genomics and selection of animals for a gene-bank
Engelsma, K.A. and Windig, J.J., Wageningen UR Livestock research, Animal Breeding and Genomics Centre, P.O. Box 65, 8200 AB Lelystad, Netherlands; jack.windig@wur.nl

We investigated the consequences for prioritization of animals for inclusion in a gene bank when genomic information is used instead of pedigree information. Diversity was compared for two recently diverged Holstein populations. Differences in overall diversity between the two groups were similar for pedigree and SNP based overall diversity. However, for some chromosome regions SNP based diversity differed from pedigree based diversity. Selection of animals for inclusion with genomic kinships resulted in a higher conserved diversity compared to pedigree kinships, but differences were small. Largest differences were found when few animals were prioritized and when pedigree errors were present. We found more differences at the chromosomal level, where selection based on genomic kinships resulted in a higher conserved diversity than pedigree based selection for most chromosomes, but for some chromosomes pedigree kinships resulted in a higher conserved diversity. To optimize conservation strategies, genomic information can help improve the selection of animals for conservation in those situations where pedigree information is unreliable or absent, or when we want to conserve diversity at specific genome regions. For maintaining diversity in a gene bank, SNP based diversity gives a more detailed picture than pedigree based diversity.

Bayes Cpi versus GBLUP, PLS regression, sparse PLS and elastic net methods for genomic selection in french dairy cattle

Colombani, C.[1], Legarra, A.[1], Croiseau, P.[2], Fritz, S.[3], Guillaume, F.[4], Ducrocq, V.[2] and Robert-Granié, C.[1], [1]INRA, Animal Genetics, UR631-SAGA, BP 52627, 31326 Castanet-Tolosan, France, [2]INRA, Animal Genetics, UMR1313-GABI, 78352 Jouy-en-Josas, France, [3]UNCEIA, 149 rue de Bercy, 75595 Paris, France, [4]Institut de l élevage, 149 rue de Bercy, 75595 Paris, France; christele.robert-granie@toulouse.inra.fr

Genomic Selection (GS) relies on genomic enhanced breeding values (GEBV) derived from dense markers information over the entire genome for the selection of young animals. GEBV predictions have to be accurate for a successful application of GS. The objective of this study was to compare the predictive ability of a Bayesian approach -Bayes Cπ- which performs SNP selection by stochastic search (including or not pedigree information) with methods already tested on French dairy cattle: Genomic BLUP (GBLUP) and dimension reduction or variable selection methods such as Partial Least Squares (PLS) regression, Sparse PLS regression and Elastic Net (EN). The prediction equation was estimated with training data sets composed of 2,976 Holstein bulls or 950 Montbéliarde bulls, genotyped for 38,462 or 39,738 SNP respectively, with a minor allele frequency of 3%. Then, phenotypes were predicted on validation data sets composed of 964 (222) young Holstein (Montbéliarde) bulls. Five milk production traits and conception rate were studied using DYD (Daughter Yield Deviation) weighted by Effective Daughter Contribution (EDC) as phenotypes. The EDC weighted correlation between GEBV and observed DYD was computed. Bayes Cπ and EN approaches gave slightly better accuracies of GEBV prediction. The convergence of Bayes Cπ was reached quickly with few samples for the Holstein data set while the prior probability that a marker has zero effect (π) had to be fixed in the Montbéliarde data set. Adding pedigree information to the BayesCπ approach did not improve the correlation nor the slope of the regression of observed DYD on estimated DYD.

Genome-wide breeding value estimation with the Bayesian elastic net

Yu, X. and Meuwissen, T.H.E., Norwegian University of Life Sciences, Animal and Aquacultural Sciences, Arboretveien 12, 1432 Ås, Norway; xijiang.yu@umb.no

Genome-wide breeding value (GWEBV) estimation is often facing a situation where the number of predictors (genetic markers) is much more larger than the number of records (phenotypes), or P>>n. Several regularization methods were proposed to reduce the predictors. The Bayesian lasso regularize the predictors with the first order function of marker effects, i.e. the sum of their absolute values, as penalty. Recently, Zou *et al.* proposed a elastic net (enet) method. In addition to the first order penalization, enet also penalizes the sum squares of marker effects. The enet can treat closely associated marker effects, which might be in a same physiological pathway, as a whole, such that it can include or exclude them from a model altogether. Enet method is currently mostly used in association studies on human diseases, where it often outperforms Bayesian lasso. In this study, we compared a Bayesian version of enet (Benet) with three other methods, including fast Bayes-B, GBLUP, and MixP, with simulated data on a real Norwegian red cattle pedigree. In Benet, the penalties on the absolute value of the marker effects is parameterized as $\pi\Lambda$, and on the squared value as $(1-\pi)\Lambda$, where Λ is a scaling parameter determining the prior variances of the marker effects, and π varies between 0 (all penalty on the squared effects) and 1 (all penalty on the absolute value). The results show that π does has little effect on the accuracy of GWEBV estimation. The accuracy goes up slowly as π increases. When π approaches 1, the results can vary a lot and are unpredictable. A value of π of 0.97 is used for final comparison, in which case the accuracy of Benet is similar to that of GBLUP, and in some cases performs slightly better. Possibly Benet may perform better in real data, where gene effects can be associated by being in the same pathway.

Genomic BLUP with additive mutational effects

Casellas, J.[1], Esquivelzeta, C.[1] and Legarra, A.[2], [1]Universitat Autònoma de Barcelona, G2R, Dep. Ciència Animal i dels Aliments, Campus UAB, 08193 Bellaterra, Spain, [2]Institut National de la Recherche Agronomique, UR631 SAGA, BP 52627, 31320 Castanet-Tolosan, France; joaquim.casellas@uab.cat

Genomic BLUP can accommodate both additive genetic (a) and mutational (m) effects by implementing y = μ + Za + Zm + e, where y stores phenotypic data, μ accounts for the population mean, Z is an incidence matrix and e is the residual term. Vectors a and m distribute under $N(0, G\sigma_g^2)$ and $N(0, M\sigma_m^2)$, respectively, G being the genomic relationship matrix. Following the original development by N. R. Wray for mutational relationships, G can be generalized to accommodate the occurrence of new mutations, i.e. the mutational genomic relationship matrix (M). Previous model was exemplified on simulated data. Simulations relied on a genome composed by 5 chromosomes (1 cM each) with 2,500 SNP and 200 QTL per chromosome. Each replicate evolved during 1000 non-overlapping generations with effective size 100. After that, generations expanded to 200 individuals and analyses were performed on three different scenarios with 3 (Scen3), 5 (Scen5) or 10 (Scen10) generations with genomic (i.e. genotypes from polymorphic SNP) and phenotypic data (h^2=0.5), and a last generation only contributing genomic data. A total of 20 replicates were analyzed for each scenario under the genomic BLUP described above (GB_m) and under a standard genomic BLUP without mutational effects (GB_0). Precision (i.e. correlation coefficient between simulated and predicted breeding values) for individuals with phenotypic data did not show differences between models GB_0 and GB_m; on the contrary, model GB_m increased precision for individuals from the last generation with advantages of 0.15±0.15% (Scen3), 0.42±0.13% (Scen5) and 1.15±0.15% (Scen10). These results evidenced that genomic BLUP can be efficiently adapted to accommodate genetic variability from new mutations, increasing the precision of the predicted breeding values for selection candidates.

Design of the reference population affects the reliability of genomic selection

Pszczola, M.[1,2,3], Strabel, T.[2], Mulder, H.A.[3] and Calus, M.P.L.[3], [1]Wageningen University, Animal Breeding and Genomics Centre, Marijkeweg 40, 6700 AH Wageningen, Netherlands, [2]Poznan University of Life Sciences, Department of Genetics and Animal Breeding, Wolynska 33, 60-637 Poznan, Poland, [3]Wageningen UR Livestock Research, Animal Breeding and Genomics Centre, P.O. Box 65, 8200 AB Lelystad, Netherlands; mbee@jay.up.poznan.pl

The objective of this study was to investigate the effect of design of the reference population in terms of relationship levels within the reference population and relationship level of evaluated animals to the reference population on reliability of direct genomic values (DGV). Data reflecting a dairy cattle population structure was simulated for a trait with heritability of 0.3. Reference populations were small and consisted of cows only, to reflect a trait that is difficult or expensive to measure, e.g. methane emission. Reference populations with different family structures were chosen: highly, moderately, lowly, and randomly related animals. Evaluated animals were chosen from one generation after the reference populations. Reliabilities of DGV were predicted deterministically using selection index theory. Randomly chosen reference populations had the lowest values of average relationship within the reference population. Family structure of the reference population strongly influenced its average relationship and had an impact on reliabilities. Reliabilities increased when the average relationship within the reference population decreased, being 0.44 for highly and 0.53 for randomly related reference populations. Average squared relationship was chosen to measure the relationship to the reference population as it was closely related to calculated reliabilities. Individual reliability strongly increased with the relationship to the reference population. Composing the most suitable reference population is an optimization problem due to a trade-off between obtaining low average relationship within the reference population and high average squared relationship to the animals in the evaluated population.

Prioritised subjects for future funding of research in breeding and reproduction.

Zeilmaker, F.[1], Neeteson, A.-M.[2] and Neuteboom, M.[1], [1]FABRE-TP Secretariat, Dreijenlaan 2, 6703HA Wageningen, Netherlands, [2]FABRE-TP Secretariat (*Aviagen Ltd per May 2011), Dreijenlaan 2, 6703HA Wageningen, Netherlands; effab@effab.info*

The Sustainable Farm Animal Breeding and Reproduction Technology Platform (FABRE-TP) is an EU recognised Technology Platform, and was launched in 2006. It brings together industry, research and other stakeholders, with a view of strategic research needs to enhance business at the short, medium and long term, and proposes EU funding topics to the EC. In 2007 the exercise of developing an Strategic Research Agenda with involvement of over 500 breeding and reproduction experts has been undertaken. This was based on reports of 34 countries and 13 expert groups. Since several big innovations have changed the landscape in breeding and reproduction, new outlines for research needs in the future have been produced by a core group of high-level experts in consultation with a large base of experts in their field from all over Europe. In addition to this, the groups have produced White Papers, showing the most evident research gaps and opportunities to contribute to the global challenges of Climate Change and Food Security. These White Papers will be presented during this session, as well as the priorities in scientific research for the next twenty years.

New Strategic Research Agenda: aquaculture breeding

Komen, J.[1], Storset, A.[2], Sonesson, A.[3], Norris, A.[4], Vandeputte, M.[5], Haffray, P.[6] and Boudry, P.[7], [1]Wageningen University, Animal Breeding and Genomics Group, P.O. Box 338, 6700 AH Wageningen, Netherlands, [2]Aqua Gen, Pb. 1240, Pir-Senteret, N-7462 Trondheim, Norway, [3]Nofima, P.O. Box 6122, NO-9291 Tromsø, Norway, [4]Marine Harvest, P.O. Box 1086 Sentrum, 0104 Oslo, Norway, [5]INRA, UMR1313 GABI, 78350 Jouy-en-Josas, France, [6]SYSAAF, Station SCRIBE/INRA, Campus de Beaulieu, 35042 Rennes, France, [7]Ifremer, 34250, Palavas-les-Flots, France; effab@effab.info

Aquaculture is an important food sector in the EU providing healthy, nutritional products of high quality. It is strategically important for Europe, especially in view of our heavy reliance on imports of seafood. Two of the major global challenges for the future of food production are climate change and food security. Aquaculture production systems and technology will have to address and contribute to these challenges and can play a major role in offering sustainable solutions. Selective breeding and reproduction are important knowledge areas for further development of the industry in a sustainable and ethical manner. There has been little development and investment in aquaculture breeding compared to terrestrial animals, partly because of the practical difficulties in doing so, and partly because of the relative newness of the industry. Without research efforts in these areas the industry will not be able to develop and compete in order to meet market demands and need for fish products in Europe. The process to prioritise important research items for EU funding resulted in a list with 11 most important topics. These are the topics that will ensure sustainability in the European aquaculture breeding sector and develop it further into the future. The aquaculture breeding research and industry community was surveyed for their opinion and their prioritisation on topics grouped into four general areas, and the results will be presented: 1) reducing the environmental impact; 2) improving fish health and welfare; 3) produce a wide range of aquatic products of high quality; 4) generic topics.

New Strategic Research Agenda: cattle breeding

Journaux, L.[1], Wickham, B.[2], Dürr, J.W.[3], Bagnato, A.[4], Swalve, H.[5] and Zeilmaker, F.[6], [1]UNCEIA, 149, Rue de Bercy, FR-75595 Paris, France, [2]ICBF, Highfield House, Shinagh, Bandon, Co. Cork, Ireland, [3]Interbull Centre, P.O. Box 7023, SE-75007 Uppsala, Sweden, [4]University of Milan, Via Celoria 10, IT-20133 Milano, Italy, [5]University of Halle, Adam-Kuckhoff-Str. 35, D-06108 Halle, Germany, [6]FABRE-TP Secretariat, Dreijenlaan 2, Office 1061, 6703HA Wageningen, Netherlands; effab@effab.info

In most European countries, in the past, animal breeding technologies have focused on increased output on a per-animal basis. Recently, breeding programmes have been modified to increasingly include functional and product quality traits. Under European conditions with high costs and high product prices, per-animal production will continue to be of high importance. However, with Europe's cattle production becoming part of a global market, breeding programmes must give attention to cost reduction and larger scale production units. From a consumer and wider society perspective, demands are expected to concentrate on environment friendly production, and an increasing awareness of food safety and quality, including nutritional and organoleptic quality. Cattle breeding programmes have become global and individual programmes find themselves in fierce competition. Major developments have occurred in the field of molecular biology and have led to an increasing awareness of the potential for the application of marker-assisted selection, gene-assisted selection and genomic selection in animal breeding. These technologies provide opportunities to meet the demands of farmers, processors, consumers and society. Challenges lie in the development of a cost-effective application of the technologies for use by farmers and breeding organizations. The role of the European research community will be to lead the way in this process by providing the tools, knowledge, and skills needed by the animal breeding sector to effectively use these new technologies. This will involve basic research, largely connected with other research activities in the field of life sciences, as well as applied research.

New Strategic Research Agenda: horse breeding

Ducro, B.[1], Janssens, S.[2] and Zeilmaker, F.[3], [1]Wageningen University, Marijkeweg 40, 6709PG Wageningen, Netherlands, [2]University of Leuven, Biosystems, Kasteelpark Arenberg 30, B-3001 Heverlee, Belgium, [3]FABRE-TP, Dreijenlaan 2, 6703HA Wageningen, Netherlands; effab@effab.info

Horses have regained popularity and the industry of sports and leisure and 'hobby' farming has been growing whereas the racing industry is still flourishing. Furthermore, horses are used in therapy or tourism, and there are niche markets for horse milk for people having allergies and for meat. Different demands of the public cause the development of a diversity of horses which fit the needs of many groups of. Insufficient standards of education of breeders and owners requests to be urgently improved to meet the same standards as in other farm animals industry. Knowledge and technology transfer between EU countries should be extended and a European network for Equines should be developed. Horse breeding industry and research in the Central and Eastern Europe have to cope with several severe structural problems and need support from EU and other European industrialised countries to reorganise their equine chain, renew education and research in equine science. If no research or development occurs in Europe, the expected impact on industry and EU citizens may be: The loss of international leadership in sport horse breeding and competition, the breeding of horses which conflict with the rising demand of citizen for e.g. hiking or agritourism, a loss of diversity of breeds and the loss of the complete horse breeding industry in some CEE countries. Furtermore reduced animal welfare, limited contribution of horses to sustainable farming and socio-cultural life, and economic losses in breeding industry are expected. For equines opportunities/needs for research are described in the short-medium and long term and include further improvement of performance traits, functional traits, disease resistance, fertility, behaviour, and a reduction of hereditary diseases. Genomics may play an important role but also the logistics of phenotyping is considered crucial.

New Strategic Research Agenda: pig breeding

Knol, E.F.[1], Bidanel, J.-P.[2], Holm, B.[3] and Higuera, M.A.[4], [1]Institute for Pig Genetics (IPG), P.O. Box 43, 6640AA Beuningen, Netherlands, [2]INRA, Domaine de Vilvert, Bat 211, F-78352 Jouy-en-Josas, France, [3]Norsvin, P.O. Box 504, NO-2304 Hamar, Norway, [4]ANPS, c/Goya 115 6-22, E-28009 Madrid, Spain; effab@effab.info

Pork production:License to produce will be critical in the years to come. Concerns on animal welfare, zoönoses, and presence of (pathogenic) micro-organisms and residues in pork are growing. The production side demands more uniformity in each aspect and the need to reduce cost price of pork, especially in terms of feed conversion remains. Pig breeding: Highly relevant is the definition of future production environment in terms of climate, housing, health and feeding. Breeding goals have evolved from production traits towards sustainability traits as survival, longevity, uniformity and ease of handling. Technology offers many opportunities in terms of data -collection, -analysis and -linkage to DNA derived genetic descriptions. Research: The sequence of the genome of the pig will be an invaluable tool for the pig breeding SME's around the world. It will add opportunities but also costs to the breeding program, EU programs need to invest in order to stay competitive. Areas to work in the years to come are 1) society aspects: welfare indicators, ban on castration, genotype-environment interactions 2) exploiting the sequence opportunities in terms of understanding the pig, marker assisted selection, genomic selection 3) trait development: maximizing feed efficiency in terms of gut health and understanding underlying nutrient efficiencies, treating human and animal health as one area of study, increased efforts in liveability, piglet survival, mothering ability and sow longevity. Research needs can be found in a) across species genomics, basic understanding of how genes operate b) across species health, basic understanding of resistance, tolerance and of host-pathogen interactions c) closer cooperation between basic research and application in breeding programs, through more cooperation and better development paragraphs in research proposals.

New Strategic Research Agenda: poultry breeding

Albers, G.[1], Burt, D.[2], Jego, Y.[3] and Kyriazakis, I.[4], [1]Hendrix Genetics, P.O. Box 114, 5830AC Boxmeer, Netherlands, [2]Roslin Institute, Easter Bush, EH259RG Roslin, Midlothian, United Kingdom, [3]Hubbard, P.O. Box 62144, FR-35220 Chateaubourg, France, [4]Newcastle University, Food and Rural Development, Claremont Rd., NE1 7RU Newcastle Upon Tyne, United Kingdom; dave.burt@roslin.ed.ac.uk

The major part of the poultry industry is at a significantly more mature development stage than other livestock Industries, both in terms of management systems and specific knowledge and application in respect of breeding. In part this is the result of basic research having been carried out in Europe over the last 40 years, which also accounts for the significant concentration of the world commercial poultry breeders in Europe. These companies and industries are significant wealth creators and must be supported to maintain their European bases. In addition there would be significant spin-off for minor breeds and other species from fundamental work carried out on chickens. For these reasons the approach and needs for the poultry sector may be viewed differently from those for other livestock species. In addition to controlling animal disease and maintaining welfare, the environmental impact and sustainability of poultry systems will be major issues to consider, with an increasing focus on genetic solutions which must be identified and understood. There is also a need to focus on management and maintenance of genetic variability and genetic diversity within poultry, the use of local resources (strains, raw materials) within the context of food security, and the economic viability (market costs and prices structure acceptability), as well as the environmental impact and resource use associated with poultry systems. The presentation sets priorities for scientific research to develop knowledge and tools for use in poultry breeding programmes.

New Strategic Research Agenda: other farm and companion animals breeding

Bienefeld, K.[1], Nielsen, V.H.[2], Boiti, C.[3], Heuven, H.[4] and Zeilmaker, F.[5], [1]Institute for Bee Research, Friedrich Engels Str. 32, 16540 Hohen Neuendorf, Germany, [2]Aarhus University, Genetics and Biotechnology, Blichers Allé 20, Postboks 50, DK-8830 Tjele, Denmark, [3]University of Perugia, Piazza Università, 1, 06100 Perugia, Italy, [4]Wageningen University, Animal Breeding and Genomics, Marijkeweg 40, 6709PG Wageningen, Netherlands, [5]FABRE-TP Secretariat, Dreijenlaan 2, Office 1061, 6703HA Wageningen, Netherlands; effab@effab.info

The species discussed in this summary (in alphabetical order: dogs, fur animals, honeybees, and rabbits) were chosen for their organized breeding and/or extensive European economic revenues. Europe is internationally leading within fur animal, honeybee, and rabbit research, breeding and production, and plays an important role globally in breeding of dogs for police/army/customs. Despite the leading role for each of the four species, it is a challenge to increase or even maintain the present European production and market share. Increased competition from non-EU countries with much lower production costs (fur animals), importation of cheaper products of lower quality from outside the EU (rabbits), indigenous species being threatened by diseases and intercrossing with imported breeds (honeybees), and the low level of organized working dog breeding in Europe, which is established elsewhere, threatens the competitive advantage Europe has now. Research is required in order to maintain the leading role. This can be divided into two main areas: 1) Study of traits and the techniques required to do so, and 2) Development of efficient breeding schemes, including improved reproductive technologies and management. Health, reproduction, behavior, biodiversity, product quality, and efficiency are the most important categories of traits. Techniques to develop or further specify these traits are needed, as well as genomics tools for understanding animal biology and for selection. For some species techniques to set up organized breeding schemes still need to be developed.

New Strategic Research Agenda: sheep and goats breeding

Astruc, J.-M.[1], Bodin, L.[2], Carta, A.[3], Conington, J.[4], Matos, C.[5] and Zeilmaker, F.[6], [1]Institute de l'Elevage, 149, Rue de Bercy, FR-75595 Paris, France, [2]INRA-SAGA, B.P. 52627, F-31326 Castanet-Tolosan, France, [3]AGRIS Sardegna, Loc. Bonassai - S.S. 291 Sassari-Fertilia Km 18,6, IT-07040 Olmedo, Italy, [4]SAC, Sir Stephen Watson Building, Bush Estate, EH26 0PH Penicuik, United Kingdom, [5]Associação de Criadores de Ovinos do Sul, Rua Cidade S. Paulo, n° 36 - Apart. 296, 7801-904 Beja, Portugal, [6]FABRE-TP Secretariat, Dreijenlaan 2, Office 1061, 6703HA Wageningen, Netherlands; effab@effab.info

Sheep and goats are among the most extensively farmed livestock species in Europe that often utilise marginal areas which are not suitable for other forms of agricultural production. Sheep and goats often provide the main source of income for these rural populations. The worrying decline in sheep numbers in the EU in recent years has had knock-on effects for the supporting infrastructure of the industry such as availability of sheep shearers, livestock hauliers and regional abattoirs. Sheep and goats maintain natural landscapes of high amenity value and benefit from having a 'green' image. However they are generally less efficient than other livestock species in the production of meat and milk because of their small size, low reproductive rate and lower yields and some commercial production systems are reliant on having a premium price. In the absence of research and development in the sheep and goat sectors in Europe and given decreasing subsidies, and if the efficiency is not improved, these industries will quickly lose out to the competition (imports and domestic alternatives) on price and quality. Loss of significant parts of the sheep and goat industries will lead to neglect of traditional landscapes and loss of income for the rural population in less-favoured areas, while biodiversity will suffer as a results of survival of only a few highly selected breeds. Development of specific genomic tools is needed to increase their efficiency and reduce their cost, for utilisation in selection schemes but also for improvement of the breeding organisation.

New Strategic Research Agenda: breeding for diversity and distinctiveness

Gandini, G.[1], Reist, S.[2], Hiemstra, S.J.[3] and Rosati, A.[4], [1]University of Milan, VSA, Via Celoria 10, 20133 Milano, Italy, [2]Bern University of Applied Sciences, Animal Science Department, Länggasse 85, 3052 Zollikofen BE, Switzerland, [3]Wageningen University, P.O. Box 65, 8200AB Lelystad, Italy, [4]EAAP-ICAR, Via Tomassetti 3 A/1, IT-00161 Rome, Italy; effab@effab.info

In Europe the consistent increases in animal production have been realized by intensification of livestock systems towards high input – high output systems. The genetic resources used for these intensive production systems are only a few breeds and lines. Continuously, many local breeds and recently developed breeds and lines are set aside from the primary food production chains. Farm animal genetic resources are sources of genetic variation of fundamental importance to ensure future genetic improvement, to satisfy possible future changes in the markets and in the production environment, and to safeguard against disasters that give an acute loss of genetic resources. In many areas local breeds adapted to harsh conditions are unique sources of income for the rural communities. The link between local breeds and the environment where they were developed makes them key components for the management and conservation the European agro-ecosystems diversity, and important elements of cultural diversity, as they reflect a history of symbiosis of relatively long periods with mankind. Diversity of breeds and their farming systems highly contribute to European food products quality and diversity. Then, farm animal genetic resources are opportunities to maintain a vital countryside. European breeding organisations are strong players in the global competition. It is crucial for Europe that the breeding sector maintains and further develops its leading role. At this respect, sustainable utilisation and conservation of the European breed (and production system) diversity can play an important role.

New Strategic Research Agenda: breeding for food quality and safety

Aumüller, R.[1], Pinard-Van Der Laan, M.-H.[2], Wall, E.[3] and Zeilmaker, F.[4], [1]GLOBALGAP, P.O. Box 190209, 50499 Cologne, Germany, [2]INRA, Genetique Animale et Biologie intégrative, Bat 211. INRA Dpt Genetique Animale, F-78352 Jouy-en-Josas, France, [3]Scottish Agricultural College, Sustainable Livestock Systems, Bush Estate, EH26 0PH Penicuik, Midlothian, United Kingdom, [4]FABRE-TP Secretariat, Dreijenlaan 2, Office 1061, 6703HA Wageningen, Netherlands; effab@effab.info

The expectations of consumers and retailers for safe foodsourced from livestock in Europe are changing very quickly. The animal breeding sector needs to be in a position to adjust selection goals consistently in order to meet these expectations. This is essential to provide new product and market opportunities through adjusted and sophisticated selection programs. We need to better understand and to exploit genetic variation in resistance to infections by current and emergening zoonotic organisms. This is one of several technologies needed to ensure food safety. The safe dissemination of genetic improvement has also an important role to play when it comes to ensure food safety. Research on food safety shall include genomics of zoonotic species to better understand host-pathogen interactions. This is to develop and to provide new and improved vaccination approaches. Without relevant research on product safety and quality which is related to primary production we will not be able within Europe to satisfy the increasing demand of consumers for diverse foods of high safety and quality. These must be produced in a transparent and traceable way. We may risk that animal breeders elsewhere in the world gain in market share. We might also miss out on the opportunity to develop differentiated value-added products from livestock production to help European farmers to compete with low-cost imports.

New Strategic Research Agenda: robustness in animal breeding

Baekbo, P.[1], Bidanel, J.[2], Ligda, C.[3] and Gengler, N.[4], [1]Pig Research Centre, Axeltorv 3, DK-1609 Copenhagen, Denmark, [2]IFIP, La Motte au Vicomte BP 35104, F-35651 Le Rheu Cedex, France, [3]NAGREF, P.O. Box 60 458, 5700 1 Thessaloniki, Greece, [4]University of Liege, Place du Vingt-Août 7, Liege, Belgium; effab@effab.info

Functional traits (health, welfare, robustness) form an increasing part of animal breeding programmes enabled by the growing knowledge in trait development, powerful computing capacity and possibilities of animal genomics to address traits too difficult or expensive to measure. Genetic antagonisms do exist, but genetic improvement in production and robustness and welfare simultaneously is now a matter of adequate selection. For thisgood data to equip selection criteria are a prerequisite:a steady flow of cost-effectively, reliably and repeatably recorded traits on thousands of individually identified pedigreed animals, managed in sophisticated database. Global food security will require efficient balanced systems, climate change also rapid adaptation to health, feed/water availability and land degradation. Proper recording methodology/ implementation need further development: e.g. indicators based on novel physiological traits, general immunity concepts, impact improved health at population level, welfare concepts moving from problem orientation to welfare as balance of the appropriate animal for a certain market/climate/circumstance with appropriate housing, feed, and management. Adapted performance recording systems creating valuable and novel phenotypic data, pooling these resources and linking them to genotypes to create critical mass are needed for genomic research; data exchange /interface development between existing data bases on a herd, regional, national and EU wide level will be important.

New Strategic Research Agenda: reproduction

Feitsma, H.[1], Hååard, M.[2], Dinnyes, A.[3] and Humblot, P.[4], [1]Institute for Pig Genetics, Schoenaker 6, 6641 SZ Beuningen, Netherlands, [2]Viking Genetics, Önsro, P.O. Box 64, SE-53221 Skara, Sweden, [3]Biotalentum, Aulich L. 26, H-2100 Godollo, Hungary, [4]SLU, P.O. Box 7054, SE 75007 Uppsala, Sweden; effab@effab.info

Reproductive techniques (RT) are indispensable for efficient animal breeding and have been used for decades to enable safe and efficient breeding and therewith guarantees for improved efficiency in food production. Research programmes on reproductive techniques must support breeding schemes which enable secure food production and food safety: high quality, healthy, affordable, diverse food products from a sustainable agriculture and aquaculture offering consumers in and beyond Europe real options for improving their quality of life. Research should enhance the competitiveness of European agriculture and aquaculture and its organisations by enabling efficient breeding in animals for sustainable and secure food production. The role of RT to enable fast and efficient implementation of genomic selection in breeding schemes to secure food availability is a major opportunity for the near future. Furthermore RT has a specific role in welfare friendly, healthy, sustainable and bio diverse agriculture and aquaculture in Europe. New opportunities are in the efficient production of transgenic farm animals generated for use in new medical approaches as models, bioreactors or organ/tissue donors. Challenges are to optimise and improve existing RT over more species and to overcome lack of knowledge, bypass anatomical and physiological limitations and develop practical applications. Priorities in RT research have been indicated by a FABRE-TP core group in order to be handed over to EU representatives.

New Strategic Research Agenda: genomics in animal breeding

Archibald, A.[1], Borchersen, S.[2], Perez-Enciso, M.[3], Velander, I.[4] and Vignal, A.[5], [1]Roslin Institute, Roslin Biocentre, EH25 9PS Roslin, Midlothian, United Kingdom, [2]Viking Genetics, Ebeltoftvej 16, Assentoft, DK-8960 Randers, Denmark, [3]Universitat Autonoma de Barcelona (ICREA), Facultat de Veterinaria, Dept. Ciencia Aniamal i dels Aliments, E-08193 Bellaterra Barcelona, Spain, [4]Pig Research Centre (Danavl), Axelborg, Axeltorv 3, DK-1609 Copenhagen, Denmark, [5]INRA Toulouse, B.P. 52627, F-31326 Castanet-Tolosan, France; effab@effab.info

The growing awareness of the finite nature of the planet's resources means that changes are needed in many economic sectors, including animal production systems, to address environmental issues and more generally sustainability. It is necessary not only to understand animals as biological systems, but also the impact of animals on their environment and vice versa. Thus, the scientific strategy for sustainable animal breeding and reproduction needs to address 'animal as systems and animals in systems'. Adapting the 21st century vision of systems biology to the needs of sustainable animal breeding and reproduction is about ensuring that the technologies and information systems customised for the target species and systems are put in place. In the next five years many new genomes will be sequenced. Significant developments in statistical and computing tools, along with improvements in capturing phenotypic data, will be required to realise the potential of genomics to inform improvements in sustainable animal production. Sequence characterisation of the genomes, and variation therein, of economically important species, populations and individuals is critical to 21st century selective animal breeding. As sequencing and genotyping costs continue to fall, effective capture of phenotypic information has become the bottleneck. Understanding the links between the genome and traits of interest is informed by data on the transcriptome, proteome and metabolome. Effective exploitation of the opportunities presented by 'omics technologies requires substantial developments in bioinformatics, statistics and computing.

New Strategic Research Agenda: genetics

Berry, D.P.[1], Boichard, D.[2], Bovenhuis, H.[3], Knap, P.[4] and Nielsen, B.[5], [1]Teagasc, Moore Research Centre, Fermoy Co. Cork, Ireland, [2]INRA, Génétique Animale et Biologie Intégrative, Domaine de Vilvert, Bat 211, F-78352 Jouy en Josas, France, [3]Wageningen University, Animal Breeding and Genomics Centre, Marijkeweg 40, 6709PG Wageningen, Netherlands, [4]PIC, Ratsteich 31, 24837 Schleswig, Germany, [5]DanBred, Axelborg, Axeltorv 3, DK-1609 København V, Denmark; effab@effab.info

Animal breeding aims at exploiting, in a sustainable manner, genetic variation within andbetween breeds to enhance competitiveness and sustainability of EU and global animal food production. Accurate estimation of genetic merit (i.e. breeding values) plays a central role in most improvement programmes. However, genetic evaluations and breeding programmes need to be re-assessed to fully exploit the rapid advancements in 'omic' technologies, especially in recent years. The usefulness of reproductive techniques, also in light of recent developments, necessitates their re-evaluation in the optimal design of both breeding programs and dissemination strategies for germplasm. The key opportunities underlying our action are the world wide recognition of European strength in the field of animal breeding, access to a wide range of genetic resources of high value, as well as access to world-class breeding infrastructure, including reproduction biotechnology centres and performance recording networks. Europe is, therefore, well positioned to rapidly transfer developments in the science of animal breeding to a more profitable and sustainable agriculture and aquaculture.

White paper on breeding for food security and climate change

Zeilmaker, F., Neuteboom, M. and Neeteson, A.M., FABRE-TP Secretariat, Dreijenlaan 2, Office 1060, 6703HA Wageningen, Netherlands; effab@effab.info

This White Paper addresses how animal breeding can contribute to the grand challenges of climate change and food security issues. Livestock production faces major challenges because of impact on, and repercussions from, climate change, shortage of feed, water and energy, and a growing global population with increasing demands for animal products. Demand for food security, combined with limited resource availability, requires increased production efficiency; we need to do more with less, and livestock will have to be adapted to less favourable conditions. Climate change and concerns about food security are likely to affect livestock housing systems through the consequences of the emergence and re-emergence of pathogens; difficulties in controlling housing conditions and the availability of feed and water resources. The latter includes competition between humans and animals for the same limited resources. In this paper it is shown that animal breeding has already contributed significantly on several levels to increase food security and at the same time reduce the impact of livestock production on climate change and the environmental footprint. In the future animal breeding can continue to play an important role in facing these challenges. The White Paper will be published on www.fabretp.info.

Practical strategies to improve the welfare of sows

Spoolder, H.A.M., Hoofs, A.I.J. and Vermeer, H.M., Wageningen UR Livestock Research, Dept. Animal Welfare, P.O. Box 65, 8200 AB, Netherlands; hans.spoolder@wur.nl

Dry sows in Europe will soon spent most of their time in group housing systems, which allow more behavioural freedom compared to stalls. For lactating sows the pressure is also on: conventional farrowing crates provide little space, and at a European level the discussion on alternatives has started. Many farmers fear a reduction in sow performance when space is increased to improve welfare. The advice provided to farmers generally focuses on details in their housing systems, but this strategy is changing as it housing provides a small part of the solution. A telephone survey of 900 Dutch pig farmers showed that pregnancy rate after first service varied widely (between 70 and 96%), but that there was no effect of housing system nor timing of entry to the group. A subsequent study on 70 farms suggested that an animal-directed approach (i.e. attention for the individual pig's needs) and attention for farm management issues, play a more important role regarding welfare and productivity than the type of dry sow system used. Strategies to improve pig welfare in the Netherlands currently focus on two points. Firstly, farmers are encouraged to participate in the design of new housing systems, aimed at providing good welfare and high performance. The Prodromi farrowing pen, currently tested at VIC Sterksel, is an example of this approach to get main stream farmers involved and change attitudes towards welfare. Secondly, farmers, chain partners and the government work together to address animal welfare not through specification of housing parameters, but through animal based performance indicators of welfare and production. This approach, currently pioneered through the Welfare Quality® methodology, addresses the management skills of the pig farmer directly, and places less emphasis on his housing system. The key to improving welfare is the farmer. Encouraging a positive attitude towards his animals, and making use of his skills and knowledge are the best strategies to improve the welfare of sows.

Genetic variation in shoulder ulcers in crossbred sows exists

Hedebro Velander, I., Nielsen, B. and Henryon, M.A., Pig Research Centre, Danish Agriculture and Food Council, Dep. of breeding and genetics, Axeltorv 3, 1609 Copenhagen, Denmark; ive@lf.dk

The aim of this study was to estimate the amount of genetic variance for the prevalence of shoulder ulcers in sows. Shoulder ulcers were observed in 17,015 lactation periods for 8,790 Landrace*Yorkshire (LY or YL) sows at nine pig production herds in Denmark. Shoulder ulcers were measured as the diameter of the ulcer, from 1 cm and larger, to capture all ulcers appearing during lactation. Sows without ulcers were recorded as 0. There were a total of 77,308 observations. Additionally, body conformation, litter size, parity number, herd and stable were recorded together with pedigree of the sows. The total number of maternal sires was 1,800 where 468 of these sires had more than 5 daughters each, representing 5,470 of all sows. Genetic variance for prevalence of shoulder ulcers was estimated by fitting a sire model and an animal model for mean ulcer size and maximum ulcer size per sow. The overall mean prevalence of shoulder ulcers in the experiment was 18.4% per parity, with a significant increasing incidence of ulcers with parity number. The mean prevalence of ulcers was 27.1% per sow. Body condition, herd and year-season at farrowing significantly affected the prevalence of ulcers. The heritabilities for size of ulcers ranged between 0.14 to 0.20 in both the sire model and the animal model. These results showed that genetic variation in shoulder ulcer exists. A selection program to reduce shoulder ulcers should include recordings from crossbred sows in production herds. However, at time of selection the information in the purebred lines is low and thereby the response to selection will be low. Introduction of genomic selection might bee useful in this case.

Practical evaluation of an indoor free farrowing system: the PigSAFE pen

Edwards, S.A.[1], Brett, M.[1], Guy, J.H.[1] and Baxter, E.M.[2], [1]Newcastle University, School of Agriculture, Food & Rural Development, Agriculture Building, Newcastle upon Tyne NE1 7RU, United Kingdom, [2]SAC, Sustainable Livestock Systems, West Main Road, Edinburgh EH9 3JG, United Kingdom; sandra.edwards@ncl.ac.uk

The farrowing crate system, whilst facilitating measures to improve piglet survival, has been shown to compromise the welfare of the sow by restriction of movement and ability to express nest building behaviour. The PigSAFE system (PS) is an alternative pen for farrowing and lactation, based on design criteria synthesised to meet both the biological needs of the animals and practical needs of the farmer. It comprises a nest area with straw and piglet protection features, a heated creep, a slatted dunging area and lockable sow feeder, in a total pen area of 7.7 m^2. To evaluate the commercial performance of this system 72 farrowings have been monitored over a 9 month period, and compared with 66 farrowings in a conventional farrowing crate system (C) taking place concurrently on the same farm, with the same staff and under the same management conditions. Sows (mean parity 4.6 sem 0.2) were allocated randomly to the two systems within each farrowing batch. Results show no significant difference in total litter size (PS=12.9, C=13.4, sem 0.35, P=0.40), stillbirth rate (with litter size covariate: PS=3.7%, C=3.8%, sem 0.77, P=0.95) or mortality of live-born piglets between birth and weaning (with litter size covariate and adjusted for net fostering, which did not differ between treatments: PS=6.4%, C=7.5%, sem 1.00, P=0.42). Piglet weight at weaning (both systems 7.3 kg sem 0.15, at 27 days sem 0.2) and sow weight and backfat loss over the lactation period have also shown no treatment difference. Similar comparative data are currently being gathered at a second site, and are now required from a wider range of commercial conditions to validate the robustness and commercial applicability of this promising alternative to the farrowing crate.

Practical strategies to improve animal welfare in growing pigs
Manteca, X.[1], Temple, D.[2] and Velarde, A.[2], [1]Universitat Autònoma de Barcelona, Animal Science, Edifici V, Campus Universitari, 08193 Bellaterra (Barcelona), Spain, [2]IRTA, Centre de Tecnologia de la Carn, 17121 Monells, Girona, Spain; xavier.manteca@uab.cat

The aim of this paper is to discuss some strategies to improve the welfare of growing pigs following the protocol of the project Welfare Quality®, which is based on four principles: good feeding, good housing, good health and appropriate behaviour. Hunger is not deliberately imposed on growing pigs, but pigs are fasted before slaughter. A prolonged fasting period causes hunger, aggressiveness and sensitivity to cold. Insufficient resting space is one of the main welfare problems related to housing. Pigs adopt a lateral recumbency position when the effective temperature is high and may experience heat stress if prevented from doing so due to lack of space. Good health can be defined as the absence of injuries, disease and pain. Tail-biting is an important cause of injuries in pigs and is a form of redirected behaviour derived from the thwarting of normal exploratory and feeding motivations. Tail biting is a multi-factorial problem involving both internal and environmental risk factors, including genetic background, sex, age, health, diet, feeding management and housing. The principle 'Appropriate behaviour' includes the expression of social and other behaviours and good human-animal relationship. Negative social interactions, such as aggression, impair animal welfare. Disruption of social groups may lead to an increase in aggressive behaviour and a reduction in positive social interactions. Housing conditions that result in increased competition for resources may heighten the number of negative social interactions. This may happen when stocking density is too high or when access to resources is insufficient, for example when feeding space is limited. A poor human-animal relationship results in the animals being fearful of the stockpersons and other humans, and the quality of stockmanship has a profound effect on the animals' welfare and productivity.

Practical strategies to improve pig welfare during transport and at slaughter
Faucitano, L., Agriculture & Agri-Food canada, Dairy and Swine Research and Development Centre, 2000 College Street, J1M1Z3 Sherbrooke, Quebec, Canada; luigi.faucitano@agr.gc.ca

The situation when the pig is in transit is considered a major stressor and may have deleterious effects on health, well-being, performance, and ultimately pork quality. Death losses during transportation in Canada may be low (0.08%), but the total loss amounts to approximately 16,000 pigs per year. In addition, other 80,000 pigs per year arrive at the plant as 'suspect' animals due to fatigue, and may need to be euthanized at the plant. We know that the rate of loss is higher during the summer months, differs with farm of origin, and with transporters. It is generally acknowledged that pot-belly (PB) trucks and some specific compartments within this truck are worse than others in terms of animal losses, but there is no real alternative truck with such large load capacity (230 pigs) and there is not universal agreement which compartments these are and the magnitude of the differences. Results arising from transportation studies run in North America over the last few years showed that the design of the PB truck, characterized by multiple and steep ramps, and 180° turns, increases the use of electrical prodding, influences pig behaviour and welfare parameters (core body temperature, heart rate and blood indicators) and reduces meat quality. Lairage and slaughter are of extreme importance for the pork chain economy, thus at this stage precautions must be then taken to insure an adequate handling and environmental control to safeguard the benefit of lairage as resting area enabling pigs to recover from the stress of transport and limit the effects of the slaughter procedure. The economic losses due animal losses, skin damage and poor meat quality depend, among others, on lairage time, on the quality of the handling and slaughter systems, and on the control of the fighting rate in mixed pigs. The objectives of these review is to show case the results of swine preslaughter research in North America.

Entire male production without mixing of unknown pigs

Rydhmer, L.[1], Eriksson, L.[1], Hansson, M.[1] and Andersson, K.[2], [1]Swedish University of Agricultural Sciences, Dept of Animal Breeding and Genetics, Box 7023, 75007 Uppsala, Sweden, [2]Swedish University of Agricultural Sciences, Dept of Animal Nutrition and Management, Box 7024, 75007 Uppsala, Sweden; Lotta.Rydhmer@slu.se

For a more sustainable pig production, castration as well as mixing of unknown pigs should be avoided and pigs of different sexes should be reared separately. Our hypothesis is that entire males that get to know each other as piglets and are reared in stable groups fight less, show less sexual behaviour and have lower levels of boar taint, compared to entire males mixed at a later age. Behaviour, performance, puberty and boar taint were studied in 96 entire males. Forty-eight males were reared in intact litters until 10 wk age. Then they were mixed with unknown males and moved to the growing-finishing unit. The other 48 males were allowed to meet piglets from another litter at 2 wk age through a small opening in the wall between the farrowing pens. Males from these litters were kept together from weaning to slaughter (i.e. no mixing when entering the growing-finishing unit at 10 wk age). The following measurements were taken: repeated weighings, repeated skin lesion records, repeated behavioural observations of general activity patterns, social (including aggressive) and sexual behaviour. After slaughter, leanness and androstenone and skatole levels in fat were recorded. Preliminary results show that piglets allowed to meet piglets from another litter at 2 wk age grew equally fast as piglets reared in intact litters, from birth to weaning and the week after weaning. On the weaning day, when the sow was removed and the opening between the farrowing pens was closed, a difference in behaviour was observed. The pigs that were used to have contact with another litter were more restless; all pigs in the pen were active (not lying down) 31% of the time as compared to 25% of the time for pigs reared in intact litters (P<0.05). Results from the growing-finishing period and at slaughter will also be presented.

The effect of group size on drinking behaviour in growing pigs

Andersen, H.M.L. and Herskin, M.S., Aarhus University, Dept. of Animal Health and Bioscience, Blichers Allé 20, 8830 Tjele, Denmark; HeidiMai.Andersen@agrsci.dk

The aim was to describe a typical visit at the water nipple with regard to number of visits, water intake, duration and diurnal variation and to investigate the effect of increased competition around the nipple on the drinking behaviour. 72 crossbreed castrated male pigs (initial live weight 20.53±1.69 kg; mean±SD) were used. The pigs were either: 3 or 10 pigs per group (four groups per treatment). All pigs were provided with a transponder ear tag. A RFID reader recorded visits at the water nipple together with at timestamp. The water flow in each pen was measured and logged each second with a timestamp. The drinking behaviour during four following days was analyzed using a linear mixed model. There was no difference between treatments in duration per visit (13.5 sec vs. 12.5 sec) or in water intake per visit (113.3 ml vs. 104.1 ml). However, there was a tendency (P=0.08) to fewer visits per day in the group with 3 pigs per water nipple (42.5±3.7 vs. 51.7±4.2 visit per day). To analyze the diurnal pattern, the day was split up into three 8-hour periods (p1: 6am to 2pm, p2: 2pm to 10pm and p3: 10pm to 6am). In the 3-pig groups, intake per visit was significantly higher in p2 and p3 compared with p1 (p1: 106.6 ml, p2: 121.8 ml and p3: 128.9 ml) whereas in the 10-pig groups, no significant difference was found between p1 and p2 but p3 was significantly higher (p1: 98.5 ml, p2: 100.8 ml and p3: 142.8 ml). The same pattern was seen for the duration per visit. In p1, there was no significant difference in number of visits between the two treatments. In p2, the 3-pig groups had fewer visits than in p1 (16.0 vs. 22.6), while there was no significant difference between p1 and p2 in the 10-pig groups (22.3 vs. 23.9). In p3 (the night), the numbers of visits were lower in both treatments compared with p1 and p2, but there was no significant difference between treatments (3.9 vs. 5.1). The result demonstrates that effects of increasing competition may be limited to specific parts of the day.

Modelling the dynamics of aggressive behaviour in groups of pigs

Doeschl-Wilson, A.B.[1], Turner, S.P.[2] and D'eath, R.B.[2], [1]The Roslin Institute and R(D)SVS, University of Edinburgh, Genetics and Genomics, Easter Bush, EH25 9RG, Midlothian, Scotland, United Kingdom, [2]Scottish Agricultural College, Sustainable Livestock Systems, Easter Bush, EH25 9RG, Midlothian, Scotland, United Kingdom; andrea.wilson@roslin.ed.ac.uk

Aggressive behaviour in pigs is a major welfare issue with detrimental effects on animal health and production. Several influencing factors (e.g. genetic predisposition, recent experience, weight and weight variation, group size and structure) have been identified. However, progress in reducing aggressive behaviour has been hampered by the poor understanding of how these factors interact in the individuals' decision making and how they influence the group dynamics. A stochastic agent-based model was developed to simulate aggressive behaviour in groups of pigs over time. Agents in this model are pigs, which, upon contact with group members, choose one of several possible actions (initiate/retaliate a physical attack/no aggressive behaviour). This choice, as well as the decision to continue or terminate an existing fight, are determined by the individuals' Resource Holding Potential (RHP), which is a combination of characteristics known to affect aggressive behavior. As the individuals' RHPs change over time, aggression profiles in the group change and dominance hierarchies develop. The model predicts individual behaviour and emerging group characteristics over time, depending on group size and structure, and on the definition of the RHP. Comparing these predictions with the same measures derived from a dataset of 24 hour continuous observations from 85 newly-mixed groups of 15 pigs, enabled us to quantify key characteristics describing aggression dynamics and the relative importance of the diverse influencing factors. Group structure, genetic predisposition as well as recent fight history were identified as key determinants of dynamic aggression patterns. The results suggest that aggression in pigs could be reduced through appropriate mixing as well as genetic selection.

Improving pig welfare and production simultaneously by breeding for social genetic effects

Camerlink, I., Bijma, P. and Bolhuis, J.E., Wageningen University, Animal Breeding and Genomics Centre and Adaptation Physiology Group, P.O. Box 338, 6700 AH Wageningen, Netherlands; irene.camerlink@wur.nl

Group housed animals affect each other's welfare, health and productivity through their social behaviour. The social effect of a pig on growth rate of its pen mates is partly genetic and can be described by a Social Breeding Value (SBV). The SBV is an individual's heritable effect on the growth of its pen mates. The mechanisms underlying SBVs in pigs are largely unknown. Here we investigate 1) the effect of social behaviours on growth rate in pigs, and 2) the relationship between social behaviours and SBVs in group housed pigs. On a commercial pig farm, 324 fattening pigs in 45 pens (8 pigs per pen) were observed at 12 w of age using 2-min instantaneous scan sampling for 6 h during daytime. Pens had a barren floor and contained a single space feeder. Estimated SBVs for growth rate were available for each pig and production data were gathered after slaughter. The ethogram distinguished between behaviours given and received. Results were analysed with SAS 9.1. by a Mixed Procedure. Pigs receiving more social nosing had a higher growth rate ($P<0.05$). For instance, pigs that received social nosing more than 1.5% of the time grew on average 30.3 ± 12.7 g/d faster than pigs that did not receive nosing from their pen mates ($P<0.05$). Pigs that received more negative behaviours, such as aggression, oral manipulation or belly nosing, tended to grow less. Pigs with a SBV above group average showed more social nosing ($P<0.05$) than pigs with a SBV below average. Our results show that positive social behaviours can increase growth performances in pigs while negative behaviours tend to decrease growth. Pigs with a high SBV for growth rate showed more positive behaviours, suggesting opportunities for breeding towards improved group productivity and animal welfare simultaneously.

Seeking a sociable swine: an interdisciplinary approach

Benard, M.[1], Camerlink, I.[2], Duijvesteijn, N.[2,3], Reimert, I.[2], Bijma, P.[2] and Bolhuis, J.E.[2], [1]Vrije Universiteit Amsterdam, De boelelaan 1085, 1081HV Amsterdam, Netherlands, [2]Wageningen University, Marijkeweg 40, 6709 PG Wageningen, Netherlands, [3]Insititute of Pig Genetics B.V., Schoenaker 6, 6641 SZ Beuningen, Netherlands; Marianne.benard@falw.vu.nl

In the Netherlands, both the government and agricultural sector aim for a sustainable livestock sector by 2023. Stricter regulations towards animal friendlier production of pig meat might, however, negatively affect other welfare aspects. For instance, a ban on castration or tail docking may lead to increased aggression and increased oral manipulation of group mates, respectively. Acknowledging the relevance of social interactions among animals may circumvent these negative effects. Current pig breeding programs, however, ignore social interactions and are purely based on individual performance. Recently, a new breeding method has been designed that includes the genetic effects that an individual has on the performance on its group mates. Potentially, these social genetic effects are associated with social interactions. In 2009, a large project has started to investigate the prospects of this selection for social effects in pigs. Aim is to study the opportunities to improve pig performance and welfare with this breeding method, in coherence with societal acceptance. Hereto, the project combines quantitative genetics, behavioural, physiological and ethics-technology studies. From the beta-side the (dis)advantages of the breeding method will be explored by genetic estimations for social effects and comprehensive animal experiments including a genotype by environment set-up. From the gamma-side a science-society dialogue will be accomplished, and future directions for welfare improvement will be formulated in collaboration with stakeholders (e.g. producers, food industry, retail, consumers, and animal welfare organisations). This interdisciplinary approach has to reveal whether adding social effects to the genetic model yields sociable and productive pigs which will be accepted by society.

Tail docking in pigs: beyond animal welfare

Stassen, E.N., Wageningen University, Animal Science, Marijkeweg 40, 6706 PG Wageningen, Netherlands; elsbeth.stassen@wur.nl

Tail docking in pigs applied at a young age is common practice in commercial pig farms. In this way tail biting behaviour is prevented. Tail biting is a serious welfare problem. The cause of tail biting outbreaks has been shown to be multifactorial. Alternative housing and management conditions have been studied in order to prevent this harmful social behaviour. Research shows that not only tail biting but also tail docking cause an impairment of pig welfare. In society tail docking and other physical animal interventions are more and more seen as proof that animals kept in intensive production systems are no longer able to adapt to their environment. It is possible to increase this adaptive capacity by genetic selection. Both physical interventions as well as genetic selection applied to address this harmful social behaviour is intuitively problematic. However, this unease cannot be dealt with solely with the help of traditional moral concepts such as animal welfare. For this, legislation concerning animal interventions will be considered in the EU. We discuss whether the concept of animal integrity is helpful to accommodate the intuition that environmental conditions should be adapted to the animal rather than vice versa.

Heritabilities and breed differences in agonistic behaviour and adaptation to an electronic sow feeder in group housed gilts

Appel, A.K.[1,2], Voß, B.[2], Tönepöhl, B.[1], König V. Borstel, U.[1] and Gauly, M.[1], [1]University of Goettingen, Department of Animal Science, Albrecht-Thaer-Weg 3, 37075 Goettingen, Germany, [2]BHZP GmbH, An der Wassermuehle 8, 21368 Dahlenburg, Germany; mgauly@gwdg.de

Due to changes in EU legislation group housing of gilts will be mandatory starting from 2013. However, mixing of unfamiliar pigs frequently leads to agonistic interactions adversely affecting welfare, longevity and productivity. Thus, strategies such as genetic selection for more docile animals are required to reduce aggression levels. A total of 613 Large White gilts (LW) and 374 Pietrain gilts (PI) were included in the dataset. Of these, 85 LW gilts and 67 PI gilts were assessed by one person for skin lesions prior to and post mixing using scores from 1 (no lesions) to 4 (little or no untouched skin). Aggressive behaviour during mixing was scored using two traits, i.e. unilateral aggression (BITE) and bilateral aggression (FIGHT). For every aggressive action (BITE, as well as FIGHT) the performing gilt got a demerit. Furthermore the adaption to a novel feeding technique (electronic sow feeder – ESF) was recorded in terms of meals not retrieved per day (NFED). A mixed model was used to test for breed differences. In addition for line LW variance components were analyzed using a multivariate linear animal model. LW gilts performed significantly more BITE than PI gilts (1.93 ± 2.19 vs. 1.05 ± 1.46) and were involved in more FIGHT actions (0.99 ± 1.29 vs. 0.53 ± 0.78). LW gilts adapted better to the unfamiliar ESF (1.01 ± 1.06 vs. 1.24 ± 1.05). Fewer skin lesions in all body sections (front, middle, back) were found in LW. Estimated heritabilities for behaviour traits in LW were $h^2 = 0.30\pm0.08$ (BITE), $h^2 = 0.17\pm0.07$ (FIGHT) and $h^2 = 0.12\pm0.06$ (NFED). These breed differences and heritabilities for agonistic behaviour indicate that integrating the evaluation of behaviour traits into commercial pig breeding programs could be a promising approach to improve animal welfare in group housed sows.

Behavioural and physiological indicators of welfare in entire male pigs raised in non-mixing versus mixing housing systems

Fàbrega, E.[1], Soler, J.[1], Tibau, J.[1], Puigvert, X.[2] and Dalmau, A.[1], [1]IRTA, Spain, [2]UdG, Spain; emma.fabrega@irta.cat

Entire male pig production may increase in the EU due to regulations on piglet castration. The objective of this study was to evaluate the effect of different entire male housing systems (HS) and slaughter strategies(SS) on physiological and behavioural responses and welfare. Ninety six entire (Large WhitexLandrace)xDuroc pigs were studied. Pigs were housed in 8 pens, 4 pens of non-mixed males (FTF) (only allowed to cross-foster at 10 days of age) and 4 of males randomly mixed twice (when starting transition and growing periods, at 28 and 65 days, MG). Two pens per housing system were slaughtered by split marketing (SM) in three days and the other 2 pens were slaughtered penwise (PW) at an individual or pen mean body weight of 120 kg, respectively. Saliva samples of all males were collected to analyse cortisol at 75, 105 and 145 days of age and skin lesions evaluated according to the Welfare Quality® protocol at 75, 90, 105, 125 and 145. Three extra saliva samples for cortisol and 2 skin evaluations were performed from the pigs of the SM treatment that remained in the pen after each slaughter batch. Behaviour was recorded by direct observation and video recording six times. Scan sampling was used to assess patterns such as activity, inactivity, exploration and feeding behaviour, and focal sampling to evaluate agonistic and sexual behaviour. No significant effect of HS on any behaviour was observed. Salivary cortisol levels did not differ between HS, but an increase over time ($P<0.05$) was observed for SM slaughtered pigs (for both HS), levels after SM being higher from those before. HS did not affect the skin lesions score, although pigs housed in the MG tended ($P<0.1$) to present more lesions at the first evaluation. After both SM batches, the skin lesions score from the remaining pigs was higher ($P<0.05$) for pigs from the MG compared to FTF. The present results suggest that slaughter strategy had a higher impact on welfare indicators than housing system.

Feet lesions of suckling piglets kept in different farrowing systems
Baumgartner, J., Animal Husbandry and Animal Welfare, University of Veterinary Medicine Vienna, Veterinaerplatz 1, 1210 Vienna, Austria; johannes.baumgartner@vetmeduni.ac.at

Keeping sows in farrowing crates causes a number of welfare problems. Apart from that the floor condition in the farrowing system has a huge impact on health and welfare of pigs and as a consequence on the economic outcome of the farm. Both crated systems and free farrowing pens are usually equipped with fully or partly slatted floor of different materials without straw bedding. Suckling piglets are highly susceptible for inadequate flooring. Different commercially available farrowing systems were tested with regard to the prevalence of feet lesions of piglets. Data collection took place in a commercial sow unit of 600 sows (LW x LR). During farrowing and lactation period pigs were kept in different commercially available farrowing systems. Four crate systems (C1-C4) and 3 free farrowing systems (F1-F3) were investigated with main focus on the claws of suckling piglets. Systems differed in dimensions, design, arrangement of the creep area and flooring. 84 litters in total were clinically inspected twice (A: day 5.4±1.3; B: day 18.7±1.8) for claw injuries. Lesions were classified for type, severity, location and quantity. PROC GLM_mult, Tukey-Test and PROC GLM_rep were used for statistical analyses (P=0.05). Almost 90% of all main digits showed heel bruising and 34% dorsal coronet erosion at inspection A, the prevalence decreased to 36% and to 1.5% respectively at inspection B. Hind limbs were more affected than front limbs. The free farrowing pen with a separated non-slatted lying area (F1) resulted in a lower prevalence of feet lesions compared to both the smaller fully slatted free farrowing pens (F2, F3) and to the crated systems (C1-C4). Plastic-coated expanded metal slats with diamond-shaped openings induced less claw lesions than slatted steel with V-shaped bars. It is concluded that both in crated and in free farrowing systems the quality of floor elements has to be improved with special emphasis to health and welfare of suckling piglets.

Effect of gender and individual characteristics on tail injuries in undocked entire male and female weaner pigs
Boyle, L.A.[1] and O'driscoll, K.M.[2], [1]Teagasc, Pig Production Department, Animal & Grassland Research & Innovation Centre, Moorepark, Fermoy, Co. Cork, Ireland, [2]Teagasc, Animal Bioscience Department, Animal & Grassland Research & Innovation Centre, Grange, Dunsany, Co. Meath, Ireland; laura.boyle@teagasc.ie

Routine tail docking is prohibited in the EU. Our aim was to determine the relationship between gender, other individual characteristics, and tail injuries in undocked pigs. At 27 days of age (d) a 'backtest score (BTS)' was determined for 448 piglets. At d28 piglets were weaned, weighed (WT1) and assigned to single sex pens balanced for BTS and WT1 in groups of 14 (entire male n=16; female n=16). At d56, pigs were inspected for tail injuries and weighed (WT2). Four tail regions were scored from 0 (normal) to 5 (severe injury/amputation) and summed to give a tail injury score (TIS). A category BITE was also created where TIS were dichotomised into bitten (TIS>0) or not (TIS=0). The effect of gender on TIS (pen = experimental unit) and BTS, WT1 and WT2 on BITE were analysed using the Mann-Whitney test in SAS (proc NPar1Way). The effect of gender on the odds of being bitten was analysed by logistic regression (proc Genmod, SAS), using General Estimating Equations to account for the pen effect. Of the 429 pigs inspected, 29.6% were bitten with more females (34.1%) affected than males (25.1%). However, taking pen into account, females were no more likely to be bitten than males (P>0.10). Furthermore, there was no effect of gender on TIS (Male: 0.30 [0-8.93] vs. Female: 0.48 [0-8.76] median score [min.-max.]; P=0.374). Finally, BTS, WT1 (7.6 kg±1.38 SD) and WT2 (17.1 kg±3.06 SD) did not differ between pigs that were bitten and those that were not (P>0.10). Neither gender nor the individual physical and behavioural characteristics measured influenced tail biting in weaner pigs.

Checking SNP and pedigree information of sibs for Mendelian inconsistencies

*Calus, M.P.L.[1], Mulder, H.A.[1], Mcparland, S.[2], Strandberg, E.[3], Wall, E.[4] and Bastiaansen, J.W.M.[5],
[1]Wageningen UR Livestock Research, Animal Breeding and Genomics Centre, P.O. Box 65, 8200 AB
Lelystad, Netherlands, [2]Teagasc, Animal & Grassland Research and Innovation Centre, Moorepark, Co.
Cork, Ireland, [3]Swedish Univ. of Agricultural Sciences, Dept. of Animal Breeding and Genetics, P.O. Box
7023, S-750 07 Uppsala, Sweden, [4]Scottish Agricultural College, Sustainable Livestock Systems Group,
Bush Estate, EH26 0PH Penicuik, Midlothian, United Kingdom, [5]Wageningen University, Animal Breeding
and Genomics Centre, P.O. Box 338, 6700 AH Wageningen, Netherlands; mario.calus@wur.nl*

Genomic selection using many SNP genotypes is becoming common practice in breeding programs.
Checking for Mendelian inconsistencies allows to identify animals whose pedigree and SNP data do
not agree. Straightforward tests compare genotypes of individual SNP genotypes between parent and
offspring, where contrasting homozygotes indicate an inconsistency. We developed two tests to identify
Mendelian inconsistencies between sib pairs, counting SNPs with contrasting homozygotes (SIBCOUNT),
or comparing pedigree and SNP based relationships (SIBREL). The algorithms were tested on a data set
of 2,078 genotyped cows and 211 genotyped sires. Theoretical expectations for distributions of the test
statistics were calculated and compared to empirically derived values. After removing 223 animals due to
parent-offspring inconsistencies, SIBCOUNT (SIBREL) identified 31 (34) additional inconsistent animals.
29 animals were identified to be inconsistent by both methods. Numbers of incorrectly deleted animals (Type
I error) were equally low for both methods. The numbers of incorrectly not deleted animals (Type II error),
was nearly twice as high for SIBREL compared to SIBCOUNT. It is concluded that counting inconsistent
SNP between a pair of sibs is slightly more precise than comparing genomic and pedigree relationships to
detect Mendelian inconsistencies between sibs.

Influence of marker editing criteria on accuracy of genomic prediction

*Edriss, V., Guldbrandtsen, B., Lund, M.S. and Su, G., Department of Genetics and Biotechnology, Faculty
of Science and Technology, Aarhus University, DK-8830 Tjele, Denmark; vahid.edriss@agrsci.dk*

We investigated the effect of marker editing criteria on the accuracy of genomic evaluations. It was found
that optimal accuracy was obtained using liberal thresholds for MAF and GC scores. The data comprised
4,429 Nordic Holstein bulls and 1,071 Danish Jersey bulls, genotyped using Illumina 50K BeadChip. The
raw SNP data (after cleaning by the laboratory) included 48,222 and 44,305 polymorphic SNP for Holstein
and Jersey, respectively. The SNP data were edited using different criteria with regard to minor allele
frequency (MAF) and GenCall (GC) scores. Approximately 20% youngest bulls were left out as validation
population and the rest were kept as reference population. Genomic breeding values for protein yield, fertility
and mastitis index were predicted using various edited SNP datasets. A Bayesian model was applied. It
captures the features of BayesA but simplifies the computing algorithm. The pseudo observations used in
the prediction were conventional EBV for Jersey and de-regressed proofs for Holstein. The results showed
that removing markers with low MAF decreased the accuracy of genomic prediction in Jersey. But not
clear trend was found in Holstein, for which the highest accuracy was obtained by including markers with
MAF above 0.01. By using more restrictive thresholds on GC scores, the accuracy of genomic prediction
decreased for all traits in both breeds.

Effect of markers density on the accuracy of GEBV

Aminafshar, M.[1], Zargarian, B.[2], Saatchi, M.[3] and Noshari, A.[4], [1]Department of Animal Science, Faculty of Agriculture and Natural Resources, Science and Research Branch, Islamic Azad university, Tehran, Iran, [2]Islamic Azad University, Faculty of Agriculture & Natural Resources, Varamin Branch, Varamin, Iran, [3]Tehran University, Department of Animal Science, Karaj, Iran, [4]Islamic Azad University, Department of Animal Science, Karaj Branch, Karaj, Iran; aminafshar@srbiau.ac.ir

It is currently possible to genotype individuals for 50,000 or more SNP with micro array DNA chip. These markers may be used to predict genomic breeding values (GEBV). It decreases the costs of animal breeding program up to 90 percent. In two designed trials, the genome consists of 3 chromosomes, each 100 cM with different marker density (100, 200, 1000 and 2,000 markers with 1, 0.5, 0.1 and 0.05 cM space) and 30 random distributed QTL were simulated. After 50 generations of random mating in a finite population (Ne=100), the population size were expanded (500 male and 500 female) to create LD. Two level of h^2 (0.1 & 0.5) were considered. Individuals of generation 51 and 52 had phenotype records and marker effects were estimated by them. Accuracy of GEBV was calculated for individuals of generations 53 to 58. Results of first trial showed, the accuracy of GEBV was increased by increasing the marker density for high heritability traits. In second trial only half of individual had a phenotype records. Results showed, the accuracy of GEBV was decreased by decreasing the number of phenotype records. In third trial, 1 chromosome with more dense markers was used (4,000 and 8,000 markers with 0.025 cM and 0.0125 cM respectively). Results showed that the accuracy of GEBV was increased when using more dense markers. In all trials, accuracy of GEBV for high heritability traits was higher than low heritability traits (with same marker density). Results of this study showed that accuracy of GEBV will decrease by passing generations due to the recombination and bias in marks effect estimation.

Reliability of direct genomic values based on imputed 50k genotypes using low density chips in Dutch Holstein cattle

Mulder, H.A.[1], Calus, M.P.L.[1], Druet, T.[2] and Schrooten, C.[3], [1]Animal Breeding and Genomics Centre, Wageningen UR Livestock Research, P.O. Box 65, 8200 AB Lelystad, Netherlands, [2]Unit of Animal Genomics, Faculty of Veterinary Medicine and Centre for Biomedical Integrative Genoproteinomics, University of Liège, B-4000, Liège, Belgium, [3]CRV, P.O. Box 454, 6800 AL Arnhem, Netherlands; herman.mulder@wur.nl

Genomic selection using 50k SNP chips has been implemented in many dairy cattle breeding programs. Cheap low density chips make genotyping of a larger number of animals cost effective. A commonly proposed strategy is to impute low density genotypes up to 50k genotypes before predicting direct genomic values (DGV). The objective was therefore to investigate the effect of using imputed 50k SNP genotypes based on a small chip on reliability of DGV. Small chips contained 384, 3,000 or 6,000 SNP. DAGPHASE, CHROMIBD and multivariate BLUP were used for imputation. Genotypes of 9378 animals were used, from which ~2350 animals had deregressed proofs. Bayesian stochastic search variable selection was used for estimating SNP effects of the 50k chip. Reliabilities of DGV were similar for DAGPHASE and CHROMIBD, but lower for multivariate BLUP. With 3,000 SNP and using CHROMIBD or DAGPHASE, 85% of the increase in DGV reliability using the 50k chip, compared to a pedigree index, was obtained. With multivariate BLUP the increase in reliability was only 40%. With 384 SNP the reliability of DGV was lower than for a pedigree index, while with 6,000 SNP about 93% of the increase in reliability of DGV based on the 50k chip was obtained when using DAGPHASE for imputation. Using genotype probabilities instead of the most likely genotype increased the reliability of DGV. The relationship between the reliability of imputation and the reliability of DGV was approximately linear across methods and SNP chips. In conclusion, DGV based on imputed SNP using a small chip with at least 3,000 SNP yields at least 85% of the increase in reliability as obtained with a 50k chip.

Impact of imputing markers from a low density chip on the reliability of genomic breeding values in Holstein populations

Dassonneville, R.[1,2], Brøndum, R.F.[3], Druet, T.[4], Fritz, S.[5], Guillaume, F.[1,2], Guldbrandtsen, B.[3], Lund, M.S.[3], Ducrocq, V.[1] and Su, G.[3], [1]INRA, GABI, 78350 Jouy-en-Josas, France, [2]Institut de l'Elevage, 149 rue de Bercy, 75595 Paris, France, [3]Aarhus University, Department of Genetics and Biotechnology, D-8830 Tjele, Denmark, [4]University of Liège, Faculty of Veterinary Medicine, Unit of Animal Genomics, B-4000 Liège, Belgium, [5]UNCEIA, 149 rue de Bercy, 75595 Paris, France; romain.dassonneville@jouy.inra.fr

The purpose of this study was to investigate the imputation error and loss of reliability of direct genomic values (DGV) or genomically enhanced breeding values (GEBV) when using genotypes imputed from a 3K single nucleotide polymorphism (SNP) panel to a 50K SNP panel. Data consisted of genotypes of 15,966 European Holstein bulls from the combined EuroGenomics reference population. Genotypes with the low density chip were created by erasing markers from 50K data. The studies were performed in the Nordic countries using a BLUP model for prediction of DGV and in France using a genomic marker assisted selection approach for prediction of GEBV. Imputation in both studies was done using a combination of the DAGPHASE 1.1 and Beagle 2.1.3 software. Traits considered were protein yield, fertility, somatic cell count and udder depth. Imputation of missing markers as well as prediction of breeding values were performed using two different reference populations in each country; either a national reference population or a combined EuroGenomics reference population. Mean imputation error rates when using national reference animals was 5.5% and 3.9% in the Nordic countries and France respectively, and 4.0% and 2.1% respectively based on the EuroGenomics reference dataset. The reliability of DGV using the imputed SNP data was 0.38 based on national reference data, and 0.48 based on EuroGenomic reference data in the Nordic validation, and the reliability of GEBV using the imputed SNP data was 0.41 based on national reference data, and 0.44 based on EuroGenomic reference data in the French validation.

Association of four SNPs with carcass and meat quality traits in Swedish young bulls of the Charolais breed

Ekerljung, M.[1], Li, X.[2], Lundström, K.[2], Lundén, A.[1], Marklund, S.[1] and Näsholm, A.[1], [1]Swedish University of Agricultural Sciences, Dept of Animal Breeding and Genetics, P.O. Box 7023, SE-750 07 Uppsala, Sweden, [2]Swedish University of Agricultural Sciences, Department of Food Science, P.O. Box 7023, SE-750 07, Uppsala, Sweden; Marie.Ekerljung@slu.se

We tested the contribution of four SNPs to the variation in carcass and meat quality in purebred Charolais bulls. Data consisted of 106 young bulls that were raised in Swedish commercial beef herds. Classification of carcass conformation was performed according to the EU system EUROP on a subjective scale with 15 levels. Similarly, fat class was subjectively assessed on a scale with 15 levels. Muscle samples were collected from M. longissimus dorsi. On day 7 post mortem, a 2 cm slice was cut out and photographed on both sides, and assessment of the intramuscular fat content based on image analysis (ratio of white to red pixels) was performed. Meat tenderness was measured as Warner Bratzler shear force. We genotyped the animals for acyl CoA:diacylglycerol acyltransferase 1 (DGAT1) K232A, stearoyl-CoA desaturase 1 (SCD1) A293V, calpain (CAPN1) C316G, and calpastatin (CAST) P52L. The DGAT1 polymorphism was associated with variation in intramuscular fat ($P<0.01$), which was in accordance with our previous subjective marbling grading performed on the same images, but showed no effect on fat class. Somewhat surprising, the CAPN1 gene was associated with fat class ($P<0.01$). The polymorphism in calpastatin was associated with carcass conformation ($P<0.05$). This may be an effect of allelic variation in calpastatin influencing the calpain's proteolytic activity on muscle fibers, where the CAST 155T allele was associated to improved EUROP conformation. Neither of the analysed genes showed any effect on shear force. The SCD1 polymorphism was not associated with variation in any of the analysed traits.

Estimating myostatin gene effect on milk performance traits using estimated gene content for a large number of non-genotyped cows

Buske, B.[1], Szydlowski, M.[2], Verkenne, C.[1] and Gengler, N.[1], [1]University of Liège, Animal Science Unit, Passage des Déportés, 2, 5030 Gembloux, Belgium, [2]Poznan University of Life Science, Department of Genetics and Animal Breeding, Wolynska 33, 60-637 Poznan, Poland; bernd.buske@ulg.ac.be

The objective of this study was to estimate the myostatin (mh) gene's effect on milk, protein and fat yield in a large heterogeneous cow population, of which only a small portion was genotyped. For this purpose, a total of 13,992,889 test-day records derived from 799,778 cows were available. The mh gene effect was estimated via BLUP using a multi-lactation, multi-trait random regression test-day model with an additional fixed regression on mh gene content. Because only 1,416 animals were genotyped, more animals of additional breeds with assumed known genotype were added to estimate the genotype (gene content) of the remaining cows more reliably. This was carried out using the conventional pedigree information between genotyped animals and their non-genotyped relatives. Applying this rule, mean estimated gene content over all cows with test-day records was 0.104, showing that most cows were homozygous +/+. In contrast, when gene content estimation was only based on genotyped animals, mean estimated gene content over all cows with test-day records was with 1.349 overestimated. Therefore, the applied method for gene content estimation in large populations needs additional genotype assumptions about additional animals representing genetic diversity when the breed composition in the complete population is heterogeneous and only a few animals from predominantly one breed are genotyped. Concerning allele substitution effects for one copy of the 'mh' gene variant, significant decreases of -76.1 kg milk, -3.6 kg fat and -2.8 kg protein per lactation were obtained on average when gene content estimation was additionally based on animals with assumed known genotype. Based on this result, knowledge of the mh genotypes and their effects has the potential to improve milk performance traits in cattle.

Analysis of subpopulation structure in Danish Jersey

Thomasen, J.R.[1,2], Sørensen, A.C.[2], Brøndum, R.F.[2], Lund, M.S.[2] and Guldbrandtsen, B.[2], [1]VikingGenetics, Ebeltoftvej 16, 8960 Randers SØ, Denmark, [2]Aarhus University, Department of Genetics and Biotechnology, P.O. Box 50, 8830 Tjele, Denmark; jrt@vikinggenetics.com

We investigate the genetic structure of a population with a heterogeneous background on parameters predicting the performance of genomic predictions. Reliabilities of genomic predictions depend on persistent associations between the marker and the QTL in the animals included in the reference population. Thus low persistence of phase between subgroups of bulls in the reference population might reduce reliabilities. The Danish Jersey (DJ) cattle population was investigated. DJ is now a mixture of 63% century-old Danish Jersey and 36% imported US Jersey. In this study 1,730 Jersey animals born in the period from 1981 to 2010 were analyzed with the Illumina 50k chip. The Structure software was used to analyze the population structure. 412 evenly spaced SNP-markers on all 29 autosomes were used to group the animals into clusters. Assuming two sub populations (DNK and US) the correlation between the probability of membership of an inferred cluster and proportion of the breed in the pedigree was 0.85 for DNK and 0.86 for US. Based on the pedigree, animals were grouped into two subgroups: 221 DNK animals with at least 75% DNK descent according to the pedigree and 171 US animals with at least 75% US descent in pedigree. The levels of LD calculated as r^2 was 0.02 higher within the two subgroups compared to the admixed group including all animals for marker spacing of 100 kb. Persistence of phase also showed divergence between the DNK and the US group with a value of 0.78 for the marker spacing interval 100-200 kb. The results confirm that the Danish Jersey population consists of two subpopulations with distinct background. A genomic model using the subpopulation structure in Danish Jersey will be validated.

Use of phenotypes from multi-trait national evaluations in genomic evaluation

Hoze, C.[1], Croiseau, P.[2], Guillaume, F.[3], Baur, A.[1], Journaux, L.[1], Boichard, D.[2], Ducrocq, V.[2] and Fritz, S.[1], [1]UNCEIA, 149 rue de Bercy, 75012 Paris, France, [2]INRA, UMR 1313 GABI, Domaine de Vilvert, 78350 Jouy-en-Josas, France, [3]Institut de l'Elevage, 149 rue de Bercy, 75012 Paris, France; chris.hoze@jouy.inra.fr

Multi-trait national evaluations using correlated traits as predictors were developed to increase the reliability of estimated breeding values (EBV) for functional traits. In genomic selection (GS), only single-trait evaluations are used so far. This study focuses on the use in GS and MA-BLUP evaluations of 'combined' phenotypes obtained from multi-trait national evaluations. Different approaches were tested in the Holstein and Montbéliarde breeds. Data consisted in 16,632 Holstein and 2,111 Montbéliard bulls genotyped on the Illumina commercial chip BovineSNP50®. Phenotypes used were daughter yield deviations (DYD) for somatic cell count and cow fertility and deregressed proof (DP) for longevity. QTL detections based on a combination of Linkage Disequilibrium and Linkage analysis and Elastic Net were performed for both direct and combined phenotypes. Then, a MA-BLUP model was tested using either direct or combined traits as phenotypes. Efficiency was measured through a validation study where the youngest bulls formed the validation population. Mean correlation between DYD or DP and EBV increased from 0.51 to 0.56 and from 0.41 to 0.50 respectively, for the Holstein and Montbéliarde breeds when phenotypes used for QTL detection were coming from the multi-trait evaluation instead of the single-trait one. Use of combined phenotypes in MA-BLUP and QTL detection allowed a gain in correlation of 0.08 in Montbéliarde but a loss of 0.06 in Holstein compared with a situation where phenotypes came from single-trait evaluation. Using more accurate phenotypes coming from multi-trait evaluations appears to increase the efficiency of GS, particularly for breeds with smaller reference population. This strategy using phenotypes from multi-trait evaluations for QTL detection is now implemented in national French genomic evaluations.

Genomic prediction in Nordic Holstein population using one-step approach

Gao, H.[1], Christensen, O.[1], Madsen, P.[1], Nielsen, U.[2], Lund, M.[1] and Su, G.[1], [1]Aarhus University,Faculty of Agricultural Sciences, Department of Genetics and Biotechnology, Blichers Allé 20, 8830 Tjele, Denmark, [2]Danish Agricultural Advisory Service, Agro Food Park 15, Skejby, 8200 Aarhus N, Denmark; Hongding.Gao@agrsci.dk

Genomic prediction has been widely implemented in dairy cattle genetic evaluation. Direct genomic breeding values are usually predicted using a reference population comprising genotyped progeny-tested bulls. However, in most cases a large proportion of bulls are not genotyped. A one-step approach allows genomic prediction using information of genotyped and non-genotyped animals simultaneously. The objective of this study was to compare a one-step model with a GBLUP model for genomic prediction of 16 complex traits in the Nordic Holstein population. The data consisted of 5,214 genotyped bulls and 9,374 non-genotyped bulls. Data were divided into reference data and validation data by birth date Oct. 1st, 2001. Thus, the validation data contained the youngest genotyped bulls, about 25%. De-regressed proof (DRP) was used as response variable for genomic prediction applying a GBLUP model and a one-step model. Using the one-step model, a set of relative weights (ranging from 0.05-0.40) were used to build the combined relationship matrix. Reliability was measured as squared correlation between genomic prediction and DRP for the validation population, and then adjusted for reliability of DRP. The results showed that when averaging over the 16 traits, reliability of genomic breeding values predicted using the one-step model was higher than those predicted using the GBLUP model by 1.9%. The relative weight used in the one-step model has a negligible effect on reliability of genomic prediction, but a significant influence on variation of genomic estimated breeding values. The results suggest that the one-step model is an appealing approach for practical genomic prediction.

Genomic predictions based on a joint reference population for Scandinavian red breeds

Heringstad, B.[1,2], Su, G.[3], Solberg, T.R.[2], Guldbrandtsen, B.[3], Svendsen, M.[2] and Lund, M.S.[3], [1]Department of Animal and Aquacultural Sciences, Norwegian University of Life Sciences, P.O. Box 5003, N-1432 Ås, Norway, [2]Geno Breeding and A.I. Association, 1432, Ås, Norway, [3]Department of Genetics and Biotechnology, Aarhus University, Tjele, DK-8830, Denmark; bjorg.heringstad@umb.no

In January 2011 the breeding organizations Geno (Norway) and Viking Genetics (Denmark, Finland, and Sweden (DFS)) agreed to create a joint reference population for Scandinavian red breeds. The joint reference population consists of more than 7,000 progeny tested bulls, 2,843 Norwegian Red, and 4,422 bulls from DFS (Swedish Red, Finnish Ayrshire and Danish Red). The aim of this study was to examine whether reliability of genomic predictions improved when using a joint reference population. Cross validation studies were carried out for Norwegian Red and DFS, respectively, comparing joint and separate reference populations. Deregressed proofs (DRP) derived from Interbull EBVs were used as response variables. Genomic breeding values were predicted using GBLUP. After editing, 46,512 SNP markers were used in the genomic analyses. Bulls born after 2001 were used as the test data (1,018 bulls for DFS and 282 for Norway). Reliability of genomic predictions was assessed as the squared correlation between genomic predictions and DRP divided by reliability of DRP for bulls in the test datasets. For DFS the average increase in reliability, when using the joint reference data was 0.018, ranging from -0.007 to 0.032. Similar results were found for Norwegian Red. In general, the largest increases in reliabilities was found for production traits, while for some of the fertility traits reliability were slightly decreased when using the joint reference data. Although the joint reference data was 60% larger than the DFS reference data and more than double the size of the Norwegian reference data, the relative gain in reliabilities were small. This might be explained by the relationships between the breeds.

Genomic Selection in dairy cattle based on genotyping subsets of bull calves: effects of pre-selection

Wensch-Dorendorf, M.[1], Yin, T.[2], König, S.[3] and Swalve, H.H.[1], [1]University of Halle, Institute of Agricultural and Nutritional Sciences, Theodor-Lieser-Str. 11, 06120 Halle, Germany, [2]University of Göttingen, Institute of Animal Breeding and Genetics, Albrecht-Thaer-Weg 3, 37075 Göttingen, Germany, [3]University of Kassel, Department of Animal Breeding, Nordbahnhofstraße 1a, 37213 Witzenhausen, Germany; monika.dorendorf@landw.uni-halle.de

Currently, genotyping for an entire population cannot be assumed. Thus, pre-selection strategies have to be developed. Here, we compare pre-selection strategies to reduce the pool of young males to be genotyped. The strategies are based either on information from bull dams (conventional BLUP EBV, GEBV, phenotype, average genetic herd level) or on random selection. Stochastic simulation was used to compare strategies according to the means of true breeding value (TBV) of selected candidates. Furthermore, the resulting young male selection candidates are analyzed with regard to inbreeding coefficients and kinship among each other. The comparison is done between pre-selection strategies, the whole population scan strategy and the conventional scheme. The number of bull calves (bull dams, respectively) is varied by increments of 1000 from 1000 to 50,000 bull calves. Simulation is done for a population size conceivable for a breeding organization with 100,000 cows. The strategies are evaluated for two heritability levels (0.10 and 0.30) and six levels of correlation between TBVs and GEBVs (r_{mg}=0.5,...,1.0). The GEBVs are generated by simple stochastic simulation using TBV. The success rate of embryo transfer is set to 1 male offspring per cow. The results indicate that for r_{mg}=0.7 or higher nearly all genomic strategies are much more successful than the conventional scheme. Several strategies produce mean TBVs comparable or even higher to the whole population scan strategy. Because of the reduced number of young males to be genotyped these strategies are strong candidates as alternatives to a whole population scan strategy.

Implications of genomic selection on production and functional traits in Fleckvieh
Neuner, S. and Götz, K.-U., Bavarian State Research Centre for Agriculture, Institute of Animal Breeding, Prof.-Dürrwaechter-Platz 1, 85586 Poing-Grub, Germany; Stefan.Neuner@lfl.bayern.de

In the recent past a variety of opportunities for adapting and designing breeding programs using genomic selection has been described. Many calculations considered just single traits, and were restricted on theoretical and optimistic assumptions about the design of the breeding program. Starting with a detailed analysis of the current breeding program several scenarios using genomic selection were examined for the dual purpose breed Fleckvieh using the deterministic simulation program ZPLAN+ in this study. The current population was modeled in its present constitution, applying realized selection intensities, applied genetic and biological parameters and current economic weights. Main characteristics of ZPLAN+ are the selection index methodology and the gene-flow matrix. All 14 traits in the total merit index for Fleckvieh were included and evaluated at the same time. The main focus was on the consequences for functional traits, because especially for lowly heritable traits strong expectations were stimulated in the beginning of the genomic selection era. Applying genomic selection just for the preselection of testbulls increases the overall genetic gain only marginally. Higher benefits could be observed if young bulls were used as sires of cows and bulls on a large scale. At first sight genetic gain per year for functional traits drops considerably for very stringent genomic breeding programs compared to the conventional scheme. However, if the accumulated genetic gains for a given genetic gain in protein yield were calculated for different selection schemes, genetic progress in functional traits was hardly affected. In consequence, the same changes in the genetic level simply will happen sooner using genomic selection. In order to make optimal use of the new technology, a total merit index can be formulated that ensures benefits for all traits compared to the status quo.

Effect of sire breed and genetic merit for carcass weight on the transcriptional regulation of the somatotropic axis in *M. longissimus dorsi* of crossbred steers
Keady, S.M., Kenny, D.A., Keane, M.G. and Waters, S.M., Teagasc, Grange, Animal and Bioscience Research Department, Dunsany, Co. Meath, Ireland; sarah.keady@teagasc.ie

The somatotropic axis plays an important role in postnatal growth and development of skeletal muscle. The aim of this study was to examine the effect of sire breed and sire expected progeny difference for carcass weight (EPD_{cwt}), on the expression of components of the somatotropic axis in M. longissimus dorsi of cattle at slaughter. Cross-bred Aberdeen Angus (AA; n=17) and Belgian Blue (BB; n=16) steers born to Holstein-Friesian dams and sired by bulls with either high (H) or low (L) EPD_{cwt} were employed in the study. Thus, there were four genetic groups viz BBH (n=8), BBL (n=8), AAH (n=8), and AAL (n=9). Blood samples were collected via jugular venipuncture at regular intervals for analysis of circulating concentrations of insulin-like growth factor-1 (IGF-1) and insulin. Total RNA was isolated from M. longissimus dorsi collected at slaughter and the mRNA expression of IGF-1, IGF-2, their receptors (IGF-1R; IGF-2R), six IGF binding proteins (IGFBPs), acid labile subunit (ALS) and growth hormone receptor (GHR) was measured by RT-qPCR. Data were analyzed using mixed models ANOVA (PROC MIXED). Gene expression of IGF-1R and IGFBP3 was upregulated in AA (P<0.001) compared to BB while IGF-1 was upregulated in H compared to L animals (P<0.01). Correlation analysis indicated moderate positive associations between gene expression of IGFBP3 and IGF-1 (P<0.001) and IGF-1R (P<0.01). In addition, mRNA expression of IGFBP3 was moderately negatively associated with M. longissimus dorsi area per kg carcass weight (P<0.05). There was no effect of either sire breed or EPD_{cwt} on concentrations of circulating IGF-1 or insulin (P>0.05). Higher transcript levels of IGFBP3 in muscle may play a role in reduced muscular growth during the finishing period. These data will contribute to a better understanding of the molecular control of muscle growth at a local tissue level in cattle.

Lactation stage dependent genome wide effects on breeding values in Holstein Friesians
Strucken, E.M. and Brockmann, G.A., Humboldt-Universität zu Berlin, Breeding Biology and Molecular Genetics, Faculty of Agricullture and Horticulture, Invalidenstrasse 42, 10115 Berlin, Germany; eva_maria_strucken@web.de

Phenotypic time dependency in milk production traits in dairy cattle has been known and analyzed over the last century. A genome wide study was conducted to investigate differences and changes in allele effects during the early lactation based on estimated breeding values (EBV). EBVs for milk yield, fat and protein yield, and fat and protein content of 2,408 German Holstein Friesian bulls were assessed in 10-day intervals for the first 60 lactation days (VIT Verden, Germany). EBVs were available for the first three lactations. The genotypic information was taken from the bovine 50k BeadChip (Illumina). 2,338 animals and 43,588 markers passed quality control (MAF <0.01, call rate <0.9, FDR <0.01, IBS<=0.95). The Bonferroni correction was applied to account for multiple testing and the λ inflation factor corrected for population stratification. Most markers were identified for fat content followed by fat yield and protein content. No significant markers were detected for protein yield. Almost all markers occurred on chromosome 14 around the DGAT1 gene. Additionally, single markers were found on chromosome 27 for fat content and chromosome 6 for protein content. Allele effects and P-values increased with later lactation days. One exception for the allele effects was fat yield where only minor differences but with a tendency to smaller effects were found for later lactation days. The most dramatic increase in allele effects and P-values was observed for protein content. Considering trend lines, a change in the slope of 10-18 units for effects on protein content was found, whereas effects on milk yield and fat content only changed for about 8-11 and 3-5 units, respectively. This study provided evidence that gene effects get more pronounced with progressing lactation, and loci with bigger effects at the beginning also tend to stronger affect the production in later lactation.

How to remove bias in genomic predictions?
Vitezica, Z.G.[1] and Legarra, A.[2], [1]Université de Toulouse, UMR 1289 TANDEM, INRA/INP-ENSAT/ ENVT, F-31326 Castanet Tolosan, France, [2]INRA, UR 631 SAGA, F-31326 Castanet Tolosan, France; zulma.vitezica@ensat.fr

Unbiased predictions are of paramount importance in selection for accurate estimates of the genetic trend and also comparison of animals across generations. A single-step method, that combines all available data jointly (pedigree and genomic information), has been recently developed. This method is based on a pedigree relationship matrix augmented with genomic information. Since the model includes the data used for selection there are in principle no problems with biases from selection. However, bias was reported in genomic prediction and in genetic parameter estimates. If observed allelic frequencies are used, genotyped animals are considered the genetic base, but this does not correctly reflect the fact that genotyped animals are more related than expected in reference to the base population, especially if they are selected. For G to be correct, base allele frequencies would be required; this is unfeasible in practice. In this work we propose a method to remove bias of genomic prediction in the single-step procedure based on the correlation between the alleles within genotyped and nongenotyped individuals relative to the whole pedigree (Wright's F_{st}). We also evaluated the effect of the correction by simulation. The corrected G is an appropriate methodological solution that takes into account the effect of nonrandom genotyping and selection on prediction. The results clearly show that a corrected G within a single-step genomic prediction approach is able to include all data and remove bias due to selection.

Genomic regions associated with natural antibody titres in milk of Dutch Holstein-Friesian cows

Wijga, S.[1], Bastiaansen, J.W.M.[1], Van Arendonk, J.A.M.[1], Ploegaert, T.C.W.[1,2] and Van Der Poel, J.J.[1], [1]Wageningen University, Animal Breeding and Genomics Centre, P.O. Box 338, 6700 AH Wageningen, Netherlands, [2]Wageningen University, Department of Cell Biology and Immunology, P.O. Box 338, 6700 AH Wageningen, Netherlands; susan.wijga@wur.nl

A genome-wide association study was performed with the aim to identify single nucleotide polymorphisms (SNP) associated with natural antibody (NAb) titres in bovine milk. Genomic associations could provide insight in the yet unknown genetic basis of NAb. Natural antibodies generally bind antigens shared by classes of pathogens. This study focused on NAb binding lipopolysaccharide (LPS), shared by gram-negative bacteria, lipoteichoic acid (LTA), shared by gram-positive bacteria, peptidoglycan (PGN), shared by gram-negative and gram-positive bacteria, and the model antigen keyhole limpet hemocyanin (KLH). Pedigree records, NAb titres and SNP genotypes (50K) were available for 1,939 first lactation cows, part of the Dutch Milk Genomics Initiative. Analyses did not show SNP associated with total NAb binding LPS, LTA, PGN and KLH at a 0.10 false discovery rate. Similar results were found for NAb isotypes immunoglobulin (Ig)G and IgA binding LTA. For NAb isotype IgM binding LTA, however, 16 SNP passed the 0.10 false discovery rate threshold. Two SNP were located on *Bos taurus* autosome (BTA)6, four on BTA18, two on BTA20, eight on BTA23 and one on BTA27. A cluster of five SNP on BTA23 was located near a gene belonging to the bovine major histocompatibility complex (MHC), mapped at about 28.6 million base pairs. Genetic control of NAb titres by genes of the MHC has recently also been found in chickens, suggesting a possible evolutionary conserved role for MHC in NAb expression. The present study showed 16 SNP associated with NAb isotype IgM titres binding LTA. Half of the associated SNP were found on BTA23 near the bovine MHC, making this a region of candidate gene(s) involved in NAb expression in dairy cows both from a functional and positional perspective.

Contributions of different sources of information to reliability of genomic prediction for Danish Jersey population

Su, G., Gao, H. and Lund, M., Aarhus University, Department of Genetics and Biotechnology, Blichers Allé 20, Postboks 50, Tjele, 8830, Denmark; Guosheng.Su@agrsci.dk

This study investigated the gain of genomic prediction via combining pedigree information and pseudo-observations of non-genotyped animals. The data comprised 1,863 Danish Jersey bulls among which 1,096 bulls were genotyped using 50k SNP chip. The whole dataset was split into a reference dataset and a validation dataset by a cut-off date (birth date of bull) such that the validation dataset consisted of approximately 20% of the genotyped bulls. Two reference datasets were created, one contained de-regressed proof (DRP) from both genotyped and non-genotyped bulls (REF$_{full}$), and the other contained DRP from genotyped bulls only (REF$_{sub}$). Genomic breeding values for 16 traits were predicted using linear mixed model based on different information sources: 1) based on REF$_{sub}$ and genomic relationship (GBLUP), 2) based on REF$_{sub}$ and the relationship combining marker and pedigree information (Onestep$_{sub}$), and 3) based on REF$_{full}$ and the relationship combining marker and pedigree information (Onestep$_{full}$). Reliability of genomic prediction was measured as squared correlation between genomic prediction and DRP, and then divided by reliability of DRP, for the bulls in the validation dataset. According to the validation analysis, reliability of genomic prediction using the GBLUP model was 4.8% higher than the reliability of pedigree index, averaged over the 16 traits. By combining pedigree information, the reliability was increased by 0.3%. By combining both pedigree information and pseudo-observations of non-genotyped animals, the reliability was increased by 1.6%. Onestep$_{full}$ led to the highest reliability of genomic prediction because of utilizing the most information and is easy to implement. Therefore, this approach could be a feasible alternative to genomic prediction in practical cattle breeding.

Genetics correlations between longevity and other traits for Slovenian populations of dairy cattle

Potočnik, K., Štepec, M., Dolinšek, B., Krsnik, J. and Gorjanc, G., University of Ljubljana, Biotechnical Faculty, Department of Animal Science, Groblje 3, 1230 Domžale, Slovenia; klemen.potocnik@bf.uni-lj.si

Longevity is an important trait in cattle breeding. In order to understand relationship between longevity and other traits we compared bull breeding values from the national official evaluation in October 2010 for the length of productive life (LPL) and other total merit index traits in Slovenian populations of Holstein (HOL), Simmental (SIM), and Brown Swiss (BSW) breeds. Survival analysis methodology was used for the evaluation of LPL while standard Gaussian linear mixed model was used for other traits. All reported correlations are among breeding values as expressed on the Slovenian scale. Somatic cell count (SCC) had the greatest positive correlation with longevity (0.29 HOL, 0.19 SIM and BSW), which implies that animals with favourable genetic merit for LPL have also favourable genetic merit for SCC. The next trait with positive correlation was hock development (0.07 HOL, 0.21 SIM and BSW) – animals with dry and clean hocks have better longevity. Positive correlations with longevity were found also for udder depth (0.05 HOL, 0.34 SIM, and 0.13 BSW) and calving interval (0.06 HOL, 0.14 SIM, and 0.09 BSW). Milking speed had positive correlation with longevity for SIM (0.22) and BSW (0.14) and slightly negative for HOL (-0.05). For all breeds negative correlations were estimated with teat thickness (-0.08 HOL, -0.22 SIM, and -0.05 BSW) and with rear leg set (-0.21 HOL, -0.07 SIM, and -0.05 BSW). Muscularity showed negative correlation for SIM (-0.25) and BSW (-0.07) and positive for HOL (0.12). Difference between these correlations can be attributed to different genetic makeup of studied breeds.

Interaction between GH gene polymorphism and expression of leptin and SCD (stearoyl-coA desaturase) in Japanese Black cattle

Sugimoto, K.[1], Kobashikawa, H.[1], Ardiyanti, A.[1], Sugita, H.[1], Hirayama, T.[1], Suzuki, K.[1], Suda, Y.[2], Yonekura, S.[3], Roh, S.-G.[1] and Katoh, K.[1], [1]GSAS, Tohoku University, Amamiyamachi, Aoba-ku, Sendai, 981-8555, Japan, [2]School of Food, Agricultural, and Environmental Science, Miyagi University, Hagurodai, Taihaku-ku, Sendai, 982-0215, Japan, [3]Faculty of Agriculture, Shinshu Universit, Minamiminowamura, Nagano, 399-4598, Japan; kato@bios.tohoku.ac.jp

Japanese Black cattle have three types of growth hormone (GH) gene single nucleotide polymorphism (SNP): Leu-Thr (allele A), Val-Thr (allele B), and Val-Met (allele C) at codon 127 and 172, respectively. We recently found that the allele C is specific for Japanese Black cattle, and cattle with the allele C show lower carcass weight but richer in intramuscular oleic acid than those with other alleles. The aims of the present study were 1) to investigate the interaction between GH SNP and plasma hormones (leptin, GH, Insulin and IGF-I) at 18 and 26 m old and mRNA levels in diaphragm tissues at 30 m old because leptin is a suppressor for feed intake and GH release, and 2) to investigate the interaction between GH SNP and mRNA expression and activities of SCD in diaphragm tissues at 30 m old because SCD is a key enzyme for synthesis of unsaturated fatty acids. Plasma concentrations of leptin and insulin, but not GH, were increased by aging. Plasma insulin concentration was lower in CC-type animals than other types (AA- and BB-types). Expression of leptin mRNA was greater in allele C than others. Expression of SCD mRNA and SCD enzyme activity were greater in allele C than others. But there is no significant linear relationship between mRNA and enzyme activities. These results imply that 1) the greater effect of GH gene SNP on leptin expression in allele C results in smaller carcass weight via suppressed feed intake and GH release, and 2) increased SCD expression and activity in allele C result in accelerated synthesis of unsaturated fatty acid compositions in beef, than those in other alleles.

Single-step genomic breeding value for milk of Holstein in the Czech Republic

Pribyl, J.[1], Boleckova, J.[1], Haman, J.[1], Kott, T.[1], Pribylova, J.[1], Simeckova, M.[1], Vostry, L.[1], Zavadilova, L.[1], Cermak, V.[2], Ruzicka, Z.[2], Splichal, J.[2], Verner, M.[2], Motycka, J.[3] and Vondrasek, L.[3], [1]Institute of Animal Science, Pratelstvi 815, 104 01 Praha - Uhrineves, Czech Republic, [2]Czech Moravian Breeding Corporation, U Topiren 860, 170 41 Praha, Czech Republic, [3]Holstein Cattle Breeders Association of the Czech Republic, Tesnov 17, 117 05 Praha, Czech Republic; vostry.lubos@vuzv.cz

Milk production of 849,693 primiparous cows with complete lactation record (including pedigree 1,643,663 animals) from the period 1995-2010 was analysed by traditional Animal Model (Breeding Value) and by Single-Step Approach of Genomic Evaluation (Genomic Breeding Value) procedures. In a Single-Step Approach, the pedigree-based relationship matrix was augmented by the genomic relationship matrix, constructed from SNP genotypes of 50K chip for 838 sires. For simplicity, only lactation model and only the first lactation were used. Breeding values were evaluated for whole dataset and for subset of cows calved until year 2005. Correlations were calculated between breeding values of groups of animals (cows/heifers/ proven bulls/young bulls, genotyped/ungenotyped), according of the data sets and procedures of predictions of breeding value. Between genotyped young bulls and their result in progeny test are propitious relations. By genotyping is partly corrected random Mendelian sampling in relationship matrix and it influences also the estimated differences of ungenotyped animals. For test of procedures were used families of programmes of BLUPF90 and DMU.

Sensitivity analyses to prior probabilities on genomic breeding values prediction

Román-Ponce, S.I.[1,2], Samorè, A.B.[1], Dolezal, M.[3], Banos, G.[4], Meuwissen, T.H.E.[2] and Bagnato, A.[1], [1]Università degli Studi di Milano, Via Celoria 10, 20134 Milano, Italy, [2]Norwegian University of Life Sciences, Ås, 1432, Norway, [3]Vetmeduni Vienna, Vienna, 1210, Austria, [4]Aristotle University of Thessaloniki, Thessaloniki, 54124, Greece; sergio.roman@unimi.it

The aim of this research was to evaluate different prior probability values, assumed for large effects markers, in the estimation of genomic breeding values (GEBV) in dairy cattle. Analyses were performed on a data set of 1,089 genotypes of Brown Swiss bulls genotyped with the Illumina Bovine 54K SNP chip. Editing on SNPs was based on genotyping rates, rate of missing genotypes and Mendelian errors, and left 39K SNP markers for the analysis. Markers on the X chromosome were not included. A total of 846 bulls born before 2001 constituted the training population while the test population included all bulls born in the period from 2001 to 2005. The effect of each SNP marker was estimated using two models: GBLUP and BayesB. Different scenarios related to the number of SNP with large effect were tested: 39 (probability 0.001), 198 (0.005), 397 (0.01), 1,985 (0.05), 3,969 (0.01) and 19,845 (0.5) markers. Estimates were derived for milk, fat and protein (yield and percent) and SCS, and ten replicates were run for each trait. GEBV were obtained by summing the single marker effects. Range of the correlation between EBV and GEBV in the test population was from 0.12 to 0.58. The estimates of markers effects were not affected ($P>0.05$) by prior assumptions. Accuracy slightly increased with BAYESB when the proportion of SNP with a large effect was assumed between 0.005 and 0.01, except for SCS that resulted in a higher accuracy with GBLUP. The prior distribution might not have affected the results because the training and evaluation animals were quite related, in which case it was difficult for BayesB to single out single SNPs with big effects.

Genetic parameters for test-day milk yields and somatic cell score in Aosta Pie Red cattle.
Cappelloni, M.[1,2], Mantovani, R.[1] and Fioretti, M.[2], [1]University of Padua, Department of Animal Science, Viale Universita,16, 35020 Legnaro (PD), Italy, [2]Italian Animal Breeders Association, Via G.Tomassetti, 9, 00161 Roma, Italy; roberto.mantovani@unipd.it

The aim of this study was to set up a test-day model for routinely EBV estimates for milk (M), fat (F), protein yields (P), and somatic cell score (SCS) in the Aosta Pie Red breed (APR). The APR is a dual purpose autochthons breed of the west alps and present mostly in the Aosta valley (80%). It is a very rustic breed that can easily adapt to the harsh mountain environment grazing at very high altitude. Data used in this study were test day (TD) M, F, P and SCS recorded on 18,113 animals enrolled in the national evaluation system and recorded over a 15 years period (1996-2010). The total number of TD used was 194,759, and they were joined with a pedigree file containing 41,620 animal effects. After a preliminary analysis, the final model adopted accounted for 52,696 levels of herd-TD by number of lactation (NL; up to 3 lactation), with a mean number of 3.7 TD per animal and lactation. Other fixed non genetic effects accounted in the model were the NL (3 levels), the class of gestation (18 levels), obtained considering the of days of gestation at subsequent TD for each animal and lactation, the class of age at parity within NL (42 classes of age), and the month of parity within NL (36 classes). A repeatability model with the EM-REML method and single trait analysis were used to estimate heritability (h^2). Results obtained were in the expected range for F (0.16 of h^2) and SCS (h^2 of 0.08), but h^2 was slightly overestimated for M and P, i.e. 0.26 for both traits. Possible overestimation were detected also analyzing repeatability, that resulted particularly high for M (0.67), although within the expected range for the remaining traits (i.e. between 0.23 and 0.40). A deeper analysis in this field seems necessary, although the TD model developed for APR could represent a suitable alternative to the present lactation model.

Proportional contributions of different breeds to Lithuanian red cattle using pedigree information when only a fraction of the population has been analyzed
Petrakova, L. and Razmaite, V., Lithuanian University of Health Sciences, Institute of Animal Science of Veterinary Academy, Animal breeding and Genetics, R. Žebenkos 12, LT-82317 Baisogala, Radviliškis district, Lithuania; Razmusv8@gmail.com

With the aim to examine the genealogical structure and statement of Lithuanian Red dairy cattle open population pedigree analysis of two large breeding herds, consisting 2,748 cows, was carried out. The data used in this study were obtained from State Enterprise the Centerof AgriculturalandRuralBusiness Informationand included pedigree records from three to five generations. Also the impact of Holstein Red breed was investigated by performing Lithuanian Red breeds lifetime milk production, lactation average and dry day period trait analysis.Analyses were performed in R 2.11.1 and Excell. High variability of genealogical structure was found within Lithuanian Red breed. More than 150 different genotypes were formed from 14 breeds, used for Lithuanian Red cattle improvement. Three groups were taken for further analysis: remaining purebred Lithuanian Red cows and cows from composite Red population. In analysis composite Lithuanian Red represented 1/2 Holstein cows and red cows without Holstein immigration. The observed results did not indicate a wide diversity to occur in dry period length. The yields of the cows in group of Lithuanian Red purebred and group without Holstein immigrationwere 1,474.99 kg and 1,113.97 kg lower, respectively then in group with 1/2 Holstein. Mean number of lactations thus in the same groups were higher (2.93 and 2.82 respectively) than in group with Holsteinimmigration (2.01).

Infrared termography to evaluate thermal comfort in Nellore cattle

Martello, L.S., Leme, P.R., Silva, S.L., Gomes, R.C., Oliveira, C.L., Canata, T.F., Souza, J.L.F. and Zillig Neto, P., Universidad de São Paulo, Av. Duque de Caxias Norte, 225, 13635900, Brazil; martello@usp.br

Rectal temperature (RT) and respiratory rate (RR) are thermoregulation traits used as indicators of thermal comfort. Infrared (IR) thermography allows measurements of surface temperature of animal's body. There is a lack of information about using IR termography in Nellore cattle as indicator of thermal comfort. Therefore the objective of this work was to evaluate the correlations between temperature measured by IR (IRT) with RT and RR at different times of the day and different anatomical regions of the animal's body. TR, FR and IRT were measured in 18 Nellore steers during ten days at 7 h, 12 h and 16 h. IRT images were taken at multiple body locations which include frontal head area (FH) and lateral head area (LH), front feet (FF), flank (FL), rump (RU), ribs (RI) eye area (EY). The RR was assessed by counting the flank movements and the RT using a thermometer inserted into the rectum of the animal. All variables were taken simultaneously. The association of different locations of IRT with RT and RR was evaluated by Pearson's correlation. Early in the morning (8 h) the RT had a positive correlation with IRT in all body locations, ranging from 0.07 with EY to 0.20 with FL. Correlations among IRT of all body locations and RR were positive and ranged from 0.04 with FH to 0.20 with RU. At 12 h correlations of RT with IRT ranged from 0.16 with FL to 0.29 with LH while correlations among RR and IRT ranged from 0.07 (RI) to 0.30 (EY). At 16 h RT and IRT were positively correlated, ranging from 0.06 with FF to 0.21 with RU and among RR and TIRT showed positive correlations with values ranging from 0.16 (EY) to 0.28 (FL). IRT are positively associated with RT and RR, indicating that increases in these temperatures are associated to the increase of RT an RR. A stronger correlation among RT and TIRT was observed at hottest time (12 h) while greater correlations between RR and TIRT were observed at the end of the day (16 h).

Small intestinal digestibility of cereal starch, protein, and fatty acids in fistulated horses

Jensen, S.K., Hymøller, L. and Dickow, M.S., Aarhus university, Institute of Animal Health and Bioscience, Research Centre Foulum, DK-8830 Tjele, Denmark; skj@djf.au.dk

Horses have a digestive tract adapted for digesting a continuous flow of grasses with high fibre and low starch contents. To meet energy requirements of performance horses they are fed energy dense cereals, a practice which can cause digestive disorders due to shifts in the environment of the digestive tract and the degradation profiles of various nutrients. To counteract detrimental effects of feeding cereals to horses different feed technological treatments can be applied which alter the physical properties of cereal kernels and change their site and course of digestion. The mobile nylon bag technique was used to study digestibilities of dry matter, starch, protein, and fatty acids of maize and barley subjected to different feed technological treatments and different types of black and yellow oats. Correlations were found between nylon bag passage time and observed digestibility of nutrients, hence all digestibilities were interpolated to digestibilities at seven hours (average passage time) prior to statistical analysis. Heat treatment, chemical treatment, and physical destruction of seed coats generally increased nutrient digestibilities of cereals. However, the effect of the feed technological treatments depended on the size of the particles entering the small intestine.

In vivo fibre digestibility and SCFA in caecum and blood plasma of Norwegian cold-blooded trotter horses fed different diets

Brøkner, C.[1], Austbø, D.[2], Næsset, J.A.[2], Bach Knudsen, K.E.[3] and Tauson, A.H.[1], [1]University of Copenhagen, Faculty of Life Sciences, Department of Basic Animal and Veterinary Science, Grønnegaardsvej 3, DK-1870 Fredriksberg C, Denmark, [2]Norwegian University of Life Sciences, Department of Animal and Aquacultural Sciences, P.O. Box 5003, N-1432 Aas, Norway, [3]Aarhus University, Faculty of Agricultural Sciences, Department of Health and Bioscience, Research Centre Foulum, Blichers Allé 20, DK-8830 Tjele, Denmark; stinne@life.ku.dk

This experiment aimed at quantifying the *in vivo* total tract digestibility of various fibre fractions and the production of short chain fatty acids (SCFA) in caecum and concentration in blood plasma. Four geldings in a Latin Square design fitted with a permanent caecal cannula were fed 4 diets: (H) timothy hay, (OB) timothy hay + whole oats and molassed sugar beet pulp (Betfor®), (BB) timothy hay + whole barley and Betfor® and (M) timothy hay + a loose chaff based concentrate. Starch did not exceed 2 g starch per kg BW per meal. Caecal content was sampled through the cannula and blood samples were taken from vena jugularis at time 0, 3 and 9 h after feeding. The horses were fitted with a harness for separate total collection of faeces and urine (Stablemaid). Caecal content and blood plasma were analysed for SCFA. Feeds and faeces were analysed for dietary fibre (DF), non-starch polysaccharides (NSP), soluble non-cellulosic polysaccharides (S-NCP) and NDF. The digestibility of organic matter in the different diets was significantly lowest (P<0.01) in H with 58% as compared to 64% in OB, 64% in BB and 63% in M. Caecal total SCFA was significantly (P<0.01) highest 3 h after feeding and ranged from 60 mmol/l in H, 67 mmol/l in OB, 76 mmol/l in BB and 70 mmol/l in M. In blood plasma total SCFA was significantly (P<0.01) highest 3 h after feeding and ranged from 812 µmol/l in H, 593 µmol/l in OB, 714 µmol/l in BB and 662 µmol/l in M. In conclusion, qualities of the diets affect the digestibility and formation of caecal SCFA.

The 13C-bicarbonate tracer technique for estimation of CO_2 production and energy expenditure in ponies

Jensen, R.B.[1], Junghans, P.[2] and Tauson, A.-H.[1], [1]University of Copenhagen, Faculty of Life Sciences, Department of Basic Animal and Veterinary Sciences, Grønnegaardsvej 7, 1870 Frederiksberg C, Denmark, [2]Leibniz Institute for Farm Animal Biology (FBN), Research Unit Nutritional Physiology 'Oskar Kellner', Wilhelm-Stahl-Allee 2, 18196 Dummerstorf, Germany; ralle@life.ku.dk

Knowledge on the energy requirements of horses is essential for optimal feeding practice, and guidelines have been given for different types of horses at different physiological stages. However, energy requirements could be given more precisely for different breeds and under different physiological conditions. The ^{13}C-bicarbonate tracer technique (^{13}C-BT) can be used for indirect determination of the CO_2 production and to estimate energy expenditure (EE). This method is based on ^{13}C kinetics in breath air after administration of ^{13}C labeled sodium bicarbonate, where the ratio between the ^{13}C and ^{12}C in samples of expired air can be used as a measure of the CO_2 production rate. The hypothesis is that the ^{13}C-BT, with oral or IV administration of the tracer in a single bolus, can be used for estimation of CO_2 production and determination of EE in equines, and hence with oral administration provide a non invasive alternative to other methods. Four Shetland ponies were used in the experiment. Measurements were performed in the stall with two administration routes of ^{13}C-bicarbonate, either oral or IV (2.5 mg $NaH^{13}CO_3$/kg body weight). Exhaled breath was collected in breath bags with a volume of approximately 1 liter by using a mask with a two-way non-rebreathing valve system. Two baseline samples were taken before and 14 samples after (5, 10, 20, 30, 40, 60, 90, 120, 150, 180, 240, 360, 640, 1440 minutes) administration of ^{13}C-bicarbonate. Two measurements were made on each administration route for each pony. The data is being processed at the moment, but will be ready for presentation at EAAP 2011. This experiment is part of a lager study where the ^{13}C-BT is validated against indirect calorimetry and used during activity.

Changes on body weight and body condition in the Lusitano broodmare

Fradinho, M.J.[1], Correia, M.J.[2], Beja, F.[2], Rosa, A.[3], Perestrello, F.[4], Bessa, R.J.B.[1], Ferreira-Dias, G.[1] and Caldeira, R.M.[1], [1]CIISA, Faculdade Medicina Veterinária, UTL, 1300-477 Lisboa, Portugal, [2]FAR, TArneiro, 7441 A. Chão, Portugal, [3]Qta Lagoalva Cima, Alpiarça, 2090 Alpiarça, Portugal, [4]Comp. Lezírias, B Prata, 2135 S Correia, Portugal; amjoafradinho@fmv.utl.pt

The main objective of this study was to characterize body weight (BW) and body condition (BC) changes along pregnancy in the Lusitano broodmare under two feeding systems: pasture plus daily supplementation (P+S) and pasture (P). BW and BC (0-5 points scale) were monthly assessed from the first month of gestation (1G) to the first month after foaling (1L) in 60 Lusitano mares over two (P+S) and three (P) breeding seasons. For both systems, data were grouped according to foaling season: Feb-Mar and Apr-May. A mixed linear model for repeated measures was used to assess the effect of foaling season, effect of gestation month and their interaction on BW and BC. Throughout pregnancy, BW and BC changes were observed. Significant increases of BW were found on the last trimester of gestation although with a smaller accrual in Feb-Mar mares on both systems. Along the study, small changes of BC were observed with maximal amplitudes of 0.22 points in P+S and 0.46 points on P. Higher BC values were registered on P+S mares for both foaling seasons. On both systems, BC of Feb-Mar mares decreased on the last trimester of gestation until 1L. The largest increase in BW and BC, between 1G and 11G, was shown by Apr-May mares. Results suggest that changes in BW and BC in the Lusitano broodmare, managed on grazing systems, are influenced by pasture production cycle, regardless of some differences in feeding regimes. These data associated with knowledge on mares' fertility and foals' growth until weaning will contribute for better decisions about the more appropriate feeding plan and foaling season, in order to achieve the higher efficiency of the production system.

3-d video morphometric measurements of conformation of the Icelandic horse: estimated means, variation and repeatability of measurements

Kristjansson, T.[1], Arnason, T.[1], Bjornsdottir, S.[2], Crevier-Denoix, N.[3] and Pourcelot, P.[3], [1]Agricultural University of Iceland, Land and animal resources, Hvanneyri, IS-311 Borgarnes, Iceland, [2]Icelandic Food and Veterinary Authority, Austurvegur 64, IS-800 Selfoss, Iceland, [3]Unité INRA BPLC Ecole Vétérinaire d'Alfort, 7 avenue du Général de Gaulle, 94704 Maisons-Alfort Cedex, France; thorvaldurk@lbhi.is

The official breeding goal for the Icelandic horse describes an ideal conformation that should facilitate multi-gaiting riding ability. The aim of this study is to quantify the conformation of the Icelandic horse in an objective way using a three dimensional video morphometric method and assess the repeatability of the measurements. 119 conformational parameters were calculated for 72 horses, which were video recorded at field tests in Iceland in the years 2008-2010. The horses are recorded in walk, using four video cameras, without the use of markers. A set of four video frames were chosen for each horse for two reference positions (forelimb/hind limb). The measurements consisted of heights, segments lengths, joint angles, inclinations and proportions. The repeatability was assessed tracking anatomical landmarks two times for 20 horses using the same frame and frames of the left and right sides. All tracking of anatomical landmarks were performed by one operator. All parameters were normally distributed. The average height at the withers in the 72 horses was 136.34±3.02 cm. The Pearson correlation r between height at the withers and height and segment length parameters was in the range of 0.02 to 0.88. The CV of the parameters was in the range of 1.98% to 12.09%. Regarding repeatability of the measurements when the same frame was tracked two times the mean repeatability of height parameters, segment lengths and angles was 0.97 (0.80-1.00), 0.83 (0.67-0.97) and 0.84 (0.67-0.99), respectively. When the left and right side of the same horse were tracked the mean repeatability of height parameters, segment lengths and angles was 0.85 (0.71-0.95), 0.82 (0.67-0.96) and 0.87 (0.80-0.96), respectively.

An alternative tool for stress assessment during competition in horses: preliminary results

Sánchez, M.J.[1], Bartolomé, E.[1], Schaefer, A.[2], Cook, N.[2], Molina, A.[3] and Valera, M.[1], [1]University of Seville., Ctra. Utrera, km1, 41013 Seville, Spain, [2]Lacombe Research Centre, 6000 C and E Trail, T4L 1W1 Lacombe, Canada, [3]University of Córdoba, Campus de Rabanales, 14071 Córdoba, Spain; v32sagum@gmail.com

The assessment of stress in horses during sport competitions has been a big deal for breeders and riders for a long time. Thus, different methods have been developed such as cortisol and heart rate variability measurement. Otherwise, these methods need direct contact with the animal, which could cause a stress response and lead to confounding results. In this preliminary study we investigated the potential of using infrared termography technology (IRT) to measure eye temperature as a means to detect stress in horses during competition and compare it with a reliable physiological method as salivary cortisol. 70 horses, ranging from 4 to 6 years old were used in this study. Maximum eye region temperature was assessed using a portable IRT camera to collect images, and cotton swabs were used to collect salivary cortisol. IRT and cortisol samples were collected during 3 competition days (warm-up exercises on day 1, classification events on day 2 and finals on day 3) of Show Jumping and Dressage competitions, and in three different moments within each competition day: 1-2 hours before competition, just after it and 3-4 hours after it. A Wilcoxon Test was made to ascertain differences among the 3 days and the 3 moments within each day of data collection. In order to highlight the existing relation among Cortisol and IRT parameters, different mathematical models were used. Significant differences ($P<0.001$) were found among most of the moments within the day for either cortisol or IRT measurements, except between previous and just after competition measurements for cortisol and between just after and 3-4 hours after measurements for IRT. Our results highlighted IRT technology as a suitable method for stress assessment in horses during competition.

Assessment of skinfold thickness as a factor related to chronic progressive lymphoedema in Belgian draught horses

De Keyser, K., Janssens, S. and Buys, N., KULeuven, Biosystems, Kasteelpark Arenberg 30, bus 2456, 3001 Heverlee, Belgium; Kirsten.DeKeyser@biw.kuleuven.be

In Belgian draught horses, disability and disfigurement of the lower limbs has been present for decades. Only in 2003 the term chronic progressive lymphoedema (CPL) was assigned to this devastating disorder, which is most practically diagnosed by clinical research. A veterinary scoring system (CPL score) has been created to define CPL condition of the limbs in draught horses. Other factors that have been related to CPL prevalence (skin condition, weight, age and sex) could confirm assigned scores and subsequently improve CPL diagnosis. Therefore, the relationship between the occurrence of CPL and these different phenotypic characteristics was investigated. Additionally, as a measure of skin condition, skinfold thickness was studied. Data are based on clinical veterinary inspections at official horse meetings (14) and farm visits (27) in Belgium (October 2009-February 2011), covering 856 horses. Skinfold thickness was measured with a Harpenden skinfold caliper. Statistical analysis was performed using a general linear model, including age, sex, body weight, height at withers and skinfold thickness. Consistent with previous studies, mares show lower mean CPL scores compared to stallions (1.98 and 2.21 respectively, on a scale from 1 = no CPL to 5 = extreme CPL). An analogous sex difference can be observed for skinfold thickness (means respectively 5.3 cm and 7.8 cm). There is a significant sex dependent relationship between CPL scores and age ($P<0.001$): males tend to develop clinical symptoms faster than females. Also skinfold thickness and CPL scores are significantly related ($P=0.0026$). Body weight and height at withers could not be related to CPL scores (n=76 individuals). CPL in Belgian draught horses is related to skinfold thickness and affected by sex and age. Skinfold thickness measurements of draught horses could confirm CPL diagnosis in the field.

Liposomes of phospholipids, a promising approach for stallion sperm freezing
Barenton, M.[1], Couty, I.[1], Labbé, C.[2], Méa-Batellier, F.[3], Duchamp, G.[4], Yvon, J.-M.[4], Desherces, S.[5], Schmitt, E.[5] and Magistrini, M.[1], [1]INRA, UMR 85 Physiologie de la Reproduction et des Comportements, 37380 Nouzilly, France, [2]INRA, UR1037 SCRIBE, Campus de Beaulieu, 35000 Rennes, France, [3]IFCE, Haras national de Blois, BP 14309, 41043 Blois Cedex, France, [4]INRA, UE PAO, 37380 Nouzilly, France, [5]IMV Technologies, ZI N°1 Est, 61300 Saint Ouen sur Iton, France; michele.magistrini@tours.inra.fr

The freezing extender is a key factor for the success of stallion sperm cryopreservation and subsequent success of artificial insemination with frozen semen. Egg yolk (EY) is an animal product used as a cryoprotective agent of sperm cells membranes in most semen cryopreservation media. In our laboratory, we previously demonstrated, *in vitro* and *in vivo*, that egg yolk can be replaced by sterilized EY plasma. Based on these results we developed a ready to use media for stallion semen cryopreservation (INRA Freeze®), composed of INRA96® supplemented with sterilized EY plasma and glycerol. In order to optimize the extender, we tested EY phospholipids (EY-PL), one of the main components present in EY plasma. We demonstrated *in vivo* that EY can be substituted by EY-PL (fertility rate per cycle: 68%(27/40) versus 55% (22/40), respectively). The present study was then conducted to confirm the efficacy of EY-PL as a substitute for EY *in vivo*. INRA Freeze® was compared to INRA96® extender supplemented with liposomes of EY-PL and glycerol (INRA96®+EY-PL). Semen from 2 pony stallions (7 ejaculates/stallion) was frozen and 52 cycles of pony mares were inseminated. Pregnancy rate per cycle showed no difference between INRA Freeze and INRA96® + EY-PL: 58% (15/26) versus 54% (14/26) respectively ($P>0.05$). We have now to precisely identify the classes of phospholipids, present in EY plasma directly implicated in the cryoprotection of spermatozoa and to optimize their ratio.

Prevention of secondary nutritional hyperparathyroidism in horses using organic minerals
Gobesso, A.A.O., Wajnsztejn, H., Gonzaga, I.V.F., Taran, F.M.P. and Moreira, C.G., Universidade de São Paulo, Faculdade de Medicina Veterinária e Zootecnia, Avenida Duque de Caxias Norte, 225 - Pirassununga/SP, 13630-700, Brazil; gobesso.fmvz@usp.br

Aiming to evaluate whether the addition of oxalic acid in the diet could induce an imbalance between calcium and phosphorus in foals, and if the diet with organic minerals, compared with diet with ionic minerals, would be able to avoid this imbalance and prevent the development of pathology, serological parameters, mineral concentrations in hair, bone mineral density and bone biopsies were analyzed. It was used 24 crossbred foals, aged between 18 and 24 months. Each treatment consisted of 6 foals (three males and three females), totaling four treatments in a completely randomized design with repeated measures on time, in a 2x2 factorial arrangement: supplementation with minerals organic or not (inorganic minerals), and presence or absence of oxalate in the diet. Sampling was conducted over a period of 150 days. The model of induced imbalance between calcium and phosphorus with addition of potassium oxalate is effective and produces the expected result. The mineral supplementation can increase bone mineral density in foals, regardless of source and sex. The creation of mineral imbalance by adding potassium oxalate decreases the concentration of calcium, phosphorus and magnesium in the bones of foals, regardless of source supplemented. Foals supplemented with organic minerals, even when challenged with the addition of potassium oxalate in the diet, maintains levels of plasma I-PTH stable, showing more resistance to the imbalance between calcium and phosphorus and avoiding the development of fibrous osteodystrophy.

Linkage of driving performance test parameters in Old Kladruber Horse
Majzlik, I., Andrejsová, L., Hofmanová, B. and Vostrý, L., CULS, Dept. of Animal Sci., Prague, 165 21, Czech Republic; majzlik@af.czu.cz

Performance in driving test of Old Kladruber Horse was studied using 15 traits in 495 horses in a time span of 1997-2008. The breed has been used as a carriage horse for centuries, which is why the performance test includes conformation parameters (Type, Body line, Fundament, Body harmony, General impression), basic riding abilities (Riding abilities, Walk, Trot, Canter, Dressage test) and driving and pulling abilities (Marathon, Driving course, 1st pulling, 2nd pulling, 3rd pulling). All parameters tested are evaluated on a 10 point scale. Data sets were analyzed by a least squares analysis (SAS) using GLM procedure; correlations between parameters under study were estimated using Pearson's coefficient for the assessment of possible linkage between them. The means of parameters varied between 6.5 – 8.7 points (min. 3, max.10 points) with C.V. 8.3-16.5%. All correlations showed positive values. Generally, the parameters of conformation indicated the highest values of correlations ($r_{p=}$0.4-0.8). The parameters of riding abilities were correlated on low to middle level. The parameters of driving and pulling abilities showed the lowest level of correlations not only among each other, but also among all parameters studied. This project was supported by grant MSM 604 607 09 01.

Oxidative status of horse feeds found on the Scandinavian market
Martinsen, T.S. and Lagatie, O., Kemin AgriFoods Europe, Toekomstlaan 42, 2200 Herentals, Belgium; tone.martinsen@kemin.com

Even though an effort has been done to investigate the dietary effects of antioxidants in horses, little focus has been directed to the oxidative status of the equine feeds and the storage stability of the raw materials. The aim of the present test was to (1) investigate the oxidative status of horse feeds fed in the Scandinavian market; (2) how do novel feed ingredients influence the oxidative balance on the feed, and (3) map the overall need for an antioxidant strategy for feed preservation. During a period of time, samples of feed found in average stables were collected and sent to Kemin's Customer Service Laboratory for oxidation stability analysis. All feeds were collected and analyzed within the shelf life period. 28 different feeds from six European countries were collected and analyzed for peroxide value (primary oxidation products) measured in meq/kg and thiobarbituric acid value (secondary oxidation product) measured as malonaldehyde in mg/kg. The results from the oxidation analysis showed that 46% of feeds in the test were high in oxidation products, meaning a TBA-value exceeding 2.0 suggesting the feeds to be fully oxidized. 50%, of the feeds were found to have a low to strong ongoing oxidation process with TBA-values ranging from 0.5-2.0. Only 4% of the feeds hada low and acceptable oxidative value (TBA value <0.5). Pelleted feeds seem to be more stable than muesli (TBA-values of 1.99±1.0 and 3.58±3.5, respectively). There does not seem to be any correlation between TBA value, fat content or starch level in the muesli-based feeds, suggesting influence from raw material. In the samples investigated, the mueslis ranged from a minimum TBA-value of 0.7 to maximum 13.8, while the pellet's values ranged from 0.6 to 3.7. More feeds are needed to be tested in order to do statistical analysis. The clinical impact on the horses fed strongly oxidized feeds and the energy balance available for the horse needs to be further investigated.

Relationship between linear conformation traits and morphological scores in the Spanish Purebred
Sánchez, M.J., Azor, P.J., Gómez, M.D., Bartolomé, E. and Valera, M., University of Seville, Ctra. Utrera km1, 41013 Seville, Spain; v32sagum@gmail.com

The conformation assesses the structure of an animal in relation to its function and its breed standard. However, in the Spanish Purebred (SPB), beauty has also become a very important economic parameter for horse breeders, getting a big relevance the score that the animal gets in morphological horse shows. In previous studies, a linear evaluation methodology has been set up together with a functional traits evaluation methodology. The aim of this scientific work was to highlight the relationship between the different classes obtained with the Linear Conformation Traits (by a linear evaluation methodology) and the beauty morphological scores (MS) obtained at the morphological horse shows. The final design included 31 linear conformation traits, collected among 2003 and 2010. The relationship between MS and the linear conformation traits was analyzed for 2485 records belonging to 1,331 horses (654 males and 667 females) born between 1990 and 2007. Breeder (540), sex (2), owner (558), coat color (9) and year of birth (17) were analyzed as environmental factors. A GLM analysis ascertained the breeder as the environmental factor that influenced MS the most. An ANOVA analysis was made with the 31 conformation traits and 12 were significant (Head length, Width of head, Length of neck, Neck-body Junction, Length of shoulder, Angle of shoulder, Lateral angle of knee, Cannon Bone Perimeter, Angle of croup. Breed quality and Harmony). In order to highlight the relationship between the different classes of the linear conformation traits and the MS, different statistical models were developed. In general, classes 6-7 achieved a higher level for MS in males whereas classes 4-5 achieved the highest for females, revealing sexual dimorphism in SPB. Our results showed that the linear conformation traits and the environmental factors influenced MS results.

Semen physiological parameter changes in different ejaculates from the same stallion
Siukscius, A., Pileckas, V., Kutra, J., Urbsys, A. and Nainiene, R., Lithuanian University of Health Sciencies, Institute of Animal Sciencies, R.Zebenkos 12, Baisogala, LT-82317, Radviliskis distr., Lithuania; arturas@lgi.lt

The purpose of the study was to determine the differences in semen physiological parameters from six different ejaculates of the same stallion both before (i.e. fresh semen) and after freezing and packaging into separate straws. Semen was collected at seven days interval from one stallion and the assessment of the main physiological parameters in fresh and subsequently cryopreserved semen was carried out. The average spermatozoa motility in fresh semen was $55\pm3.65\%$. Out of six evaluated ejaculations only 50% met the requirements for cryopreservation. pH value of the semen was always stable 7.25 ± 0.04. Ejaculates produced on the average 156 ± 31.1 straws of cryopreserved semen. Morphological parameters of fresh semen ranged widely. The number of intact spermatozoa per ejaculation ranged from 44 to 70%. Tail pathology was most common and accounted for $16.0\pm2.27\%$. The number of spermatozoa with head pathology reached 12% in one of the ejaculates, i.e. their number exceeded the total average more than twice. Neck pathology was found in $3.2\pm0.87\%$ of spermatozoa. Cryopreservation had no significant influence on the morphological parameters of semen: the number of intact spermatozoa was lower yet insignificantly from 6 0.0 ± 3.97 in the fresh semen to $51.7\pm3.07\%$ ($P<0.05$) in the frozen semen. The changes in the spermatozoa motility in different straws with the frozen semen from the same ejaculate were significant, i.e. from 5.9 to 8.7% in ejaculate 1 (mean $7.8\pm0.93\%$) ($P>0.5$), from 9.1 to 16.7% in ejaculate 2 (mean $11.8\pm2.47\%$), from 18.7 to 24.5% in ejaculate 3 (mean $21.7\pm1.68\%$), from 10.3 to 13.1% in ejaculate 4 (mean $11.4\pm0.87\%$), from 7.4 to 17.9% in ejaculate 5 (mean $12.3\pm3.06\%$) and from 20 to 23.3% in ejaculate 6 (mean $21.9\pm0.98\%$). Qualitative parameters of the thawed semen differ in different ejaculates and in different straws of the same ejaculate.

Effect of including ricinoleic acid from castor oil (*Ricinus communis* L.) in the diet of horses: hematological and biochemical parameters
Nunes Gil, P.C., Centini, T.N., Françoso, R., Gandra, J.R. and Gobesso, A.A.O., College of Veterinary Medicine and Animal Science, University of São Paulo, Brazil, Nutrition and Animal Science, Av Duque de Caxias Norte, 225, 13635000, Brazil; cateto@usp.br

Castor oil contains 90% ricinoleic fatty acid, which gives important characteristics to animal production: controlling of pathogens by the antimicrobial activity, having an antioxidant activity, improving digestion by stimulating enzyme activity. The aim of this study was to evaluate the effects of different levels of ricinoleic acid in the diet of horses on their hematological and biochemical parameters. The trial was carried in the OuroFino Agronegócios LTDA research center, Brazil. Eight adult horses were used of which four were mares and four geldings, weighing on average 361.8 ± 23.6 kg. The horses were divided to four treatments: (T1) 1 g, (T2) 2 g, (T3) 4 g and (T4) 8 g ricinoleic acid per day. The experiment lasted 12 days, and the horses were fed at a maintenance level, together with the dose of ricinoleic acid. For measuring hematological and biochemical parameters, blood samples were collected every morning at 0, 24, 48, 144, 192 and 240 hours of experimental period. Data obtained at time 0 were used as covariates in statistical models. The means were adjusted by LS-means and analyzed using PROC MIXED as repeated measures in time. Statistically significant ($P>0.05$) effect of ricinoleic acid supplementation was observed on the hemoglobin value and mean corpuscular hemoglobin and absolute neutrophil and lymphocyte counts. Also the effect of the sampling time on the concentration of red blood cells, hematocrit, leukocytes and platelets was statistically significant. In the case of the biochemical parameters, significant interaction between time and ricinoleic acid supplementation was found on creatinine, urea and gamma glutamyltransferase concentrations. In conclusion, the addition of ricinoleic acid influenced the hematological and biochemical values of horses.

Fatty acid composition of Mongolian mare milk fat
Minjigdorj, N., Haug, A. and Austbø, D., Norwegian University of Life Sciences, Department of Animal and Aquacultural Sciences, P.O. Box 5003, N-1432 Aas, Norway; naidankhuu.minjigdorj@umb.no

This study aimed to investigate the fat content and the fatty acid composition of milk from Mongolian mares. Milk samples from a total of 12 mares were collected from the two different regions Steppe and Forest-steppe in the summer and in autumn. The mares were kept outdoors all year and grazed on natural pastures only without any form of concentrate supplementation. The milk fat content, as well as the concentration of total saturated (SFA) and total unsaturated fatty acids (TUFA) was similar for the milking seasons and the two regions. The average total milk fat content was 1.9 g/100 g milk which is slightly higher compared to other commercial mare milk products, but lower than bovine and human milk. The sum of TUFA was about 55% of total fatty acids. The sum of PUFA represented about 28% of total fatty acids, with ALA and LA as the main PUFA accounting for 19.5% and 8.1% of total fatty acids, respectively. The concentration of ALA, a fatty acid regarded as favorable in human nutrition, was much higher than observed in most other mare milk, bovine milk and human milk. The ratio linoleic to α-linolenic acid (LA/ALA) was low (0.5), indicating that Mongolian mare milk could be considered a desirable nutritional factor in human diets. Moreover, the saturated/unsaturated fatty acid ratio could also be considered a desirable nutritional factor in the human diet. The high concentration of gamma-linolenic acid (GLA) in milk from mares grazing on the Mongolian steppe region, especially in the summer (1.34% of total fatty acids) is an interesting finding that may also be of relevance to human nutrition and health. Compared to bovine milk, the Mongolian mare milk had higher concentrations of polyunsaturated fatty acids (PUFA), especially ALA, similar concentration of monounsaturated fatty acids (MUFA), especially palmitoleic (16:1) and oleic acid (18:1), and a lower concentration of SFA.

Evaluation of polyherbal supplementation on methane emission from dairy goats
Mirzaei, F.[1], Prasad, S.[2] and Mohini, M.[2], [1]AnimalScience Research Institute of Iran, Livestock Production Management, Beheshti Avenue, 31585,Karaj, Iran, [2]National Dairy Research Institute,Karnal, Livestock Production Management,Nutrition Division, Karnal, 132001,Karnal, India; fmirzaei@gmail.com

It is accepted that one of the most important environmental constraint to agricultural development in the present century is climate change. The question of the relationship between animal farming and the environment was recently highlighted by the FAO report 'Livestock's Long Shadow. The aim of the present work was to monitor the effects of polyherbal supplementation on cross bred does, starting from the last month of pregnancy to weaning, on methane production. Thirty does were divided into three treatments: low level supplementation (LS), high level supplementation (HS) and not supplemented treatment (NS). The study was carried out in October, 2008 till to June, 2009. Overall means of CH_4 for low level polyherbal supplementation was slightly lower than other groups. It seemed that polyherbal supplementation with lower dose tended to improve CH_4 reduction, when it expressed as g/kg digestible dry matter (DDM) and miM/kg DMI the similar trend was noticed. Literature on effect of polyherbal supplementation on Methane gas (CH_4) emission in goats is not available.

Prion protein gene (PRNP) genetic diversity in local Greek and Skopelos goat breeds
Kanata, E.[1], Panagiotidis, C.H.[1], Ligda, C.[2], Georgoudis, A.[3] and Sklaviadis, T.[1], [1]Aristotle University of Thessaloniki, School of Health Sciences, Pharmaceutical Sciences, Aristotle University Campus, 54 006 Thessaloniki, Greece, [2]National Agricultural Research Foundation, P.O. Box 60 458, 57 001 Thessaloniki, Greece, [3]Aristotle University of Thessaloniki, School of Agriculture, Animal Production, Aristotle University Campus, 54 006 Thessaloniki, Greece; chligda@otenet.gr

Aim of the present study was to determine the PRNP variability in Greek goat breeds, with emphasis on scrapie resistance associated variants. The breeds analysed were the local Greek goat raised in the whole country and the Skopelos breed, mainly raised in Sporades and Magnissia. Blood samples from 34 goats (13 local Greek, 21 Skopelos) were collected from farms spread over the traditional breeding region of each breed. Genomic DNA was extracted by commercial kits. The whole caprine PRNP ORF was amplified and subjected to sequencing analysis on both strands using the dye terminator chemistry and a 3730 ABI Genetic Analyzer. Raw sequencing data were analyzed with the Variant Reporter Software. Ten previously described SNPs were detected. Eight resulted in amino acid substitutions (G37V, W102G, I142M, R151H, N146S, P168Q, Q222K, S240P) and two correspond to silent changes (codons 42, 138). As expected, codon 240 accounted for the greater degree of variation. Allele P240 was the most frequent in both breeds (61.5% local Greek, 80.9% Skopelos), suggesting that for Greek goats P240 is the wild type allele instead of S240. Also remarkable, K222 which confers resistance to scrapie, was found in both breeds and in relatively high frequencies (11.5% local Greek, 4.7% Skopelos). 9 PRNP SNPs were detected in local Greek goats and 5 in Skopelos (including silent changes). This may be the result of the Skopelos breed geographical isolation. Our results suggest differential PRNP variation among the local Greek and Skopelos goat breeds and a high degree of variation in the local breed with important representation of scrapie-resistance associated alleles.

Comparision of lifetime production in Bovec and Improved Bovec sheep
Kompan, D. and Gorjanc, G., University of Ljubljana, Biotechnical Faculty, Department of Animal Science, Groblje 3, 1230 Domzale, Slovenia; drago.kompan@bf.uni-lj.si

Production data of 1,734 Bovec and 520 Improved (with East-Friesian) Bovec sheep born between years 1989 and 2006 were used for the analysis of culling dynamics and lifetime production for lambs born, weight of weaned lambs and milk. A linear model with approximate modeling of culling dynamics was used for statistical analysis. Results have shown that in the period ≤1996 Bovec breed had on average 4.4 lactation in their lifetime, which was almost one lactation more than in Improved Bovec breed. In the period after the year 1996 the highest proportion of animals were culled in the first lactation (as expected due to culling dynamics), with slightly higher proportion in Improved Bovec breed than in Bovec breed. Lifetime number of born, liveborn and weaned lambs was with 7.8, 7.4, and 6.4 higher for about one lamb in Improved Bovec breed. Lifetime lamb weaning weight was around 85 kg and did not differ between breeds, due to shorter suckling period in Improved Bovec breed. During lifetime, Improved Bovec sheep produced 820 kg of milk, which was 200 kg (~34%) more than in Bovec sheep. The difference for milk fat was about 12 kg (~32%) and about 10 kg (~34%) for milk protein. However, if the average metabolic weight of breeds is taken into account the difference in milk production amounts to only ~6%. These results show that Bovec breed has satisfactory milk production, but too low body weight for intensive milk production. On the other hand, low body weight makes this breed suitable for extensive farming in hill and mountain areas.

Effect of phytogenic feed additives on performance parameters of fattening lambs
Ringdorfer, F., ARAC Raumberg-Gumpenstein, Sheep and Goats, Raumberg 38, 8952 Irdnig, Austria; ferdinand.ringdorfer@raumberg-gumpenstein.at

In order to evaluate the use of phytogenic feed additives (PFA) in fattening lambs a trial with weaned male and female lambs (20 kg) until their slaughter (40-45 kg) was carried out where the performance parameters like feed intake, daily weight gain and feed conversion was recorded. After slaughter the carcasses was classified by the EUROP-System and undergo further analyses. There were 4 groups, one control group (D) fed with standard concentrate, and 3 trial groups A, B and C fed with standard concentrate plus phytogenic feed additive A, B and C. The animals of all groups had access to water and concentrate *ad libitum* during the whole trial period. The amount of hay was depending on bodyweight. Until 25 kg they get 200 g, from 25 to 30 kg 250 g, from 30 to 35 kg 300 g, from 35 to 40 kg 350 g and more than 40 kg 400 g hey. There were 14 animals per group where feed intake was recorded daily and body weight weekly. There was no effect found on feed intake, daily gains and feed conversion. Average DMI was 1.38 kg in each group (P=0.9989), daily gain was 397, 410, 420 and 409 g in group A, B, C and D (P=0.7234) and feed conversion was 40.3, 38.9, 38.4 and 38.9 MJ ME/kg BW gain in group A, B, C and D (P=0.6588). Average DMI was 1.44 kg in male and 1.32 kg in female lambs (P<0.001). Daily gain was 453 and 365 g for males and females (P<0.001) and feed conversion was 36.6 and 41.6 MJ ME/kg BW gain for males and females (P<0.001). Also carcass characteristics were not affected by PFA. Dressing percentage was 47.8, 47.7, 48.1 and 47.6% for group A, B, C and D, respectively (P=0.9197). Carcass conformation was 2.8, 3.0, 2.7 and 2.8 for group A, B, C and D (P=0.2001), whereas 1=E, 2=U, 3=R, 4=O and 5=P. The average value of fatness was 3.13, 3.03, 3.03 and 3.24 for group A, B, C and D (P=0.6801). Female lambs had fatter carcasses than male 3.29 vs. 2.83 (P=0.0003). The portion of muscle in sirloin was 52.8% resp. 49.4% for male and females (P=0.0022) and for fat 25.8% resp. 31.1% (P=<0.001).

Effects of feed location, roof and weather factors on sheep behaviour in outdoor yards
Jørgensen, G.H.M.[1] and Bøe, K.E.[2], [1]Bioforsk Norwegian Institute for agricultural and environmental research, Nord Tjøtta, P.O. Box 34, 8860 Tjøtta, Norway, [2]Norwegian University of life sciences, Department of animal and aquacultural sciences, P.O. Box 5003, 1432 Ås, Norway; grete.jorgensen@bioforsk.no

The aim of this experiment was to investigate the effect of roof cover and location of feed on sheep's use of an outdoor yard under different weather conditions. A 2 x 2 factorial experiment was conducted with roof covering of outdoor yard (yes or no) and location of feed (indoors or outdoors) in four different pens. Groups were rotated systematically between treatments every week. Twenty adult ewes of the Norwegian White breed were randomly allotted to 4 groups with 5 animals. Weather parameters were automatically recorded. The following behavioural parameters were scored using instantaneous sampling every 15 minutes throughout 24 hour video recordings: location (indoors or outdoors) and general behaviours (stand/walk, resting, feeding). Weather factors did not seem to have any large influence on sheep behaviour. A roof covering the outdoor yard increased time spent in the yard, had no effect on feeding time, a limited effect on resting time but increased the time spent resting outdoors. Locating the feed outdoors increased time spent in the yard, but also increased the time spent resting indoors, indicating that if a dry and comfortable resting area is offered indoors, the feed should be located in the outdoor yard.

Milk protein variants and their associations to milk performance traits in East Friesian Dairy sheep
Giambra, I.J., Brandt, H. and Erhardt, G., Justus-Liebig-University, Department of Animal Breeding and Genetics, Ludwigstr. 21b, 35390 Giessen, Germany; Isabella.J.Giambra@agrar.uni-giessen.de

Milk protein polymorphisms of 235 East Friesian Dairy sheep, hold in four different flocks in Germany (G) and the Netherlands (NL), were analysed by isoelectric focusing. For association studies between milk protein variants and milk production traits, 150-day lactation data (n=130) of 82 animals (flock 1 to 3, G) and 240-day lactation data (n=260) of 76 animals (flock 4, NL) were available. Isoelectric focusing lead to the simultaneous identification of α_{s1}-casein (CN) alleles A, C, and H, α_{s2}-CN A, B, and C, and β-lactoglobulin (LG) A and B, whereas κ-CN and α-lactalbumin were monomorph. Highest frequencies showed α_{s1}-CN allele C, α_{s2}-CN A, and β-LG A. Casein haplotype analyses revealed highest frequency of α_{s1}-/α_{s2}-CN-haplotype CA and a linkage between these casein loci. Significant positive effects of α_{s1}-CN CH on fat percentage and yield, of α_{s2}-CN AB on protein yield, and of β-LG AA on fat and protein percentage were identified concerning the 240-day lactation data. Furthermore, α_{s1}-/α_{s2}-CN-haplotype CB/HB was associated with highest fat and protein percentage and yield, respectively. Association studies including the 150-day lactation data showed no association between α_{s1}-CN and milk performance traits, but significant positive effects of β-LG BB on milk, protein, and fat yield, and of β-LG AA on protein content. Additionally, α_{s2}-CN BB was associated with highest protein percentage, also leading to the significant positive influence of α_{s1}-/α_{s2}-CN-haplotype CB/CB on protein content. The discrepancies in the observed associations are probably due to flock effects and different standardised lactation lengths comparing the German and Dutch data. However, these results show potential for future consideration of milk protein variability in dairy sheep breeding, due to economic importance.

Lamb traceability evaluation by visual ear tags, electronic boluses and retinal imaging

Rojas-Olivares, M.A.[1], Caja, G.[1], Carné, S.[1], Costa-Castro, A.[1], Salama, A.K.K.[1,2] and Rovai, M.[1], [1]Grup de Recerca en Remugants (G2R), Ciencia Animal i dels Aliments, Universitat Autònoma de Barcelona, 08193 Bellaterra, Spain, [2]Animal Production Research Institut, Sheep & Goat, 4 Nadi El-Said, 123311 Dokki, Giza, Egypt; maristela.rovai@gmail.com

Spanish Recental lamb (23 to 25 kg BW) traceability using different identification (ID) devices was studied under farm and slaughterhouse conditions. Lamb primary ID was done at birth with temporary official visual ear tags (V1; 2.8 g, 40 x 15 mm; n=241). Lamb secondary ID was done at weaning with permanent official visual ear tags (V2; 5.2 g, 38 x 39 mm; n=104) and with electronic mini-boluses (MB; ceramic, 19 g, 56 x 12 mm). Moreover, 81 lambs were ID with glass encapsulated transponders s.c. injected (IT) in the left armpit for tracing carcasses through slaughter. Electronic ID by MB and IT used 32-mm half-duplex transponders. Retinal images of live lambs (n=98) were taken at 80 d of age (both eyes) for auditing lamb ID. Head position was compared (normal, n=67; reversed, n=31) after slaughter. On-farm traceability did not vary according to ID device ranging from 98.6 to 100%; P>0.05). The V1 and V2 were removed at beheading and MB at evisceration, enabling carcass ID that was assumed to be the same as the slaughtering order. Although only 78.8% IT were retained after slaughter, they proved that carcass order was altered at weighing, reducing carcass traceability to 68.3%. All retinal images matched in live lambs, but live vs. slaughtered image matching markedly decreased in the normal vs. reversed head position (56.4 vs. 75.0%; P<0.05). In conclusion, V1, V2, MB and IT were efficient devices for individually tracing live lambs but all of them failed for tracing carcasses efficiently. Retinal images efficiently audited live lambs and most of carcasses. Individual tracing from farm to carcass using radiofrequency ID devices would be possible if carcass order is maintained in the slaughterhouse.

Effect of cereal grain source in flushing diet on reproductive efficiency and concentration of blood parameters in Shaal ewes

Sadeghipanah, H.[1], Zare Shahneh, A.[2], Pahlevan Afshar, K.[3], Asadzadeh, N.[1], Aliverdinasab, R.[1], Javaheri Barfourooshi, H.[1], Aghashahi, A.[1] and Khaki, M.[1], [1]Animal Science Research Institute of Iran, Department of Biotechnology, 3146618361 Karaj, Iran, [2]University of Tehran, Department of Animal Science, Karaj, Iran, [3]Islamic Azad University, Abhar Branch, Department of Animal Science, Abhar, Iran; hassansadeghipanah@yahoo.com

In order to investigate the effects of cereal grain source in flushing ration on reproductive efficiency and serum concentrations of steroid hormones and metabolites, in reproductive season, 88 Shaal ewes were blocked by age then allocated into two groups on the basis of ration: 1: 300 gr corn grain plus basal diet (corn group); and 2: 300 gr barley grain plus basal diet (barley group). Ram introducing was performed 14 days after start of flushing. From each group, four 4-year-old ewes were selected for blood sampling. Both on day 14 (ram introducing day) and on day 35 (end of flushing), blood urea concentrations tend to decrease (P<0.10) in corn group (respectively 54.6 and 27.5 mg/dl) in comparison with barley group (respectively 64.1 and 32.6 mg/dl). Differences in other metabolites between two groups were not significant; but blood concentrations of glucose, triglycerides, cholesterol and proteins in corn group were quantitatively higher than in barley group. On day 14, serum progesterone concentration in corn group (7.71 ng/ml) tend to increase (P=0.07) in comparison with barley group (2.95 ng/ml). Source of cereal grain in diet does not significantly affect on parameters relating to reproductive efficiency, but it quantitatively improved most of these parameters like pregnancy rate, fecundity and lamb crop. Totally, these results show that use of corn grain instead of barley grain in flushing diet of sheep, improves balance of metabolites and hormones relating to reproduction and thereby it maybe improves reproductive efficiency in Shaal ewes. So that, each ewe in corn group produced 2.8 kg weaned lamp more than each ewe in barley group.

Grazing saltbush (*Atriplex* spp.) during summer improves vitamin E concentration in muscle and the colour stability of retail meat cuts

Fancote, C.R.[1,2,3], Norman, H.C.[2,3], Vercoe, P.E.[1,2], Pearce, K.L.[4] and Williams, I.H.[1], [1]University of Western Australia, Faculty of Natural and Agricultural Sciences, Crawley, 6009, Australia, [2]Future Farm Industries Cooperative Research Centre, Crawley, 6009, Australia, [3]CSIRO, Centre for Environment and Life Sciences, Wembley, 6913, Australia, [4]Murdoch University, Division of Veterinary and Biomedical Sciences, Murdoch, 6150, Australia; fancoc01@student.uwa.edu.au

Vitamin E is an antioxidant that is important for meat colour because it slows the browning of meat caused by the oxidation of the muscle pigment myoglobin, from oxymyoglobin (red), to metmyoglobin (brown). In southern Australian farming systems the lack of green feed for livestock during summer can lead to vitamin E deficiency and retailers have reported reduced shelf life of meat as a result. Saltbush shrubs (Atriplex spp.) used for revegetation of saline land can be used as feed for livestock during these summer months and have been found to contain high levels of vitamin E (140 mg/kg). We tested the hypothesis that grazing saltbush during summer would increase muscle vitamin E concentration and improve the colour stability of retail meat cuts. Crossbred lambs aged 10-months (n=48) grazed 8 adjacent plots planted to saltbush or control pastures for 64 d during summer. After the grazing period they were fed a low vitamin E diet (7.5 mg/kg DM vitamin E) for 38 d before slaughter. Lambs that grazed saltbush over summer had higher concentrations of vitamin E in the m. longissimus lumborum (LL) than controls (1.6±0.08 mg/kg compared to 0.8±0.08 mg/kg: P<0.001) which led to improved colour stability of meat during retail display (P<0.05), determined over 96 h by the ratio of oxymyoglobin to metmyoglobin. Meat from animals that grazed saltbush was also redder than controls over the 96 h display (P<0.05). Our findings indicate that meat from animals fed saltbush during summer may be more acceptable to consumers because of increased redness and have a longer shelf life as a result of improved of colour stability.

Healthier goats: a project for eradicating contagious diseases in norwegian goats

Lindheim, D.[1], Leine, N.[1] and Sølverød, L.[1,2], [1]TINE Norwegian Dairies SA, Project Healthier Goats, Deparment of Goat Health Services, Tine rådgiving, Postboks 58, N-1431 Ås, Norway, [2]TINE Norwegian Dairies SA, Deparment of Mastitis Laboratory, Postboks 2038, N-6402 Molde, Norway; dag.lindheim@tine.no

'Healthier goats' is a project with the purpose of eradicating CAE, CLA and Johne's disease in the Norwegian goat population. The project started in 2001. By February 2011, 383 farmers have applied to join the project. The project is intended to last until 2013. The goal is to include all the Norwegian goat flocks; 450 flocks of milking goats and about 620 other flocks. For eradication of the different pathogens, the technique of 'snatching' kids has to be used. The kids are taken away from the mother and the 'infected' barn as quickly as possible after birth. The kids are housed in a clean barn, given cow colostrum and raised separated from older animals. The old goats are kept until the lactating period is finished and are then slaughtered. Thereafter, the barn and the near surroundings are cleaned and disinfected. The kids will start to produce milk at the same time of the year as usual for the farm. There is an extensive testing. Some flocks have all the three diseases, while others only have CAE or CLA. The first test for detection of CAE-antibodies is done before the kids are 5 weeks old. The next test is within a year. Bulk milk testing is then done twice a year. An ELISA-test is applied for both blood and milk. CLA is tested by an ELISA, but the detection relies mainly upon clinical examination in the initial phase of the infection. Johne's disease is tested by interferongamma-test, ELISA, culturing of faeces and by pathological examination. The test results after eradication show very good results, when the recommended procedures are followed. So far, less then 2% CAE-positive animals have been detected. One case of Johne's disease has occurred where the procedures were not followed during eradication. CLA have been confirmed in 8 flocks, and represent a major challenge because of its very contagious nature.

Studies on major milk proteins polymorphism in Carpathian local goat breed and in F1 hybrids (Carpathian x Saanen)

Lazar, C., Pelmus, R., Marin, D. and Ghita, E., National Research Development Institute for Animal Biology and Nutrition, Laboratory of Animal Biology, Calea Bucuresti, no. 1, Balotesti, Ilfov, 077015, Balotesti, Romania; cristina_lazar17@yahoo.com

The purpose of the paper was to determine the protein concentration (Bradford method) and the protein polymorphism in the milk from Carpathian goats and from F1 Carpathian x Saanen hybrids. Milk samples were collected from 12 Carpathian goats and from 15 F1 Carpathian x Saanen hybrids reared at ICDCOC Palas-Constanța. The average mean value for milk protein concentration in the samples of Carpathian goat milk was 1.3%, which is below the normal limits reported by the literature (2.9-6% total protein). The average milk protein concentration in the samples of F1 (Carpathian x Saanen) hybrids was 5%, with values which are a little bit higher than normal limits reported by the literature (average 2.7% for Saanen and 3.3 for Carpathian). This shows that the cross of the two breeds produced hybrids which yielded a higher concentration of milk protein. These results revealed that it was obtained a higher percentage of protein (3.3-6%) in goat than in cow milk (3.3 to 3.5%), and the other components are relatively equal, and slightly lower than in the sheep milk (5.2-6.5%). Our results also show a higher polymorphism of these proteins. The following casein proteins were determined by electrophoresis (SDS PAGE) in the milk samples from Carpathian goats and from the Carpathian x Saanen hybrids: alpha s2 and alpha s1-casein, beta-casein, kapa-casein, and beta-lactoglobulin. Milk quality was improved, because of the individuals with a strong expression of alpha-LA, which can be used in reproduction to obtain a good quantitative milk production. To verify the results, real time PCR analysis of the polymorphic genes of the milk proteins will be performed at a subsequent stage, to determine the genetic variants from the locus of each protein and to notice the differences between the Carpathian breed and F1 hybrids (Carpathian x Saanen).

Short study of major milk protein polymorphism in local sheep breed

Pelmus, R.S., Lazar, C., Pistol, G., Marin, D., Gras, M. and Ghita, E., National Research and Development Institute for Animal Biology and Nutrition, Animal Biology Laboratory, Calea Bucuresti, no 1, 077015, Balotesti, Romania; pelmus_rodica_stefania@yahoo.com

There is an international interest for preservation and improvement of local animal breeds, due to their superior biological traits: rusticity, resistance and adaptability to very different local environment. The local Romanian Blackhead Teleorman sheep breed fits very well the current economic demands, such as milk production and prolificacy. The purpose of our study was to determine milk quality indices as well as milk protein content and milk protein polymorphism in the local Blackhead Teleorman sheep breed, using 24 milk samples. In milk samples were determined the total percentage of milk solids non-fat, fat and protein content. Total milk proteins were determined using Bradford method. The types of different milk proteins were identified by SDS-PAGE. After gel visualisation, the results were analysed with specialized software. The test day milk yield and the chemical composition assays performed during the milking period of the Blackhead Teleorman sheep showed that the yield of milk fat and protein of these local sheep ranged within the quality indices specific to the breed (6.56% fat and 5.9% protein), described in the literature. Milk samples were further analyzed for milk protein polymorphism. The electrophoretic pattern of milk samples of Teleorman Blackhead sheep showed the presence of four major caseins variants (alpha s1-, alpha s2-, beta- and k-casein) and two whey proteins, beta-lactoglobulin and alpha-lactalbumin. Our study revealed that milk samples from local sheep are characterised by medium expression level of alpha s1-, beta-, k-casein and beta-lactoglobulin and by low expression level of alpha s2-casein and alpha-lactalbumin. For a higher certitude of these results we will subsequently perform a real time PCR analysis of the milk protein gene polymorphisms, which will give information regarding the genetic variants from the locus of each protein.

Effect of the buffer system on fluorescence yield when evaluating membrane integrity of ram sperm
Yániz, J.L., Mateos, J.A. and Santolaria, P., Universidad de Zaragoza, Producción Animal y Ciencia de los Alimentos, Ctra. Cuarte s/n, Huesca, 22071, Spain; jyaniz@unizar.es

This study was designed to evaluate the effect of various buffers on the fluorescence yield of two fluorchromes (IP and CFDA) when used to assess the membrane integrity of ram sperm. Second ejaculates from six adult males were collected using an artificial vagina, pre-diluted in a milk-based extender and stored at 15 °C until analysis. Within the first 24 h of storage the semen samples were carefully mixed and diluted to 50 x 10^6sperm/ml with either HEPES, MOPS, TES, TRIS, citrate or phosphate-based extenders, immediately before the analysis of fluorescence yield during the assessment of sperm plasma membrane integrity. Sperm were labeled with carboxyfluorescein diacetate and propidium iodide, for the assessment of sperm membrane integrity. Sperm were photographed and the digital images processed. The fluorescence intensity (FI) in the sperm head (200 sperm/fluorochrome and sample) and the fluorescence background noise (FBN) were determined quantitatively using the Image J processing open software. The FI and FBN of the samples were related to the intensity measured for a fluorescence intensity standard. The fluorescence contrast (FC) was defined as FI- FBN. Differences in sperm FI, FBN, and FC between extenders for each fluorochrome were examined through analysis of variance. If the F value was significant, a Tukey test was used for a posteriori multiple comparisons between the diluents. Significantly higher sperm fluorescence intensity and contrast were recorded when TRIS diluent was used, rather than the other diluents, both in the propidium- and in the fluorescein-labeled cells and that should be considered when evaluating their membrane integrity. Acknowledgements: This work was supported by the spanish MICINN (IPT--010000-2010-33) and Fundación ARAID (OTRI 2010-0464).

Effect of drying and fixation on sperm nucleus morphometry in the ram
Yániz, J.L., Vicente-Fiel, S. and Santolaria, P., Universidad de Zaragoza, Producción Animal y Ciencia de los Alimentos, Ctra. Cuarte s/n, Huesca, 22071, Spain; jyaniz@unizar.es

This work was designed to study separately the effect of two stages of conventional sample processing (drying and fixation) on the parameters of sperm nucleus morphometry. Ejaculates from 15 adult males were collected using an artificial vagina, pre-diluted in a milk-based extender and stored at 15 °C until analysis. Within the first 3 h of storage the semen samples were carefully mixed and diluted to 100 x 106 sperm/ml with a citrate-based solution, immediately before processing for sperm head morphometry analysis. For each sample an aliquot was labelled directly with Hoechst (FRESH), and six semen smears were prepared, allowed to air dry. Two slides were directly labelled with Hoechst without further processing (DRIED), two fixed with 2% glutaraldehyde and two fixed with 50% methanol (MET). All samples were labelled with the fluorochrome Hoechst 33342 and photographed with a digital camera. The images were processed with using Image J open software to obtain the morphometry parameters (area, perimeter, length and width). Differences in sperm morphometry parameters between groups were examined through analysis of variance. If the F value was significant, a Tukey test was used for a posteriori multiple comparisons between the groups. There was a significant decrease in head length, width, area and perimeter of air-dried sperm samples with respect to fresh sperm. On average, this decrease was of 4.3% in length, 4.3% in width, 9.2% in area, and 2.8% in perimeter. Between semen smears, fixation significantly increased head sperm dimensions with GLUT> MET> DRIED, although this increase was not enough to recover the size of fresh sperm heads. Acknowledgements: This work was supported by the spanish MICINN (IPT--010000-2010-33) and Fundación ARAID (OTRI 2010-0464).

Genetic parameters for chosen udder morphology traits in sheep

Margetín, M.[1], Čapistrák, A.[1], Apolen, D.[1], Milerski, M.[2] and Oravcová, M.[1], [1]Animal Production Research Centre Nitra, Hlohovecká 2, 951 41 Lužianky, Slovakia (Slovak Republic), [2]Institute of Animal Science, Přátelství 815, 104 00 Praha Uhříněves, Czech Republic; margetin@cvzv.sk

Udder morphology traits were assessed throughout milking period using linear scores (LS) and exact udder measurements (EUM) in ewes of an experimental sheep flock. LS assessment was based on a nine-point scale and included the following traits: udder depth (UD), cistern depth (CD), teat position (TP), teat size (TS), udder cleft (UC), udder attachment (UA) and udder shape (US). EUM were taken in the same ewes and included the following traits: udder length (UL), udder width (UW), rear udder depth (RUDx), cistern depth (CDEx), teat length (TL) and teat angle (TA). Genetic parameters were estimated using non-transformed data (1275 LS for each trait in 381 ewes and 1,185 EUM for each trait in 355 ewes). Multiple-trait models (REMLF90, VCE programs) were used for estimation. In addition to random additive genetic effect of animal and permanent effect of ewe, the models involved fixed effects of control year (7 levels), lactation stage (4 levels), breed group (9 levels) and parity (3 levels). With LS, the highest h^2 were found for CD (0.294), TS (0.275) and TP (0.242). Genetic correlation coefficient between CD and TP was 0.980. This value indicates that the same genes determine both traits. From breeder's point of view, negative genetic correlation coefficient between TP and TS (-0.381) is an important finding. With EUM, mostly higher heritability and genetic correlation coefficients were found. The highest h^2 were found for RUDx (0.448), UL (0.338) and TA (0.295). Moderate or high genetic correlation coefficients between the traits of udder size were found (0.525 between UL and UV; 0.923 between UL and RUDx). High genetic correlation coefficients between EUM and LS traits were found ($r_g >0.8$) when common genetic evaluation was done. Genetic correlation coefficient between TS and TL was 0.937, between CD and RUDx was 0.932, between UD and RUDx was 0.855.

Partitioning of milk accumulation in the udder of Tsigai, improved Valachian and Lacaune ewes

Mačuhová, L.[1], Uhrinčať, M.[1], Mačuhová, J.[2] and Tančin, V.[1,3], [1]Animal Production Research Centre Nitra, Hlohovecká 2, 951 41 Lužianky, Slovakia (Slovak Republic), [2]Institute for Agricultural Engineering and Animal Husbandry, Prof.-Dürrwaechter-Platz 2, 85586 Poing, Germany, [3]Slovak University of Agriculture in Nitra, Tr. A. Hlinku 2, 949 76 Nitra, Slovakia (Slovak Republic); macuhova@cvzv.sk

The aim of this investigation was to evaluate the partitioning of milk accumulation in the udder of Tsigai (TS), Improved Valachian (IV), and Lacaune (LC) (n=16/breed). Ewes were routinely milked twice a day in 1 x 24 milking parlour. Milk flow type occurrence was evaluated at morning milking (control milkings) during three successive days in the middle of two months (June, July). After the fourth morning milking (OT milkings) at both months, oxytocin was injected i.v. at a dose of 2 UI. Afterwards the ewes were milked again in order to remove residual milk (RM). During milkings, an actual milk yield was recorded in one-second intervals using graduated electronic milk collection jars. The milk flow curves were classified into four types: 1 peak (1P), 2 peaks (2P), plateau I (maximal milk flow over 0.4 l/min (PLI)), plateau II (maximal milk flow less than 0.4 l/min (PLII)). Values of fractional milk (MY:MS:RM) were 62:21.5:16.5%; 61:29:10% and 66:18:16% for TS, LC and IV, resp. LC ewes had the lowest amount of RM (0.054±0.006 l) and RM/TMY (9.86±1.16%). It can be supposed that the ewes of LC retained less milk in the ductual system of the udder than TS (0.088±0.008 l; 16.47±1.44%) and IV (0.069±0.010 l; 15.99±1.78%). The highest occurrence of 2P milk flow type was in LC as compared with TS and IV. It can be conclude that LC had higher milkability and better aptitude to machine milking than IV and TS.

Influence of stocking density on behavior of feedlot lambs
Leme, T.M.C., Titto, C.G., Amadeu, C.C.B., Fantinato Neto, P., Alves Vilela, R., Geraldo, A.C.A.P.M. and Titto, E.A.L., University of Sao Paulo - Animal Science and Food Engineering Faculty, Animal Science Department - Biometeorology and Ethology Laboratory, Av. Duque de Caxias Norte, 225, 13635-900 -Pirassununga - São Paulo, Brazil; titto@usp.br

The aim of this study was to verify the influence of the animal density on the behavior of confined lambs. We used 86 animals confined after weaning in 23 pens of two lambs (double pens) or 10 pens of ten animals (collective pens). During confinement, behavioral patterns were recorded instantly and continuously using the focal and interval sampling methods every 30 minutes, from 06:00 to 18:00 hours, for 4 days. The behavioral variables were: posture (standing or lying), activity (eating, ruminating, leisure, drinking water, and grooming) and events (nid-nodding, pushing, picking up, bellowing, mounting, defecating, or urinating). For the evaluation of behavioral variables, the analysis of variance and multiple comparison procedure by Student t test was used. For the posture 'standing', there was a significant effect at 08:30 ($P<0.05$), 11:30 ($P<0.01$), 14:30 ($P<0.01$), 16:30 ($P<0.01$), and 17:30 ($P<0.01$), where a higher percentage of animals housed in double pens remained standing compared to the animals housed in collective pens, which remained 'lying' in those times. For the activity 'eating', a significant effect between treatments at 08:30 ($P<0.05$) and at 16:30 h ($P<0.05$) was observed, in which most animals in the double pens remained in this activity. No statistical difference was found for the other activities and events between the treatments. The number of animals per group influenced the behavior of confined lambs, changing the pattern of food intake.

Influence of pre-slaughter management on cortisol level in lambs
Titto, E.A.L.[1], Leme, T.M.C.[1], Titto, C.G.[1], Amadeu, C.C.B.[1], Fantinato Neto, P.[1], Vilela, R.A.[1] and Pereira, A.M.F.[2], [1]University of Sao Paulo - Animal Science and Food Engineering Faculty, Animal Science Department - Biometeorology and Ethology Laboratory, Av. Duque de Caxias Norte, 225, 13.635-900 - Pirassununga – Sao Paulo, Brazil, [2]Universidade de Évora – Instituto de Ciências Agrárias e Ambientais Mediterrânicas, Animal Science Department - Biometeorology and Animal Wellfare, Apartado,94, 7000-554 – Évora, Portugal; titto@usp.br

This study aimed to verify the influence of the transport in open or closed compartments, followed by two resting periods (1 and 3 hours) for the slaughter process on the levels of cortisol as a indicative of stress level in lambs. The slaughterhouse was located 85 km away from the place of confinement and the transportation of the lambs was carried out in a cage truck type, each cage had a divider in the middle making the front without external visual access to the environment, and the back portion of the cages with visual access to the road. At the slaughterhouse, blood samples were taken from 86 lambs after the transport and before slaughter (1 or 3 hours of resting) for plasma cortisol analysis. Variables were evaluated through the PROC GLM procedure from the Statistical Analysis System ©, version 9.1.3 software. The method of transport influenced in the cortisol concentration ($P<0.01$), the animals transported in the closed compartment had a lower level. After the resting period in the slaughterhouse, there was a decline in the plasmatic cortisol concentration, with the animals subjected to three hours of rest presenting the lower average cortisol value ($P<0.05$). It can be inferred that the lambs that remained three hours in standby before slaughter had more time to recover from the stress of the transportation than those that waited just one hour. Visual access to the external environment during the transport of the lambs is a stressful factor changing the level of plasmatic cortisol, and the resting period before slaughter was effective in lowering stress, reducing the plasmatic cortisol in the lambs.

The effect of dietary crude protein different levels and probiotic on fattening performance and blood metabolites of Kordish three monthes male lambs

Delkhorooshan, A., Forughi, A.R., Soleimani, A., Vakili, R. and Vosooghi Poostin Doz, V., Azad university of Kashmar, Animal sience, mashhad- street dr.beheshti between 36 and 38 number 2.132, 9175843811, Iran; hooman1350@yahoo.com

This investigation was carried out to study of the effect of dietary crude protein different levels and probiotic on fattening performance and blood metabolites of Kordish three months male lambs. The experiment was conducted using 24 Kordish three monthes male lamb with initial live weight of 29.38 ± 3.1 kg and 90 ± 6 days aged. This experiment was conducted in a 2×2 factoriel design that contained 4 experimental diet during 60 days. The treatment were: 1)14.5% CP without probiotic, 2)14.5% CP with 2 gr probiotic per lamb per day, 3)16.5% CP with 2 gr probiotic per lamb per day, 4) 16.5% CP without probiotic. Lambs were weighed every 15 days before morning feeding. For measuring blood metabolites(glocuse and BUN) on 13 and 41 days 4 lambs were randomly selected in each treatment and blood sample were collected 3 hours after morning feeding. Result showed that between experimental diets for average daily feed consumption, average daily gain, feed efficiency and feed conversion were not significant. Also result indicated that between experimental diets for blood serum glocuse and BUN were significant ($P<0.05$).

Absorptive mucosa of goat kids fed lyophilized bovine colostrum

Moretti, D.B., Nordi, W.M., Lima, A.L., Pauletti, P. and Machado-Neto, R., ESALQ/Universidade de São Paulo, Zootecnia, Av. Pádua Dias, 11., 13.418-900, Piracicaba, SP, Brazil; raul.machado@esalq.usp.br

Colostrum intake in newborn goat kids is essential for the acquisition of immunoglobulins (Ig) influencing also metabolism and nutritional status. It was studied the morphological characteristics of the small intestine of goat kids fed lyophilized bovine colostrum (LBC), an alternative source of Ig, or goat colostrum (GC). At 0, 7 and 14 hr of life 15 male newborns received 5% of body weight of LBC and 14 GC, both with 55 mg/ml of IgG. Samples of duodenum, medium jejunum and ileum were collected at 18, 36 and 96 hr of life. Three animals were sampled at birth, without colostrum intake (0 hr). The enteric tissues were analyzed by scanning electron microscopy. The morphological characteristics were not different between LBC and GC in all segments. There were no morphological differences in the duodenal and jejunal villi in different sampling time. Duodenal villi were finger-thick, short, with different heights and with frequent anastomoses. Leaf-shaped villi and duodenal folds could also be verified. In the jejunum, fingerlike villi, thin and thick, of different heights were observed as well as leaf-shaped. Vacuoles with colostrum were observed in the jejunum of goats sampled at 18 hr of life. In the ileum, animals showed fingerlike villi, with different widths. The villi were the highest at 0 hr and the lowest at 18 and 36 hr. The structure of absorptive mucosa of ileum showed an intermediate height and few anastomoses at 96 hr. At all sampling time, frequently cell extrusion processes were observed at the apex of the ileum villi with a group of cells and along the villi with isolated cells. Use of lyophilized bovine colostrum did not influence the small intestine morphological characteristics ensuring this source of Ig as a possible substitute of goat colostrum.

Effect of feeding lyophilized bovine and goat colostrum on serum antibodies fluctuation in newborn goat kids
Lima, A.L., Moretti, D.B., Nordi, W.M., Pauletti, P., Susin, I. and Machado-Neto, R., ESALQ/Universidade de São Paulo, Zootecnia, Av. Pádua Dias, 11, 13.418-900, Piracicaba, SP, Brazil; raul.machado@esalq.usp.br

Colostrum that constitutes the only source of antibodies for newborns goat kids, is very often a vehicle of Caprine Arthritis Encephalitis virus. It was evaluated the effect of initial levels of passive protection acquired from goat and bovine colostrum and the viability of using lyophilized bovine colostrum to newborns goat kids. Twenty-five newborns female goat kids were randomly allocated to five treatments, two fed goat (GCA) and bovine (BCA) colostrum, with 45 to 55 mg/ml of IgG, another two, goat (GCB) and bovine (BCB) colostrum, with 15 to 25 mg/ml of IgG, and a fifth treatment fed lyophilized bovine colostrum (LBC), with 45 to 55 mg/ml of IgG. The animals received 5% of body weight of colostrum at 0, 12, 24 hr of life and, after, cow's milk twice a day and concentrate *ad libitum* until 60 days of life. Blood samples were collected at 0, ½, 1, 2, 5, 10, 15, 20, 25, 30, 35, 40, 50 and 60 days of age and analyzed by electrophoretic fractionation of serum proteins to determine gamma globulin (CELMGEL). Serum gamma globulin concentration showed significant effect of period (P<0.05) and treatment (P<0.05). The levels of serum gamma globulin at birth were the smallest, with a mean concentration of 0.24±0.04 g/dl, and at 12 hr (½ day) the values were lower than those found between 30 to 60 days. The mean serum Ig at 60 days was 7.1±0.04 mg/ml. Animals from LBC, with a mean concentration of 6.7±0.04 mg/ml, differed from GCB and BCB (4.7±0.04 and 4.7±0.05 mg/ml, respectively). Gamma globulin serum concentration found in animals from GCA and BCA, 6.2±0.04 and 5.1±0.04 mg/ml, respectively, were not different from concentrations found in animals from GCB and BCB, that have received colostrum with lower concentration of antibodies. The results indicate that lyophilized bovine colostrum can be used to feed newborns goat kids as an alternative source for the acquisition of initial protection.

Performance of lambs fed diets with different proportions of physic nut meal (*Jatropha curcas*)
Oliveira, P.B.[1], Lima, P.M.T.[1], Campeche, A.L.[2], Mendonça, S.[3], Cabral Filho, S.L.S.[1], Mcmanus, C.[4], Abdalla, A.L.[2] and Louvandini, H.[2], [1]Universidade de Brasília - UnB, Brasília/DF, 70910-900, Brazil, [2]Centro de Energia Nuclear na Agricultura - USP, Piracicaba/SP, 13400-970, Brazil, [3]Embrapa Agroenergia, Brasília/DF, 70770-901, Brazil, [4]Universidade Federal do Rio Grande do Sul, Porto Alegre/RS, 900040-060, Brazil; pedrobatelli@gmail.com

Although it has high protein value (23%), the physic nut is not utilized in animal feed due to its toxicity, resulting from the presence of the phorbol ester. Studies have shown it is possible to detoxify physic nut bran and development of varieties of physic nut without the phorbol ester, representing a potential use in ruminant feeding. The objective of this work was to evaluate the performance of Santa Inês lambs fed diets with physic nut meal (Jatropha curcas) with zero concentration of phorbol ester. Twenty four intact lambs, 120 days of age and 21+1.88 kg of body weight (BW) were housed during 60 days and divided in four treatments according to the percentage of physic nut meal in the concentrate (0, 20, 40 and 60%). They were fed with Tifton (Cynodon dactylon) hay *ad libitum* and a concentrate isoenergectic and isoproteic mixture (corn, soybean meal and plus mineral mixture). Feed intake and BW were measured three times a week and every 15 days respectively. The statistical analyses were carried out utilizing the statistical software Statistical Analysis System (SAS) and an analysis of variance and Tukey's test at probability level of 5% was carried out. The daily weight gain and total weight gain showed no difference between the four groups. The average daily weight gain was 137+27.2; 122+9.0; 129+32.3; and 116+16.4 g/animal/day and the total weight gain was 8.2+1.63; 7.3+0.54; 7.7+1.94; and 7.0+0.99 kg for 0, 20, 40, 60% proportion treatment respectively. Therefore, the physic nut meal with zero concentration of phorbol ester had shown to be a viable alternative source of feed for ruminants.

Biosecurity practices in goats of the region of Coquimbo, Chile

Maino, M., Lazcano, C., Soto, A., Duchens, M., Perez, P. and Aguilar, F., Faculty of Veterinary Medicine University of Chili, Department of Animal Production Development, Av. Santa Rosa 11735, La Pintana, Santiago, Chile, 8820808, Chile; mmaino@uchile.cl

Biosecurity is a set of preventive measures designed to reduce the entry of pathogens to the livestock, as well as avoid or reduce its dispersal. The biosecurity contributes to competitivity through (1) to improve the productivity of the herds (2) to improve the quality of the products and (3) to improve the conditions for commercialization the products. Additionally, the biosecurity improves public health lowering zoonosis risks and food borne diseases, animal productivity and trade. The objective of the present study was to identify biosecurity level of caprine livestock in the Chilean region of Coquimbo, which is the area that gathers 60% of the Chilean caprine flock, and correlate farms biosecurity level with a group of social and productive attributes. In order to reach these objectives we defined biosecurity protocol's technical specifications. In order to define the protocol technical specifications, an expert panel worked over a bibliographic review of the main foreign biosecurity measures for goats. The resulting biosecurity protocol was applied as a check-list by means of a direct interview to 55 goat farmers of the region. The relations between the fulfillment of biosecurity measures and their social and productive characteristics were analyzed using Spearman correlation; the analysis was performed through InfoStat version 2004 software. The 14.9% of the assessed measures are fulfilled by stockbreeders. The most performed measures are feed storage (94.5%), health products storage and disposal (62.5%), manure removal and disposal (56.4%), milking practices (51%), and water supply (49.1%). No relations between biosecurity measures and social or productive characteristics were found.

***Ad libitum* feed intake patterns of sheep are repeatable**

Bickell, S.L.[1,2], Toovey, A.F.[2,3], Revell, D.R.[3] and Vercoe, P.E.[1,2], [1]The University of Western Australia, School of Animal Biology, 35 Stirling Highway, Crawley, WA, 6009, Australia, [2]Sheep Cooperative Research Centre, Armidale, NSW, 2351, Australia, [3]CSIRO Livestock Industries, Private bag 5, Wembley, WA, 6914, Australia; samantha.bickell@uwa.edu.au

Patterns of daily methane production (DMP) have been associated with diurnal grazing patterns of ruminants, where peaks in DMP correspond with peaks in eating activity. Due to this diurnal grazing pattern, some studies of DMP of sheep in respiration chambers have been designed to simulate grazing patterns by offering feed in equal morning and evening portions. However, in other DMP studies animals are offered their ration once or are fed *ad libitum* and very little is known about the animal's feeding pattern in this artificial environment. Our aim was to investigate the pattern of feed intake of sheep fed an *ad libitum* diet in methane respiration chambers. We also investigated the repeatability of the pattern of feed intake over time as eating rates of animals have been shown to be repeatable. Hourly amounts of food consumed by 20 sheep while in a methane respiration chamber for 23 h were recorded on two separate occasions 4 weeks apart. Hourly food consumption was determined from digital scales remotely logging the weight of each animal's feed container every 5 min. Individual sheep had different eating patterns and their hourly feed intake, measured 4 weeks apart, was highly repeatable with a correlation of 0.82. Feed intake increased for all animals as expected but feed intake increased at a constant linear rate for some animals, while others showed distinct bouts of eating and non-eating. Individual sheep have repeatable, but distinct patterns of feed intake when fed *ad libitum* in methane respiration chambers. An animal's pattern of feed intake may be an important factor to consider when assessing how well chamber studies reflect an animal grazing in the field. We are now investigating how patterns of feed intake influence the pattern and amount of methane produced in the chambers.

Fattening and carcass quality characteristics and ultrasound measurements of heavy lambs of Tsigai sheep in Slovakia

Polák, P., Tomka, J., Krupová, Z., Krupa, E. and Oravcová, M., Animal Production Research Centre Nitra, Deperment of Animal Breeding and Product Quality, Hlohovecká 2, 951 41 Lužianky, Slovakia (Slovak Republic); polak@cvzv.sk

Tsigai is a dual purpose breed traditionally kept in milk production scheme with selling of light lambs before Easters. Production of heavy lambs by suckling mothers in pastures is not traditional for the breed in the region. Fattening carcass quality characteristics of 20 heavy lambs of Tsigai sheep were obtained at the average age of 137 days. Sonograms of transversal cut of musculus longissimus thoracis et lumborum (MLTL) were obtained one day before slaughter on last thoracic vertebra by echocamera Aloka PS 2 and ultrasound probe UST 5820 – 5. Muscle width, thickness and area and fat thickness were measured on digitalised sonograms in special software for video image analysis NIS – elements. Relatively high variability, more than 10%, were found for dressing percentage, weight of carcass, weight of meat in carcass, weight of valuable cuts in carcass, average daily gain, lean meat production per day and backfat thickness. The highest variability was found out for fat proportion in carcass. The width of MLTL had the higher coefficients of correlation with carcass quality variables among ultrasonic measurements (0.35 for weight of carcass, 0.36 for weight of meat in carcass and 0.32 for weight of valuable cuts in carcass). Muscle thickness and muscle area had negative small (close to zero) correlation coefficients. Ultrasound measurements had coefficient of correlations with average daily gain between 0.26 and 0.33. Despite of limited number of analysed animals in this preliminary study, the findings indicate that ultrasound can be successfully utilized in prediction of carcass quality characteristics.

Relationship between electrical condutivity and milking fraction of milk of Murciano-Granadina goats

Romero, G.[1], Pantoja, J.C.[2], Sendra, E.[1], Alejandro, M.[1], Peris, C.[3] and Diaz, J.R.[1], [1]Universidad Miguel Hernández (UMH), Dep Tecnología Agroalimentaria., Ctra. Beniel, km. 3,2, 03312 Orihuela, Spain, [2] University Wisconsin-Madison, Dairy Science, Observatory Drive, Madison (WI), USA, [3]Universidad Politécnica Valencia, Instituto Ciencia Animal, Camino de Vera s/n, 46022-Valencia, Spain; gemaromero@umh.es

The aim was to study the relationship between electrical conductivity (EC) and milking fraction (MF) of goat milk from glands previous to development of algorithms for goat mastitis detection. Additionally it was studied the relationship of EC with milk composition. 57 goats were enrolled in the study (28 healthy and 29 with unilateral mastitis infection) from the experimental farm of the UMH. A single sampling day was done in the fourth month of lactation. The relationship between EC and milking fraction (MF, 3 levels; F1: first 100 ml, F2: machine milk, F3: last 100 ml), parity (NP, 2 levels; primiparous or multiparous) and health status of the mammary gland (HS, 2 levels: healthy or infected) was assessed using a mixed linear model. EC was transformed to logarithm base 10 (LEC). All the variables considered were significant. From all possible interactions, only MF with HS was significant. The progress of milking caused a decrease of LEC, regardless of the NP and HS. Greatest differences between healthy and infected glands were obtained at F3 fraction (5.01 vs 5.14 mS/cm), although not high enough to be detected by commercial conductimeters. The relationship between EC with milk macro-composition (fat, casein, whey protein, lactose and ash) was assessed using a linear regression model. EC was negatively correlated (P<0.001) with fat (R^2=0.27), lactose (R^2=0.39) and whey protein (R^2=0.08) and positively with ash (R^2=0.02), and was not related to the casein content. All fractions studied may provide useful on-line EC measurements for mastitis detection. However, further studies are needed on automatic daily measurements of EC to determine highly optimized algorithms for mastitis detection.

Longissimus muscle fatty acid profile of lambs fed a high-concentrate diet containing canola, sunflower or castor oil

Maia, M.O., Ferreira, E.M., Nolli, C.P., Gentil, R.S., Polizel, D.M., Petrini, J., Pires, A.V., Mourão, G.B. and Susin, I., University of São Paulo/ESALQ, Av. Padua Dias, Piracicaba, SP, 13418-900, Brazil; ivasusin@esalq.usp.br

Lamb meat is a good protein source; however, it's high in saturated fatty acids. Thirty six Longissimus muscle of crossbred Dorper x Santa Inês lambs (BW at slaughter=36.7±1.5 kg and 120 d old) were used to determine the effects of diets with 3% of vegetable oil on meat fatty acid composition. Lambs were assigned to a randomized complete block design and fed the experimental diets during 8 weeks. The control (CON) diet contained 90% concentrate and 10% hay in a DM basis.In the remaining treatments, diets were added with 3% canola oil (CAN), 3% sunflower oil (SUN) or 3% castor oil (CAS). Total lipids were extracted, esterified and methylated. Methyl esters were separated by gas chromatography (Agilent/Model 7890) using a 100 m capillary column. Data were analyzed using the MIXED procedure of SAS and means were compared by Tukey Test. CON and CAS diets increased (P=0.01) proportions of C16:1. SUN diet increased concentrations of C18:1 trans (4.12, 4.00, 6.57, 3.88% for CON, CAN, SUN and CAS, respectively) and decreased (P=0.02) C18:1n9 compared to CAS (35.53, 39.99% for SUN and CAS, respectively). CAS decreased (P=0.02) C18:2 n-6 (7.68, 8.45, 9.03 and 6.33% for CONT, CAN, SUN and CAS, respectively). CAN diet increased the concentration of C18:3 n-3 (0.21, 0.36, 0.21 and 0.20% for CON, CAN, SUN and CAS, respectively). The Inclusion of vegetable oils did not affect MUFA:SFA ratio. However, CAN and SUN diets increased PUFA:SFA (0.34 and 0.35%, respectively) in comparison with CAS (0.25%). The C18:1-OH was present only in CAS (0.21%) diet because the composition of the castor oil. The proportion of C18:2 cis-9, trans-11 (CLA) was higher when CAN (0.59%) and SUN (0.44%) were included in the diet, in contrast with CON and CAS (0.31 and 0.33%, respectively). Canola and sunflower oils changed Longissimus muscle fatty acid composition of crossbred Dorper x Santa Inês lambs, showing a higher CLA content.

Response to a divergent selection based on somatic cell counts in Alpine dairy goats

Caillat, H.[1], Bouvier, F.[2], Guéry, E.[3], Martin, P.[4], Rainard, P.[5] and Rupp, R.[1], [1]INRA, UR631 SAGA, F-31326 Castanet-Tolosan, France, [2]INRA, UE332 Bourges, F-18390 Osmoy, France, [3]Laboratoire Départemental d'Analyses, 216 Rue Louis Mallet, F-18020 Bourges, France, [4]Capgenes, 2135 RTE CHAUVIGNY, F-86550 Mignaloux-Beauvoir, France, [5]INRA, UR1282 IASP, F-37380 Nouzilly, France; rachel.rupp@toulouse.inra.fr

Milk somatic cell count (SCC) is routinely collected in the French Alpine and Saanen dairy breeds and its heritability has been estimated to be around 0.20.Accordingly, it is possible to consider improving the mastitis resistance in goat by selection on SCC. However, efficiency of such a selection raises some concerns, especially because non infectious factors of variation have a large effect on goat milk SCC. Our study therefore consisted in evaluating the consequences of SCC-based selection on intra-mammary infections. Using 13 progeny-tested AI bucks selected for extreme breeding values for somatic cell scores (SCS), we created two groups of28 High SCS and 25 Low SCS goats which were raised at the INRA experimental facility of Bourges. Milk bacteriological analyses and SCC of half-udders were performed at the kidding date and at 7monthly points in first lactation. Milk production was similar in the 2 lines. The mean SCC was 1,709,000 cells/ml and 592,000cells/ml for the High and Low SCS goats, respectively, with a significant (P<0.05) difference of 1.4 point in SCS. Regarding milk bacteriology results, 35% of samples were positive. The bacterial types found most frequently were Coagulase Negative Staphylococci (67%), S. xylosus being the most prevalent (19.5%). Frequency of positive samples was significantly higher in the High SCS (43%±2) than in the Low SCS line (25%±2). Additionally, the mean SCS was significantly higher for the positive than for negative milk samples (4.8±0.2 vs 3.8±0.1). This difference was also significant within the goat line. Preliminary results in the High and Low SCS lines therefore gave evidence that SCS-based selection in goat will lead to decreased intra-mammary infection prevalence and mean SCS.

Induction and synchronization of ovulation by combining the male effect, intravaginal progesterone capsules and cloprostenol, in goats during the non-breeding season
Gomez-Brunet, A.[1], Toledano, A.[1], Coloma, M.A.[1], Velazquez, R.[1], Castaño, C.[1], Carrizosa, J.A.[2], Urrutia, B.[2], Santiago, J.[1] and López, A.[1], [1]INIA, Animal Reproduction, Avd. Puerta de Hierro km 5.9, 28040, Spain, [2]IMIDA, Animal Production, La Alberca, 30150 Murcia, Spain; gomez@inia.es

Looking for alternative methodologies, focusing on limiting the use of progestagens to estrus synchronization in small ruminants, recently, we have developed an effective method, the IMA-PRO2®, based on male exposure, progesterone (P4) administered by i.m injection and cloprostenol 9 days later. Because the EU legislation forbidding i.m P4, a first experiment was conducted to examine the potential value of using intravaginal P4 capsules to replace i.m injection at buck introduction, on estrous behavior and LH secretion. In a second experiment, pregnancy rates were determined, in goats inseminated after cloprostenol and previously treated with either i.m or intravaginal. In Experiment 1, goats were divided into three groups. Does in groups 1 and 2 were treated, respectively, with intravaginal P4 capsules (120 mg in palm oil) or with 25 mg progesterone i.m in olive oil at the time of introduction of bucks, whereas does in group 3 were treated with an intravaginal progestagen-FGA sponges during 30 h after bucks introduction. There were no treatment differences neither in the time of estrus behavior or in the preovulatory LH surge. The preovulatory LH surge occurred 58,0±3,9 h in group 1, 69.0±2.5 h in group 2 and 65.0±1.3 h in group 3. In Experiment 2, goats in the two groups were inseminated intravaginall 50 h after cloprostenol. Pregnancy rates were similar in inseminated does treated with the IMA-PRO2® (60.6%, n=61) and in those treated with the intravaginal P4 capsules (58.5%, n=62). In conclusion, our results show that the use of intravaginal capsules at the time of buck introduction followed by an i.m cloprostenol, 9 days later, constitutes an effective method to synchronize estrous and ovulations during seasonal anestrous.

Environmental impact associated with freshwater use along the life cycle of animal products
De Boer, I.J.M.[1], Hoving, I.E.[2] and Vellinga, T.V.[2], [1]Wageningen University, Animal Production Systems Group, P.O. Box 338, 6700 AH Wageningen, Netherlands, [2]Wageningen UR Livestock Research, P.O. Box 65, 8200 AB Lelystad, Netherlands; Imke.deBoer@wur.nl

Freshwater is essential for human well-being and ecosystem quality. Awareness of preserving freshwater as a resource is increasing. The water footprint of a product measures the amount of water that is required along the life cycle of that product, and does not assess the environmental impact associated with this water use, such as the impact on human health (HH), ecosystem quality (EQ) or resource depletion (RD). Recently, factors were published that determine the impact of freshwater use in a region or country on HH, EQ and RD. We used this knowledge in order to develop an approach to assess the regional impact of freshwater use on HH, EQ and RD along the life cycle of an animal product. The approach developed identifies hotspots of animal chains in terms of regional impacts associated with freshwater use, and, therefore, identifies were water-saving may have the greatest benefits. Production of one kg of Dutch fat-and-protein corrected milk (FPCM) on a dairy farm with intensive irrigation, for example, required 61 kg of consumptive water. About 75% this water was required to irrigate on-farm roughage production, 15% for production of concentrates and 5% for drinking and cleaning water. Consumptive water use related to use of fossil energy, fertilizer and transport along the milk chain was 5% only. Production of one kg of FPCM resulted in an impact on HH of 0.82×10^{-9} DALY (disability adjusted life years), on EQ of 0.011 m^2.yr and on RD of 6.7 KJ. The impact of producing this kg of FPCM on RD was caused mainly by cultivation of concentrate ingredients, and appeared lower than the average impact on RD of production of one kg of broccoli in Spain or one kg of cotton in Egypt. Results highlight, therefore, the need for life cycle assessment to go beyond simple water volume accounting when the focus is on freshwater scarcity.

Life cycle assessment of milk at the farm gate

Kristensen, T.[1], Mogensen, L.[1], Knudsen, M.T.[1], Flysjo, A.[2] and Hermansen, J.E.[1], [1]University of Aarhus, Faculty of Agricultural Sciences, Department of Agroecology and Environment, P.O. Box 50, 8830 Tjele, Denmark, [2]Arla Foods amba, Sønderhøj 14, 8260 Viby J, Denmark; troels.kristensen@agrsci.dk

Life cycle assessment (LCA) represents a comprehensive method to estimate the accumulated impact on the environment of a production process or a product. During recent years, several papers have reported the green house gas (GHG) emission per kg milk produced by use of the LCA concept. The aim of this paper is firstly to introduce the life cycle methodology and illustrate the impact of some methodological choices, secondly to give typical figures, including hot spots, for GHG emission as a result of the production process until the farm gate and thirdly to identify and quantify biological and management options of significant importance for the variation in the emission and thereby identify possible mitigations options. Important methodological choices include whether to use a steady state approach (attributional) or a marginal approach (consequential), how to include aspects of land use changes as a result of the dairy production system and how to handle allocation of emissions on multiple products like milk and meat. In a typical EU milk production system, the GHG emission from the farm process accounts for 80-90% of the total emission in the whole chain ending at the consumer. On the farm CH_4 is the major source (40-60%), followed by N_2O (30-40%) and with use of fossil energy (CO_2) only accounting for 15-20%. Variation in on farm emission and emission from imported products is one of the major differences between production systems, which illustrate the importance of including the emission related to production before the farm. Variation in feed conversion rate, calculated as milk yield from dry matter intake (DMI) at herd level, highly affects GHG emission per kg milk on Danish dairy farms, followed by variation in stocking density.

Carbon footprinting of New Zealand lamb from an exporting nation's perspective

Ledgard, S.F.[1], Lieffering, M.[2], Coup, D.[3] and O'brien, B.[4], [1]AgResearch, Ruakura Research Centre, Private Bag 3123, Hamilton, New Zealand, [2]Ag Research, Grasslands Research Centre, Palmerston North, New Zealand, [3]Meat Industry Association, Wellington, P.O. Box 345, New Zealand, [4]Beef and LambNZ, Wellington, P.O. Box 121, New Zealand; stewart.ledgard@agresearch.co.nz

New Zealand (NZ) is the world's largest exporter of lamb. Lamb from NZ is exported to distant markets and this has implications for transportation, 'food-miles', greenhouse gas (GHG) emissions and final product costs. The NZ agricultural sector has a strong focus on efficiency of production and development of low-cost farming systems that rely on a temperate climate for all-year grazing of perennial pastures and little use of conserved feed. Two major drivers for this project have been; i. demands for information by customers (primarily supermarket chains) accentuated by NZ's distance from markets and related transport emissions, and ii.NZ government policy on limiting GHG emissions via an Emissions Trading Scheme which includes agriculture. The NZ meat industry (including processors) initiated a national project to determine the carbon footprint of lamb using a Life Cycle Assessment approach. Emphasis was on the whole life cycle and applying methods that complied with ISO14044 and the PAS2050, particularly since the UK is a major market for NZ lamb. Primary data was used for lamb production and meat processing, while secondary data was used for retailer/consumer/waste stages. The carbon footprint averaged 19 kg CO_2-equivalent/kg lamb meat, with 80% from the cradle-to-farm-gate (mainly animal CH_4 and N_2O emissions), 3% from processing, 5% from all transportation stages (mainly shipping), and 12% from retailer/consumer/waste stages (mainly retail storage, home cooking). Inclusion of consumer transport to purchase the meat added up to 7% to the total carbon footprint. Sensitivity analysis was carried out for all life cycle stages. Findings are being used by the NZ meat industry to determine the most effective areas in the life cycle for carbon footprint reduction and improved system efficiency.

A life cycle assessment of seasonal grass-based and confinement total mixed ration dairy farms

O'brien, D.[1,2], Shalloo, L.[1], Patton, J.[1], Buckley, F.[1], Grainger, C.[1] and Wallace, M.[2], [1]Animal & Grassland Research and Innovation Centre, Livestock Systems Research Department, Teagasc, Moorepark, Fermoy, Co.Cork, Ireland, [2]University College Dublin, School of Agriculture, Food Science and Veterinary Medicine, Belfield, Dublin 4, Ireland; daniel.obrien1@ucdconnect.ie

The purpose of this study was to compare the resource use and environmental impact of a seasonal pasture-based and a confinement total mixed ration (TMR) dairy farm operated in Ireland using life cycle assessment (LCA). The method was applied to assess the whole life cycle for the production of raw milk, from the production of inputs to products being sold from the farm. Biological data required to assess both systems was obtained from previously published work. The impact categories included in the analysis were; global warming, eutrophication, acidification, land use and non-renewable energy use. The study found that the grass-based system required 10% less land and used 40% less non-renewable energy per unit of milk. The grass-based system performed better in all environmental impacts categories relative to the confinement TMR system reducing greenhouse gas emissions, acidification potential and eutrophication potential per unit of milk by 14%, 49% and 31%, respectively. Concentrate feed was the dominant contributor to environmental impacts in the confinement TMR system. On-farm pasture production was a major cause of all impacts in the grass-based system. Inorganic and organic fertilizers and fossil fuels use were major sources of impacts for both grass and concentrate production. The environmental impact of producing concentrate was greater than the production of on-farm forages. Consequently, the confinement TMR system required more resources and had a greater environmental impact for a given level of milk compared to the seasonal grass-based system. These finding indicate that a shift away from grazing-based dairy systems will not increase the sustainability of dairy production.

Is feeding more maize silage to dairy cows a good strategy to reduce greenhouse gas emissions?

Van Middelaar, C.E.[1], Dijkstra, J.[2], Berentsen, P.B.M.[3] and De Boer, I.J.M.[1], [1]Wageningen University, Animal Production Systems Group, P.O. Box 338, 6700 AH Wageningen, Netherlands, [2]Wageningen University, Animal Nutrition Group, P.O. Box 338, 6700 AH Wageningen, Netherlands, [3]Wageningen University, Business Economics Group, P.O. Box 8130, 6700 EW Wageningen, Netherlands; corina.vanmiddelaar@wur.nl

The dairy sector contributes to climate change through emission of greenhouse gases (GHGs): carbon dioxide (CO_2), methane (CH_4), and nitrous oxide (N_2O). From an animal perspective, a feeding strategy with high potential to reduce enteric CH_4 emission is replacing grass silage with maize silage. Increasing maize silage in the ration, however, affects the farm plan (e.g. ploughing grassland for maize land, type of concentrate), and consequently GHG emissions along the milk production chain. This study evaluated the effect of replacing grass silage with maize silage on GHG emission from a life cycle perspective. Life cycle assessment, linear programming (economic optimization) and dynamic system modelling were combined to define a reference farm (based on an average Dutch dairy farm on sandy soil) and to evaluate GHG emissions along its production chain. Subsequently, maize silage was increased by 1 kg DM/milking cow/day at the expense of grass silage, and the model was used again to evaluate the new optimal farm plan for the farm with this strategy, and GHGs emitted. For each ton of milk produced, increasing maize silage reduced annual emissions by 18 kg CO_2-eq. compared to the reference situation. Lower use of concentrates due to higher nutritional value of maize compared to grass, and lower enteric CH_4 emission were the main contributors to this reduction (respectively 50% and 25%). Ploughing grassland for maize land, however, leads to loss of soil carbon and resulted in non-recurrent CO_2 and N_2O emissions of 652 kg CO_2-eq. in total. In conclusion, it takes up to 36 years before annual emission reductions by feeding more maize silage compensates for emissions related to ploughing grassland for maize land.

Comparing strategies to aggregate environmental performances of individual farms into a dairy sector score

Dolman, M.A.[1,2], Vrolijk, H.C.J.[1] and De Boer, I.J.M.[2], [1]LEI, part of Wageningen UR, Alexanderveld 5, 2585DB The Hague, Netherlands, [2]Wageningen UR, Animal Production Systems Group, Marijkeweg 40, 6709PG Wageningen, Netherlands; mark.dolman@wur.nl

The environmental performance of an agricultural sector can be assessed based on input-output data at sector level (top-down) or on aggregating environmental scores of individual farms (bottom-up). A bottom-up approach has the advantage that variation among farms is assessed, and it can be used to deduce mitigation options. Our objective is to compare strategies to aggregate environmental scores of Dutch dairy farms into a sector score, using the Dutch Farm Accountancy Data Network (FADN). Life cycle assessment (LCA) was used to quantify the environmental performance of milk production at individual dairy farms (i.e. land use, energy use, climate change, eutrophication and acidification). The functional unit is one kg of fat-and-protein corrected milk. Two aggregation strategies are compared: 1) a conventional FADN weighting method based on a stratification of economic farm size and farm type only, which does not consider other farm characteristics that determine environmental impacts, such as milk production per ha or soil type; 2) a strategy based on statistical matching using additional farm characteristics to compute a weighting factor per farm. Aggregated environmental scores are validated by a leave-one-out cross validation in which observed and estimated values for individual farms are compared. With respect to soil type, economic size, milk production per ha and the ratio of maize silage, results per region were more representative using statistical matching. The variance of the weighting factor was higher, resulting in a higher sensitivity of the weighting scheme applied. The choice of weighting factor highly affected environmental performance (range 10-20%). Statistical matching was identified as a better method to aggregate environmental scores of individual farms into a sector score for environmental impacts.

Greenhouse gas emissions of Spanish sheep farming systems: allocating between meat production and ecosystem services

Ripoll-Bosch, R.[1], De Boer, I.J.M.[2], Bernués, A.[1] and Vellinga, T.[3], [1]CITA de Aragón, Avda Montañana, 930, 50059 Zaragoza, Spain, [2]Animal Production Systems Group, Wageningen University and Research Centre, P.O. Box 338, 6700 AH Wageningen, Netherlands, [3]Wageningen UR Livestock Research, Wageningen University and Research Centre, P.O. Box 65, 8200 AB Lelystad, Netherlands; rripoll@aragon.es

Sheep farming systems (SFS) in Spain are considered pasture-based and low-input, but large differences in input utilization, land use and intensification level exist, and their environmental impacts, therefore, are expected to differ. We used life cycle assessment (LCA) to evaluate and compare greenhouse gas (GHG) emissions of three contrasting SFS: (1) grazing (G): located in alpine mountains, with 1 lambing per year and free ranging; (2) mixed sheep-cereal (M): located in mid-altitude ranges, with 3 lambings in 2 years and guided grazing; (3) zero-grazing (Z): located in low altitude semi-arid conditions, with 5 lambings in 3 years and no grazing. The functional unit (FU) was 1 kg of lamb live-weight leaving the farm. Emissions of GHGs from on-farm processes and farm inputs were computed according to IPCC guidelines (Tier 2 level). Per FU, GHG emissions were highest for G (28.4 CO_2-eq), intermediate for M (24.3 CO_2-eq) and lowest for Z (19.5 CO_2-eq). Besides meat, however, these, SFS also provide ecosystem services to society (e.g. biodiversity and landscape conservation). We valued these services for each SFS based on agri-environmental subsidies of the EU, and used economic allocation to distinguish GHG emissions of SFS between meat and ecosystem services. Correcting for multifunctionality of SFS, GHG emission per kg live-weight changed, i.e. lowest for G (15.2 CO_2-eq), intermediate for M (18.0 CO_2-eq) and highest for Z (19.5 CO_2-eq). A comparison of GHG emissions among SFS should account for the multifunctionality of these systems.

Evaluation of the environmental sustainability of different European pig production systems using life cycle assessment

Dourmad, J.Y.[1], Ryschawy, J.[1], Trousson, T.[1], Gonzalez, J.[2], Houwers, H.W.J.[3], Hviid, M.[4], Nguyen, T.L.T.[5] and Morgensen, L.[5], [1]INRA Agrocampus Ouest, UMR1079 SENAH, 35590 Saint-Gilles, France, [2]IRTA, Finca Camps i Armet, 17121 Monells, Spain, [3]Wageningen UR Livestock Research, P.O. Box 65, 8200 AB Lelystad, Netherlands, [4]DMRI, Maglegaardsvej 2, DK-4000 Roskilde, Denmark, [5]DJF, Univ. of Aarhus, 8830 Tjele, Denmark; jean-yves.dourmad@rennes.inra.fr

The environmental sustainability of 12 European pig production systems has been evaluated within the EU Q-PorkChains project, using life cycle assessment (LCA). One conventional and two differentiated systems were evaluated for each of four countries: Denmark, Netherlands, Spain and France. The information needed for the calculations was obtained from an enquiry conducted on 10 farms from each system. The environmental impacts were calculated at farm gate, including the inputs, and expressed per kg live pig and per ha land use. For the conventional systems, the impact per kg pig produced on climate change, eutrophication, acidification, energy, and land use were 2.01 kg CO_2-eq, 41.5 g SO_2-eq, 26.3 g eq PO_4-eq, 15.2 MJ and 4.0 m^2, respectively. The corresponding values for the differentiated systems were on average 40, 22, 49, 28 and 80% higher, but with large variations between systems. Conversely, when expressed per ha of land use, the impacts were lower for the differentiated systems, by 10 to 20% on average, depending on the impact category, due to higher land occupation per kg pig produced. The use of litter bedding tended to increase climate change impact per kg pig. The use of traditional local breeds, with lower productivity and feed efficiency, resulted in higher impacts per kg pig produced, for all categories. Differentiated systems with extensive outdoor raising of pigs resulted in markedly reduced impact per ha land use. The results indicated that the conventional systems were generally better for global impacts, expressed per kg pig, whereas differentiated systems were often better for local impacts, expressed per ha land use.

Measuring water footprints in dairy production worldwide in a climate change scenarios

Sultana, M.N., Uddin, M.M., Ndambi, O.A. and Hemme, T., International Farm Comparison Network (IFCN) Dairy Research Center, Germany, Department of Agricultural Economics, University of Kiel, Schauenburger Str. 116, 24118 Kiel, Germany, Germany; torsten.hemme@ifcndairy.org

The decreasing water availability is a risk to food security. Water footprints has emerged as an important indicator for water use in agriculture and food production which is sensitive to climate change and can also benchmark water use efficiency in dairying. This study aims at developing a method for calculating water footprints in dairying. An extended version of the TIPI-CAL (Technology Impact Policy Impact Calculations model) of International Farm Comparison Network (IFCN) was used for this analysis. The underlying farm data set for this study are the typical farms of the IFCN. The method was tested on 12 typical dairy farms from six developed countries: Canada, Germany, New Zealand, Spain, Switzerland and USA and six developing countries: Argentina, Bangladesh, China, Czech Republic, Jordan, and Pakistan. The results show that cows have their highest water requirement during lactation period which varied from 66% of their total requirement in Bangladesh to 97% in Jordan. Water use during dry period was highest in Bangladesh (33%) due to very long dry period. The water footprint per kg milk varies from 430 l in USA to 2,400 l in Pakistan due to variability in milk yield and management system. The water used for drinking and servicing ranged between 3.5 and 56.0 liters for Germany and Pakistan respectively. It was concluded that feed production is the major driver for water footprints and the greatest challenges were in obtaining co-efficient for water input from feed production. The measuring of water footprints in dairying is a step towards achieving efficient water use which will augment food security and enhance climate change adaptability.

Nitrogen and mineral content in broilers

Schlegel, P.[1] and Menzi, H.[2], [1]Agroscope Liebefeld-Posieux, Tioleyre 4, 1725 Posieux, Switzerland, [2]Swiss College of Agriculture, Länggasse 85, 3052 Zollikofen, Switzerland; patrick.schlegel@alp.admin.ch

The nitrogen and mineral content of broilers were analyzed to update the coefficients used for nutrient export within the nutrient cycle assessment of broiler farms. Twenty seven ready-to-slaughter birds were collected on 14 farms. They represented 4 broiler organizations and 4 production programs (coquelets, short, normal and extensive). Extensively produced birds were Hubbard and Isa JA genetics, others were all Ross genetics. The processing of the killed animals were: weighing, coarsely grinding, freezing, lyophilisation, weighing, finely grinding followed by the chemical analysis. The average body weight (BW) for coquelets, short, normal and extensive categories were respectively 0.84±0.05, 1.47±0.22, 2.09±0.20 and 2.19±0.27 kg. Average body contents per kg dry matter were 82 g N, 82 g ash, 19.8 g Ca, 16.4 g P, 236 mg Fe, 9.7 mg Mn and 60.5 mg Zn. Magnesium, K, Na and Cu contents decreased (P<0.05) with increasing BW (Mg: 1.1 g – 0.044*BW; K: 8.4 g – 0.517*BW; Na: 3.87 g – 0.424*BW; Cu: 6.73 mg – 0.94*BW). The dry matter content increased with increasing BW (311 (P<0.001) + 23 (P<0.001) * BW). Average body contents per kg fresh matter were 29.1 g N, 29.1 g ash, 7.09 g Ca, 5.83 g P, 0.35 g Mg, 2.62 g K, 1.08 g Na, 1.75 mg Cu, 83.7 mg Fe, 3.36 mg Mn and 21.3 mg Zn. The present data are 12%, 12% and 8% higher for respectively N, P and K when compared with the values actually used in Switzerland to calculate nutrient cycles in broiler production.

Carbon cycles, pyruvate carboxylase, and the potential for chaos in liver of dairy cows during the transition to lactation

Donkin, S.S., Purdue University, Animal Sciences, 915 W State Street, West Lafayette, IN, 47907, USA; sdonkin@purdue.edu

Meeting the nutritional requirements of the transition dairy cow greatly impacts productivity, longevity, and animal well-being and metabolic disorders are often observed when gluconeogenic capacity fails to adapt to the increased demands for glucose to support lactose synthesis and mammary metabolism. Coordinated adaptations in hepatic energy production and the synthesis of glucose are necessary to support mammary lactose synthesis and other tissue functions. Experiments from our laboratory and others have identified the importance of changes in expression of genes for pyruvate carboxylase (PC) and cytosolic phosphoenolpyruvate carboxykinase (PEPCK-C) as part of the molecular adaptations necessary for a smooth transition to lactation. The mRNA abundance and activity of these enzymes is closely linked to gluconeogenic capacity in bovine liver. Expression of PC is elevated 5- to 6-fold with the onset of calving and PEPCK expression is elevated 2-to 3-fold after lactation is established. The bovine PC gene contains 3 separate promoters and the proximal promoter (promoter 1) of bovine PC is uniquely expressed in lipogenic and gluconeogenic tissues. Expression of bovine PC promoter 1 is activated in response to the PPAR-α agonist Wy14643 and with serum from feed restricted cows. Data indicate that regulation of PC expression in liver is due to changes in the level and profile of fatty acids released from adipose tissue. In addition, shifting the endproducts of rumen fermentation towards more propionate prior to calving increases hepatic PEPCK-C mRNA expression in transition cows. *In vitro* and *in vivo* studies indicate an activation of PEPCK-C promoter activity by propionate that suggests a feed-forward mechanism of metabolic control in ruminants. The available data link molecular control of gluconeogenesis with the profile of rumen fermentation end products and mobilization of adipose tissue to coordinate capacity for gluconeogenesis in transition dairy cows.

Contribution of amino acids to liver gluconeogenesis in transition dairy cows
Larsen, M. and Kristensen, N.B., Aarhus University, Department of Animal Health and Bioscience, Blichers allé D20, DK-8830 Tjele, Denmark; Mogens.Larsen@agrsci.dk

The transition from pregnancy to lactation represents a great challenge to the adaptation of nutrient metabolism in the liver and other tissues in order support the dramatically increased need for glucose and amino acids in the dairy cow. The liver release of glucose increases abruptly after calving, but the propionate availability from ruminal fermentation of feed is limited due to delayed increases in feed intake. Thus, the relative importance of glucogenic carbon recycling via lactate, alanine and glycerol is greater in *post partum* transition cows. Further, it is generally believed that the limited propionate availability leads to greater contribution of glucogenic amino acids for liver gluconeogenesis. We have measured the net liver release of glucose and net liver removal of glucogenic precursors in three experiments with totally 23 periparturient dairy cows catheterised in the hepatic portal vein, the hepatic vein, and an artery. A complete randomised design was used and dietary treatments were initiated at the day of calving. The contribution of precursors to liver gluconeogenesis was estimated as maximal contributions by dividing the net removal of each precursor with the release of glucose and using a stoichiometric relation of 2:1. This does not take other metabolic pathways utilising precursors into account; thus, estimated contributions are maximal contributions. Our results question the dogma of glucogenic amino acids being major precursors for liver gluconeogenesis in *post partum* transition cows. Alanine was the only amino acid that likely contributed substantially to liver gluconeogenesis in *post partum* transition cows, in line with alanine being an interorgan transporter of amino groups from peripheral protein reserves. These findings imply that amino acids are prioritised for milk protein synthesis and vital body functions over gluconeogenesis in the *post partum* transition cow. The potential benefits of increasing protein supply in this period need further investigation.

The activity of glycosidase enzymes in serum of dairy cows in their first lactation
Marchewka, J., Grzybek, W., Horbańczuk, K., Pyzel, B., Bagnicka, E., Strzałkowska, N., Krzyżewski, J., Poławska, E., Jóźwik, A. and Horbańczuk, J.O., Institute of Genetics and Animal Breeding of the Polish Academy of Sciences, Animal Sciences, Postępu 1, Jastrzębiec, 05-552, Poland; j.krzyzewski@ighz.pl

For the dairy cows in middle/peak of lactation the deficiency of energy is covered by energy from fat (fatty acids). This process results in higher glycerol synthesis in both liver and mammary gland of the cow. Glycerol as a good glycogenic substrate can be involved in gluconeogenesis process. Therefore the aim of the study was to determine the activity of glycosidase enzymes: AcP, BGAL, BGLU, NAG and MAN in serum of dairy cows. Samples were collected from 48 cows in the 60th and 200 day of their first lactation (24 higher producing – H and 24 lower producing – L cows) fed on TMR diet. The results of the study showed that in 60th day of lactation the activities of BGAL, BGLU, NAG and MAN were higher for both H and L cows (on average 5.74, 0.82, 29.05 and 15.22 nmol/mg of protein/hour, respectively) compared to day 200 (on average 4.44, 0.43, 19.24 and 8.45 nMol/mg of protein/hour, respectively). The activity of AcP was higher in 200 than in 60th day for both H and L cows (on average 7.45 vs 5.11 nmol/mg of protein/hour). Comparing serum samples of H and L cows independently of stage of lactation only the activity of AcP showed difference (on average 6.93 vs 5.80 nmol/mg of protein/hour). The results of the study confirm that glycosidase activities can help higher producing cows to cope to maintain optimal energy level. Research was realised within the project 'BIOŻYWNOŚĆ – innowacyjne, funkcjonalne produkty pochodzenia zwierzęcego' (BIOFOOD – innovative, functional products of animal origin) no. POIG.01.01.02-014-090/09 co-financed by the European Union from the European Regional Development Fund within the Innovative Economy Operational Programme 2007-2013.

Gluconeogenesis and mammary metabolism and their links with milk production in lactating dairy cows

Lemosquet, S.[1], Lapierre, H.[2], Galindo, C.E.[2] and Guinard-Flament, J.[3], [1]INRA, UMR1080, Dairy Producrtion, 35590 Saint Gilles, France, [2]Agriculture and Agri-Food Canada, Sherbrooke, QC, JM1Z3, Canada, [3]Agrocampus Ouest, UMR1080, Dairy Production, 35062 Rennes, France; Sophie.Lemosquet@rennes.inra.fr

In dairy cows, whole body (WB) glucose availability, measured as WB glucose rate of appearance (WBGRa), largely depends on gluconeogenesis or more precisely on WB glucose production, representing at least 62% of WBGRa. Glucose is mainly taken up by the mammary gland and plays an important role in regulating milk volume through lactose synthesis. However, the relationships between WBGRa, mammary glucose utilization, and milk volume are not clear. Neither lactose yield nor mammary glucose uptake represent a fixed proportion of WBGRa and varied between 39% to 59% and 59% to 84% of WBGRa, respectively, in mid lactating dairy cows. Increasing supply of glucogenic nutrients increased WBGRa indicating that glucose production responds to the push system. The apparent conversion of a single nutrient towards glucose production, however, does not appear to be constant. For example, a relative low apparent efficiency of conversion of propionate to glucose (30% to 40%) was observed when its infusion in the rumen increased its molar proportion above 17%. This variable efficiency of conversion of glucogenic nutrients to glucose could be explained if the demand for glucose utilisation is another driving force than the push system to regulate glucose production. Indeed, in cows receiving phlorizin which increased urinary glucose output, WBGRa increased probably to sustain milk yield that did not decrease. On the reverse, lactose yield and milk volume did not increase in parallel to WBGRa in response to increasing intestinal supply of non essential amino acids probably because mammary glucose uptake was not limited by WBGRa. In conclusion, glucose production and mammary glucose utilization for lactose synthesis could depend on the balance between glucogenic nutrient availability (push system) and mammary metabolic demand (pull system).

Holstein cows in early lactation: milk and plasma fatty acids contents along with plasma metabolites and hormones as influenced by days in milk, parity and yield

Knapp, E., Dotreppe, O., Hornick, J.L., Istasse, L. and Dufrasne, I., Nutrition Unit, Veterinary Faculty, University of Liege, Bd Colonster 20, 4000 Liege, Belgium; eknapp@ulg.ac.be

Physiological factors such as days in milk, parity and yield can affect the energy metabolism in high producing dairy cattle. There are however difficulties for an accurate monitoring. Blood and milk samples along with a gynecological examination were obtained on 32 cows from 5 private farms on 4 occasions monthly, starting from calving. When the data were grouped according to 3 days in milk periods (0-50, 51-99 and $\geq$100 d), there were increases in the C18:1 and C18:0 fatty acids contents in milk (P<0.001) and decreases in the C4-14 contents (P<0.001) with no changes in fat content in period 1 as compared to periods 2 and 3. Plasma cholesterol, triglycerides, insulin, IGF1 and progesterone concentrations increased with time while β-hydroxybutyrate (BHB) and non esterified fatty acids (NEFA) decreased with advancing lactation periods (P<0.05, 0.01 or 0.001). With days in milk, in the NEFA fraction, the C18:1 decreased greatly (P<0.001) and the C18:0 increased (P<0.05). So NEFA C18:1 could be considered as a reliable marker of fat mobilization. When the primiparous cows were compared with the multiparous cows, the short chain fatty acids concentration were lower and the long chains fatty acids higher in milk (P<0.05, 0.01 or 0.001) indicating large mobilization with the younger animals even when multiparous cows had a higher yield. In the plasma, the pluriparous cows were characterized by lower triglycerides (P<0.05) and IGF1 (P<0.001) concentration and higher BHB (P<0.001) concentration than primiparous cows. These differences have to be associated to a faster adaptation of the liver to the negative energy balance with older cows. When milk yield classes were compared (<29, 29-39 and >39 kg) there were no significant changes between classes for most of the measured parameters. This suggested that cows with either a very high or with a lower yield were able to adapt their energy metabolism.

Effect of supplementing linseed on liver, adipose and mammary gland metabolism in periparturient dairy cows

Mach, N.[1], Zom, R.L.G.[1], Widjaja, A.[1], Van Wikselaar, P.[1], Weurding, R.E.[2], Goselink, R.M.A.[1], Van Baal, J.[1], Smits, M.A.[1] and Van Vuuren, A.M.[1], [1]Wageningen UR Livestock Research, Edelhertweg 15, 8219 PH Lelystad, Netherlands, [2]Agrifirm BV, Boogschutterstraat 1A, 7940 KA Meppel, Netherlands; nuria.mach@wur.nl

The effects of dietary omega-3 fatty acids (FA) on liver, adipose and mammary gland metabolism were studied in 14 periparturient dairy cows that were randomly assigned to control or linseed supplementation. Milk samples and biopsies of liver and adipose tissue were taken in wk-3, wk1, wk4 and wk6. Liver and adipose mRNA abundance of key gluconeogenic and lipogenic regulator genes were measured by qPCR. Additionally, mammary gland biopsies were taken to study genome-wide differences in gene expression on Affymettrix Bovine Genome Arrays in wk6. Linseed did not modify dry matter intake, but increased milk yield and decreased milk fat concentration, which coincided with lower FA synthesized de novo and higher proportions of stearic acid, conjugated linoleic acid and linolenic acid in milk fat. Although linseed did not result in significant transcriptional changes in adipose and liver tissue, changes occurred in the mammary gland tissue. In the mammary gland, gene sets related to cell proliferation and remodeling, as well as immune system response were predominantly up-regulated by linseed supplementation, whereas those involved with nutrient metabolism were reduced. The functional analysis indicated that linseed reduced the expression of pathways associated with lipid and carbohydrate metabolism. The proportion of short-chain FA in milk was mainly positively correlated to most of the lipogenic genes, whereas that of long-chain FA was negatively correlated, suggesting a regulatory role for these components at the transcriptional level of mammary gland lipid metabolism. Such molecular knowledge may provide the basis for more detailed functional studies to improve the dairy cow's energy metabolism and health, as well as quality aspects of dairy cows products.

Pre-calving liver activity and post-calving relationships among biochemical and haematological profiles of dairy cows

Abeni, F.[1], Cavassini, P.[2] and Petrera, F.[1], [1]CRA-FLC Centro di Ricerca per le Produzioni Foraggere e Lattiero-Casearie, Sede distaccata di Cremona, Via Porcellasco 7, 26100 Cremona, Italy, [2]Ascor Chimici s.r.l., Via Piana 265, 47032 Bertinoro (FC), Italy; fabiopalmiro.abeni@entecra.it

The aim of this paper was to evaluate the effect of a combined index of pre-calving liver activity (PLA) on post-calving biochemical and haematological profiles. Data were obtained from a trial to assess the effect of rumen protected choline (RPC; 50 g of commercial product, with 50% choline as choline chloride) supplementation in transition cows; however, this paper discusses only results where PLA did not interacted with RPC supplementation. Twenty-two pluriparous Italian Friesian cows were randomly assigned to 2 experimental groups to be supplemented with RPC from -3 to +5 wk around calving, or to consume basal diet only (CON). Jugular blood samples were collected weekly. The PLA was calculated from normalized albumin and cholesterol concentrations in plasma 4 wk before calving, to stratify the cows within both treatment groups treatment in lower (L) and upper (U) PLA value. Data were analyzed as a randomized block design, with diet supplementation, PLA, wk of trial, and their interactions as main factors, with cow repeated in time. Correlations among biochemical and haematological features were analyzed within each wk after calving. At +1 wk from calving, cows with L PLA had lower plasma cholesterol (P<0.05), higher (P<0.05) circulating leukocytes (WBC), neutrophil count (NEU), and NEU% of WBC than U cows. Cows in U group exhibited slower decline in erythrocyte volume (MCV) after calving, when compared with L cows. In the first wk after calving, positive correlations were evidenced for plasma beta-hydroxy-butyrate (BHB) with WBC (r=0.48, P=0.02), and with NEU (r=0.57, P=0.006), whereas a negative correlation was reported with lymphocyte % of WBC (r=-0.66, P<0.001). These results suggest further efforts to better understand the relationships between liver activity and cow ability to cope with inflammatory conditions just after calving.

Effects of an induced hypoglycemia for 48 hours on metabolism in lactating dairy cows
Van Dorland, H.A., Kreipe, L., Vernay, M.C.M.B., Oppliger, A., Wellnitz, O. and Bruckmaier, R.M., Veterinary Physiology, Vetsuisse Faculty, University of Bern, Bremgartenstrasse 109A, 3001, Bern, Switzerland; anette.vandorland@vetsuisse.unibe.ch

Effects of an induced hypoglycemia over 48 hours on metabolic parameters in plasma and liver of mid-lactating cows were studied. Eighteen dairy cows were randomly assigned to one of three infusion treatments for 48 h (each n=6), including a hyperinsulinaemic hypoglycaemic clamp (HypoG), to obtain a glucose concentration of 2.5 mmol/l, a hyperinsulinaemic euglycaemic clamp (EuG) in which the effect of insulin was studied, and a control treatment with a 0.9% saline solution (NaCl). Blood was collected for glucose, insulin, NEFA, and BHBA analysis. Liver tissue was taken before and after the treatment for mRNA abundance measurement of genes encoding enzymes involved in metabolism. Plasma BHBA level decreased in response to treatment in EuG cows (P<0.05), and was lower (0.41±0.04 mmol/l) on day 2 of the treatment compared to HypoG and NaCl cows (on average 0.61±0.03 mmol/l, P<0.01). In liver, only differences between treatments for their effects were observed for mitochondrial phosphoenolpyruvate carboxykinase (PEPCKm) and glucose-6-phosphatase (G6PC). In HypoG, mRNA abundance of PEPCKm was up-regulated, whereas it was down-regulated in EuG and NaCl cows (P<0.05). The EuG treatment down-regulated mRNA expression of G6PC (P<0.05), which was a marked effect (P<0.05) compared to the unchanged transcript expression in NaCl. No significant treatment differences were observed for genes related to lipid metabolism, and plasma NEFA remained unaffected, likely due to the infusion of insulin. In conclusion, low glucose concentrations in dairy cows affect liver metabolism at a molecular level, through upregulation of PEPCKm mRNA abundance. The results suggest that metabolic regulatory events in early lactation are directed, apart from hormones, by a drain and excess supply of substrates, such as glucose and fatty acids.

Effect of abomasal casein infusion in *post partum* transition dairy cows
Larsen, M. and Kristensen, N.B., Aarhus University, Department of Animal Health and Bioscience, Blichers allé D20, DK-8830 Tjele, Denmark; Mogens.Larsen@agrsci.dk

Previous studies based on splanchnic amino acid fluxes and milk protein output showed that dairy cows mobilise close to 5 kg of essential amino acids in the first month *post partum*. The present study aimed at investigating the effects of alleviating the protein deficiency in *post partum* transition dairy cows. Eight Holstein cows (second lactation) were used in a complete randomised design with repeated measurements at 4, 15 and 29 days in milk (DIM). At the day of calving, cows were randomly assigned to 1 of 2 treatments: continuous abomasal infusion of casein (CAS) or water (CTRL) via a ruminal cannula. All cows were fed the same diet. Abomasal casein infusion followed this profile: 360 g/d at 1 DIM, 720 g/d at 2 DIM, followed by daily reductions of 19.5 g/d ending at 194 g/d at 29 DIM. Plasma volume was determined by 125I-albumin dilution. Data were analysed using PROC MIXED in SAS with treatment, DIM and their interaction as fixed effects. Cow was considered as random effect and DIM within cow as repeated measurement. Dry matter intake was unaffected (P=0.36) by treatment. Milk yield increased more rapidly after calving with CAS (P<0.01) and averaged 43.8±1.0 kg/d with CAS and 36.6±1.0 kg/d with CTRL. Milk protein yield with CAS was 1,664±39 g/d at 4 DIM compared with 1,212±86 g/d for CTRL, whereas milk protein yield did not differ at 29 DIM (1,383±48 g/d; interaction, P=0.02). The calculated utilisation of abomasally infused casein to milk protein was 64 and 60% at 4 and 29 DIM, respectively. Plasma volume with CAS was 50±1 ml/kg at 29 DIM compared with 47±1 ml/kg for CTRL, but did not differ at 4 DIM (40±1 ml/kg; interaction, P=0.05). In conclusion, these results suggest that extra amino acid supplied to *post partum* transition cows is mainly prioritised for milk production. However, the faster restore of normal plasma levels in cows with casein infusion indicate positive effects on regaining normal body functions after calving.

Prospects for a regional feed centre
Galama, P.J., Wageningen UR, Livestock Research, Edelhertweg 15, 8219 PH Lelystad, Netherlands; paul.galama@wur.nl

A feed centre buys crops from dairy farmers and arable farmers, processes these with simple feedstuffs and minerals into total rations and sells them to a number of dairy farmers, to be delivered at the feeding fence. The operations of the feed centre can also include the supply of rations as well as nutritional and crop management support. A feed centre offers many advantages. The dairy farmer can contract out part of his work. From several farms, on-farm feed storage is no longer necessary, as feed is stored centrally. The farmyard will be less cluttered and appearances will be improved. A poor farm lay-out (land far away from the farm) is also less problematical as crops are brought to a central place. Separating the production and storage of feedstuffs from milk production allows it to optimise the cultivation and feed supply regionally instead of on a farm level. The dairy farmer may grow the grass, nature grass, maize, protein crops (such as lucerne or lupin) or concentrate feed replacers (e.g. cereals) regionally on the best spot. He can specialise in milk production, which will also enable him to grow crops on his fields. He will also be able to contract out part of the latter activity to arable farmers. The feed centre then acts as a link between feed producers and dairy farms. This makes dairy farming land-related on a regional level. A feed centre can have a major impact on the income growth of feed producers and dairy farmers. A side-effect is that there will be more traffic movements. The sustainability of a regional feed centre is assessed on the basis of economics, traffic movements, energy consumption and landscape. Calculations are made for the situation with and without a feed centre for several dairy farm systems. The additional energy consumption (MJ per 100 kg of milk) due to the higher transport requirements has been weighed against energy saving by growing more feedstuffs in the region. The experiences in several countries of feed centre will be shown.

---

e-Cow: a web-based animal model that predicts herbage intake, milk yield and live weight change in dairy cows at grazing
Baudracco, J.[1], Lopez-Villalobos, N.[1], Holmes, C.W.[1], Comeron, E.A.[2], Macdonald, K.A.[3] and Barry, T.N.[1], [1]Massey University, IVABS, Private Bag 11-222, Palmerston North, 5301, New Zealand, [2]INTA Rafaela, Ruta 34 Km 227, 2300, Argentina, [3]DairyNZ, Private Bag 3221, Hamilton, 3240, New Zealand; jbaudracco@yahoo.com

An animal model was developed in a web-based version. The model, named e-Cow, simulates a single dairy cow at grazing, with and without supplementary feeding, predicting herbage dry matter (DM) intake, milk yield, milksolids (fat plus protein), and changes in body lipid reserves and live weight, for the whole-lactation on a daily basis. The model combines physical, metabolic and ingestive constraints in the prediction of herbage DM intake, accounts for homeostatic and homeorhetic drives and is able to predict the performance of Holstein Friesian (HF) cows of differing genetic merit. Inputs required are quantity and quality of herbage and supplements, type of pasture (ryegrass or lucerne), pregnancy date, lactation number, body condition score (BCS) and live weight at calving, strain of HF cow (North American, NA, or New Zealand, NZ) and potential yields of milk, fat and protein. In validation analysis, the model showed satisfactory accuracy of prediction, with concordance correlation coefficients over 0.80 for herbage DM intake, over 0.70 for milk yield and over 0.60 for live weight change. The performance of cows grazing ryegrass-based pastures was predicted for a situation with high pasture allowance (25 kg DM/cow/day, expressed at 4 cm above ground level) and high quality pasture (11 MJ ME/kg DM) over a 305-day lactation. The model predicted productions of 443 and 446 kg milksolids per cow and final BCS (scale 1-5) of 2.8 and 3.1 for NA and NZ HF cows, respectively, when no supplements were offered. It predicted 611 and 553 kg milksolids per cow and final BCS of 3.0 and 3.5 for NA and NZ HF cows, respectively, when 6 kg DM of concentrate were offered. The web-based version of the e-cow model is a valuable tool for applied research, teaching and extension purposes.

Including rapeseed in the concentrate of grazing dairy cows improves nutritional quality of milk
Prestløkken, E., Storlien, T.M., Helberg, A., Galmeus, D. and Harstad, O.M., Norwegian University of Life Sciences, Department of Animal and Aquacultural Scienses, P.O. Box 5003, 1432 Ås, Norway; egil.prestlokken@umb.no

Fat has the potential to reduce enteric methane emissions from dairy cows. However, it is essential to ensure that the fat additive does not influence negatively the nutritional quality of milk. In Norway, rapeseed is an actual source of fat, and the objective of this experiment was to study its effect on nutritional quality of milk from grazing dairy cows. Eight multiparous dairy cows of the Norwegian Red breed weighing (mean ± SD) 574±39 kg, were 34.1±13.5 DIM and producing 33.5±3.1 kg milk/d at the start of the experiment were used. The cows (4 with and 4 without rumen fistula) were assigned to two treatments in a crossover design. Cows were pastured day and night and supplemented with 9 kg/d concentrate in the milking parlor. The experimental concentrate (RSC) constituted 100 g/kg ground rapeseed (62 g fat/kg DM), whereas the control concentrate (CC) was without rapeseed (44 g fat/kg DM). Milk yield was measured on three consecutive days in each period, and representative individual samples within period were analysed for chemical composition. Daily milk yield were 26.4 and 26.9 (SEM=0.73, P=0.28) EKM/d for RSC and CC, respectively. Milk fat content was 32.0 and 37.4 g/kg milk (P<0.05, SEM=2.3) for RSC and CC, respectively. However, including rapeseed in the diet significantly (P<0.05) reduced the proportion of palmitic acid (23.6 vs. 21.9 g/100 g FA) and increased the proportion of oleic acid (25.6 vs. 36.1 g/100 g FA) in the milk. We conclude that including rapeseed in the diet improved nutritional quality of milk without negatively effecting milk yield, although milk fat content was decreased.

Including rapeseed in the concentrate of grazing dairy cows lowers enteric methane production
Storlien, T.M.[1], Beauchemin, K.[2], Mcallister, T.[2], Prestløkken, E.[1], Helberg, A.[1], Galmeus, D.[1], Risdal, M.[1] and Harstad, O.M.[1], [1]Norwegian University of Life Sciences, Department of Animal and Aquacultural Sciences, P.O. Box 5003, 1432 Ås, Norway, [2]Agriculture and Agri-Food Canada Research Centre, P.O. Box 3000, Lethbridge, Alberta, Canada T1J 4B1, Canada; tonje.storlien@umb.no

The objective of this experiment was to study whether providing rapeseed to dairy cows on pasture lowers enteric methane (CH4) emissions. Eight multiparous dairy cows of the Norwegian Red breed weighing (mean ± SD) 574±39 kg, 34.1±13.5 DIM, and producing 33.5±3.1 kg milk/d at the start of the experiment were used. The cows (4 with and 4 without rumen fistula) were assigned to two treatments in a crossover design. Cows were pastured day and night and supplemented with 9 kg/d concentrate in the milking parlor. The control concentrate without rapeseed (CC) contained 44 g fat/kg DM, whereas the experimental concentrate (RSC) constituted 100 g/kg of ground rapeseed and had a fat content of 62 g/kg DM. Enteric CH4 emissions were measured over 5 d in each period using the sulfur hexafluoride tracer technique. Milk yield was measured on three consecutive days and its composition determined to calculate energy corrected milk (ECM). Milk yield (ECM, kg/d) was similar for RSC and CC (26.4 and 26.9 respectively; SEM=0.73; P=0.28). However, RSC lowered enteric CH4 emissions compared with CC (P<0.05) expressed as g/cow/d (269 vs. 295, SEM=15.8) and as g/kg ECM (10.1 vs. 11.3, SEM = 0.58). We conclude that including rapeseed in the concentrate is a strategy to reduce enteric CH4 emissions from dairy cows on pasture.

Effect of the feed allowance level on the growth performance of boars vaccinated against gonadotrophin-releasing factor, using ImprovacTM, and consequence on nutritional requirements after the second injection

Quiniou, N.[1], Monziols, M.[1], Colin, F.[2], Goues, T.[1] and Courboulay, V.[1], [1]IFIP-Institut du Porc, BP 35104, 35650 Le Rheu, France, [2]Pfizer - Division Santé Animale, 23-25 avenue du Dr Lannelongue, 75668 Paris Cedex 14, France; nathalie.quiniou@ifip.asso.fr

An over-consumption of feed is reported in male pigs vaccinated against boar taint (Improvac™) after the second vaccination (V2). The increase in feed intake may result in increased body fatness and decreased feed efficiency, when compared to boars. The effect of feed restriction on these criteria and on pig behavior was investigated over the 22 to 115 kg BW range in a batch of 120 group-housed boars (five pigs/pen). The male pigs were first vaccinated at 62 days of age and a second time (V2) at 130 days of age and then slaughtered 4 to 5 weeks later. They were either fed *ad libitum* (AL treatment) or received a maximum daily feed allowance of 2.5 (R2.5 treatment) or 2.75 kg/pig (R2.75 treatment). Behavioral observations and lesion scoring were conducted on V2-6, V2+7 and V2+21 days. Between V2 and slaughter, feed allowance in R2.75 and R2.5 treatments was 15 to 22% below average AL feed intake (3.20 kg/d), respectively. The statistical analysis was performed with the BW at V2 or at slaughter introduced as a covariate so that results were adjusted over a similar BW range. Feed restriction was associated with a reduced ADG after V2 (871, 941 and 1017 g/d in R2.5, R2.75 and AL groups, P<0.001) but had no effect on FCR (3.00 kg/kg on average, P>0.10) or backfat thickness at slaughter (13 mm, on average, P=0.08). A slight increase in negative behavior was observed at the beginning of the restriction period resulting in higher lesion scores in R2.5 and R2.75 pigs than in AL ones (P<0.05). According to our results, immunocastrated pigs should not be restrictively fed during the late finishing period. Additional calculations are planned in order to model the nutrient requirements after V2 using the InraPorc software.

Effect of two alternative feed supplements on ruminal physicochemical parameters in dairy cows

Bayourthe, C.[1], Monteils, V.[1], Julien, C.[1], Aubert, T.[2] and Arturo-Schaan, M.[2], [1]Université de Toulouse INRA UMR 1289 INRA INPT ENVT TANDEM, avenue agrobiopole, 31326 castanet tolosan, France, [2]CCPA, Zone d'activités Nord Est du Bois de Teillay, 35150 Janze, France; bayourthe@ensat.fr

Three Holstein dairy cows, fitted with ruminal cannulas, were allocated in a 3×3 Latin square design. They were given a total mixed ration as control diet (CD) supplemented with 100 g/d of malate or plant extract, plus inactivated yeast (MY and PEY, respectively) during a 28-d experimental period (10 d of diet adaptation, 3 consecutive days for measurement and sampling, and 15 d of transition). Ruminal pH and redox potential (Eh) were recorded hourly over a 9-h period from 1 h before to 8 h after the morning meal, using the *ex vivo* method of Marden *et al.* The Clark's Exponent (rH) was calculated by integrating both pH and Eh values in the Nernst's equation: rH = Eh (mV)/30 × 2 pH. Samples of ruminal fluid were taken at 0, 2, 4, 6, and 8 h after feeding for VFA, NH3-N, and total lactate determinations. All data were analysed using the SPSS software, using a repeated-measures model that included as main plot the effects of cow, treatment, and period. Mean pH tended (P=0.077) to be greater with MY (6.36) and PEY (6.31) compared with CD (6.20). Mean Eh and rH were significantly lower with PEY (-192 mV and 6.20) and with MY (-188 mV and 6.43) than with CD (-158 mV and 7.12). Total VFA remained constant between treatments. Acetate concentration tended (P=0.08) to be greater in MY and PEY diets compared with CD (54 vs 49 mM). Others VFA, lactate and ammonia concentrations did not differ among treatments. In conclusion, the stabilization of pH with MY and PEY was not associated with a lower lactate concentration. Both supplements had a pH stabilization effect, probably via their capacity to neutralize protons, and strengthened the reducing power of the milieu. All these better conditions of ruminal environment could favor the activity of cellulolytic bacteria.

Milk fatty acids content in buffalo fed two ryegrass diets

Chiariotti, A., Pace, V., Carfi, F. and Tripaldi, C., CRA-PCM, Animal Production, Via Salaria 31, 00015 Monterotondo (RM), Italy; vilma.pace@entecra.it

To extend buffalo-based agriculture in dry areas, where maize cultivation is difficult, ryegrass was studied as potential animal food, due to its reduced water requirements. At this purpose ten multiparous lactating Mediterranean buffaloes were fed two isoenergetic (0.92 MilkFU/kg DM) and isoproteic (P=15%) diets based both on ryegrass hay (group H) or ryegrass silage (group S). Milk fatty acids content was scored on samples taken at 60 and 90 days in milking (DIM) extracted and methylated according to ISO/FDIS 14156-2001 and ISO/15884-2002 procedures for milk and milk products in general. Fatty acids methyl-esters were analyzed by Gas Liquid Chromatography either for SCFA (Short Chain Fatty Acid C4:C8), MCFA (Medium Chain Fatty Acid C10:C14), LCFA (Low Chain Fatty Acid >C: 18, oleic acid omitted) and CLA (C18:2 cis-9, trans-11 isomer) content recorded as %/100 mg extracted fat. Data were analyzed by ANOVA and statistical difference was discussed at P<0.01 level. Silage treatments did positively affect milk fat yields (group S: 5.70 Vs group H: 5.13 kg/head/d), mozzarella production (group S:1.43 Vs group H:1.26 kg/head/d) and LCFA (group S:2,48 Vs group H:2.05). As lactation progressed concentration of LCFA decreased in both groups (H:1.89 VS 1.68 and S:2.71 Vs 2.31) confirming the tendency reported in literature in cows. Considering that LCFA are from the diet or body fat depots the time effect could be caused by change in nutrient intake and partitioning of energy between milk and body reserves. As lactation progresses, body fat mobilization is reduced and eventually body reserves increase leading to a less LCFA content. Results suggest that feeding ryegrass to animals is a valid option in arid areas to avoid intensive crop irrigation and the silage treatment has a positive effect on milk quality.

The effects of combining sodium hydroxide treatment with fungal cellulases on the *in vitro* fermentation characteristics of wheat straw

Adebiyi, A.O. and Chikunya, S., Writtle College, Chelmsford, CM1 3RR, United Kingdom; sife.chikunya@writtle.ac.uk

A few studies have attempted to evaluate the potential synergistic effects of combining alkalis with exogenous fibrolytic enzymes for upgrading straws. In this study, wheat straw was treated with either water (STR) or with 5% sodium hydroxide (NaOH) and subsequently treated with cellulases (40 IU /g DM) from either Trichoderma reesei (TR) or Aspergillus niger (AN). The 6 treatments were: STR, STR+AN, STR+TR, NaOH, NaOH+AN and NaOH+TR. Substrates were incubated using the *in vitro* gas production procedure. Measurements of were taken after 3, 6, 12, 24, 48 and 72 hrs. Cumulative total and rate of gas production differed significantly among the treatments at all incubation times (P<0.05), both the effects of NaOH and enzyme type were significant. After 72 hrs, NaOH+AN produced the most gas whilst STR produced the least gas, with means of 336 and 177 ml/g OM respectively. At all times, treating with NaOH reduced NH_3-N compared to not treating with NaOH (P<0.05). The effects of using the two cellulases on NH_3-N were variable. Trichloroacetic acid precipitable-nitrogen increased in NaOH-treated compared to non-alkali treated samples (P<0.001). After 12, 48, and 72 hrs incubation, neutral detergent fibre digestibility (NDFD) was increased by both alkali treatment (P<0.001) and by the addition of fungal cellulases (P<0.001). However, NaOH alone tended to be more effective at improving NDFD than using fungal cellulases alone or in combination with NaOH. At 72 hrs NDFD values were: 854, 825, 721 for NaOH, NaOH+AN and NaOH+TR, compared to 584, 552 and 545 g/kg DM for STR, STR+AN, and STR+TR respectively. As in previous studies, these results demonstrate efficacy of NaOH as means for upgrading straws. Although using fungal cellulases improved gas production with or without alkali treatment; it did not elicit any further improvements on NDFD. It suggests that NaOH treatment is effective on its own and the benefits of combining it with exogenous fungal cellulases are marginal.

High-yielding dairy cows and contet of non-structural carbohydrates in the nutrition during a transition period

Mudrik, Z.[1], Hucko, B.[1], Kodes, A.[1], Polakova, K.[2] and Kudrna, V.[2], [1] Czech University of Life Science, Prague, Department of Microbiology, Nutrition and Dietetics, Kamýcká 129, Prague 6 - Suchdol, 165 21, Czech Republic, [2]Institute of Animal Science, Prague-Uhříněves, Přátelství 815, Prague - Uhříněves, 104 00, Czech Republic; Mudrik@af.czu.cz

The aim of this study was to evaluate the effect of the high content of non-structural carbohydrates (NFC) in high-yielding dairy rations before and after parturition (21days before and 50 days after carving) on dry matter intake and milk production. Thirty-six high-yielding dairy cows were allocated into one of the three well-balanced groups (C, E 1 and E 2). Rations in each group differed in the content of non-structural carbohydrates. Control group – C received diet containing 33.7% NFC/kg dry matter, group E received diet with content of 40.8% NFC/kg dry matter, E 2 group have diet with 47.6% of NFC/kg dry matter. From the first day *post partum*, all cows received the same ration TMR *ad libitum*. The cows were observed dry matter intake, milk yield, live weight, BSC, some values in blood and rumen fluid. Average daily dry matter intake before calving was highest in group C – 14.32 kg/head. Differences between groups were statistically significant (P<0.05). Dry matter intake after calving was stabilized. The highest milk yield is in group E 1 (43.71 kg/head/day), but the fatness of the total production of milk and milk fat were lowest. The highest fat content (4.10%) and fat corrected milk production (44.03 kg/head/day) were recorded in group E 2, while the highest concentration of milk protein was detected in the milk group E 2. The concentration of NFC in the ration of dairy cows before parturition did not affect dry matter intake *post partum* or milk yield, milk composition, live weight, either as measured parameters in blood serum, and the value of rumen fluid after calving.

Effects of different levels of sunflower residue silage replacing with alfalfa hay and corn silage on nutrient digestibility of Mohabadi dairy goats

Pirmohammadi, R., Urmia university, animal science, Urmia university-Urmia city, 0098441, Iran; r.pirmohammadi@urmia.ac.ir

There is a shortage of feedstuffs in many countries. The use of agricultural by-products is often a useful way of overcoming this problem. Sunflower residuals are one of these by-products. Using such by-products for animal feeding is a means of recycling which otherwise, if accumulated, might cause environmental pollution. *In vivo* digestibility data on sunflower silage (SRS) that can be used in the formulation of diets for animals are limited. This experiment was conducted to study the effects of replacing alfalfa hay and corn silage with different levels of sunflower residual silage on feed nutrient digestibility of Mohabadi goats. Eight goats with 60±5 kg of BW fed experimental diets in a Latin square design 4×4 with change over design arrangement. Four treatments (1, 2, 3 and 4) included 0% (control), 30%, 60% and 90% replacement of alfalfa hay and corn silage with SRS, respectively. The results showed that dry matter intake (DMI) of experimental goats was significantly different between the treatments (P<0.05). The highest and the lowest DMI were shown in the treatments 2 and 4, (1.45 and1.062 kg), respectively. DM, OM, CP and NDF digestibility was significantly different between the treatments (P<0.05). It may conclude that alfalfa hay and corn silage can be substituted with SRS at up to 60% level of substitution with no negative effects on nutrient digestibility of Mohabadi goats.

Potential for conservation of local livestock breeds through delivery of ecosystem services
Small, R.W., Liverpool John Moores University, Natural Sciences and Psychology, Byrom Street, Liverpool L3 3AF, United Kingdom; r.w.small@ljmu.ac.uk

Many valued terrestrial habitats in Europe are not natural but are the product of centuries of agriculture, including the grazing of livestock. Their conservation frequently requires management, particularly to prevent succession to woodland. Where grazing was the traditional management which was instrumental in creating the habitat it is often now the preferred means of maintaining the conservation interest. Although there are alternatives, such 'conservation grazing' is seen as a natural means of management that increases both plant and animal diversity. Conservation grazing also provides opportunities for local and possibly endangered breeds, which are seen as both an element of the cultural heritage and as adapted to local environmental conditions. Thus conservation grazing can also contribute to the maintenance of farm animal genetic resources. In addition, many European countries have support measures for such breeds which encourage breeders to participate in approved breeding programmes. The U.K. takes a different approach by providing support only through agri-environment measures, with no requirement for farmers to participate in breeding programmes. As well as clear benefits for the conservation of farm animal genetic resources this approach also entails risks, as the livestock may be perceived solely as a means of delivering ecosystem services, rather than for their intrinsic societal value. This review explores the relationships between grazing and ecosystems biodiversity, the contribution conservation grazing can make to the conservation of farm animal genetic resources and the value of approved breeding programmes.

Traditional low producing dairy breeds used as suckler cows, a threat or a future oriented solution?
Sæther, N. and Rehnberg, A.C., Norwegian Genetic Resources Centre, P.O. Box 115, N-1431 Ås, Norway; nhs@skogoglandskap.no

During the last ten years the total number of cattle in Norway has been reduced with 10% whereas the number of suckler cows has increased with 152%. As no beef cattle breeds are native to Norway, most of the suckler cows are of imported breeds. However, an increasing number of herds with the native and endangered cattle breeds are being kept as suckler cows, thus the breeds are traditionally dairy breeds. Figures from the The Norwegian Genetic Resources Centre show that up to 85% of the cows of one of these six breeds are now kept as suckler cows. This particular breed has a yearly milk yield of 3,000 kg whereas the other five native and endangered dairy breeds have an annual milk yield of 4,000 kg. The average annual milk yield in Norway is 7,000 kg. These figures might explain why the endangered breeds are preferred used as suckler cows to dairy cows. However, some farmers dislike this development as they fear that using endangered and traditionally dairy breeds as suckler cows will destroy the milk and udder traits, and that the traditional knowledge on dairy production based on these breeds will vanish. The farmers with suckler cows correspond to this concern that breeding work and utilisation of farm animals always have been a dynamic process and without finding new market for the endangered breeds they are much more threatened by extinction. The study will present more figures on the development of how utilisation of the endangered breeds have developed and changed during the last 10 years. According to the Convention on Biological Diversity (CBD) *in situ* conservation is the fundamental requirement for the conservation of biological diversity. Furthermore The Global Plan of Action for Animal Genetic Resources states that the cultural importance of animals is a key factor in *in situ* conservation. Both these documents stress the importance of *in situ* conservation, but there is no guidance on how to handle changes in markets and thereby utilisation of breeds.

The Norwegian Coastal Goat, the queen of the coastal alpine heaths in Norway
Stubsjøen, L. and Sæther, N., Norwegian Genetic Resources Centre, P.O. Box 115, 1431 Ås, Norway;
nina.sather@skogoglandskap.no

As a result of land-use change the cultural landscape in Europe has gone through big changes during the last decades. The area reduction by 80% of the coastal heathlands, which represents a 4-5,000 year old cultivated landscape type, illustrates how dramatic this alteration has been for some landscape categories, and explains why this ecosystem is classified as greatly endangered under the EC Habitats Directive. This landscape borders the Atlantic coast from Northern Norway to Portugal and is a result of adaptation and interaction between humans and their farm animals; mainly native and extensive sheep breeds. However, at the north-west part of the Norwegian coastline, characterized by steep alpine mountains, the topography might be too tough even for these adapted sheep breeds. In this region the main traditional browser is the Coastal Goat. Even if this livestock production is part of a several hundred year old tradition, the Coastal Goat breed was more or less unknown to Norwegian authorities up to the middle of the 1990s. During the last 15 years a conservation programme for the breed has been carried out. The breed is still under serious threat as the female population consists of 300 goats and only 15-20 herds. The main product from this breed is four year old castrates, illustrating how extensive this production is. The main challenges in the conservation programme have been lack of interested local farmers and restrictions on moving animals across borders set by the animal health authorities. Bucks bought for semin production have shown signs of inbreeding depression by poor semin quality, hardly usable for ai. The inbreeding depression might be solved by exchanging animals from flocks at different sides of the mentioned animal health border, requiring dispensation by the animal health authorities. Improved local respectability is needed to achieve renewed interest. A new chapter in the conservation programme is initiated by inviting a wide range of stakeholders to a joint effort to secure this unique breed.

The development of population structure and inbreeding parameters in dairy cattle breeds kept for conservation and grazing purposes
Rehnberg, A.C. and Sæther, N., Norwegian Genetic Resources Centre, P.O. Box 115, 1431 Ås, Norway;
anna.rehnberg@skogoglandskap.no

Since 1990 five of the six endangered dairy cattle breeds in Norway have been recorded in electronic herd books. With population sizes around 100 cows, the main breeding goal for the breeds since 1990 has been to increase the population sizes and breed as many males and females as possible. The farmers have been strongly advised to keep high focus on kinship, and some reduced focus on milk yield and growth rate in their breeding practice. This has been a challenge, as farmers in general have a high focus on the importance of improving traditional production traits. However, since these breeds have significantly lower milk production than the dominating dairy breed in Norway, Norwegian Red, 4,000 kg/yr and 7,000 kg/yr, respectively, the endangered breeds are mainly kept as suckler cows where annual milk yield is of less importance. The electronic herd books have offered advice to the farmers on where to find pure bred and relatively unrelated bulls for their cows, either as live breeding bulls or as ai-bulls from the gene bank established and continuously extended by Geno in cooperation with Norwegian Genetic Resources Centre. The study on the development of the population structure presents data on annual numbers of breeding females, breeding sires and born calves during two decades, from 1990 to 2009 for five native and endangered dairy cattle breeds. The potential breeding female population (> two yrs) for all breeds increased by between 180% and 280% during the decade from 1990 to 2000, whereas the increase in females in the following decade varied between 2% and 350%. The population data for the breeds in 2009 varied between 186 and 785 females. The annual numbers of breeding sires per born calve varies between two and seven calves/father/ year for the whole period and over all breeds, indicating that the farmers have succeeded in using a high share of breeding males in the breeding work. More parameters on inbreeding, such as ΔF, will be presented.

Harmonization and joint evaluation in Interbull MACE as basis for genomic evaluation
Wickham, B., ICBF & Sheep Ireland, Highfield House, Shinagh, Bandon Co. Cork, Ireland;
bwickham@icbf.com

Over the last forty years dairy cattle breeding has become increasingly international while phenotype recording, the very basis of genetic improvement, remains a local and largely national activity. A key development has been the formation of Interbull and the establishment of the Interbull Centre, located at SLU in Uppsala Sweden, to provide international genetic evaluation services. The development of this important infrastructure is the product of a unique international partnership between organisations involved in; performance recording, operating national cattle breeding databases, conducting genetic evaluations and undertaking cattle breeding research. It provides dairy farmers and commercial cattle breeding organisations world-wide with information that has enabled rapid genetic change. The key steps in creating this powerful and widely used system include; establishing international identification standards, development of MACE, development of standards for national evaluations to be accepted as inputs to international evaluations and the development of an agreement for data sharing between evaluation units world-wide. Interbull's regular meetings and workshops have become one of the most important international forums for the sharing of knowledge and research findings relevant to cattle breeding. This infrastructure has enabled dairy breeders to quickly exploit the benefits of genomic selection and has great potential to deliver further benefits in the future. In the immediate future the key role for Interbull is to ensure the on-going availability of MACE evaluations free of biases due to genomic pre-selection and facilitating international genomic evaluations consolidating information from several countries.

Opportunities to optimize the role of functional traits in dairy breeding goals using genomic information
Strandberg, E., Swedish University of Agricultural Sciences, Department of Animal Breeding and Genetics, P.O. Box 7023, 75007 Uppsala, Sweden; Erling.Strandberg@slu.se

Functional traits have received renewed interest in recent years in several countries. Part of the reason for that is that previous selection and breeding programs have overemphasized production traits, resulting in a correlated decrease in functional traits. However, even if fertility and health traits are included in the breeding goal, genetic response in these traits may be low, unless sufficient number of daughters are tested for each sire. The relative accuracy of, say, milk yield ($h^2=0.3$) vs fertility ($h^2=0.05$) is 2.1 if 50 daughters are tested, but 1.4 with 150 daughters. Thus, even with large daughter groups, relative response is different from that indicated by the breeding goal weights. One of the potential advantages of genomic selection is that the 'phenotypes' of the bulls in the reference population are daughter averages (or EBVs) based on several hundreds of daughters, making the accuracy of the measure, and thus of the genomic BVs, almost equal for all traits. This would give a genetic progress in breeding goal traits in good agreement with the breeding goal weights. Given that the reference population bulls are sufficiently old, their daughters will also have observations on traits measured later in life, making it possible to select for, say, longevity more efficiently than in progeny testing. However, it is possible that in extending the reference population, e.g. by including data from several countries, mainly information on production will be included, because these traits are always recorded and in a more standardized way. There might also be more genotype by environment interaction for functional traits than for production traits. Another opportunity is to measure traits that are expensive to record, in experimental farms or cooperator herds. Cooperation across countries and universities can create a reference populations of enough number of cows with, say, progesterone measures, to give reasonably accurate GBVs for selection of bulls.

Animal breeding in the genomics era: challenges and opportunities for the maintenance of genetic diversity

Simianer, H., Chen, J. and Erbe, M., Georg-August-University, Department of Animal Sciences, Albrecht-Thaer-Weg 3, 37075 Göttingen, Germany; hsimian@gwdg.de

Genomic tools are being rapidly adopted in farm animal selection programs, with dairy cattle taking the lead. While an increased rate of genetic gain per year is expected with genomic selection, it remains to be studied which impact this has on the maintenance of genetic diversity. Genomic selection requires large calibration sets, which limits the possibilities to implement it in small breeds, and may widen the gap of the achievable genetic progress in large breeds (like Holstein) versus many small breeds, thus posing an additional threat on small breeds. Potential strategies to overcome this problem, such as across breed calibration studies, are discussed. It is further argued that genomic selection within a given breed has a built-in property of selecting a more diverse set of candidates, and, by this, maintaining diversity. This is most relevant for traits of small heritability, where traditional BLUP-based approaches result in family selection, i.e. in recruiting groups of related candidates. Genomic selection removes this dependency of the extent of family selection on the degree of heritability, which will be illustrated both with simulated and empirical data. Additionally, the composition of the calibration set is shown to have a measurable impact on the representation of sire families in the selected proportion, which raises the general question of the optimal composition of the calibration set. Further, approaches of actively balancing genetic progress and loss of genetic diversity using genomic tools are addressed. While there is a good chance to reduce the loss of diversity per generation in genomic selection programs, the loss of diversity per year (or per unit of genetic progress) will likely be increased. It might be taken into consideration to invest a proportion of the achievable short-term genetic progress into the maintenance of genetic diversity and fitness, resulting in an increased potential long-term genetic progress.

Application potential of genomics beyond elite animals' evaluations

Winters, M.[1] and Coffey, M.P.[2], [1]DairyCo, AHDB, Stoneleigh Park, Kenilworth, Warwickshire, CV8 2TL, United Kingdom, [2]SAC, SLS, Roslin Institute Building, Easter Bush, Midlothian, EH25 9RG, United Kingdom; Marco.Winters@dairyco.ahdb.org.uk

Since the availability of low cost genotyping, many countries have introduced genomic evaluations for dairy cattle. Application to date has mainly been restricted to the screening and evaluation of elite males and females to accelerate genetic gain and reduce costs of progeny testing. However the potential of genomic data beyond this specific application is large in the wider population, and is set to become even more informative with the availability of increasing numbers of whole-genome sequences and low density (and lower cost) genotyping combined with imputation. Immediate applications being proposed with the lower density SNP chips are (1) animal identification and parentage validation using around 100 SNPs, and (2) female young stock screening using around 3,000 SNPs. SNP data is being used to test for known genetic defects, most of which are patented; however the ability to detect new genetic defects is becoming easier and with accelerated rates of gain is increasing in importance. Future applications for herd management will be developed, and might be applied to areas such as vaccine or drug sensitivity and specificity, feeding regimes tailored to the genetic profile, product traceability, individual mate selection. Critical to the success of these will be the collection of existing and new phenotypes combined with international standardization of trait definitions. For farmers to widely engage with genomics, efficient and cost effective new services will be needed to facilitate the genotyping of the most informative animals in each herd, utilizing the repository of genotypes in each country (or population). Maximum long term exploitation potential will be achieved by international sharing of data and coordinated international monitoring of inbreeding/SNP homozygosity. In order to ensure widespread application, it is important to enable unrestricted usage of the SNP data, free from patent right restrictions.

Implementation of genomics in dairy cattle breeding schemes and uptake by the farmers: North American perspective
Funk, D., Genus, 1525 River Road, DeForest, Wisconsin, 53532, USA; dfunk@absglobal.com

Genomic predictions of genetic merit for dairy cattle became official January 2009 in the US and August 2009 in Canada. The breeding industry has widely incorporated genomic predictions into breeding programs, especially AI companies and breeders of elite genetic merit animals. Currently genomic evaluations are publicly available for females, older bulls, and only selected unproven bulls; for other bulls <2 years of age evaluations are provided only to AI companies. Number of young bulls progeny tested has been reduced, although reduction in numbers is not universal among AI companies. Usage of unproven bulls as sires of sons has increased dramatically, estimated at 50% for current contract matings. There is increased ownership of elite genetic-merit females by AI companies, especially heifers. Multiple genotyping panels are now available; more than 40k females have been genotyped as of March 2011, with nearly 50% genotyped using the Illumina 3k panel. To date there has been limited genotyping of entire herds when the herd owner pays for genotyping. Few animals have been genotyped with higher density panels (>600k SNPs) if previously genotyped with lower density options. International collaborations likely will result in adding several thousand dense genotyped (>600k SNPs) Holsteins, mostly daughter proven bulls, to the genomic database. Farmers have much greater access to semen from genomic young sires than previously, although marketing efforts differ by AI companies. Semen prices for most genomic young sires are substantially lower than for proven bulls of similar genetic merit. The inflow of performance data from traditional progeny test programs may decline with genomics, especially conformation data. In the US, a Dairy Data Alliance (DDA) responsible for stewardship of national data resources is being established. This Alliance will be governed and funded by the industry and will identify data needs of the future and calculate and distribute genomic evaluations.

Implementation of genomics in dairy cattle breeding schemes and uptake by the farmers: European perspective
Reents, R., IT Solutions for Animal Production (vit), Heideweg 1, 27283 Verden, Germany; reinhard.reents@vit.de

Traditional selection schemes in dairy cattle have been based on BLUP estimated breeding values (EBV) from progeny testing schemes. Selection response has been high but progeny testing breeding programs are rather costly. Genotyping of single nucleotide polymorphisms (SNP) markers has given animal breeders a powerful, cost effective new tool. Key factors for a successful implementation are highly reliable conventional EBVs on a considerably high number of bulls that have been genotyped. It has been shown that combining bull data from several countries can effectively increase the reliability of genomic enhanced EBVs(GEBV). Availability of MACE EBVs through the international genetic evaluation service of Interbull is the input to combined genomic evaluation systems. Collaboration in Europe has happened either in bilateral swapping of bull genotypes, or collaboration in consortia like EuroGenomics (Holstein) or Intergenomics for the Brown Swiss populations. The latter have implemented a genomic evaluation system in the framework of Interbull with the Interbull centre as service provider. Reliabilities of GEBVs from large reference populations are nearly as high as reliabilities from progeny tested bull EBVs. A recently introduced validation procedure by Interbull/ICAR assures that a given genomic evaluation system yields unbiased GEBVs. Because of the cost effective way of calculating young bull GEBVs the selection intensity has increased dramatically. Some countries report that ten times more young bulls are now genomically evaluated compared to the number of bulls that were selected for entering a progeny test before. Farmers are using these highly selected bulls. High selection intensity along with a shortened generation interval can lead to a significant increase in inbreeding with a given time frame and puts an increased responsibility for monitoring this to AI centres and breeding organisations.

Genomic selection from a farmer's point of view
Gunnarrson, L.I., Viking Genetics, Box 64, 532 21 Skara, Sweden; lars.inge@skottorpssateri.se

As farmer we have about 1,150 ha with a milk quota of 7,000 tons divided on 530 Holstein cows and 120 SRB cows. Average milk production is 11,900 kg/cow. In addition to being chairman of VikingGenetics, I am board member of the Swedish Dairy Association and chairman of Nordic Genetic Cattle Evaluation, NAV. The goal for working with genomic selection is to increase genetic progress. By genomic selection we can achieve a more balanced genetic progress than by traditional progeny test – differences in reliability between functional and production traits is less for genomic breeding values than for traditional breeding values based on large progeny groups. In a Nordic context, genomic selection in particular gives better results for functional traits with low heritability like mastitis resistance, resistance against other diseases, calving ease and female fertility for young animals. The Nordic Total Merit Index, NTM, has high weight on these economical important functional traits, giving the farmers full benefits from using genomic selection. Dairy farmers in Finland, Sweden and Denmark can now order genomic test on females, and all young bulls used for AI have been genomic tested. Genomic breeding values are today used as a strong selection tool for young bulls and females for flushing. VikingGenetics expect that the genetic trend gradually will increase along with improved genomic prediction in coming years and the genetic progress will probably be about fifty percent higher than in the traditional progeny testing scheme. Implementation of HD Chip in addition with sequencing will give possibilities to build up the future reference bulls and open up for higher efficiency in exploring new traits like female production efficiency and milk components such as fatty acids and differences in protein quality. Therefore the Milk Genomic project is also an important area for having future progress in milk production. Genomic breeding values will certainly strengthen the selection procedure and possibilities for using more outcross sires.

Interbull's role in the era of genomics
Philipsson, J., SLU, Dept of Animal Breeding and Genetics, Box 7023, S-75007 Uppsala, Sweden; Jan.Philipsson@slu.se

Effects of animal breeding have historically been boosted at certain times by important technical discoveries or theoretical developments. The globalization of cattle breeding has been possible due to the possibility of using frozen semen of individual bulls across the world. The issue of identifying which are the best bulls to be used got a boost by the development of the mixed-model procedures (BLUP) supported by the enormous computer capacity developments. Further developments (MACE) have enabled the opportunity to evaluate practically all AI bulls across countries. And now we are in the midst of a technical break-through where technology developments including sequencing of the cattle genome combined with use of quantitative genetic methods form the basis for genomic selection. During the past 20 years an enormous development has taken place in the major dairy breeds. Interbull has played an important role for this development as facilitator through delivery of genetic evaluations to be used across countries, so that all bulls are ranked correctly for the predominant environment of each country or region – a win-win situation for both importers and exporters of semen. The achievements, so far reaching 73 populations representing six breeds in 30 countries and more than 40 traits, have been supported by the collaborative research conducted by Interbull Center and its partners attracting top scientists around the world. The regularly arranged open seminars and technical workshops have laid the basis for a spirit of cooperation and freely sharing of research results and practical experiences. In the era of genomics experiences already show an increased need of cooperation and sharing of experiences for the industry to fully benefit from adoption of this new fast developing technology. Additional activities at the Interbull Center might include sharing of information on genotyped or sequenced bulls, estimation of SNP effects, and an increasingly important task of monitoring genetic trends, genetic diversity and inbreeding.

Contribution of rumen ciliates to host digestive system: usefulness of specific genes sequences

Belanche, A.[1] and Balcells, J.[2], [1]Aberystwyth University, Institute of Biological, Environmental and Rural Sciences, Penglais, Aberystwyth SY23•3AL, United Kingdom, [2]ETSEA, Departament de Producció Animal, Av. Alcalde Rovira Roure 191, LLeida, Spain; balcells@prodan.udl.cat

Microbial protein is the main source of protein for ruminants. Therefore the ability to distinguish between microbial N and non-microbial N represents a crucial point in the ruminant's nutrition. By combining adequate digesta flow and microbial markers microbial yield can be estimated. However, no marker has proven completely satisfactory. The existence of specific microbial rDNA sequences (DS) may help to overcome these limitations. Several studies were carried out to validate the use of DS as microbial markers, initially survival of microbial DS throughout gastric digestion and factors involved were analysed. Gastric acidity had a significant effect on DS integrity; pH 1.2 hydrolyzed almost all incubated DS although pepsin had a limited effect and the presence of fibre mitigated the negative effect of the acidity on the DS. Under simulated abomasal conditions almost all the bacterial DS and around 78% of protozoal DS maintained their molecular integrity and were detected after the incubation. *In vivo* survival of microbial DS was studied in two groups of five lambs sacrificed at weaning and fattening stage. If microbial DS are able to by-pass acidic digestion then they can be use as novel and specific microbial marker. This point was evaluated and its efficacy compared with purine bases (PB) as a microbial marker. The survival of microbial DS was determined in three digestive sites (omasum, abomasum and duodenum) but no differences were observed, suggesting that DS maintains the molecular stability along the digestive tract. Bacterial 16S and protozoal 18S rDNA sequences persisted through the gastric digestive tract and their utilization as a high-specific microbial marker must be considered. In conclusion, DS can be considered as potential microbial markers to estimate microbial protein synthesis.

Effect of different sources of non-protein nitrogen on the ciliated protozoa concentration in Nellore steers

Corte, R.R.P.S.[1], Nogueira Filho, J.C.[1], Britto, F.O.[1], Leme, P.R.[1] and Manella, M.[2], [1]Faculdade de Zootecnia e Engenharia de Alimentos, Av Duque de Caxias Norte, 225, 13635900, Brazil, [2]Alltech do Brasil, Rua Curió, 312, Araucária, 83705552, Brazil; jocamano@usp.br

To evaluate the effect of replacement of soybean meal (SBM) by different non-protein nitrogen sources (NPN) on ruminal ciliated protozoa, four rumen-fistulated Nellore steers were fed four diets: 1) CTL (Control diet): 12% SBM and 1% urea, 2) O: 6% SBM and 1.8% slow-release urea (Optigen®), 3) U: 6% SBM and 1.7% urea, and 4) UO: 6% SBM, 1.0% urea and 0.7% Optigen®. Diets had 78.5% concentrate and were isonitrogenous (15.5% CP), isocaloric (77.4% TDN) and similar in rumen degradable protein (10.4%). Treatments were tested in a 4x4 Latin square and data were analyzed by SAS software. On d2 and d5 following 14-d adaptation, rumen fluid samples were taken before and at 4 h after feeding. In order to identify the ciliate protozoa a Sedgwick-Rafter counter cell and an optic microscope with slide area of 0.4362 mm^2 were used. The number of ciliated protozoa was higher (P<0.001) at 4 h after feeding (48.3x10^5/ml) than before feeding (46.8 x 10^5/ml). Compared to CTL, NPN sources presented higher numbers (P<0.001, for all genus) of Entodinium (CTL=28.8, O=34.8, U=35,3 and UO=40.2 x 10^5/ml), Diplodinium (CTL=1.9, O=3.0, U=3.3 and UO=3.4 x 10^5/ml), Epidinium (CTL=1.7, O=2.7, U=2.4 and UO=3.1 x 10^5/ml), Isotricha (CTL=2.1, O=3.1, U=2.7 and UO=3.6 x 10^5/ml), Dasytricha (CTL=1.8, O=2.7, U=2.5 and UO=3.5 x 10^5/ml), Ostradinium (CTL=0.9, O=1.0, U=1.1 and UO=1.3 x 10^5/ml), Euliplodinium (CTL=0.7, O=0.8, U=1.0 and UO=1.3 x 10^5/ml), and total ciliated protozoa (CTL=37.9, O=48.0, U=47.9 and UO=56.3 x 10^5/ml). The combination of NPN sources (UO diet) showed superior numbers when compared with O diet for all ciliated protozoa (P<0.001). Ciliated protozoa were increased by nitrogen source supplementation.

Effect of forage source and a supplementary methionine hydroxy analogue on rumen fermentation parameters
Whelan, S.J., Mulligan, F.J., Callan, J.J. and Pierce, K.M., University College Dublin, Agriculture, Food Science and Veterinary Medicine, Animal Nutrition Building, UCD Lyons Research Farm, Newcastle, Dublin, 0, Ireland; stephen.c.whelan@ucd.ie

Optimising rumen function is essential for the efficient capture of dietary nutrients and maintenance of animal health. This trial investigates the effect of (1) forage source (grass silage (GS) vs. maize silage (MS)) and (2) supplementary methionine hydroxy analogue (HMBi) on rumen pH, VFA and NH_3-N over time in dairy cows offered low crude protein (133 g kg DM^{-1}) total mixed rations (TMR). Four ruminally cannulated dairy cows were offered 1 of 4 dietary treatments in a 2*2 factorial, Latin square design. The diets were GS based TMR, GS based TMR + HMBi, MS based TMR and MS based TMR + HMBi. Diets were iso-energetic (0.96±0.013 UFL kg DM^{-1}) and iso-nitrogenous (98±2 g PDI kg DM^{-1}). Diets contained 440 and 360 g kg DM^{-1} concentrate for GS and MS based diets respectively. Animals were housed in metabolic stalls to facilitate collection of rumen fluid every 2 h for 48 h. Data was analysed using PROC MIXED of the SAS institute. There were no effects of HMBi supplementation on measurements. Rumen pH (6.2±0.07) was unaffected (P>0.05) by diet. There were strong effects (P<0.01) of time on pH with lowest pH (5.9±0.07) approx. 14 hr post feeding. Rumen VFA concentrations (mmol l^{-1} ± s.e.d.) were higher (P<0.05) in cows offered GS (114±2) vs. MS (108±2) based diets and peaked (125±3) approx. 14 hr post feeding (P<0.01). Rumen ammonia nitrogen (NH_3-N) (mg ml^{-1} ± s.e.d.) was lower (P<0.05) in cows offered MS (56±1.8) vs. GS (62±1.8) based diets. Peak NH_3-N (91±4.1) occurred approx. 2-4 h post feeding (P<0.05). The reduction in rumen pH was related to increased VFA concentrations. However, at no time was the pH low enough to be classified as sub acute ruminal acidosis. Rumen NH_3-N concentrations were optimum for microbial growth. These data sugest that HMBi exhibits a good degree of rumen protection.

Nutrients digestibility and blood constituents of dairy buffaloes fed diet supplemented with flaxseed
Kholif, S.M.[1], Morsy, T.A.[1], Abedo, A.A.[2], El-Bordeny, N.E.[3], Abd El-Aziz, M.[1] and Abdo, M.M.[1], [1]National Research Center, Dairy Science Dept., Dokki, Giza, 12622, Egypt, [2]National Research Center, Animal Production Dept., Dokki, Giza, 12622, Egypt, [3]Faculty of Agric., Ain Shams Univ., Animal Production Dept., Cairo,Egypt, 68, Egypt; abedoaa@hotmail.com

Twenty one multiparous lactating buffaloes averaging 561 kg body weight (BW) were allotted at calving to three groups of seven buffaloes blocked for similar calving dates to determine effects of feeding crushed flaxseed on nutrients digestibility and concentration of some blood metabolites. Buffaloes within each block were assigned to one of 3 dietary treatments: C: Control (total mixed ration consisting of soybean meal, cottonseed meal, sunflower meal, yellow corn, wheat bran and berseem clover containing no flaxseed, CF17: Control with 17 g/kg DM crushed flaxseed, and CF34: Control with 34 g/kg DM crushed flaxseed as a source of linolenic acid. Diets were fed for ad-libitum intake from calving to week 16 of lactation. Crushed flaxseed supplementation had no effect on BW and dry matter intake (12.65, 12.68 and 12.63 kg/day, for C, CF17 and CF34, respectively). Flaxseed supplementation improved nutrients digestibility. Digestibility of OM was 69.0, 73.8 and 73.0% for C, CF17 and CF34, respectively. Diets' nutritive value (TDN) was 49.4, 63.3 and 62.9% and DCP value was 12.0, 12.9 and 12.8% for C, CF17 and CF34, respectively. Blood serum concentrations of total protein (66, 69 and 76 g/l), globulin (32, 32 and 41 g/l) and total lipids (2.68, 2.83 and 2.92 g/l, for C, CF17 and CF34, respectively) were increased at the highest level of flaxseed, while cholesterol (1.46, 1.33 and 1.25 g/l for C, CF17 and CF34, respectively) was decreased with flaxseed supplementation.

Milk urea nitrogen (MUN) as a tool to assess the nutritional status of buffaloes in Iran

Golghasemgharehbagh, A.[1], Pirmohammadi, R.[2], Ghoreishi, S.F.[3] and Rezaei, S.A.[4], [1]Payame Noor University, Department of Agriculture Science, 19395-4697 Tehran, 90441 Urmia, Iran, [2]Urmia University, Department of Animal Science, 365 Urmia, 90441 Iran, [3]Jihad Agriculture Association, Animal Science, Jihad Agriculture Association, West Azarbaijan, 90441 Urmia, Iran, [4]Urmia University, Department of Pathobilogy, 365 Urmia, 90441, Iran; A_GH_gharebagh@yahoo.com

In recent years, an adequate supply of protein and energy to dairy animals has been increasingly constrained. A higher demand for milk protein relative to other milk constituents puts more emphasis on the adjustment of available protein and energy supply to stimulate milk protein synthesis. Milk urea nitrogen (MUN) may be used as a management tool to improve dairy herd nutrition and monitor the nutritional status of lactating dairy herds. Buffalo breeding in the western Azerbaijan province of Iran is under rural conditions which according to climatic conditions, breeding systems and buffalo population, can be divided into three regions (North, Central and South). To monitor MUN, samples of morning and evening milkings were collected monthly from the buffaloes for 3 months. The MUN was estimated by according to the segmented flow method as described by De Jong *et al.* Data were subjected to ANOVA statistical analysis and mean values were compared with Duncan test. Results showed that average MUN concentration in the South (10.8 mg/dl) was significantly ($P<0.05$) higher than in the Central (8.8 mg/dl) region. The MUN concentration between North (10.2 mg/dl) and South was not significantly different. Moreover, in all areas, MUN in morning samples was lower than in evening samples. This may be due to the difference between feeding at morning and at evening. It was concluded that buffaloes under these rural conditions receive insufficient dietary protein.

Nutritional regulation of mammary protein synthesis

Hanigan, M. and Arriola, S., Virginia Tech, Dept. of Dairy Science, Blacksburg, VA 24061, USA; mhanigan@vt.edu

Milk protein yield is relatively plastic responding dramatically to changes in nutrient supply to the animal, particularly when supply is limited. Changes in milk protein synthesis can be mediated by alterations in rates of gene transcription and mRNA translation. Factors regulating these processes are thus of great interest. Initiation of lactation results in reduced expression of a number of genes involved with beta-oxidation of lipids, mRNA splicing, protein transport, ubiquitin-dependent protein degradation, and immunologic function. At the same time, expression of genes encoding milk proteins is enhanced. These changes appear to be at least partially regulated by prolactin via Jak/Stat signalling and glucocorticoids via glucocorticoid response elements in gene promoters. Translation is regulated through the mammalian target of rapamycin (mTOR) and General Amino Acid Response (GAAR) signalling pathways. The mTOR pathway integrates signals arising from insulin and growth factor binding to receptors, amino acid (AA) supply in the cell, and energy status of the cell. Low concentrations of insulin, depletion of intracellular amino acids, or inadequate ATP supply depresses rate of translation initiation and elongation resulting in dramatic reductions in protein synthesis rates. The GAAR pathway also affects the expression of many genes, some of which encode translational machinery, and thus affect rates of translation. During AA depletion, translation rates and transcription of activating transcription factor 4 are increased. This transcription factor heterodimerises with CCAAT-enhancer binding protein β and binds to AA response elements to induce expression of target genes. Genes affected by AA deprivation are involved in regulation of transcription, AA transport and metabolism, and protein translation. Amino acid depletion in the cell also affects expression of genes involved in lipid and carbohydrates metabolism. Mechanisms have been studied primarily in other species and tissues, although some extension to mammary tissue has occurred.

Characterizing individual differences in performance responses to a nutritional challenge

Schmidely, P.[1,2], Duvaux-Ponter, C.[1,2], Laporte-Broux, B.[1,2], Tessier, J.[1,2] and Friggens, N.C.[1,2], [1]INRA, UMR 791 Modélisation Systémique Appliquée aux Ruminants, 16 rue Claude Bernard, F-75005, Paris, France, [2]AgroParisTech, UMR 791 Modélisation Systémique Appliquée aux Ruminants, 16 rue Claude Bernard, F-75005, Paris, France; nicolas.friggens@agroparistech.fr

Being able to characterise robustness at the level of the individual animal would be valuable for refining management and selection strategies. However, there is little agreement on which are the key biological components, this presents a difficulty for characterising robustness. Recent work suggests that multivariate approaches are needed and that key information about the coping ability of an animal can be derived from the dynamic of its response to a challenge. Here we examined coping ability with respect to a nutritional challenge. Measurements of behavior, performance, and metabolites were made in 16 dairy goats exposed to a 2-day nutritional challenge, at 2 different stages of lactation. Each challenge consisted of a 1-week control period with standard TMR, 2 days of straw feeding, and a 1-week recovery period on the TMR. All feeds were offered *ad libitum*. As the focus of this work was to examine individual differences in responses we are interested in the portion of the variation that is usually assigned to the error term in standard ANOVA approaches. As expected the challenge resulted in a large drop in intake and milk yield with increases in milk fat content and milk protein content. Using a mixed models approach with random regression, individual deviations from the average response trajectory were isolated as offsets and slopes. Using principal components analysis these individual deviations were characterized. This revealed different response strategies between individuals with regard to energy partitioning in milk. In conclusion, there is significant variation between individuals in response to a nutritional challenge and this can be used to develop a quantitative description of coping ability.

Effect of rumen protected methionine on nitrogen metabolism in lactating mediterranean buffaloes fed a reduced-protein diet

Pace, V., Chiariotti, A., Contò, G., Carfì, F., Di Giovanni, S., Mazzi, M. and Terramoccia, S., CRA-Research Centre for Animal Production, via Salaria 31, 00015, Monterotondo (Rome), Italy; vilma.pace@entecra.it

The aim of the research was to evaluate the effect of supplementing rumen protected methionine (RPM) on nitrogen metabolism and rumen microbial growth in lactating buffaloes fed a reduced-protein diet. Sixteen multiparous Mediterranean buffaloes were divided in two groups (A and B) and fed for 120 days on two isoenergetic diets (0.90 MilkFU/kg DM) containing, on DM basis, 44% corn silage,13% soybean meal, 15% corn meal, 26% lucerne hay (group A; CP=15.5%) and 44% corn silage, 9.5% soybean meal, 18,5% corn meal, 26% lucerne hay, 12 g/head/d RPM (group B; CP=14.2%). Blood and urine samples from each animal were collected every two weeks to determine urea and insulin in plasma, and total-N, urea-N and creatinine in urine. Samples of whole rumen content were collected from four cannulated animals, fed the same diets, one hour before the morning feed, to determine total and cellulolytic bacteria, fungi and protozoa. The differences between groups were tested using a monofactorial model (ANOVA). Urea in plasma in group A was significantly higher than in group B: 7.4 vs 6.2 mM/l ($P<0.01$). Urea-N in urine was also higher in group A than in group B: 154 vs 132 g/head/d ($P<0.05$). The amount of total-N excreted through urine was 208 in group A vs 179 g/head/d in B ($P<0.01$). With respect to rumen microbial biomass, no differences were detected in bacteria and fungi, but the number of protozoa in group A was significantly higher than in group B (1.7×10^8 vs 0.7×10^8, $P<0.01$). The reduction of dietary protein level combined with RPM supplementation apparently reduced the growth of some rumen microorganisms, that are unable to utilize RPM, but exerted a positive effect in lowering urinary N excretion, thereby reducing the impact of buffalo herds on the environment.

Effect of anaerobic enzyme matrix on the digestibility, rumen parameters and lambs growth performance

Gado, H.[1] and Borhami, B.E.[2], [1]Faculty of Agriculture, Ain Shams University, Animal Production, Shoubra Al-Kheima, Cairo, 11424, Egypt, [2]Faculty of Agriculture (El-Shatby), Alexandria University, Animal Production, 11306 El-Shatby, Alexandria, Egypt; gado@link.net

One hundred lambs (initial body weight 18.4 kg±0.12) were used to evaluate the effect of ZADO® (an enzyme mixture of cellulase, xylanase, protease and alpha amylase) on feed intake, animal growth performance, apparent digestibility, and rumen concentrations of volatile fatty acids (VFA) and ammonia nitrogen (NH_3-N). Lambs were randomly divided into two groups of 50 animals each. Basal diet with or without 5 g of ZADO®/animal/d was fed. A growth performance trial of 60-days, ending by a digestibility trial of 21 days for 3 animals within each group, was carried out. Including ZADO® increased feed intake (1.72 vs. 1.32 kg/d; P<0.04), average daily gain (0.38 vs. 0.25 kg/d; P<0.05) and feed efficiency (5.0 vs. 5.1 kg DM/kg live weight gain; P<0.05) and decreased calculated net energy required for 1 kg live weight gain (4.8 vs. 5.3 MJ; P<0.05). Including ZADO® increased nutrients digestibility coefficients and total digestible nutrients (74.1 vs. 68.3%; P<0.06). Including ZADO® increased the digestibility of neutral detergent fiber (57.3 vs. 49.2%; P<0.001) and ruminal concentrations of VFA (meq/l) at 3 h from 75 to 91 and at 6 h from 73 to 82 and NH_3-N at 3 h after feeding from 159 to 175 mg/l (P<0.05). The improved animal performance by inclusion of ZADO® was as a consequence of the improved digestibility and the increase in VFA. In conclusion, ZADO® as anaerobic enzyme matrix improved the nutritive value the ration.

Effect of dry period length, and lipogenic vs. glucogenic diets on dry matter intake, milk production and composition in Holstein dairy cows

Soleimani, A.[1], Moussavi, A.H.[2], Tahmasbi, A.[2], Danesh Mesgaran, M.[2] and Golian, A.[2], [1]Islamic Azad University – Kashmar Branch, Department of Animal Science, Seyyed morteza blvd, 91888-17717, Iran, [2]Ferdowsi University of Mashhad, Department of Animal Science and Excellence Center for Animal Science, Mashhad - Khorasan Razavi Province, 91775-1163, Iran; ak.soleimani@iaukashmar.ac.ir

The aim of this study was to determine the effects of dry period length (35 vs. 20 days) and dietary strategy (lipogenic vs. glucogenic diet) on dry matter intake (DMI), milk production and milk composition in early lactating Holstein cows using a 2x2 factorial arrangement. Twenty-four Holstein dairy cows were housed individually and allocated to 4 experimental treatments: G35: a glucogenic diet with a 35-d dry period; L35: a lipogenic diet with a 35-d dry period; G20: a glucogenic diet with a 20-d dry period and L20: a lipogenic diet with a 20-d dry period. Dry matter intake was recorded daily until 8 week *post partum*. After parturition, cows were milked 3 times per day at 04h00, 12h00 and 20h00. Milk yield was recorded and milk was sampled during 8 weeks *post partum*. Data were statistically analyzed using MIXED procedure of SAS. No significant difference was observed in prepartum and *post partum* DMI among treatments. A significant difference (P<0.05) was found between treatments G35 and G20 regarding milk production and daily yields of lactose, milk protein and milk solids non-fat (38.4 vs. 30.6, 1.79 vs. 1.43, 1.19 vs. 0.96 and 3.29 vs. 2.63 kg/d, for G35 and G20, respectively). Milk fat concentration for G35 was significantly (P<0.05) lower than for L35 (3.66 vs. 4.08%). Week of lactation had a significant (P<0.01) effect on DMI, milk production and milk composition. These results suggest that the mammary gland would require a dry period length of more than 20 days to allow a high milk secretion.

Food Security: Policy perspectives for industry and research

Vriesekoop, P.W.J.[1], Williams, J.[2], Tubb, R.[3] and Jensen, J.[4], [1]Wageningen UR Livestock Research, P.O. Box 65, 8200AB Lelystad, Netherlands, [2]INRA Phase, INRA Research Centre, F-37380 Nouzilly, France, [3]MTT Agrifood Research Finland, Humppilantie 9, FI-31600 Jokioinen, Finland, [4]Aarhus University, P.O. Box 50, DK-8830 Tjele, Denmark; info@animaltaskforce.eu

This paper from a working group of the Animal Task Force focuses on the future challenges of food security for the animal production sector (incl. aquaculture). Four areas for consideration where new initiatives are needed are: 1) Improvement of resource efficiency. Increasing global population, and increasing demands per individual from large population segments, will intensify competition for food production resources. Genetic improvement of animals and crop plants, improved control of processes, (e.g. through precision farming), and reduction of waste throughout the production chain, will improve efficiency and mitigate negative environmental impacts. Production systems with low emissions and low carbon footprints must be developed. 2) Development of animal products contributing to improved human health and the reduction of obesity and related metabolic and skeletal disorders. Research is also required on (a) prevention of disease transmission (zoonoses) (b) combating antimicrobial resistance, and (c) understanding how toxic substances enter the food chain. We must also support the integration of veterinary and human medicine.3) Animal production systems that are sustainable on local and aggregated levels, adapted to diverse economic and geobiological conditions, and incorporating different requirements on land use, must be developed. This includes improving the logistics of production and distribution, optimising the scale of production systems and better integration of crop and animal production. 4) Adaptation of animal production systems to changing climate. Both regional differences in climatic conditions and the frequency of extreme weather events are expected to increase. Animal production systems will need to be more adaptable to changing climate and extreme weather conditions.

High animal production efficiency for less environmental footprints: opportunities and challenges

Zebeli, Q. and Klevenhusen, F., Institute of Animal Nutrition, Vetmeduni Vienna, Veterinaerplatz 1, 1210 Vienna, Austria; qendrim.zebeli@vetmeduni.ac.at

It is commonly known that livestock has major impact on the environment and contributes to global warming due to greenhouse gas (GHG) emissions. In livestock husbandry, CO_2 is produced mainly during concentrate production and transport, and N_2O predominantly results from manure management and fertilizing grass- and croplands. Methane derives to great parts from ruminants due to enteric fermentation and many researchers are searching for ways to mitigate enteric CH_4 production. This presentation will start with a brief overview of achievements in the use of diverse feed supplements, which have already been tested for their CH_4 mitigating potential. Despite encouraging results in some *in vitro* trials, the *in vivo* studies conducted so far showed that their effects on CH_4 mitigation are either minor or they impair ruminal digestion and jeopardize animal performance. One strategy that has gained increasing interest in lowering GHG emissions from livestock, in a long term, is the improvement of the animal production efficiency, which results in less GHG per unit edible product. In this respect, we will elaborate different digestive physiological approaches, which can lead to increased levels of milk and meat yields from fewer nutrients and less animal waste. Improving animal production efficiency can lower the environmental footprint by reducing animal numbers while gaining at least the same amount of edible product. Among others we will discuss options to include highly digestible feedstuffs in the diet, promote animal health, and prolong animal's productive lifespan. As well, opportunities to increase animals' feed intake potential, enhance nutrient uptake, and 'dilute' relative maintenance requirements will be described as potential long-term solutions to mitigate GHG. However, although sound in theory, applications of certain strategies in practice may be more difficult. In this regard, the presentation will conclude with limitations, future challenges, and suggestions for further research.

Chasing multiple breeding goals – can genetics deliver multiple public-private good outcomes?
Wall, E. and Moran, D., SAC, West Mains Road, Edinburgh, EH9 3JG, United Kingdom;
eileen.wall@sac.ac.uk

Contemporary livestock production systems have a dual role combining food production and public good objectives. The livestock industry will have a number of drivers in the development of future breeding goals due to the challenges of global change (climate change, reduction of environmental impact, food security). The private breeding goal for producers, and the industry as a whole, will be profit driven and therefore a financial based breeding goal is appropriate. But this needs to be augmented with metrics to accommodate and measure other public good objectives including challenges of reducing greenhouse gases (GHG), improving resources efficiency as well as wider social/public goals such as improved quality and animal welfare. Some of these public goals have a partial private economic benefit that can be incorporated in the private breeding goal. However, public targets and private targets may not always be compatible, and could be in conflict. This will be a challenge for livestock producers in developing the next generation of breeding goals. This paper will explore methods for the inclusion of public good environmental efficiency targets in private breeding goals, quantifying the private (financial) and public (GHG emissions, animal welfare) co-benefits and trade-offs. The results will show that genetic improvement tools provide a useful and cost-effective mechanism to help livestock agriculture meet the challenges of the reducing GHG emissions. We will explore the wider and future challenges that face livestock production and how they could be reflected in, and potentially limit future breeding goals.

Methane, animal breeding and beef and sheep farmers' readiness for uptake
Bruce, A., University of Edinburgh, High School Yards, Edinburgh EH1 1LZ, United Kingdom;
ann.bruce@ed.ac.uk

Methane emissions from cattle and sheep have increasing profile in the context of climate change. A range of breeding technologies have been suggested as providing ways of reducing these emissions. However, sheep and beef cattle producers have historically been slow at adopting breeding technologies to improve productivity for reasons that are not well understood. The objectives of this project were to understand why the adoption of breeding technologies has been slow and uneven and how the context of methane emissions might influence this situation. This project sought to analyse wider system factors critical to innovation. The research consisted of 40 semi-structured interviews with a range of sheep farmers, beef producers and relevant associated stakeholders around the UK. The aim was to sample a diversity of stakeholders systematically. Statistical analysis This is qualitative research that does not purport to provide statistically significant samples but to unravel some of the complex processes involved in technology adoption. Analysis was undertaken with inductive coding using Nvivo9 software and causal mapping using Banxia Decision Explorer™ software. EBVs are not trusted, even among those who use this information. Environmental policies may limit the ability of farmers to respond to other concerns. Breeding has a strong cultural component. Economic signals can emphasise conformation. In the public sphere, methane production from cattle and sheep may be presented as an important cause of climate change but to livestock farmers this is a natural product that has always existed and they see few ways in which methane production can be reduced. No one, single barrier to adoption of breeding techniques was found. Instead there is a complex array of cultural, economic and other barriers to adoption, many of them inherent in current production systems.

Mitigation options of greenhouse gas emissions for Finnish beef production

Bouquet, P., Hietala, P. and Juga, J., University of Helsinki, Department of Agricultural Sciences, Latokartanonkaari 7, 00014 Helsingin yliopisto, Finland; peggy.bouquet@helsinki.fi

In Finland, 9% of the greenhouse gas (GHG) emissions came from Agriculture in 2009. Milk and beef production, which contribute to a large part of them, are tightly linked together since 85% of beef comes from dairy animals. The left 15% comes from specialized beef production. The latter has been shown to produce more GHG emissions per kg of meat than dairy herds. Following the objective of a more environmentally efficient beef production, this study evaluated GHG emissions per meat kg in different scenarios of intensified beef production from dairy herds. We studied the effects of increasing insemination of dairy cows by beef bulls to produce cross-bred slaughter animals, using of sexed semen for cow replacement and cross-bred calf production, decreasing replacement rate by increasing cow herd life and eventually increasing growth rate of slaughter animals. In these calculations changes in meat amount from culled dairy cows were not accounted for. When a breeding strategy in dairy herd increased the meat production by 1% (around 500 tons of beef) the net reduction in GHG emissions by replacing specialized meat production with meat from dairy herds would be between 0.35 and 0.4% (around 3,500-4,000 tons of CO_2 eq.). Reducing the replacement rate from 35% to 30% allowed for a 3% increase in beef production from dairy animals. A similar increase could be achieved if the annual growth rate of slaughter animals would be increased by 5%. Increasing the percentage of beef insemination in dairy herds by 5% would increase the total meat production only by 0.2%. Using sexed semen was even less efficient: if 40% of cows were inseminated by sexed semen the increase in meat production would be only 0.1%. It is possible, even though limited, to reduce GHG emissions per kg of meat produced in a system promoting dairy rather than specialized beef production. Increasing the herd life of dairy cows is the most effective breeding strategy to achieve this.

Breeding for resource efficiency in grazing animals on resource-poor rangelands

Rauw, W.M., INIA, Departamento de Mejora Genética Animal, Crta de la Coruña km 7.5, 28040 Madrid, Spain; rauw.wendy@inia.es

Grazing animals are often subject to recurrent periods of undernutrition in which large amounts of body tissue may be catabolized. Variation in the ability to graze selectively may offer the opportunity to breed for better adaptation to poor-quality rangelands, resulting in healthier animals and improved production. However, feed intake in grazing animals cannot be accurately recorded without sophisticated methods that are unlikely to be used in a selection criterion. An alternative method is proposed based on changes in body weight of animals after a period of grazing. Of 905 ewes grazing approximately 2.5 months on poor-quality Nevada (USA) rangelands, 94% lost body weight. Body weight change was moderately heritable at $h^2=0.29$, which indicates that selection against loss of body weight will result in a positive selection response. The following model can be used for estimation of grazing ability in terms of estimated energy units consumed: $EGI_i = FI_i - b_0 - e_i = (b_1 \times BW_i^{0.75}) + (b_2 \times BWG_i) + (b_3 \times PROD_i)$, were EGI_i = estimated grazing intake, FI_i = actual feed intake, $BW_i^{0.75}$ = metabolic body weight, BWG_i = body weight gain, $PROD_i$ = level of production (kg milk or wool, etc), b_0 = population intercept, b_1, b_2 and b_3 = partial regression coefficients representing maintenance requirements, feed requirements for growth, and feed requirements for (re)production, respectively, and e_i = the error term, representing residual feed intake. This model can be finalized by including estimates of the b-values. This method can be used to assess the nutritional adequacy of rangelands and determine if additional supplementation is necessary. Grazing ability can be compared between animals living in different environments or different species living in the same environment. The method can be used to evaluate the grazing potential or load of a flock on a given ecosystem. Selection for grazing ability would foremost result in healthier animals that can produce offspring without compensating their welfare.

Novel estimation of residual energy intake and their quantitative trait loci in growing-finishing pigs
Shirali, M.[1,2], Doeschl-Wilson, A.[3], Knap, P.W.[4], Duthie, C.[1], Kanis, E.[2], Van Arendonk, J.A.M.[2] and Roehe, R.[1], [1]Sustainable Livestock Systems Group, SAC, West Mains Road, Edinburgh, EH9 3JG, United Kingdom, [2]Animal Breeding and Genomics Centre, Wageningen UR, P.O. Box 338, 6700 AH Wageningen, Netherlands, [3]The Roslin Institute, University of Edinburgh, Easter Bush, Roslin, Midlothian, EH25 9RG, United Kingdom, [4]PIC International Group, Ratsteich 31, 24837 Schleswig, Germany; mahmoud.shirali@sac.ac.uk

The aims of this study were to find the best model for estimation of residual metabolisable energy intake (REI) and to identify QTL for REI at different stages of growth and the entire growing-finishing period of pigs. Data were available on 315 F_2 pigs from a three-generation full-sib design developed by crossing Pietrain sires with a commercial dam line. REI was estimated, as the difference between actual metabolisable energy intake and predicted using three different models that included 1) non-adjusted average daily protein (APD) and lipid deposition (ALD) (REI1); 2) besides other fixed effects adjusted average daily gain and backfat, a commonly used method (REI2) and 3) besides other fixed effects adjusted APD and ALD, as a novel method (REI3). The coefficients of determination for the three models showed that REI3 fitted the data better than the others (R^2=0.63, 0.74 and 0.76 for REI1, REI2 and REI3 respectively). Three significant QTL were identified for REI3 on SSC14, SSC2 and SSC8, during growth from 60-90 kg, 90-120 kg and 120-140 kg, respectively. In addition, a QTL was identified on SSC14 for REI3 during the 60-140 kg period, which did not differ from QTL for 60-90 kg. These QTL explained 4.6%, 3.6%, 4.2% and 4.7% of the variance in REI during 60-90 kg, 90-120 kg, 120-140 kg stages of growth and the 60-140 kg, respectively. No significant QTL were detected for REI2. This study showed that accurate estimation of REI by adjusting for traits such as APD and ALD, which are directly related to energy requirements, is of high importance for detecting QTL and for efficient breeding for REI.

Genetic parameters for residual energy intake and its relationships with production and other energy efficiency traits in Nordic Red dairy cattle
Liinamo, A.-E.[1], Mäntysaari, P.[2] and Mäntysaari, E.A.[1], [1]MTT Agrifood Research Finland, Biotechnology and Food Research, Biometrical Genetics Group, FIN-31600 Jokioinen, Finland, [2]MTT Agrifood Research Finland, Animal Production Research, FIN-31600 Jokioinen, Finland; anna-elisa.liinamo@mtt.fi

The aim of this paper was to study the genetic parameters of residual energy intake and its relationships with milk production, dry matter intake, body weight, body condition score and energy balance in Nordic Red dairy cattle from an experimental data set. The data was collected at the MTT Agrifood Research Finland Rehtijärvi experimental farm in four feeding trials between 1998 and 2008, and included lactation weeks 2-30 for 291 Nordic Red nucleus heifers descending from 72 different sires. Residual energy intake (REI) was defined as the difference between total energy intake of each animal, and the energy required for milk, maintenance and body weight change based on the animal's observed lactation performance, body weight and body weight change. REI was expressed as weekly averages in ME MJ/d. The other studied traits included the weekly averages for energy corrected milk yield kg/d (ECM), dry matter intake kg/d (DMI), body weight kg (BW), body condition score 1-5 (BCS), and energy balance ME MJ/d (EB). The data were analyzed with random regression models. The highest heritability estimates for REI were obtained in the beginning of the lactation and they were moderate (0.20). The heritability estimates of REI then dropped almost to zero between lactation weeks 10 to 20 before starting to rise again. The genetic correlations of REI in different lactation weeks suggest that REI in early and mid-lactation periods are partially different traits. The genetic correlations between REI and the other studied traits were high and positive for DMI and EB, and moderate and positive for BW and BSC. The genetic correlation of REI with ECM was negative between lactation weeks 2 and 5, but changed into moderate to low and positive after that period.

Application of random regression models for Gaussian and binary traits to estimate genetic parameters in low input dairy cattle herds

Yin, T.[1,2], Bapst, B.[3], König-Vb., U.[1], Simianer, H.[1] and König, S.[2], [1]Dept. of Animal Breeding and Genetics, Albrecht-Thaer-Weg 3, 37075 Göttingen, Germany, [2]Dept. of Animal Breeding, Nordbahnhofstraße 1a, 37213 Witzenhausen, Germany, [3]Schweizer Braunviehzuchtverband, Chamerstrasse 56, 6300 Zug, Switzerland; tyin@gwdg.de

Due to restrictions in feeding and husbandry, low input farming is substantially different from conventional production systems. Such differences suggest the implementation of an own breeding program. A prerequisite for the latter objective is the estimation of (co)variance components of traits to be included in the overall breeding goal. Data for production and functional traits were available from 1283 Brown Swiss cows kept in 54 low input farms in Switzerland. Different random regression models have been applied to the data to infer genetic parameters for two types of time dependent covariates, i.e. days in milk (DIM) and parity of the cow. For Gaussian distributed test-day records in milk yield (MY), fat percentage (Fat%), protein percentage (Pro%), lactose percentage (Lac%), SCS, and milk urea nitrogen (MUN), a multivariate random regression animal model was applied. Daily heritabilities followed the same pattern as found for high input production systems. Expected genetic antagonisms were found between MY and Pro% for all DIM, and also between MY and Fat% apart from the earliest stage of lactation. Antagonistic relationships between MY and SCS were only found for the first third of lactation. Threshold random regression sire models were applied to estimate daily correlations between MY and a binary distributed fertility trait having limited observations within DIM. Pronounced antagonistic relationships between MY and conception rate ranged between -0.7 to -0.8. Interval traits describing the cow's ability for recovering after calving had higher heritabilities compared to traits including the component of a successful insemination. In conclusion, random regression methodology can be applied for data from small herds and for limited categorical observations per animal.

Application of lactation efficiency in pig breeding programs

Bergsma, R. and Knol, E.F., IPG, Institute for Pig Genetics, R&D, P.O. Box 43, 6640 AA Beuningen, Netherlands; rob.bergsma@ipg.nl

Sow fertility, in piglets per sow per year, keeps increasing, reducing the amount of sow feed invested per piglet. Milk production should keep pace with the increased piglet production. The objective of this study was to investigate whether sows could facilitate increased milk production by increased efficiency rather than by increased feed intake or more weight loss during lactation. Lactation efficiency was defined as the energy efficiency of sows during lactation. It is an equivalent trait to feed conversion or residual feed intake in finishers. Estimates on 2,202 litters and 7,188 finishers yielded low heritabilities for lactation efficiency and of residual feed intake as a grower-finisher (0.12 and 0.11 respectively), and a moderate favorable genetic correlation (-0.51) between the two. There are no disadvantages of including lactation efficiency in the breeding objective in dam lines, other than reduced feed intake capacity. The 10% most energy efficient lactations in our dataset had, compared to the mean, a decreased feed intake, which was not fully compensated by the increased protein mobilization during lactation. The consequence of this reduced availability of protein was that milk yield stayed behind. This demonstrates that increasing energy efficiency should be accompanied by reformulation of diets. It also demonstrates that current diets limit milk yield of the most efficient sows in a sow herd. Genetic variation in lactation efficiency exists and is positively correlated with feed efficiency in finishers. Maximizing energy efficiency will have consequences for formulation of diets for future feeding, but also for current diets in order to properly recognize the most efficient animals in the existing breeding herds. Better tuning of the demands of animal production in the diets offered is one of the pillars behind reduced ecological foot print of animal production. In that way, farmers as well as the environment can benefit of genetically improved energy efficiency to the maximum.

Genetic selection for lower predicted methane emissions in dairy cattle

De Haas, Y.[1], Windig, J.J.[1], Calus, M.P.L.[1], Bannink, A.[1], De Haan, M.[1], Dijkstra, J.[2] and Veerkamp, R.F.[1],
[1]Wageningen UR Livestock Research, P.O. Box 65, 8200 AB Lelystad, Netherlands, [2]Wageningen University,
P.O. Box 338, 6700 AH Wageningen, Netherlands; Yvette.deHaas@wur.nl

Mitigation of ruminant methane (CH_4) emission has become an important area of research because accumulation of CH_4 has been linked to global warming. Little information is yet known on opportunities for mitigation via animal genetics. Measuring CH_4 production directly from animals is difficult and hinders direct selection on reduced CH_4 emissions. However, improvements can be made through selection on traits that are correlated to CH_4 emissions (e.g. RFI) or through selection on CH_4 predicted from feed intake and diet composition (e.g. the International Panel on Climate Change Tier-2). The aim of this study was to estimate phenotypic and genetic associations between residual feed intake (RFI) and predicted CH_4 emission. Data was used from an experimental farm. Genotypes, daily feed intake records, weekly live weights and weekly milk productions were available from 588 heifers. RFI (MJ/d) was calculated as the difference between net energy intake and calculated energy requirements for milk, fat, and protein yields, and maintenance costs as a function of metabolic live weight. Predicted CH_4 emission (gram/day) was calculated with the IPCC-method and is 6% of gross energy intake. Estimated heritability for predicted CH_4 emission was 50%. Both phenotypic and genetic correlation between RFI and predicted CH_4 emission were on average approximately 0.60, showing that it is possible to decrease the predicted CH_4 emission by selecting more efficient cows, with the assumption that diet digestibility does not differ between efficient and non-efficient cows. Depending on lactation stage, the correlation between predicted CH_4 emission and milk production varied between +0.5 and -0.6. In late lactation more food was consumed than required for milk production. Hence, simply increasing yield is not always the answer to reducing CH_4 emission.

Animal variation in methane emission from breath of Danish Red, Holstein and Danish Jersey

Lassen, J.[1], Løvendahl, P.[1] and Madsen, J.[2], [1]Aarhus University, Department of Genetics and Biotechnology,
Blichers Allé, 8830 Tjele, Denmark, [2]University of Copenhagen, Department of Large Animal Science,
Grønnegårdsvej 6, 1870 Frederiksberg, Denmark; jan.lassen@agrsci.dk

On a total of 155 dairy cows in an automatic milking system (AMS) a Fourier Transformed Infrared (FTIR) measuring unit was used to make large scale individual methane emission records. The cows were 62 Red Danish, 50 Holsteins and 43 Jerseys. The FTIR unit was put in the front part of an AMS close to the cows head for 3 days. The unit made an air analysis every 20 second. We tested the capability of the FTIR unit to make individual measurements by estimating repeatability for the median. The idea of the measurements was to use the cow's carbon dioxide concentration in its breath as a tracer gas and from that be able to quantify the production of methane. The repeatability of the methane to carbon dioxide ratio was 0.30 for Red Danish, 0.37 for Holsteins and 0.33 for Jerseys. These estimates are higher than earlier results from other studies where SF_6 were used as tracer gas, and on a smaller number of animals than in this study. For Holstein and Jersey the model used to analyze the data contained an effect of three weeks average daily yield and roughage intake and both had a small but significant effect on methane emission. Milk production level only had a significant effect in the Red Danish experiment and days in milk and lactation number were non-significant. The results from this pilot study suggest individual cow's methane emission can be measured using FTIR equipment. The data can be used for both management and genetic analysis and opens for future studies on other gasses from breath. The results show that even after correcting for a number of effects that are reasonable to believe have an effect on methane emission we still observe a considerable amount of individual variation. This variation may well be under genetic control.

Monitoring and status of Lithuanian farm animal breeds
Sveistiene, R.[1,2], Razmaite, V.[1,2] and Muzikevicius, A.[2], [1]Lithuanian University of Health Sciences, Institute of Animal Science, Animal Breedig and Genetics, R.Zebenkos 12, LT-82317, Lithuania, [2]Lithuanian Endangered Farm Animal Breeders Association, R.Zebenkos 12, LT-82317, Lithuania; ruta@lgi.lt

The aim of the study was to analyse farm animal monitoring data and estimate the status and effective population size of Lithuanian farm animal breeds (LFAB).The principles of conservation of animal genetic resources and evaluation of breed status are based on the experience of animal breeding in small conserved herds and on the criteria and global strategy of FAO for the management of farm animal genetic resources. The status of LFAB was evaluated by their monitoring. LFAB genetic resources include old and new breeds. The total numbers of some protected breeds are decreasing, and separate breeds have grown only due to the 5 year obligation to increase herds by participating in the Rural Development Programme. The Ne size of LFAB is low and could be characterized as the endangered and critical. The genetic status of a population-inbreeding can be minimized by having a larger Ne>50 and by using within family selection to as large an extent as possible.The numbers of animals from rare breeds were stabilized and even have increased for some breeds, the numbers of sires should be enlarged, and higher requirements for pure breeding. The farmers seek for economic benefit. The breeding of animal from rare breeds and income from their production is not competitive compared with industrial breeds. It could be defined that nowadays rare breed animals are kept not for commercial purposes but for breed restoration and herd stability maintenance by preserving biodiversity for future generations. Therefore, compensatory payments have been helped to conserve the genetic resources of LFAB by reproducing new herds or animals and follow the special mating rules and schemes in order to delay inbreeding and prevent single individuals from getting extreme levels of inbreeding.

Genetic correlations between traits scored with detailed and simplified exterior systems in Norsvin Landrace
Aasmundstad, T.[1,2], Grindflek, E.[1] and Vangen, O.[2], [1]Norsvin, P.B. 504, 2304 Hamar, Norway, [2]Norwegian University of Life Sciences, Department of Aquaculture and Animal Science, P.B. 5003, 1432 Aas, Norway; Torunn.Aasmundstad@norsvin.no

Welfare and production parameters indicate that increased effort on improved longevity and robustness in Norwegian pig populations are needed. In 2009, non-functional leg/exterior caused 12.5% of all sow-culling. In 1996, Norsvin implemented a detailed system for exterior scoring (S1) consisting of 21 traits. The system was implemented for purebreed boars at 100 kg and for gilts prior to first farrowing. However, intensive pre-selection of gilts was performed by farmers based on the quality of the exterior traits, resulting in inaccurate estimation of breeding values and reduced selection pressure for all other traits. To avoid this problem Norsvin wanted to perform exterior evaluation on all gilts in a litter (approx. 45,000/year) and to enable this a simpler evaluation system was needed (S2). In S2 the traits are grouped in 6 traits; fore leg, hind leg, motorics, standing under, hoofs and top-line. In both systems traits were scored on a linear scale. The aim of this study was to estimate genetic correlations between traits scored with S1 and S2. Since S2 was implemented in end of 2010 and few semen boars therefore are represented as sires, only preliminary results are presented. The data included information on 47,000 gilts for S1 traits and 7,000 gilts for S2 traits. The traits were analysed in bivariat animal models. The preliminary genetic correlations ± standard errors are: S1.front leg knee: S2.front leg= 0.10 ± 0.25, S1.front leg pastern: S2.front leg=no convergence, S1.rear leg pastern: S2.hind leg=0.62 ± 0.20, S1.waddling hind: S2.motorics=0.51 ± 0.27. We conclude that positive correlations indicate that our simplified system capture some of the same variation as the detailed system. The large standard errors could be due to the genetic structure of data, or the complexity of the new combined traits.

Breeding values for production efficiency in the Swiss dairy breeding programs

Cutullic, E.[1], Bigler, A.[2], Schnyder, U.[3] and Flury, C.[1], [1]Swiss College of Agriculture, Länggasse 85, 3052 Zollikofen, Switzerland, [2]Swissherdbook, Schützenstrasse 10, 3052 Zollikofen, Switzerland, [3]Arbeitsgemeinschaft Schweizerischer Rinderzüchter, Schützenstrasse 10, 3052 Zollikofen, Switzerland; erwan.cutullic@bfh.ch

Production efficiency is a trait of growing interest in dairy breeding. In Switzerland, live weights are not routinely collected and production efficiency was not considered in breeding programs so far. In the present study we investigated the relationships between linear description measurements evaluated by experts and live weights, recorded between 30 and 270 days in milk, in 84 primiparous and 192 multiparous cows. Dairy breeds were Holstein or Red Holstein (n=167). Dual purpose breeds were Swiss Fleckvieh (37) and Simmental (72). Relationships were analysed by multiple regressions, simplified stepwise. The final model (R^2=78%, r.s.e.=38 kg) included log-transformed lactation rank (P<0.001), days in milk (P=0.099) and 4 linear description parameters: heart girth (P<0.001), rump length (P<0.001), muscularity (P<0.001) and body condition score (P=0.008). Heart girth and muscularity could be replaced by rump width, chest width and stature with moderate precision loss (R^2=71%, r.s.e.=44 kg). Replacing muscularity by distinguishing Simmental from other breeds lead to a comparable precision (R^2=80%, r.s.e.=37 kg). These first results suggest that a generic equation could be used. However, a validation on New-Zealand type Holstein cows revealed an overestimation for such light cows. Similarly, reported equations for primiparous Brown Swiss cows underestimate live weights of our dataset. A generic equation would thus require a larger dataset with a better coverage of different parities and cow types. Further variance components, production efficiency breeding values and correlations with other breeding values will be presented. Other measures to define production efficiency will be discussed and an outlook for the implementation of production efficiency into Swiss dairy breeding programs will be given.

Genetic and economic evaluation of different Sahiwal cattle breeding programmes in Kenya

Ilatsia, E.D.[1,2], Roessler, R.[1], Kahi, A.K.[3] and Valle Zárate, A.[1], [1]Hohenheim University, Institute of Animal Production in the Tropics and Sub tropics (480a), Garbenstrasse 17, 70593, Stuttgart, Germany, [2]Kenya Agricultural Research Institute, National Animal Husbandry Research Centre, P.O. Box 25, 20117, Naivasha, Kenya, [3]Egerton University, Animal Sciences, P.O. Box 536, 20115, Egerton, Kenya; evansilatsia@yahoo.com

The Sahiwal cattle breeding programme in Kenya has been operational on interim basis with no efforts undertaken to evaluate its suitability compared to other alternative programmes. The objective of this study was to evaluate the genetic and economic success of the different setups for a Sahiwal cattle breeding programme in Kenya. The breeding programmes examined were the current closed nucleus breeding programme with two breeding strategies: a purebreeding (CN_{PURE}) and a crossbreeding system (CN_{CROSS}) involving Sahiwal sires and East African Zebu dams. An open nucleus where a certain proportion of pastoral-born Sahiwal bulls are used to produce cows in the nucleus was simulated as an alternative breeding programme. In this programme only a purebreeding strategy (ON_{PURE}) was considered. The breeding strategies were evaluated under two breeding objective scenarios that addressed traditional markets where animals are sold on body size/weight basis (BSWB) and the Kenya Meat Commission where payment is based on carcass characteristics (CSCB). The genetic gain and profit per cow for all investigated breeding programmes varied within breeding objectives. The CN_{PURE} was the most attractive economically but less competitive in regard to genetic superiority compared to either CN_{CROSS} or ON_{PURE}. Returns and profits were generally higher for the CSCB compared to the BSWB for all evaluated breeding strategies. The CN_{CROSS} plays a complimentary role of facilitating the exploitation of trade-offs that exist between the Sahiwal and the EAZ, it also represents an intermediate phase in the up-grading programme.

How to keep and feed horses in groups?

Søndergaard, E., AgroTech, AgroFoodPark 15, 8200 Aarhus N, Denmark; evs@agrotech.dk

Horses are social animals by nature; thus keeping horses in groups is keeping horses naturally. Despite this fact most riding horses are kept singly. When horse owners are asked why they keep their horses singly the risk of injuries and the problems of feeding the horses individually are mentioned as the main reasons. Another reason may be that most housing systems for horses are designed for keeping horses individually thus just letting the horses loose without reconsidering the system may create problems. Keeping horses in groups requires considerations about group size and composition, about handling the horses within and outside the group and about feeding them according to their needs. All has to be considered in relation to the risk of injuries and human safety. For ease of feeding, homogenous horse group are preferable, but for e.g. the behavioural development of young horses mixed groups are more suitable. Handling horses in larger groups may be more difficult than handling horses in smaller groups but it may be easier for horses to cope in larger groups, especially for horses low in hierarchy. Feeding horses individually when they are kept in groups may be a challenge especially considering that horses prefer to synchronise their behaviour. Also designing systems to allow all horses to rest and relax together is a challenge. Designing systems for keeping and feeding horses in groups is an ongoing process and thus this presentation does not aim to give the final solution, but just some suggestions according to our present knowledge.

Housing of trotters in Norway – a survey

Bøe, K.E.[1] and Jørgensen, G.H.M.[2], [1]Norwegian University of Life Sciences, Department of Animal- and Aquacultural Sciences, P.O. Box 5003, 1432, Norway, [2]BioForsk, BioForsk Nord, Tjøtta, 8860, Norway; knut.boe@umb.no

The aim of this survey was to investigate type of housing systems for trotters in Norway. In cooperation with the Norwegian Trotting Association a questionnaire with 32 different questions concerning housing, ventilation and outdoor paddocks were sent to 442 stables with trotters. 275 (61%) responded and 268 stables with altogether >2,400 horses were included in the final dataset; 38 small stables (1-2 horses), 164 medium-sized (3-10 horses) and 60 large stables (11-90 horses). The majority of the buildings (56%) had earlier been used and built for other purposes, often dairy production. 83.6% of the stables were insulated, warm buildings, 10.1% non-insulated, cold buildings and 1.1% simple sheds with three walls. Mechanical ventilation systems were present in 50% of the stables. Keeping the horses in single boxes inside was the dominant way of housing (84%). 80% the medium-sized stables kept two or more horse together in the outdoor paddocks, whereas the corresponding number for large stables was 90%. The majority of the horses seemed to have daily access to the outdoor paddocks, both during winter (96%) and summer (98%).

Impact of unrelated adults on the behaviour of weanlings and young horses.
Henry, S.[1], Bourjade, M.[2] and Hausberger, M.[1], [1]UMR CNRS 6552 EthoS, Campus de Beaulieu, Université Rennes 1, 263 avenue Général Leclerc, 35042 Rennes, France, [2]Max Plank Institute, Leipzig, 04303 Leipzig, Germany; severine.henry@univ-rennes1.fr

Young horses normally live in small year-round stable groups including one stallion, their mothers, a few other mares, their siblings and unrelated peers. On the contrary, most of young domestic horses are generally maintained in same-age and same-sex groups from weaning until training. One has to consider that the absence of adult partners during ontogeny may be a source of behavioral disorders. In a first study, we focused on social conditions at weaning. While it is well known that presence of peers is of high importance to alleviate weaning stress, we investigated here the effects of the introduction of unrelated adult mares in groups of weanlings. Results showed that signs of stress were less pronounced and shorter in time in weanlings housed with adult mares than in weanlings kept in same-age groups (e.g. distress vocalizations: $P<0.05$; salivary cortisol: $P<0.05$). Besides, only foals deprived of adult presence exhibited increased aggressiveness towards peers ($P<0.05$) and abnormal behaviors ($P<0.05$). In conclusion, the presence of two unrelated adults in groups of weanlings not only alleviated weaning stress, but also favored positive social behavior and limited the emergence of abnormal behaviors. In a second study, we examined the impact of the temporary presence of adult horses on the behavior of 1- and 2-year-old horses. Results showed that young horses reared in homogeneous groups had a reduced behavioral repertoire, no real preferred partner and displayed many agonistic interactions compared to domestic horses reared under more natural conditions. Interestingly, after the introduction of adults, young horses expressed new behaviors (e.g. snapping, lying recumbent), preferential social associations emerged ($P<0.05$) and positive social behavior increased ($P<0.05$). Taken together, these results have important implications in terms of husbandry, indicating the importance of keeping young horses with adults.

Enrichment items during turnout – effect on horse behaviour?
Bøe, K.E.[1] and Jørgensen, G.H.M.[2], [1]Norwegian University of Life Sciences, Department of Animal- and Aquacultural Sciences, P.O. Box 5003, 1432, Norway, [2]BioForsk, BioForsk Nord, Tjøtta, 8860, Norway; grete.jorgensen@bioforsk.no

The aim of this study was to investigate the use of enrichment items intended to provide enrichment during turnout, both for individual and group kept horses in an attempt to reduce the amount of passive behaviours. The study was divided into two parts, where study 1 involved eight horses rotated through eight individual paddocks, each containing one of seven enrichment items and one paddock being kept without item, functioning as a control. The horses' item-directed behaviours; passive behaviours or other non-item related activities were scored using instantaneous sampling, every minute for one hour at the beginning and the end of the turnout period. Study 2 involved six horse groups (3-6 horses) exposed to four different items that the horses interacted the most with during study 1 (straw STRA, ball filled with concentrates CBALL, branches BRAN and scratching pole POLE). Both horses kept individually and in groups performed significantly more item-directed behaviours towards edible items like STRA and CBALL than other objects. There was however no overall relation between the numbers of item-directed behaviours and the number of passive behaviours observed, indicating that the enrichment items did not alone reduce the amount of passive behaviours during turnout periods. Such a reduction was only apparent when horses spent more time eating green leaves growing on the paddock surface (R=-0.97 study 1, R=-0.67 study 2, $P<0.0001$). Access to STRA in group kept horses also seemed to reduce the amount of agonistic behaviours. In conclusion, if grass is not available in paddocks, the provision of roughage reduces the amount of passive behaviours in singly kept horses and it also reduces the risk of agonistic interactions between horses kept in group.

Relations between management, emotionality and cognitive abilities in riding school horses

Lesimple, C., Fureix, C., Richard-Yris, M.A. and Hausberger, M., Laboratoire EthoS, 263 avenue du gal Leclerc, 35042 Rennes Cedex, France; clemence.lesimple@univ-rennes1.fr

Previous studies showed that housing conditions have a major impact on young horses' behaviour: horses kept in group in paddock are less prone to express undesirable behaviours at weaning and at work than stalled ones. A study highlighted that the time spent stabled was associated with an increase of aggressiveness in horses. In riding schools, horse behaviour is crucial regarding users' and professionals' safety. Our study aimed to investigate how human management could impact on riding school horses' behaviour. 3 emotionality tests (Arena, Novel object & Bridge tests), and 1 learning test (Chest test) were performed on 184 horses from 22 riding schools involved in the same general activity (teaching, beginner to moderate level) and differing in terms of housing conditions. Factorial correspondence and multivariate analyses were used to assess the effect of each factor on horses' behaviour. Non parametric tests (Kruskall-Wallis & Mann-Whitney) were used to compare horses' responses between categories of schools. The results show that riding schools could be separated into 4 categories according to their horses' behaviour. The MANOVA revealed a strong impact of breed (Wilks'λ=0.49,F(60, 646)=2.14,P<0.001) and housing conditions (Wilks'λ=0.33 ,F(4,179)=7.56,P<0.001) on horses' behaviour. In particular, horses from riding schools with box housing reacted more strongly in the novel object test (MW, X_{box}=21.11±2.37, X_{padd}=9.65±1.51,U=9,P<0.005) and showed more active locomotion (MW, X_{box}=42.66±5.98% of horses/school, X_{padd}=6.56±3.99% of horses/school, U=5.5,P=0.0005). This study highlights the importance for riding school owners to take into account horses individual characteristics, as well as to have a more general reflection on how management (in particular housing) may impact on safety.

Injury recording in horses kept in groups

Mejdell, C.M.[1], Jørgensen, G.H.M.[2], Rehn, T.[3], Keeling, L.[3] and Bøe, K.E.[4], [1]Norwegian Veterinary Institute, P.O.Box 750 Sentrum, 0106, Norway, [2]Bioforsk, P.O.Box 34, 8860 Tjøtta, Norway, [3]Swedish University of Agricultural Sciences, Box 7068, 750 07 Uppsala, Sweden, [4]Norwegian University of Life Sciences, P.O.Box 5003, 1432 Ås, Norway; cecilie.mejdell@vetinst.no

Horses are social animals and group housing has obvious advantages for their well-being. However, the risk of injuries is a common concern among horse owners and many horses are therefore kept singly outdoors as well as indoors. In order to investigate the number and severity of injuries in horses kept in groups, we developed a system for injury recording consisting of six well defined categories from no injury (=0) to very serious injury (=5). The scoring system was tested for observer agreement using 43 agricultural students who after a training session classified 40 photographs of injuries presented to them in random order twice, so that all images were scored 86 times. Attribute agreement analysis was performed using Kendall's coefficient of concordance (Kendall's W), Kendall's correlation coefficient (Kendall's τ) and Fleiss' kappa. Intra- and inter-observer agreement as well as agreement with the 'gold standard' was good to excellent. The scoring system was used when examining 100 riding horses kept in groups. Number of injuries, category and body location were recorded for all horses. A total of 308 injuries were found, whereof 79% in category 1 (hair loss only), 17% in category 2, 4% in category 3 (minor laceration) and none in categories 4 and 5. Rump and barrel had the highest number of injuries but category 3 injuries were typically found on head and legs. 27 horses had no injuries at all, the median number was 1, and one horse had 28 injuries. In conclusion, severe injuries were not recorded. This is in accordance with data from 67 groups of horses in the Nordic project 'Group housing of horses under Nordic conditions: strategies to improve horse welfare and human safety', where the same protocol was used.

From grass to food – utilization of Norway rangeland for meat production
Holand, Ø. and Steinheim, G., Norwegian University of Life Sciences, Dept. of Animal and Aquacultural Sciences, P.O.Box 5003, N-1432 Ås, Norway; oystein.holand@umb.no

Pre-WW1 the Norway's livestock industry used rangeland pastures intensively. Wild ruminant populations were small. Continuing industrialization and new technologies led to reduced use of pastures and by mid century cattle grazing was strongly reduced, but partly being replaced by increased sheep and (later) wild ruminant populations. Hence, the decrease in total grazing was moderate, but the shift in grazer/browser ratio was large. Mountain and forest grazing is feasible on roughly 2/3 of Norway's area. Grazing is the only way to produce proteins from rangelands. Indeed, ruminants are keystone species, producing ecological services. Today ~2 million sheep graze rangelands in summer, together with ~300,000 other livestock, including ~200,000 cattle, harvesting ~320 million feed units. About 200,000 semi-domestic reindeer (year round grazing) produce ~2,000 tons meat/yr. The wild biomass is mainly moose (~100,000) and red deer (~200,000) and ~5500 tons of meat are harvested annually. Rangeland meat production competes with other industries as well as the large protected predators, for spatial resources. Predators cause increased mortality and reduced growth and preventive measures are expensive and inefficient. Large predators and grazers should therefore be spatially managed in an integrated way. Low profitability leads to fewer and bigger livestock units and thus a more clumped pasture use, calling for new technologies and revised management and legislation schemes. Global change may cause food scarcity making us rediscover the importance of food security. It will affect livestock and wild cervids' pastures quality and quantity. But direction and magnitude is hard to predict, as effects will vary greatly with altitude and latitude. Ruminant diversity may thus become increasingly important for efficient meat production. The rangeland resources have potential for increased production and should be utilized using management schemes that optimize the efficiency of our 'tools', the domestic and wild ruminants.

Sustainable rangeland utilization for sheep and goat production in Iceland
Dýrmundsson, Ó.R., Farmers Association of Iceland, Bændahöllin,Hagatorg, IS-107 Reykjavík, Iceland; ord@bondi.is

Iceland, an island of 103,000 km^2 bordering on the Arctic Circle, is better know internationally for its volcanic activity, glaciers and mountainous landscapes than for its grassland-based agriculture. However, natural rangeland pastures are amongst the most valuable resources of the country providing summer grazing for 1 million sheep and a small goat population, as well as sustaining considerable horse, cattle and reindeer grazing. The short-tailed Iceland sheep and the Iceland goat, both of Nordic origin, are well adapted to the traditional, extensive grazing system applied, which is characterized by free-range browsing for 5-6 months (June-November) followed by housing with silage/hay feeding for 6-7 months (November-May). Thus the economic contribution of rangeland utilization is substantial, especially to the sheep sector, amounting to approximately half of the annual roughage consumption of sheep and goats. The diverse vegetation of the rangelands is charcterized by hardy grasses (Gramineae) and sedges (Cyperaceae) as well as by dwarf shrub heaths and mosses. There is also a great variation, partly seasonal, in plant quality, selection, intake and nutritive value as well as in vegetation cover and soil condition. These sustainable pastures, which do not receive any fertilizers, thus have extremely variable land carrying capacity. Expressed as ewes with lambs (mainly twins) per hectare this could range from 0.4-2.0 on lowlands and 0.4-0.8 on uplands, at a moderate grazing pressure. In the Quality Controlled Sheep Production Scheme, now including over 90% of the sheep population, the guidelines expressed as hectares needed per ewe with lambs aim for 2.5-8.0 on rangelands below 400 m above sea level, assuming stocking rates which ensure light or moderate grazing pressure. Lambs and kids raised under such rangeland conditions are free from parasitic infection and are generally healthy. They grow fast, losses during the summer are minor and the meat is tender and palatable with an excellent consumer image.

Investigation of the present situation and trends in the Cyprus sheep and goat sector

Papachristoforou, C.[1], Hadjipavlou, G.[2], Miltiadou, D.[1] and Tzamaloukas, O.[1], [1]Cyprus University of Technology, Agricultural Sciences, Biotechnology & Food Science, P.O.Box 50329, 3036 Lemesos, Cyprus, [2]Agricultural Research Institute, Animal Science, P.O.Box 22016, 1516 Lefkosia, Cyprus; georgiah@arinet.ari.gov.cy

In Cyprus, meat and milk from sheep and goats (S&G), contribute about 18% to the total value of animal production. The sheep population seems to have stabilized around 300,000 after a 30-year record low observed between 1995 and 2000. The goat population of 280,000 in 2009, showed a downward trend over the last 5 years, following a 30-year record high between 1999 and 2003. These changes seem to be associated with the management of scrapie disease in the two species. About 21% of farms keep only sheep, 46% only goats and 33% both species, the respective average flock size being 150, 110 and 198 (106 sheep + 92 goats). Concerning farmers' age, only 15% are younger than 40 years, while 35% are older than 60 years. Farm income from meat and milk is, respectively, 48 and 52% for sheep, and 55 and 45% for goats. Only 20% of milk produced is processed on farm, the rest is sold to dairies and used for production of the local haloumi cheese and for yogurt. Farm-gate prices are similar for lamb and kid meat but the price of sheep milk is 60% higher than that of goat milk. Goat milk is undervalued as it fetches only 10% higher price than cow milk. Good prospects for the S&G sector would be foreseen in case the haloumi cheese were recognized as a PDO product. Most (65%) lambs and kids slaughtered, are between 3 and 6 months of age, giving carcasses of 16-17 kg on average. The goat population consists of the locally adapted Damascus type (28%), crosses between the Damascus and local breeds (55%) and local breeds (17%). The majority of sheep (58%) are crosses between the Chios and the Cyprus fat-tailed breeds, while 38% belong to the Cyprus-Chios type. Using molecular techniques, a project is being implemented for the improvement of Chios milk. The project is funded by the Cyprus Research Promotion Foundation.

Sustainable goat husbandry in Turkey

Olfaz, M., Onder, H. and Tozlu, H., Ondokuz Mayis University, Animal Science, Ondokuz Mayis University, Agricultural Faculty, Dep of Animal Science, 55139, Turkey; hasanonder@gmail.com

Turkey is the prominent country in the World in terms of goat production. In this sense, goat production has an important margin for exportation in addition to compensate the necessity of internal market. Besides, reduction of the number of goats in last 20 years caused the important retreatment on milk, meat and leather production of goat. This situation is the most efficient factor for depletion of rural people taking the margin of employment into consideration. To increase the income of goat producers in Turkey, it is essential to put economical measures into action immediately. The first is to make technically is reorganizing the breeding. However, bring to a successful conclusion of organizing the breeding, it is needed to economic incentives for short, medium and long dated periods. In short-dated period, required backstopping should be done in favor of producers by price formation of goat products. For medium and long dated periods, some precautions could be taken as become cooperatives of producers and enlarging of small farms. The aim of this article is to discus the precautions which should be taken for sustainable goat production in Turkey.

The use of GPS, pedometry and acoustics to infer activity of grazing animals
Ungar, E.D., ARO - the Volcani Center, Department of Agronomy and Natural Resources, Institute of Crop Sciences, P.O.B. 6, Bet Dagan 50250, Israel; eugene@volcani.agri.gov.il

The activity states of a grazing animal are most simply classified as graze, rest, and travel. Knowledge of the activity timeline can help improve our understanding of grazing systems and has the potential to improve grazing management decisions. Activity needs to be inferred from information provided by some kind of sensor placed on the animal. Three technologies are discussed: GPS, pedometry, and acoustics. Using GPS collars, activity can, in principle, be inferred from inter-point distance: short, intermediate and long distances correspond to rest, graze and travel activities, respectively. The main problem tends to be confusion of rest and graze. Lotek GPS collars incorporate motion sensors, and data from these can be used to improve classification. The IceTag pedometer produced by IceRobotics provides a detailed timeline of the number of steps taken, and the time allocation between active, standing and lying states. This data can be used to infer the activity timeline but confusion of rest and graze activities remains an issue. Inference methods using GPS or pedometer data include discriminant analysis and partition analysis. More advanced analytic techniques which may further reduce the misclassification rate remain to be explored. Acoustic analysis is an experimental technique based on attaching an inward-facing microphone to a hard tissue in the region of the head. The recording enables sound-producing jaw movements to be identified readily, and these can be classified as bites or chews. This is a promising methodology for monitoring ingestive behaviour and indentifying graze activity. Primary challenges are automation of sound analysis and construction of robust equipment for long periods of deployment. If successful, acoustic monitoring would provide a wealth of information beyond the completely accurate identification of periods of active grazing.

Effects of sheep grazing on sandy soils and turf quality
Pistoia, A.[1], Bondi, G.[1], Casarosa, L.[1], Poli, P.[1] and Masciandaro, G.[2], [1]University of Pisa, DAGA, Via S. Michele degli Scalzi, 2, 56124 Pisa, Italy, [2]CNR, ISE, Via Moruzzi, 1, 56124 Pisa, Italy; gbondi@hotmail.it

Overgrazing causes a worsening of soil and vegetation quality. Sandy soils are more resistant to compaction and allow greater infiltration of water, but it causes a worsening of roots anchorage to soil, so that vegetation is less resistant to ovine browsing. The aim of this study was to verify sheep overgrazing damages on sandy soils and turfs at different intensities of grazing. The experimental period was divided into two phases: spring and autumn; 15 Massese sheep introduced to grazing about eight hours per day. In a spontaneous turf fenced area of 150 m[2], never used for grazing, was carried out a grazing trial, with high stocking density (0.1 head/m[2]). Sampling focused on: the turf at the starting of the trial and at autumnal sprout, to evaluate floristic changes by botanical analysis; the soil, sampled after 100 hours and 250 hours of grazing for two seasons respectively. Soil samples were subjected to major physical, chemical and biochemical analysis. These parameters were analyzed by ANOVA one way, within each treatment. The significant levels reported (P<0.05) are based on the Pearson coefficients. Floristic composition, evaluated at the end of the first phase of grazing, shows Leguminosae disappearance (25% vs 0%) and Graminaceae increase (40% vs 47%) and especially the Other Species (35% vs 53%) results more resistant to grazing sheep. Soil bulk density does no change because the sandy texture is resistant to compaction. Stability of OM parameters, important for soil fertility, don't underline the prevalence of mineralization processes, but a balance with humification process. The β-glucosidase) increased trend of enzyme activities (dehydrogenase and after the first phase (spring), while in the second phase (autumn) decreased. Overgrazing by sheep on sandy soils pasture, does not cause substantial changes on soil characteristics, but it worsens turf quality, with negative effects on environment.

Fatty acid profile of typical Norwegian lambs sustained entirely on fresh mountain pastures

Mushi, D.E.[1], Steinheim, G.[2], Johannessen, R.[2], Thomassen, M.[2] and Eik, L.O.[3], [1]Sokoine University of Agriculture, Department of Animal Science and Production, P.O. Box 3004, Morogoro, Tanzania, [2]Norwegian University of Life Sciences (UMB), Department of Animal and Aquacultural Sciences, P.O. Box 5025, N-1432 Ås, Norway, [3]UMB, Department of International Environment and Development Studies, P.O. Box 5025, N-1432 Ås, Norway

A study was conducted to map the fatty acid composition of typical Norwegian slaughter lambs from high-altitude areas. The lambs studied were in six different flocks and grazed mountain pastures in the core of southern Norway. These pastures are among the highest mountain pastures used for livestock grazing in Scandinavia, and the sheep flocks mainly grazed above 1000 m. a. s. l. From each flock, 10 animals were randomly selected for fatty acid profiling. Mean age of lambs at slaughter was 136 d with mean carcass weight of 21 kg. Meat from the lambs was highly unsaturated with mean iodine number of 53. Female lambs had overall higher ($P<0.05$) proportions of polyunsaturated fatty acids (PUFA) while male lambs had higher ($P<0.05$) proportions of n-6 PUFA, mainly due to the higher proportions of C18:2 n-6. The n-6:n-3 PUFA ratio of the lambs ranged from 1 to 1.6, which is far below the maximum recommended level (<4). Meat from lambs grazed on Norwegian mountain pastures has a salubrious fatty acid profile and this may be used when promoting its consumption.

Fat tailed sheep from Southern Africa: a special reference to the Damara

Almeida, A.M., IICT & CIISA, Centro de Veterinária e Zootecnia, Av. Univ. Técnica, 1300-477 Lisboa, Portugal; aalmeida@fmv.utl.pt

Sheep production is one of the most important activities in animal production in tropical regions. Southern Africa (SA) fat tailed sheep such as the Afrikaner, Namaqua Afrikaner, Nguni, Persian Black Head, Tswana, Pedi, Sabi and particularly the Damara are important animal genetic resources very well adapted to their harsh home environments. In this review, autochthonous sheep genetic resources of SA are reviewed regarding history and productive performance. A special focus is made to the most internationalized of such breeds: the Damara. The following aspects are reviewed: history, standards, productive performance in Southern Africa and importation to Australia. Finally, a reference is made to the influence of fat tailed sheep formation of composite breeds such as the Dorper or the Meatmaster.

Milk quality of goats at mountain pasture grazing

Dønnem, I., Randby, Å.T. and Haug, A., Norwegian University of Life Sciences, Department of animal and aquacultural sciences, pb 5003, 1432 Ås, Norway; ingjerd.donnem@umb.no

Problems with milk fat lipolysis and poor sensory milk quality of Norwegian goat milk are prominent during the mountain grazing period, which normally is in mid lactation. However, other milk constituents, which may be beneficial for human health, may increase on mountain pasture compared to milk produced on silage and concentrates. In this study the composition of goats' milk during a mountain grazing period was evaluated and compared with milk obtained by the same animals when offered high quality grass silage and concentrate indoors earlier in the same lactation. The indoor period lasted from lactation week 2 to 18, while the mountain pasture period lasted from lactation week 23 to 34. Milk yield and milk fat concentration decreased when goats were let out on mountain pasture. The concentration of milk free fatty acids (FFA) increased while sensory milk taste quality did not change. Grazing on mountain pasture increased the concentration of α- tocopherol in milk, and the milk fat proportion of monounsaturated and polyunsaturated fatty acids (MUFA and PUFA) and decreased the milk fat proportion of saturated fatty acids (SFA). This study confirmed the problem with high FFA concentrations in milk on mountain pasture, and the benefits of pasture concerning milk fatty acid composition.

An application of data envelopment analysis to measure technical efficiency on a sample of Irish dairy farms

Kelly, E.[1,2], Shalloo, L.[1], Geary, U.[1] and Wallace, M.[2], [1]Teagasc, Livestock Systems Research Department, Moorepark, Fermoy, Co. Cork, Ireland, [2]School of Agriculture, Food Science and Veterinary Medicine, University College Dublin, Belfield, Dublin 4, Ireland; eoin.kelly@teagasc.ie

Since joining the EU Irish milk production has been regulated by Common Agricultural Policy (CAP) through market support and the milk quota regime. However milk quota is set to be abolished by 2015 allowing potential to expand at farm level. In order for dairy farmers to prosper post quota they must become more efficient. The aim of this study was to determine levels of technical efficiency on a sample of Irish dairy farms and quantify the key management and production characteristics that differ between efficient and inefficient producers. Data Envelopment Analysis (DEA) was used in this study to generate technical efficiency scores under constant returns to scale (CRS) and variable returns to scale (VRS) assumptions and to identify best practise production. DEA efficiency scores range between 0 and 1. The inputs used were land area, cow numbers, labour, concentrate, fertilizer and other direct and overhead costs. The output variable was expressed as kg of milk solids and included milk solids produced and a milk equivalent of other farm output from subsidiary farm enterprises. On average Irish dairy farmers were technically inefficient with technical efficiency scores of 0.785 under CRS and 0.833 under VRS. Key production characteristics of efficient and inefficient producers were compared using an analysis of variance procedure in SAS (PROC GLM). Technically efficient producers with a score of 1 were compared to technically inefficient producers with a score of less than 1. Efficient producers used less input per unit of output, had higher production per cow and per hectare, had a longer grazing season, higher milk quality, participated in milk recording and had greater land quality compared to inefficient farms.

Effect of pre-grazing herbage mass on milk production, grass dry matter intake and grazing behaviour of dairy cows

Tuñon, G.[1], Kennedy, E.[2], Hennessy, D.[2], Kemp, P.[1], Lopez Villalobos, N.[1] and O'Donovan, M.[2], [1]Massey University, Palmerston North, -, New Zealand, [2]Animal & Grassland Research and Innovation Centre, Teagasc, Moorepark, Fermoy, Co. Cork, -, Ireland; g.tunon@massey.ac.nz

The experiment explored the effect of pre-grazing herbage mass (HM) on milk production, grass dry matter intake (DMI) and grazing behaviour of dairy cattle. From 24 April to 17 October 2010, 45 Holstein-Friesian dairy cows (mcan calving date 23 March 2010) were randomly assigned to one of three treatments – low (L), medium (M) or high (H) pre-grazing HM (targets 800, 1,500 or 2,200 kg DM/ha >3.5 cm, respectively). Paddocks were rotationally grazed to a post-grazing sward height of 4 cm. Pre-grazing HM was determined twice weekly. Milk production was recorded daily and milk composition weekly. Grass DMI was measured twice during the experiment using the n-alkane technique. Cow grazing behaviour was monitored during the first intake measurement using IGER behaviour recorders. The experimental period was divided into Period 1- summer (24 April to 25 July) and Period 2 -autumn (26 July to 17 October). In Period 1, milk solids (MS) production was similar across treatments (1.87 kg MS/cow/day). In Period 2, cows grazing L and M swards had significantly ($P<0.05$) higher MS yields (1.33 and 1.37 kg/day, respectively) compared to cows grazing H swards (1.24 kg MS/day). There was no difference in grass DMI between treatments in Period 1 (16 kg DM/cow/day). In Period 2 there was a trend ($P<0.09$) towards higher DMI for M cows compared with the other two groups (L-15.2, M-16.5, H-15.7 kg DM/cow/day). In Period 1 cows offered low HM swards grazed for 1.5 hours more than M and H cows. Within this time, cows grazing L swards had 15% more grazing bites than the other two treatments. There were two main conclusions from the study: (1) cows grazing high HM swards had lower MS production during the autumn; (2) cows grazing low HM swards spent 1.5 more hours per day grazing to achieve slightly less intake and MS production than the other two treatments.

Behavioral aspects of Caracu and Red Angus cattle breeds in a pasture with shade and water immersion

Geraldo, A.C.A.P.M.[1], Pereira, A.M.F.[2], Nogueira Filho, J.C.M.[1] and Titto, E.A.L.[1], [1]Animal Science and Food Engineering Faculty - University of São Paulo, Av.Duque Caxias Norte,225, 13635-900 Pirassununga–SP, Brazil, [2]University of Évora, Ap.94, 7000-554 Évora, Portugal; ana.de.mira.geraldo@gmail.com

The shade is an important resource in the extensive production systems in tropical regions. The aim of this study is to understand through behavioral analysis, the preferences of animals for resources for environmental protection. The experiment was conducted in the Biometeorology and Ethology Laboratory of FZEA-USP. Six male of Caracu and Red Angus cattle breeds were used. The animals were submitted to 2 different treatments: availability of artificial shade and water for immersion and availability of water for immersion. The observations of the behavioral patterns were recorded using the focal sampling method every 15 minutes (12 h/day). The observed positions were: in the sun, under the shade and in the water. The posture observed were: standing, lying down and behavioral activities were grazing, ruminating and rest. The data concerning to the time spent in different behaviors and different positions were analyzed by the multifactorial variance (ANOVA-GLM). The fixed factors used were the breed and treatment. The results had shown that grazing activity was the behavior in which both breeds had spending more time (especially in the cooler periods), followed by ruminating in the Caracu and by resting in the Red Angus. The results also had shown that shade can be used as a shelter against solar radiation as well against rain. The Caracu had presented a clear preference for the shade, particularly in the hottest hours. However this was not always observed in Red Angus, who sometimes chose to remain in the water. In hot climates, resources for defense against heat load, as shade and water for immersion can really improve the welfare of the cattle.

Effect of limited vs. *ad libitum* concentrate feeding on the performance and carcass and meat quality of Parda de Montaña bulls finished on pasture

Casasús, I., Joy, M., Albertí, P. and Blanco, M., CITA-Aragón, Avda Montañana 930, 50059 Zaragoza, Spain; icasasus@aragon.es

Supplementation is a common practise to improve performance and meat quality of grazing ruminants. When delivered on *ad libitum* basis to reduce workforce, concentrate intake might exceed the maximum 40% of the daily diet imposed by organic farming regulations. The aim of this study was to compare the performance of grazing young bulls with restricted vs. *ad libitum* concentrates. Parda de Montaña autumn-born bull calves (n=16, aged 6.5 mo, 237 kg) were rotationally grazed on natural meadows. Eight bulls received a fattening concentrate (13.7% CP, 18.8% NDF) on *ad libitum* basis (ADLIB) and the other 8 were fed daily 3 kg per head (3KG). Grass and concentrate intakes were estimated weekly by group, in the case of grass by the herbage regrowth and disappearance method. At the end of the summer (12 mo), bulls were slaughtered and carcass and Longissimus thoracis meat characteristics were assessed, and analyses of variance were performed. Growth rate was higher in ADLIB than in 3KG bulls (1.50 vs. 1.27 kg/d, respectively, P<0.05), due to a greater concentrate intake (6.3 vs. 2.7 kg DM/d, respectively), although grass intake was slightly lower than that of the 3KG group (5.9 vs. 7.2 kg DM/d, respectively). Slaughter weight was not different (480 vs. 448 kg in ADLIB and 3KG, P=0.12) but ADLIB bulls had heavier carcasses (292 vs. 253 kg, P<0.01), and greater dressing percentage (60.8 vs. 56.6%, P<0.001). ADLIB carcasses had slightly better conformation score (11 vs. 10 in a 15-point scale, P=0.06) and backfat thickness (1.31 vs. 0.76 mm, P=0.07), but similar subcutaneous fat colour. Meat shear force and colour were similar for both groups and evolved similarly through the ageing period. Meat chemical composition was similar in both treatments, and there were only slight differences in individual fatty acid contents. In conclusion, the different feeding management resulted in different gains and carcass weights but did not influence meat quality.

Automated activity monitoring as a management tool in dairy production

Munksgaard, L., Aarhus University, Department of Animal Health and Bioscience, Blichers Allé 20, P.O.Box 50, 8830 Tjele, Denmark; lene.munksgaard@agrsci.dk

The structural development in dairy farming has lead to increased farm size, where each co-worker is responsible for monitoring an increasing number of animals. At the same time there is an increasing focus on animal health and welfare. Thus there is a need for new tools for monitoring and troubleshooting at dairy farms. The first symptom of a disease is very often a change in behavioural patterns. The most well known changes in behaviour in response to disease are reduced activity, decreased feed intake and increased time spend lying. Development of devices for automatic registration of behaviour can contribute considerably to on-farm assessment of animal welfare as well as a tool for consultancy, and can be used as documentation for a given standard of animal welfare. During the last years devices for wireless measurements of three-dimensional acceleration has been used to estimate important aspects of dairy cow activity. There is some documentation that accelerometer technology can be used to estimate lying, standing and walking as well as the number of steps taken when placed on the leg of cows. There are also some preliminary results suggesting that when placed on the neck of cows the technology may be useful to estimate eating behavior indoors as well as on pasture. Commercially available activity loggers and 3-dimensional accelerometers attached to the legs have been used to identify cows in heat and different algorithms have been developed for identification of lame cows. Furthermore, there is some evidence that automated recording of activity may be useful for detection of other diseases. Accuracy in automatic recording of different activities of dairy cows will be discussed as well as the value of the automated recordings of varies activities of dairy cows in relation to prediction of the status of the animal.

Prediction of cow pregnancy status using conventional and novel mid infrared predicted milk traits

Hammami, H.[1,2], Bastin, C.[1], Gillon, A.[1,3], Soyeurt, H.[1,2] and Gengler, N.[1,2], [1]University of Liège, Gembloux Agro-Bio Tech, Animal Science Unit, Passage des Déportés, 2, B-5030 Gembloux, Belgium, [2]National Fund for Scientific Research, Rue d'Egmont 5, B-1000 Brussels, Belgium, [3]Walloon Breeding Association, Research and Development, B-5590 Ciney, Belgium; hedi.hammami@ulg.ac.be

The objective of this study was to determine the ability of conventional milk cow characteristics and novel traits predicted by mid infrared (MIR) obtained from milk recording to predict the pregnancy status once the cow was inseminated. Conventional milk recording, spectral, and reproductive data collected in Luxembourg Hoslteins between 2008 and 2010 were used. Cows were defined as pregnant if they were positively checked and calved between 267 and 295 d later after the last AI or if they had calved between the later intervals when no checks were recorded. Pregnant or not within 3 intervals after last AI (<=35 d, 45-60 d, and 60-90 d) was modeled using logistic regression models firstly as a function of conventional cow milk characteristics and extended to fatty acids as novel traits predicted by MIR in a second step. The lactation curve characteristics for milk, fat, protein, and lactose yields were estimated using modified best prediction method. Test-day fatty acid contents were estimated using an appropriate calibration equation. Two third proportion and one third of the whole data set were randomly selected for calibration and validation models respectively. The relation between the predicted and observed probabilities of cow pregnancy was approximately linear for calibration and validation models. The sensitivity-specificity combination for cow pregnancy increased when fatty acids were added to conventional milk characteristics as inputs to the different models (from 78 to 85% for sensitivity and from 40 to 52% for specificity). Results based on those models showed that it would be possible to help breeders to manage cow fertility using such tool implemented in the milk recording organizations.

Development and validation of a automatic optical systems for the control of the body condition score of dairy cows

Salau, J.[1], Junge, W.[1], Harms, J.[2] and Suhr, O.[3], [1]CAU, Institute of Animal Breeding & Husbandry, Olshausenstraße 40, 24098 Kiel, Germany, [2]Bavarian State Research Center for Agriculture, Institute of Agricultural Engineering & Animal Husbandry, Prof.-Dürrwaechter-Platz 1, 85588 Poing, Germany, [3]GEA Farm Technologies GmbH, Siemensstraße 25-27, 59199 Bönen, Germany; jsalau@tierzucht.uni-kiel.de

Studies on the body condition classification have shown, that the results essentially depend on the person who classifies. The aim is a more objective jugdement of the cows body condition using digital image processing on the images of a Time-Of-Flight-camera. These cameras send out infrared light modulated at 20 MHz which is reflected by the object. The distance between object and sensor is then calculated out of the angular phase shift. The camera is positioned on the research farm Karkendamm of Kiel University. We get three dimensional diagrams which show the differences of height on the cows back. Firstly we can draw line profiles of the cow out of the diagrams by cutting at right angle to the backbone. Thus we can measure the angle with which the cows surface decrements from the backbone to the side or how gaunt the cow has become by determining the curvature or convexity of the line profile. Secondly we can measure distances on the cows back like the width between the hips. Using mathematical tools for surface examination we might thirdly get the possibility to examine the changes in time of the fat tissues thickness in two dimensional areas on the cows back. We might also be able to measure the roundness at protruding knucklebones like the cows ischial tuberosities. These ideas could be used to get ratios from the image, which could reveal insight about the cows body condition without touching the animal or locking it into position. But first the images have to be sorted according to if one can see a useful part of the cow on the image or not, and additionally the cow has to be cut out and some important body parts (backbone, tail, hips, ischial tuberosities) have to be detected automatically.

SpermVital – field trials with this new semen processing approach
Kommisrud, E.[1], Grevle, I.S.[1], Sunde, J.[1], Garmo, R.T.[1] and Klinkenberg, G.[2], [1]BioKapital, Holsetgata 22, 2317 Hamar, Norway, [2]Sintef, Pb 4760 Sluppen, 7465 Trondheim, Norway; ek@biokapital.no

To make artificial insemination (AI) an even more valuable breeding method, by improving fertility results, an immobilization technique for spermatozoa was developed. The SpermVital technology intends to prolong the shelf life of spermatozoa both *in vitro* prior to AI, and after AI in utero, and is described in patent application PCT/NO07/00256. Spermatozoa are immobilized within a solid gel-network made of calcium alginate gel. The technology has successfully been combined with cryopreservation, a necessity for AI in cattle. Both lab-trials and smaller AI trials have been integrated while developing this technology, to verify the progress. Several recent AI trials in cattle have indicated that the SpermVital semen is competitive with, or superior to, traditionally preserved semen, and that the timing of AI is less critical. In a recent trial ejaculates were split-sampled, one half processed by ordinary cryopreservation procedure, the other half cryopreserved in combination with the SpermVital technology, and used for AI in cows and heifers in 3 experimental groups. A control group was inseminated with ordinary semen at optimal time in estrus. The 2nd group was inseminated at optimal time using SpermVital semen, while the 3rd group received SpermVital semen with AI estimated 24 h before optimal time. Pregnancy rates were equal in all groups. A large randomized field trial was initiated autumn 2010, using semen from 16 unproven Norwegian Red bulls. Ejaculates were split-sampled and produced by ordinary procedures and with SpermVital technology, as described earlier. Totally, 32,000 AI doses were produced, set up in goblets of 10 straws, 5 of each treatment, and distributed randomly across Norway. Statistical analyses are performed in a logistic regression model with 56 d non-return rate as outcome. Data on first AI, previous calving date, parity, season, AI personnel, and data on herd level are obtained from the Norwegian Dairy Herd Recording System.

The use of computerized tomography in pig breeding
Kongsro, J., Norsvin, P.O. Box 5003, N-1432 Aas, Norway; jorgen.kongsro@norsvin.no

Computed Tomography (CT) is recognized as one of the most powerful tools in medical diagnostics. Since the development of the first generation EMI scanner in the 1960s, the technology has evolved into modern slip ring designs and helical CT processing. The first use of CT in animal science was made by H. Skjervold and colleagues at the Norwegian University of Life Sciences in the early 1980s. The main goal of this work was to predict the body composition of farm animals. CT is based on the attenuation of X-rays through a body related to its density, using a slip-ring and helical scanning. There are a number of opportunities for capturing different phenotypes on live animals using CT, provided there is variation in image pixels or texture which can be sampled from the attenuation data. In general, there are three areas where CT can contribute to accurate phenotyping on live animals: body composition, meat quality and diagnostic imaging. For body composition, research over the last few decades has shown CT can be used to accurately predict the body composition of live farm animals. Meat quality traits like intramuscular fat (IMF) and fatty acid composition (FA) can be measured by CT on a macro scale (whole animals and cuts), and CT has recently been shown to provide accurate measurements, as long as there is detectable variation in attenuation or texture that can be related back to chemical IMF or FA reference data. The next step for CT in animal breeding is to apply the use of diagnostic imaging for animal health and robustness, by utilizing the differences in X-ray attenuation in diseases like leg weakness, osteochondrosis and scapular morphology in relation to shoulder sores. For animal breeding it is critical to develop large scale recordings of image diagnostic. This requires bringing together veterinary medicine with image analysis, computer programming and mathematical morphology, to provide large scale recording that can be utilized to produce a healthier and more robust breeding animal.

Differences in body composition between pigs crossbreds of 30 kg measured *in vivo* by computed tomography

Carabús, A., Gispert, M., Rodriguez, P., Brun, A., Francàs, C., Soler, J. and Font I Furnols, M., IRTA, Irta monells granja Camps i Armet, 17121 Monells, Spain; anna.carabus@irta.cat

The aim of this experiment was to prove if there were body's composition differences between genetics types at the early weight of 30 kg of live pigs by analyzing Computed Tomography (CT) images. Fifty eight pigs of different crossbred Duroc x (Landrace x Large White) (DUx(LDxLW)), Pietrain x (Landrace x Large White) (PIx(LDxLW)) and (Landrace x Large White) (LDxLW) were evaluated *in vivo* at 30 kg of weight. All pigs were scanned (after anaesthesia and immobilization), every 7 mm, from the cranial to caudal position by CT device following acquisition protocols previously established (140 kW, 145 mA, axial 1s, 7 mm thick). CT images were studied with the VisualPork program to obtain measures of fat thickness, muscle depth, body length, areas and perimeter at different levels. After CT evaluation, 15 of the 58 animals (5 of each genetic type) were slaughtered and carcass cutting and full dissection were performed. Dissection was used to obtain results of real parameters such as ham, loin and carcass weight and lean content, and then these parameters were compared with the results of prediction equation obtained by CT images variables. GLM procedure of SAS determined differences among genetics with live weight as a covariate. LDxLW crossbred presented the lowest subcutaneous fat thickness above the middle of the vertebral column compared with DUx(LDxLW) and PIx(LDxLW) (7.1 mm vs. 8.9 and 8.4 mm). It was also relevant how PIx(LDxLW) hams were bigger (height, area and perimeter) than those of the other two crossbreds studied already at the weight of 30 kg.

Use of loin intramuscular fat content predicted with ultrasound technology in the Canadian swine improvement program

Maignel, L.[1], Daigle, J.P.[2], Groves, J.[1], Wyss, S.[1], Fortin, F.[2] and Sullivan, B.[1], [1]Canadian Centre for Swine Improvement inc., Central Experimental Farm, Building 54, 960 Carling Avenue, Ottawa, Ontario, K1A0C6, Canada, [2]Centre de développement du porc du Québec inc, Place de la Cité, tour Belle Cour, 2590, boul. Laurier, bureau 450, Quebec, Quebec, G1V 4M6, Canada; laurence@ccsi.ca

Intramuscular fat predicted on live pigs has been recently included in the Canadian accreditation program for ultrasonic technicians. The scanning technique and image analysis tools based on an Aloka 500 device and developed by Biotronics have improved over the recent years, but require a special training and follow-up. Image collection and analysis are carried out separately, and require a specific data management framework to ensure results are available quickly for genetic evaluation and selection decisions. Data collected across Canada in the past three years have brought new knowledge for the development of innovative selection tools on meat quality. A large research project involving 6,000 Duroc pigs scanned across Canada was designed to enlarge the live IMF database and confirm genetic parameters estimated in a previous study. Simulation studies are used to optimize the use of this new data in selection indices along with other EBVs in order to maintain or increase marbling levels in Canadian Durocs. Data collected in the project are also used for genetic evaluation. Another component of the project will provide valuable information about the effect of genetics and management (especially feeding) on the marbling level in pork loins. Boars with either low or high IMF EBVs will be selected to produce commercial pigs fed with a standard feeding program or a specific feeding program formulated to increase marbling. The main outcome of this project will be new tools for the Canadian producers to customize marbling levels in their hogs using a combination of genetics and feeding.

Modelling the dynamics of feed intake in growing pigs; interest for modelling populations of pigs

Vautier, B.[1,2], Quiniou, N.[2], Van Milgen, J.[1] and Brossard, L.[1], [1]INRA – Agrocampus Ouest, UMR 1079 SENAH, F-35590 Saint-Gilles, France, [2]IFIP-Institut du Porc, BP 35104, F-35651 Le Rheu Cedex, France; bertrand.vautier@rennes.inra.fr

Modelling approaches are more and more used as an alternative to animal experimentation. Growth models are most often based on the prediction of feed intake, the efficiency of nutrient utilization, and the phenotypic potential of the animal. The drawback of these models is that feed intake is described by an empirical function of body weight and that variation between pigs is ignored. The aim of the present study was to compare four functions describing feed intake (linear, power, exponential or the gamma function) and to characterize the relation between parameters describing growth in a population. Seven batches of 144 growing barrows and gilts each, originating from crossbred LWxLD sows and eight purebred or crossbred sire lines were used. The pigs were weighed at 10, 13, 16, 19, 21 and 23 weeks of age and their individual feed intake was recorded daily through an automatic feed dispenser. For each equation, models were fitted individually to the data and the quality of the prediction was evaluated by the root mean square error of prediction. The gamma function provided the most precise fit of the data irrespective of genotype and gender. The gamma function is the 'least-empirical' one and expresses feed intake relative to the maintenance energy requirement, implying that animals eat for maintenance once they are mature and non-productive. Based on this model, eight parameters define an individual profile. Some relations between parameters have been shown, while some others are considered independent. This will allow the generation of virtual batches built on a realistic variation structure. These virtual batches will allow later stochastic simulations, taking into account individual variation in the response of the batch.

An integrated automatic traceability system for the salami swine meat chain based on RFID technology

Beretta, E.[1], Nava, S.[1], Montanari, U.[2] and Lazzari, M.[1], [1]University of Milan, Department of Veterinary Science and Technology for Food Safety, Via Celoria 10, 20133 Milano, Italy, [2]Hi.Pro. Solution, Via Quasimodo 55, 40013 Castel Maggiore, Italy; ernesto.beretta@unimi.it

In Europe, regulations do not impose the individual recognizing of swine. In Italy, a relevant part of the meet industry is evolving toward transformed products of high quality that require a secure certification of the meet origin. For this reason traceability system based on the individual swine recognizing are needed. In this frame a system based on RFID technology has been studied and developed also starting from previous experience conducted on meet production chain. E-tags were used to identify: animals, hooks in the slaughtering line and meat cuts. RFID readers were installed in the farm (livestock gates), slaughterhouse (entrance, evisceration and carcass weighting areas), carcass cutting working stations, and meat shops (weighing place). The whole system was controlled by specific software integrated by 4 main components: livestock, slaughtering, storage/shop and web-oriented DB. Critical readings for tracing were: ear tag reading in the farm, in the slaughterhouse (before the stunning box), combined ear tag and hook tag reading at the evisceration area, carcass labeling and recording at the weighing area, registration of carcass entry in the cutting area, and using read-and-write e-tags for adding meat cuts in the transformed salami. Selling actions in the store/shop area were also recorded. The system has been tested successfully in the slaughterhouse of Cà Lumaco (Zocca, Italy) were its different components proved to be functional and able to integrate all the subsystems. All the products (salami, hams, etc.) are identified using RFID or 2D barcode in this way all the traceability data are available. In the pigs nursery cameras are installed to allow the farmer, using a smartphone or PC, to switch on/off the lights of the room and to see the pregnant sows and, in case, to intervene to help the swine.

Modeling and monitoring sows' activity types in the farrowing house using 3D acceleration data
Cornou, C.[1], Lundbye-Christensen, S.[2] and Kristensen, A.R.[1], [1]University of Copenhagen, Faculty of Life Sciences, Department of Large Animal Science, Grønnegårdsvej 2, 1870 Frederiksberg, Denmark, [2]Århus University Hospital, Aalborg Hospital, Department of Cardiology, Center for Cardiovascular Research, Sdr. Skovvej 15, 9000 Aalborg, Denmark; cec@life.ku.dk

The objectives are to develop a method for classifying sows' activity types performed in farrowing house and to monitor sows' behaviour around the onset of farrowing. In a first step, five types of activity are modelled using multivariate dynamic linear models: High active (HA), Medium active (MA), Lying laterally on one side (L1), Lying laterally on the other side (L2) and Lying sternally (LS). The classification method is based on a Multi Process Kalman Filter (MPKF) of class I. The activities are modelled using a Learning data set and the performance of the method is validated using a Test data set. Results of activity classification appear satisfying: 75 to 100% of series are correctly classified within their activity type. When collapsing activity types into active (HA and MA) vs. passive (L1, L2, LS) categories, results range from 96 to 100%. In a second step, the suggested method is applied on series collected for 19 sows around the onset of farrowing, including 9 sows that received bedding materials (57 sow days in total) and 10 sows that received no bedding material (61 sow days in total). Results indicate that there is a marked (1) increase of active behaviours (HA and MA, $P<0.001$) and (2) decrease of lying laterally (L1 and L2) behaviours starting 20 to 16 hours before the onset of farrowing; during the last 24 hours before parturition, the averaged time spent lying laterally in a row decreases and the number of changes of activity types for HA and MA increases. These behavioural changes occur for sows both with and without bedding material, but are more marked when bedding material is provided. Straightforward perspective for application of this classification method for monitoring activity types is e.g. automatic detection of farrowing.

Detecting sows' parturition using 3D acceleration measurements
Cornou, C.[1] and Lundbye-Christensen, S.[2], [1]University of Copenhagen, Faculty of Life Sciences, Department of Large Animal Science, Grønnegårdsvej 2, 1870 Frederiksberg, Denmark, [2]Århus University Hospital, Aalborg Hospital, Department of Cardiology, Center for Cardiovascular Research, Sdr. Skovvej 15, 9000 Aalborg, Denmark; cec@life.ku.dk

The objective is to develop and assess methods for detecting sows parturition using series of 3-dimensions acceleration measurements previously classified into activity types. Two groups of sows are monitored: a first group (n=9) provided with straw (S), and a second group (n=10) where no straw is provided (NS); two types of activity are taken into account: High Active behaviour (corresponding to feeding, rooting and nest building behaviours) and Total Active behaviour (including any active activity type). Two different methods are suggested. The first method suggests modelling sows' diurnal pattern of activity using a saw-tooth function for the probability of being active and monitoring the series using a Dynamic Generalized Linear Model (DGLM). The second method is based on a cumulative sum (CUSUM) of hourly differences of activity, from day-today. Both methods use a threshold value, optimized for each group, to detect the onset of farrowing. Best results in terms of sensitivity and specificity are observed for the CUSUM method, using individual variance and monitoring High Active (sensitivity = 100%; specificity = 100%) and Total Active behaviours (sensitivity = 100%; specificity = 95%) behaviours. Results of the DGLM method indicate a sensitivity of 100% and a specificity of 89% for both group S and NS. Observing the occurrence of alarm times, the DGLM method allows (1) earlier detection of farrowing: 15 hours before the onset of farrowing, for both groups, as compared to 9-12 for the other suggested methods; and (2) a better distribution of alarms, i.e. most alarms occur in the interval 6-18 hours before farrowing.

Genetic analyses of carcass, meat and fat quality traits measured by computed tomography and near-infrared spectroscopy in Norwegian Landrace and Duroc pigs

Gjerlaug-Enger, E.[1,2], Kongsro, J.[1], Olsen, D.[1], Aass, L.[2], Ødegård, J.[3] and Vangen, O.[2], [1]Norsvin, P.O. Box 504, 2304 Hamar, Norway, [2]UMB, P.O. Box 5003, 1432 Ås, Norway, [3]Nofima Marin, P.O. Box 5010, 1432 Ås, Norway; eli.gjerlaug@norsvin.no

In this study, computed tomography (CT) technology was used in large-scale measurement to calculate lean meat percentage (LMP) on live pigs for breeding purpose. Norwegian Landrace (L; n=4,525) and Duroc (D; n=3,624) boars, selections candidates to be elite boars, were CT-scanned between 2008 and 2011 as part of an ongoing testing programme at Norsvin's boar test station. Meat chops (*M. longissimus dorsi*) and subcutaneous fat from carcasses (L; n=3,220-6,565, D; n=1,764-4,345) was also sampled to investigate the sources of variation for intramuscular fat (IMF), drip loss and fatty acid composition (FAC). Near-infrared spectroscopy (NIRS) was applied for prediction of the IMF and FAC. A multi-trait animal model using AI-REML methodology was used for the genetic analyses. The r_g to more traditional production traits such as average daily gain (ADG) and feed conversion ratio (FCR) from 25-100 kg were also estimated. Among the production and carcass traits studied here, the ADG, FCR and LMP had unfavourable r_g to most of the meat and fat quality traits. The highest unfavourable r_g was estimated between LMP and IMF. Selection for increased LMP yields more C18:2n-6 in the fat, while C18:1n-9 showed lower r_g to the LMP. The r_g between ADG and meat and fat quality traits was low. A decreased FCR was unfavourable for all of the meat and fat quality traits. The r_g between meat and fat quality trait were favourable. In order to have a sustainable breeding for carcass and production traits, the meat and fat quality must be taken into consideration. Many meat and fat quality traits measured with rapid methods demonstrate a high genetic variation that makes it possible to select for such traits. The use of new technologies makes it possible to select for efficient, lean pigs with improved sensory, technological and nutritional qualities.

The contribution of social genetic effects to the heritable variation of growth and feed intake of Swiss Large White pigs from a testing station

Canario, L.[1] and Hofer, A.[2], [1]INRA, Animal Genetics, UMR 444 ENVT Génétique Cellulaire, BP 52627, F-31320 Castanet-Tolosan, France, [2]SUISAG, Allmend, CH-6204 Sempach, Switzerland; laurianne.canario@toulouse.inra.fr

Feed intake during and pig body weight at the end of fattening period were recorded on 16,041 pigs born from 7,757 litters of the Swiss Large White dam population. At around 26 kg of body weight, animals were moved to the testing facility in which pens were equipped with electronic feeders. They were raised in groups of 8 to 10 pigs. Groups were formed by mixing of females and castrate males. Pigs were fed *ad libitum*. The testing period was 30 to 103 kg live weight. Traits analyzed were average daily gain (ADG), average feed consumption per day (AFC) and feed conversion ratio (FCR) corrected for weight at slaughter. The statistical models of analysis included the fixed effects of number of group mates, sex and a year-season combination, and the random effects of pen, group, litter, dam permanent environment, and direct (Ad) and social (As) additive genetic effects. For the 3 traits under study, the effect of the litter of origin was significant whereas the dam permanent environment was negligible and not accounted for in the selected model. Social genetic effects were highly significant for ADG and AFC, but negligible for FCR. The heritability value accounting for social effects T^2 was 0.53 (SE=0.06) for ADG and 0.72 (SE=0.07) for AFC. Social genetic effects contributed 28% and 31% of total heritable variance of ADG and AFC, respectively. Genetic cooperation was found for feed intake (r Ad,As = 0.53 (SE=0.22)). Response to social selection of growth and feed intake is investigated. The results suggest that social genetic models are of interest to genetically improve growth and feed intake, and this type of selection would be more efficient if applied on traits recorded on test rather than on the derived trait feed conversion ratio.

Estimation of genetic variation in macro- and micro-environmental sensitivity

Mulder, H.A.[1], Rönnegård, L.[2,3], Fikse, W.F.[2], Veerkamp, R.F.[1] and Strandberg, E.[2], [1]Animal Breeding and Genomics Centre, Wageningen UR Livestock Research, P.O. Box 65, 8200 AB Lelystad, Netherlands, [2]Swedish University of Agricultural Sciences, Department of Animal Breeding and Genetics, Box 7023, SE-75007 Uppsala, Sweden, [3]School of Technology and Business Studies, Högskolan Dalarna, SE-79188 Falun, Sweden; herman.mulder@wur.nl

Genetic variation in environmental sensitivity means that animals genetically differ in their response to environmental factors. Some factors (e.g. temperature) are known and called macro-environment, whereas other factors are unknown and called micro-environment. We assume that genetic variation in sensitivity on macro- and micro-environmental level is expressed as genetic variation of the slope of a linear reaction norm and in environmental variance, respectively. The objectives of this study were to develop a statistical method to estimate genetic parameters for macro- and micro-environmental sensitivity and to investigate bias and precision of the estimated genetic parameters. A reaction norm model to estimate genetic variance in macro-environmental sensitivity was combined with a structural model for residual variance to estimate genetic variance in micro-environmental sensitivity using a double hierarchical generalized linear model in ASREML. A dairy cattle population was simulated and results indicated that designs with 100 bulls with at least 50-100 daughters are required for estimation of all genetic parameters with sufficient precision. When the number of daughters increased, the standard deviations of estimates across 100 replicates decreased substantially, especially for the genetic variance of the slope of the reaction norm and the genetic variance in environmental variance. The standard deviations of estimated genetic correlations were quite large (0.1-0.4). No bias was observed in all parameters. It is concluded that genetic parameters for macro- and micro-environmental sensitivity can be estimated with sufficient precision in designs with 100 bulls with 50-100 daughters each.

Inbreeding effective population size in composite beef cattle breeds

Cantet, R.J.C., Facultad de Agronomía Universidad de Buenos Aires & CONICET, Av.San Martín 4453, C1417DSQ, Buenos Aires, Argentina; rcantet@agro.uba.ar

Composite breeding is popular in countries with large numbers of beef cattle. Characteristics of certain composites breeds are: 1) breeding and mating schemes usually consist of up to five generations of grading-up, where cows in the earlier generation are mated to bulls from the same or more advanced ones; 2) an open policy for registration, which results in a large portion of animals having very little pedigree information beyond the identification of sire and dam; 3) the predominant use of embryo transfer and artificial insemination in advanced generations induces a low ratio bull:cow when compared with other beef breeds. As a consequence, the observed inbreeding on recorded animals from advanced generations is higher in composites than in the founder breeds. Thus, when evaluating the genetic variability of composites, the lack of complete pedigree information sets the need for a demographic measure such as the effective population number (N_e). Previous research has pointed out that calculation of N_e in composite breeds requires taking into account the existence of nucleus and multiplier tiers with different mating systems functioning as subdivided populations, and the use of demographic information such as (co)variances of family sizes. A formula for N_e has been obtained that accounts for these effects by adapting an inbreeding – subdivided population approach. First, equations were obtained for the inbreeding developed in the advanced generations, and for the average coancestry between individuals from the nucleus and multiplier herds. These equations depend on probabilities that two genes chosen at random from two different offspring of one sex are derived from the same parent, and other demographic parameters. The resulting recursive system is solved for a small set of parameters and N_e is obtained from there. Although the resulting formula is complex, it allows monitoring the variability in composites and evaluating the impact of different mating policies.

Effect of the gene diacylglycerol-O-transferase 1 (DGAT1) polymorphism on the global expression pattern of genes in the mammary gland tissue of dairy cows

Mach, N.[1], Blum, Y.[2,3], Causeur, D.[4], Lagarrigue, S.[2,3] and Smits, M.[1], [1]Wageningen UR Livestock Research, Edelhertweg 15, 8219 PH Lelystad, Netherlands, [2]INRA, Rue de Saint-Brieuc, 65, 35000 Rennes, France, [3]Agrocampus Ouest, UMR598, Animal Genetics, Rue de Saint-Brieuc, 65, 35000 Rennes, France, [4]Agrocampus Ouest, UMR598, Applied Mathematics, Rue de Saint-Brieuc, 65, 35000 Rennes, France; nuria.mach@wur.nl

It is well documented that the non-conservative alanine to lysine amino acid change (K232A) in exon VIII of the diacylglycerol-O-transferase 1 (DGAT1) gene affects the milk composition from dairy cows. However, little is known about the possible molecular basis changes in the mammary gland tissue. DNA microarray analysis was used in the present study to determine the most affected genes and pathways in the mammary tissue of dairy cows in response to the DGAT1 polymorphism. The genotype at the locus was designated AA, AK, or KK for homozygous Ala, heterozygous, or homozygous Lys, respectively. Mammary gland biopsies of 9 AA cows, 13 AK cows, and 4 KK cows were taken. The Factor Analysis for Multiple Testing methodology was used to associate the 7,486 expressed genes on Affymettrix Bovine Genome Arrays and the DGAT1 polymorphism. DGAT1 polymorphism did not modify dry matter intake, and milk yield, although changes occurred in the C14:0, C16:1cis-9, and unsaturated fatty acids in milk. DGAT1 polymorphism resulted in 652 differentially expressed genes. Within this set of genes, we uncovered significantly enriched pathways related to cell growth, proliferation and development, remodelling, immune system response, as well as lipid biosynthetic processes. When comparing AA genotype relative to KK genotype, the most inhibited pathways were related to lipid biosynthesis, which likely reflected counter mechanisms of mammary tissue to respond to changes in milk FA composition. Such molecular knowledge on the physiology of the mammary gland might provide the basis for further functional research on dairy cows.

Direct and maternal genetic relationships between calving ease, fertility, production and somatic cell count in UK Holstein-Friesian heifers

Eaglen, S.A.E.[1], Coffey, M.P.[1], Woolliams, J.A.[2] and Wall, E.[1], [1]SAC, Sustainable Livestock Systems, Easter Bush, EH25 9RG, Midlothian, United Kingdom, [2]The Roslin Institute and R(D)SVS, University of Edinburgh, Easter Bush, EH25 9RG, Roslin, United Kingdom; sophie.eaglen@sac.ac.uk

Focus in cattle breeding is shifting from traits that increase income, towards traits that reduce costs. As a result, an increasing number of functional traits, of which calving ease (CE) is an example, are included in national breeding indices. Yet, knowledge of genetic relationships between CE and other traits of interest is scarce. The objective of this study was to estimate these relationships, using a national dataset of 27,845 primiparous cow performance records. Traits of interest are calving interval (CI), days to first service (DFS), number of inseminations per conception (NRINS), non-return after 56 days (NR56), milk yield at day 110 (110MY) and accumulated 305 day milk yield (305MY) and the lactation average somatic cell count (SCC). CE (4-grades, ascending in difficulty) is affected by both a direct genetic effect (DCE, ease of birth) and a maternal genetic effect (MCE, ease of calving). Therefore, to account for DCE, a random sire of the calf effect was included in the multi-trait linear sire models fitted using ASREML v3.0. Significant results show that DCE has a negative relationship with CI (-0.60±0.25), 110MY (-0.31±0.15), 305MY (-0.44±0.15) and NR56 (0.67±0.27). MCE showed to be positively correlated to CI (0.61±0.25) and NRINS (0.62±0.22). CE was found not to be related to DFS and SCC, the genetic direct-maternal correlation was negative but non-significant and fertility and production traits were unfavourably correlated. A mirror image of reversed signs was displayed between DCE and MCE relationships, which demonstrates that CE is a complex trait and needs careful consideration when analyzed and interpreted. Further study is needed and caution is vital when selection decisions are being made, as response to selection is not straightforward.

Genetic and environmental variability of goat milk FTIR spectra

Dagnachew, B.S. and Ådnøy, T., Norwegian University of Life Sciences, Department of Animal and Aquacultural Sciences, P.O.Box 5003, 1432, Ås, Norway; binyam.dagnachew@umb.no

In the current dairy animal breeding programs, milk components are often analyzed by infrared spectra. Fourier Transform Infrared (FTIR) spectrometry is the most common technique used today because of its multiplex capability. Until now genetic variability of goat milk FTIR spectra are only known indirectly as they contribute to the major milk components. The aim of this study is to investigate genetic and environmental variability of goat milk FTIR spectra directly. A dataset containing 74,858 milk FTIR spectral observations which belong to 28,260 goats of 271 herds was used for the analysis. Principal component analysis (PCA) was applied to the spectral data, and new traits (PC traits) were defined because the whole spectra could not be analyzed for variance components. Genetic and environmental variance components were estimated for these PC traits using multi-trait mixed model. The estimated variances were back transformed to the original spectral variables. The PCA resulted in 8 components (PCs) which explained almost 99% of the spectral variation. Genetic variance ratios of these 8 PC traits ranged from 0.011 to 0.285 of the total phenotypic PC trait variation. Heritabilities of the spectral variables ranged from 0.018 to 0.408 and variance ratios of the permanent environmental effects ranged from 0.002 to 0.184 of the phenotypic spectral variation. High to moderate heritabilites were observed in particular in some spectral regions: between 1,030 cm^{-1} and 1,300 cm^{-1}, 1,500-1,600 cm^{-1}, 1,700-1,800 cm^{-1}, and 2,800-3,000 cm^{-1}. This could be related to the genetic variability of major milk components (fat, lactose and protein). Our results confirmed that there is a substantial amount of genetic variation in goat milk FTIR spectra, however, not all FTIR regions are of breeding interest. The observed environmental variability of the spectra implies that the FTIR spectra should be considered in milk quality improvement programs including feeding advice.

Genetic response of milk coagulation properties in Italian Holstein Friesian dairy cattle population using different sets of genetic parameters

Pretto, D.[1], López-Villalobos, N.[2] and Cassandro, M.[1], [1]Department of Animal Science, University of Padova, Viale dell'Università 16, 35020 Legnaro (PD), Estonia, [2]Institute of Veterinary, Animal and Biomedical Sciences, Massey University, Private Bag 11-222, 4442 Palmerston North, New Zealand; denis.pretto@unipd.it

Aim of this study was to analyze genetic response of milk coagulation properties (MCP) under current selection index in Italian Holstein Friesian (IHF) dairy cattle population using different sets of genetic parameters. The selection index analysed was a sub-index (59% relative weight of total index) of current selection index applied in IHF population and included protein and fat yield, protein and fat percentage, somatic cell score. Genetic response after 10 years of selection, for all the traits in selection index and correlated response for MCP, milk yield and milk acidity (expressed as pH or titratable acidity), was estimated using selection index theory. Genetic parameters for all traits were assumed from different sources and validated for positive definite matrix. Genetic response was estimated assumed three different sets of genetic correlation parameters for MCP traits. Four pathways of selection were used. Selection intensities and generation intervals for the respective pathways were in according with current selection scheme. Bulls were assumed to have been evaluated on 100 daughters in separate herds, and cows were evaluated from own records (1 or 2 lactations records depend on the pathway). Expected genetic change in 10 years for rennet coagulation time (RCT) was -0.07, -0.03 and +0.12 genetic standard deviation (σ_g) for set A, B and C, respectively; while for curd firmness was +0.38, +0.12 and +0.15 σ_g for set A, B and C, respectively. Current selection index with two set of correlation seems not to deteriorate MCP traits, while with set C an increase in RCT is expected.

Genetic analysis of twinning rate and milk yield using a multiple-trait threshold-linear model in Japanese Holsteins

Masuda, Y., Baba, T., Kaneko, H. and Suzuki, M., Obihiro University of Agriculture and Veterinary Medicine, Department of Life Science and Agriculture, Inada W2-9, 0808555, Obihiro, Hokkaido, Japan; masuday@obihiro.ac.jp

The objective of this study was to estimate genetic parameters for twinning rate in the first three parities (TR1, TR2 and TR3) and milk yield in the first lactation (MY) using a 4-trait threshold-linear animal model in Japanese Holsteins. Twinning and 305-d milk production for 200,027 cows calving between 1990 and 2007 were analyzed. Twinning was treated as a binary character i.e. single (1) or twin/triplet (2). Individual milk yields were calculated using a multiple trait prediction for cows with at least 8 test-day records. Herd (TR1) and herd-year of calving (TR2 and TR3) were included in the mathematical model as uncorrelated random effects to avoid extremely category problems. (Co)variance components were estimated with THRGIBBS1F90 computer program via Gibbs sampling. Posterior means and standard deviations (SD) for parameter estimates were calculated using 80,000 samples after burn-in. Reported twinning rate for first, second, and third parity were 0.70%, 2.87% and 3.73%, respectively. Posterior means (SD) of heritabilities for T1, T2 and T3 were 0.12 (0.01), 0.15 (0.01) and 0.15 (0.02), respectively. Genetic correlations between parities ranged from 0.85 (T1 and T3) to 0.95 (T2 and T3). Genetic correlations of MY with twinning rate were 0.07 or less and their 95% highest probability density interval (HPD95) contained zero. Multiple births at different parities were considered as genetically same traits in spite of the low frequency of reported twinning in younger compared to older cows. No genetic relationship between twinning and milk yield suggests that the selection for twinning rate may be feasible without reducing milk yield.

Additive and dominance genetic effects for litter size components in Pannon White rabbits

Nagy, I.[1], Gorjanc, G.[2], Čurik, I.[3], Farkas, J.[1] and Szendrő, Z.[1], [1]Kaposvár University, Guba S. 40, 7400 Kaposvár, Hungary, [2]University of Ljubljana, Groblje 3, 1230 Domzale, Slovenia, [3]University of Zagreb, Svetošimunska 25, 10000 Zagreb, Slovenia; gregor.gorjanc@bf.uni-lj.si

Quantitative genetic parameters were estimated for the number of rabbits born alive (NBA) and dead (NBD) in a synthetic Pannon White rabbit population. The data collected between 1992 and 2009 consisted of 18,398 records from 3,883 rabbit does and pedigree with 4,804 members. Four repeatability models, all including inbreeding of a doe and litter as covariates, were used: a) with additive genetic effect (A model), b) A model and permanent environment effect (AP model), c) AP model and dominance genetic effect (ADP model), and d) A model with dominance genetic effect (AD model). With the A model estimated heritability (h^2) was 0.120 ± 0.009 for NBA and 0.031 ± 0.005 for NBD. With the AP model estimated h^2 decreased to 0.057 ± 0.010 for NBA and 0.022 ± 0.006 for NBD with the estimated proportion of permanent environment variation (p^2) being equal to 0.061 ± 0.010 for NBA and 0.015 ± 0.007 for NBD. With the ADP model estimated h^2 further slightly decreased to 0.055 ± 0.011 for NBA and 0.019 ± 0.006 for NBD while p^2 decreased to 0.051 ± 0.010 for NBA and 0.004 ± 0.008 for NBD with estimated dominance variation (d^2) being equal to 0.048 ± 0.006 for NBA and 0.053 ± 0.006 for NBD. With the AD model estimated h^2 were intermediate (0.089 ± 0.010 for NBA and 0.020 ± 0.006 for NBD), while estimated d^2 was markedly higher for NBA (0.118 ± 0.006) and similar for NBD (0.059 ± 0.005) in comparison to the ADP model. Correlations between solutions for A, D, and P effects were high (>0.966) for all models and both traits. Estimates of regression on inbreeding obtained with ADP model were -0.14 and -0.41 for NBA and 0.18 and 0.06 for NBD per 10% increase in inbreeding of a doe and litter, respectively, and did not differ substantially between models. Results show importance of permanent environment and dominance effects for the estimation of variance components for litter size components.

Genetic assessment of fighting ability in Valdostana cattle breeds

Sartori, C. and Mantovani, R., University of Padua, Deparment of Animal Science, Viale Universita, 16, 35020 Legnaro (PD), Italy; cristina.sartori@unipd.it

Fighting ability is of main importance in social species when dominance relationships are built up to regulate the access to resource. A bellicose temperament at pasture is well known in Aosta Black Pied and Aosta Chestnut cattle breeds, and it revives in traditional tournaments (i.e. 'Batailles de Reines') annually performed in Aosta Valley. Despite the lack of planned breeding programs, an empirical selection on fighting ability (FA) has been performed for decades, but in recent years a genetic evaluation has been considered. The aim of the present work is to overview the studies on fighting ability as regard: (1) behavioral analysis of FA; (2) genetic assessment of FA; (3) effect of an inclusion of competitors, either within the phenotype or as IGE; (4) incidence of inbreeding on FA. Behavior analysis on 188 video recorded battles indicates that cows embrace different behaviors of increasing intensity, as defense of resource, exhibition and physical fight. Battles resulted affected by asymmetries in weight, age, fights experienced and genetic differences in FA. About genetic analysis on FA (n=23,999 fights by 8,259 cows), a score mainly based on the position reached at battle and on the size of the tournament (Placement Score) resulted a suitable phenotype for FA, and an EM-REML variance component estimates on a classical quantitative model allowed to assess a h^2 of about 0.08. Including the opponent within the phenotype the heritability of the trait dramatically decrease (h^2=0.04), whereas accounting for the competitor as IGE slightly enhances the estimate (h^2=0.09), however improving the goodness of fit. Inbreeding slightly affects the genetic evaluation for FA (0.02 of variation in h^2), and linear regression analysis carried out on the 33 main lineages to which most of participants belonged (n=6,087) revealed an overall negative trend of EBVs compared to the increase of F (b=-21.3). Thus, studies on FA revealed that also behavior may be considered for breeding.

Possibilities for organic cattle breeding

Nauta, W.J.[1] and Roep, D.[2], [1]Louis Bolk Institute, Hoofdstraat 24, 3972LA, Netherlands, [2]Wageningen University, Rural Sociology Group, Hollandseweg 1, 6706 KN, Netherlands; w.nauta@louisbolk.nl

Organic dairy farmers should use organic breeding and breeding stock that suit their production systems. However, organic farmers are still using breeding stock from companies focused on conventional production and which are produced with embryo transfer (ET). This conflicts with the rules and principles of organic production. Options for suitable organic breeding systems were studied in PhD research (2000-2009). Farmers' visions on breeding and breeding goals were explored, genotype by environment (GxE) effects on production traits estimated, and the various options for organic breeding compared and contrasted. Farmers' visions indicated a wish for more organic based breeding. An examination of GxE effects on production traits did not yield convincing results for starting an organic breeding program. Farmers set greater store by functional traits. However, populations are relatively small for effective testing schemes, and the diversity of (cross)breeds also poses problems. But animals with high genetic merit for production and the use of ET technologies in conventional breeding schemes force the organic sector to find new ways for organic breeding systems. The issue is complex. Many stakeholders are involved and have their opinion on it. After many decades of being supplied with breeding stock through AI, it is not easy for farmers to switch to AI bulls from organic farms or to natural mating. Three possible scenarios for development were identified: natural farm-based breeding, an organic breeding plan and the use of specific ET-free bulls from conventional programs. Knowledge is lacking among farmers and other stakeholders, and pilot projects for practical learning are being launched. Old and new knowledge can help farmers take steps towards organic breeding practices. Modern breeding technologies like genomics and sperm separation, but also the use of native and cross breeds are possibilities for organic breeding towards a fully closed organic production chain.

Estimation of genetic parameters for test day milk yield in a seasonal calving system

Mc Carthy, J.[1] and Veerkamp, R.F.[2], [1]Irish Cattle Breeding Federation, Bandon, Co Cork, Ireland, [2]Animal Breeding and Genomics Centre, Wageningen UR Livestock Research, Lelystad 8200AB, Netherlands; jmccarthy@icbf.com

Variance components for test day milk yield were estimated for primiparous animals in a seasonal calving system, where 80% of calvings occur within a 5 month period in the Spring. Month in Milk (MIM) and Test Month (TM) were used to fit both the permanent environment effect and additive genetic effect. Estimates of heritability (0.13-0.27) were found to be highest during early and late lactation for all calving months. Genetic variance for all calving months was found to be highest at the beginning of the lactation, with a plateau towards the middle of the lactation. Genetic correlations between corresponding stages of lactation were highest between consecutive calving months, but decreased when intervals between calving months increased, however remained above 0.96 in all cases. The peak of heritability was later in lactation for later calving months. Additive genetic variance was found to be highest for animals calving early in the season. Differences in heritability between animals at the same stage of lactation but with different calving months were less that 0.01 between adjacent calving months.

mtDNA diversity of some Turkish native goat breeds

Çinar Kul, B. and Ertugrul, O., Ankara University, Faculty of Veterinary Medicine, Department of Genetics, Diskapi, 06110 Ankara, Turkey; bkul@veterinary.ankara.edu.tr

Domestication of the animals, considered as one of the most important milestones of modern human life, began about 11,000 years ago in the Neolithic age on the lands known as 'Fertile Crescent'. The four livestock species were given priority in the domestication course: cattle, pig, sheep and goat. In comprehensive studies, which are mainly focused on the goat domestication, cytochrome b and D-loop regions located on mitochondrial DNA (mtDNA) have been frequently used. In this study, the mtDNA control region was analyzed in Angora (Ankara), Honamli, Kilis, Hair (Kil) and Norduz goat breeds to reveal diversity of mitochondrial DNA and differentiation of goat breeds and cytochrome b analyzed for estimating the molecular clock and the Time to Most Recent Common Ancestor. In terms of nucleotide and haplotype diversity, except for Norduz goat breeds, all groups show high values as compared with mainly world goat breeds. This finding indicates a central localization for Turkey in the goat domestication course. A, D and G haplogroups have been determined in 5 breeds studied. Additionally, our results indicated that Haplogroup G is the most ancient haplogroup for Turkish native goat breeds and oldest haplogroup used in the domestication course was Haplogroup A. Through this study, an inexpensive novel method based on PCR-RFLP was introduced to determine G haplotype, a unique haplotype to Fertile Crescent regions. Inbreeding and mating with culture breeds, yield to vanishing of the native breeds. Despite their low production profile, the loss of native breeds, given their adaptability features acquired throughout thousands of years via natural selection, is of quite importance to genetics. This data showed that the studied 4 goat breeds preserved their mtDNA diversities. However, these results should be further analyzed using alternative molecular markers and comparisons, in order to construct new preservation strategies accordingly. This study was supported by the project TURKHAYGEN–1 (KAMAG-106G005).

Optimizing cryopreservation – determining the core set of the semen bank
Nikkonen, T., Strandén, I. and Mäki-Tanila, A., MTT Agrifood Research Finland, Biotechnology and Food Research, Myllytie 1, 31600 Jokioinen, Finland; terhi.nikkonen@mtt.fi

In an animal genetic resources programme the general aim of the semen bank is to store as much ancestral genetic variation as possible. The bank is a back-up in case of a disaster, e.g. large parts of the population are lost due to disease outbreak or environmental disaster, or the population has inbreeding problems. To optimize the cryopreservation and also to minimize the costs for the long-term preservation, the core set needs to be found. It is unfeasible to preserve several doses from every AI bull in the long term storage from economical and storage space point of view. The core set should have enough doses and ancestral variation so that it can be used to replenish the variation or altogether re-establish a breed. In this study the GENCONT software was used to optimize the contributions from the stored bulls so that the coancestry measure was minimized. We started from a chosen base set and considered the contributions of the other candidates existing in the semen bank to this set. We used data from the Eastern Finncattle. The pedigree, which is used to set up the relationship matrix, had 9 913 animals. Within the analysed time frame up to 2006 the semen bank consisted of 61,707 doses from 42 different bulls. Aim of long-term cryopreservation according to the Finnish National Action Plan for AnGR is 25 different bulls and 200 doses per bull, in total 5,000 doses. The oldest bulls with only few doses and the bulls with no parents were the base set. Doses from the other bulls were added by optimum contribution principles until we had the core set with 5,000 doses. The optimum contribution method provides a useful framework to assess the state of cryo storage and to find guidelines in determining the core set in the bank.

The genetic architecture of quantitative traits
Goddard, M., University of Melbourne, LFR, University of Melbourne, Royal Pde, Parkville, 3010, Australia; meg@unimelb.edu.au

Genome wide association studies and genomic selection experiments find that almost every trait is influenced by very many locations in the genome and that nearly all of these have very small effects. Typically even the larger loci explain only 0.001 of the variance of the trait. For many traits there are no loci with large effects but some traits have loci that explain up to half the genetic variance. These are usually mutations that would have once been detrimental but have recently been favoured by selection so that they have increased in frequency and now explain a large part of the variance.

Estimation of the effective number of genes underlying quantitative traits based on chromosomal partitioning of the genomic variance

Simianer, H.[1], Erbe, M.[1], König, S.[2] and Pimentel, E.C.G.[1,2], [1]Georg-August-University, Department of Animal Sciences, Albrecht-Thaer-Weg 3, 37075 Göttingen, Germany, [2]University Kassel, Department of Animal Breeding, Steinstr. 19, 37213 Witzenhausen, Germany; hsimian@gwdg.de

Partitioning the genomic variance of complex traits into chromosomal contributions often reveals that the contribution of each chromosome to the total genetic variance is highly correlated to the physical chromosome length. While this observation is not suited to formally prove that an infinitesimal model of inheritance is underlying the respective traits, the observed results are close to what would be expected if an infinitesimal model was the true state of nature. Based on this argument, we suggest to define the effective number of genes underlying a quantitative trait, N, to be the number of randomly distributed independent genes of equal genetic variance, which produces the same reliability of the association of chromosome length and genetic contribution as empirically observed. This approach was applied to a dataset of 2,294 progeny tested Holstein bulls genotyped for 39,557 SNPs on the 29 autosomes. Reliabilities of the association between the proportion of the genetic variance per chromosome and the physical length (after correction for DGAT1) were 0.82 for milk yield (Mkg), 0.64 for fat percentage (F%), 0.77 for protein percentage (P%), and 0.79 for somatic cell score (SCS). The resulting estimated effective number of underlying genes ranged from N=400 (for F%) over N=700 (for P%) and N=800 (for SCS) to N=900 (for Mkg). 95 per cent confidence intervals are substantial (e.g. 600 < N < 1,500 for Mkg). It is argued that these numbers are conservative lower bound estimates, suggesting that presumably >1000 genes are involved in the inheritance of the traits studied, which is well in line with recent findings suggesting a highly polygenic background of production traits in dairy cattle.

Empirical determination of the number of independent chromosome segments based on cross-validated data

Erbe, M.[1], Reinhardt, F.[2] and Simianer, H.[1], [1]Georg-August-University, Department of Animal Sciences, Animal Breeding and Genetics Group, Albrecht-Thaer-Weg 3, 37075 Göttingen, Germany, [2]Vereinigte Informationssysteme Tierhaltung w.V. (VIT), Heideweg 1, 27283 Verden/Aller, Germany; merbe@gwdg.de

In dairy cattle, prediction of genomic breeding values has become a basis for selecting young bulls which are not yet progeny tested. In many cases it is useful to derive the expected accuracy of prediction in advance of a study to find the best design regarding e.g. the required size of the training set for a specific level of accuracy for a given trait. Hence, deterministic equations have been suggested which use information of the number of animals in the training set, the heritability, and assumptions regarding the effective population size and the number of independent chromosome segments (M_e) for predicting the accuracy. The aim of our study was to use one of these deterministic predictions backwards for deriving M_e empirically based on cross-validated data. A data set of 5,698 Holstein Friesian bulls genotyped with the Illumina 50K SNP chip was available. All bulls had estimated breeding values based on progeny testing for milk yield and somatic cell score with an accuracy >0.84. Different k-fold (k = 2-10, 15, 20) cross-validation scenarios were performed using a genomic BLUP approach. Each scenario was repeated 50 times while animals were assigned to the folds randomly. Afterwards, a maximum likelihood approach was used to estimate the value of M_e assuming that the accuracies obtained with the cross-validated data in each scenario were normally distributed with a mean equal to the expected accuracy derived from the deterministic equation. The highest likelihood was obtained with M_e of 1,239 and 1,046 for somatic cell score and milk yield, respectively, when using a modified form of the deterministic equation assuming that the maximum accuracy is less than one. This corresponds to an effective population size $9 < N_e < 216$, which is in good agreement with the usual estimates of Ne in the Holstein population.

Understanding dominance
Wellmann, R. and Bennewitz, J., University of Hohenheim, Department of Animal Husbandry and Animal Breeding, Garbenstraße 17, 70599 Stuttgart, Germany; r.wellmann@uni-hohenheim.de

Knowledge of the genetic architecture of a quantitative trait is useful to adjust methods for the prediction of genomic breeding values. The inclusion of dominance effects into these models could increase the accuracy of the predictions and predicted dominance effects could be used to find mating pairs with good combining ability. Additive variance, dominance variance and inbreeding depression have been estimated for some populations and traits. But it is widely unknown which architectures are consistent with available estimates, and how the genetic architecture of a trait does affect these quantities. Formulas are derived that can be used to estimate the number of QTL that affect a quantitative trait from variance components and inbreeding depression. It is shown that a lower bound for the number of QTL depends on the ratio of squared inbreeding depression to dominance variance. That is, high inbreeding depression must be due to a sufficient number of QTL because otherwise the dominance variance would exceed the true value. Parameters of the distribution of additive and dominance effects can also be estimated from variance components. Moreover, the second moment of the dominance coefficient depends only on the ratio of dominance variance to additive variance and on the dependency between additive effects and dominance coefficients. This has implications on the relative frequency of overdominant alleles. The formulas are applied to milk yield and productive life in Holstein cattle. Plausible estimates were obtained for PL only if the model assumes that dominance effects diminish rather than increase the additive variance. This questions the common practice to sample additive effects independent from dominance coefficients in simulation studies. Our findings support the hypothesis of a large number of QTL that affect this trait, with large QTL being close to recessivity. However, the formulas should be applied with caution as dominance variance is difficult to estimate.

Application of spike and slab variable selection for the genome-wide estimation of genetic effects and their complexity
Wittenburg, D. and Reinsch, N., Leibniz Institute for Farm Animal Biology, Genetics and Biometry, Wilhelm-Stahl-Allee 2, 18196 Dummerstorf, Germany; wittenburg@fbn-dummerstorf.de

Quantitative traits are typically influenced by a myriad of genes with their additive, dominance and epistatic effects, whereat their number and kind of interplay are hardly known. Although additive genetic effects are of major interest in breeding applications, the knowledge about non-additive genetic effects is important for the understanding of phenomena like heterosis or inbreeding depression as well as for an appraisal of broad sense heritability. When analyses of genome-wide marker information try to make allowance for non-additive effects, the separation of non-zero effects from unimportant ones is an issue e.g. for the improvement of genetic value prediction and deeper insights into the relative importance of additive and non-additive effects for genetic variation. A previously published spike and slab variable selection approach was adapted to this problem by specifying a complexity parameter for each kind of genetic effect included in a statistical model. These parameters represent the proportion of non-zero effects within each source of genetic variation. With aid of the complexity parameters, an empirical selection procedure was appended to identify the significant non-zero genetic effects a posteriori. The suitability of this approach was verified via simulations with either a small number or a multitude of QTL with additive and dominance effect. The accuracy of genetic value prediction was always at a high level in both scenarios. Despite that most of the genetic variation was due to additive effects, the contribution of dominance effects could be assessed. The spike and slab approach had, however, difficulties not only to determine the amount of dominance variation but also the number of dominance effects; its proportion was underestimated to a larger extent than the proportion of additive effects.

Using whole genome sequence data to predict phenotypic traits in *Drosophila melanogaster*

Ober, U.[1], Stone, E.A.[2], Rollmann, S.[3], Mackay, T.[2], Ayroles, J.[2,4], Richards, S.[5], Gibbs, R.[5], Schlather, M.[6] and Simianer, H.[1], [1]Georg-August-University Göttingen, Animal Breeding and Genetics Group, Albrecht-Thaer-Weg 3, 37075 Göttingen, Germany, [2]NC State University, Department of Genetics, Raleigh, NC 27695-7614, USA, [3]University of Cincinnati, Department of Biological Sciences, Cincinnati, OH 45221-0006, USA, [4]Harvard University, Department of Organismic and Evolutionary Biology, Cambridge, MA 02138, USA, [5]Baylor College of Medicine, Human Genome Sequencing Center, Houston, TX 77030, USA, [6]Georg-August-University, Centre for Statistics, Goldschmidtstraße 7, 37073 Göttingen, Germany; uober@math.uni-goettingen.de

Predicting geno- and phenotypes plays an important role in many areas of life sciences. So far, genomic prediction was based on genome-wide SNP sets of variable density. Ultimately, prediction based on complete genome sequences will be the way to go. We report results of a first attempt to implement sequence-based prediction in the model organism Drosophila melanogaster. The recently generated 'Drosophila Genetic Reference Panel (DGRP)' comprises 192 inbred lines which are fully sequenced and phenotyped for various quantitative traits. We predicted the phenotypes of starvation resistance for 157 DGRP-lines using whole-genome sequence data. After filtering, 1.8 million SNPs were used to construct the genomic relationship matrix according to VanRaden, which was then used in a genomic BLUP model. Accuracy of prediction was assessed as the correlation between true and predicted phenotypes by cross-validation (CV). With 5-fold CV an accuracy of 0.26±0.01 was achieved. Applying the approach of Daetwyler et al., this suggests the presence of 1,600 independently segregating genome segments. Even after accounting for the specific genomic properties of D. melanogaster, this value does not coincide well with what is expected from estimates of N_e derived from the decay of LD structure with distance and from LD of non-syntenic loci. Possible reasons for this discrepancy will be discussed.

Understanding the genetics of complex traits in milk using RNA sequencing

Medrano, J.F., Wickramasinghe, S., Islas-Trejo, A. and Rincon, G., University of California, Davis, Dept of Animal Science, Davis, California, 95616, USA; jfmedrano@ucdavis.edu

A comprehensive expression profiling of genes expressed in bovine milk somatic cells at different stages of lactation was performed to study the genetic regulation of milk composition traits. Our main focus has been on genes coding for enzymes involved in milk oligosaccharide metabolism. Milk oligosaccharides are complex sugars that form the third largest component in milk. Because of their complex structures and presence as free and conjugated oligosaccharides, little is known of their genetic determination. Milk samples were collected from six Holstein and six Jersey cows at days 15 (transition) and 250 (late) of lactation and RNA was extracted from the pelleted milk cells. Expression analysis was performed by RNA sequencing (RNAseq) using the Illumina GAII analyzer. Reads were assembled, annotated and analyzed in CLC Genomics Workbench. Data was normalized by calculating 'reads per kilo base per million mapped reads' (RPKM) for each gene and annotated with Ensembl (24,580 unique genes). T-tests were performed to identify genes with significant changes in expression between the two lactation stages. From RNAseq analysis large numbers of coding SNP were identified for use in phenotype-genotype association studies. Results revealed that 48-51% of annotated genes are expressed in bovine milk. No significant differences were observed between breeds in either oligosaccharide profiles or expressions of 92 glycosylation-related genes expressed in milk. Most of the genes exhibited higher expression at day 250 indicating increases in net glycosylation-related metabolism in spite of decreases in free milk oligosaccharides in late lactation milk. Based on the 92 glycosylation-related genes identified, metabolic pathways were constructed to identify key regulators of oligosaccharide metabolism in milk. These genes together with coding SNP will guide the design of a targeted breeding strategy to optimize the content of beneficial oligosaccharides in bovine milk.

Building gene networks underlying traits routinely recorded in dairy cattle

Szyda, J.[1], Kamiński, S.[2] and Żarnecki, A.[3], [1]Wrocław University of Environmental and Life Sciences, Institute of Animal Genetics, Kozuchowska 7, 51-631 Wroclaw, Poland, [2]University of Warmia and Mazury, Department of Animal Genetics, Oczapowskiego 5, 10-718 Olsztyn, Poland, [3]National Research Institute of Animal Production, Institute of Animal Breeding and Genetics, Krakowska 1, 32-083 Krakow, Poland; joanna.szyda@up.wroc.pl

As a result of national genomic selection programmes estimates of SNP effects densely distributed along the bovine genome are now available for many quantitative traits routinely recorded in dairy cattle. Moreover, the bovine genome sequence data base contains information on many genes. The goal of the current study is to combine the two sources of information in order to identify the causal mutations underlying the most significant SNP estimates and to build networks between them. The available material are estimates of SNP effects obtained from the Polish genomic selection programme, based on a data set of 2,309 Holstein-Frisian bulls genotyped using the Illumina 50K bead chip. The SNP effects were estimated by a mixed linear model with deregressed national proofs for production, somatic cell score, fertility as well as type and conformation traits as dependent variables and a random SNP effect. The position of significant SNPs was matched with the Btau4.0 sequence information in order to identify underlying genes. Finally, for the selected genes a regulatory network is created.

Mapping of fertility traits in the combined Nordic red dairy cattle population by genome-wide association

Schulman, N.F.[1,2], Sahana, G.[2], Iso-Touru, T.[1], Tiirikka, T.[1], Lund, M.S.[2] and Vilkki, J.H.[1], [1]MTT, Agrifood Research Finland, Myllytie 1, 31600 Jokioinen, Finland, [2]Aarhus University, Research Centre Foulum, 8830 Tjele, Denmark; nina.schulman@mtt.fi

The Nordic red cattle is an admixed population with the original breeds being Finnish Ayrshire (AY), Danish red (RDM), and Swedish red (SRB). The objective was to combine data from the three breeds in order to increase power and precision of QTL mapping compared to analyzing the breeds separately. We performed single marker association mapping of female fertility traits in the combined Nordic red dairy cattle population. The analyzed data included 4,115 AI bulls (AY 1,734, RDM 835, and SRB 1,546). Each individual is assigned to a breed based on country of origin but the actual breed proportions within each individual differ. Genotypes were obtained with the Illumina BovineSNP50 panel and a total of 38,388 informative, high-quality SNP markers were used. We performed the association analysis using a mixed model approach which fitted a fixed effect of the SNP and a random polygenic effect. The breed effect was accounted for either by adding breed or the breed proportions estimated by the software STRUCTURE as fixed effect. Five significant QTL areas were detected after Bonferroni correction. The significance and precision of the detected QTL areas were increased compare to individual breed analyses. This shows that there is an advantage in combining the Nordic red population in association analyses.

Dissection of genetic variation in the neighbourhood of the SIGLEC5 and DGAT1 genes underlying traits routinely evaluated in dairy cattle

Suchocki, T.[1], Szyda, J.[1], Kamiński, S.[2], Ruść, A.[2] and Żarnecki, A.[3], [1]Wrocław University of Environmental and Life Sciences, Institute of Animal Genetics, Kozuchowska 7, 51-631 Wroclaw, Poland, [2]University of Warmia and Mazury, Department of Animal Genetics, Oczapowskiego 5, 10-718 Olsztyn, Poland, [3]National Research Institute of Animal Production, Institute of Animal Breeding and Genetics, Krakowska 1, 32-083 Krakow, Poland; tomasz.suchocki@gmail.com

The large number of genotypes and individuals which is now available through national genomic selection programmes allows for a more precise insight into the genetic variation underlying traits, which are routinely recorded and evaluated in dairy cattle. Bulls selected from the training data set of the Polish genomic selection programme were additionally genotyped for DGAT1 (502 bulls) and SIGLEC5 (462 bulls). Within SIGLEC5 gene a new missence mutation was found. An association analysis with an without LD information, as well as a linkage analysis, all based on a mixed linear models, were applied to the data set in which 21 SNPs located in the proximity of DGAT1 on BTA14 and 21 SNPs located in the proximity of SIGLEC5 on BTA18 were considered. All needed variance-covariance parameters in mixed models were obtained using the Expectation Maximization algorithm. The main goal of the study was to check which of the polymorphisms considered on BTA14 and BTA18 are responsible for the highest variation of the quantitative traits routinely evaluated in dairy cattle and to compare estimates obtained by the three different statistical models considered. As a phenotypic values deregressed proofs of 3 production, 21 type, one somatic cell and 4 fertility traits were used. The significant effects of DGAT1 gene ware found for production traits and for SIGLEC gene for fertility and type traits. Additionally, the SIGLEC gene is a one of the few known genes, which may potentially affect the fertility traits.

Integrating QTL controlling fatness, lipid metabolites and gene expressions to genetically dissect the adiposity complex trait in a meat chicken cross

Blum, Y.[1,2], Demeure, O.[1], Désert, C.[1], Guillou, H.[3], Bertrand-Michel, J.[4], Filangi, O.[1], Le Roy, P.[1], Causeur, D.[2] and Lagarrigue, S.[1], [1]Agrocampus Ouest/INRA, UMR598 Animal Genetics, 35042 Rennes, France, [2]Agrocampus Ouest, Applied Mathematics Department, 35042 Rennes, France, [3]INRA, UR66 Pharmacology and Toxicology Laboratory, 31027 Toulouse, France, [4]INSERM, MetaToul, 31024 Toulouse, France; yuna.blum@rennes.inra.fr

Background: Many studies have been performed to identify QTL affecting fat deposition in chicken. Most of them used a single QTL model. We propose in this study focused on abdominal fatness (AF) trait in a meat chicken cross to use multiple QTL models. In addition, we will analyze the co-localization between the QTL controlling AF, the mQTL controlling different hepatic lipid components and the eQTL controlling gene expressions to better functionally characterize the AF QTL regions through these elementary phenotypes that they control. Methods: A total of 177 offspring in 4 sire families originating from crosses between two lines divergently selected for AF were analyzed for AF, hepatic lipid composition (about 30 variables) and hepatic gene expression (28,743 probes). Results: Using single-QTL and multi-QTL models, 4 and 2 QTL for AF were detected on five chromosomes. For each QTL region, co-localized eQTL were identified with a hotspot on GGA2 (339 genes, including 1 cis-eQTL, P<0.01). We analyze this hotspot gene list using functional GO, KEGG and IPA terms information and constructing a gene network. Functional terms related to lipid metabolism have been identified. The mapping of mQTL controlling hepatic lipid profiles is currently performed and should also be used to characterize the QTL region. Conclusion: This large-scale information at different levels of observation can bring precious information about the 'function' of a QTL region of interest and then about the causative genes. As far as we know, it is the first time in a study on livestock species that metabolite measurements and gene expressions are included in a QTL analysis.

Complex Inheritance of melanoma and pigmentation of coat and skin in grey horses

Sölkner, J.[1], Druml, T.[1], Seltenhammer, M.H.[1], Sundström, E.[2], Pielberg, G.R.[2], Andersson, L.[2] and Curik, I.[3], [1]BOKU - University of Natural Resources and Life Sciences Vienna, Gregor Mendel Str. 33, A-1180 Vienna, Austria, [2]Uppsala University, P.O.Box 597, SE-751 24 Uppsala, Sweden, [3]Univiersity of Zagreb, Svetosimunska, HR-10000, Croatia; johann.soelkner@boku.ac.at

Melanoma, speed of greying of the coat, vitiligo and amount of speckling in grey horses do not follow simple inheritance patterns. To dissect their complex inheritance we estimated narrow sense heritability (h^2) which was further decomposed to polygenic heritability (h^2_u), grey mutation, i.e. 4.6-kb duplication at syntaxin 17 (STX17) heritability (h^2_{STX17}) and agouti gene (ASIP) heritability (h^2_{ASIP}). We also estimated genetic correlations (r_u) among the traits studied. The analysis was based on repeated measurements on 500-600 horses, respectively. Heritabilities were high for most traits invesitgated, h^2 values were 0.79 for greying, 0.63 for vitiligo and 0.66 for speckling, while lower values (h^2=0.37) were obtained for melanoma. A large proportion of phenotypic variation was caused by STX17 (for greying h^2_{STX17}=0.22; for melanoma h^2_{STX17}=0.18; for speckling h^2_{STX17}=0.55 and for vitiligo h^2_{STX17d}=0.23), the agouti gene only affected melanoma (h^2_{ASIP}=0.02). Moderate to high genetic correlations among traits were estimated. However, after adjusting for gene effects the genetic correlation was reduced substantially, with exception of the correlation of greying and vitiligo. We present a case of strong pleiotropic gene action, where a single gene substantially affects several traits and their correlations.

QTL for female fertility traits in the Nordic Holstein cattle using genome-wide association mapping

Höglund, J.K.[1,2,3], Sahana, G.[1], Guldbrandtsen, B.[1] and Lund, M.S.[1], [1]Aarhus University, Faculty of Agricultural Sciences, Department of Genetics and Biotechnology, Blichers allee 20, DK-8830 Tjele, Denmark, [2]Swedish University of Agricultural Sciences, Faculty of Veterinary Medicine and Animal Science, Department of Animal Breeding and Genetics, P.O. Box 7070, 750 07 Uppsala, Sweden, [3]VikingGenetics, Ebeltoftvej 16, Assentoft, DK-8960 Randers SØ, Denmark; johanna.hoglund@agrsci.dk

Dairy production has experienced a trend towards poorer female fertility, especially for the Holstein breed. Female fertility can be improved by the use of genetic markers in breeding schemes. However, detected QTL need to be confirmed before information about them is included in breeding schemes. The objective of this study was to detect and confirm QTL for female fertility in the Nordic Holstein population. A total of 3,475 Danish, Swedish and Finnish Holstein bulls with breeding values for eight female fertility traits were genotyped with the 54K bovine SNP array. An association study using a single marker approach identified a total of 108 QTL at a nominal 5% genome wide significance level using a Bonferroni correction (threshold = 1.3 10-6). We identified four QTL for number of inseminations per conception in heifers, 5 QTL for number of inseminations per conception in cows, 18 QTL for the interval from calving to first insemination, 10 QTL for days from first to last insemination in heifers, 15 QTL for days from first to last insemination in cows, 17 QTL for 56-day non-return rate in heifers, 27 QTL for 56-day non-return rate in cows, and 12 QTL for the fertility index. Twelve of these QTL overlapped with the QTL detected in a previous study in Nordic Holstein. They are therefore considered to be confirmed. This study shows that there is convincing evidence for the presence of multiple QTL with effects on female fertility traits and many of the results presented here confirm previous findings, both within Holstein Friesian cattle and in red dairy breeds. However, our study also revealed strong signals for QTL not previously detected.

A QTL landscape of swine dry-cured ham sensory attributes and their association to intramuscular fat content and composition

Pena, R.N.[1], Gallardo, D.[2], Guàrdia, M.D.[3], Reixach, J.[4], Arnau, J.[3], Ramírez, O.[2], Amills, M.[2] and Quintanilla, R.[1], [1]IRTA, Genètica i Millora Animal, Av. Rovira Roure 191, 25198 Lleida, Spain, [2]UAB, Ciència Animal i dels Aliments, Campus UAB, 08193 Bellaterra, Spain, [3]IRTA, Tecnologia dels Aliments, Finca Camps i Armet, 17121 Monells, Spain, [4]Selección Batallé S.A., Av Segadors s/n, 17421 Riudarenes, Spain; romi.pena@prodan.udl.cat

Herewith, we have carried out a QTL scan for sensory attributes of dry-cured hams from a commercial Duroc population. Flavour and texture descriptors and the overall liking of 300 dry-cured hams were evaluated by a panel of trained tasters. A correlation study confirmed the tight association of these sensory attributes with intramuscular fat content (IMF) and fatty acid (FA) profile of raw material. IMF was positively correlated to the most favourable descriptors (flavour: aged and matured; texture: fat melting and crumbliness), and negatively linked to non-desirable ones (flavour: bitterness, metallic and piquantness; texture: pastiness and adhesiveness). Consistently, IMF content showed a positive correlation (0.23 to 0.31) with the overall liking. Also, both monounsaturated and saturated FA content were associated with desirable properties of dry-cured hams, whereas polyunsaturated FA content displayed opposite associations. A panel of 116 microsatellites covering the whole genome was used to perform the QTL scan. Genome-wide significant QTL were mapped on SSC 3, 6 and 12 for texture (hardness, adhesiveness and pastiness), and on SSC 4, 7, 11 and 17 for flavour (aged, matured, sweetness and umami) descriptors. Additionally, a number of chromosome-wide significant QTL were detected across the porcine genome. Considering the QTL landscape as a whole, the most relevant results were observed on SSC 1, 3, 6, 7 and 12. These chromosomes harboured several QTL for sensory attributes co-localizing with QTL for fat deposition and IMF traits. These results represent a first contribution to elucidate which genetic factors modulate the sensory properties of dry-cured hams.

Ultimate pH at 24 hours post mortem differential gene expression in the Longissimus thoracis muscle of Iberian×Landrace back-crossed pigs

Noguera, J.L.[1], Casellas, J.[2], Pena, R.[1], Folch, J.M.[2], Fernández, A.I.[3] and Ibañez-Escriche, N.[1], [1]IRTA-UdL, Av. Rovira Roure, 191, 25198 Lleida, Spain, [2]UAB, Campus UAB, 08913 Bellaterra, Spain, [3]SGIT-INIA, crta. Coruña km 7,5, 28040 Madrid, Spain; joseluis.noguera@irta.es

Ultimate pH at 24 hours post mortem (pH_{24}) is one of the most accurate determinants of meat quality in pigs. Moreover, pH_{24} is genetically determined. Moderate heritabilities (~ 0.2) are reported in the literature. Given these outcomes, the aim of this study was to identify differential expression of pH_{24} related genes in the Longissimus thoracis (Lt) muscle of finished pigs. A total of 108 Iberian (25%) × Landrace (75%) back-crossed pigs were fattened under standard management conditions, and slaughtered at an average age of 179.9±2.6 d. The pH_{24} was measured directly on Lt muscle. For all individuals, a 50 g sample of the Lt was collected in the slaughter line and snap frozen in liquid nitrogen. The GeneChip Porcine Genome array (Affymetrix) was used to profile gene expression in each Lt sample, and we applied the gcRMA algorithm of the Bioconductor package for normalizing data. After that, gene expression data from 5,637 probes was analyzed with the GEAMM software, under a multivariate mixed linear model accounting for the systematic effect of each array (n=108) as well as four sources of variation, modeled under normal priors: probe, sex, fattening batch, and pH_{24} (continuous effect). The Bayesian analysis launched a unique Monte Carlo Markov chain with 60,000 iterations, the first 10,000 of them being discarded as burn-in. Focusing on the link between pH_{24} and gene expression, 34 probes reached the significance threshold after correcting for multiple testing with FDR 0.1 (P<0.000754). Additionally, Ingenuity Pathway Analysis was used to explore the associated pathways. The analyses revealed that differences in pH_{24} could be related to the ubiquitin-proteasome system (DNAJB9 and HSPA4),beta-catenin intercellular signalling (LRP6 and TCF7L1), and protein-kinase (TCF7L1).

Differential gene expression linked to fatty acid profile in the Longissimus thoracis muscle of Iberian×Landrace back-crossed pigs

Ibanez-Escriche, N.[1], Noguera, J.L.[1], Pena, R.N.[1], Folch, J.M.[2], Fernandez, A.I.[3] and Casellas, J.[1,2], [1]IRTA-UdL, Av. Alcalde Rovira Roure, 191, 25198 Lleida, Spain, [2]UAB, Campus UAB, 08913 Bellaterra. Barcelona, Spain, [3]SGIT-INIA, crta. Coruña km 7,5, 28040 Madrid, Spain; joaquim.casellas@uab.cat

The aim of this study was to identify differential gene expression conditional on the fatty acid profile (FAP, C:12-C:22 interval) in the Longissimus thoracis (Lt) muscle of finished pigs. A total of 110 Iberian (25%) × Landrace (75%) back-crossed pigs were reared under standard management conditions. From these, 108 Lt muscle samples were collected at slaughter, snap frozen and used for FAP and total RNA isolation. Each RNA sample was individually hybridized in GeneChip Porcine Genome arrays (Affymetrix, Santa Clara, CA). In order to obtain two extreme groups of comparison for FAP, a principal component analysis, that allows the synthesis of the global FAP variability in our population, was performed by means of the princomp function of R (http://www.R-project.org). The first principal component accounted for 36% of the global FAP variance and was used to rank the animals, selecting extreme pigs to create the HIGH (n=20) and LOW (n=22) groups. After normalizing raw data with the gcRMA algorithm from the Bioconductor package, gene expression values from 5,681 probes were analyzed with the BRB software under an analysis of variance with a fixed effects model accounting for: sex (male or female), fattening batch (three levels) and FAP group (high, low). The results from the FAP group showed 39 genes differential expressed with a False Discovery Rate smaller than 0.05. Additionally, Ingenuity Pathway Analysis (IPA, http://www.ingenuity.com) was used to explore the associated networks enriched in the resulting gene list. A total of 5 top networks were identified, of which, lipid metabolism was one of the most relevant networks represented by 9 of the 39 differentially expressed genes.

Targeted genetical genomics to detect causative mutations

Nadaf, J.[1,2], Berri, C.M.[1], Godet, E.[1], Dunn, I.C.[2], Le Bihan-Duval, E.[1] and De Koning, D.J.[2,3], [1]INRA, Recherches Avicoles, UR83, Nouzilly, 37380, France, [2]University of Edinburgh, Roslin Institute and R(D) SVS, Easter Bush, Midlothian, EH25 9RH, United Kingdom, [3]Swedish University of Agricultural Sciences, Animal Breeding and Genetics, Box 7023, 750 07 Uppsala, Sweden; dj.de-koning@slu.se

Genetical genomics has been shown to be a promising approach for dissecting complex traits. However, high cost and therefore low sample size has been a problem. One strategy to overcome this problem is to focus on only one or a few identified QTL and select homozygous genotypes in the segregating population for the gene expression study (targeted genetical genomics). In this study, we focused on a QTL affecting the initial pH of meat on chicken chromosome1. Meat pH is one of the most important indicators of chicken meat technological quality, and can be indicative of changes in meat colour and meat drip loss. From 698 F2 chickens used for QTL mapping, gene expression profiles of 24 birds (12 from each homozygous genotype at the QTL) were investigated. We identified up to 16 differentially expressed genes in the QTL area (potential cis-eQTL). These were much more significant than any genes outside the QTL region. Several of these local signals corresponded to unknown genomic elements, while others corresponded to known genes, which could contribute to variation of the trait due to the QTL. Combining these results with analysis of downstream effects of the QTL (using gene network analysis) suggests that the QTL is involved in the pH variation by controlling oxidative stress. The eQTL results were reproduced with only 4 arrays of pooled samples (with lower significance level). Selected genes have been confirmed using qRT-PCR. The results clearly suggest that this cost effective approach is promising for the functional characterization of QTL. We are now in the process of doing targeted re-sequencing of the QTL region in birds with different QTL genotypes to identify the causal genetic variation(s) and mechanism(s) controlling gene expression in this region.

Cytoplasmic line effects for birth weight and preweaning growth traits in the Asturiana de los Valles beef cattle breed.

Pun, A.[1], Goyache, F.[2], Cervantes, I.[1] and Gutiérrez, J.P.[1], [1]University Complutense of Madrid, Producción Animal, Avda. Puerta de Hierro s/n, E-28040 Madrid, Spain, [2]SERIDA-Deva, Camino de Rioseco 1225, E-33394 Gijón, Spain; gutgar@vet.ucm.es

Biological mechanisms explaining maternal effects may be partially related to mitochondrial DNA, which is almost exclusively maternally inherited. The aim of this research was to quantify the contribution of cytoplasmic line effects to phenotypic variance of preweaning growth traits in a sample of Asturiana de los Valles beef cattle breed. Production data and pedigree information were obtained from an existing performance recording database. The analysed database included a total of 22,852 animals. Database was formed by 15,645 records of birth weight (BrW), weaning weight (WW) and preweaning average daily gain (ADG) from 6,055 cows and 2,121 dam lines. All runs were carried out using the TM program. The fitted models included the herd-year of calving, the calving season, the sex of calf, the age of the dam at calving as fixed effects, and for WW and ADG also the effect of creep feeding and the age of the calf at weaning. Four different models were defined using different combinations of the additive genetic, the maternal genetic, the maternal permanent environmental and the cytoplasmic line effects. The estimates of the proportion of phenotypic variance accounted by the cytoplasmic line were rather low ranging from 0.03±0.005 for WW to 0.01±0.003 for BW. Those estimates were higher when both cytoplasmic line and maternal permanent environmental effect were considered in the model, suggesting that these effects are due to different causes. Cytoplasmic line may have a marginal effect on those traits and seems to be more important for preweaning growth traits.

Improving environmental sustainability of the dairy cow

Capper, J.L., Washington State University, Department of Animal Sciences, P.O. Box 646310, Pullman, WA 99164-6310, USA; capper@wsu.edu

The global dairy industry faces the challenge of providing sufficient animal protein to fulfill requirements of the growing population while reducing environmental impact per unit of dairy product. A deterministic model based on the nutrient requirements and metabolism of dairy cows was used to assess the environmental impact (resource use and waste output) per kg of milk produced by the U.S. dairy industry in 1944 compared to 2007. Advances in nutrition, genetics and management facilitated an increase in annual milk yield from 2,074 kg to 9,193 kg over this time period, resulting in 21% of the animals, 23% of the feed, 35% of the water and 10% of the land being required to produce one kg of milk in 2007 compared to 1944. Greenhouse gas (GHG) emissions were reduced by 63% per kg milk. A similar model evaluating the use of recombinant bovine somatrotopin (rbST) to increase milk yield per cow by an average of 4.5 kg/d demonstrated that milk could be produced with a 9% decrease in overall environmental impact. It is clear that improved productivity provides a means to reduce environmental impact through the dilution of maintenance effect, whereby the proportion of the total daily nutrient requirement attributed to maintenance is reduced. Strategies to reduce the daily maintenance cost by using smaller bodyweight animals would also be predicted to mitigate environmental impact providing that productivity was not unduly affected. Deterministic modeling of Cheddar cheese production from Jersey cows (454 kg bodyweight, 20.9 kg milk at 4.8% fat and 3.7% protein) compared to Holstein cows (680 kg bodyweight, 29.1 kg milk at 3.8% fat and 3.1% protein) demonstrated that the interaction between bodyweight and milk composition compensated for a reduction in milk yield, with reductions of 11% land and 32% water and 20% GHG emissions per kg cheese produced. To improve the environmental sustainability of dairy cows it is crucial to consider animal productivity and efficiency metrics, rather than focusing on productivity alone.

Interactions between dairy cow nutrition and environmental burdens
Sebek, L.B.J., Wur Livestock Research, Animal Nutrition, Edelhertweg 15, 8219 PH, Netherlands;
leon.sebek@wur.nl

European dairy farms encounter more and more environmental challenges. Not only nitrogen (N) and phosphorus (P) losses have to be minimized, but also greenhouse gas emissions. Sometimes these different goals can be achieved by the same measurements, but not always. To prevent uncontrolled swapping between environmental goals it is important to have a farm scale evaluation of the effects of measurements. However, farm scale results are the combined effect of the results of the different farm components. How important is the interaction between animal nutrition and farm scale environmental burdens? The results of the project 'Cows and Opportunities', including experimental dairy farm 'De Marke', show that the herd is an important farm component in reducing environmental burdens. It also shows that decreasing the amount of nutrients within the farm cycles could be limited by the interaction between the quality of home grown feeds and animal performance. If animal performance decreases due to feed quality, the farm scale effect of decreasing nutrient inputs can be contra productive. Hence, environmental goals are to strict and should be adapted. Another approach is to adapt the nutritional qualities of the dairy cow and/or the nutritive value of home grown feeds.

The best dairy cow system for future in terms of environment, economics and social criteria - worldwide
Hemme, T., IFCN Dairy Research Center, Schauenburgerstrasse, 24118 Kiel, Germany;
torsten.hemme@ifcndairy.org

In the dairy chain the major share of (a) the costs, (b) resources used, (c) emissions created and d) the political challenges come from producing the milk itself. This means the dairy farming system has a significant impact on the overall performance of the dairy chain. That's the reason why the IFCN – International Farm Comparison Network – is analysing and comparing various dairy farming systems world wide. Using the cost of milk production as an economic indicator we sees small scale farms in Africa and partly Asia as low cost producers and stantion barn farms in Western Europe as the farms with the highest costs. Using carbon emissions per kg milk as environment indicator we see farming systems with high milk yields and high share of compound feed having the lowest emissions. In contrast farming systems with low milk yields (1000-2,000 kg per cow) using mainly crop residues and feed base have 3-4 times emissions per kg milk. Using the employment per kg milk as a social indicator we see the picture that dairy farming systems in developing countries create 100 times more employment per kg milk than larger farms in developing countries. Outlook: Defining fitting future dairy farming systems is a very complex task as a dairy farming system by itshelf is is very complex. Moreover the key drivers like milk/beef prices, input prices (feed, land, labour, etc) change dramatically and new requirements like carbon footprints arise. Defining the best fitting dairy cow system is a continuous task and needs in each dairy region the connexion of different research disciplines interacting with dairy farmers and all players of the dairy chain.

High milk production and good reproduction: is it possible?
Zeron, Y.[1], Galon, N.[2] and Ezra, E.[3], [1]Sion-Israeli Company for AI and Breeding Ltd., M.P. Shikmim, 79800, Israel, [2]Hachaklait Veterinary Services Ltd., Habareket 20, Caesaria, 38900, Israel, [3]Israel Cattle Breeders Association, Habareket 20, Caesaria, 38900, Israel; yoel@sion-israel.com

Recently, it has been stated that as milk production increases fertility decreases. Genetic, nutritional and management factors were suggested to explain this phenomenon. Israel has close to 1000 herds with about 120,000 Holstein dairy cows, with exceptionally high milk production. This study aimed to identify and measure some of the factors affecting the reproduction in high producing dairy herds. Herd size ranged from 100-1000 milking cows. A multi-herd study was done on data from the Israeli Herdbook, from the year 2000, using a multivariate linear regression model. Risk factors included: herd, parity, year, season, service number, A.I. technician, waiting period, genetic breeding values and interactions between the major factors. The mean of 305 d milk production per cow increased from 11,930 kg in 2005 to 12,254 kg in 2009. The simple mean CR of the first 5 A.I. dropped from 33.9% in 2005 to 32.9% in 2009. Cows and herds with higher milk production also had a better CR, compared to those with low milk production. The shortening of the waiting period by a few days led to a correlated drop in CR. The economically driven shift from winter to summer inseminations caused a 0.9% drop in CR from 2005-2009. The inter-calving interval has remained almost unchanged and the mean of open days has gotten shorter. The genetic breeding value for both milk production and daughter fertility has increased in the last decade. Summer heat stress control had a corrective affect on CR. The high and increased milk production did not cause a major drop in fertility parameters. Farmers' agenda is profitability and not fertility. Management improvements can minimize the decline in fertility. Quality data combined with critical analysis of the contributing risk factors, enable making the necessary corrections.

Impact of breeding strategies using genomic information on fitness and health
Egger-Danner, C.[1], Willam, A.[2], Fuerst, C.[1], Schwarzenbacher, H.[1] and Fuerst-Waltl, B.[2], [1]ZuchtData GmbH, Genetics, Dresdner Str. 89/19, A-1200 Vienna, Austria, [2]University of Natural Resources and Applied Life Sciences, Sust. Agric. Sys., Gregor-Mendel-Str. 33, A-1180 Vienna, Austria; egger-danner@zuchtdata.at

A complex deterministic model (computer program ZPLAN) was used to model the breeding goal and breeding structure for the Austrian Fleckvieh (dual purpose Simmental) breed. The breeding goal is described by the Total Merit Index (TMI). Presently fat and protein yield have a relative weight of 37.9%, beef traits of 16.5%, fitness traits of 43.7% and milkability of 2.0%. The breeding program is characterized by 250,000 cows under performance recording, 1,500 bull dams, 140 test bulls with a test capacity of 25%, 25 proven bulls and 8 bull sires. The Annual Monetary Genetic Gain (AMGG) is mainly achieved by fat and protein yield (84.3%), and only by 10.1%, 4.4% and 1.2% by beef traits, fitness traits and milkability, respectively. By including economic weights for the direct health traits early fertility disorders, cystic ovaries and mastitis, the relative AMGG of fitness traits increased up to 8.3% turning the presently slightly negative AMGG for the fertility and udder health index in a positive direction. The impact of strategies using genomic information was also analyzed. A change in the breeding program with a reduction to 100 test bulls, the increase of the percentage of inseminations with young bulls to 40% and a proportion of candidates with GEBVs and selected young bulls from 10:1 increased the AMGG by 19.1% with a slight relative shift to fitness traits. The increase of the percentage of inseminations with young bulls from 25 to 70% increased the AMGG by 7 percent points, the relative AMGG for fitness traits by 2 percent points. A moderate positive natural genetic gain per year for fertility and udder health can only be achieved by increasing their economic weights. An increase by 50% resulted in a relative AMGG for fitness traits of 20%, doubling the weights led to an increased relative AMGG of 34.9%.

Genetic analysis of calf vitality in Dutch dairy cows
Van Pelt, M.L. and De Jong, G., CRV, AEU, P.O.Box 454, 6800 AL Wageningen, Netherlands;
mathijs.van.pelt@crv4all.com

For Holstein heifers, the percentage of calves born alive decreased from 89% to 84% during the last two decades. For Holstein cows, the percentage of life-born calves remained stable at 95% in that same period. The aim of this study was to estimate genetic parameters of calf vitality and to investigate what factors are causing the deterioration of heifer calf vitality. The data set for the parameter estimation comprised 266,578 calvings of heifers and 656,801 calvings of cows with at least 75% Holstein blood. A still-born calf was recorded as a calf that died before, during or within 24 hours after birth. Variance components were estimated with a bivariate sire-maternal grandsire model including a direct and maternal effect. The genetic parameters were used in a breeding value estimation on the complete national data set and solutions for all effects in the model were averaged per year x month. For heifers, the heritability for direct calf vitality was 0.038 and for maternal calf vitality was 0.086. For cows, the heritability was very low with 0.005 for both direct and maternal calf vitality. Genetic correlations between heifers and cows were 0.57 for direct calf vitality and 0.52 for maternal calf vitality. Therefore calf vitality for heifers and cows are considered as different traits. The decrease of calf vitality for heifers is caused by genetics and management. The average solution per herd x year decreased over the years for the maternal effect and explained half of the decline of calf vitality for heifers. The other half of the decline of calf vitality is caused by changes in management.

Future breeding goals in beef cattle: effect of changed production conditions on economic values
Åby, B.A.[1], Vangen, O.[1], Aass, L.[1] and Sehested, E.[2], [1]Norwegian University of Life Sciences, Department of Animal and Aquacultural Sciences, P.O. Box 5003, 1432 Ås, Norway, [2]GENO Breeding and AI Association, P.O. Box 5003, 1432 Ås, Norway; bente.aby@umb.no

The intensification of the livestock sector during the last decades has led to a continuously increasing amount of concentrates in animal diets worldwide. Human population growth and climate change may however reduce the amount of grain available for this use in the future. A change towards more extensive roughage based production systems influences both animal performance and economy, and thus the effect on economic values is of interest. The aim of this study was to estimate the effect of more extensive production conditions on economic values for production and functional traits in beef cattle. A bio-economic model was developed simulating the lifetime production of suckler cows of Continental and British breeds kept under semi-intensive production conditions (intensive indoor feeding of fattening animals, extensive cow-suckler cow enterprice). The model follows a suckler cow from first calving to slaughter, and includes six production traits and eight functional traits. Surplus heifers and bulls are kept for fattening on farm. Economic values were estimated as the increase in profit from a 0.1% increase in the mean of the trait considered. Based on a completely roughage based feeding system, the effect of three different roughage qualities were investigated. Results suggest that functional traits were more important than production traits for Continental breeds (66 vs. 31%), while production traits were more important for British breeds fed low quality roughage (60 vs. 40%). A decreasing roughage quality increases the economic importance of production traits. A sensitivity analysis to estimate the effect of increased roughage price was conducted, indicating increased importance of production traits. Other altenative production scenarios (e.g minimal use of concentrates) will be investigated.

Economic selection indexes for various breeds under different farming and production systems in Slovenia

Klopcic, M.[1], Veerkamp, R.F.[2], Zgur, S.[1], Kuipers, A.[3], Dillon, P.[4] and De Haas, Y.[2], [1]University of Ljubljana, Biotechnical Faculty, Dept. of Animal Science, Groblje 3, 1230 Domžale, Slovenia, [2]Wageningen UR Livestock Research, P.O. Box 65, 8200 AB Lelystad, Netherlands, [3]Wageningen UR, Expertise Centre for Farm Management and Knowledge Transfer, Wageningen, Netherlands, [4]Teagasc, Moorepark, Production Research Centre, Fermoy, Co. Cork, Ireland; Marija.Klopcic@bf.uni-lj.si

In recent times the economic perspective of the breeding programme is used more frequently in deciding the breeding objectives. Therefore, the aim of the study was to calculate economic selection indices for one dairy (Holstein) and two dual-purpose breeds (Simmental and Brown) in Slovenia that could be easily used by the farmers. Prices are becoming more difficult to predict with increased fluctuations in all prices which adds a level of complexity to the inclusion of economics into the selection indices. Various farming systems were considered. Economic values were calculated for milk, fat and protein yields, survival, calving interval for all three breeds and growth for dual-purpose breeds, using a farm economic model (Moorepark Dairy Systems Model), adapted to the Slovenian situation. The constructed economic indices ranked bulls significantly different compared to the ranking based on the currently applied Total Merit Indices. The economic indices were rather robust towards sensitivity in prices, only changes in milk price affected the economic weights significantly. Sensitivity analyses showed that values of animal sale price, beef price and production level affected the ranking of bulls a bit, as well as the values belonging to the organic farming system. Values that represent low-cost farms and farming in hilly areas gave no re-ranking of sires. The effect of growth as factor in the analysis depended on the assumptions made.

The relationship between diet characteristics, milk urea, nitrogen excretion and ammonia emissions in dairy cows

Bracher, A.[1,2], Münger, A.[1], Stoll, W.[1], Schlegel, P.[1] and Menzi, H.[2], [1]Research Station Agroscope Liebefeld-Posieux, P.O. Box 64, rte de la Tioleyre 4, 1725 Posieux, Switzerland, [2]Swiss College of Agriculture, Länggasse 85, 3052 Zollikofen, Switzerland; annelies.bracher@alp.admin.ch

In Switzerland, several cantons launched programs focused on efficient nitrogen use and reduction of ammonia emissions. 79% of ammonia emission from livestock stem from cattle. Various programs foresee feeding measures in dairy cows as an option to reduce N excretion and thus emissions. Reliable indicators on which to base incentives would be highly appreciated. In the present study, feeding measures with a potential effect on ammonia emission and possible indicators were evaluated based on a meta analysis of data from Swiss feeding and nitrogen balance trials with dairy cows across winter and summer rations. Single and multiple regressions were used to establish relationships. In a second step, relations were implemented in a model to simulate N excretion over a whole lactation. N intake in the balance trials varied between 160 and 800 g/day and urinary and total N excretion ranged between 30-430 and 100-600 g/day, respectively. Milk production varied between 3 and 49 kg/day. Among diet characteristics, crude protein content, CP/NEL-ratio, N/DOM-ratio and ruminal N balance correlated with milk urea and urinary N excretion. Fecal N is more closely related with feed intake and whole tract digestibility and varies independent of urea turnover. A curvilinear relationship between N digestibility and CP content of the diet ranging between 80 and 260 g/kg DM was found (r^2=0.88). On the assumption of a known N intake, the corresponding fecal and urinary N excretion can thus be estimated. Both, milk urea and ruminal N balance (g/day) show a high correlation with urinary N excretion and are potential indicators for ammonia emissions. However, residuals in simple regressions indicate further influencing factors such as protein supply over protein requirements, energy balance and diet type.

Genetic relationships between energy balance, fat protein ratio, body condition score and disease traits in Holstein Friesians

Buttchereit, N.[1], Stamer, E.[2], Junge, W.[1] and Thaller, G.[1], [1]Institute of Animal Breeding and Husbandry, Hermann-Rodewald-Str.6, 24098 Kiel, Germany, [2]TiDa GmbH, Bosseer Str.4c, 24259 Westensee, Germany; nbuttchereit@tierzucht.uni-kiel.de

Various health problems in dairy cows have been related to the extent and duration of the energy deficit *post partum*. Energy balance indicator traits like milk fat protein ratio (FPR) and body condition score (BCS) could be used in selection programs to help predicting breeding values for health traits, but currently there is a lack of appropriate genetic parameters. Therefore, the genetic relationship of energy balance, FPR, and BCS to mastitis, metabolic disorders, and claw and leg diseases was investigated using fixed regression and threshold models on data from 1,693 primiparous Holstein Friesian cows recorded within the first 180 days in milk. Average daily energy balance, FPR, and BCS were 8 MJ NEL, 1.13, and 2.94, respectively. Disease frequencies were 24.6% for mastitis, 9.7% for metabolic disorders, and 28.2% for claw and leg diseases. Heritability of energy balance was low (h^2=0.06), whereas heritability estimates for FPR and BCS were moderate (h^2=0.30 and 0.34, respectively). For the disease traits, heritability estimates ranged from 0.04 to 0.15. The genetic correlations were, in general, associated with large standard errors, but, although not significant, the results suggest that an improvement of overall health can be expected if energy balance traits are included into future breeding programs. A reduction of FPR might have favorable effects on metabolic stability and health of claw and legs. For BCS and mastitis, a significant negative correlation of -0.40 was estimated. The study provides a new insight into the role energy balance traits can play as auxiliary traits for robustness. It was concluded that both, FPR and BCS, are potential variables to differentiate between cows that can or cannot adapt to the challenge of early lactation.

Fat-to-protein-ratio in early lactation as an indicator of herdlife for first lactation dairy cows

Bergk, N. and Swalve, H.H., Martin-Luther-University Halle-Wittenberg, Institute of Agricultural and Nutritional Sciences, Theodor-Lieser-Str. 11, 06120 Halle, Germany; nadine.bergk@landw.uni-halle.de

The ratio of fat to protein in the milk is a common parameter used in herd management since this parameter has the distinct advantage that it is easily derived from milk recording data and allows for inferences on the energy balance of the cow. Based on test day records of 21 large contract herds from Mecklenburg-Western Pomerania, alternative definitions of the fat-to-protein-ratio were derived for the early lactation. Data was augmented by information on culling of cows resulting in different definitions of survival up to 100, 300, and 450 days *post partum* of the first calving. The three data sets used comprised 46,630, 45,115, and 42,976 cows in first lactation. An analysis on the phenotypic level revealed that the fat-to-protein-ratio of the first 50 days significantly influenced milk yield, average SCS, and survival rates. Substantial effects were found for animals with a fat-to-protein-ratio deviant from norm values. Compared to cows with a normal fat-to-protein-ratio at the second test day and using the definition of survival to 300 days, for cows with extreme high fat-to-protein-ratio culling rates increased from around 10% to 22%. An estimation of variance components using a sire model did not yield clear conclusions with respect to the association of the fat-to-protein-ratio and milk production as well as survival rates. Substantial differences exist between different definitions of the fat-to-protein-ratio and furthermore, associations between fat-to-protein-ratio and survival appear to be non-linear.

Genetic associations between feed efficiency and fertility in beef cattle

Crowley, J.J.[1,2,3], Evans, R.D.[4] and Berry, D.P.[1], [1]Teagasc, Animal & Grassland Research & Innovation Centre, Fermoy, Co. Cork, 0, Ireland, [2]Teagasc, Animal & Grassland Research & Innovation Centre, Dunsany, Co. Meath, 0, Ireland, [3]University College Dublin, School of Agriculture, Food Science & Veterinary Medicine, Belfield, Dublin4, 0, Ireland, [4]Irish Cattle Breeding Federation, Bandon, Co. Cork, 0, Ireland; john.crowley@live.ie

While feed efficiency in young beef animals is heralded as the primary focus of research in this area, the contribution of the cow herd to overall profitability of beef production systems cannot be ignored. The objective of this study, using a large dataset, was to quantify the genetic covariances between feed efficiency in growing animals measured in a performance test station, and beef cow fertility in commercial herds. Feed efficiency data were available on 2,605 bulls from one test station. Records on cow performance were available on up to 94,936 beef cows. Animal and animal-dam linear mixed models were used to estimate genetic covariances. None of the feed efficiency traits were correlated with calving interval, calving to first service, calving difficulty or perinatal mortality. However, genetic correlations estimated between age at first calving and FCR (-0.55±0.14), Kleiber ratio (KR; 0.33±0.15), residual feed intake (RFI; -0.29±0.14), RG (0.36±0.15) and relative growth rate (RGR; 0.37±0.15) all suggest that selection for improved efficiency may delay the age at first calving. We speculate, using information from other studies that, this may be due to a delay in the onset of puberty. Results from this study, as indicated from the estimated genetic correlations, suggest that selection for improved feed efficiency will have no deleterious effect on cow fertility or calving performance with the exception of possibly delaying the onset of puberty.

Efficiency of milk production in farms with a high share of grasslands

Litwinczuk, Z., Teter, W., Chabuz, W. and Stanek, P., University of Life Sciences in Lublin, Department of Breeding and Genetic Resources Conservation of Cattle, Akademicka 13, 20-950 Lublin, Poland; zygmunt.litwinczuk@up.lublin.pl

The aim of this study was to determine the gross profit of farms having a high share of grasslands and producing milk. The study included 28 farms keeping 10 and more Simmental cows. The farms were divided into two groups depending on the model of nutrition. In group I the fodder from grasslands were only used, and in group II also applied the corn silage. There were identified: the stocking (LSU/100 ha), annual milk yield per cow (kg), annual cost of cow's maintenance (€), cost and gross profit per 1 tonne of milk and amount of agricultural subsidies. Calculation of unitary costs was performed according to the modified 'resolution-organic' method. In the statistical study the one-way analysis of variance was used. The significance of differences was verified by Duncan test. The area of grasslands was nearly 29 ha in group I and approximately 66 ha in group II, and the animal stocking (LSU/100 ha) – from 88 in group II to 100 in group I. The cost of cow's keeping were higher in group I, i.e. approximately 700 €/cow (nearly 150 € more than in group I). However, taking into consideration the higher cow's productivity (on average 5,700 kg of milk per year in group II, and 5,000 kg in I) and a smaller share of indirect cost (at 2.6%), the lower cost of milk production (at 5 €/t) ultimately achieved in group II. Gross profit from milk production, which also included the payment for the work of the farmer (obtained after deduction of actual cost), was 107.5 €/t in group I and 127.5 €/t in II. In conclusion, it was stated that farms from group II achieved the higher gross profit per cow. The obtained extra charge increased the incomes from milk production at about 225-250 €/cow. The share of subsidies in the gross incomes was higher in group I, i.e. in farms using solely the fodder from grasslands (37% compared to 28% in group II).

Heart rate variability measured at milking in primiparous and multiparous cows
Speroni, M. and Federici, C., Agricultural Research Council (CRA), Fodder and Dairy Productions Research Centre (FLC), via Porcellasco 7, 26100 Cremona, Italy; marisanna.speroni@entecra.it

The interest in measuring heart rate variability (HRV) in animal welfare studies is increasing because a lot of information can be inferred from it. Healthy individuals show a high degree of HRV, that is a good indicator of emotional and physical adaptability to different situations; total HRV is mainly determined by autonomic innervations; a reduction of the total variance is usually considered indicative of a reduced vagal modulation; increased HR and reduced HRV following a particular challenge can indicate that welfare is threatened. However in interpreting HRV as welfare indicator it is important to take account not only of the environmental context, but also of animal factors as age, physiological state and metabolic condition including nutritive and activity level. Early lactation may be a critical period for primiparous cows in intensive farming since they are generally low ranked and have to cope with high productive performances. The current study wanted to verify the effect of parity on HRV and was conducted as part of a larger research aiming to use HRV and behavioral measures as tools to phenotype copying style in dairy cattle. Seventeen dairy cows were included in the study: 9 primiparous (P) and 8 multiparous (M); average days in milk during the experimental period were 38 (range 32-48, sd=3.6); HRV was measured at milking using a commercial heart rate monitor (Polar®). HRV data concerning the first 5 minutes of milking was analysed by Kubios software (Dep. of Physics, Univ. of Kuopio, Finland). Primiparous cows showed higher HR and lower RR (R-wave to R-wave interval), SD1(short-term component of HRV derived from a quantitative analysis of Poincaré plot), RMSSD (square root of the mean of the sum of the squares of differences between successive inter-beats intervals) than M (P<0.05). These results indicated a lower vagal tone at milking in P in comparison with M.

Motivating factors in Swedish young cattle production
Bostad, E.[1], Swensson, C.[1] and Pinzke, S.[2], [1]Swedish University of Agricultural Sciences, Department of Rural Buildings and Animal Husbandry, P.O. Box 59, 230 53 ALNARP, Sweden, [2]Swedish University of Agricultural Sciences, Dept. of Work Science, Business Economics and Environmental Psychology, P.O. Box 88, 230 53 ALNARP, Sweden; elise.bostad@ltj.slu.se

Swedish young cattle (red veal and young bulls) producers' goals and attitudes were studied as a complement to a larger study of the labour input and working environment in Swedish beef cattle production. Previous research has found that attitudes and goals are major determinants of the farmers' behavior such as the way of managing the farm and the productiveness of the enterprise. This study attempted to investigate if different motivation of farming could be identified, and if the identified factors could be related to 1) 'labour efficiency' measured as work time/animal and 2) 'farm size' measured number of cattle per unit and 3) physical work environmental factors. Through questionnaire studies of ~50% of Swedish young cattle producers anually finishing 100-1500 young cattle the farmers further left their opinions about the physical and psychosocial environment and which factors that are more or less justifying in their work. Factor analysis and Principal Component Analysis (PCA) was used in order to identify common factors for the farmers' driving forces in the beef production, and through stepwise regression factors of goals and attitudes to the actions of the farmer were identified. The study is not providing a complete understanding of farmers' motivation and behavior but is aiming to be a supplementary dimension of the working conditions and possible correlations to underlying goals and attitudes of the farmer.

Use of an electronic system for automatic calving detection in dairy cattle: technical and economical preliminary results

Marchesi, G.[1], Tangorra, F.M.[1], Pofcher, E.J.[2] and Beretta, E.[1], [1]Università degli Studi di Milano, via Celoria,10, 20133, Milan, Italy, [2]Universidad Nacional de La Plata, Casilla de Correo 296, 1900, La Plata, Argentina; gabriele.marchesi@unimi.it

Dystocia is a common problem in dairy cattle farms with rates of about 10% in cows and up to 30% in heifers. An inadequate monitoring of pregnancies may lead to a prolongation of the expulsive phase, and an increase in calving complications. Such aspects have a negative economic impact, reducing the animals reproductive performances. In the past several protocols were proposed to detect exactly when the calving process begins. Recently an electronic system that, once attached to the vulva some days before the expected calving, gets activated by expulsion of fetal membranes generating a SMS (Short Message Service) to the farmer was proposed on the market. Previous tests carried out during 2007-09 by other researchers using this technology on dairy cows in an experimental farm of Central Italy highlighted a strong reduction of dystocia (-90%) and calving interval. To validate these results, we carried out a test (2010) on 27 dairy cows equipped with the automatic calving detection system, and housed in free-stall barns of a commercial farm of Northern Italy. Preliminary results obtained confirm the usefulness of the system in dystocia control. To evaluate the potential economic benefits of the introduction of this technology a model was developed using literature and commercial data. The model considers the benefits deriving from the decrease of dystocia problems at calving and the reduction of calving intervals from 420 to 380 days. Considering only the effects related to the first aspect, in a case study involving 100 lactating dairy cows a 4 years time of investment return has been calculated. If also the second aspect is considered, the time of investment return decreases. In the most favorable scenarios it could be also less than 1 year. The work are still in progress and more data are needed to confirm our preliminary results.

Changes of metabolic conditions around calving affect oxidative stress and acute phase protein in pluriparous dairy cows

Sgorlon, S.[1], Guiatti, D.[1], Trevisi, E.[2], Ferrari, A.[2] and Bertoni, G.[2], [1]University of Udine, Dept. Agricultural and Environmental Science, via delle Scienze 208, 33100 Udine, Italy, [2]Università Cattolica del Sacro Cuore, Istituto di Zootecnica, Via Emilia Parmense 84, 29122 Piacenza, Italy; erminio.trevisi@unicatt.it

The study aimed to evaluate the relationships between the variation of metabolism during the *post partum* and the inflammatory conditions and oxidative stress in dairy cows. For this purpose, two commercial farms were selected and 30 pluriparous Friesian cows for each farm were identified on the basis of expected date of calving. Blood was sampled from cows every fortnight starting from 14 days before calving and until 70 days *post partum* (7 samples). At the same time of blood sampling, fat thickness of pelvic region was measured with an ultrasound scanner equipped with a 5 MHz transducer. Blood was analysed for acute phase response (ceruloplasmin, haptoglobin, albumin and zinc), metabolic (b-OHB, glucose, NEFA, cholesterol, urea, haemoglobin, total proteins and cortisol) and oxidoreductive status (paraoxonase, SOD, GPx and glutathione). Data were analysed with a factorial model (GLM) with fixed effects for time (7 levels), farm (2 levels) and their interaction. Correlation coefficients between the analysed parameters and the levels of significance were calculated with 2 tails Pearson test. Statistical analysis allowed to separate the markers influenced by time of sampling from that influenced by farm, and to select the best predictors of metabolic and oxidative stress around parturition, i.e. total proteins, urea, haptoglobin, cholesterol, BOHB, NEFA, paraoxonase and SOD. The present work also allowed to underline the relationships between stress *post partum* and oxidoreductive status. Animals presenting impaired body conditions at the beginning of lactation showed an highest probability to develop oxidative stress. In particular the correlation analysis showed significant relationship between fat thickness and albumin, ceruloplasmin, cholesterol, GPx and glutathione.

Genetic evaluations, genetic trends and inbreeding in Scandinavian trotter populations
Arnason, T., International Horse Breeding Consultant, Knubbovagen 34, S-744 94, Sweden; ihbcab@bahnhof.se

For two decades multiple trait animal model BLUP methods have been used in practice for genetic evaluation of Standardbred trotters in Sweden (SST) and Nordic (cold-blood) trotters (NT) in Norway and Sweden. The genetic evaluations are based on annually summarized racing results in the age range of 3-5 years in SST and 3-6 years in NT. The traits evaluated in a multivariate framework are transformed functions of ranks, earnings and racing time in addition to the all-none trait racing status. The pedigree files included (in 2010) 236,059 animals in SST and 102,923 in NT. The number of horses with records on racing status were 143,216 (85,848 with racing results) in SST and correspondingly 57,046 (25,669) in NT. For the last 10 years the annual genetic gain has been 6.2% and 3.5% of the standard deviation in the aggregate phenotype for SST and NT, respectively. The average inbreeding coefficient is currently about 8% in SST and 6% in NT. The recent rate of inbreeding per generation is 1.3% in SST and 1.0% in NT corresponding to N_e of 40 and 50, respectively. Quotas for maximum number of mares that can be mated to each stallion are applied as a vague attempt to restrict inbreeding in NT, but other more effective methods for balancing inbreeding and genetic progress have been investigated in both breeds, but have not been put into practice as yet. Genetic analysis of 2.5 million records on individual races in SST in years 1995-2010 have been initiated. Genetic evaluation of ranks in races based on Bayesian Thurstonian models will be compared with the use of random regression models for evaluation of traits with repeated measurements and with the present system of genetic evaluation in the SST breed.

Comparison of three different methodologies for the genetic evaluation of performance traits in the trotter horses
Gomez, M.D.[1], Molina, A.[1], Menendez-Buxadera, A.[1], Varona, L.[2] and Valera, M.[3], [1]Univ. Cordoba, C.U.Rabanales. Ed.Gregor Mendel., 14071 Cordoba, Spain, [2]Univ. Zaragoza., C, Miguel Servet 177, 50013 Zaragoza, Spain, [3]Univ. Seville, Ctra. Utrera Km1, 41013 Seville, Spain; pottokamdg@gmail.com

The Spanish Trotter Horse is a breed of increasing importance in Spain. Their breeding program includes the improvement of performance results on national and international races as main breeding goal. The aim of this analysis is the comparison between three methodologies tested in this population (BLUP animal model, Random Regression model -RRM- and Thurstonian model -Thurs-) in order to select this with better results and more advantages for the Trotter population. Therefore, the same model was performed with each methodology with the same database in order to compare them. Ranking trait was analyzed including: sex (2 levels), age (3) and race (3,920) as fixed effects in the model; and trainer-jockey, permanent environment and animal effects as random ones. Genetic parameters and breeding values were estimated. The RRM was applied throughout the trajectory of distances (from 2,000 to 2,250). Heritability values are low with all the tested methodologies, being 0.05 with BLUP and 0.09 with Thurs. The RRM shows that heritability decreases when race distance increases (from 0.15 to 0,11), being 0.12 the average value. Genetic correlations between race distances, estimated with RRM, range between 0.82 and 0.99, being the higher values between the adjacent distances. Differences in the genetic ranking by the breeding values are shown using the percentage of coincidence for the top and bottom 5% of animals in the genetic ranking, taking BLUP methodology as reference. The percentage of coincidence between BLUP and RRM was 89.04% and 88.85% for the top and bottom 5%, respectively. The values between BLUP and THURS was 86.35% and 89.42% for the top and bottom 5%, respectively. Finally between RRM and THURS was 82.69% and 83.65%, respectively.

The use of a Tobit-like-classification in genetic evaluation of German trotters
Bugislaus, A.-E.[1], Stamer, E.[2] and Reinsch, N.[3], [1]University of Rostock, Justus-von-Liebig-Weg 8, 18059 Rostock, Germany, [2]TiDa, Bosseer Str. 4c, 24259 Westensee, Germany, [3]FBN, Wilhelm-Stahl-Allee 2, 18196 Dummerstorf, Germany; antke-elsabe.bugislaus@uni-rostock.de

Evaluation of breeding values in German trotters is based on a multiple trait BLUP animal model utilizing individual results from all racing trotters. Only the best one third of participants in races receives earnings. This could lead to the assumption that only results from trotters with earnings are meaningful for genetic evaluation because these trotters show their real racing potential. The objective of this study was to apply a Tobit-like-classification of observations for genetic evaluation in German trotters. Data consisted of 111,822 performance observations from 8,095 trotters. In pedigree four generations were considered (25,862 animals). In many races the first five ranks obtained earnings. For this reason the classified trait placing status was applied in a problem-oriented manner and genetically analyzed by using a univariate animal model and REML. The first, second, third, fourth and fifth ranked trotters obtained a 5, 4, 3, 2 and 1, respectively, for the placing status; all other observations were lumped together into a common class with trait value of 0, thereby treating them as censored observations with the only information of having been slower than their competitors with higher scores. Significant fixed effects were sex, age, year-season, condition of race track, distance, driver and individual race. Estimated heritability for the trait placing status was 0.12. In comparison, variance components were estimated for the trait racing time per km including results from all starting trotters in races. Heritability for the trait racing time per km was with 0.11 a little bit lower. Rankings of breeding values for the traits placing status and racing time per km showed for raced stallions a low correlation (r=0.20). So, a Tobit-like-classification resulted in a different ranking of breeding values. Further analysis with a real Tobit-model are recommended.

Recommendations for implementation of optimal contribution selection of the Norwegian and the North-Swedish cold-blooded trotter
Olsen, H.F., Klemetsdal, G. and Meuwissen, T.H.E., University of Life Sciences, Dept. of animal and aquacultural sciences, P.O. Box 5003, 1432 Aas, Norway; hanne.fjerdingby@umb.no

The aim of this study was to apply optimal contribution selection (OCS), and from the results obtained, to recommend how OCS could be best used in future breeding of the Norwegian and the North-Swedish cold-blooded trotter. OCS was implemented using the software Gencont with overlapping generations, with assumptions for number of mated mares per sire, rate of inbreeding per generation and on the selection candidates. We concluded that OCS can be implemented in the Norwegian and the North-Swedish cold-blooded trotter, as a dynamic tool to both recruit young stallions as well as to assign annual breeding permit for earlier approved stallions. For number of mares per sire, a constraint in accordance with the maximum that a sire can mate naturally is recommended. Mare candidates can well be those that were mated the previous year, while the number of sire candidates need to be restricted from continuous recording of reliable information on death and gelding, as well as only considering as first-time candidates those between 3 and 10 years of age, with phenotype above average within a year class. The restriction on rate of inbreeding only had a minor effect on the selected animals.

Utilizing information from related populations for estimation of breeding values in a small warmblood riding horse population
Furre, S.[1], Heringstad, B.[1], Philipsson, J.[2], Viklund, Å.[2] and Vangen, O.[1], [1]Norwegian University of Life Sciences, Department of Animal and Aquacultural Sciences, P.O.Box 5003, N-1432 Ås, Norway, [2]Swedish University of Agricultural Sciences, Department of Animal Breeding and Genetics, P.O. Box 7070, S-750 07 Uppsala, Sweden; siri.furre@umb.no

There is an extensive exchange of genetic material between sport horse populations. However most published breeding values on horses today only include information on a national level. Recent studies have shown high genetic connectedness, and correlations between traits tested in different studbooks. Consequently joint genetic evaluation between populations is possible. The aim of this study was to investigate if it is beneficial to include information from other populations in the Norwegian Warmblood breeding value estimation using Swedish Warmblood as an example. Three different analyses were undertaken; records from young horse tests (YHT) in 1) Norway or 2) Sweden, or 3) both countries combined. Testing schemes in both countries have been almost identical during the test-period. Our objective was to examine how these three different approaches would affect ranking of stallions. A total of 18,116 horses with YHT records in Sweden (17,462) and Norway (654) were included in the study. The number of stallions with offspring test-results in both countries was 240. For 58 stallions approved in Norway, breeding values and rank correlations on an individual level were estimated. Heritabilities were 0.12 and 0.17 for conformation, 0.26 and 0.29 for walk, 0.34 and 0.37 for trot and 0.41 and 0.34 for canter from analyses 1 and 3 respectively. Rank correlations between BLUP-values for stallions with information from 1 and 2 were 0.13, 0.24, and 0.19 for walk, trot and canter. When information came from analyses 1 and 3, the rank correlation was 0.80, 0.78 and 0.67 for the three traits respectively. Further studies will include more traits and information from Finnish and Danish warmblood.

Improved understanding of breeding values of Swedish Warmblood horses
Viklund, Å., Fikse, W.F. and Philipsson, J., Swedish University of Agricultural Sciences, Animal Breeding and Genetics, P.O. Box 7023, 75007 Uppsala, Sweden; asa.viklund@slu.se

For estimated breeding values (EBVs) to be an effective tool in horse breeding, the EBVs have to be easily understood by the breeders. Breeding values are based on complex calculations and breeders' questions typically relate to the (perceived lack of) connection between actual performances and estimated breeding values. EBVs for Swedish Warmblood (SWB) horses are estimated with multivariate BLUP animal models based on information from two young horse performance tests and competition data. EBVs are published for the goal traits show jumping and dressage, and nine sub traits, i.e. individual gaits, jumping and conformation traits. In this study the EBVs were partitioned to highlight the contribution from three information sources: parents, own performance and progeny. The contribution from the parents was calculated as the average EBV of the parents. The contribution from own performance was expressed as regressed observations adjusted for all fixed effects in the model and parent average. As such, it represents the estimated Mendelian sampling deviation based on own performance. The contribution due to progeny information was calculated as the weighted average of the EBV of the progeny adjusted for the mates. The standard deviations (SD) of contribution from own performance were 0.15 (jumping traits) and 0.13 (dressage traits), where the EBVs were expressed with a genetic SD of 0.60 (jumping traits) and 0.51 (dressage traits). The SD of progeny contributions were 0.11 and 0.10, respectively, indicating that own performance was a more important information source than progeny information for the bulk of horses. The partitioning of EBVs will be helpful in explaining the EBV of individual horses, for example the change in breeding values from two subsequent genetic evaluations. Improved understanding may lead to greater acceptance for the EBVs among breeders. Altogether, a more effective use of EBVs in SWB breeding will result in increased genetic progress.

Comparison of models for genetic evaluation of Icelandic horses

Albertsdóttir, E.[1], Árnason, T.[1], Eriksson, S.[2], Sigurdsson, Á.[1] and Fikse, W.F.[2], [1]Agricultural University of Iceland, Faculty of Land and Animal Resources, IS-311 Borgarnes, Iceland, [2]Swedish University of Agricultural Science, Department of Animal Breeding and Genetics, P.O. Box 7023, SE-75007 Uppsala, Sweden; Freddy.Fikse@slu.se

The current genetic evaluation of Icelandic horses by multiple-trait animal model is based on riding ability and conformation traits assessed at breeding field tests. Inclusion of competition traits in the genetic evaluation would better reflect the breeding goal. Horses participating in field tests or competitions are not a random sample of the population. Therefore, an all-or-none test status trait, defined as attendance of horses born in Iceland at breeding field tests and/or in competition was created. Our aim was to study the effect of integrating either or both competition traits and test status trait into the genetic evaluation. Four sets of multi-trait models for genetic evaluations were compared using five methods: estimation bias, predictive ability, accuracy, correlations between breeding values, and ranking of sires. Breeding field test data included 19,954 records for 15 traits from horses assessed in 11 countries during 1994-2008. Competition data included 44,160 records for 4 traits from 7,687 horses competing in Iceland and Sweden during 1998-2008. In general, the differences in estimation bias and predictive ability between different genetic evaluation models were very small. Strong correlations were found between breeding values from different models for three overall traits: total conformation score, total riding ability score and total score. Accuracy increased for most traits when competition traits and/or test status were added to the model, and the ranking of sires was altered. We concluded that competition traits should be integrated into the genetic evaluation of Icelandic horses, and that further studies of genetic parameters for test status and its relationship with the other traits are needed.

Genetic analyses of movement traits in German Warmblood horses

Becker, A.-C.[1], Stock, K.F.[1,2] and Distl, O.[1], [1]University of Veterinary Medicine Hannover, Institute for Animal Breeding and Genetics, Buenteweg 17, 30559 Hannover, Germany, [2]Vereinigte Informationssysteme Tierhaltung w.V., Heideweg 1, 27283 Verden / Aller, Germany; kathrin-friederike.stock@tiho-hannover.de

Movement characteristics are important selection criteria for the Warmblood riding horse, and scores from gait evaluations during free movement and under rider are considered in current breeding programs. However, trait definition is rather general and does not reflect specific motion patterns of the individual horses. Therefore, in a pilot study new movement traits were defined and recorded during routine inspections of mares and foals of the Oldenburg Society in 2009 and 2010. In addition, scores from studbook inspections (SBI; 1,994 mares presented in 2009) and mare performance tests (MPT; 2,749 mares tested in 2000-2008) were considered for genetic analyses which were performed in linear animal models with REML. Genetic correlation analyses were based on univariately predicted breeding values for selected movement traits from foal and mare evaluations, SBI and MPT. Detailed judgments of movement were available for 3,203 foals and 2,822 mares and included binary records for indications of imbalance, irregular tail posture, and irregular motion pattern in hind legs. Foal data were found to be more suitable for variance component estimations than mare data, with heritability estimates of 0.03-0.06 on the original scale and 0.14-0.23 using threshold model transformation. Significant correlations were found between breeding values for indications of imbalance in foals and selected traits from SBI (correctness of gaits, impetus and elasticity, walk at hand) and MPT (walk, trot and canter during free movement and under rider, rideability). According to our results, foal evaluations may be included in future breeding programs as an early source of information to improve movement characteristics in the Warmblood horse.

Heritability and repeatability of young stallions jumping parameters

Lewczuk, D.[1] and Ducro, B.[2], [1]Institute of Genetics and Animal Breeding PAS Jastrzebiec, ul.Postepu 1, 05-552 Wolka Kosowska, Poland, [2]Wageningen Institute of Animal Sciences, P.O. Box 338, 6700 AH Wageningen, Netherlands; d.lewczuk@ighz.pl

In the years 2003-2008 a group of 438 stallions (1,266 jumps) were filmed by digital camera (operated 25 frames per second) during the free jumping tests on the Polish official performance tests. Jumping parameters performed on the doublebarre obstacle were measured using video image analysis program. Two parameters of the length of the jump were investigated – taking off and landing distances. The height of the jump was described by heights of elevation of the body measured in the croup, withers and head in the highest position of the bascule. The position of the head and lifting of legs were measured for every limb above the highest point of the obstacle. The time of jumps were divided into phases and measured. Analysis of the data was done using ASREML. The statistical model included random effect of the horse, fixed effects of the heights of the obstacle, data of the test as the group of evaluation and age of the horse (three and four years old).The repeatabilities of jumping parameters were calculated between 0.13-0.75 and the heritabilities reached the values from 0 to 0.41. Data for four parameters were not estimable. The highest values of heritability were achieved for heights of hind limbs lifting (0.24-0.41) and position of the head (0.31). Time of the jumping phases was heritable on the level about 0.25. The highest repeatabilities were estimated for landing distance, elevation of bascule points, position of the head in the highest frame of the bascule and time of landing.

Genetic analyses of the Franches-Montagnes horse breed with genome-wide SNP data

Hasler, H.[1], Flury, C.[1], Haase, B.[2], Burger, D.[3], Stricker, C.[4], Simianer, H.[5], Leeb, T.[2] and Rieder, S.[1], [1]Berne University of Applied Sciences, Swiss College of Agriculture, Laenggasse 85, 3052 Zollikofen, Switzerland, [2]University of Bern, Vetsuisse Faculty, Institute of Genetics, PO-Box 8466, 3001 Bern, Switzerland, [3]Swiss National Stud Farm, Les Longs Prés, PO-Box 191, 1580 Avenches, Switzerland, [4]agn Genetics GmbH, Boertjistrasse 8b, 7260 Davos, Switzerland, [5]Georg-August-University, Goettingen, Department of Animal Sciences, Animal Breeding and Genetics Group, Albrecht-Thaer-Weg 3, 37075 Goettingen, Germany; heidi.hasler@bfh.ch

The Franches-Montagnes (FM) is the only indigenous Swiss horse breed. The population is estimated to be about 21,000 animals with a total of around 2,500 foalings per year. The running breeding program includes the yearly estimation of BLUP breeding values for 43 different traits. In the present project 1,151 FM horse samples were genotyped with the Illumina Equine SNP50 BeadChip. The sample set of selected FM horses represents about one-third of the present breeding population. These horses were chosen based on the number of offspring and the distribution and accuracy of different breeding values. After data preparation and plausibility check 40,200 SNPs were kept for further analysis. Based on the mentioned SNP data linkage disequilibrium (LD) was determined. The resulting pair-wise r^2-values were used to estimate effective population size N_e. This was done for different sample sets to proof consistency of results. Additionally, marker-based relationship was compared with pedigree-based relationship. Coherence between estimations was high ($R^2 \sim 0.65$). The SNP data enabled the detection of pedigree inconsistencies and to deduce relationships between individuals. In addition we started to map traits from the running breeding program and to estimate allele effects for those. The latter were used to deduce genomic breeding values. The accuracies of these first genomic breeding values were found within a range comparable to those in dairy cattle in Switzerland.

Genome-wide association study for osteochondrosis in warmblood horses

Distl, O., Lampe, V. and Dierks, C., University of Veterinary Medicine Hannover, Institute for Animal Breeding and Genetics, Buenteweg 17p, 30559 Hannover, Germany; ottmar.distl@tiho-hannover.de

The objective of this work was a genome-wide scan for equine osteochondrosis (OC) using the Illumina Equine 50K Beadchip and to perform further large scale genotypings for verification of significant quantitative trait loci (QTL). OC-QTL have been identified in Hanoverians employing a whole genome linkage analysis and chromosome scans using very dense microsatellite sets. Fine-mapping was done on horse chromosomes (ECA) 2, 4, 5, 16, 18 and 21. We performed a genome-wide association study employing the Illumina Equine 50K Beadchip. From a large sample of Hanoverian warmblood horses, 244 most distantly related animals were chosen to avoid stratification of data by families. One hundred and fifty six animals were affected by osteochondrosis in fetlock and/or hock joints and 88 were free from any signs of OC in each limb joint. Phenotypic traits were OC in fetlock and/or hock joints, osteochondrosis dissecans in fetlock and/or hock joints. In addition, these OC conditions were also analysed separately for fetlock and hock joints. For genome-wide mapping we performed association analyses (MAF>0.01) controlling for cryptic structure of the genotypic data, non-genetic fixed effects and genomic relationships among animals. Associated loci explained more than 60% of the phenotypic variance and for all these loci candidate genes were identified. The most significant hits for OC in fetlock and/or hock joints were on ECA1, 2, 5, 8, 9, 10, 12, 14, 16, 18, 20, 23 and 26. QTL for OC in fetlock joints were on ECA2, 8, 11, 14, 16, 18 and 26. For OC in hock joints, ECA1, 5, 10, 14, 16, 20 and 23 contained significant QTL. These results are useful for validation studies in other horse breeds and fine mapping of candidate regions as well as development of genomic breeding values for horses.

Remaining questions about the inheritance of Roan coat color in horses

Langlois, B.[1], Valbonesi, A.[2] and Renieri, C.[2], [1]INRA, GABI-BIGE, CRJ - Domaine de Vilvert, 78 350 Jouy-en-Josas, France, [2]School of environmental Science, University of Camerino, via Circonvallazione 93, 62024 Matelica, Italy; bertrand.langlois@jouy.inra.fr

Since Hintz and Van Vleck the genetic determinism of horses Roan coat color has been attributed to a dominant allele that is supposed to be lethal in the homozygous state. This was later confirmed by Marklund *et al.* at the molecular level. However, breeders and Bowling contest the generality of these results. We propose here to analyse the segregation of this character in three French draft breeds the 'Ardennais' the 'Trait du Nord' and the 'Auxois'. Respectively 10,033, 3,322 and 2,798 crossings of non affected parents, affected by non affected, and affected by affected, produced 0.4, 39 and 79% of affected offsprings. Therefore the segregation data do not support the monofactorial hypothesis. Other alleles or genes should interfere with the first elucidated mechanism.

Genetic analysis of Romanian Lipizzaner from Sambata de Jos stud farm: reproductive isolation and age structure

Maftei, M.[1], Popa, R.[1], Popa, D.[1], Marginean, G.[1], Vidu, L.[1], Vlad, I.[1], Pogurschi, E.[1] and Girlea, M.[2], [1]University of Agricultural Sciences and Veterinary Medicine, Bucharest, Faculty of Animal Sciences, Horse breeding, Bld. Marasti, no. 59, sector 1, 011464, Bucharest, Romania, [2]Veterinary and Food Safety Department, Ilfov County, Food Safety, Bld. Ion Ionescu de la Brad, no. 8, sector 1, 013811, Romania; mariusmaftei@yahoo.com

This study is a part of an ample research concerning the genetic analysis (history) of Lipizzaner horses from Sambata de Jos stud farm. The genetic analysis studies are a part of Animal Genetic Resources Management because just start of them we elaborate the strategies for inbreeding management. This study has as purpose to present two important aspects of genetic analysis: reproductive isolation level and age structure. This parameters has a capital importance in animal breeding because there has a directly influence in animal population evolution. The reproductive isolation situation was quantified using the relation elaborated by S. Wright in 1921. The age structure situation is based on the age distribution histogram. The analysis showed that the Lipizzaner horse from Sambata de Jos stud farm is a reproductively isolated population and have its own evolutionary path. The import of biological material from other studs has not disrupted the development of this distinctive population. Age structure is not balanced with negative repercussions on generation interval.

Partial results regarding the genetic analysis of Romanian Pure Arabian horses from Mangalia stud farm: reproductive isolation and age structure

Maftei, M.[1], Popa, R.[1], Popa, D.[1], Marginean, G.[1], Vidu, L.[1], Vlad, I.[1], Pogurschi, E.[1] and Girlea, M.[2], [1]University of Agricultural Sciences and Veterinary Medicine, Bucharest, Faculty of Animal Sciences, Horse breeding, Bld. Marasti, no. 59, sector 1, 011464, Bucharest, Romania, [2]Sanitary Veterinary and Food Safety Department, Ilfov County, Food Safety, Bld. Ion Ionescu de la Brad, no. 8, sector 1, 013811, Romania; mariusmaftei@yahoo.com

This study is a part of an ample research (research contract no 154/2010 U.E.F.I.S.C.S.U.) concerning the genetic analysis of all horse breeds from Romanian national stud farms. The main reason is to elaborate an inbreeding management program for horse breeds starting from its statute. The genetic analysis studies are a part of Animal Genetic Resources Management because just start of them we elaborate the strategies for inbreeding management. One of the most important horse breed in Romania (from economically and genetically point of view) is the Pure Arabian, bred in Mangalia stud farm, in a special environmental conditions. This study has as purpose to present two important aspects of genetic analysis: reproductive isolation level and age structure. These parameters have a capital importance in animal breeding because they have a direct influence in animal population evolution. The reproductive isolation situation was quantified using the relation elaborated by S. Wright in 1921. The age structure situation is based on the age distribution histogram. The analysis showed that the Romanian Pure Arabian horse from Mangalia stud farm is still a population with its own evolutionary path despite of stallion imports from same breed but from other populations. Age structure is not balanced with negative repercussions on generation interval but that can be improve especially at the mares level.

Influence of foreign trotter populations in the Spanish trotters horses' breeding values

Gómez, M.D.[1], Cervantes, I.[2], Valera, M.[3] and Molina, A.[1], [1]Univ. of Cordoba, C.U. Rabanales. Ed. Gregor Mendel, 14071 Córdoba, Spain, [2]Univ. Complutense of Madrid, Avda. Puerta del Hierro s/n, 28040 Madrid, Spain, [3]Univ. of Seville, Ctra. Utrera km 1, 41013 Seville, Spain; pottokamdg@gmail.com

The Studbook of Spanish Trotter Horses (STH) was created in 1979 to register the trotter horses born in Spain. Nowadays, it has 17,859 horses registered till 2007. This Studbook remains open to include animals from other trotter populations all around the world, artificial insemination beeing the most frequent reproductive practice. Therefore, a 37.3% of the registered horses are imported from foreign trotter populations. USA and France are the most represented countries of origin in the pedigree (23.3% and 18.7% of the horses have more than 75% of influence of USA and French founders, respectively). The aim of this study was to analyze the influence of the genetic contribution of foreign populations on the breeding values for 4 performance traits analyzed in the STH: annual earnings (AE), percentage of first placing in a year (PFP), time per kilometer (TPK) and best annual racing time per hippodrome and type of start (BRTHS). The breeding values were estimated using the BLUP animal model used in the systematic genetic evaluation of this population following Gómez *et al.* Data included 285,538 racing records from 5086 horses. A response surface regression analysis was performed for each trait using the breeding values for those traits as dependent variables; the genetic contribution of USA, France, Italy and Spanish founders were used as independent variables, with linear and quadratic adjustment. Also the interactions between effects were included. For TPK, France and USA percentages were statistically significant, whereas for BRTHS, only the genetic contribution of USA founders was significative, being no significative for the other traits. The interaction effect between France and USA was statistically significant in AE, TPK and BRTHS. Finally, the best combination to maximize the breeding value was determined.

The genoype for coat colour genes as a conservation criterion to preserve the genetic diversity in Menorca horses

Solé, M.[1], Azor, P.J.[1], Valera, M.[2] and Gómez, M.D.[1], [1]University of Córdoba, Department of Genetics, C.U.Rabanales. Ed. Mendel, pl. baja., Córdoba, 14071, Spain, [2]University of Seville, Department of Agroforestry Science, Crta.Utrera, Km 1, Seville, 41013, Spain; ge2sobem@uco.es

The Menorca Horse is an endangered Spanish horse breed, mainly located in the Balearic Island. All the animals registered in the Studbook are black coated. The aim of this study is the quantification of genetic diversity in the population and the calculation of loss of genetic diversity when the heterozygous horses for MC1R gene (Ee), responsible of the chestnut coat color, are excluded. The study of genetic variability has been assessed using the genotypes of 1,215 registered horses for 16 microsatellite markers, analyzed in the Central Laboratory of Veterinary (Algete), belonging to the Spanish Ministry of Environment and Rural and Marine Affairs. For the estimation of the loss of genetic diversity excluding the heterozygous horses, 213 individuals were genotyped for the gen MC1r in the Laboratory of Genetic Veterinary Diagnostic of MERAGEM Research Group (Córdoba). The level of heterozygosis, the average number of alleles per locus and the loss of genetic diversity for the homozygous and heterozygous horses (EE: 155 individuals and Ee: 58 individuals) was estimated for these 213 individuals. The heterozygosis, taking only the homozygous horses (EE), is reduced from 0.71 to 0.70. Average number of alleles per locus and the effective allele size has been also reduced from 7.31 to 7.19 and from 3.73 to 3.65, respectively. Therefore, when we exclude the heterozygous horses (Ee), a genetic diversity loss occurs in this population. These values show high levels of genetic variability despite of being an endangered breed, but a loss of variability occurs when only the homozygous horses (EE) are included. So, coat color could be a good criterion to maintenance the genetic diversity within the conservation program of this breed.

Integration of foreign breeding values for stallions into the Belgian genetic evaluation for jumping horses

Vandenplas, J.[1], Janssens, S.[2] and Gengler, N.[1,3], [1]University of Liège - Gembloux Agro Bio-Tech, Animal Science Unit, Passage des déportés, 2, 5030 Gembloux, Belgium, [2]KULeuven, Department Biosystems, Livestock Genetics, Kasteelpark Arenberg 30 - bus 2456, 3001 Heverlee, Belgium, [3]National Fund for Scientific Research, rue d'Egmont, 5, 1000 Bruxelles, Belgium; jvandenplas@ulg.ac.be

The aim of this study was to test the integration of foreign estimated breeding values (EBV) for stallions into the Belgian genetic evaluation for jumping horses. Belgian breeders import horses from neighbouring countries for which foreign information is needed as prior to estimate a more accurate EBV. The Belgian model is a bivariate repeatability BLUP animal model. For the year 2009, pedigree and data files contained 101,382 horses and 712,212 performances. 98 French and 67 Dutch stallions were selected and their foreign EBVs were converted into Belgian national trait. Associated reliabilities were also estimated. A Bayesian approach was applied to integrate this prior information into the Belgian evaluation. It led to a slight modification of the average EBV and of the standard deviation for the whole population. It also led to a new Belgian ranking of the foreign stallions more similar to foreign rankings. However, the adequacy of the Belgian model was not damaged. With regards to prediction ability, Bayesian evaluations using conversion equation estimated by Weighted Least Squares procedure predicted the best traditional EBVs of the year 2009 for the French stallions. For the Dutch ones, it were the evaluations associated to the conversion equation based on Wilmink *et al.* For both countries, Bayesian evaluations using conversion equation based on Goddard improved the most the stability of EBVs. Finally, integration of French and Dutch information improved reliabilities of the Bayesian EBVs of at least 5% and 2% for French and Dutch stallions, respectively. These results confirm the interest to integrate foreign information into the Belgian evaluation for jumping horses.

Crossbreed genetic performance study in the eventing competition using the Spanish Sport Horse as example

Cervantes, I.[1], Bartolomé, E.[2], Valera, M.[2], Gutiérrez, J.P.[1] and Molina, A.[3], [1]Dept. of Animal Production. University Complutense of Madrid, Av Puerta de Hierro, 28040, Spain, [2]Dept. of Agro-forestry Sciences. University of Sevilla, Ctra Utrera 1, 41013, Spain, [3]Dept. of Genetics.University of Córdoba, CU Rabanales, 14071, Spain; icervantes@vet.ucm.es

The Spanish Sport Horse (SSH) is a composite breed. Since the Foundational Registry was closed, an animal is considered SSH either when both parents were registered as SSH or only when one of them was SSH, allowing the other parent to be from some specific breeds. The eventing competition is a very heterogeneous equestrian discipline that combines dressage, jumping, cross exercises and a conformation evaluation. The aim of this study was to ascertain the influence of the parental breeds on the performance by linking their paternal genetic contribution with the individual breeding values of the SSH animals for five performance traits. Breeding values were obtained using BLUP animal model that included age, sex, judge, competition, training and stress as fixed effects and rider-horse interaction, animal and error as random factors. Data included 2,548 records, 52% of them belonging to 210 SSH animals. A response surface regression analysis was performed for each trait using the breeding values as dependent variables and the genetic contribution of Spanish Purebreed (SPB), Arab Horse (A), Thoroughbred (TB), Selle Français (SF), German breeds (G) and Netherlands breeds (N) as independent variables, using a linear and quadratic adjustment. The interactions between effects were also included. For dressage breeding values SPB, SF, G and N were significantly different. For jumping TB, SF, G and N were significant. Regarding cross, SPB and SF were significant. For the conformation trait, A, SF, G and N were significant and for cross aptitude only SPB was different. These results are discussed on a joint frame of parental breed and heterozygosis influences. Further analyses are needed to fit the best crossbreed combinations performance in eventing competitions.

Effect of year and season of birth, sex, sire and breeder on ossification of the distal epiphysial cartilage of the radial bone in Thoroughbred horses
Łuszczyński, J. and Pieszka, M., Agricultural University, Horse Breeding Department, Al. Mickiewicza 24/28, 30-059 Kraków, Poland; mpieszka@ar.krakow.pl

Alarming frequency of different limb injuries is observed among young Thoroughbred horses as a result of their early usability. It seems significant to evaluate the optimal time of training initiation in the way the intensive exercise will not negatively influence the organism condition. The aim of this study was to evaluate the effect of chosen factors on ossification of the distal epiphysial cartilage of the radial bone in Thoroughbred horses. The study was carried out on 452 two-year-old Thoroughbred horses that were under race training at Warsaw Race Track between 2000 and 2007. X-ray tests were conducted and interpreted by a veterinarian using X-ray apparatus Orange 8016 HF (54 kV, 1 mAs) every month starting at 18 month of life. In the statistical model the year of birth, sex, birth season, sire and breeder were treated as constant effects. The significant differences were estimated using Tukey's test. Significant influences of year of foal birth, seasons of the year, sire and breeder on ossification of the distal epiphysial cartilage of the radial bone were stated in the study. Horses born in 2000 and in 2005 were characterised by the earliest time of ossification of the epiphysial cartilage (781.8 and 784.2 days, respectively). The average ossification time of horses born in spring time was 766.0 days which was significantly shorter than the time for horses born in winter time (810.3 days). The earliest stage of ossification of the epiphysial cartilage of the radial bone was noted for foals by sire Professional (770.0 days). Widzów Stud was shown as the breeding center of the Thoroughbred horses producing foals reaching their somatic maturity in the shortest time (770.4 days). The results could have a practical application for breeders and trainers in the planning of training, properly chosen for the stage of maturity of young Thoroughbred horses.

The effects of season on plasma insulin-like growth factor (IGH-F) concentration in young horses
Łuszczyński, J. and Pieszka, M., Agricultural University, Horse Breeding Department, Al. Mickiewicza 24/28, 30-059 Kraków, Poland; mpieszka@ar.krakow.pl

Insulin-like growth factor could be a useful marker in the horse for diagnostic, selection or forensic purposes, provided its physiological regulation is well understood. The objective of this study was to investigate the effect of season on blood plasma IGF-I in normal, healthy, young horses. The experiment used 81 young horses comprising of four breeds: Thoroughbred (10 colts and 9 fillies), Arabian (10 colts and 10 fillies); Anglo-Arabians (10 colts and 11 fillies) and Hucul (10 colts and 11 fillies). All studied horses were fed using meadow hay, oat seeds, red carrot and pasture grass, according to their age, sex and year season. Blood samples from all studied horses aged between 19-27 months were collected in winter (December) and summer (June). Venous blood was sampled from jugular vein to heparinised polystyrene test-tubes always between 10.00 and 12.00 a.m. Plasma samples were frozen and then only thawed immediately prior to radioimmunoassay (IRMA IGF-I A15729, Immunotech a Beckman Coulter Company, France). The concentration of plasma IGF-I factor was compared between winter and summer season using one-factor variance Anova and Tukey's test. After the analysis the significant influence of year season on blood plasma IGF-I concentration was shown – in most studied breeds (Arabian, Anglo-Arabian and Hucul) the IGF-I concentration was higher in summer season comparing to winter season but only for Huculs the differences were significant (summer: 327,78 ng/ml and winter: 206,25 ng/ml; $P \leq 0,05$). Hucul are a primitive breed of horses and it seems that because of difficult living conditions they evolved specific traits, which allow them to react fast to environmental changes. Thoroughbred horses were characterized by opposite tendency – concentration of IGF-I in winter was higher than in summer for 12,99 ng/ml but the differences were not significant. This work was supported by a grant N N311 315935.

Status quo of the personality trait evaluation in German horse breeding
Pasing, S.[1], Christmann, L.[2], Gauly, M.[1] and König V. Borstel, U.[1], [1]University of Göttingen, Albrecht-T.W. 3, 37075 Göttingen, Germany, [2]Hannoveraner Verband, LindhooperStr. 92, 27283 Verden, Germany; koenigvb@gwdg.de

The aim of the present work was to analyse the status quo of the personality trait evaluation in German horse breeding and to identify possibilities for improvement. All performance test judges and test riders officially registered with the German Equestrian Federation were interviewed (response rate: 24%) regarding the criteria they use to evaluate the traits temperament, character, willingness to work, constitution and rideability. All breeding experts agreed that personality traits are important (27%) or very important (73%). The majority (93%) also agreed that the evaluation of personality traits during performance tests should be continued, but at the same time 82% see room for improvement. Criticism included that scoring is often based on experience while concrete criteria are lacking. Existing traits were described, among others, by the following attributes: Character: Preparedness to work, behaviour during handling and in the barn; Temperament: Attention, behaviour towards novel objects, nervousness; Willingness to work: Braveness, jumping manner, nervousness, preparedness to work. Few (14%) of the experts were familiar with the official guidelines for personality trait evaluation, and when presented with the official guidelines' description of score 7 (4) out of 10 for rideability, only 47% (46%) assigned the correct score. The others exceeded the correct score by one score (score 8 instead of 7) and exceeded (39%) or went below (16%) score 4 by up to three scores. Overall, the existing guidelines were applied insufficiently, resulting in considerable differences of trait definitions by individual experts within traits as well as overlap among traits. Including concrete behaviour traits into new evaluation guidelines and calibrating judges' use of the guidelines could contribute to a more objective personality trait evaluation.

Pilot study on the OC(D) disease in Polish sport horse breeding population
Lewczuk, D.[1], Bereznowski, A.[2], Hecold, M.[2] and Kłos, Z.[2], [1]Institute of Genetics and Animal Breeding PAS Jastrzebiec, ul.Postepu 1, 05-552 Wolka Kossowska, Poland, [2]Warsaw University, Department of Veterinary Medicine; Clinic for Horses, ul.Nowoursynowska 100, 02-797, Poland; d.lewczuk@ighz.pl

In the years 2009-2010 a group of 114 mares and 87 stallions were examined twice for OCD- in the beginning and in the end of the training on young horses performances. Ten digital X-rays for fetlock, stifle and hock were made. Images were judged in the scale 0-3 where 0 was considered as no changes and 1,2,3 were considered as signs of OC(D) of different degree. The results were given in the binary scale as well (0 – no OCD and 1 – OCD with the highest degree meaning osteochondrosis dissecans). The data was transformed according to the threshold scale theory, then the analysis of variance was performed using SAS program procedure GLM. The statistical model included random effect of the horse, fixed effects of the training, breed and sex, as well as regressions on the age (in days), basic measurements and conformation. No statistical changes were found between sex groups, breeds and investigations (before and after training). The frequencies of OCD free mares reached 74% before training and 73% after training. For stallions the frequencies were adequately 71% and 69%. Some differences between breeds and sex group were observed for single images and wider scale used. The conformation and age regressions were statistically significant for 0-1 scale, but no clear connections were found between other regressions and OCD observations using 0-3 scale.

Development of a genetic evaluation for conformation traits for the Austrian Noriker draught horse

Fuerst, C.[1] and Fuerst-Waltl, B.[2], [1]ZuchtData EDV-Dienstleistungen GmbH, Dresdner Strasse 89/19, A-1200 Vienna, Austria, [2]University of Natural Resources and Applied Life Sciences Vienna, Department of Sustainable Agricultural Systems, Gregor Mendel Strasse 33, A-1180 Vienna, Austria; birgit.fuerst-waltl@boku.ac.at

The Austrian Noriker draught horse is presumably the oldest autochthonous draught horse breed in Central Europe and is considered as an endangered breed. Conformation plays an important role in the breeding program of the Noriker population, but was exclusively based on phenotypic information so far. The goal of this study was to estimate genetic parameters and to develop and implement a genetic evaluation for conformation traits. Scores of 11 conformation traits (scores from 1 to 10) and 5 measures are routinely collected by breeding commissions mainly between the age of 2.5 yrs (stallions) and 3.5 yrs (mares) when entering the studbook. Applying multivariate linear animal models heritabilities and genetic correlations for the 16 traits were estimated. Effects being taken into account were the fixed effects sex, age and classification place*day (or breeding area*half-year) as well as the random permanent environmental effect and the random genetic effect of the stallion/mare. The dataset consisted of 260 and 10,827 records of stallions and mares, respectively. Heritabilities ranged from 0.15 (hind legs) to 0.47 (type) for the 11 conformation scores and from 0.33 (caliber index) to 0.70 (height at withers) for the measures. Breeding values were estimated in three separate runs (11 conformation scores, 4 measures, caliber index) using the same models. EBVs for active stallions are published with a mean of 100 and a genetic standard deviation of 20 using a rolling base. Genetic trends were positive for all traits except for caliber index, where a rather stable trend was observed. Routine genetic evaluation was introduced in February 2011 and will be carried out once a year.

Genetic correlation between jumping and dressage performances of Belgian Warmblood Horses

Brebels, M., Buys, N. and Janssens, S., K.U.Leuven, Department of Biosystems, Kasteelpark Arenberg 30, 3001 Leuven (Heverlee), Belgium; machteld.brebels@biw.kuleuven.be

The Belgian Warmblood studbook (BWP) approves stallions for either jumping and dressage ability. In order to assess the need of specialized lines within the population of Belgian warmblood horses, an analysis of the genetic correlation between jumping and dressage performances seemed appropriate. Data were based on competition data provided by the Belgian Equestrian Federation (KBRSF-FRBSE) which consisted of 350,566 jumping performances and 117,810 dressage performances for the period 1991-2009. A BLUP animal model was applied using a pedigree file of 272,857 individuals. The repeatability model for dressage included the fixed effects sex, age, level and participation and the random effects event, rider and animal. Dressage performances were transformed according to the level. Hence, the dressage performance was expressed as 'percentage' and three transformations; 'low level', 'medium level' and 'high level'. The jumping model was based on the current model of the genetic jumping index (GSI) with transformed ranking as the trait. Variance and covariance components for jumping and dressage performances were estimated with linear models using REML-techniques. For dressage, the heritability ranges from 0.223 ± 0.004 to 0.544 ± 0.004. The effect of the rider explains an important part of the variance. The genetic correlation between jumping and dressage is low negative and varies with the applied model and trait; from -0.154 ± 0.083 to -0.044 ± 0.077. The transformation of the original percentage amplifies the negative correlation. This underlines the importance of the choice of the transformation (i.e. the weight given to 'high level' performance).

Intra-line and inter-line genetic diversity in sire lines of the Old Kladruber horse based on pedigree information

Hofmanova, B.[1], Vostry, L.[2] and Majzlik, I.[1], [1]Czech University of Life Sciences Prague, Animal Science and Ethology, Kamycka 129, 165 21 Prague 6, Czech Republic, [2]Czech University of Life Sciences Prague, Genetics and Breeding, Kamycka 129, 165 21 Prague 6, Czech Republic; hofmanova@af.czu.cz

The Old Kladruber horse is the original Czech horse breed included among the genetic resources of the Czech Republic, currently kept in two color varieties – grey and black. Pedigree records are available from 18th century. The population is closed from 2002, so there is a concern about the loss of genetic variation. According to effective population size (Ne=204) the Old Kladruber horse belongs to the group of endangered breeds. The population is divided into 9 sire lines. The genetic diversity within and amongst sire lines was evaluated using pedigree information. The analysis included a total of 324 individuals with fully informative pedigrees of the five generations of ancestors. The procedure INBREED of SAS was employed for inbreeding coefficient (F_X) and coefficient of relationship (R_{XY}) calculation. The average inbreeding coefficient of 0.076 ± 0.038 was estimated. Differences between sire lines were found. Lower mean values of coefficients of relationship were estimated between sire lines of different color varieties (grey x black) compared to values estimated within each color variety. Genetic diversity was assessed by cluster analysis – procedure VARCLUS of SAS. These results are useful for the development of breeding strategies, as well as for the preservation of genetic diversity at the population. This work was supported by the project MSM 6046070901.

Phenotypic and genetic correlations to asses morphological patterns for 'Menorca dressage' in Menorca horses

Solé, M.[1], Gómez, M.D.[1,2], Valera, M.[3] and Molina, A.[1], [1]University of Córdoba, Department of Genetics, C.U.Rabanales. Ed. Mendel, pl. baja., Córdoba, 14071, Spain, [2]A.C.P.Caballos de Raza Menorquina, C, Bijuters 17, Ciutadella de Menorca, 07760, Spain, [3]University of Seville, Department of Agro-forestry Science, Crta.Utrera, Km 1, Seville, 41013, Spain; ge2sobem@uco.es

The Menorca Horse is an endangered breed mainly located in the Balearic Island. Conformation traits and functional performance in Classic and Menorca Dressage (a special type of Dressage from Menorca) are the main selection objectives included in their breeding program. The aim of this study is to estimate the phenotypic (P) and genetic (G) correlations between 10 conformation traits (objective body measurements) and two functional traits (dressage punctuation: final reprise score and menorca movements), in order to determine the conformation traits with higher influence on performance results for Menorca Dressage. A total of 425 dressage results from 93 males were included in the analysis. All these animals were analyzed for the conformation traits. P and G correlations were estimated with a REML methodology (animal model). Rider (46 classes), judge (10 classes) and reprise level (4 classes) were included as fixed effects. The age was included as a covariable effect and the animal and the permanent environmental effect as random. P and G correlations ranged between 0.02 and 0.74, being the highest for hock angle with final reprise score (P=0.49; G=0.70), cranial view of the posterior aplomb with menorca movements (P=0.43; G=0.74) and shoulder angle with menorca movements (P=0.33; G=-0.71). In the other hand, the lowest correlations were for group angle with final reprise score (P=-0.02; G=-0.03) and withers height with final reprise score (P=0.02; G=0.05). So, this values show high correlations between most of the posterior traits with menorca movements, which determine a major easiness for this type of exercises.

Genome-wide association study of insect bite hypersensitivity in Dutch Shetland pony mares

Schurink, A.[1], Ducro, B.J.[1], Bastiaansen, J.W.M.[1], Frankena, K.[2] and Van Arendonk, J.A.M.[1], [1]Wageningen University, Animal Breeding and Genomics Centre, P.O. Box 338, 6700 AH Wageningen, Netherlands, [2]Wageningen University, Quantitative Veterinary Epidemiology Group, P.O. Box 338, 6700 AH Wageningen, Netherlands; Anouk3.Schurink@wur.nl

Insect bite hypersensitivity (IBH) is a common allergic disease in horses found throughout the world and caused by bites of Culicoides spp. The most prominent clinical symptom is severe itch, which results in self-mutilation. Welfare of affected horses is therefore seriously impaired. IBH is a multifactorial disease. A genetic background has been confirmed in various horse breeds using phenotypes and pedigree information. However, research on genomic level is limited and genes associated with IBH have not yet been reported. The aim of our study was therefore to detect DNA regions associated with IBH in Dutch Shetland pony mares. Most mares were acquired through the routine inspection system of the studbook. Mares were selected according to a matched case-control design and were matched on residence, withers height category, coat color, age and genealogy. Phenotypes were confirmed by a veterinarian and owners were asked about the history and management of IBH. After quality control, phenotypes and genotypes (SNP) of 188 mares were analyzed using genome-wide association. Logistic regression was used and data were corrected for stratification using the EIGENSTRAT method implemented in the R package GenABEL. Analyses indicated that SNP on eight chromosomes passed the threshold (p-value = 0.0005). Three chromosomes seemed most promising. Applying knowledge about genomic information on IBH will result in a more efficient breeding program to reduce IBH incidence compared to traditional breeding programs based on progeny testing.

Genetic of melanoma in Old Kladruber horse

Vostry, L.[1], Hofmanova, B.[1], Vostra Vydrova, H.[2], Majzlik, I.[1] and Pribyl, J.[3], [1]Czech University of Life Sciences Prague, Faculty of Agrobiology, Food and Natural Resources, Kamycka 129, 165 21 Prague, Czech Republic, [2]Czech University of Life Sciences Prague, Faculty of Economics and Management, Kamycka 129, 165 21 Prague, Czech Republic, [3]Institute of Animal Science, Pratelstvi 815, 10401 Prague - Uhrineves, Czech Republic; vostry@af.czu.cz

The aim of this study was to asses the prevalence of melanoma to investigate a possible genetic variation of this trait in the Old Kladruber horse. An overall of 702 Gray variety of Old Kladruber horse, 238 stallions and 326 mares were analysed. Melanoma status was recorded in for different degree. Two different analyses were conducted: a linear animal model (LM) with melanoma classified into five categories, threshold animal model (TM) with melanoma classified into five categories. All models included the fixed effects of year of evaluation, age, line, sex, greying level, vitiligo A and vitiligo F status and random direct genetic effect and the effect of animal's permanent environment. Heritability for occurrence of melanoma was been estimated for LM – 0.09 and for TM – 0.24. The coefficient of repeatability was been estimated for LM – 0.68 and for TM – 0.67. The values of Pearson's correlation coefficient and Spearmen's rank correlation coefficient among breeding values estimated by models LM and TM were 0.86 and 0.84 for data with pedigree information and between 0.78 and 0.78 for subset of animals with measurements. Results suggest that additive genetic variation of melanoma occurrence seems large enough in Old Kladruber horse to be exploited in specific breeding program. Supported by the project No. MSM 6046070901, and by project no. MZE 0002701404.

Genetic parameters of insect bite hypersensitivity in Belgian Warmblood Horses

Peeters, L.M., Brebels, M., Buys, N. and Janssens, S., KULeuven, Department of Biosystems, Kasteelpark Arenberg 30, BE-3001 Heverlee, Belgium; lies.peeters@biw.kuleuven.be

Insect bite hypersensitivity (IBH) is a seasonal recurrent allergic reaction of horses to the bites of certain Culicoides spp. The etiology of IBH is multifactorial in origin and involves environmental and genetic factors. The aim of this study was to estimate the effect of environmental factors and the heritability of IBH in the Belgian Warmblood Horse population in Belgium. A total of 1205 IBH scores with pedigree information could be collected using a questionnaire at 30 horse competitions during the summer of 2009. IBH prevalence was 9.4% (n=113) and the association with the following factors was tested: 'coat color' (P=0.238), 'sex' (P=0.747), 'soil humidity' (P=0.194), 'management foal' (P=0.852), 'housing system' (P=0.614), 'management horse' during summer (P=0.107) or winter (P=0.265),'frecuency deworming' (P=0.875), 'body condition' (P=0.0551), 'age' (P=0.0002) and 'vegetation' (P=0.0003). The prevalence of IBH seems to be higher when the vegetation of the pasture is woody (15.15%) compared to when the vegetation is forested (5.79%), open land (8.06%) or when the horse is always stabled (8.06%). If the body condition of the horse is 'normal' (8.84%) the risk of developing IBH seems to be lower compared to when the horse is 'skinny' (20%) or 'overweight' (15.38%). Horses younger than 1 year did not show clinical signs of IBH and the prevalence of IBH is much lower for younger horses. In a first preliminary study, using a linear animal model with these three fixed factors, a heritability of 0.287±0.061 was calculated. However, several other models (ex. threshold models) will be compared.

Mastitis in meat sheep

Green, L.E.[1], Huntley, S.[1], Cooper, S.[1], Monaghan, E.[1], Smith, E.M.[1], Bradley, A.J.[2] and Purdy, K.[1], [1]University of Warwick, School of Life Sciences, Coventry, CV4 7AL, United Kingdom, [2]University of Nottingham, School of Veterinary Sciences and Medicine, Sutton Bonington, LE12, United Kingdom; laura.green@warwick.ac.uk

Estimates of the occurrence of clinical mastitis in meat sheep during lactation range from 0-5%; the consequences of clinical mastitis are severe; affected sheep usually lose the diseased udder half and a proportion of sheep die. At weaning many more sheep have palpable lesions in their udders and on many farms 20- 30% ewes are culled with this chronic mastitis. Sheep farmers consider mastitis an important disease issue and one that is frustratingly unmanageable. The occurrence of clinical mastitis in meat sheep is considerably lower than the current incidence of mastitis in high yielding dairy cow herds (50-70 cases/100 cows/year) but more similar to that seen in dairy cows in low yielding herds with contagious mastitis, where low grade infections are common but clinical mastitis more rare. We have been testing the hypothesis that many sheep have an intramammary infection with few displaying detectable clinical signs. Our hypothesis is that with each lactation the udder is at risk of infection and with repeated lactations infections accrue. If this hypothesis is correct then we can start to develop a management programme for mastitis in meat sheep. Currently we have evidence that many sheep do have subclinical infections based on somatic cell counts (SCC) and bacterial culture. Somatic cell counts are higher in older ewes, particularly those in low body condition and ewes with poor udder conformation but there is no association with teat lesions. Lambs reared by ewes with high SCC grow more slowly. Lamb growth rate is also lower when ewes are older, in low body condition or have teat lesions.

Mastitis in sows - current knowledge and opinions

Preissler, R.[1,2] and Kemper, N.[2], [1]Institute of Animal Breeding and Husbandry, Christian-Albrechts-University Kiel, Olshausenstraße 40, 24098 Kiel, Germany, [2]Institute of Agricultural and Nutritional Science, Martin-Luther-University Halle-Wittenberg, Theodor-Lieser-Straße 11, 06120 Halle (Saale), Germany; rpreissler@tierzucht.uni-kiel.de

Mastitis in sows is most frequently observed *post partum* and affects both sows' and piglets' health and welfare. The insufficient intake of colostrum by the piglets can cause apathy, secondary infections, diarrhea or even death. Puerperal mastitis in sows is of complex nature, and known world-wide under several names, with Mastitis-Metritis-Agalactia- Syndrome (MMA), *Post partum*-Dysgalactia- Syndrome (PDS) or Coliform Mastitis as the important ones. Main clinical signs are mastitis, dysgalactia and fever above 39.5 °C. However, physiological hyperthermia is often observed in postparturient sows, specifically gilts, leading to misinterpretations. Additional clinical examinations of mammary glands, and in particular of behavioral changes in the sows and the piglets, allow more precise diagnoses. As a multifactorial disease, several influences contribute to the clinical picture and can be attributed to the causing pathogens, the environment or the host. Many different pathogens, especially coliforms, have been isolated from the milk of affected sows and environmental factors affect the clinical course of this disease. The host, including the genetic variation and individual factors as well as parity number or birth condition, is an important factor, too. A detailed analysis of phenotypic and genetic variation of puerperal mastitis with a holistic approach was carried out in the recent research project 'geMMA' at our Institute. Clinical examinations and detailed bacteriological analyses of milk samples were carried out in order to characterize phenotypic variation. Genetic variation was analyzed through high-throughput genotyping to identify possible candidate genes, and results can be used for subsequent biological analyses of possible disease mechanisms.

Incidence of mastitis and its effect on the productive performances in Rasa Aragonesa ewes

Ruz, J.M.[1], Marco, J.[2], Folch, J.[3] and Fantova, E.[1], [1]UPRA-Grupo Pastores, Mercazaragoza, 50014 Zaragoza, Spain, [2]Analitica veterinaria, Aritzbidea 18 bajo, 48100 Mungia, Vizcaia, Spain, [3]Cita de Aragon, Unidad de Tecnología en Producción Animal, Avenida Montañana 930, 50050 Zaragoza, Spain; jfolch@aragon.es

Rasa Aragonesa is an autochthonous breed of sheep oriented exclusively to meat production, reared in semi-extensive conditions in a 3 lambing/2 years reproductive system. In a first work, the incidence of mastitis has been checked in 106 ewes from 11 flocks. The presence of subclinical mastitis (positives to California Mastitis Test – CMT), lost udder and clinical mastitis (diagnosed by palpation) has been detected in 40.0%; 8.6% and 7.5% ewes, respectively. Bacteriological analysis detected the presence of intra-mammmary infection in 44% of the checked ewes with high differences between flocks (from 10% to 70%). In a second work, the number of lambs born and their growth rate have been recorded in 62 ewes showing mastitis, (either positive at CMT or presenting pathological damages in the udder at clinical exploration), comparing with 130 apparently healthy ewes from the same flocks. Affected ewes presented a higher perinatal mortality (14.1 vs. 7.06%) and lower growth rate between birth and weaning at 45 days (-17.1%)

Relevance of somatic cell count and bacteria for udder health in sheep
Kern, G.[1], Traulsen, I.[1], Kemper, N.[2] and Krieter, J.[1], [1]Institute of Animal Breeding and Husbandry, Christian-Albrechts-University, Oshausenstraße 40, D-24098 Kiel, Germany, [2]Institute of Agriculture and Nutritional Science, University Halle-Wittenberg, Theodor-Lieser-Straße 11, D-06120 Halle (Saale), Germany; gkern@tierzucht.uni-kiel.de

Mastitis is one of the most important diseases in sheep husbandry. The objective of this study was to determine the somatic cell count (SCS) and to detect the different bacteria in milk to estimate their impact on udder health in sheep. A dataset with milk-samples of 592 udder halves of sheep from 20 farms of Northern Germany was analysed. The dataset included information about husbandry (dairy, meat, extensive), age of the ewes, breed, date of lambing, number of lambing, udder traits, milk score, somatic cell count, percentages of protein, fat, lactose and isolated bacteria. Bacteria were detected in 68.2% of the right udder halves (n=296) and in 69.3% of the left udder halves (n=296). In 51.4% of the ewes bacteria were isolated in both udder halves. Bacteria of the family Stapylococcaceae and Streptococcaceae were identified most frequently. A mixed model was applied to determine the systematic effects on SCS of each udder half. Husbandry, milk-score and week showed a significant effect on SCS ($P<0.05$). The SCS was significant lower ($P<0.05$) in extensive husbandry (left: 3.06, right: 3.45) than in meat (l: 4.78, r: 4.94) and dairy systems (l: 4.34, r: 4.35). The SCS was significant higher assuming low milk-score (4.34) than a high milk-score (3.78). Detected bacteria showed no significance on SCS ($P<0.1$). Bacteriaincreased the SCS from 3.82 (negative) to 4.30 (positive) on the left side and from 3.97 (neg.) to 4.52 (pos.) on the right. Due to the fact that only a few ewes showed a clinical mastitis a relation between SCS and bacteria concerning udder health could not be determined. In further investigations additional milk-samples of ewes are collected to increase the sample size.

Behavior of electrical conductivity and yield ARIMA models for mastitis detection on dairy goats
Romero, G.[1], Roca, A.[1], Peris, C.[2] and Díaz, J.R.[1], [1]Universidad Miguel Hernández (UMH), Tecnología Agroalimentaria, Ctra. de Beniel km. 3,2, 03312- Orihuela, Spain, [2]Universidad Politécnica de Valencia, Instituto de Ciencia Animal, Camino de Vera s/n, 46022-Valencia, Spain; gemaromero@umh.es

The aim of the study was to evaluate the behavior of electrical conductivity (EC) and yield (YI) of milk for the mastitis detection on dairy goats using ARIMA models. The study was carried out at the experimental farm of small ruminants of UMH. EC and YI were measured daily by gland in 8 primiparous and 10 multiparous Murciano-Granadina goats, whose glands were free of intramammary infection. After a period of 3.5 weeks controlling sanitary status of the glands, various situations adverse to health of the glands were simulated. Confirmed the establishment of intramammary infection, HS analyzes and daily measuring of EC and YI continued. It was studied the behavior of the EC and YI, separately, for the detection of mastitis (sensitivity: SENS, specificity: SPEC, positive predictive value: PPV and negative predictive value: NPV) using the ARIMA model (Proc ARIMA, SAS 9.1., 2002). All data prior to infection were included in the model, and it was checked deviations of the variables outside the range predicted by the model. A case of 'positive' occurred if any of the values measured during the 5 days after the establishment of infection was diverted above the upper threshold predicted by EC model or below the lower threshold of YI, separately. A total of 12 glands were infected, 10 subclinical and 2 clinical in a total of 9 multiparous goats (6 unilateral and 3 bilateral). The results (SENS, SPEC, PPV, NPV) obtained by EC (30%, 96%, 80%, 68%, respectively) were better than those obtained by YI (30%, 80%, 44%, 68%, respectively). 100% of clinical mastitis cases (n =2) were detected by EC. The high SPEC and SENS obtained by EC for clinical mastitis have the advantage of its use would not result in unnecessary treatment and is capable of detecting mastitis that causes high glandular damage, respectively.

Genetic resistance to natural gastrointestinal nematode infections in German sheep breeds

Idris, A., Moors, E., König, S. and Gauly, M., Livestock Production Systems, Department of Animal Science, Albrecht-Thaer-Weg 3, 37075, Göttingen, Germany; mgauly@gwdg.de

The objective of this study was to evaluate the possibility of breeding for resistance against gastrointestinal nematode infections in German sheep breeds based on faecal egg counts (FEC), dag score (DS), faecal consistency (FCS), body condition (BCS) or FAMACHA© score (FS). Genetic analyses were carried out on 3,023 lambs (from 170 sires) of four sheep breeds and various farms under the conditions of a natural parasite infection. Individual faecal samples were taken once during the grazing period. At the time of sampling, also DS, FCS, BCS, and FS were determined. Heritabilities and genetic correlations were estimated within an animal model using REML-methodology. The model included the age of lambs as covariate and fixed effects of farms, sex, birth type and season nested within years. Additive genetic and residual were included as random effects. Estimated heritabilities for FEC ranged from 0.15 to 0.60 among breeds. The heritabilities estimated for the potential infection indicator traits varied also between the breed. Heritabilities ranged from 0.11-0.80 for DS, 0.09-0.23 for FCS, 0.01-0.55 for BCS, and 0.02-0.26 for FAMACHA©. The phenotypic correlations between FEC and the potential infection traits were generally low. However, only BCS showed significantly negative correlations (-0.20, -0.21; P≤0.001) to FEC in three of the four breeds. FEC did not show a consistence genetic correlation with the other traits and in most of the cases the estimated values had high standard errors. DS was moderate to highly genetically correlated (0.49 to 0.97) with FCS. It can be concluded that the use of FEC as the selection trait, when breeding for parasite resistance, is feasible in Merinoland, Merino Long-wool and Texel sheep. An indirect selection for nematode resistance using DS, FCS, BCS or FS seem to be unreliable.

Is the establishment rate and fecundity of Haemonchus contortus related to body or abomasal measurements in sheep?

Idris, A.[1], Moors, E.[1], Erhardt, G.[2] and Gauly, M.[1], [1]Livestock Production Systems, Animal Science, Albrecht-Thaer-Weg 3, 37075 Göttingen, Germany, [2]Animal Breeding and Genetics, Ludwigstr. 21b, 35390 Giessen, Germany; mgauly@gwdg.de

The relationship among parasitological parameters, abomasal size and body size measurements were investigated in lambs following an experimental infection with Haemonchus contortus. In total 100 lambs (German Merino (GM), Texel × GM, Suffolk x GM, German Blackhead Mutton x GM and Ile de France × GM) were experimentally infected with 5000 infective 3rd stage larvae of H. contortus at 12 weeks of age. Four and six weeks after infection (p.i.), individual faecal samples were collected to estimate the faecal egg counts (FEC). Furthermore wither height, shoulder width, heart girth, loin girth, and body length were taken at 18 weeks of life. Lambs were slaughtered and necropsied seven weeks p.i., and worm counts, abomasal volume and surface area were determined. Positive correlations (P<0.05) were found between different body size parameters, body weights and abomasal sizes. FEC and worm counts were not significantly correlated with any of the measured parameters(P≥0.05). The mean worm burden in GM lambs was higher (P<0.05) than in crossbred lambs. There was no significant difference in abomasal size between GM and crossbred lambs. In addition, the Texel crossbred lambs had a bigger abomasal volume in comparison to the other crossbreds. No significant differences between the crossbreds could be observed in the parasitological parameters. The results suggest that there is no relation between worm burden following an experimental infection with H. contortus and body or abomasal size parameters. Therefore, other factors, including genetic based differences in resistance, must cause the variation of the parasite establishment between and within breeds.

The effect of age of lambs on response to natural *A.phagocytophilum* infection

Grøva, L.[1,2], Olesen, I.[2,3], Steinshamn, H.[1] and Stuen, S.[4], [1]Bioforsk Organic Food and Farming Division, Gunnarsvei, 6630 Tingvoll, Norway, [2]University of LIfe Sciences, Animal and Aquacultural Sciences, POBox 5003, 1432 Ås, Norway, [3]Nofima Marin, POBox 5010, 1432 Ås, Norway, [4]Norwegian School of Veterinary Science, Kyrkjevegen 332/334, 4326 Sandnes, Norway; lise.grova@bioforsk.no

A main scourge in Norwegian sheep farming is tick-borne fever (TBF) caused by the bacteria Anaplasma phagocytophilum and transmitted by the tick Ixodes ricinus. It is shown, through infection studies, that the clinical response to A.phagocytophilum is less severe in young lambs compared with older lambs. Lambs get colostrum with immunoglobulines from mother after birth, and this passive immunity contribute to defeat infection and develop their own immunity to a certain level. The objective of this study was to reveal effects of early pasturing on tick-infested pastures on the performance of lambs. A field study was conducted on two sheep farms in tick-infested areas in 2008 and 2009. On each farm and year about 45 ewes and 90 lambs were assigned to two treatments with respect to age at pasturing: 1) Young lambs; 3-7 days old at start of pasturing and 2) Old lambs; >3 weeks old at start of pasturing. Body weight recordings, rectal temperature, tick-counts, observation of clinical signs of disease and blood sampling of the lambs were conducted during the spring grazing period (average 56 days long). Comparison of Gompertz growth curve parameters of lambs infected with A.phagocytophilum when they were young compared to old was done using the PROC MIXED procedure in SAS. Lambs that were young when let on to tick-infested pasture, and hence young when naturally infected with A.phagocytophilum, obtained a 38.5 g/day higher ($P > 0.0001$) maximum daily weight gain compared to lambs that were old at start of pasturing. On tick-infested pastures, early pasturing can therefore be recommended as a preventive measure to reduce indirect losses to TBF.

Reduced feeding space for ewes: effect on behaviour and feed intake

Bøe, K.E. and Andersen, I.L., Norwegian University of Life Sciences, Department of Animal- and Aquacultural Sciences, P.O. Box 5003, 1432, Norway; knut.boe@umb.no

Reducing the number of feeding places may have negative effects on both feed intake and behaviour even if ruminants are fed roughage *ad libitum*. The aim of this experiment was to investigate effects of reduced feeding space on competition, activity budget and feed intake of ewes fed *ad libitum* on two types of roughages. A 3 x 2 factorial experiment was conducted with number of animals per feeding place (1:1, 2:1 or 3:1) and type of roughage (grass silage or hay) as main factors. A total of 48 pregnant, adult ewes of the Dala breed were randomly divided into eight groups of six animals; four of these groups were fed grass silage whereas the other four groups were fed hay. The ewes were video recorded for 24 h at the end of each experimental period. Daily intake of silage was not affected by reducing the feeding space from 1:1 to 3:1, but for ewes fed on hay, the daily intake was significantly reduced by 6.8% ($P < 0.05$). However, time spent eating hay was significantly longer than time spent eating silage ($P < 0.05$). Irrespectively of type of roughage, time spent eating was significantly reduced ($P < 0.0001$) whereas CV for eating was increased ($P < 0.05$) when feeding space was reduced. Overall, the competition level was higher when the ewes were fed on hay than on silage (queuing: $P < 0.01$; displacements: $P < 0.01$). In the hay treatment, the time spent queuing increased from 0.3% in the 1:1 treatment to 5.3% in the 3:1 treatment, but this was not significant in the silage treatment ($P < 0.001$). Number of displacements was low at one ewe per feeding place, but increased significantly when feeding space was reduced in the hay treatment ($P < 0.01$).

Light treated bucks induce a well synchronized estrus and LH peak during anestrous season by male effect in north Moroccan goats

Chentouf, M.[1] and Bister, J.L.[2], [1]Institut National de la Recherche Agronomique, Centre de Tanger, Unité de Recherche sur les Productions Animales, 78, Bd Sid Mohamed Ben abdallah, Tanger, 90010, Morocco, [2]University of Namur, 51, Rue de Bruxelles, Namur, B-5000, Belgium; mouad.chentouf@gmail.com

The efficiency of buck effect for the induction and synchronization of reproductive activity in north Moroccan goats was studied during seasonal anestrus. Immediately before joining bucks on April 16, 21 non cyclic does treated for 11 days with intra-vaginal sponge impregnated with 45 mg of Fluorogestone acetate, were randomly assigned to the three groups LTB (Light treated buck; n=7), LMTB (Light and melatonin treated buck; n=7) and NTB (non treated buck; n=7). Treated male were subjected to artificially long days between November 1 and January 15 followed by natural day light. Long days were provided using the flash method in open barns and artificial light was given from 06:00 to 09:00 and from 22:00 to 24:00. At the buck introduction, estrus was checked every 4 hours and blood sampled every 2 hours during 48 hours from heat detection for the determination of plasmatic level of LH by ELISA kit. LTB and LMTB groups showed estrus respectively 44.8 ± 13.0 hours and 33.3 ± 10.3 hours after teasing. Concerning LH secretion, 86% of LTB and LMTB groups displayed LH peak respectively 61.3 ± 12.3 hours and 55.0 ± 13.4 hours after joining. Furthermore, no reproductive activity was observed in NTB group. It can be concluded that bucks treated with light only can induce well synchronized estrus and ovulation in anestrous female by male effect in north Moroccan goats.

European Partnership: a stop to piglet castration in the EU

Bonafos, L., European Commission, DG Sanco - Animal Welfare, Rue Froissart 101, 1040 Brussels, Belgium; laurence.bonafos@ec.europa.eu

The European Commission (EC) has prepared a European Partnership around the castration of piglets. This is a partnership between parties representing the pig and pork production sector and EC services, committed to develop tools to implement the end of surgical castration. Those parties include: associations and federations of pig producers, abattoirs, processors, retailers, consumers, feed industry, housing industry, breeding, veterinary medicine, veterinarians. Plus academia and animal welfare NGOs.The EC services involve 6 directorates: Health & Consumers, AGRI, JRC-IRMM, RTD, ENV, Enterprise & Industry. The principles of a European Partnership include: (1) Voluntary participation, (2) Based on scientific and economic data, (3) Consensus-based approach, (4) Pragmatic mechanisms and a workable structure (steering committee, specialised working groups), (5) Dialogue and transparency towards stakeholders and interested parties. This should ensure a lean and manageable structure with strong political visibility (EC, European Parliament, Member States,Stakeholders, NGOs), with focused communication, and with dedicated human and financial resources. Late 2010, a pilot group prepared a 'Declaration on alternatives to surgical castration of pigs', which is now open for EU-wide ratification. The main points are: (1) Ensure the acceptance of products from not surgically castrated pigs by authorities and consumers in the EU but also in Third Countries, (2) Agree on a common understanding of boar taint, (3) Perform or coordinate R&D on mutually recognised methods for the assessment of boar taint, on a European recognised reference method for the measurement of boar taint compounds, and on rapid inline detection methods, (4) Develop information and training of farmers and other members of the pork chain, (5) Do a cost-benefit analysis on the consequences of the end of surgical castration, and (6) Publish an annual public progress and cost report. The main objective is to end surgical castration EU-wide by 1 january 2018.

Consumer attitude and acceptance of boar taint

Tacken, G.[1], Font-I-Furnols, M.[2], Panella-Riera, N.[2], Blanch, M.[2], Olivier, M.[2], Chevillon, P.[3], De Roest, K.[4], Kallas, Z.[5] and Gil, J.[5], [1]LEI - Wageningen UR, Consumer and Behavior, P.O. Box 35, 6700 AA Wageningen, Netherlands, [2]IRTA, Finca Camps i Armet s/n, E-17121 Monells (Girona), Spain, [3]IFPRI, BP 35104, 35651 LE RHEU Cedex, France, [4]Centro Ricerche Produzioni Animali - C.R.P.A. S.p.A., Corso Garibaldi, 42, Reggio Emilia, Italy, [5]CREDA-UPC-IRTA, Edifici ESAB, Avinguda del Canal Olimpic s/n, 08860-Castelldefels (Barcelona), Spain; gemma.tacken@wur.nl

From societal viewpoint, castration of pig meat is no longer acceptable in some countries in the EU. The goal of the ALCASDE project was to gain insight into (1) consumers' attitudes towards pig castration (2) consumer acceptance of meat with boar taint and (3) the potential impact of this acceptance on the behavioral intentions of pork. In 6 countries was tested: the Netherlands, Germany, United Kingdom, France, Italy and Spain. In each country a representative sample of pork consumers was selected (n=130). Regular boar meat was used and qualified by the Androstenone and Skatol levels of neck fat. Using a questionnaire, color, odor, freshness and shelf life were found to be much more important buying factors than brand, origin, package type and the type of production. Respondents considered themselves not well informed on pig welfare. Within welfare, castration was less important than housing and feed. A sensory test revealed that only in France, Italy and the Netherlands significant differences were found between the consumer perception of tainted boar pork loin and gilt pork loin. The relatively low androstenone and skatole levels in Spain and United Kingdom could explain this. After odor perception during preparation, respondents filled in a questionnaire on behavioral intention. Despite perceived abnormal odor, still a lot of respondents in Italy, the Netherlands and Spain, would serve the meat to family members. 50% of the respondents that smell an abnormal taste, indicate to stop buying pork for a while. 1/3 to 3/4 of the respondents would not visit the a store again, that sells this meat.

Consumer studies of boar taint: what can we conclude?

Font I Furnols, M., IRTA-Food Technology, Granja Camps i Armet, 17121 Monells, Spain; maria.font@irta.es

Boar taint is a sensory defect of pork. Consumers are the last step in the production chain, and consequently those that may suffer this problem. Different consumer studies have been carried out to evaluate boar taint. However, there are differences among studies regarding type of muscle or sample evaluated, criteria of classification of samples (boar taint level, androstenone and/or skatole content, sex), methodology used to prepare the samples and to analyze the boar taint compounds, information provided to the consumer, sensitivity of consumers to androstenone, etc. and results seem to point into different directions. An EU project carried out at the end of the 90s in 7 countries using the same methodology, showed differences among countries and it was concluded that skatole had a more important effect than androstenone in the odour while both compounds contributed in the flavour scores of the meat. Recently, another EU work with consumer studies in 6 countries also confirmed differences among countries and the effect of high levels of androstenone on meat acceptability. In Sweden, two studies showed differences in meat acceptability depending on the presence of boar taint or the sex while in Denmark and in the Netherlands two studies showed the importance of skatole levels in the acceptability of pork, being androstenone not important. In a Norwegian work, sensitive consumers had lower acceptability of odour of meat with androstenone while frying but not when they tasted it. In Spain and France various studies showed less acceptability of pork from entire males with respect to castrates or gilts. In the majority of the studies carried out in United Kingdom, Ireland, Canada and United States, consumers did not find important differences in meat depending on the sex of the animals. The aim of the present review is to understand the differences between several studies and to try to find out a way to summarize the existing literature and to have conclusions.

Androstenone cut-off levels for acceptability of entire male pork by French consumers
Chevillon, P.[1], Nassy, G.[1], Gault, E.[1], Lhommeau, T.[1] and Bonneau, M.[2], [1]IFIP, La Motte au Vicomte, 35650 Le Rheu, France, [2]INRA, UMR INRA-Agrocampus Ouest 1079 SENAH, 35590 St Gilles, France; michel.bonneau@rennes.inra.fr

Three surveys were conducted in order to assess the acceptability of fresh pork from entire males by French consumers with a particular emphasis on trying to establish cut-off levels for androstenone. Each survey involved 140 consumers who assessed fried cutlets during cooking and slices of cooked roasts at eating. The ability of the consumers to perceive the smell of pure androstenone was measured. Androstenone and skatole levels were measured using HPLC and expressed as µg pure fat. Survey 1, conducted within the EU project ALCASDE, demonstrated that (1) French consumers readily accepted pork from entire males with androstenone and skatole levels similar to those observed in castrates and gilts and (2) consumers which perceived the odour of pure androstenone as unpleasant reacted more negatively to tainted meat than consumers that were anosmic to androstenone or perceived it as pleasant. In surveys 2 and 3 consumer perception of cooking odour, flavour and taste did not differ between entire male pork with very low androstenone and skatole levels (similar to those observed in castrates and gilts) and entire male pork with very low skatole levels and androstenone levels in the range of 1.5-1.8 µg pure fat (surveys 2 and 3) or 2.1-3.4 µg pure fat (survey 3). Taken together, these results suggest that for entire male fresh pork with very low skatole levels, androstenone cut-off levels for consumer acceptability are higher than 2.5 µg pure fat.

Imbalances in supply and consumption of sustainable food products - recent Dutch findings
Backus, G., Wageningen University and LEI Agricultural Economics Research Institute, Postbus 8130, 6700EW Wageningen, Netherlands; Ge.Backus@wur.nl

There is a gap between the current and the potential market for sustainable food products in the Netherlands. Consumers are interested in sustainable food products and seem to care about animal welfare and environmental aspects besides more traditional ones like price, taste, convenience, health and safety. However the social environment and product choice do not stimulate them enough to influence their buying decisions. Retailers, out-of-home food providers and food processors acknowledge that they have a role to play in supplying more sustainable food products. However their actions do not always fit into the consumer demand. They are relatively often focussed on internal sustainability programmes like energy saving, while sustainable procurement and animal welfare are the issues of specific relevance for consumers. The vulnerability to 'naming and shaming' by NGO's and media keeps food processors and retailers away from communicating their progress in more sustainable production. These findings are based on consumer surveys and face to face interviews with CEO's in retail, out-of-home and food processors, in the framework of a large project for the Dutch government, called 'Food balance 2011'.

Consumer awareness and acceptance of the method of surgical castration and the use of vaccination (Improvac®) to control boar taint

Schmoll, F., Sattler, T. and Jäger, J., Leipzig University, An den Tierkliniken 11, 04103 Leipzig, Germany; schmoll@vetmed.uni-leipzig.de

Surgical castration of male piglets is routinely performed in most European countries to prevent the occurrence of boar taint in pig carcasses. From an animal welfare and production efficiency point of view vaccination of male fattening pigs against gonadotropin-releasing hormone (GnRH) with Improvac® is beneficial. However, its successful commercial introduction to market requires the agreement of the pork consumers, too. The objective of this study was to evaluate the awareness and acceptance of the method of surgical castration and the preference of the vaccine method relative to castration. The study was performed in cooperation with two leading market research companies. In Austria 1000 on-line interviews were conducted by meinungsraum.at. In Germany 1786 face-to-face interviews by 420 interviewers (Institute for Public Opinion Research Allensbach) were done. The majority of consumers in Austria (67%) and in Germany (63%) had never heard about boar taint. The level of knowledge of the existing practice of surgical castration to avoid boar taint is similar low in Austria (27%) and Germany (24%). Not surprisingly, vaccination of fattening boars to control boar taint is even less known in both countries (6%). Participants were informed about the method of surgical castration and the vaccine method. Thereafter (61%) preferred the vaccine method over surgical castration (27%) in Austria (2% gave no statement). In Germany vaccination was preferred (41%) over the practice of castration (19%), too. Interestingly 40% of the german participants were not able to form an opinion which method to prefer. Overall, it can be concluded that the vaccination approach appears to be acceptable to consumers, if they are informed about the issues. Though the thematic of the piglet castration is now discussed in public, the results of this study are still similar to those of the study published three years ago by Allison *et al.*

Facing new EU policies towards animal welfare improvement: the relative importance of pig castration

Kallas, Z.[1], Gil, J.M.[1], Panella-Riera, N.[2], Blanch, M.[2], Tacken, G.[3], Chevillon, P.[4], De Roest, K.[5] and Oliver, M.A.[2], [1]Center for Agro-food Economy and Development -Polytechnic University of Catalonia (CREDA-UPC), Esteve Terradas, 8, 08860 - Castelldefels (Barcelona), Spain, [2]Institut de Recerca i Tecnològica Agroalimentaria, Product Quality, Finca Camps i Armet, S/N, 17121 Monells (Girona), Spain, [3]Agricultural Economics Research Institute, Consumer and Behaviour, Afdeling Dier Postbus 35, 6700 AA Wageningen, Netherlands, [4]IFIP Institut du Porc, Viandes Fraîches et Produits Transformés, 35651 Le Rheu Cedex, BP 35104 Le Rheu Cedex, France, [5]Centro Ricerche Produzioni Animali, Ufficio Economia, Corso Garibaldi, 42, Reggio Emilia, Italy; nuria.panella@irta.cat

Our research studied the relative importance of animal welfare (pig castration) and boar taint relevance among different attributes of the fresh pork meat within six EU countries (United Kingdom, France, Italy, The Netherlands, Germany and Spain). The empirical analysis uses consumer-level questionnaires to elicit information regarding consumer attitudes toward pig welfare. We use the Analytical Hierarchy Process (AHP) as a multi-criteria decision-supporting method aiming to decompose consumers' buying decision of fresh pork meat. Our results show a lag of information among countries about 'gender of the animal' and as a consequence about pig castration. Effective communication campaign about the relationship between pig welfare and castration is needed in order to help policy makers in addressing the real social interests. Thus, policy authorities will be in a position to design appropriate policy instrument concerning castration.

Acceptability of meat with different levels of boar taint compounds for Spanish and English consumers
Panella-Riera, N.[1], Blanch, M.[1], Kallas, Z.[2], Gil, J.M.[2], Gil, M.[1], Oliver, M.A.[1] and Font I Furnols, M.[1], [1]IRTA, Monells, 17121, Spain, [2]CREDA, Castelldefels, 08860, Spain; nuria.panella@irta.es

Boar taint is an abnormal sensory trait of pork mainly due to androstenone (AND) and skatole (SKA). The aim of this work was to study consumers' acceptance of pork with different levels of boar taint in Spain (ES, N=133) and United Kingdom (UK, N=146). Three types of samples were used: loins from Females (FE) and loins from entire male pigs with two levels of boar taint according to and and SKA levels: BT020 had 0.20±0.07 µg/g of and and 0.06±0.02 µg/g of SKA; BT107 had 1.07±0.40 µg/g of and and 0.18±0.07 µg/g of SKA, on pure fat basis. Loins were cooked on a cooking plate at 180 °C and served warm to consumers, and the following attributes were assessed: 'Delicious', 'Odour' and 'Taste' (rated on a Likert scale going from 1: 'dislike very much' to 9:'like very much'), and 'Strength of odour', 'Abnormal odour', and 'Abnormal taste' (scored between 1:'low perception' to 9:'strong perception'); avoiding always the intermediate level (5). Consumers answered information about their behaviour when buying, cooking and consumption of pork (i.e. eat pork with or without fat) as well as socio-demographic questions (age and gender). The SAS Freq procedure and the Mixed procedure were used for data analysis. Age affected the acceptability in ES (higher as age increased), while only abnormal odour was affected in UK (26-40 age group scored higher than 18-25 group). Regarding the gender, only in ES women scored better delicious and odour than men. No significant differences were found on the attributes among consumers classified according to their habits. Under these experimental conditions, main differences were found due to the type of samples and age of consumer. Gender and consumers' behaviour did not affect the acceptability of the meat. Further research is needed to understand why meat with an average of 1.07 µg/g and on pure fat basis was accepted at the same level than meat from gilts, while meat with lower and levels was less accepted.

Consumer perception of meat from entire male pigs as affected by labeling and malodorous compounds
Trautmann, J.[1], Meier-Dinkel, L.[1], Gardemin, C.[1], Frieden, L.[2], Tholen, E.[2], Knorr, C.[1], Wicke, M.[1] and Mörlein, D.[1], [1]Georg-August-University of Göttingen, Department of Animal Sciences, Albrecht-Thaer-Weg 3, D-37075 Göttingen, Germany, [2]University of Bonn, Institute of Animal Science, Endenicher Allee 15, D-53115 Bonn, Germany; daniel.moerlein@agr.uni-goettingen.de

Surgical castration is performed to prevent the so called 'boar taint'. However, for animal welfare reasons, castration was recently declared to be stopped within the EU by 2018. While the relative contribution of the compounds skatole, indole and androstenonewas investigated repeatedly, little is known about the effects of specific top down processes on sensory evaluation of boar meat. Previously, consumer perception of pork labeled 'free range' or 'organic' was shown to be positively shifted compared to 'conventional pork'. We hypothesized an opposite shift when labeling pork as 'boar' meat dependent on prior knowledge and individual experience of the consumers. 145 people were invited to a central location sensory test. In a 2x2 factorial experiment, each participant evaluated 4 samples of pork loin. While the actual meat was from entire male pigs (boar) or from castrates/gilts (control), it was labeled either 'pig meat' or 'young boar's meat'. Pigs were raised on performance testing stations. Back fat samples were analysed for skatole, indole (HPLC-FD) and androstenone (GC-MS). Loins for sensory evaluation were selected with respect to the level of odorous compounds, i.e. 0.5 up to 2.5 µg/g androstenone and skatole up to 0.2 µg/g (melted fat). After the sensory evaluation, consumers were asked to answer a questionnaire and to perform a smell test to evaluate perceived intensity, liking and familiarity of skatole and androstenone. According to ANOVA, neither label nor actual meat type affected hedonic ratings for taste and overall acceptability. Subsequent smell tests did not explain consumer reactions, nor did the answers of the questionnaire concerning prior experience or knowledge of castration.

Description of EU pork consumers: a survey carried out in 6 countries

Panella-Riera, N.[1], Blanch, M.[1], Font I Furnols, M.[1], Kallas, Z.[2], Gil, J.M.[2], Tacken, G.[3], Chevillon, P.[4], De Roest, K.[5], Gil, M.[1] and Oliver, M.A.[1], [1]IRTA, Monells, 17121, Spain, [2]CREDA, Castelldefels, 08860, Spain, [3]LEI, Wageningen, 6700AA, Netherlands, [4]IFIP, Le Rheu, 35651, France, [5]CRPA, Reggio Emilia, 42100, Italy; marta.gil@irta.es

The aim of this work was to characterize the European consumer of pig meat (within ALCASDE project). The survey was carried out Germany (DE n=132), Spain (ES n=133), France (FR n=139), Italy (IT n=140), Netherlands (NL n=132) and United Kingdom (UK n=146). All 822 respondents were selected for consuming pork >1 time/month and stratified by age and gender, within each country profile. Respondents answered socio-demographic questions and frequency of consumption of different pork products, the most common purchasing place for fresh pork meat, if they were responsible for buying fresh pork at home, if they were responsible for cooking at home, and if they usually eat the pork with the fat. Data was analysed with FREQ procedure of SAS software. In general, over ninety percent of consumers ate fresh pork >2 times/week (DE 96.2%; ES 95.5%; IT 92.9%; NL 93.9%; UK 97.3%) except for FR (34.8%). The most consumed product was the sausage in DE, dry cured ham in ES and IT; cooked ham in FR, mince meat in NL and sliced bacon in UK. The supermarket was the most common purchasing place of fresh pork with the exception of NL, where it was the traditional market. In general, the percentage of respondents responsible for buying fresh pork in their household was 91.0%. In all countries, women were more responsible for buying fresh pork than men, and they were mostly between 41-60 years old. Ninety-one percent of respondents were partially responsible for cooking at home. Women were more responsible for cooking at home than men. France was an exception, where 49.6% women and 50.4% men cooked at home. Considering all respondents, 44.5% ate the pork with the fat in all the countries (35.4% of women and 54.6% of men). The study showed differences among countries regarding respondents' traits.

The effect of long-term under/over feeding on milk and plasma fatty acids profile and on insulin and leptin concentrations of sheep

Tsiplakou, E.[1], Chadio, S.[2] and Zervas, G.[1], [1]Agricultural University of Athens, Nutritional Physiology anf Feeding, Iera Odos 75, Athens, GR11855, Greece, [2]Agricultural University of Athens, Anatomy and Physiology of Farm Animals, Iera Odos 75, Athens, GR11855, Greece; eltsiplakou@aua.gr

The objective of this study was to determine the effects of long-term under/overfeeding on sheep milk fatty acids (FA) profile and on blood plasma insulin and leptin concentrations. Twenty-four lactating sheep divided in three sub-groups and fed the same ration in quantities covered 70% (underfeeding), 100% (control) and 130% (overfeeding) of their energy and crude protein requirements. The results showed that the concentrations of C18:0, C18:1 and monounsaturated FA in milk fat were found to be significantly higher, while those of C6:0, C8:0, C10:0, C11:0, C12:0, C14:0, short chain FA (SCFA), and medium chain FA lower in the group of underfed sheep compared to controls and overfed. The overfed sheep had significantly higher C6:0, C8:0, C10:0, trans-11 C18:1, C18:2n6c, C18:3n3, SCFA and polyunsaturated FA and lower C14:0, C14:1, C16:1 and long chain FA concentrations compared to controls. In the milk of the underfed sheep were observed significantly lower concentrations of trans-11 C18:1, C18:2n6c, C18:3n3 compared to the overfed. In sheep blood plasma underfeeding caused a significant increase on C18:0, C18:1, C20:3n3 and C24:0 and a decrease on C14:0, C18:2n6c and C18:3n6 concentrations compared to control and overfeeding. Overfed sheep had significantly higher leptin and insulin concentrations compared to underfed. In conclusion, the long term under/overfeeding modified in sheep the milk, plasma FA profile and the hormones concentrations.

Association between measurements of assessing selenium status in sheep

Van Ryssen, J.B.J., Coertse, R.J. and Smith, M.F., University of Pretoria, Animal & Wildlife Sciences, Hatfield, 0002 Pretoria, South Africa; jvryssen@up.ac.za

Concentration of selenium (Se) in whole blood, plasma/serum and the liver of animals is widely measured to assess the Se status of animals. Such values have been published in tables and guidelines using criteria such as deficient, marginally deficient, sufficient and toxic ranges. Associations between Se concentrations in these fluids and the liver were calculated from data obtained from 12 independent trials. The Se status of the sheep ranged from deficient to toxic. In 8 trials the sheep (post-weaning) were on feedlot diets and 4 on Se deficient grazing. The duration of the trials ranged from 69 to 150 days. In a meta-analysis the means of treatments were used to estimate the log (base 10) liver Se values from log blood and log plasma Se, as well as log plasma Se from log blood Se. Random coefficient regression analysis using the Linear Mixed Models method was applied to predict for instance the dependent log 10 liver Se from the independent log 10 blood Se. Blood Se levels ranged from 8.5 to 1817 µg/l. Plasma Se concentration was ca. 43% that of blood, suggesting that plasma Se is as reliable as blood Se to indicate Se status. As Se concentration in livers increased both blood and plasma Se concentrations constituted an increasingly smaller proportion of liver Se levels, suggesting that liver Se levels responded more to changes in Se intake than blood and plasma Se. In feedlot sheep: At 16 µg Se/l blood, liver concentration was 36 µg/kg DM and plasma, 6.8 µg/l. At 100 µg Se/l blood, liver was 555 µg/kg DM and plasma, 43 µg/kg. At 400 µg Se/l blood, the corresponding liver concentration was 4335 µg/kg DM and plasma, 270 µg/l. These results suggest that the associations between liver Se and blood and plasma Se are affected by relative Se status of the animal.

Effect of mannan-oligosaccharide supplementation in calf milk replacer on zootechnical and health performances of veal calves

Bertrand, G.[1], Martineau, C.[1], Andrieu, S.[2] and Warren, H.[2], [1]Agesem, Monvoisin, 35652 Le Rheu, France, [2]Alltech Biotechnology Centre, Summerhill Road, Dunboyne, Co. Meath, Ireland; hwarren@alltech.com

Research into mannan-oligosacchride (MOS) supplementation in young ruminant feeds has shown several benefits to performance and health parameters. A trial was conducted using male veal calves to investigate the effect of including a MOS (Bio-Mos™, Alltech Inc, KY) in calf milk replacer (CMR). Sixty male Holstein calves (8-10 d, ~48 kg live weight (LW)) were allocated to one of two dietary treatments for 161 d according to LW and haematocrit: control (CMR) or MOS (control + 4 g MOS/h/d). Additionally, a compound starter feed (fibrous pelleted feed, based on mixed cereals) was offered manually via a bucket once a day. Faecal consistency, feed intake, LW, growth rate (ADG), health status and carcass parameters were assessed. Data were analysed using a General Linear Model. Faecal consistency, feed intake and carcass parameters were unaffected by treatment. Calves from the MOS group had numerically higher LW at day 161 compared with control. From d0 to d22, calves receiving the supplemented CMR had significantly higher ADG (P<0.01) compared with control animals. There was a strong trend (P=0.06) towards higher ADG for MOS calves at d64. Incidence of respiratory disorders was numerically more frequent in the control group from d64 to d161 (finishing period). The average carcass weight for the calves receiving MOS was numerically 4 kg higher than control group (P=0.22). Supplementing veal calf CMR with MOS resulted in a tendency for increased growth and carcass weight and significant improvements in early growth.

Effect of dietary threonine on weaned piglets susceptible or not to *Escherichia coli* K88, under *E. coli* K88 challenge

Trevisi, P.[1], Simongiovanni, A.[2], Casini, L.[1], Mazzoni, M.[1], Messori, S.[1], Priori, D.[1], Corrent, E.[2] and Bosi, P.[1], [1]University of Bologna, Via F.lli Rosselli 107, 42123, Italy, [2]Ajinomoto Eurolysine S.A.S., rue de Courcelles 153, 75 817 Paris Cedex 17, France; paolo.bosi@unibo.it

Threonine is an important component of mucin and immunoglobulins which can be affected by *E. coli* infection. The effect of the susceptibility to enterotoxigenic *E. coli* K88 (ETEC) on the Thr requirement of ETEC-challenged pigs was studied by a 2×2 factorial design. Forty-two weaned pigs were divided into 2 groups using the MUC4 gene as a marker for ETEC-susceptibility (2 MUC4$^{-/-}$ and 2 MUC4$^{+/+}$ pigs per litter). Within genotype, pigs were fed 2 diets differing in the standardized ileal digestible Thr:Lys ratio: 65% (Thr-) vs. 70% (Thr+). Pigs were orally challenged with 1.5×10^{10} CFU ETEC on d7 and slaughtered on d12 or 13. Data were subjected to an analysis of variance with diet, genotype, their interaction, and litter effects. In the 1st week, the Thr+ group consumed more feed (P<0.05). The Thr+ diet alleviated the loss of gain to feed induced by challenge (P=0.08) and increased growth in the overall trial (P=0.08) compared to the Thr- diet. Before challenge, the Thr+ group excreted less *E. coli* in feces (P<0.05) while in the post-challenge period the diet did not affect the number of days with diarrhea and the fecal shedding of ETEC. The MUC4$^{+/+}$ pigs responded to the challenge with a raise of anti-K88 IgA in blood and jejunal secretum (P<0.001). Among them the Thr+ group had a higher increase of anti-K88 IgA values (P<0.05 in blood and P=0.07 in secretum) than the Thr- group. The diet did not affect the morphometry of jejunal villus and crypts. The trend of improved growth in the whole trial may result from the combination of different partial effects of Thr: a better initial feed intake, an improved immune response and a better control of gut microbiota. It was concluded that a Thr:Lys ratio at 70% is advisable for pigs in the first two weeks after weaning, whatever the genotype for ETEC susceptibility.

Effect of valine on performance of weaning piglets

Dissler, L.[1], Häberli, M.[1], Probst, S.[2] and Spring, P.[1], [1]Swiss College of Agriculture, Animal Sciences, Laenggasse 85, 3052 Zollikofen, Switzerland, [2]Egli Muehlen AG, Schuermatte 4, 6244 Nebikon, Switzerland; peter.spring@bfh.ch

The authorization of valine as a feed additive in the EU offers new possibilities in the formulation of diets for monogastric animals. The aims of the present trials were to investigate the effects of different dietary crude protein and amino acid concentrations as well as valine:lysine ratios on piglet performance and health. Two 4-week trials with 90 weaned piglets each were conducted. Both trials were set up as a block design with 3 treatments and 6 replicates per treatment (a total of 18 pens with 5 piglets each). Piglets were weaned at 4 weeks of age with an average weight of 7.65 and 8.12 kg in trial 1 and 2, respectively. Diets were based on barley, corn, rice, oat flake, wheat starch, lactose, potato protein and soy and were offered *ad libitum* as meal. The following treatments were tested: Trial 1: (1) 14.0 MJ DE, 180 g CP, 12.5 g Lysine, 8.6 g Valine; (2) 14.0 MJ DE, 165 g CP, 11.5 g lysine, 7.9 g valine; (3) 14.0 MJ DE, 165 g CP, 11.5 g lysine, 8.5 g valine. Trial 2: (1) 14.0 MJ DE, 180 CP, 12.4 g lysine, 8.6 g valine; (2) 14.0 MJ DE, 165 g CP, 12.4 g lysine, 7.9 g valine; (3) 14.0 MJ DE, 165 g CP, 12.4 g lysine, 8.6 g valine. Feed intake, weight gain and FCR were recorded for 2 two-week periods. Animal health and fecal scores were recorded daily. Data were analyzed with ANOVA and treatment means were compared with the test of Tukey-Kramer. In trial 1, the higher crude protein and amino acid concentrations led to a significant improvement in FCR (1.43a, vs 1.53b and 1.51b kg/kg; P<0.05). Weight gain also tended to be higher with higher dietary crude protein and amino acid concentrations. Increasing the amino acid concentrations in the treatment with reduce CP concentration allowed to maintain performance at the level of the standard diet in trial 2. No significant differences were recorded in trial 2 for the entire trial period. The supplementation of the essential amino acid valine did not affect performance or animal health in both trials.

Effect of phosphorus level and phytase inclusion on performance, bone mineral concentration and mineral balance in finisher pigs

Varley, P.F., Callan, J.J. and O'doherty, J.V., University College Dublin, Animal Nutrition, Lyons Research Farm, Newcastle, Co. Dublin, Ireland; patrick.varley@ucd.ie

Two experiments were conducted to investigate the interaction of phosphorus (P) level and phytase (PHY) inclusion in the diet of finisher pigs, with the objective of investigating the effects of dietary treatment on growth performance, bone mineralisation and phosphorus and calcium (Ca) balance. In Experiment 1, the growth performance and bone analysis experiment, pigs (n=6; initial body weight (BW)=45.2 kg) were allocated to one of six dietary treatments in a 3×2 factorial arrangement: (T1) 1.5 g available (a)P/kg (low P diet); (T2) T1 and 500 PHY units (FTU)/kg; (T3) 2.0 g aP/kg (medium P diet); (T4) T3 and 500 FTU/kg; (T5) 2.5 g aP/kg (high P diet); (T6) T5 and 500 FTU/kg. Experiment 2 consisted of a digestibility and P and Ca balance study. These pigs (n=6; BW=67.3 kg) were offered identical diets to those offered in Experiment 1. There was an interaction between dietary P level and PHY inclusion for overall average daily gain (ADG) and carcass weight (CW; P<0.05). Pigs offered the low P diet supplemented with PHY had a higher overall ADG and CW compared to pigs offered the non-PHY, low P diet. However, there was no effect (P>0.05) of PHY inclusion in the medium and high P diet on overall ADG and CW. A higher concentration of bone ash, P and Ca concentrations were noted in pigs fed medium and high P (P<0.001) and PHY (P<0.01) diets compared to pigs offered low P and non-PHY diets. Pigs offered PHY diets had lower faecal P output (P<0.01) and a higher P digestibility (P<0.001) and P retention (P<0.05) compared to pigs offered non-PHY diets. In conclusion, supplementation with PHY in a low P finisher pig diet resulted in pigs having a similar carcass weight, but weaker bones than pigs offered a medium and high P diet.

Soluble and insoluble fibre from sugar beet pulp enhance intestinal mucosa morphology in young rabbits

El Abed, N., Delgado, R., Abad, R., Romero, C., Villamide, M.J., Menoyo, D., Carabaño, R. and García, J., Universidad Politécnica de Madrid, Dpto. Producción Animal, ETSI Agrónomos. Ciudad Universitaria s/n, 28040 Madrid, Spain; rosa.carabano@upm.es

The effect of soluble and insoluble fibre from sugar beet pulp (SBP) on mucosa morphology was investigated. Four diets were formulated with similar NDF (33% DM) and protein (16% DM) level. Control diet (C) was based on wheat starch, casein and wheat straw and sunflower hulls as fibre sources (36, 15.4, 18 and 18%, respectively) and contained 3.7% DM of soluble fibre (SF). Fibrous sources of C diet were partially substituted (40%) for SBP (9.7% SF on DM), or the insoluble fibre of SBP (3.1% DM SF, iSBP diet), respectively. Another diet was obtained substituting 6% of starch by pectins from SBP in the Control diet (PEC diet, 8.4% SF). Fifty six weaned rabbits (25 d, 460±77 g SD) were assigned to experimental diets, and fed *ad libitum* for 11 d after weaning. Then they were slaughtered and 6-cm from the middle part of jejunum collected to record villous height, crypt depth and goblet cell number. Means were compared using a protected t-test. Rabbits fed C diet (low SF and high insoluble non fermentable fibre) showed the shortest villous height and deepest crypt depth (506 and 133 μm) compared to rabbits fed the other three diets, resulting in the lowest villous height/crypt depth ratio (3.83, P<0.05). Rabbits fed SBP diet (high SF and high insoluble fermentable fibre) showed similar villous height than those fed PEC and iSBP diets (649 μm), but higher villous height/crypt depth ratio (6.76 vs. 6.02, P<0.05) derived from the lower crypt depth found in rabbits fed SBP and iSBP diets respect to those fed PEC diet (96 and 102 vs. 114 μm., P<0.05). Goblet cells number per villous increased from rabbits fed C, iSBP and PEC diets (11.8, 14.0, 15.8, respectively, P<0.05), being those fed SBP diet intermediate between iSBP and PEC diets (15.3). These results indicate a positive effect of SF and insoluble fermentable fibre of SBP, and specially the combination of both, on mucosa morphology.

Forage quality in ewe diets determines fatty acid profile and lipogenic gene expression in *Longisimus dorsi* of suckling lambs

Dervishi, E.[1], Joy, M.[1], Alvarez-Rodriguez, J.[1], Serrano, M.[2] and Calvo, J.H.[1], [1]CITA, Tecnología en producción animal, Avda. Montañana 930, 50059, Spain, [2]Instituto Nacional de Investigaciones Agrarias INIA, Mejora genetica animal, Crta. de la Coruña, km. 7,5, 28040 - Madrid, Spain; edervishi@aragon.es

The aim of this study was to investigate whether forage quality in ewe diets affects IMF fatty acid profile and the expression of genes related with fat metabolism in l. dorsi muscle from suckling lambs of Churra Tensina sheep breed. The effect of ewe diet on the expression of some lipogenic (LPL, ACACA, FASN, FABP4, DGAT1, SCD, PRKAA2), and transcription factors genes (SREBP1, PPARG, PPARA and CEBPB) was also studied. Twenty-four lambs were used for IMF fatty acid profile determination and gene expression studies: HAY group lambs (n=12) and GRE group lambs (n=12) were raised by ewes receiving meadow hay and grazing green forage, respectively. When lambs reach 10-12 kg live weight were slaughtered and sample of l. dorsi was used for IMF fatty acid profile and gene expression studies. The fatty acids profile and gene expression levels were determined. Statistical analysis were carried out using the SPSS 15.0, using GLM. The relationship between gene expression and FA indicators was determined using stepwise linear regression analysis. GRE lambs promoted the formation and deposition of vaccenic (C18:1 n- 7), CLA and PUFA n-3 in L. dorsi from their suckling lambs (P<0.05). Significant statistical differences were found in SCD gene expression (P=0.04), and CEBPB was at the limit of significance (P=0.05). Relative gene expression of SCD was 0.22 lower in lambs from GRE group compared to HAY group. While CEBPB gene expression was 1.31- fold higher in GRE group compared to HAY group. Regression analysis showed that SCD and CEBPB gene expression in suckling lambs are modulated by PUFA n-6/ n-3 ratio. Higher levels of n-6/n-3 stimulate SCD expression and inhibited CEBPB gene expression in HAY group lambs.

Effects of mannanoligosaccharide–β glucan or antibiotics on health and performance of dairy calves

Dabiri, N.[1], Nargeskhani, A.[2] and Esmaeilkhaniani, S.[3], [1]Animal Science Department, Islamic Azad University, Karaj Branch, Karaj, 3187644511, Iran, [2]Animal Science Department, Ramin Agricultural and Natural Resources University, Ahvaz, 81785199, Iran, [3]Isfahan Research Center for Agriculture and Natural Resources, Isfahan, 81785199, Iran; Najdabiri@hotmail.com

Twenty four newborn Holstein calves (initial body weight at birth = 40±3.0 kg) were used to study the effects of mannanoligosaccharide–β glucan (MOS-β glucan) or antibiotics on health and performance. Calves were assigned randomly to one of three treatments. Treatments included: whole milk with no additives (control), whole milk containing MOS-β glucan at 4 g per meal, whole milk containing antibiotic at 500 mg of oxytetracycline/d. Calves received whole milk twice daily. Water and starter were offered for ad lib intake throughout the trial of 56 days. Body weight was measured at birth and thereafter weekly till 8 wk of age. Starter intake was measured daily. Fecal scores were monitored 3 times per week. Blood samples were collected at 3, 7, 15, 30, 45 and 56 days, and analyzed for total protein, blood urea N, albumin and ratio albumin:globulin. Analysis of weekly dry matter intake (DMI) revealed no significant difference from week 1 to 7 among the treatments. The calves fed MOS-β glucan or antibiotic had greater (P<0.05) DMI than control at week 8. Furthermore, at week 8 average daily gain (ADG) was significantly greater (P<0.01, SE 36 g/d) for MOS-β glucan (1024 g/d) and antibiotic groups (1,014 g/d) compared with control group (791 g/d). When the entire study period was evaluated, it was observed that calves fed MOS-β glucan had lower (P<0.05, SE 0.03) fecal score (1.35) than control (1.54) and antibiotic (1.54) treatments. No treatment differences in feed efficiency and blood samples were detected during the trial. This study showed that MOS-β glucan can be a substitute for antibiotic as a supplement for calf growth.

Increasing milk selenium concentration by nutritional manipulation

Yosef, E., Ben-Ghedalia, D., Nikbahat, M. and Miron, J., Institute of Animal Science, The Volcani Center - ARO, Bet-Dagan, 50250, Israel; edithyos@agri.gov.il

Selenium is an essential element for human and animal physiology, found in seleno-protein compounds with antioxidant activity that protects cell membranes. Milk of cows could be a potential supplement of selenium for human nutrition. In a survey of selenium content in common forages grown in Israel, it was found that corn, wheat and sorghum silages contain 38, 66 and 85 ppb selenium, respectively, while vetch and clover hays contain about 123 ppb. Selenium contents in the total mixed rations (TMR) fed to dairy cows were in the range of 134 to 539 ppb. The aim of this study was to increase the selenium concentration in cow milk by nutritional manipulations. Two experiments were conducted in a research dairy farm, equipped with individual tie-stalls. Trial 1 checked the possibility to increase milk selenium concentration by addition of organic selenium to the TMR. Two groups of 19 milking cows each (similar in initial performances) were assigned to the two dietary treatments. The control group fed TMR containing 0.3 ppm inorganic selenium (Na_2SeO_3 used in most Israeli TMRs), while the experimental group fed TMR containing 0.3 ppm Na_2SeO_3 + 0.3 ppm organic selenium (Se-yeast). Data was analyzed according to the repeated measurement Proc mixed model of SAS. The addition of selenium yeast to the TMR, increased milk selenium concentration by 65% (from 77.7 to 128.2 ppb, P<0.05), and increased milk production from 44.7 kg/day in the control group to 46.0 kg milk/day in the experimental group (P<0.05). In trial 2, 26 lactating cows were divided into two groups (of 13 cows each) similar in initial performance. The control group fed TMR containing 0.3 ppm selenium as Na_2SeO_3 while the experimental group fed TMR containing 0.6 ppm selenium as Se-yeast. The experimental TMR increased milk selenium concentration, from 89.6 ppm in the control group to 185.2 ppm in the experimental group (P<0.05).

Livestock farming with care

Scholten, M.C.T.[1], Gremmen, H.G.J.[2] and Vriesekoop, P.W.J.[3], [1]Wageningen UR, ASG, PO 65, 8200 AB Lelystad, Netherlands, [2]Wageningen UR, META, PO 8130, 6700 EW Wageningen, Netherlands, [3]Wageningen UR, LR, PO 65, 8200 AB Lelystad, Netherlands; Martin.Scholten@wur.nl

The Netherlands has a reputation for modern, efficient livestock farming. But it is also a sector that is a topic of public debate, with farming practice, human health, animal welfare and the environment as central issues. So how justified are these concerns? Wageningen UR researchers compiled a kaleidoscopic review on a sustainable intensification of livestock farming based on care ethics. Driven by rising global demand, livestock farming will continue to grow, with expansion virtually unstoppable. So how can we accommodate this trend in a responsible way within the carrying capacity of the planet earth? Organisations like the FAO, CGIAR and the Global Research Alliance call for knowledge sharing, best practices, best ecological means and best animal health means. And from the global perspective the question arises to what extent the Netherlands is able and willing to play a leading role in a further modernization of livestock farming practices. As it can be argued that a smart, modern and strictly managed intensified animal farming provides the best opportunity to resolve negative trade offs. However, in intensified livestock farming, the caring about the wellness of the animal, in broadest sense, is of utmost importance. Using the principles of care ethics we have formulated a principle of livestock framing with care. A modern and smart designed farming, that guarantees a devoted (careful) treatment of animals that are entrusted to the farmer, while (s)he stands for a conscious (careful) compliance of professional merits to earn an economic fair share. It translate into 4 axioms: one health safety, animals integrity, zero emissions and modern entrepreneurship. Values like responsibility, trust, commitment and operating carefully are held in high regard. These values are important for the whole chain from producer to consumer, and in turn add value to the economic value of the livestock industry.

Towards a framework based on contributions from various disciplines to study adaptive capacities of livestock farming systems

Douhard, F.[1,2], Friggens, N.C.[1] and Tichit, M.[2], [1]INRA, UMR 791 MoSAR, 16 rue Claude Bernard, 75231 Paris, France, [2]INRA, UMR 1048 Sadapt, 16 rue Claude Bernard, 75231 Paris, France; frederic.douhard@agroparistech.fr

Achieving ecological intensification of livestock farming systems (LFS) requires improvements to strengthen LFS ability to deal with exogenous uncertainties. Such improvements concern both 1) the biological capacities of animals to maintain their performance in unpredictable environment and 2) farmer's abilities to apply adaptive management such that animals can express their potential. Starting from a socio-ecological viewpoint, we reviewed the concept of resilience and contributions from others sciences and developed some propositions to model the adaptive capacities of LFS. We illustrate our propositions with several case studies in dairy production. Within such complex adaptive systems, farmers have to manage several objectives and adjust dynamically their herd structure or their production project to steer production processes. We show that these adjustments fit well with the homeostasis concept from biology and with the notion of 'regime' in ecology. When experimenting or anticipating uncertainties, LFS can also transform and evolve toward another type of management. This change of 'homeorhetic trajectory' or 'regime shift' defines a new set of LFS configurations linking a new production project with a new herd structure. In this respect, the tensions between the production project and biological fitness trajectories of animals will enable us to assess dynamically LFS ability to co-evolve with their environment. Given in-herd individual variability, we suggest that there exists some flexibility, i.e. different ways, to manage these tensions while remaining in the desired regime. To conclude, we highlight that this flexibility should be evaluated in a multi-criteria framework (e.g. environmental, economical) to support different efficiency perspectives.

Crops and fodder for sustainable organic low input dairy systems

Marie, M.[1,2] and Bacchin, M.[1], [1]INRA ASTER, Av. Louis Buffet, 88500 Mirecourt, France, [2]ENSAIA, INPL-Nancy, Nancy Université, B.P. 172, 54505 Vandœuvre lès Nancy, France; michel.marie@mirecourt.inra.fr

In a long-term experiment involving a mixed crop-livestock dairy system, four rotation modalities (6 or 8 years, with spring or winter cereals and three years of temporary pastures) have been conducted. Twelve environmental indicators have been computed at field and system levels: nitrate leaching, phosphorus losses, pesticides losses in surface and groundwater, pesticide loss in the air, ammonia volatilization, nitrous oxide, methane and carbon dioxide emission, soil organic matter content, phosphorus availability and non-renewable energy use. Globally, the impact of the mixed crop-livestock system on abiotic resources is medium to low, as well as for water pollution, soil quality and energy consumption. Air pollution presented the worst performance (medium to high impact), while other indicators yielded a score above the tolerance value (7), except for nitrate leaching (6.9) and ammonia volatilization (6.9). Winter cereals not preceded by a cover crop result in low score for nitrate leaching, as well as some over-fertilization of permanent meadows by manure. The organic matter content was low for plots which received no fertilization or for which no residues wee incorporated in the soil, and high for spelt, triticale, and temporary meadows or alfalfa plus dactyls. Spreading of slurry, poor in phosphorus, reduces phosphorus availability, as also observed for crops not preceded by fertilization. Ammonia volatilization was lower in 8 years – spring cereals rotation system, due to the fact that it includes alfalfa plus dactyle, which is not fertilized. On the other hand, mixed cereals and proteaginous induced more ammonia volatilization. Nitrous oxide emission was higher in six years rotation systems because half of their plots are temporary meadows receiving manure. Such a monitoring of environmental performance of production systems is a useful tool to design sustainable and efficient cropping systems and livestock management.

Effect of uncertainty on GHG emissions and economic performance for increasing milk yields in dairy farming

Zehetmeier, M. and Gandorfer, M., Technische Universität München, Alte Akademie 14, 85350 Freising, Germany; monika.zehetmeier@tum.de

Increasing milk yield per cow is proposed as a strategy to reduce greenhouse gas (GHG) emissions and to improve economic performance of dairy farms. However, most studies ignore the effects of uncertainty in their analysis and the fact that milk and beef production are closely interlinked with dual purpose cows as a key component. Therefore, the aim of this study was to model the effect of uncertainty on GHG emissions and economic performance considering two different system boundaries. The first system boundary was set at the dairy farm gate allocating GHG emissions between milk and co-products while the second boundary took into account dairy farming as well as beef fattening systems assuming different levels of beef demand. A bio-economic model was set up incorporating indoor dairy cow production systems differing in milk yield (6,000 to 12,000 kg cow^{-1} year^{-1}) and breed (dual purpose and dairy breeds), bull, heifer and calf fattening systems as well as beef cow production systems. GHG emissions were calculated considering on and off farm emissions of CO_2, N_2O and CH_4. To model price risk and uncertainty of GHG emissions we used Monte Carlo simulation based on market and published data. Stopping at the dairy farm gate (system boundary 1), results showed that with increasing milk yield GHG emissions (per kg milk) decreased while dairy farm profit increased at each level of probability. Results indicated that high milk yields were stochastic dominant first degree over lower ones in terms of GHG emissions and profit. However, if both milk and beef production were considered (system boundary 2), this convenient and often drawn conclusion did not hold in all cases (e.g. when a high amount of beef needs to be replaced by beef cows). Further research should also consider variability in milk output and replacement rate as a function of milk yield.

Livestock farming systems and ecosystem services: from trade-offs to synergies

Magda, D.[1], Tichit, M.[2], Durant, D.[3], Lauvie, A.[4], Lécrivain, E.[5], Martel, G.[6], Roche, B.[6], Sabatier, R.[7], De Sainte Marie, C.[5] and Teillard, F.[2], [1]INRA-SAD, UMR 1248 AGIR, chemin de Borde Rouge, F-31320 Castanet-Tolosan, France, [2]INRA-SAD, UMR 1048 SADAPT, 16 rue Claude Bernard, F-75005 Paris, France, [3]INRA-SAD, UE 57, route Bois Maché, F-17450 Fouras, France, [4]INRA-SAD, UR 0045 LRDE, quartier Grossetti, F-20250 Corte, France, [5]INRA-SAD, UR 767 Ecodev, Saint Paul Site AgroParc, F-84914 Avignon, France, [6]INRA-SAD, UMR 980 SAD Paysage, route de Saint Brieuc, F-35042 Rennes, France, [7]University of Göttingen, Büsgenweg 4, 37077 Göttingen, Germany; magda@toulouse.inra.fr

New stake for livestock farming systems (LFS) is to reconcile production with natural resource management and the provision of multiple ecosystem services. Such stake invites the animal science community and other disciplines to analyze biodiversity in a twofold perspective: (1) as a key resource for LFS; (2) as a product derived from processes operating in LFS. Starting from a biotechnical viewpoint, we reviewed contributions from different disciplines (animal production, ecology, management science and economy) and developed a new conceptual framework for analyzing LFS/biodiversity relationships. We illustrate our propositions with several case studies of natural grassland based livestock systems. We showed that it is necessary to reframe criteria for biodiversity characterization. It is also necessary to reconsider key organizational levels in relation with the socio-technical framework. Based on these results, we examined key concepts for the development of a new research agenda. Such agenda highlights the need for interdisciplinary and participatory research in order to reveal adaptive capacities of LFS. It also points ways to move from trade-offs to synergies between LFS and biodiversity. It is concluded that scientific breakthroughs must occur in animal production and management science to move toward an agro-ecological perspective of livestock farming systems.

It's all about livestock ecology!

Meerburg, B.G. and Vriesekoop, P.W.J., Wageningen University & Research Centre, Livestock Research, P.O. Box 65, 8200 AB Lelystad, Netherlands; Bastiaan.Meerburg@wur.nl

Livestock ecology is the scientific study of the interactions of domestic animals with their production environment. Such interactions have effects on livestock numbers, densities and their distribution. In modern agriculture, individual farmers should pay attention to a number of issues that may influence their farm management directly, such as animal welfare, local biodiversity, soil quality, carbon storage and greenhouse gas emissions. The changes in the Common Agricultural Policy (CAP) by the European Commission demonstrate the need for farmers to pay attention to such aspects. However, many of such aspects go further than the operating level of the individual farmer and often cannot be easily optimized at this low level of scale. For example, the implications of mineral flows at regional and national level go far beyond the boundaries of the individual farm. The current global in-balance in phosphorous (P) may serve as an example of a strong trade-off between economy and environment. Many challenges lay ahead where the farm-level and the higher system levels need to be better connected. How can farmers contribute with their farm management to global ecological challenges? What does this mean for their land use? What is the opportunity for biofuels, which might generate value to many residuals of agriculture and provide a source of farm income? And at the regional level, how can modern farms be embedded in the rural landscape? Concluding, we should pay more attention to robust farming systems that not only optimize their farm management around the animal, but also focus on the potential trade-offs of this optimization in terms of ecology and economy.

Ecological intensification of livestock farming systems: landscape heterogeneity matters

Tichit, M.[1], Sabatier, R.[2] and Doyen, L.[3], [1]INRA, umr 1048 SADAPT, 16 rue Claude Bernard, 75005 Paris, France, [2]University of Göttingen, Büsgenweg 4, 37077 Göttingen, Germany, [3]MNHN, umr 7204 CERSP, 55 rue Buffon, 75005 Paris, France; muriel.tichit@agroparistech.fr

Taking advantage of heterogeneity at different scales of livestock farming systems (LFS) is a new challenge for ecological intensification. Empirical evidence suggests that landscape heterogeneity affects the long-term dynamics of wildlife and the production/biodiversity trade-off. In this paper, we examined ecological intensification in a large range of grassland landscapes where agro-ecological processes are driven by a variety of livestock farms. Objective was to test whether landscape heterogeneity improved the production/biodiversity trade-off. We developed a spatially explicit dynamic model linking grass dynamics in grazed and mown fields to bird population dynamics at landscape scale. We simulated contrasting landscapes composed of extensive or intensive farms which implement different land uses having either compensatory or complementary effects on birds. Farm types influenced the landscape composition (i.e. land use proportion). Farmer's land use rules determined landscape heterogeneity (i.e. land use spatial arrangement). Our results showed that land use proportion was the main driver of bird dynamics in landscapes made of compensatory land uses. However, in landscape made of complementary land uses, landscape heterogeneity was an important, although secondary, driver of bird dynamics. In both cases, land use proportion had a strong impact on LFS productive performance. Finally, in more complex landscapes combining compensatory and complementary land uses, simulations showed that landscape heterogeneity improved the production/biodiversity trade-off. This improvement was 15% higher in landscapes made of intensive farms. We discuss the importance of heterogeneity in a practical viewpoint. Increasing heterogeneity requires coordination tools to facilitate cooperation and synergies between livestock farms.

Endotoxemia as a model for evaluating naturally occurring and nutritionally-induced variations in the stress and innate immune responses of cattle and swine

Carroll, J.A.[1], Burdick, N.C.[1], Randel, R.D.[2], Welsh, Jr., T.H.[3], Chase, Jr., C.C.[4], Coleman, S.W.[4] and Sartin, J.L.[5], [1]USDA-ARS, Livestock Issues Res. Unit, Lubbock, TX, 79403, USA, [2]Texas AgriLife Research, Overton, TX, 75684, USA, [3]Texas AgriLife Research, College Station, TX, 77843, USA, [4]USDA-ARS, SubTropical Agri. Res. Station, Brooksville, FL, 34601, USA, [5]Auburn University, Anatomy, Physiology & Pharmacology, Auburn, AL, 36849, USA; jeff.carroll@ars.usda.gov

While the generic innate immune response is a highly conserved and essential immunological response necessary for life, natural and nutritionally-induced variations exist that can have a significant impact on an animal's health and productivity. Within our laboratory, we have developed reliable endotoxemia models utilizing an *E. coli*-derived lipopolysaccharide (LPS) that can be effectively utilized to discern these individual variations in cattle and swine. Utilizing this model, we characterized variations in the stress and innate immune responses following exposure to LPS in different breeds of *Bos taurus* cattle that are considered to be either heat-tolerant or heat-sensitive. Recently, our group also demonstrated natural variations in the stress and innate immune responses of *Bos indicus* cattle following administration of LPS. These studies revealed sexually dimorphic variations following exposure to LPS, as well as variations directly linked to the disposition of the animal. In swine, we have utilized our LPS model to evaluate the ability of non-antibiotic supplements to provide a degree of immunological protection. We have demonstrated that the inclusion of fish oil and yeast products can significantly alter the stress and innate immune responses in a manner that appears to be beneficial to the animals' overall health and productivity. Elucidating these naturally occurring and nutritionally-induced variations in the stress and innate immune responses is essential to developing new management strategies that will improve the overall health, productivity, and well-being of livestock.

Heat stress in farrowing sows under piglet-friendly thermal environments

Malmkvist, J. and Pedersen, L.J., Aarhus University, Animal Health and Bioscience, Blichers alle 20, DK-8830 Tjele, Denmark; jens.malmkvist@agrsci.dk

Piglets face a considerable thermal challenge at delivery under the temperatures typically used in the in-door housing of farrowing and lactating sows. Consequently, provision of additional on-site floor heating (34 °C) at birth and early in life has proven favorable for the re-gaining of normal rectal temperature, the initiation of suckling and for the survival of piglets. For the sow, however, this additional heating is above the optimal 18 to 23 °C comfort temperature range; but only little is known about heat stress in sows around parturition. Therefore, we studied farrowing sows during and after high temperatures in several experiments, using LxY sows loose housed in farrowing pens to sample behavioural as well as physiological stress responses. In comparison to an unheated control group, heating (34 °C until 48 h after birth of first piglet) in the whole pen-floor acted as a stressor since it resulted in elevated plasma concentrations of cortisol and ACTH in sows around parturition. There were, however, no concurrent changes in plasma oxytocin, farrowing duration, piglet inter-birth intervals, lactate in umbilical cord blood or behavioural activity. Comparing groups with part of the pen floor heated or not, we found that the sows neither preferred nor avoided the heated floor as farrowing site, but used the heated area increasingly during the first 3 days post-partum, as the piglets actively sought the heated floor. Additionally, we quantified effects of floor heating (34 °C) in part of the pen with different durations of floor heating (12 vs. 48 h after birth of first piglet) and different room temperatures (15, 20, 25 °C). Results on birth problems, sow BW changes, feed intake, thermoregulation and behaviour were sampled. In general, long-term indicators of thermal stress were unaffected, probably because the sows in the experimental pens are able to perform thermoregulatory behaviour and successfully adapt to the thermal challenges during the three weeks of lactation.

Underlying infections enhance the physiological responses of cattle to endotoxemia

Sartin, J.L.[1], Walz, P.[1], Givens, D.[1] and Elsasser, T.H.[2], [1]Auburn University, Anatomy, Physiology and Pharmacology; and Animal Health Research, College of Veterinary Medicine, Auburn, AL 36849, USA, [2]USDA/ARS, Mastitis Research Group, Beltsville, MD, USA; sartijl@auburn.edu

Many disease models focus on the effects of single disease entities. However, in nature, it is not uncommon to find concurrent diseases of varying severity. These mixed disease models may present with atypical symptoms and may produce altered responses in the host compared to single disease models. This review will first focus on studies of underlying, low level parasitic infection followed by endotoxin (LPS) administration. Here the asymptomatic presence of a prior parasitic infection resulted in enhanced insulin release in response to LPS though plasma glucose was unchanged, suggestive of insulin resistance. Plasma nitrate/nitrite levels (indicative of NO production) were elevated more in the coinfection model, suggesting an enhanced responsiveness to LPS in the dual infection model. This effect of increased NO was also demonstrated to result in increased tissue nitration signals in the lungs and pancreas. In a second model, studies have examined infections with a mildly virulent strain of bovine viral diarrhea virus followed by LPS. In the viral-LPS compared to the parasite-LPS model, similar enhanced responses were observed for key hormones and metabolites. For example, there were increased cortisol, insulin, free fatty acid, and glucose responses in the BVDV – LPS group compared to either the BVDV or LPS groups. Also, body temperatures were depressed in BVDV + LPS cattle compared to BVDV alone. These exaggerated metabolic and endocrine responses to coinfections suggest that complex disease interactions may exist and should be examined further to determine the metabolic cost to the animal, the altered expression of symptoms in mixed disease entities as well as determine whether new strategies are required to both diagnose and treat these animals.

Impacts and mechanisms underlying stress-associated post-translational nitration of protein structure and function in signaling pathways

Elsasser, T.[1], Ischiropoulos, H.[2], Collier, R.[3] and Sartin, J.[4], [1]USDA-ARS, BARC, Beltsville MD 20705, USA, [2]U of Pennsylvania, CHOPS, Philadelphia,PA 19104, USA, [3]U of Arizona, ARC, Tucson Az 85719, USA, [4]Auburn Univ, CVM, Auburn, Al 36849, USA; theodore.elsasser@ars.usda.gov

Production stresses commingle across a myriad of situations ranging from the natural processes of birth and parturition, to disease incidences, to climatic influences. Analysis of signaling mediators has revealed that many factors are common to the progression of several of these stresses and ultimately impact animal health. Recently discovered, a highly reactive nitrogen anion (peroxynitrite, $ONOO^-$) is generated in intracellular compartments by the reaction of nitric oxide (NO) and superoxide anion ($O_2^-\cdot$) during responses to proinflammatory stress. Evidence suggests that oxidative stress processes generate similar nitrating reactants through interactions between H_2O_2 and NO_2^- and catalyzed by peroxidases These nitrating species are (1) generated preferentially in subcellular compartments that enable the needed spatial and temporal elaboration of the reactants, (2) reactively outcompete enzyme-driven reactions targeted to the impacted protein epitope, (3) highly reactive towards phenolic groups (especially those of protein tyrosines) and (4) defined by the local amino acid sequence, orientation, and hydrophobic interactions of favored residues. The consequences of this post-translational modification include altered spatial conformation that reduces and in some instances results in gain of biological function. Our laboratories have evaluated the generation of protein nitration in association with hormone-directed signal transduction processes and mitochondrial function in cattle and regional changes in blood flow in the neonatal pig as models of proinflammatory production stress. Our data suggest that nutrition-based intervention strategies can be developed that effectively reduce the generation of nitrated proteins thereby facilitating faster recovery from these stresses.

Changes in expression of TGF-β, CTGF, collagen and elastin associated with seasonal time of sampling in bovine claw laminar tissue

La Manna, V.[1,2] and Galbraith, H.[1,2], [1]University of Aberdeen, School of Biological Sciences, 23 St Machar Drive, Aberdeen AB24 3RY, United Kingdom, [2]University of Camerino, Environmental and Natural Sciences, via Pontoni 5, 62032 Camerino, Italy; h.galbraith@abdn.ac.uk

Lameness in cattle continues to be a major, internationally endemic, welfare problem with particular problems in dairy cow production. Particular issues relate to lesions in 'soft' tissues of the sole of claw which may extend into keratinised horn. Such lesion formation has been ascribed to inefficiencies of body-weight-bearing by the suspensory system of the laminar region dermis and epidermis. Molecular signalling systems for regulation of cellular and extracellular synthesis and breakdown which are central to responses of these tissues to adverse conditions are poorly understood. The aim of this work was to examine, in female cross-bred cattle (n=64), expression of two important growth factors, transforming growth factor β1 (TGF-β1) and connective tissue growth factor (CTGF) and two structural proteins, collagen type I and elastin in post mortem laminar tissue collected in late winter (February) and early summer (May) at 57° latitude (Aberdeen). Semiquantitative PCR analysis showed variability of expression in relation to house keeping gene of reference (18S ribosomal RNA) with statistically significantly (P<0.05) smaller mean values for winter compared with summer (for TGF-β1: 0.568±0.425 vs. 0.887±0.694 and for CTGF: 0.09±0.11 vs. 0.3±0.32). No differences were recorded for relative expression of collagen and elastin. The results suggest that the relative expression of these growth factors differed according to conditions at the two times of sampling and may be considered of potential importance in regulation of biology of laminar suspensory tissues. The absence of difference in the expression of the two structural proteins may indicate that the growth factor responses did not produce changes in relative expression of the connective tissue molecules studied and which appeared insensitive to season.

Combining elevated total somatic cell count in first parity cows

Heuven, H.C.M.[1,2], Xue, P.[1], Van Arendonk, J.A.M.[1] and Van Den Berg, S.M.[3], [1]Wageningen University, Animal breeding and genomics centre, Marijkeweg 40, 6709 PG Wageningen, Netherlands, [2]Utrecht University, Faculty of Veterinary Medicine, Yalelaan 104, 3584 CM Utrecht, Netherlands, [3]University of Twente, Faculty of Behavioral Sciences, P.O. Box 217, 7500 AE Enschede, Netherlands; henri.heuven@wur.nl

Mastitis in dairy cows is associated with elevated total somatic cell count (TSCC). Since the total number of cells in milk is virtual constant during lactation in absence of pathogens the elevated TSCC can be identified. The mean of the TSCC, excluding the elevated TSCC are a predictor of the 'normal' level. The goal of this research is to summarize the elevated TSCC using Item Response Theory during first lactation. Data, milk yield and SCC, were collected on 2,094 cows in 417 herd. In total 61,262 data points from first lactation cows were available. Each TSCC measurement during lactation is designated as 'normal' or elevated (0/1) based on an algorithm developed by Heuven. This ignores the size of an elevation because it depends to a large extent on the pathogen involved and day post infection and to a lesser extent on the ability of a cow to react. Patterns of elevated TSCC over lactation will be shown as well as how the combination of the elevated TSCC in a latent variable per cow is correlated with the occurrence of clinical mastitis for this cow.

A cure survival model for Taura syndrome resistance in Pacific white shrimp (*P. vannamei*)

Ødegård, J.[1,2], Gitterle, T.[3,4], Madsen, P.[5], Meuwissen, T.H.E.[2], Yazdi, M.H.[4], Gjerde, B.[1,2], Pulgarin, C.[3] and Rye, M.[4], [1]Nofima, P.O. Box 5010, NO-1432 Ås, Norway, [2]Norwegian University of Life Sciences, Departement of Animal and Aquacultural Sciences, P.O. Box 5003, NO-1432 Ås, Norway, [3]CENIACUA, Cra 9 C No. 114 - 60, Bogota, Colombia, [4]Akvforsk Genetics Center AS, Sjølseng, NO-6600 Sunndalsøra, Norway, [5]Aarhus University, Faculty of Agricultural Sciences, Tjele, DK-8830, Denmark; jorgen.odegard@nofima.no

In some cases, failure time data may come from a population with a fraction non-susceptible ('cured') individuals, causing excessive censoring. Hence, standard survival analysis may be inappropriate, and a Bayesian mixed cure survival model is proposed. The observed survival time was assumed to be a result of two underlying liability traits; susceptibility (whether at risk or not) and endurance (individual hazard for susceptible individuals). The model was applied to survival-time data of Pacific white shrimp (*Penaeus vannamei*) challenge-tested with the Taura syndrome virus. In total, 15,261 shrimp from 513 full-sib families of three generations were challenge-tested in 21 separate tests. The overall mortality (across generations) was 28%, while the estimated fraction susceptible shrimp in the cure model was 38%. The estimated underlying heritability was high for susceptibility (0.41 ± 0.07), but low for endurance (0.07 ± 0.03). Furthermore, the estimated genetic correlation between the to underlying traits was low (0.22 ± 0.25). A classical survival model (ignoring the cure fraction) prodused EBV closely correlated with the cure model EBV for susceptibility, but less correlated with EBV for endurance. Seemingly, susceptibility status dominates genetic variation in observed survival time. However, earlier termination of the challenge-test may distort this relationship, and shift the focus of a classical survival model towards endurance rather than susceptibility. In most cases, improvement of the latter trait is of preference, as improved susceptibility will prevent mortality, while improved endurance is more likely to delay mortality.

Estimating diagnostic accuracy of the tuberculin skin test and abattoir meat inspection from bovine tuberculosis surveillance data

Bermingham, M.L.[1], Handel, I.G.[1], Glass, E.J.[1], Woolliams, J.A.[1], De Clare Bronsvoort, M.B.[1], Skuce, R.A.[2,3], Allen, A.R.[2], Mc Dowell, S.W.J.[2], Mc Bride, S.H.[2] and Bishop, S.C.[1,2], [1]The Roslin Institute & (D) SVS, University of Edinburgh, Midlothian, EH25 9RG, United Kingdom, [2]Agri-Food and Biosciences Institute Stormont, Belfast, BT4 3SD, United Kingdom, [3]The Queen's University of Belfast, Dept. of Vet. Science, Belfast, BT4 3SD, United Kingdom; mairead.bermingham@roslin.ed.ac.uk

Bovine tuberculosis (TB) is an infectious disease caused by Mycobacterium bovis. Bovine TB control in the United Kingdom(UK) and Republic of Ireland (ROI) is based on diagnosis using the single intradermal comparative tuberculin test (SICTT), supported by active abattoir surveillance. The heritability of responsiveness to the SICTT has been estimated at 0.14 and 0.16 in ROI and Great Britain (GB) Holstein-Friesian dairy cow populations respectively. However, imperfect diagnostic performance may have resulted in underestimation of SICTT heritability estimates. The aim of this study was to extend the Bayesian formulation of the Hui-Walter latent class model to estimate diagnostic test parameters and true prevalence of bovine TB from surveillance data. Data from 73,003 Holstein-Friesian dairy cows across 409 TB herd breakdowns in Northern Ireland between 1995 and 2010 were included in the analysis. Both diagnostic tests were highly specific, although abattoir inspection had poorer sensitivity than the SICTT. The estimates of true prevalence were higher than the apparent prevalence within herd breakdowns. This study provides an extended Hui-Walter latent class model that allows for the unbiased estimation of diagnostic test parameters and true prevalence of bovine TB from surveillance data. Furthermore, correcting the heritability estimates using the diagnostic parameter estimates from this study, led to predicted true heritability estimates for SICTT responsiveness in the ROI and GB studies of 0.16 and 0.19 respectively; indicating that true genetic variation is likely to be greater than estimated from surveillence data.

Cellular and extracellular biology of bovine claw laminar tissue
La Manna, V.[1,2], Di Luca, A.[1] and Galbraith, H.[1,2], [1]University of Aberdeen, School of Biological Sciences, 23 St Machar Drive, Aberdeen, AB24 3RY, United Kingdom, [2]University of Camerino, Environmental and Natural Sciences, via Pontoni 5, 62032 Camerino, Italy; h.galbraith@abdn.ac.uk

Lameness in cattle continues to be a major welfare problem which has diminished little in the last decade. There are particular problems in dairy cow production. Typical issues relate to lesions arising from interaction of the 'hard' keratinised and 'soft' tissues of the claw and adverse physical and socially interactive environments of cow husbandry. The aim of the present study was to investigate cellular and extracellular morphology of bovine claw tissue. Sections were prepared in transverse and longitudinal orientations from the claw laminar region and examined histologically. Focus was given to the presence (1) in the epidermis, of coronary wall horn and development and keratinisation of cap horn and interlaminar horn and (2) in the dermis (laminar and reticular), the presence of blood vessels, fibroblasts and expression of connective tissue and adhesion proteins. Laminar basal and suprabasal keratinocytes were identified in terms of denucleation and keratinisation, which increased with increasing axial to frontal abaxial perspectives, in the formation of cellular clusters. The presence of blood vessels, important in inflammatory events and elastin fibres in parallel and perpendicular orientations was demonstrated in dermis. Immunohistochemical investigation showed the presence, in keratinocytes, of β-actin, cytokeratins, integrin α6, and cadherins. Signals in dermal tissue were recorded for vimentin, indicative of fibroblast cells, fibronectin, and collagens I, IV and VII. The presence of certain molecules of cytoskeletal structure and adhesion, particularly in cells and extracellular tissue located at the extended basement membrane separating epidermis and dermis were noted. These are considered of particular importance for maintenance of structural integrity and prevention of injury to internal soft tissues giving rise to lameness.

Possible involvement of rodents in spread and transmission of *Coxiella burnetii*, causative agent of Q-fever, to humans and livestock
Meerburg, B.G.[1] and Reusken, C.B.E.M.[2], [1]Wageningen University & Research Centre, Livestock Research, P.O. Box 65, 8200 AB Lelystad, Netherlands, [2]National Institute for Public Health & Environment (RIVM), Laboratory for Zoonoses & Environmental Microbiology, P.O. Box 1, 3720 BA Bilthoven, Netherlands; Bastiaan.Meerburg@wur.nl

Elucidation of the role of rodents in Q-fever epidemiology is necessary for adequate risk-management. Several indications exist that wild rodents persistently disseminate Coxiella burnetii from the endemic sylvatic cycle to the domestic cycle. We reviewed currently available scientific literature on C. burnetii and the role of rodents in its epidemiology. Goats, sheep and cattle are the primary reservoirs but many domesticated and wild animals (mammals, fish, birds, reptiles, arthropods) can be carriers. Ruminants are often infected subclinically or asymptomatically. Urine, feces and birth material of infected animals are sources for environmental C. burnetii contamination. The bacterium can survive for years. Aerosolized, it can travel downwind, which may cause human morbidity after inhalation. Direct contact with infected livestock by-products such as wool, hides or straw, or consumption of raw products such as milk and meat may also result in infection. Natural infections were demonstrated in over 30 different rodent species on 4 continents: Asia, Africa, Europe and North-America. However, the contribution to Q-fever epidemiology may vary substantially based on species-specific characteristics like host status, habitat preferences, pathogenesis and behavior. From US and UK studies it is known that rodents show evidence of exposure relative to their interaction with ruminants. Species that have more contact have a higher seroprevalence, probably caused by spread of aerosolized bacteria amongst small ruminants. Studies are needed that look specifically examine the significance of rodents in the transmission of C. burnetii to livestock both within and between herds, and in direct transmission to humans.

Diagnosis of paratuberculosis: evaluation of an ELISA in milk in relation to serum and bacterial culture of tissue in goats

Van Hulzen, K.J.E.[1], Heuven, H.C.M.[2], Nielen, M.[1] and Koets, A.P.[1,3], [1]Utrecht University, Farm Animal Health, Yalelaan 7, 3584 CL Utrecht, Netherlands, [2]Utrecht University, Clinical Sciences of Companion Animals, Yalelaan 108, 3584 CM Utrecht, Netherlands, [3]Utrecht University, Department of Infectious Diseases and Immunology, Yalelaan 1, 3584 CL Utrecht, Netherlands; k.j.e.vanhulzen@uu.nl

Diagnosis of paratuberculosis by culture is labour intensive, time consuming and expensive. The objective of this study was to evaluate the commercially available paratuberculosis ELISA (Institut Pourquier, Montpellier, France). Test results for milk were compared with results for counterpart serum samples generated with the same ELISA, and results from bacterial culture of tissue samples in goats with and without Mycobacterium avium subspecies paratuberculosis (MAP) infection. Paired blood and milk samples from 62 goats were collected to compare ELISA results for milk in comparison with serum. Goats were selected based on advanced age and/or suspicion of MAP infection by the farmer. Fifty-four animals tested negative for both milk and serum, six animals tested positive for milk and serum, one animal tested positive for milk and negative in serum, and one animal tested negative for milk and positive for serum. Non-randomly, 57 goats were subjected to bacterial culture of tissue samples to evaluate the ELISA for milk in relation to bacterial culture. Thirty-three goats tested positive in the bacterial culture. Of these, 10 goats tested positive and 23 goats tested negative in the milk ELISA. Twenty-four goats tested negative in the bacterial culture. Of these, 22 goats tested negative and two tested positive in the milk ELISA. In this study, the milk ELISA was equally accurate compared with the serum ELISA. The proportion of positive results for ELISA in milk and bacterial culture in tissue samples was different, but similar to observations in cattle. Results of this initial validation study indicate that the milk ELISA has poor sensitivity but good specificity and may be a useful tool in MAP management for dairy goat farmers.

Effect of farming system characteristics on the effectiveness of paratuberculosis control in a dairy herd

Fourichon, C.[1], Marcé, C.[1,2], Seegers, H.[1], Pfeiffer, D.U.[2] and Ezanno, P.[1], [1]Oniris, INRA, UMR1300 BioEpar, Atlanpole Chantrerie, F-44307 Nantes, France, [2]Royal Veterinary College, Veterinary Epidemiology & Public Health Division, University of London, Hatfield AL9 7TA, Hertfordshire, United Kingdom; christine.fourichon@oniris-nantes.fr

Paratuberculosis control is a major concern in major dairy producing areas. In most European countries, no vaccine is available and control relies on reducing within-herd transmission through improved biosecurity for young calves, and on testing and culling infected animals. Effectiveness of control in dairy farms is highly variable, and farming system characteristics are assumed to play an important role in the observed differences. A paratuberculosis transmission model and a herd simulation model were combined to investigate effects of calf housing, hygiene, and occurrence of involuntary culling due to other disorders (mastitis and infertility) on the effectiveness and the efficiency of paratuberculosis control. The herd was managed under a quota constraint. Several surveillance systems and control actions for paratuberculosis were combined and compared. Effects of the control programmes can be compared in terms of prevalence of infectious animals, replacement rate, resulting herd size and quota achievement. The type of calf housing had a minor effect contrary to hygiene management and time-to-cull infectious animals. Whatever the surveillance in place to trigger a control programme (surveillance based on either clinical case, or on systematic testing with tests of limited sensitivity), hygiene and involuntary culling interacted to a large extent on the results of the control programme. Implementing systematic test-and-cull was more efficient than clinical surveillance or doing nothing after 7 to 13 years. Return on investment of systematic surveillance was higher and occurred earlier when hygiene was impaired. The differences between control scenarios were small when hygiene was improved. Modelling is a valuable tool to account for farm characteristics.

Characterisation of the contact network in the pig supply chain
Büttner, K., Krieter, J. and Traulsen, I., Institute of Animal Breeding and Husbandry, Christian-Albrechts-University, Olshausenstr. 40, 24098 Kiel, Germany; kbuettner@tierzucht.uni-kiel.de

The structure of contact networks can be characterised by applying network analysis. Thus the spread of infectious diseases can be predicted and finally prevented. To identify premises with a central position in the network of pig production, e.g. premises with a high number of trade connections, parameters describing these properties are measured (in-degree, out-degree and degree distribution. In an observation period from 2006 to 2009 contact data from a producer community in Northern Germany were analysed. Supplier, purchaser, number and type of delivered livestock were recorded. The data contains information on 15,372 animal movements between 658 premises. To apply static network analysis repeated trade connections between two premises were aggregated to a single one. Thus, the number of trade contacts decreased to 2,018. In this directed network, i.e. each edge has a certain direction, a node represents a premise (multiplier, farrowing, finishing, farrow-to-finishing farm or abattoir) and an edge denotes a trade contact (piglet, fattening pig, sow or boar). The number of trade partners which deliver animals to a specific premise ranges between 0 and 206 (in-degree). The number of trade partners which receive animals from a specific farm (out-degree) is smaller (0-108). 75% of the premises show an in- or out-degree of 3 or less. Due to their position in the network, different farm types reveal various degrees. The median out-degree of multipliers is 5. Abattoirs show a median in-degree of 40. The results indicate that the degree distribution follows the so-called power-law distribution. Networks with such a degree distribution are considered to be highly resistant concerning the random removal of nodes. However, by strategic removal of highest-degree premises, e.g. selective vaccination or culling, the network structure changes and can decompose into fragments. By this means, the chain of infections is interrupted and further disease spread can be prevented.

Impact of infection or vaccination for Bluetongue virus serotype 8 on reproduction in dairy cows
Nusinovici, S., Seegers, H., Joly, A., Beaudeau, F. and Fourichon, C., Oniris, INRA, UMR1300 BioEpar, Atlanpole Chantrerie, F-44307 Nantes, France; christine.fourichon@oniris-nantes.fr

Serotype 8 of the bluetongue virus (BTV8) was introduced in northern Europe in 2006 and resulted in a large outbreak in France in 2007 and 2008. Production losses were reported but not precisely quantified. Inactivated virus vaccines have been widely used to control BTV8. To evaluate BTV vaccination programs, an appraisal of both reduction in production losses and possible side effects is needed. The objectives of this study were to quantify (1) detrimental effect of infection in naïve herds and (2) a possible side effect of vaccination against BTV-8 using inactivated vaccines on the fertility of dairy cows, in field conditions. Fertility was assessed by return-to-service following artificial insemination (AI). The study was performed on dairy herds either (1) exposed to BTV8 in 2007 or (2) not exposed to BTV. Statistical multivariate models were performed to quantify effects on conception failure, early and late embryonic death. Cows inseminated from 10 weeks prior to disease detection in their herd to 4 weeks after had an increased risk of return-to service. Average estimated risk varied according to the spatiotemporal dynamics of the disease (hazard ratio between 1.2 and 1.5). Due to uncertainty on the true date of onset of infection in the herds, the mechanism underlying the effect shown could not differentiate between possible causes of return-to-service. Only cows receiving a second vaccine injection between 2 and 7 days after AI had a significantly higher risk of 3-week-return-to-service (relative risk 1.19). No later effect was found. This low side effect could be due to an increase of early embryonic death. When accounting for reproduction performances, detrimental effects of the BTV8 infection are much higher than a slight side effect of vaccination only in the week following AI.

Impact of Bluetongue outbreaks on the number of calves born in French beef cattle farms

Mounaix, B., Ribaud, D. and David, V., Institut de l Elevage, 149 Rue de Bercy, 75000 Paris, France; beatrice.mounaix@inst-elevage.asso.fr

In 2009, a significant decrease in the number of calves born in French beef cattle farms was observed. This study, based on data analysis, was set up to get a better understanding of the role of the bluetongue disease in this decrease. The National identification database was analyzed to define relevant calving periods and to calculate an indicator of the herd productivity in calves from the assessment of the number of the mothers-to-be which were present at the beginning of each calving period. Variations of this indicator between 2005 and 2009 were compared between seven epidemiological areas for three beef cattle breeds. For each breed, mixed regression analysis was used to measure significant differences between four calving periods and seven areas, taking into account variability among farms. A total of 23,039 one-breed herds were included in the analysis. In each area, the productivity indicator decreased significantly ($P<0.0001$) within the calving period immediately following the bluetongue outbreak. In the East of France, where bluetongue spread in 2007, a decrease of the productivity indicator was observed during the 2007-2008 calving period. In both central basins which were infected in 2008 the decrease happened during the 2008-2009 calving period. In the Brittany area, where the viral outbreak was very low in 2008, no significant variation of the productivity indicator was observed. These results confirmed the significant impact of bluetongue on beef cattle reproduction: a decrease in the number of calves born and late calving. In France, this impact was moderate: 2.7% fewer calves in 2009 in calving units, and limited to the period following the outbreak. Then, this impact was observable at the national level and could have an economic impacts. Also, it could be important for farmers to take into account the possible impacts of the bluetongue without vaccine protection when deciding about voluntary vaccination.

Using network properties to predict disease transmission in the pig supply chain

Traulsen, I., Büttner, K. and Krieter, J., Institute of Animal Breeding and Husbandry, Christian-Albrechts-University, Olshausenstr. 40, 24098 Kiel, Germany; kbuettner@tierzucht.uni-kiel.de

Pork production is characterized by movement of animals between farms from farrowing to slaughter. Infected but clinically unsuspicious animals can easily spread infectious diseases. A data set including15,372 pig movements (piglet, fattening pig, sow or boar) between 658 premises (multiplier, farrowing, finishing, farrow-to-finishing farm or abattoir) from 2006 to 2009 on a weekly basis (156 weeks) were analysed by methods of network theory. Network properties of the premises indicated a homogeneous network structure, although premises participating at the contacts changed over time. To assess the influence of the structure on disease transmission, epidemics were simulated using an SIR-epidemiological model. Scenarios included temporal changes in trading relations (random order of weeks) and randomly chosen initially infected premises. Results indicated a significant influence ($P<0.05$) of both, the initially infected premises' type and the number of premises that received animals from the initially infected premise in the first week of the epidemic. Multipliers resulted in largest epidemics (6.9 infected premises on average), while farrowing farms and farrow-to-finishing farms revealed epidemics with an average size of 2.7 and 2.1, respectively. Starting an epidemic on a farm that delivered animals to only one premise, however resulted in average epidemics with 2.2 infected premises. An increase in the number of trading partners (3, 6, 14) indicated an increase in epidemic size (9.9, 17.6, 21.5). The impact of the order of the weeks changed the epidemic size only slightly (4.5 to 4.8 infected premises). In conclusion, type and number of trading partners of the initially infected premise was more important for the size of an epidemic than trading relations of the following weeks. In case of a disease outbreak, risk assessment of further disease spread could be improved using properties of network theory describing trading relations.

Effect of region and stocking density on performance of farm ostriches
Bouyeh, M., Islamic Azad University, Animal science Department, Islamic Azad University Rasht branch, 4193963115, Iran; mbouyeh@gmail.com

Although industrial production of ostriches has existed for only a few decades in the world, within this short period it has developed considerably. The aim of this study was to evaluate the performance of farm ostriches which are reared outdoors and fed concentrate in different climates and densities. In this research, production statistics recorded by owners over 5 years on 78 farms, with a mean number of 63 birds per farm, distributed in different regions were studied, examining the effects of treatments including: climate of the region i.e.hot and dry, mild and humid, or alpine; and density i.e. lower than 100, 100-300, or higher than 300 m^2 space per bird. The data were statistically analyzed using by SPSS and the Duncan test was used for comparison of means. Results showed that In climate had no significant effect on egg production, egg weight, fertility, hatchability and weight of day-old-chick among the three climates overall but in the alpine climate there was a lower performance compared with other climate zones. However, the mild and humid climate was significantly (P<0.05) associated with lower maturity age of males and females and longer duration of production season. Stocking density did not have a significant impact on egg production, hatchability, maturity age of males and females or duration of production season. However in relation to egg weight and day-old-chicks, an area more than 300 m^2 resulted in lower performance compared with an area of 100-300 m^2 or less than 100 m^2. Regarding fertility, an area less than 100 m^2 had a lower performance compared with the other two groups (P<0.05). The results of this work showed that ostriches have a higher performance in hot and dry, and mild and humid climates compared with alpine climates; and in regard to stocking density, allocation of 100-300 m^2 area per breeder bird in is better than a higher or lower density.

Correlated random effects in survival analysis
Mészáros, G.[1], Sölkner, J.[1] and Ducrocq, V.[2], [1]University of Natural Resources and Life Sciences, Division of Livestock Sciences, Gregor-Mendel-Str. 33, A-1180 Vienna, Austria, [2] UMR 1313 Génétique Animale et Biologie Intégrative INRA, Domaine Vilvert, 78352 Jouy-en-Josas, France; gabor.meszaros@boku.ac.at

Survival analysis is a popular methodology when analyzing the occurrence of an event as a function of time. It allows to use both time independent and time dependent explanatory variables, as well as to include unobserved random heterogeneity into survival data models. The Survival Kit software package is frequently used for survival analysis in animal breeding, as it allows handling very large databases with a possibility to account for the relationship between animals. From version 6 on, it is possible to jointly estimate variances for two normally distributed random effects, assuming that they are independent from each other. In certain situations however one might suspect some connection between the two random effects. This is the case for genetic effects which vary with time (e.g. the sire effect on culling of his daughters in early vs later lactations) or for survival traits influenced by direct and maternal genetic effects (e.g. survival from birth to weaning). We present an extension of the Survival Kit where it is possible to include correlated random effects, accounting for their relatedness. The correlation coefficient can be estimated or fixed, in case it is already known. The methodology is demonstrated using a practical example.

Combining survival analysis and a linear animal model to estimate genetic parameters for social effects on survival in layers

Ellen, E.D.[1], Ducrocq, V.[2], Ducro, B.J.[1], Veerkamp, R.F.[3] and Bijma, P.[1], [1]Wageningen UR, ABGC, P.O. Box 338, 6700 AH Wageningen, Netherlands, [2]INRA, UMR 1313 GABI, 78352 Jouy-en-Josas, France, [3]Wageningen UR Livestock Research, ABGC, P.O. Box 65, 8200 AB Lelystad, Netherlands; Esther.Ellen@wur.nl

Mortality due to cannibalism in layers is a difficult trait to improve genetically, because censoring is high (animals still alive at the end of study period) and mortality depends on both the individual itself and behaviour of its cage mates. To estimate genetic parameters for social effects on survival time, we compared a two-step approach (2STEP) that combines survival analysis and a linear animal model including social effects (LAM) with LAM applied directly to observed survival time. LAM, therefore, ignored censoring. Survival data of three purebred White Leghorn layer lines from Institut de Sélection Animale B.V., a Hendrix Genetics company, were used in this study. In total, 16,780 hens with intact beaks were kept in four-bird cages. Cross validation was used to compare 2STEP with LAM. Using 2STEP, total heritable variance, including both direct and social genetic effects, expressed as proportion of phenotypic variance, ranged from 32% to 64%. Though this was substantially larger than heritability with LAM (6% to 19%), those values cannot be compared directly. Cross validation showed that 2STEP and LAM performed equally well in predicting records that were set to missing, and gave approximately the same response to selection when animals were selected based on predicted records. Further investigation showed that this occurred because all records were censored at the same survival time point. When censoring was at different time points, 2STEP was superior to LAM. Cross validation showed that selection based on both direct and social genetic effects, using either 2STEP or LAM, gave largest response to selection. It can be concluded that 2STEP is useful to estimate genetic parameters for social effects on survival time in layers when individuals are censored at different time points.

Threshold and linear models for the genetic analysis of bull fertility in the Italian Brown Swiss population

Tiezzi, F.[1], Penasa, M.[1], Cecchinato, A.[1], Maltecca, C.[2] and Bittante, G.[1], [1]University of Padua, Department of Animal Science, Viale dell'Università 16, 35020, Legnaro, Italy, [2]North Carolina State University, Department of Animal Science, Campus Box 7621, 27695, Raleigh, NC, USA; francesco.tiezzi@unipd.it

The objective of this study was to estimate genetic parameters for bull fertility in the Brown Swiss Population. Bull fertility traits were non-return rate at 56 days (NR56) and pregnancy rate (PR). Data included 124,206 single inseminations on heifers and cows reared in Bolzano-Bozen province (northeast Italian Alps) from 1999 to 2008. Observations were restricted to matings with AI service bulls with known ancestors and a minimum of 100 observations. Bayesian linear (LM) and threshold (TM) models were implemented via Gibbs sampling for both traits. All models accounted for the 'fixed' effects of year-season of insemination, days in milk within parity of the cow, and status of the service bull at breeding (young or proven), and the 'random' effects of herd and technician. In addition, three additional random effects were included sequentially: the additive genetic effect of the service sire, the permanent environmental effect of the cow, and the additive genetic effect of the sire of the cow. Marginal posterior means of heritability ranged from 0.02 to 0.03 for NR56 and from 0.04 to 0.07 for PR, and all models led to similar ranking of sires for both traits.

Rationale for estimating genealogical coancestry from molecular markers

Toro, M.A.[1], Garcia-Cortes, L.A.[2] and Legarra, A.[3], [1]UPM, Ciudad Universitaria, 28040 Madrid, Spain, [2]INIA, Carretera La Coruña km 7, 28040 Madrid, Spain, [3]INRA, UR 631 SAGA, F-31326 Castanet Tolosan, France; miguel.toro@upm.es

Genetic relatedness or similarity between individuals is a key concept in population, quantitative and conservation genetics. When a genealogy is available, genetic relatedness between individuals is measured by the coancestry coefficient that assumes a founder population where genealogy starts. With molecular markers there are two basic ways of calculating empirically the genetic similarity between individuals: the molecular coancestry and the molecular covariance. Here we derive the expected values of these empirical measures of similarity as a function of the genealogical coancestry if the individuals are linked by genealogical pathways. From these formulas it is easy to derive estimators of genealogical coancestry from molecular data. However, the estimators are severely biased if the distribution of gene frequencies in the founder population is unknown. This is illustrated with some simulation examples. A real data example in dairy cattle is also shown. Estimators of genealogical coancestry from molecular data are easy to derive. If gene frequencies in the founder population are unknown some consequences and alternatives of this limitation are briefly discussed.

Mapping of calving traits in dairy cattle using a genealogy-based mixed model approach

Sahana, G., Guldbrandtsen, B. and Lund, M.S., Aarhus University, Genetics and Biotechnology, Blichers Allé 20, P.O. Box 50, 8830 Tjele, Denmark; Goutam.Sahana@agrsci.dk

Among the existing methods for genome-wide association mapping, tree-based association mapping methods show obvious advantages over single marker-based and haplotype-based methods because they incorporate information about the evolutionary history (genealogy) of the genome in the analysis. Local genealogies are genealogies for the longest chromosome region around a marker that do not require recurrent mutation or recombination. We have developed a genealogy-based mixed model (GENMIX) for association mapping for quantitative traits in population with complex pedigree. Using simulated data, we have shown that GENMIX has more power compared to linear mixed model approach in association mapping. We apply the method to map QTL for calving traits in the Danish, Swedish and Finnish Holstein cattle population. The data analyzed included 4,258 AI bulls. Genotypes were obtained with the Illumina BovineSNP50 panel. A total of 38,545 informative, high-quality SNP markers were used for the association analysis. The traits analyzed were calving ease, calf size and stillbirth. Additionally, two combined indices, birth index and calving index, were analyzed. The chromosomal regions habouring QTL for these calving traits were identified.

Runs of homozygosity and levels of inbreeding in cattle breeds
Ferencakovic, M.[1], Hamzic, E.[2], Fuerst, C.[3], Schwarzenbacher, H.[3], Gredler, B.[2], Solberg, T.R.[4], Curik, I.[1] and Sölkner, J.[2], [1]University of Zagreb, Svetosimunska 25, HR-10000, Croatia, [2]BOKU – University of Natural Resources and Life Sciences Vienna, Gregor Mendel Str. 33, A-1180 Vienna, Austria, [3]Zuchtdata EDV Dienstleistungen GmbH, Dresdner Straße 89/19, A-1200 Vienna, Austria, [4]Geno, Holsetgata 22, 2317 Hamar, Norway; mferencakovic@agr.hr

Inbreeding results in identity by descent of maternally and paternally inherited alleles. Due to the process of recombination that combines maternal and paternal DNA in very large chunks during meiosis, inbreeding to a recent ancestor results in relatively long homozygous segments of the genome, 'runs of homozygosity'. High density genotyping renders the search for such homozygous segments relatively easy. Runs of homozygosity that are several megabases (Mb) long are due to inbreeding to a recent ancestor. We have searched for runs of homozygosity that are longer than 2 Mb, 4 Mb, 6 Mb, 8 Mb and 10 Mb, respectively, and expressed these segments as proportions of the total length of the autozygous genome, based on SNP positions available. Very long segments most likely result from recent inbreeding while shorter segments may have been caused by inbreeding to ancestors further in the past. The breeds involved in the study are Fleckvieh (dual purpose Simmental), Brown Swiss, Norwegian Red, 500 animals each, and Tyrol Grey, 213 animals. On long and informative pedigrees, inbreeding coefficients were calculated for the full pedigree available and for subsets including a maximum of 5 generations for each individual. Results indicate a high correlation (>0.80) of individual runs of homozygosity with pedigree inbreeding coefficients of animals when calculating correlations across breeds. Within breeds, correlations were smaller (~0.70) because of differences in inbreeding levels in these breeds. We conclude that the proportion of the genome arranged in long homozygous segments provides a good indication of level of inbreeding of an animal that is more informative than the probability of autozygosity derived from from pedigree data.

Prediction of breed composition in an admixed cattle population
Frkonja, A.[1], Gredler, B.[1], Schnyder, U.[2], Curik, I.[3] and Sölkner, J.[1], [1]BOKU - University of Natural Resources and Life Sciences Vienna, Gregor Mendel Str. 33, A-1180 Vienna, Austria, [2]swissherdbook, Schützenstrasse 10, CH-3052 Zollikofen, Switzerland, [3]University of Zagreb, Svetosimunska 25, HR-10000, Croatia; anamarija.frkonja@boku.ac.at

Swiss Fleckvieh has been established in 1970 as a cross/composite of Simmental and Red Holstein Friesian cattle. The current population includes ~50,000 herdbook cows. The composition of the population involves ~30% Simmental and ~70% Holstein Friesian ancestry, with a wide range. For the current analysis, Illumina Bovine SNP50 Beadchip data were available for 100 pure Simmental, 100 pure Holstein Friesian and 300 admixed bulls. The scope of the study was to compare the performance of hidden Markov models, as implemented in STRUCTURE software with methods used in genomic selection (Bayes B, partial least squares regression, LASSO variable selection). In the latter case, the training population consisted of the sets of purebred animals with 'phenotypes' 0 for Simmental and 1 for Holstein Friesian, crossbred animals were the test population. We compared the algorithms for the full set of 40,492 SNP and for subsets of 50%, 20%, 10%, 5% and 1% evenly distributed SNP from this set. We also selected subsets of SNP based on difference in allele frequencies in the pure populations, using F_{ST} as an indicator. SNP with F_{ST} higher than 25% (5635), 30% (3904) and 35% (2,620 SNP) were used. Additionally 48 and 96 SNP with the highest F_{ST} values were extracted. Results are correlations of admixture levels estimated with admixture levels based on pedigree information. For the full set of SNP used, PLS, Bayes B and STRUCTURE are performing very similar (correlations of 0.97), while the correlation of LASSO and pedigree admixture was lower (0.93). With decreasing number of SNP, correlations only decreased substantially when 5% or 1% of all SNP were used. With SNPs chosen according to F_{ST}, results were very similar to using the full set. Only when using 96 and 48 SNP with the highest F_{ST}, correlations dropped to 0.92 and 0.90, respectively.

Ignoring half-sib family structure leads to biased estimates of linkage disequilibrium

Gomez-Raya, L. and Rauw, W.M., Instituto Nacional de Investigación y Tecnología Agraria y Alimentaria - INIA, Departamento de Mejora Genética Animal, Crta. de la Coruña, km. 7,5, 28040 Madrid, Spain; rauw.wendy@inia.es

Maximum likelihood methods for the estimation of linkage disequilibrium (LD) between SNPs in half-sib families were developed (LD-HS). Three situations were considered: sire is double homozygote, sire is homo-heterozygote, and sire is double heterozygote. Current methods to estimate LD ignoring half-sib family structure are the method of Excoffier and Slatkin (LD-EX) for unrelated individuals, and informative maternal haplotypes (LD-MH). Using algebra, we showed 1) method of Excoffier and Slatkin applied in half-sib families is biased for homo-heterozygote and double heterozygote sires, and 2) the use of informative maternal haplotypes in progeny from a double heterozygote sire leads to severe upwards bias estimation of LD if SNPs are closely linked. Monte-Carlo computer simulations were carried out for a variety of scenarios regarding sire SNPs genotypes, linkage disequilibrium and recombination fractions. The allele frequencies in the SNPs were set at 0.50. Each simulation set was analyzed with LD-HS, LD-EX, and LD-MH. Results for estimating linkage disequilibrium in half-sib progeny from a double heterozygote sire for a simulated LD of 0.00 and recombination fractions of 0.00, 0.25, and 0.50 were 0.11, 0.06, and 0.00 for the method LD-EX, and 0.25, 0.13, and 0.00 for the method of LD-MH, respectively. Results for simulated LD of 0.20 were 0.22, 0.16, and 0.09 for the method LD-EX, and 0.25, 0.23, and 0.20 for the method of LD-MH, respectively. LD-HS was always asymptotically unbiased. The general conclusion is that ignoring half-sib family structure might lead to upward biased estimates of linkage disequilibria at short genetic distances and downward biased estimates of linkage disequilibria at long genetic distances. Inferences on population structure and evolution of cattle or sheep should be based on linkage disequilibria estimated after accommodating existing half-sib family structure.

Characterization of linkage disequilibrium in a Danish, Swedish and Finnish Red Breed cattle population

Rius-Vilarrasa, E.[1], Iso-Touru, T.[2], Strandén, I.[2], Schulman, N.[2], Guldbrandtsen, B.[3], Strandberg, E.[1], Lund, M.S.[3], Vilkki, J.[2] and Fikse, W.F.[1], [1]Swedish University of Agricultural Science, Undervisningsplan 4A, 750 07 Uppsala, Sweden, [2]MTT Agrifood Research Finland, Biotechnology and Food Research, 31600 Jokioinen, Finland, [3]Aarhus University, Faculty of Agricultural Sciences, Blichers Allé 20, 8830 Tjele, Denmark; Elisenda.Rius-Vilarrasa@slu.se

Differences in LD patterns exist between populations. This will reduce the benefit of genomic selection in multi-breed populations. For the Nordic Red dairy cattle populations, the benefit from using a combined reference population is influenced by the genetic ties between breeds. The purpose of this study was to investigate the extent of LD in and between Danish Red (DR), Swedish Red (SR) and Finnish Ayrshire (FAY). Genotype information for 6,107 bulls and 38,647 SNP markers were available after editing. Marker pairs less than 500 kilobases (kb) apart were used to calculate the r^2 measure of LD within breed. At distances <5 kb the average r^2 within breed was similar for all three populations (0.58-0.59). LD decreased with distance. For all longer distances r^2 was highest in FAY and lowest in DR. The difference in average r^2 between populations was highest when markers were 80-100 kb apart. Persistence of phase across populations was assessed by the correlation between breeds of r computed within breed. Correlations of r were 0.80, 0.86 and 0.94 for DR-FAY, DR-SR and SR-FAY. Differences in the extent and patterns of LD were found between populations, where the largest variation was observed between DR and FAY. These results indicate possible disadvantages of genomic selection in a combined reference population for the Nordic Red dairy cattle.

NIMBUS: an open library to implement random forest in a genome-wide prediction context
Gonzalez-Recio, O.[1], Bazán, J.J.[1,2] and Forni, S.[3], [1]INIA, Mejora Genética Animal, Ctra Coruña km 7.5, 28040 Madrid, Spain, [2]Universidad Autónoma de Madrid, Física Teórica, Cantoblanco, 28049 Cantoblanco, Madrid, Spain, [3]Genus Plc, 100 Bluegrass Commons Blvd. Ste 2200., Hendersonville, TN, USA; gonzalez.oscar@inia.es

This abstract presents a software library named Nimbus that implements the random forest algorithm to analyze categorical or continuous traits using genomic information. It can be used for genome-wide association studies or genomic-assisted prediction of breeding values. Random Forest is a massively non-parametric machine-learning algorithm, robust to over-fitting. The algorithm is able to capture complex interaction structures in the data without a large number of parameters that other parametric counterparts may need. The reduction of parameters may alleviate several problems of analyzing genome-wide data. This library has been designed to predict the genetic merit for complex traits of individuals and yet-to-be observed records of animals without phenotypes. It also estimates the relative importance of markers for the expression of the trait analyzed. Nimbus allows a friendly implementation of random forest to analyze genomic data. The software can build classification and regression trees using bootstrapped samples of the data set. Two versions are publicly available for implementation, either on regression or classification problems. The predictive ability performance of random forest was shown using different data sets and was compared to linear regression commonly used in similar genome-wide prediction scenarios.

Improving the fit of non-linear regression models in SNP genotype association studies
Pollott, G.E., Royal Veterinary College, Royal College Street, London NW1 0TU, United Kingdom; gpollott@rvc.ac.uk

Many quantitative traits are measured repeatedly over time on the same animals. Genetic analyses of such traits often involve fitting a linear mixed model, including a term to account for the change in the trait over time. Such models can suffer from three major drawbacks; the best-fit overall relationship between the trait and age may be impossible to linearise, the estimates of fixed effects may vary with age and the variance of the trait may vary with age. All three drawbacks will produce estimates of genetic parameters or SNP genotype effects which are less than optimal. This paper demonstrates some solutions to these problems and illustrates them with a SNP genotype association analysis of cattle liveweight which improved the results from non-significant to highly significant. The solutions involve using the best-fit overall relationship between age and liveweight and the change in SD of liveweight with age to preadjust the liveweight data before analysis with a suitable mixed model. Using the TFAM SNP as an example, preadjusting the data for weight-for-age changed the effect of SNP genotype from non-significant to significant ($P<0.001$). By further adjusting the data by dividing by the SD-for-age improved the model once again. One additional improvement was tried by fitting a within-genotype regression on age to the data adjusted for both weight-for age and SD. This reduced the residual variance of the model further and allowed the identification of changes in the relative effect of the SNP genotypes with age. In cases where a linearised mixed model may be less than optimal for analysing a trait repeatedly measured over time, preadjustment of the data can reveal relationships in the data which were hidden in the original analysis. Although this method is illustrated using growth as an example it is more generally applicable to any trait measured repeatedly over some linear scale.

Associations between DGAT1 and SCD1 polymorphisms and production, conformation and functional traits in dairy cattle

Bovenhuis, H.[1], Visker, M.H.P.W.[1], Mullaart, E.[2] and Van Arendonk, J.A.M.[1], [1]Wageningen University, Animal Breeding and Genomics Centre, Marijkeweg 40, 6709 PG Wageningen, Netherlands, [2]CRV, P.O. Box 454, 6800 AL Arnhem, Netherlands; Henk.Bovenhuis@wur.nl

Diacylglycerol acyltransferase 1 (DGAT1) and stearoyl-CoA desaturase 1 (SCD1) are genes involved in fat metabolism. Both genes are polymorphic in Holstein Friesian dairy cattle and have major effects on milk production and milk fat composition. The aim of the present study was to estimate effects of the DGAT1 K232A and SCD1 A293V polymorphisms on conformation and functional traits. For this purpose data of 4426 progeny tested Holstein Friesian bulls born between 1985 and 2005 were available. The frequency of the DGAT1 A allele was 0.66 and the frequency of the SCD1 V allele was 0.32. Highly significant effects of DGAT1 on milk production traits were found: differences between AA and KK genotypes were +708 kg of milk, -29 kg of fat, +11 kg of protein, -0.74% of fat and -0.18% of protein. Additionally, DGAT1 showed significant effects on maturity and gestation length: the AA genotype was associated with lower scores for maturity (which is a measurement for increase in production from lactation 1 to lactation 3) and a shorter gestation period than the KK genotype. The SCD1 polymorphism showed significant effects on milk-, fat- and protein yield: differences between VV and AA genotypes were +100 kg of milk, +7 kg of fat and +3 kg of protein. Further, significant associations of the SCD1 polymorphism were found for non-return rate at 56 days, gestation length and angularity: the VV genotype was associated with a 1% lower non-return on 56 days, a shorter gestation period and a higher score for angularity than the AA genotype. These results illustrate that polymorphisms in DGAT1 and SCD1 have effects beyond milk production and fat composition.

Identification of haplotypes contributing to androstenone on SSC5 in Duroc

Van Son, M.M.[1], Lien, S.[2,3], Agarwal, R.[2,3], Kent, M.[2,3] and Grindflek, E.[1], [1]Norsvin, Pb 504, 2304 Hamar, Norway, [2]Norwegian University of Life Sciences, Centre for Integrative Genetics (CIGENE), Pb 5003, 1432 Ås, Norway, [3]Norwegian University of Life Sciences, Department of Animal and Aquacultural Sciences, Pb 5003, 1432 Ås, Norway; maren.moe@umb.no

Boar taint gives an unpleasant odour and flavour to the meat from some uncastrated male pigs. It is caused by elevated levels of androstenone and/or skatole in the adipose tissue of boars, with high levels of androstenone being the most serious issue in Norwegian pig populations. One way of solving the problem of boar taint would be to identify genetic markers for use in breeding, however high and unfavourable correlations between androstenone and other sex steroids make genetic selection a challenge as we do not want to reduce animal fertility. A QTL scan previously performed on 840 Duroc and 1,150 Landrace boars using the pig 60K SNP chip identified several QTL for androstenone in adipose tissue. Only three of the QTL were significantly affecting levels of androstenone without affecting any of the other sex steroids studied (testosterone, 17β-estradiol and estrone sulphate in plasma). One highly significant genome-wide Duroc-specific QTL affecting androstenone only was detected on SSC5. Examination of linkage disequilibrium (LD) within this region was performed using Haploview and a haploblock of 11 SNPs with considerable LD was identified. An association study was conducted using ASReml, and the entire haploblock was found to significantly affect levels of androstenone, explaining 4.5% of the total variance. Using Beagle we found six unique haplotypes present in this block with the two most frequent haplotypes reducing and increasing the level of androstenone respectively. Whole-genome resequencing of two extreme androstenone boars of each breed generated an average of 4X coverage and detected a number of new SNPs within the SSC5 QTL region. These SNPs allow fine mapping of the QTL region and will potentially allow the identification of a functional mutation.

Fine-mapping of a QTL segregating on pig chromosome 2 highlighted epistasis
Tortereau, F.[1], Sanchez, M.P.[2], Billon, Y.[3], Burgaud, G.[3], Bonnet, M.[1], Bidanel, J.P.[2], Milan, D.[1], Gilbert, H.[1,2] and Riquet, J.[1], [1]INRA Laboratoire de Génétique Cellulaire, Chemin de Borde Rouge, 31326 Castanet-Tolosan, France, [2]INRA GABI, Domaine de Vilvert, 78352 Jouy-en-Josas, France, [3]INRA GEPA Le Magneraud, St Pierre d Amilly, 17700 Surgeres, France; flavie.tortereau@toulouse.inra.fr

A QTL underlying fatness traits was described between 30 and 50 cM on pig chromosome 2 in F2 pedigrees involving Large White (LW) and Meishan (MS) breeds. A recombinant progeny testing has been initiated for the fine mapping of this QTL. One F1 sire heterozygous LW/MS was mated to LW sows to produce recombinant offspring on SSC2 in the initial confidence interval. All the sons carrying distinguishable recombinant haplotypes were progeny tested, and the location of the QTL was then refined relative to the recombinant points and the results of segregation analyses. Among the different sires progeny tested, four of them were full-sibs: in the SSC2 QTL interval, all of them inherited the same maternal haplotype and a recombinant paternal haplotype differing only on the position of the recombination point. Highly significant results were obtained for two of them, but no likely localisation of the QTL could be deduced from combining the MS chromosomal portions of these four sires and their estimated genotype at the QTL. To explain discrepancies obtained, it was hypothesized that another locus could interact with this SSC2 QTL. Genotypes at another locus in the genome may influence progeny results estimated for SSC2 QTL. A genome scan was performed on those four sires and finally three regions (on SSC3, SSC8 and SSC13) were retained as candidate epistatic region. Interactions between these regions and SSC2 QTL were first evaluated with the four recombinant BC families, and then epistatic effect observed between SSC2 QTL and SSC13 was confirmed by the analysis of the F2 pedigree from which originated the different BC families.

The genetics of wool shedding in sheep breeds known for their wool shedding characteristics
Pollott, G.E., Royal Veterinary College, Veterinary Basic Sciences, London, NW1 0TU, United Kingdom; gpollott@rvc.ac.uk

In some countries, the economics of sheep production are such that wool shearing has a negative effect on profitability and so attention is turning to breeds that shed their wool naturally. Data were available from two groups of sheep breeders interested in wool shedding; one developing a composite (Sheep Improved Genetics Ltd.; SIG) and the other a purebred shedding breed (Wiltshire Horn). Wool shedding was scored on a 5-point scale (1 to 5; No shedding to completely shed) once a year in early summer (adults) or late summer (lambs). The SIG animals were a wool-shedding type based on using Easycare, Katahdin, Dorper and Wiltshire Horn rams (all shedders) on non-shedding ewes (Suffolk, Texel, Lleyn and Friesland). Systematic crossing resulted in a population of shedders. Shedding score data from 1,467 SIG first cross (F_1) and first backcross (BC_1) of F_1 to shedding rams facilitated a genetic analysis. The model of inheritance for wool shedding was developed in these sheep. Animals were divided into two groups; shedders, having a shedding score >1 at least once in their life, and non-shedders, only having a shedding score of 1. The four common modes of inheritance for Mendelian traits were tested on the F_1 and BC_1 animals against their expected ratios and tested using a chi-squared test. Only the autosomal dominant mode of inheritance was found to be significant (P<0.001). Within the group of sheep carrying this gene, genetic variation of the shedding score was found with a heritability of 0.55±0.07 in lambs and 0.22±0.06 in adults. Not all lambs, known to be shedders as an adult, shed in their 1st year. The heritability of shedding score in the Wiltshire Horn group, measured at 14 months of age, was found to be 0.41±0.20. Thus wool shedding in these breeds operates under a dual model of inheritance, an initial dominant switch gene but a polygenic trait determining the extent of wool loss amongst shedders.

Genomia: across-Pyrenees genomic selection for dairy sheep

Beltran De Heredia, I.[1], Ugarte, E.[1], Aguerre, X.[2], Soulas, C.[2], Arrese, F.[3], Mintegi, L.[3], Astruc, J.M.[4], Maeztu, F.[5], Lasarte, M.[6], Legarra, A.[7] and Barillet, F.[7], [1]NEIKER, Granja Modelo de Arkaute, Apdo 46, 01080, Vitoria-Gasteiz, Spain, [2]CDEO, Quartier Ahetzia, 64130 Ordiarp, France, [3]ARDIEKIN, Granja Modelo de Arkaute, Apdo 46, 01080, Vitoria-Gasteiz, Alava, Spain, [4]Institut de l Elevage, BP42118, 31321 Castanet Tolosan, France, [5]ITG Ganadero, Avda Serapio Huici 22, 31610 Villava, Navarra, Spain, [6]ASLANA, C/ Aintziburu s/n, 31170 Iza, Navarra, Spain, [7]INRA, UR 631, BP 52627, 31326 Castanet Tolosan, France; andres.legarra@toulouse.inra.fr

Dairy sheep breeds Latxa (in Spain) and Manech (in France), divided into blond and black strains, are the object of 5 independent selection programs since the 80's/70's. These populations, comprising 164,000 females and a few hundred AI males on the whole, originate in the same ancestral across-borderline breed; frequent Latxa-Manech exchanges occur, in particular in the Blond Face strain, albeit not in a systematic way. The advent of genomic selection has raised the interest of a systematic sharing of resources of these breeds. European Union FEDER funds, through POCTEFA (http://www.poctefa.eu/) finances a project called GENOMIA, whose aim is to investigate the possibilities of a joint, international, breeding scheme. Particular focus is given to: description and analysis of the existent schemes; exchange of genetic material; study of sanitary issues for systematic exchange of AI doses; genomic selection. Current similar points involve traits of interest (milk yield) and methodologies (data recording, AI-based selection, BLUP-based evaluation). Current disagreements involve different recording systems for milk composition and type traits, and emphasis on resistance to Scrapie, that has been heavily selected for in Manech whereas in Latxa it is not a major concern. Concerning genomic selection, plans are genotyping ~1400 progeny-tested rams by country, in order to assess, by cross-validation, the accuracy of across-country genomic evaluations and the interest of using both populations together.

Heritability estimates for *Mycobacterium avium* ssp. paratuberculosis in fecal tested German Holstein cows

Küpper, J.[1], Brandt, H.[1], Donat, K.[2] and Erhardt, G.[1], [1]Justus-Liebig-University, Department of animal breeding and genetics, Ludwigstr. 21b, 35390 Giessen, Germany, [2]Thüringer Tierseuchenkasse, Victor-Goerttler-Str. 4, 07745 Jena, Germany; Julia.d.kuepper@agrar.uni-giessen.de

Paratuberculosis (ParaTB) or Johne's disease caused by Mycobacterium avium ssp. paratuberculosis (MAP) is a chronic enteritis in cows which leads to high economic losses. The world wide classical control programs are based on management arrangements and culling of infected animals. Breeding for disease resistance will be an additional useful tool to inhibit the distribution of ParaTB. Therefore, we estimated the heritability for occurrence of ParaTB based on fecal culture. On the basis of 11,285 German Holstein herd book cows originated from 15 farms we determined heritabilities between 0.157 and 0.228. This lead to the conclusion that the MAP detection by fecal culture is from interesting to control the disease as well as an appropriate feature for further genetic analyses to detect MAP associated chromosome regions.

The impact of missing or erroneous SNP on the accuracy of genomic breeding values
Huisman, A.E.[1], Visscher, J.[2] and Van Haandel, B.[3], [1]Hendrix Genetics, Research and Technology Centre, P.O. Box 114, 5830 AC Boxmeer, Netherlands, [2]Hendrix Genetics, Institut de Sélection Animale, P.O. Box 114, 5830 AC Boxmeer, Netherlands, [3]Hendrix Genetics, Hypor, P.O. Box 30, 5830 AA Boxmeer, Netherlands; abe.huisman@hendrix-genetics.com

Genomic selection has gained a lot of attention, the main goal of genomic selection is to maximise the predictive accuracy of breeding values using dense marker information, and by doing so increase genetic gain. Several methods have been proposed to estimate these so called 'genomic breeding values', however we can safely assume that not all genotyping work is 100% accurate, especially when smaller panels of single nucleotide polymorphism (SNP) are used to impute genotyped individuals towards larger SNP panels or even whole genome sequence. Main objective is to investigate what the effect of missing or erroneous genotype information is on the accuracy of genomic breeding values, secondary objectives included the method of analysis (G-BLUP or BayesB), the quantity of missing or erroneous SNP genotypes (10% or 40%) present, and the heritability of the trait of interest (0.15 or 0.40). Several datasets, including genotypes and phenotypes, were simulated to study the aforementioned scenarios, and results were averaged. In all cases the genome consisted of 5 chromosome of 1 Morgan, with 1001 evenly spaced SNP markers placed on them, in total 5,005 SNP markers. Each genome harboured 50 quantitative trait loci that, together, explained 80% of the genetic variance. Twelve generations of matings were simulated, starting in generation 1 with 100 males and 1000 females and expanding with 20 males and 200 females every subsequent generation. Each mating resulted in on average 3.5 offspring. Selection and culling decisions where made based on conventional EBV for a trait that was only recorded on the females. All animals with an accuracy of 70% or higher for their conventional EBV were put in the reference panel, the test panel consisted of males from the top 25% full sib families in generation 12.

Parent-of-origin effect found in the associative genetic effect for survival in crossbred laying hens
Peeters, K., Eppink, T.T., Ellen, E.D., Visscher, J. and Bijma, P.; katrijn.peeters@wur.nl

Mortality due to cannibalism and feather pecking is undesirable from a welfare and economic perspective. So far, genetics of survival has been primarily studied in purebreds. To study survival in crossbreds, the Institut de Sélection Animale B.V. provided data on 7,668 W1xWB and 7,344 WBxW1 laying hens. Birds, with untrimmed beaks, of the same line were kept in groups of four. As survival in laying hens is affected by social interactions, a bivariate direct-associative animal model was used to estimate the genetic parameters within and between crosses. The average survival rate in the parental (W1 and WB) lines was higher than in the crossbreds, indicating negative heterosis. Moreover, the average survival rate differed between crosses. The associative effects were large and strongly significant in both crosses, contributing 87% and 72% to the total heritable variance of W1xWB and WBxW1, respectively. Within crosses, the genetic correlation between the direct and associative effect was moderately negative for W1xWB (-0.37±0.17), but highly negative for WBxW1 (-0.83±0.10). Between crosses, the genetic correlation between the direct effects was high (0.95±0.23), while the genetic correlation between the associative effects was only moderate (0.42±0.25). This indicates a parent-of-origin effect for the associative effect, as it matters which parental line provides the sire and which provides the dam. This can be due to maternal effects, imprinting or sex-linked effects. Since the cross with the highest survival received the paternal chromosome from the pure line with the highest survival, and vice versa, the effect appears to be paternally transmitted. This indicates the presence of maternal imprinting or a Z-chromosome-linked effect. A sex-linked effect would support a previous study which found associative QTLs on the Z-chromosome.

Sporadic interspersed LD defines long range haplotype blocks in cattle and chicken

Lipkin, E.[1], Fulton, J.[2], Arango, J.[2], O'sullivan, N.P.[2], Bagnato, A.[3], Dolezal, M.[4], Rossoni, A.[5] and Soller, M.[1], [1]The Hebrew University of Jerusalem, Genetics, Dept. of Genetics, The Hebrew University of Jerusalem, Jerusalem, 91904, Israel, [2]Hy-Line Int, Dallas Center, IA, 50063, USA, [3]Universita` degli Studi di Milano, Veterinary Sci. and Technologies for Food Safety, Milano, Italy, 20133, Italy, [4]Vetmeduni Vienna, Institut für Populationsgenetik, Vienna, Austria, 1210, Austria, [5]ANARB, Italian Brown Cattle Breeders Association, Loc. Ferlina Bussolengo (VR), 204 - 37012, Italy; lipkin@vms.huji.ac.il

Bovine and chicken haplotype block structure was studied in Brown Swiss dairy cattle sires (BSS), and in chicken sires from a Brown-egg layer line (BES) selected for production traits while maximizing retention of genetic variation. BSS and BES were genotyped by Illumina 50K and 42K arrays, respectively. LD r^2 values were obtained from the BSS non-redundant haplotypes dataset and all BES haplotypes. Both populations gave similar results: general absence of high LD (>0.5) between adjacent markers, but large blocks characterized by long-range, sporadic, low level LD. The markers in these blocks showed moderately low but highly significant LD (>0.25) with a small number of markers in the block, along with very low LD (<0.10) with most markers in the block. The number of major and minor haplotypes (MAF>0.05) across these blocks was far below the number expected without LD. Thus, it may be possible to capture much of the LD in the block by use of tag haplotypes. The results imply that a QTL will be in moderately low but significant LD with a subset of markers in its vicinity. This should enable effective GWS based on these arrays, depending on the proportion of the genome covered by the blocks: almost complete for BEL, but only about half for BSS. Large scale high resolution GWA mapping by LD to single markers may not be possible due to overall paucity of high LD. GWA mapping by haplotype-to-QTL LD may be more informative in identifying QTL, but because of large extent of the haplotype blocks may not provide high-resolution mapping.

Association of melatonin receptor 1A gene polymorphisms with production and reproduction traits in Zandi sheep

Hatami, M., Rahimi-Mianji, G. and Farhadi, A., Laboratory for Molecular Genetics and Animal Biotechnology, Department of Animal Science, Faculty of Animal Science and Fisheries, Sari Agricultural Sciences and Natural Resources Universiy, 578, Iran; ayyoob_farhadi@yahoo.com

Melatonin regulates some major physiological processes such as maturation and function of reproductive system, pubertal development and seasonal reproduction and adaptation. The activation of melatonin hormone is mediated by melatonin receptor. Previous studies showed melatonin receptor 1A gene (MTNR1A) is highly polymorphic in seasonally breeding species. The aim of the present study was to detection of MTNR1A gene polymorphism and its association with production (body weights at birth, 1, 3, 6, 9 and 12 month of age) and reproduction (litter size) traits in Zandi sheep using PCR-RFLP methodology. Hundred blood samples were randomly collected and genomic DNA was isolated using modified salting out method. A part of exon 2 of MTNR1A gene was amplified using specific primer pairs. After amplification, the PCR product was digested with MnII endonuclease. Restricted digestion allowed the determination of two alleles (M, m) and two genotypes (MM, Mn) with frequencies of 0.91, 0.09 and 0.82, 0.18 for MTNR1A marker site in zandi sheep population. Least square means showed that MM individuals had higher body weight at one month of age than Mn individuals (P<0.05).

Using microsatellites for fine-mapping of recessive traits: an obsolete principle?
Buitkamp, J., Edel, C. and Goetz, K.-U., Bavarian State Research Center for Agriculture, Dep. of Animal Genetics, Prof.-Duerrwaechter-Platz 1, 85586 Poing, Germany; johannes.buitkamp@lfl.bayern.de

Since the first description of high-density SNP panel-based mapping of bovine recessive diseases by the group of Michel Georges the use of DNA Chip technology is considered as the most efficient method for homozygosity mapping in farm animals. Nevertheless, fine mapping can be hindered by unusual long recombination blocks in intensively selected cattle populations. Microsatellites and SNPs mutate at different rates and may therefore provide complementary results in fine-mapping of recessive traits. We used both, dense SNP data and microsatellite genotypes to track down the mutation causing the arachnomelia syndrome in Simmental cattle. The minimal region without recombination was determined by haplotype analysis using both, high density SNP and microsatellite sets. The results were very similar, but in the case of arachnomelia syndrome one microsatellite genotype allowed the detection of a single recombination event, shaping the region of interest from 3.1 to 1.8 Mb. We present the comparison of both fine mapping experiments as well as a general discussion on the effective use of microsatellites.

Bovine macrophages and mastitis: *in vitro* **gene response to** *Staphylococcus aureus* **exposure**
Lewandowska-Sabat, A.M.[1], Boman, G.M.[1], Sodeland, M.[2], Heringstad, B.[3], Storset, A.[1], Lien, S.[2] and Olsaker, I.[1], [1]Norwegian School of Veterinary Science, P.O. Box 8146 Dep., NO-0033 Oslo, Norway, [2]Centre for Integrative Genetics, Norwegian University of Life Sciences, P.O. Box 5003, NO-1432 Ås, Norway, [3]Geno Breeding and A.I. Association, P.O. Box 5003, NO-1432 Ås, Norway; anna.sabat@nvh.no

In this study, we integrated genomic data from genome wide association (GWA) mapping in cattle, with transcriptomic data for *in vitro* gene response to Staphylococcus aureus exposure of bovine macrophages. Macrophages are members of the innate immune system, and are widely distributed in all tissues. We have used these cells as a simple *in vitro* model of bovine mastitis. About 420 of the 17,000 genes on the ARK-Genomics bovine cDNA array responded as a result of exposure to S. aureus. Approximately 70% of the responding genes had a known identity (Entrez Gene ID) and were used in cluster analysis and identification of regulated pathways. Identified pathways include Toll-like receptor signalling, MAPK signalling, Th1/Th2 differentiation and apoptosis. The microarray results were verified by qRT-PCR analysis of selected genes in biological replicates of the *in vitro* macrophage exposure experiments. Further comparison of genomic location of differentially expressed genes with the QTL markers position revealed 31 candidates. Among these, several significant pathways and biological functions were found. Of specific interest is the role of cytokines. Cytokines mediate communication between immune cells and are responsible for the recruitment and migration of inflammatory cells into tissue during inflammation and infection. The expression of candidate genes is currently being tested in Norwegian Red sires with low and high breeding values for mastitis. Results from these analyses may emphasize further functional studies for identification of factors contributing to resistance to mastitis pathogens in cattle.

Characterization of the ovine αs1-casein (CSN1S1) gene and association studies with milk protein content in Manchega sheep breed

Calvo, J.H.[1], Serrano, M.[2], Sarto, P.[1], Berzal, B.[1] and Joy, M.[1], [1]CITA, Tecnología en Producción animal, Avda. Montañana 930, 50059- Zaragoza, Spain, [2]INIA, Mejora Genética Animal, Ctra. La Coruña km 7,5, 28040- Madrid, Spain; jhcalvo@aragon.es

There are evidences showing that polymorphism of the αs1-casein (CSN1S1) gene have been associated with an effect on milk protein, casein and fat content as well as on cheese yield in ruminants. This work focuses on the characterization and evaluation of the ovine αs1-casein (CSN1S1) as a candidate gene related to protein content. Primers were designed from ovine cDNA, bovine sequences and the Sheep Genome Assembly vs 1.0. Genomic DNA from animals with extreme values for protein content (n=8, Manchega and Assaf sheep breeds) was used to search polymorphisms. BLAST software was used to confirm gene identities. Subsequently, we used ClustalW software to align and identify polymorphisms. Studies of putative regulatory elements within the promoter and potential target sites for miRNA within the 3' UTR regions were performed using TF Search and microinspector soft wares. The CSN1S1 complete genomic DNA sequence was determined (18,427-bp), including promoter and UTRs regions. Exons were identified by comparison with ovine and bovine sequences. Sequencing studies revealed 61 polymorphisms: 5 polyT, 1 poly A, 1 GT microsatellite, 2 indels and 52 SNPs. Two polymorphisms detected in the 5' flanking region were located within possible trans-acting factor binding sites, modifying a putative CdxA and GATA-1 consensus sites. The 2 SNPs located in coding regions were synonymous substitutions (exon 14 and 17). Finally, a SNP located in exon18 modify a putative target site for a miRNA (bta-miR-631). This SNP was genotyped using RFLP-PCR. No associations were found between polymorphism in exon 18 and EBVs for protein content in a daughter design comprising 13 sire families in Manchega sheep breed. Further studies are being carried out in order to test the effect of CSN1S1 promoter polymorphisms in protein content.

Genetic parameters of test day milk yield in Brazilian Girolando cattle using an autoregressive multiple lactation animal model

Costa, C.N.[1], Freitas, A.[1], Barbosa, M.V.[1], Paiva, L.[2], Thompson, G.[3,4] and Carvalheira, J.[3,4], [1]Embrapa Gado de Leite, Juiz de Fora, MG, 36038-330, Brazil, [2]Girolando Breeders Association, Uberaba, MG, 38040-280, Brazil, [3]ICBAS, Universidade do Porto, Porto, 4099-003, Portugal, [4]ICETA/CIBIO, Universidade do Porto, Vairão, 4485-661, Portugal; cnc1@mail.icav.up.pt

The Girolando is a dairy cattle formed by crossing the Holstein and the Gir breeds. It is the most predominant cattle in dairy farming in Brazil. In the late nineties the Girolando Breeders Association (GBA) started running an AI progeny test of crossbred young sires. Genetic evaluation is currently based on fitting a lactation model. This study was aimed to estimate variance components and genetic parameters for test day milk yield of Girolando cattle, using an autoregressive test day multiple lactations (AR) animal model. Data consisted of test day (TD) records produced by Girolando cows under milk recording supervised by the GBA. After editing, 108,218 TD records from the first three lactations of 9,119 cows, sired by 1,285 bulls and calving from 1992 to 2008 in 214 herds were used to fit the AR model that included the fixed effects of herd, year-season of calving, days in milk within lactation order, regressions on age at calving (linear and quadratic), additive direct, dominance and recombination effects. The random effects were animal, short and long term environmental effects (fitted with autoregressive covariance structures) and the residuals (accounting for heterogeneity of variance by parity number). Medium heritability estimates, ranging from 0.17 to 0.27 of milk TD yield indicate opportunities for genetic gain by selection. Results from this study confirm the potential of using TD yields to replace the lactation model to estimate breeding values of Girolando sires and cows in Brazil. Further studies are needed to compare the resulting predictions of breeding values from these two models and the impact on selection decisions made by dairy farmers and on the expected genetic gain in milk yield of the Girolando cattle.

Comparison of residual variance effect on breeding values estimated by random regression models
Takma, Ç.[1], Akbaş, Y.[1], Orhan, H.[2] and Özsoy, A.N.[2], [1]Ege University, Department of Animal Science, Faculty of Agriculture, 35100 Bornova-IZMIR, Turkey, [2]Süleyman Demirel University, Department of Animal Science, Faculty of Agriculture, Çünür 32260-Isparta, Turkey; cigdem.takma@ege.edu.tr

Heterogeneous residual variance effect on breeding values was examined for test day milk yields of Turkish Holsteins. A third order random regression model including the fixed, random additive genetic, permanent environmental and residual effects were used. Lactation stages in months are assumed to have two residual variance schemes (RV1 and RV10). The predicted error variance was found 5.7302 with RV1 model and this variance was changed between 4.70 and 10.33 with RV10 model. Additive genetic variances (2.10-7.52 for RV1 and 1.5-6.7 for RV10) and permanent environmental variances (4.87-12.63 for RV1 and 5.2-12.71 for RV10) were higher at the beginning and late part of lactation, but lower at the middle of the lactation. However, heritability values (0.39-0.15 for RV1 and 0.3-0.11 for RV10) were decreased during the late part of lactation. Pearson product-moment, Spearman rank and Kendall rank correlations within cows (0.90, 0.87, 0.76) were found higher than within sires (0.86, 0.85 and 0.74). These correlations indicate that breeding values estimated from RV1 and RV10 models were highly correlated. The difference between the probability in same order EBV's and different orders EBV's was also highly correlated.

Estimation of the genetic diversity level in local populations of Turano-Mongolian horse breeds
Voronkova, V. and Sulimova, G., Vavilov Institute of General Genetics of Russian Academy of Sciencies, Laboratory of comparative animal genetics, Gubkin Str 3, 119991 Moscow, Russian Federation; valery.voronkova@gmail.com

Turano-Mongolian horses are ancient horse breeds inhabiting Altaian, Tuvinian, Transbaikalian and Mongolian steppe that are the regions believed to be the origins of horse domestication. To determine the genetic diversity and phylogenetic relationships among 13 populations (n=531) of 6 indigenous horse breeds (Mongolian, Tuvinian, Altaian, Transbaikalian, Buryat, Astrakhan) we employed two types of molecular markers: mtDNA polymorphism and ISSR fingerprinting. Sequences of the control region (D-loop) of mtDNAs (15,341-15,837) were made for 25 samples of 6 populations (Northern Mongolian, Gobi desert, Central Mongolian, Buryat and Transbaikalian). The 398 bp long fragments (15,399-15,796) of the mtDNA were taken after alignment for further analysis. We found 41 variable sites where 39 transitions and 2 transversions were observed. Phylogenetic tree was built using software MEGA 4.1 by a Neighbor-joining method. Three northern populations (Northern Mongolian, Transbaikalian and Tuvinian) with a high bootstrap value (90) were placed in one cluster. Their natural habitats are closely located and migration between the populations may occur. Buryat, Central and Southern Mongolian populations are detached; Gobi desert population is the most distinct one. For the ISSR analysis two primers were chosen out of 6 primers for the final study as the most informative marker systems. A total of 47 ISSR bands was obtained with two primers – $(ACC)_6G$ and $(GAG)_6C$ – among which 46 (97,9%) were polymorphic ones. High genetic diversity level was shown for all the studied populations in comparison with thoroughbred horses. Two Altaian populations had the highest genetic diversity level what can indicate a gene flow or crossbreeding. Genetic distances were calculated and analyzed with a method of principal components. Population distribution was in accordance with their geographical affiliations.

**Characterization of growth and egg production of two Portuguese autochthonous chicken breeds:
Preta Lusitânica and Amarela**

*Soares, M.L.[1], Lopes, J.C.[1], Brito, N.V.[1] and Carvalheira, J.[2,3], [1]IPVC, Escola Superior Agrária, Ponte de
Lima, 4990-706, Portugal, [2]University of Porto, ICBAS, Porto, 4099-003, Portugal, [3]University of Porto,
ICETA/CIBIO, Vairão, 4485-661, Portugal; laurasoares@esa.ipvc.pt*

The aim of this study was to analyze female growth and egg production of 2 Portuguese indigenous chicken
breeds, Preta Lusitânica (PL) and Amarela (AM). These breeds are classified as endangered with less than
2,000 females per breed. The growth period was from birth to the end of the first laying period and included
2 lots per breed: PL1 (AM1) born in September (November) of 2008 and PL2 (AM2) born in March (April)
of 2009. The birds were fed a commercial starter diet from 0 to 20 d and then were allowed to graze freely
with a daily supplement of corn in a free range production system. Body weights (BW) were recorded every
15 d. The Gompertz function was chosen to describe the growth curve. The adult BW was estimated for
PL1 (PL2) as 2,642±45.5 g (2,433±17.6 g) and for AM1 (AM2) 2,027±43.6 g (1,902±12.7 g), respectively.
The maximal growth rate was estimated by taking the derivative of the Gompertz equation and was 20.3
g/d (17.9 g/d) at 65 d (78 d) for PL1 (PL2) and 14.9 g/d (18.2 g/d) at 62 d (71 d) for AM1 (AM2). Egg
laying started at 152 d (199 d) of age for PL1 (PL2) with an average BW of 2,247 g (2,226 g), averaging 65
eggs/bird/yr (40 eggs/bird/yr). AM chickens started laying at 149 d (154 d) of age for AM1 (AM2) with an
average BW of 1,703 g (1,696 g), averaging 120 eggs/bird/yr (99 eggs/bird/yr). The average egg weight in
the different flocks was 53.5±6.3 g for PL1, 56.9±4.7 g for PL2, 54.8±19.1 g for AM1 and 54.8±2.6 g for
AM2. Differences within breed are related with month of birth. PL1 and AM1 which began laying in March
and April (longer days length) had better results than those initiated in January. Increase in light exposure
triggers hens to begin laying eggs. Differences between breeds for these traits will provide information for
a better definition of the chicken's type of production.

Genetic analysis of fertility in Swedish beef breeds

*Näsholm, A., Swedish University of Agricultural Sciences, Animal Breeding and Genetics, P.O. Box 7023,
SE 750 07 Uppsala, Sweden; anna.nasholm@slu.se*

Traits of importance for fertility were studied in Swedish beef breeds. The overall aim was to enable
inclusion of fertility in the genetic evaluation. Traits studied were age at first calving (age1) and interval
between first and second (int12), second and third (int23), and third and fourth (int34) calving. Data included
in total 33.622 observations for Charolais cows and 17.130 observations for Hereford. Linear uni- and
multivariate animal models were used for the analyses. Average for age1 was 26.6 and 25.9 months for
Charolais and Hereford, respectively. Average calving intervals varied for Charolais between 370 and
383 days and for Hereford between 369 and 376 days. The values decreased with calving number. Direct
heritabilities for age1 were medium high (0.25 to 0.38) and maternal heritbilites were low (0.06 to 0.12).
A moderately negative genetic correlation (-0.58 to -0.74) was found between direct and maternal effects.
Heritabilities for int12 were low (0.05 to 0.12) and even lower for int23 and int34 (0 to 0.03). Genetic
correlations between age1 and int12 were low (0.14 to 0.23) whereas between int12 and int23 moderately
high to high genetic correlations (0.60 to 0.92) were estimated. Low negative genetic correlations (-0.33 to
-0.07) were found between age1 and 365 d-weight of the heifer. It is concluded that age1 and int12 could
be used for genetic evaluation of fertility in Swedish beef breeds. Maternal genetic effects on age1 should
then be considered.

Variance components and heritability estimations of milk yield from partial test day milk records using random regression model

Akbaş, Y.[1], Takma, Ç.[1], Orhan, H.[2] and Özsoy, A.N.[2], [1]Ege University, Department of Animal Science, Faculty of Agriculture, 35100 Bornova Izmir, Turkey, [2]Süleyman Demirel University, Department of Animal Science, Faculty of Agriculture, 32260 Çünür Isparta, Turkey; yavuz.akbas@ege.edu.tr

In this study variance components and heritability of milk yield were estimated from partial test day milk records using random regression model. Partial records were including 3, 6 or 10 monthly test day milk yields. The random regression model was used to estimate variance components and heritability of milk yields. The third order random regression model included fixed, random additive genetic, permanent environmental and residual effects. Data set was consisting of 5,616, 11,260 and 18,788 records with 3, 6 or 10 TD of 1901 Holstein Friesians from 1998 through 2009 in Turkey, respectively. Constant residual variances were 6.19, 6.48 and 7.03 using 3, 6 or 10 TD, respectively. Random additive genetic, permanent environmental variances and heritabilities were highest at the beginning of lactation and decreased towards to last part of lactation. Average heritability estimates of TD milk yields were 0.24, 0.28 and 0.34, respectively. Heritability was slightly underestimated by using partial records.

Multiple trait genetic analysis of growth in the pirenaica beef cattle breed

González-Rodríguez, A., Altarriba, J., Moreno, C. and Varona, L., University of Zaragoza, Animal Breeding and Genetic Quantitative, Miguel Servet 177, 50013, Spain; aldemarango69@yahoo.es

One of the main objectives of selection for beef cattle is weight at slaughter. Live weights along growth are currently used as selection criteria under the hypothesis of a positive genetic correlation with weight at slaughter. In this study we have used data for weight at 120 (W120) and 210 (W210) days and at slaughter (WS) and the gain of weight between them (Δ120-210 and Δ210-WS) in the Pirenaica Beef Cattle Breed to increase our knowledge about the genetic and environmental relationships between different phases of growth. The posterior mean (and standard deviation) estimates of heritability were 0.343 (0.016), 0.347 (0.019), 0.349 (0.022), 0.292 (0.030), and 0.334 (0.058) for W120, W210, WS, Δ120-210 and Δ210-WS, respectively. Further, genetic correlations were high and positive between weights at different ages (W120 and W210), but more moderate when related with weight at slaughter (WS). These results question the efficiency on the current scheme of selection. Genetic and residual correlations among gains of weight and immediately posterior weights (Δ120-210 and W210 and Δ210-WS and WS) were positive and very high. However, genetic correlations between gains of weight and previous weights were low or null and residual correlations between them were negative (-0.30). The presence of these negative residual correlations suggests the existence of compensatory growth that is not considered in regular strategies of genetic evaluation.

Characterization of the gene Melatonin Receptor 1A (MTNR1A) in the Rasa Aragonesa sheep breed: association with reproductive seasonality

Martinez-Royo, A.[1], Lahoz, B.[1], Alabart, J.L.[1], Folch, J.[1] and Calvo, J.H.[2], [1]CITA, Producción Animal, Av Montañana 930, 59059 Zaragoza, Spain, [2]ARAID, Fundación, 50004 Zaragoza, Spain; amartinezroyo@aragon.es

This work focuses on the characterization and evaluation of MTNR1A as a candidate gene related to reproductive seasonality in the Rasa Aragonesa sheep breed. MTNR1A has shown influence on reproductive seasonality in other breeds. Cyclic and non cyclic ewes for a given month were classified as a dichotomous variable coded with '1' and '0', respectively. The percentage of cyclic ewes between February and July 2009 was analyzed using a generalized linear model for categorical variables with repeated measures by the CATMOD procedure of SAS. The model used was FC(ik) = μ + gene(i) + month(k) + GxM(ik), where FC(ik) is the frequency of cyclic ewes with genotype(i) and month (k); μ is the overall mean percentage of cyclic ewes from the total set; gene(i) is the effect of genotype(i) nested to the ram; month (k) is the effect of the month, treated as a dichotomous repeated variable cyclic/non cyclic within 6 months of the experimental tests; GxM (ik) is the interaction effect. A significant effect was found between SNP606 of the MTNR1A gene and spontaneous out of season estrus behaviour. The T allele was associated with cyclicity in the Rasa Aragonesa breed. This finding, along with the fact that this polymorphism does not result in an amino-acid substitution, suggest that SNP606 may act in linkage equilibrium with a mutation in other genes responsible for out of season breeding. New polymorphisms (11 SNPs) in the coding region of the gene MTNR1A did not show any association for reproductive cyclicity during anestrus in the Rasa Aragonesa breed. Association between polymorphisms (17 SNPs) found in the promoter region and cyclicity is expected to be completed soon.

Detection of QTL affecting serum cholesterol, LDL, HDL, and triglyceride in pigs

Cinar, M.U., Uddin, M.J., Duy, D.N., Tesfaye, D., Tholen, E., Juengst, H., Looft, C. and Schellander, K., Institute of Animal Science, University of Bonn, Animal Breeding and Husbandry, Endenicher Allee 15, 53115 Bonn, Germany; ucin@itw.uni-bonn.de

The objective of this research was to investigate the chromosomal regions influencing the serum level of the total cholesterol (CT), triglyceride (TG), high density protein cholesterol (HDL) and low density protein cholesterol (LDL) in pigs. For this purpose, a total of 330 animals of the Duroc × Pietrain F2 population were phenotyped for serum lipids using ELISA and were genotyped by using 122 microsatellite markers covering all porcine autosomes for QTL study in QTL Express. Blood sampling were performed at approximately 175 days just before slaughter of the pig. Most of the traits were correlated with each other and were influenced by average daily weight gain, slaughter date and age at slaughter of the animals. A total of 18 QTL including three QTL with imprinting effect were identified on 11 different porcine autosomes. Most of the QTL reached to 5% chromosome-wide (CW) level significance including a QTL at 5% experiment-wide (GW) and a QTL at 1% GW level significance. Of these QTL, four were identified for both the CT and LDL, whereas two QTL were identified for both the TG and LDL. Moreover, three chromosomal regions were detected for the HDL/LDL ratio in this study. A QTL for HDL on SSC2 and two QTL for TG on SSC11 and 17 were detected with imprinting effect. The highly significant QTL (1% GW) was detected for LDL at 82 cM on SSC1, whereas significant QTL (5% GW) was identified for HDL/LDL on SSC1 at 87 cM. Chromosomal regions with pleiotropic effects were detected for correlated traits on SSC1, 7 and 12. Two novel QTL on SSC16 for HDL and HDL/LDL ratio and an imprinted QTL on SSS17 for TG were detected in the pig for the first time. The results of this work shed additional light on the genetic background of serum lipid concentrations and these findings will be helpful to identify candidate genes in these QTL regions related to lipid metabolism and serum lipid concentrations in pigs.

Fine mapping of quantitative trait loci for androstenone and skatole levels in a French commercial Large White pig population

Le Mignon, G.[1], Robic, A.[2], Shumbusho, F.[1], Iannuccelli, N.[2], Billon, Y.[3], Bidanel, J.-P.[1] and Larzul, C.[1], [1]INRA, UMR1313 GABI, F-78350 Jouy-en-Josas, France, [2]INRA, UMR444 LGC, F-31326 Castanet-Tolosan, France, [3]INRA, UE967 GEPA, F-17700 Surgères, France; jean-pierre.bidanel@jouy.inra.fr

The objective of this study was to map boar taint QTL in a French commercial population of 480 Large White pigs using the Illumina PorcineSNP60 BeadChip whole genome single nucleotide polymorphism (SNP) assay. Pigs were slaughtered at 110 kg liveweight and backfat sample was taken for skatole and androstenone measurements by HPLC and GC-MS, respectively. We first performed a genome wide association study (GWAS) using a family-based association tests for quantitative traits to detect SNP associations; then we combined linkage disequilibrium with linkage analysis (LDLA) using the same whole data set. We detected four loci with the first method and 14 new QTL regions with the second one. For these 18 boar taint QTL (9 for androstenone and 9 for skatole), we performed haplotype association tests and characterized four highly significant haplotypes specifically associated with either the levels of either androstenone or skatole. The three androstenone haplotypes and the skatole haplotype explained 39.6% and 11% of trait genetic variance, respectively. The most significant skatole QTL was located on SSC8 and was detected by the three strategies. The most promising haplotype to reduce androstenone in French LW pigs was located in the CYP11A1 region.

Polymorphism evaluation of microsatellite DNA markers and genetic structure of Baluchi sheep

Dashab, G.R.[1], Aslaminejad, A.A.[1], Nassiri, M.R.[1], Esmailizadeh Koshkoih, A.[2] and Saghi, D.A.[1], [1]Ferdowsi University of Mashhad, Animal science, Azadi square, 9177948974, Mashhad, Iran, [2]Shahid Bahonar University, Animal science, 22 Bahman Blvd,Kerman, 76169-133, Kerman, Iran; golamr.dashab@agrsci.dk

Population structure, diversity and genetic variation in two flock of Baluchi sheep breed was investigated using seven microsatellite loci. In all, 503 individuals belonging to two flocks were genotyped. The total number of alleles were 38 on the seven studied loci and ranged from 4 in two loci bm1853 and bms1714 to 7 in mcm200 and rm0006 loci. The seven tested loci were all polymorphic in two flocks of Baluchi sheep population. The average direct count of heterozygosity for overall loci in each flock was less than the expected heterozygosity. Tests of genotype frequencies for deviation from Hardy-Weinberg equilibrium (HWE), at each locus in the whole population, revealed significant departure from HWE. A slightly low rate of inbreeding within two flocks was noticed (Fis=0.05). Low genetic differentiation was detected by estimation of Fst index between all populations. The Fst values showed that about 2.8% of the total genetic variation was explained by the population differences and 97.2% corresponded to differences among individuals. The mean of polymorphism information content (PIC) for all loci in Baluchi population was 0.65, which indicated that all the studied loci were highly polymorphic and could be very efficient for QTL mapping.

Estimation of direct and maternal (co)variance components for growth traits in Moghani sheep
Savar Sofla, S.[1], Abbasi, M.A.[2], Nejati Javaremi, A.[1], Vaez Torshizi, R.[1] and Chamani, M.[1], [1]Department of Animal Science,Faculty of Agri.,Science and Research Branch,Islamic Azad University, Tehran, 0098-21, Iran, [2]Animal Science Research Institute, Karaj, 0098-261, Iran; simasavar@gmail.com

The objective of this study was to estimate genetic and non-genetic parameters for birth weight (BW), weaning weight (WW), 6-month weight (6MW), average daily gain from Birth to Weaning (ADG_{0-3}), and average daily gain from Weaning to 6 months of age (ADG_{3-6}), Kleiber ratio from Birth to Weaning (KR_{0-3}) and Kleiber ratio from weaning to 6 months of age (KR_{3-6}). Data were collected at Moghani Sheep Breeding Station at Jafarabad, Iran, during 1995 to 2008. (Co)variance components and genetic parameters were estimated by univariate and multivariate animal models using Restricted Maximum Likelihood procedure. The log likelihood ratio test (LRT) was carried out to determine the most suitable model for each trait. The effects of year and season of birth, birth type, and sex of lamb were significant for all studied traits. The age of dam was significant for BW, WW, ADG_{0-3}, ADG_{3-6}, KR_{0-3} and KR_{3-6}. Direct heritability estimates for BW, WW, 6MW, ADG_{0-3}, ADG_{3-6}, KR_{0-3} and KR_{3-6} based on most appropriate models were 0.104 ± 0.03, 0.098 ± 0.01, 0.085 ± 0.02, 0.212 ± 0.04, 0.023 ± 0.04, 0.132 ± 0.04 and 0.014 ± 0.01, respectively. The estimate of maternal heritability varied from 0.017 ± 0.01 for KR_{0-3} to 0.127 ± 0.03 for ADG_{0-3}. The estimates of maternal permanent environmental effect were 0.099 ± 0.01, 0.046 ± 0.02 and 0.068 ± 0.05 for 6MW, ADG_{0-3} and KR_{0-3}, respectively. Direct genetic correlations among studied traits ranged from -0.70 for BW and ADG_{3-6} to 0.98 for WW and ADG_{0-3}. The low estimates of direct heritabilities for growth traits obtained in this study indicated that selection for growth traits would result in slow genetic improvement. Maternal effects were significant on weights in different ages of Moghani sheep and should be taken account in any selection program to improve the efficiency of this breed.

Genetic and phenotypic trends estimation for growth traits in Moghani sheep
Savar Sofla, S.[1], Nejati Javaremi, A.[1], Abbasi, M.A.[2], Vaez Torshizi, R.[1] and Chamani, M.[1], [1]Department of Animal Science,Faculty of Agri.,Science and Research Branch,Islamic Azad University, Tehran, 0098-21, Iran, [2]Animal Science Research Institute, Karaj, 0098-261, Iran; simasavar@gmail.com

Data of Moghani sheep were used to estimate genetic and phenotypic trends during 1994-2008 for growth traits. Animal's breeding values were predicted by multi-trait animal models based on best linear unbiased prediction (BLUP) techniques. Genetic and phenotypic trends were estimated using regressing of mean of breeding value or phenotypic on year of birth, respectively. Genetic trends of birth weight (BW), weaning weight (WW), six month of age (6MW), average daily gain from birth to weaning (ADGa), average daily gain from weaning to six month of age (ADGb), klieber ratio from birth to weaning (KRa) and kleiber ratio from weaning to six month of age (KRb) were 0.007 kg, 0.082 kg, 0.084 kg, 0.807 gr, 0.392 gr, 0.032 gr and 0.014 gr per year, respectively. Phenotypic trend was 0.031, 0.528, 0.036 kg per year for BW, WW, 6MW and 4.258, 0.144, 0.165 and -0.235 gr per year for ADG0-3, ADG3-6, KR0-3 and KR3-6, respectively. The genetic trend was positive and significant for all studied traits (P<0.01) and phenotypic trend was positive and significant for BW (P<0.01), WW and KR3-6 (P<0.05). It confirms that selective breeding can lead to significant genetic improvement in Moghani sheep. The relatively low annual progress reflects actual lack of consistent directional selection for defined selection goals. During the 14-year period, total genetic improvement was 0.101, 1.005 and 1.144 kg for BW, WW and 6MW, and was 9.702, 4.640, 0.379 and 0.171 gr for ADGa, ADGb, KRa and KRb, respectively. Generally, the result of our study showed low genetic improvement in traits during this period. Although it confirmed that selective breeding could result in significant genetic improvement in performance of Moghani sheep. The main reason could be the absence of comprehensive selection programs or well defined criteria.

Use of antimicrobial drugs in Europe

Van Geijlswijk, I.M.[1], Bondt, N.[2], Puister-Jansen, L.[2] and Mevius, D.J.[1], [1]Utrecht Universtity, fac of Veterinary Medicine, Pharmacy, Yalelaan 106, 3584 CM Utrecht, Netherlands, [2]LEI, part of Wageningen UR, Markets & Chains, P.O. Box 35, 6700 AA Wageningen, Netherlands; i.m.vangeijlswijk@uu.nl

Veterinary antimicrobial drug use is usually reported in tonnes of active substance. A method to convert tonnes to daily dosages per animal year (present livestock) is presented, and the resulting data for 9 European countries and the USA, for the years 2001-2009 are shown. A comparison of Sweden, Norway, Finland, Denmark, Germany (2003 and 2005 data only), Switzerland, The Netherlands, UK, France and the USA revealed the highest use in the USA and The Netherlands, while the lowest use was in Norway and Sweden. The veterinary use of antimicrobials seems to be higher in countries with higher density of livestock. A higher density could imply an increased risk for transmission, what could lead to more treatments. In humans ambulatory use no relation with density of inhabitants is found. Remarkable is the big difference between human and veterinary use in the Netherlands, where France displays relatively high use in both animals and humans and Norway shows overall low usage.

An overview of veterinary antimicrobial use policies, legislation, and regulations: background, scientific data, and the basis for discussions in the USA

Hoang, C., American Veterinary Medical Association, 1931 N. Meacham Rd., Schaumburg, IL 60173-4360, USA; CHoang@avma.org

For more than 40 years, antibiotics have been used in the United States to protect our food supply and improve animal health and welfare. Some estimates show that up to 95% of antibiotics used for livestock are devoted to therapeutic uses (defined as treatment, control or prevention) for disease conditions. Others have suggested a much greater proportion than 5% of antibiotics in livestock used for production purposes (feed efficiency, improved rate of weight gain). These drugs are often referred to as 'growth promoters'. Some countries have instituted bans on the use of antibiotic growth promoters (AGPs), yet there has not been clear evidence of a significant reduction in human antibiotic resistance patterns as a result. The data suggests that the bans did, however, initially result in increased death and disease among animals. Greater amounts of antibiotics continue to be used to treat and prevent disease in animals, and little evidence to suggest that antibiotic resistance in humans has stabilized or improved. Discussions on veterinary antibiotic use in the United States often revolve around disagreements on interpretations of data, differing perceptions and definitions, and proposals of how to proceed in the future. It is important to note that veterinarians are the only health professionals that routinely operate at the interface of human and animal health. This makes veterinarians uniquely qualified as stewards of proper antibiotic usage by adhering to judicious therapeutic use guidelines to ensure safe animal husbandry and food production. Presently, the United States has several layers of protection in place to ensure antibiotics used to keep animals healthy do not harm public health – including FDA approval, post-approval monitoring, judicious therapeutic use guidelines, and various monitoring and surveillance systems. Yet, many feel that those are not enough and there is still a need for additional policies, legislation, and regulation.

Antibiotic replacement strategies in intensive livestock production
Den Hartog, L.A.[1,2] *and Sijtsma, S.R.*[1], [1]*Nutreco, R&D and Quality Affairs, P.O. Box 220, 5830 AE Boxmeer, Netherlands,* [2]*Wageningen University, Animal Nutrition Group, P.O. Box 338, 6700 AH Wageningen, Netherlands; leo.den.hartog@nutreco.com*

The past success of antibiotics to prevent and treat infectious animal diseases threatens its future utility. The selection for antibiotic-resistant bacteria in intensive livestock production and its impact on animal and human health has become a major concern. As a result the development of alternative strategies receives high priority. Most alternatives focus on optimising the health status of the animal and prevention of disease. There are immunological, ecological and bactericidal strategies. The immunological strategy of vaccines has favourable attributes such as low cost, administration ease, multiple agent efficacy and safety. Ecological and bactericidal strategies seem to be less well developed. Ecological strategies usually focus on maintaining gut health and include feeding measures, such as the use of prebiotics, probiotics and enzymes. Pre and probiotics improve the intestinal microbial balance. In addition, enzymes improve gut morphology and nutrient absorption. Ecological strategies could be successful and cost-effective, if the knowledge on the gastrointestinal ecosystem and microbial-host interactions is increased. Bactericidal strategies focus on elimination of microorganisms in or outside the GIT. Such strategies include bacteriophages, bacteriocins, organic acids, and a wide range of natural products such as etheric oils. Bacteriophages and bacteriocins are attractive alternatives to antibiotics due to their specificity, mode of action and safety. However, unless new production techniques are introduced, costs are too high. Economic feasibility seems to be better for organic acids and medium chain fatty acids, in particular when taking into account the proven efficacy. In conclusion: both ecological and bactericidal strategies could be successful and cost effective alternatives. However, total replacement of antibiotics is not yet possible with the current products.

Sustainable use of antimicrobials in animal health: an industry view
Simon, A.J., Pfizer Animal Health, 23-25 Ave du Dr Lannelongue, 75688 Paris, France; tony.simon@pfizer.com

The effective management of disease in animals is critically important for supply of sufficient affordable and wholesome food and for the maintenance of animal welfare, including companion animals. Antimicrobials have helped revolutionise agriculture since their introduction in the late 1940s. Initially widely used, but concerns are now growing over antimicrobial resistance. The focus of regulatory agencies, veterinarians and others is now more on risk management and sustainable use. Concerns were initially limited to target pathogens but currently are more about the movement of resistance bacteria or genes between bacteria in animals and human pathogens. For a new antimicrobial, risk management starts with a decision to invest € 10 Ms. If successful, development ends with formal review of the data generated by regulatory agencies, including environmental and antimicrobial safety. The final decision on whether the antimicrobial is approved, for what indications and with what restrictions or warnings, also rests with the regulatory agencies. Following approval, antimicrobials can only be used in Europe on veterinary prescription. This means that they should only be prescribed for their indicated uses or, where no suitable antimicrobial is available in accordance with the European prescribing Cascade: the veterinarian assesses likely efficacy and safety before making an appropriate choice. Veterinary use should follow prudent use guidelines and, in addition, post authorisation monitoring of resistance status is often required by the regulatory agencies. Antimicrobials will remain critically important for agricultural and companion animals for the foreseeable future and new classes will be necessary. Continued awareness of the wider environmental implications, including possible effects on human medicine, is important. Decisions must be based on good science with a strong evidence base and not on broad brush measures: a realistic assessment of the real impact of any changes should be undertaken before they are introduced.

Usage of antibiotics in organic and conventional livestock production in Denmark

Sørensen, J.T., Aarhus Faculty of Science and Technology, Department of Animal Health and Bioscience, Blichers Alle 20, DK-8830 Tjele, Denmark; jantind.sorensen@agrsci.dk

The increased usage of antibiotics in Danish livestock production is a major public concern. The increase in medication is seen as a symptom of an unsustainable industrialised animal production with a compromised animal welfare. Further the increased usage of antibiotics increase the risk of resistance problems in humane medicine, such as Methicillin-resistant Staphylococcus aureus (MRSA) which may origin from pig herds. Studies on the usage of antibiotics in Danish livestock herds show that there are a large variation in usage in antibiotics between herds, which is not reflected in a similar variation in animal health. Thus, there seems to be room for a reduction in usage of antibiotics though improved management. The usage of antibiotics in organic slaughter pig production has been found to be only 10-20% compared to the level used in conventional slaughter pig production. The number treatments with antibiotics in organic dairy production has been found to be 64% of what is found in conventional dairy herds. The substantial lower usage of antibiotics in organic livestock production may be an inspiration for conventional livestock production. The differences in management between the two production systems, which may relate to disease level and usage of antibiotics are further investigated in this paper and the possibilities for developing a sustainable livestock production with a high level of animal welfare combined with a low level of usage of antibiotics is discussed.

Cow health and medicine use as part of dairy farm management in The Netherlands

Kuipers, A.[1] and Wemmenhove, H.[2], [1]Expertise Centre for Farm Management and Knowledge Transfer Wageningen UR, P.O. Box 35, 6700 AA Wageningen, Netherlands, [2]Livestock Research Wageningen UR, P.O. Box 65, 8200 AB Lelystad, Netherlands; abele.kuipers@wur.nl

Antibiotics use has become a hot public issue in The Netherlands, because of the growing resistance against antibiotics in human medical treatment. In animal husbandry registration of medicines on farm level by farmers is an obligatory part of quality assurance schemes. However, the utilization of these data for management purposes has not been developed yet. On 75 conventional farms from 8 veterinary practices the medicine use was administered in detail with goal to make such data useful for management purposes. In this report we restrict ourselves to antibiotics. Data were collected over the period 2005-2009. The antibiotic use on farm level is expressed in daily dosages per cow per year. The average over the 5 year period was 9.2 dosages per cow/year with a spread from 3-12. The farms were pictured according to the criteria: below/above average and decreasing/increasing over years. Daily dosages were split up in contributions by dry-off treatments (45%), mastitis (22%) and other diseases (33%). Farm and farmer characteristics were studied affecting antibiotics use, practicing a step-wise regression procedure. Number of total daily dosages are explained significantly by health status (+), average cell count herd (-) and quota size (+). Number of daily dosages mastitis are affected by herd size (+), level of milk production/cow (+) and access to pasture (when outside than more use). Number of daily dosages other diseases by quota size (+), young stock/10 cows (+), cell count (-) and % cows removed from herd (-). Number of daily dosages dry-off are strongly negatively correlated with cell count, expressing that there is a trade off between level of cell count and level of antibiotics use for dry off treatment. Question arises: which of both is more important in the long run to be reduced for the dairy sector?

Impression about cow health and medicine use in Germany using data from research farms
Thaller, G. and Junge, W., Institut für Tierzucht und Tierhaltung, Christian-Albrechts-Universität zu Kiel, Hermann-Rodewald-Strasse 6, 24118 Kiel, Germany; Georg.Thaller@tierzucht.uni-kiel.de

Constitution of cows and maintenance of health is of increasing interest in German cattle populations. When considering longevity however, it seems that the functionality steadily decreases and that there is some need to improve health traits. Application of medicine is seen more and more critical by consumers and the focus should be set on breeding a more robust cow which is not in need to be treated with drugs. Drug use on 3 experimental farms in Germany was analyzed. Results of this analysis will be presented and placed in a wider perspective.

Animal health and medicine use in Slovenia and some other Balkan countries
Klopcic, M.[1] and Stokovic, I.[2], [1]University of Ljubljana, Biotechnical Faculty, Dept. of Animal Science, Groblje 3, 1230 Domzale, Slovenia, [2]Veterinary Faculty, Zagreb, Croatia; Marija.Klopcic@bf.uni-lj.si

Antibiotics use in animal husbandry is an issue of growing public concern. However, practices and intensity of antibiotics use will vary over Europe. In this contribution veterinary practices in Slovenia and Croatia are described: how do veterinaries work together with farmers and institutions, and which guidelines are followed to apply medicines for cattle. An orientation took place about frequency and amounts of antibiotics used on dairy farms. In Slovenia, different health categories are being linked to antibiotics use to gain more insight. Also general practices towards drying-off therapy are inventoried. In Europe drying-off represents the major input of antibiotics in the udder besides mastitis treatments. Moreover, farmers are being questioned concerning their practices of applying antibiotics. What influences their behavior towards the level of application of medicines?

Antibiotics, hormones and sustainability of the US dairy and beef industries
Capper, J.L., Washington State University, Department of Animal Sciences, P.O. Box 646310, Pullman, WA 99164-6310, USA; capper@wsu.edu

Sustainability encompasses environmental, economic and social issues. As the global population increases, livestock industries face a challenge in producing animal protein that is economically affordable, has a low environmental impact and meets consumers' social expectations. US consumers often perceive sustainable agriculture as being confined to extensive low-input:low-output systems and thus regard conventional agricultural systems as being less sustainable. Animal proteins are considered staple foods, yet concern over the perceived sustainability of conventional animal production may threaten future social license to operate. Antibiotics, ionophores and hormones may be used within conventional US beef and dairy systems to improve growth, feed efficiency, milk production and reproductive efficiency. Improved efficiencies bestowed by these productivity-enhancing technologies (PET) confer financial advantages to the producer in terms of profit over expenses that can be passed onto the consumer, thus improving economic sustainability. Environmental sustainability is also improved by PET use – results from a deterministic model based on beef cattle nutrition and metabolism showed that energy use was reduced by 15.0%, land use by 18.4%, water use by 15.4% and the carbon footprint by 15.0% per kg of beef in conventional systems using PET. Nonetheless, the social sustainability issues relating to PET use are considerable. The majority of US consumers purchase food based on price, convenience and taste, however a growing group of consumers regard PET as having significant human health implications ranging from endocrine imbalances to antibiotic resistance. These concerns have led to retailers demanding the withdrawal of PET from their supply chains in order to gain market advantage. The contribution of PET to two of the three pillars of sustainability (economic and environmental) is without question, however, overcoming the social issues involved with PET use within the livestock industry remains a significant concern.

Reducing antibiotics in organic dairy farming
Nauta, W.J., Van Veluw, K. and Wagenaar, J.P., Louis Bolk Institute, animal production, Hoofdstraat 24, NL-3972LA, Netherlands; w.nauta@hccnet.nl

Policy makers in the Netherlands have set the task for livestock production to reduce the use of AB by 50% in the next 3 years to limit the risks of the spreading of resistant bacteria. Knowledge and experience of organic farmers can be used to help conventional farmers with this important task. Organic dairy farmers are only allowed to use antibiotics (AB) in specific circumstances. Preventive use of AB is prohibited and treatments with AB are only allowed to prevent animals from serious suffering. Milk of animals that have been treated more then two times a year, can not be sold as certified organic product. Many farmers struggle to find new ways to keep their herds healthy. In a survey among 182 Dutch organic dairy farmers information was gathered on farming styles, the use of AB and alternatives. Relations between farms, breeds and health treatments were studied. Also practical information of 15 AB-free farms was documented. Preliminary results show that there is a large variation in the use of health treatments and medications. A growing number of Dutch organic farmers (now about 20%) does not use AB at all. They indicate that converting from conventional to organic farming was easier than converting to AB free milk production. First they run into more health problems, mainly concerning udder health. Cows have to get used to the AB free environment. It is seen as part of a whole farming concept that has to be in balance. Other farmers try to limit the use of AB and seek for knowledge and information on alternative treatments. AB is mainly used for udder problems. The choice of the breed and housing system seem to have influence too. More results are coming up and the information will be used to transfer knowledge to researchers, advisors, policy makers and farmers involved in animal health projects, consultancy and study groups. It can be concluded that the use of AB can be limited, even to zero, however, more information is needed about the impact and best practices at farm level.

Consumption of antibiotics in Austrian cattle production

Obritzhauser, W.[1], Fuchs, K.[2], Kopacka, I.[2] and Egger-Danner, C.[3], [1]Institute for Veterinary Public Health, University of Veterinary Medicine, Vienna, Austria, [2]Austrian Agency for Health and Food Safety, Data, Statistics & Risk Assessment, Graz, Austria, [3]ZuchtData EDV-Dienstleistungen GmbH, Dresdner Straße 89/19, 1200 Vienna, Austria; egger-danner@zuchtdata.at

There are major concerns about the impact of antibiotics used in veterinary medicine on the spread of resistance because this causes difficulties to treat human diseases. The prudent use of antibiotics is therefore crucial in animal production. In order to evaluate the effect of antibiotics on antimicrobial resistance in cattle, data about the usage of antimicrobial substances for the treatment of cattle diseases are a prerequisite. Records of diagnoses and prescriptions from 10 veterinary practices dealing with cattle were used as a spot test to project the consumption of antibiotics. The amount of antimicrobials used was linked to a key ofdiagnoses, which wasgenerated for the Austrian health monitoring system for cattle. As a unit of measurement for the consumption of antibiotics the number of prescribed daily doses (PDD) per 500 kg bodyweight (BW) per year was estimated by Monte Carlo simulation techniques. A median number of 2.82 PDD/500 kg BW was estimated for the overall consumption per year in cattle. Tetracyclines, beta-lactam antibacterials and combinations of antibiotics were most frequently used. The biggest quantities of antimicrobials were applied for the metaphylaxis of crowding diseases and the therapy of pneumonia in fattening cattle. penicillins, cephalosporins and combinations of antibacterials were most often used for the treatment of mastitis and for drying off in dairy cattle. Antimicrobials classified as prioritized critically important (cephalosporins and quinolones) added up to approx. 12% of all prescribed doses. Most often these substances are used in mastitis and diseases of the respiratory tract. Related to a method to estimate the consumption of antimicrobials, recombining consumption data and diagnosis data gives an insight into the intensity of treatment.

'Therapy evaluation': a tool for sustainable antibiotic use

Huijps, K., Van Hoorne, J., De Hoog, J. and Koenen, E., CRV, P.O. Box 454, 6800 AL Arnhem, Netherlands; erwin.koenen@crv4all.com

Animal health is no longer an issue for the veterinarian alone. Political and social developments take place at high speed where sustainable antibiotic is a major issue. The veterinarian will play an important role by advising farmers for a sustainable use of antibiotics. To give this advice, it is of great importance for a veterinarian to have as many farm specific data on medicine use, dosage, frequency, and duration as possible. Other relevant information is a sudden increase in use of antibiotics in a certain period and a comparison with other farmers. In the Netherlands 'Therapy-evaluation' is available now for veterinarians by means of PiR-DAP, a platform for data exchange between veterinarians and farmers. PiR-DAP is the first platform world-wide in this context and contains modules regarding therapy-evaluation on udder health, metabolic disorders, fertility, claw health, drying off, vaccinations, young stock, and other health issues. Next to a farm overview of treatments, individual treatments and cow information can be looked up and it is possible to indicate farm specific goals. By means of these overviews it is possible for a veterinarian to advise the farmer about sustainable antibiotic use. Furthermore, a veterinarian can set up treatment protocols for a specific farmer or for multiple farmers at the same time to stimulate working according to a protocol. The overviews also help a farmer to get a quick and complete overview of his farm situation. A pilot is conducted with 5 veterinary practices (36 farmers). Goal of this pilot was to develop and test the techniques, educate veterinarians and farmers, and prepare for national implementation. Within the pilot group, a decision support tool was set up including analyses of the effectiveness of treatment, analyses of incidences of different health issues, insight in farm goals, national averages and cow information. By means of 'Therapy-evaluation' decision support for sustainable antibiotic use and improved animal health is made possible for all veterinarians and farmers.

Management of ewe lamb replacements: effects on breeding performance and lifetime

Hanrahan, J.P., Teagasc, Athenry, Co. Galway, Ireland; Seamus.Hanrahan@teagasc.ie

The context for this paper is pastoral meat production systems. The efficiency of such systems can be enhanced by exploiting the potential of ewes to lamb at ~13 months of age. This requires attainment of puberty plus high conception rate during a joining period that usually starts 3 to 4 weeks later than for adults. The literature indicates that ewe lambs intended for mating should be ≥60% of mature weight by joining and live-weight gain should continue through the joining period. The 'ram effect' can be used to advance first oestrus; better results are achieved when exposure to teaser males is for 17 rather than 8 days prior to joining. Advancement of oestrus will minimise effects on pregnancy rate due to the fact that fertilised ewe-lamb oocytes are inherently less competent than those of adults. Ewe lambs and adults should not be mixed for joining since rams preferentially mate with adults, and duration of oestrus is shorter in ewe lambs. Furthermore, pregnancy rate in ewe lambs is higher if mature rams, rather than ram lambs, are used in paddock mating. Shearing pre-joining has been suggested to advance date of first oestrus but this is doubtful. However, there is good evidence that in many circumstances shearing pre-joining increases final pregnancy rate. This effect can be especially large in systems where ewes are in winter housing prior to joining. Shorn ewe lambs produce lambs that are heavier at birth. Nutrition during pregnancy needs particular attention as high feeding levels can impact negatively on lamb birth weight due to an associated switch of nutrients from the foetus to maternal growth. High management standards are essential at lambing and during lactation; yearling ewes and their lambs need high quality pasture, and concentrate supplement is often advisable, to ensure adequate performance of both categories. Risks from gastrointestinal parasites need particular attention. It is well established that lifetime production is 10 to 20% higher when ewes are joined to lamb at 1 year and there is no evidence for higher wastage/culling rates.

Dairy goats' replacement strategies and does's breeding

Lefrileux, L.Y.[1] and Bocquier, B.F.[2], [1]IE, Le pradel, 07170 Mirabel, France, [2]INRA, UMR-SELMET (868), 2 place Pierre Viala, 34060 Montpellier, France; yves.lefrileux@inst-elevage.asso.fr

Dairy goats' replacement strategies and does's breeding In France, dairy goat's farming systems have widely changed during the last few years so that the goat's farmers are by now highly specialized, and have to manage large flocks. Concerning the kids' breeding, the farmers' priority is to control the health risks to have high reproductive results and thanks to a sufficient growth rate that allows the kids to be pregnant during their first year. Concerning the kids' breeding, their technical options have to be efficient because these animals are determining the future production of the flock. Nowadays the technical recommendations, made from the birth to the first kidding, are based on many experiments lead either in farms or on research stations that have been published. On the sanitary plan, the recommendations focus primary on preventive management practices to avoid CAEV occurrence and actions to control the contaminations by cryptosporidium, coccidian and, in grazing system, the gastro-intestinal parasites. Concerning animals' feeding strategies, several programs are proposed, based on evolution of animals' requirements (calculated by INRA); they all aim at getting a high growth rate during the suckling phase, at minimizing the stress at weaning and promoting rumen's growth by feeding fibrous forages and adapted levels of concentrates. For the reproduction, the natural mating is largely in use during natural seasonal estrus or during artificial heats (induced by appropriate lighting treatments). The artificial insemination is only little used on kids, because of the low results which can be mainly explained by the fact that the protocols are not strictly respected and implemented. In the future, new research programs could be design to consolidate the actual technical recommendations to be used on farms, new questions that address animals' welfare and use of more fibrous diets' to improve feed autonomy and reduce environmental impacts are to be developed.

Anti-Müllerian hormone (AMH) plasma concentration and eCG-induced ovulation rate in prepubertal ewe lambs as predictors of fertility at the first mating

Lahoz, B.[1], Alabart, J.L.[1], Monniaux, D.[2], Mermillod, P.[2] and Folch, J.[1], [1]CITA, Av. de Montañana 930, 50059 Zaragoza, Spain, [2]INRA UMR 6175 - CNRS - Univ. Tours - Haras Nationaux, PRC, 37380 Nouzilly, France; blahozc@aragon.es

In sheep, delays in the age at first lambing give rise to unproductive periods and important economic losses. The objective of this work was to determine if the AMH plasma concentrations and the ovarian response to eCG before puberty is related to sexual precocity in sheep. Plasma samples were taken from 76 Rasa Aragonesa ewe lambs aged 109 ± 18 days (Mean$\pm$SD) for AMH determination by ELISA. At the same time 600 IU of eCG was applied and ovulation rate (OR) recorded by laparoscopy 6 days later. Ewes were first joined to rams at 312 ± 18 days. Correlations among OR, AMH, age and weight were assessed (Spearman's; ρ). Differences in AMH were tested by ANOVA and percentages by generalised linear models for categorical variables. A logistic model was fitted and the ROC curve analysed to evaluate AMH as a predictor of ovulation. AMH plasma concentrations were highly correlated with OR ($\rho=0.37$; $P<0.001$), but not with age or weight. Differences in AMH were highly significant between non-ovulating ewe lambs (43.0 pg/ml) and ewes presenting an OR of either 1-2 (111.9 pg/ml) or >2 (163.1 pg/ml; $P<0.0005$ for both). Ewes with AMH concentrations before puberty higher than 23 pg/ml (optimum cut-off point to predict ovulation) displayed higher fertility at the first mating opportunity than ewes with AMH levels of ≤23 pg/ml (+27.9%, $P<0.005$). Similarly, fertility was higher in ovulating than in non-ovulating prepubertal ewes (+31.5%, $P<0.0001$). The present data suggest that both the ovarian response to eCG and the AMH plasma concentration could be reliable markers of the ovarian maturity status in ewe lambs. These results may be useful in the selection of replacement ewes at a very precocious age in terms of predicted fertility at the first mating. Financed by MICINN (PET 2008-76). B. Lahoz was supported by an INIA fellowship.

The management of replacement ewe and ram lambs for breeding in Iceland

Dýrmundsson, Ó.R. and Jónmundsson, J.V., Farmers Association of Iceland, Bændahöllin,Hagatorg, IS-107 Reykjavík, Iceland; ord@bondi.is

In Iceland approximately 12-16% of the weaned lambs are kept for flock replacement and these lambs constitute nearly 20% of the national winterfed sheep population of 470.000. They are all of the Nordic short-tailed Iceland bred, both females and males being early sexually maturing, however, normally not used for breeding in former times. Improved winterfeeding after the middle of the 20th century enabled sheep farmers to include the lambs in their breeding flocks. There was a considerable variation in the application of this practice, both between flocks and regions, until it became adopted on most farms some 20 years ago. This coincided with the general adoption of pre-mating shearing in autumn, known to improve fertility in mated ewe lambs. After weaning in September, when most of the lambs are 120-130 days of age, they are normally kept on enclosed rangelands or cultivated pastures close to the farm buildings until indoor feeding begins, often 2-3 weeks before the adults ewes. As a rule ewe lambs and ram lambs, respectively, are kept separately from other sheep during the winter. Ram lambs are used for both ewes and ewe lambs. For ewe lambs polled rather than horned tups are often used. Most of the lambs weigh 40-50 kg when mating takes place. Although both lambing and prolificacy rates are lower in ewe lambs than ewes, good progress has been made on most farms, both due to better management and genetic selection for improved reproductive performance. Ultrasonic scanning has become a common practice. The best results seem to be achieved when the lambs are maintaining good body condition, both before and during the winterfeeding period, with special attention to the nutritional requirements of pregnant ewe lambs throughout gestation. Breeding of properly managed 7 month old lambs does not compromise their adult reproductive performance.

Genetic relationship between ewe lamb growth and adult ewes liveweight

Francois, D.[1], Leboeuf, M.[1], Bourdillon, Y.[2], Fassier, T.[2] and Bouix, J.[1], [1]INRA, Génétique Animale, UR 631 SAGA, 31326 Castanet Tolosan, France, [2]INRA, Génétique Animale, UE 332 Domaine de la Sapinière, 18390 Osmoy, France; Dominique.Francois@toulouse.inra.fr

Data of the experimental flock of la Sapinière (Romane INRA-401 breed) concerning the liveweight at weaning at 70 days, at the selection for replacement at 120 days and, adult liveweight at mating, data collected on 9671 ewes. Heritability of Average Daily gain between 70 and 120 days was 0.24, heritability of Adult Liveweight was 0.53. Genetic correlation between Average Daily gain between 70 and 120 days, and Adult Liveweight was 0.83. Since growth is a selection criteria in this nucleus, a tendency to increase Adult Liveweight is expected. Increase of liveweight means increase of maintenance nutrition requirements and is not wanted by breeders. To prevent increase of liveweight in such a context, management of ewes (adult & replacement) must tend to reduce environmental effect in particular nutrition. Analysis of ewes liveweights data for 20 years (1990-2009) for this flock showed a decrease from 66.5 kg in 1990 to 56 kg in 2009 meaning a straight reduction in environmental effect due to more focused feeding all along the year.

Reproductive management of the ewe lambs and their impact on productive life

Beltrán De Heredia, I., Bataille, G., Amenabar, M.E., Arranz, J. and Ugarte, E., Neiker-Tecnalia, Animal Production, P.O. Box 46, 01080 Vitoria, Spain; ibeltran@neiker.net

The productivity of a sheep flock depends to a great extent on the reproductive efficiency, as the fixed costs are roughly the same, no matter whether sheep lamb or not. The reproductive management of the Latxa dairy sheep is very traditional, with one lambing season per year during autumn and winter. Annual replacement rates vary between 15 and 25% of the flock size. The average fertility of ewes lambs ranges between 36 and 54%, although some flocks pursue the first lambing at the age of 2 years old, affecting therefore the productivity of the system. The aim of this study was to evaluate the influence of the reproductive strategy regarding replacement lambs on their growth and productive life in the Latxa dairy breed. The data were collected within the experimental flock of Neiker-Tecnalia in Arkaute Center. The productive data corresponding to the 339 replacement lambs born between the years 2000 and 2009 were analysed depending on whether they lambed or not around the first year of life. The reproductive and productive efficiency of the females lambing at the age of 8.0 to 8.5 months, depends on their weight at mating. The reproductive and productive efficiency does not affect the growth of the female, or their subsequent reproductive performance or milk production throughout their life. Contrary to what expected, an increase in the final output obtained per animal was observed. In addition, there was and unquestionable advantage from the point of view of the breeding scheme: as their genetic index can be estimated at the end of their first lactation when they were 1.5 years-old, it allows selecting the best mothers of future generations for the following mating season. Then, there is the possibility of getting an additional generation per ewe, since the current average life expectancy of sheep in the existing commercial flocks is around 4-5 years.

Relationship between maternal traits of the dam and growth of dairy replacement heifers
Bergk, N., Swalve, H.H. and Wensch-Dorendorf, M., Martin-Luther-University Halle-Wittenberg, Institute of Agricultural and Nutritional Sciences, Theodor-Lieser-Str. 11, 06120 Halle, Germany; nadine.bergk@landw.uni-halle.de

Body weight at first service of virgin heifers was measured in contract herds of a dairy cattle breeding organization in Northeast Germany. Daily gain from birth up to first insemination was computed and calving information of the dam was added. Traits of the dam were: gestation length, dystocia, and twin parity. The dataset included 10,040 heifers. A proportion of heifers had records on their birth weight. (Co)variance components for the traits age and body weight and average daily gain from birth up to first insemination were estimated. All three traits are heritable on a moderate level. Parity of dam, gestation length, twin parity and birth weight significantly influenced age and body weight at first service as well as average daily gain. Heifers from first parity cows grow slower and have the lowest average daily gain. Replacement heifers out of second parity cows have an optimal development and show the highest body weight gain. A decline of daily gain from birth up to first insemination started at third parity of the dam. The development of twin calves in comparison with normal singletons is delayed. A positive phenotypic relationship exists between gestation length of the dam and growth of replacement heifers. A negative influence of gestation length on the age at first breeding is also demonstrated. Body weight at first breeding and average daily gain showed a moderately positive genetic correlation with gestation length. Birth weight and age at first breeding are moderately negative correlated. The correlation of birth weight with body weight at first breeding and average daily gain is positive on a moderate level. This applies to the phenotypic as well as the genetic level.

A descriptive study of rearing procedures in western dairy French herds
Le Cozler, Y.[1], Récoursé, O.[2], Ganche, E.[1], Giraud, D.[1], Lacombe, D.[1], Bertin, M.[3] and Brunschwig, P.[4], [1]Agrocampus Ouest, Animal Science, 65 rue de St-Brieuc, 35000 Rennes, France, Metropolitan, [2]CLASEL, 141 Boulevard des Loges, 53942 Saint Berthevin, France, Metropolitan, [3]Les Contrôles de Performances, rue de la Géraudière, 44939 Nantes, France, Metropolitan, [4]Institut de Elevage, 9 rue Andre Brouard, 49105 Angers, France, Metropolitan; yannick.lecozler@agrocampus-ouest.fr

Impact of dairy heifers rearing conditions on short-, medium-, or long-term performance has been widely published. However, very few information is available regarding on-farm procedures used by French breeders. Present study aimed at collecting such information and accessing if and how dairy farmers use pre-set rearing targets and data monitoring for the evaluation of rearing results. An inquiry was performed on 449 herds, mainly located in Pays de la Loire (Western part of France). The questionnaire was full-filled in farms. Economical aspects were not considered and a complementary quantitative analysis was performed on 286 representative herds from the 449 initials farms. It showed that size (46.6 cows/herd) and milk production (7,953 kg milk/cow/lactation) were closed to National values (47.3 cows/herd and 8,109 kg / cow/lactation, respectively). As in many other countries, this inquiry confirmed that growth during rearing is not precisely monitored in most herds, although more than 80% of farmers claimed they had precise targets for age and weight at 1st calving. Rearing conditions differed between farms, but similar priorities and/or interests were found: age or body weight at weaning or service for example, use of non-marketable milk for calves, visual methods for heat detection… Finally, farmers estimated that they spent 12 to 15 h on average per reared heifer. Rearing heifers was considered as a necessity, a pleasure or a chore by 67, 36 and 2% of farmers respectively. As already noticed in other dairy countries, heifer-rearing management in Western part of France could be improved considerably.

How do different amounts of solid feed in the diet affect the behaviour and welfare of veal calves?
Webb, L.E.[1], Bokkers, E.A.M.[1], Engel, B.[2] and Van Reenen, C.G.[3], [1]WUR, Animal Production Systems, P.O. Box 338, 6700 AH Wageningen, Netherlands, [2]WUR, Biometris, P.O. Box 100, 6700 AC Wageningen, Netherlands, [3]WUR, Livestock Research, P.O. Box 65, 8200 AB Lelystad, Netherlands; Laura.Webb@wur.nl

The current EU solid feed requirement for veal calves is insufficient to prevent the development of behaviours indicative of poor welfare. This study investigated the effect of different amounts of solid feed in the diet on the behaviour and welfare of veal calves. 48 bull calves were fed four diets (A, B, C, D) with solid feed (0, 9, 18, 27 g DM/kg metabolic weight/day, with diet D roughly equal to 6 times the EU minimum requirement) and milk replacer provided for similar growth performance between diets. Abnormal oral behaviours (AOB) were recorded once a week around feeding (continuous), and chew/ruminate (CRB) and play behaviours were recorded throughout the day (scan sampling). To study motivation to orally interact with novel objects, we presented calves with a ball (study month 1) and an overall (months 2, 3, 4) for 3 min. Only diet D calves showed less AOB compared with diet A calves (P=0.003). In month 1, CRB were more frequent in C and D than in A (P<0.05) and more frequent in D than in B (P=0.004). In month 2, only D calves performed more CRB than A calves (P=0.019) and CRB differences between diets disappeared in months 3 and 4. Play behaviours were too infrequent for analysis. At the end of month 1, diet D calves touched the ball slower than diet A calves (P=0.013). Across months 2,3,4 diet D calves spent less time orally manipulating the overall than calves in diets A and B (P<0.05). Only diet D provided a welfare benefit by reducing AOB. Diet D only showed increased CRB in the first two months, most likely indicative of an increasing chewing/ruminating efficiency. Motivation to orally interact with novel objects was lower in diet D compared with A, pointing to a frustrated drive that is the cause for the development of AOB in veal calves fed little solid feed.

Examination of the bovine leukocyte environment using immunogenetic biomarkers to assess immunocompetence following exposure to weaning stress
O'loughlin, A.[1,2], Earley, B.[1], Mcgee, M.[1] and Doyle, S.[2], [1]Teagasc, Animal and Bioscience Research Department, Animal and Grassland Research and Innovation Centre, Dunsany, Co. Meath, Ireland, [2]National University of Ireland, Maynooth, Department of Biology and National Institute for Cellular Biotechnology, Maynooth, Co. Kildare, Ireland; aran.oloughlin@teagasc.ie

The study objective was to characterise the physiological and immunogenetic responses to weaning stress in beef calves. Twenty-eight calves were assigned to one of 4 treatments where calves were weaned and either penned adjacent to the dam (1) female calves (mean age (s.d) 227 (17.9) d); (2) male calves (mean age (s.d) 197 (38.1) d) or away from the dam (3) female calves (mean age (s.d) 224 (25.1) d); (4), male calves (mean age (s.d) 209 (37.3) d). On d -3, 0, 1, 2, 3, 7, and 11 relative to weaning (d 0), blood samples were collected for haematology and gene expression profiling. Haematological data and relative gene expression were analysed (2×2 factorial design) using the PROC MIXED procedure of SAS. Following weaning, neutrophil number increased (P<0.001) in all calves and was greater (P<0.001) in calves weaned away from the dam. The expression of pro-inflammatory cytokine genes in leukocytes, including IL-1β, IL-8, IFN-γ and TNFα, were up-regulated (P<0.01) on d 1 following weaning. There was increased (P<0.001) expression of the glucocorticoid receptor, GRα, the pro-apoptotic gene, Fas, and the Gram-negative pattern recognition receptor TLR4, post-weaning. The earlier and more profound increase in neutrophil number and N:L ratio together with reduced lymphocyte number in calves penned away, compared with calves penned near their dams post-weaning, suggests that calves are sensitive to maternal separation at weaning. Additionally, this study has characterised the inflammatory responses of calves to weaning stress and identified a number of novel immunogenetic biomarkers of weaning stress.

Nutrition and welfare: role of inflammation and its modulation

Bertoni, G., Piccioli-Cappelli, F., Cogrossi, S., Grossi, P. and Trevisi, E., Istituto di Zootecnica, Università Cattolica del Sacro Cuore, via E. Parmense, 84, 29122 Piacenza, Italy; giuseppe.bertoni@unicatt.it

In humans, the tremendous changes occurring during diseases (e.g. fever, anorexia, depression, then reduction of welfare) are mediated by pro-inflammatory cytokines (IL-1, IL-6, TNFα). These can be modulated by nutrition that is therefore able to decrease the effect or the risk of inflammation in both disease and healthy situations. Aim of this paper was the supply Ascophyllum nodosum (AN, 50 g/capo/d) to dairy cows around calving (±2 wk) to reduce inflammations often occurring in this period. Eight cows (4 treated and 4 untreated = CTR) were frequently checked around calving (±4 wk) for health status, feed intake, rectal temperature, BCS, milk yield and composition and metabolic profile. Weekly energy balance (EB) and the net energy efficiency (NEE) were also evaluated. Statistical analysis was based on repeated measures variance test, using a model that includes treatment, days and their interaction. During peripartum cows of both groups suffered of some mild diseases, but AN group showed a lower somatic cell count till 28th day in milk (DIM; 129±126 vs 205±206 n/μl of CTR) and a lower rectal temperature before calving, but not in the wk after calving, likely for 2 cases of retained placenta in the AN group. AN exhibited higher values of haptoglobin and myeloperoxidase till 14th DIM, suggesting a marked acute phase reaction and, possibly, an abundance of neutrophil granulocytes into the blood. Nevertheless, AN vs CTR showed a quicker raise of plasma cholesterol and similar levels of other negative acute phase proteins. Moreover, AN had a slightly higher feed intake; a similar milk yield, a sensibly lower fat content and a reduced lipomobilization (e.g. higher glucose and lower level of BHB: 0.8±0.4 vs 1.4±0.9 of CTR till 28th DIM). Altogether, these last changes have determined a better EB in AN (+2 Mcal/cow/d after calving) and similar NEE. On the whole, AN seems to exert a positive effect on welfare of periparturient dairy cows.

Grass/red clover silage to growing/finishing pigs - influence on performance and carcass quality

Høøk Presto, M.[1], Rundgren, M.[1] and Wallenbeck, A.[2], [1]SLU, Dept. of Animal Nutrition and Management, Box 7024, SE- 75007 Uppsala, Sweden, [2]SLU, Dept. of Animal Breeding and Genetics, Box 7023, SE- 75007 Uppsala, Sweden; Magdalena.Presto@slu.se

Production of grass and red clover are important for the crop rotation and has good potential for nutrient supply and can improve pig welfare by offering increased possibilities for foraging behaviour. However, pigs' ability to consume roughage may vary according to nutrient properties and straw length. This study aimed to evaluate how grass/red clover silage affected pig performance and carcass quality. Sixty-four growing/finishing pigs (Y x H) was fed either commercial feed + chopped silage mixed (SM), commercial feed + intact silage fed separately (SS), commercial feed + grinded silage, mixed and pelleted (SP) or commercial feed alone (C). The pigs were fed twice daily according to the Swedish energy recommendations. Silage constituted of 20% on energy-basis in the silage treatments. Performance and carcass quality was analyzed with mixed GLM in SAS. Preliminary results showed that SM and SS pigs had lower weight gain than SP pigs followed by C pigs (742[a], 779[a], 842[b] and 892[c] g/day, P≤0.001). Silage pigs had a lower carcass weight compared with C pigs (P≤0.001). SM and SS pigs had on average a lower daily energy intake compared with SP and C pigs (27.0, 26.8, 28.6 and 29.0 MJ ME, respectively), probably due to silage refusals, which might explain the lower weight gain. Feed conversion ratio (MJ ME/kg weight gain) was however higher in SM pigs compared with the others (36.8[a] vs. 34.6[b], 34.2[b] and 32.9[b], P=0.001). Estimated daily lean meat growth was lowest for SM followed by SS pigs, and highest for SP and C pigs (318[a], 340[b], 352[c] and 378[c], P≤0.001), whereas lean meat % was higher in SS and SM pigs compared with SP pigs (58.9[a], 58.0[a] vs. 56.6[b], P=0.005) and C pigs were intermediate (57.5[ab]). In conclusion, pigs with dietary inclusion of grass/red clover silage have the potential to perform as optimal as conventionally fed pigs when silage is fed in pelleted form.

The effect of phosphorus restriction during the weaner-grower phase on serum osteocalcin and bone mineralisation in gilts

Varley, P.F., Sweeney, T., Ryan, M.T. and O'doherty, J.V., University College Dublin, Animal Nutrition, Lyons Research Farm, Newcastle, Co. Dublin, Ireland, Co. Dublin, Ireland; patrick.varley@ucd.ie

Ninety-six female pigs with an initial bodyweight (BW) of 10.0±1.6 kg were assigned to 4 dietary treatments to determine the effects of restricting dietary phosphorus (P) level during the weaner-grower [approximately 10 to 50 kg BW; day (d) 0 to 59] and finisher (approximately 50 to 100 kg BW; d 59 to 131) period on serum osteocalcin concentration and bone mineralisation. The dietary treatments were: (1) 4.0 g total P (tP)/kg from d 0 to 131 (LL); (2) 4.0 g tP/kg from d 0 to d 59 and 6.0 g tP/kg from d 59 to 131 (LH); (3) 6.0 g tP/kg from d 0 to 131 (HH) and (4) 6.0 g tP/kg from d 0 to 59 and 4.0 g tP/kg from d 59 to 131 (HL). Bone data were analysed as a completely randomised design using the General Linear Model Procedure of SAS. For serum osteocalcin analysis, the data was analysed by repeated measures analysis using the Proc Mixed procedure of SAS. During the weaner-grower period (d 0 to 59), pigs offered high P diets had higher bone ash (P<0.05) and serum osteocalcin concentration (P<0.05) compared to pigs offered low-P diets. Pigs offered LL diet had a lower bone ash (P<0.05), bone P (P<0.01) and bone calcium (P<0.05) concentration than pigs offered LH, HL and HH diets on d 131. Pigs offered LH diet had higher concentration of osteocalcin compared to pigs offered LL (P<0.01), HH (P<0.05) and HL (P<0.05) diets on d 88 and 108. In conclusion, compensatory effect occurred in bone mineral concentration at the termination of the finisher stage when high P was introduced at the initiation of the finisher stage, following consumption of a low-P diet during the weaner-grower stage. However, pigs offered LH diet did not surpass the level of bone mineralisation achieved by pigs offered HH diet.

Effects of probiotic, prebiotic and synbiotic on perfomance and humoral immune response of female suckling calves

Mohamadi, P. and Dabiri, N., Islamic Azad University, Karaj Branch, Animal Science, Natural Resource and Agriculture Azad University of Karaj, Bolvar Eram, Mehrshahr, Karaj, Tehran, 3187644511, Iran; najdabiri@hotmail.com

Thirty two Holstein female calves (initial body weight = 38±4.0 kg) were assigned randomly to one of four treatments to examine the effect of probiotic, prebiotic and synbiotic on performance and humoral immune response of female dairy calves. treatments include whole milk with no additives (control), whole milk containing probiotic at 1 g protexin (multi-strain probiotic contain 7 bacteria strains and 2 yeast strains with 2×10^9 cfu/g) per day, whole milk containing prebiotic at 4 g Tipax (polysaccharides of saccharomyces cereviciae cell wall) per day and whole milk containing 1 g probiotic and 4 g prebiotic (synbiotic) per day. Calves were fed by colostrum for 3 days then were separated from their dams and housed in individual pen. Calves received whole milk twice daily at 0700 and 1600 h. calf starter and water were offered ad lib throughout the trial of 60 days. Starter intake was recorded daily throughout the trial. BW were measured at birth and thereafter weekly up to 60 d of age. To evaluate the effects on humoral immune response calves were vaccinated with ovalbumin on d 35 and 49. Blood was collected via jugular venipuncture on d 35, 42, 49 and 56 and serum was analyzed for IgG titer to ovalbumin using an ELISA. Dry Matter Intake was greater significantly in calves fed synbiotic (P≤0.02). Average daily gain was greater in calves fed prebiotic and synbiotic than probiotic and control treatments (P≤0.05). Calves fed probiotic, prebiotic and synbiotic had higher feed efficiency than control calves (P≤0.05). No significant differences were detected in serum IgG titer to ovalbumin on d 35, 42 or 49; however on d56 a linear increase (P≤0.06) in serum IgG titer was noted with calves fed synbiotic. Probiotic, prebiotic and especially synbiotic had beneficial effects on performance; however, effects on humoral immune response warrant further research.

Effects of dietary garlic on broiler performance and blood parameters
Mirza Aghazadeh, A. and Tahayazdi, M., Urmia University, Animal Science, Urmia, 57153-165, Iran; a.aghazadeh@urmia.ac.ir

In developed societies arteriosclerosis is strongly related to the dietary intake of cholesterol and saturated fatty acids. Use of garlic as a spice and herbal remedy in human is widespread in all parts of the world. In recent years, several clinical reports have described the hypercholesterolemic effect of garlic in human. The objective of this study was to determine the effects of sun-dried garlic on performance, carcass characteristics and some blood parameters in broilers. Two-hundred-day-old mixed sex Ross (308) broiler chicks were randomly allocated into 5 dietary treatments each whit four replicates of 10 chicks for a 6-wk feeding trial, including 7 to 21 d and 21 to 42 d periods. Experimental diets were: 1) basal diet (control=BAS), 2) BAS with 0.4% sun-dried garlic (SDG), 3) BAS with 0.8% SDG, 4) BAS with 1.2% SDG, and 5) BAS with 1.6% SDG. At the end of the experiment, 8 broilers of each treatment were killed by cervical dislocation for blood collection and carcass analysis. Data were analyzed by one-way ANOVA procedure of SAS software. Means were separated by using Duncan's multiple test range test whenever the F-test was significant ($P<0.05$). The effects of garlic on feed intake and body weight gain improved FCR slightly, but this effects were not statistically significant ($P>0.05$). Relative edible parts (breast and thigh) weights and carcass yield were not significantly affected by dietary garlic levels. However, supplementing basal diet with 1.6% garlic significantly decreased abdominal fat pad content relative to the control. Feeding garlic (1.2 and 1.6%) resulted in significant reduced levels of plasma cholesterol and triglycerides. In conclusion, the results of this experiment confirm the cholesterol lowering effects of garlic without altering growth of the chicks and feed efficiency.

Effect of butyric acid on cecal microbial population of broilers fed wheat-corn based diets
Mirza Aghazadeh, A., Sepahvand, M. and Jafarian, O., Urmia University, Animal Science, Urmia, 57153-165, Iran; a.aghazadeh@urmia.ac.ir

Since the 2006, duo to the ban of the use of some antibiotics as growth promoters, the interest for use of organic acids as an alternative for antibiotics has gained increasing interest in animal nutrition. The objectives of this study were to evaluate the effect of butyric acid administration through the drinking water on the performance and cecal microflora of broiler chickens. Two hundred and forty day-old (Ross 308) broiler chicks were randomly distributed into four groups with six replicates each. Four groups were fed with basal diet based on wheat- corn, and soybean meal (control) or diets supplemented with butyric acid through drinking water at the levels of 0.25, 0.35 and 0.5%. Water and feed were provided adlibitum. Microbial populations were determined by serial dilution (10-1 to 10-10) of ceceal samples in anaerobic diluent. Total aerobic counts (TCA) and coliforms of ceceal samples were evaluated by MHA and EMB grown media and aerobic plate count (APC), and surface plate count (SPC) methods. Immediately after weighing at 21 and 42 days of age, one bird (close to the (mean body weight) from each pen killed and processed to determine the internal organ weights and bacterial counts. Data from performance and microbial counts were analyzed by one-way ANOVA procedure of SAS software. Means were separated by using Duncan multiple range test whenever the F-test was significant ($P<0.01$). No significant difference between butyric acid and control supplemented groups were found in live weight gain, feed intake and FCR at 0-42 days of age. Total aerobic counts and coliforms in cecal content were not affected by dietary treatments.

The pH dynamics of the goats and cows rumen and abomasums

Birģele, E., Ilgaža, A. and Keidāne, D., Latvian University of Agriculture Faculty of Veterinary Medicine, Preclinical institute, Helmaņa 8, 3002 Jelgava, Latvia; aija.ilgaza@llu.lv

There were compared the dynamics of the intrarumenal and intraabomasal pH in adult goats and cows one hour before and seven hours after unlimited hay and concentrated mixed feed based on barley, wheat and sunflower meal (goats – 0.5 kg and bulls – 2 kg) feeding. In this study five bulls (n=5) in the age of 1.5 and one year old goats (n=5) were included. To investigate the changes of reaction of abomasum and rumen in animals before and after feeding, the uninterrupted long-lasting intra gastric pH measurement method was applied together with chronic fistula method. For intrarumenal and intraabomasal pH measurements special flexible pH-probe with two antimony electrodes (12 cm distance between each other) and one calomel electrode at the end of the probe was used In adult bulls it was established that intrarumenal pH before morning feeding were within the range from 6.9 to 7.9, but in goats – from pH 7.2 to 7.5. The pH values in the rumen of bulls rumen practically did not change after the feeding, but in goats in the first two hours pH significantly decreased, gaining the level of 6.0±0.4 (P<0.05), pH below 7.0 remained for further three hours. Average of the bulls intraabomasal pH, independently from feeding, were in the level of pH 3.2-3.9. In goats abomasum pH before morning feeding was comparatively higher – 4.2-4.5, but in the two hours after feeding pH in abomasum radically decreased, reaching the lowest pH level 3.2-3.3 (P<0.05). In conclusion, it was established that in adult bulls pH in rumen and abomasum did not significantly changed after the feeding of the concentrated mixed feed and hay, but in goats after the similar feeding pH level in the rumen and abomasum significantly decreased during the two hours period. Thereby these functional processes in rumen and abomasum are different in these ruminants.

Bovine neonatal survival - is improvement possible?

Mee, J.F., Teagasc, Animal and Bioscience Research Department, Moorepark, Fermoy, Co. Cork, Ireland; john.mee@teagasc.ie

The answer to this rhetorical question is – in theory, yes, but the inconvenient truth is in practice often, no. This dichotomous answer hints at the enigmatic discord between what is theoretically possible and what actually occurs in practice. Though this view may conflict with received thinking, evidence for this divergence can be found in the disparity between bovine neonatal survival rates (in the first two days of life) internationally and between results from research studies and farm-level data. For example, dairy calf neonatal survival rates in some countries, e.g. Norway, are amongst the highest in the world; in contrast to those in many North American Holstein-Friesian-dominated dairy industries. Whereas experimental and observational studies have identified critical risk factors for improved neonatal survival the results from such studies are not always replicated at farm level. In fact the reverse has occurred in recent years, with a decrease in bovine neonatal survival rates reported in the peer-review literature from many countries around the world. Unfortunately this decline in neonatal survival has not attracted the same degree of interest or research funding as the well documented decline in dairy cow 'fertility', of which it is an adjacent problem. The reason for the lack of improvements in neonatal survival stems from de-prioritisation of the issue relative to other animal health and welfare concerns. This has resulted in less funding of research work with consequent downstream atrophy of knowledge metastasis through extension and implementation programmes. While there are knowledge gaps constraining progress towards improved neonatal survival requiring more transdisciplinary research including the 'omic' technologies, re-prioritisation of neonatal survival as an important welfare deficiency signal and better communication of existing knowledge are of greater importance in reversing current trends.

Effect of acid-base disturbances on perinatal mortality (PM) in dairy cattle
Szenci, O., Szent Istvan University, Faculty of Veterinary Science, Clinic for Large Animals, Dora major, H-2225 Ullo, Hungary; szenci.otto@aotk.szie.hu

Profitability of cattle breeding is greatly influenced by the rate of calves born alive and reared to adulthood. In spite of the speedy developments of animal breeding, PM is still very high (4-12% especially in heifers) and constitutes approximately half of the total calf losses. Direct and indirect asphyxia was suggested as the cause of death because in 73-75% of the calves that died in the perinatal period no pathological changes were detected. As a result of the disturbances in the utero-placental circulation occurring during calving due to the rupture of foetal membranes and uterine contractions, all calf foetuses develop more or less severe hypoxia and consequently acidosis. The foetus responds to hypoxia by an oxygen-conserving adaptation of its circulation. This means that all organs, that are not essential for intra-uterine life (lungs, spleen, thymus, muscles and skin, GI, liver and kidneys) are supplied with minimum blood. O_2 is spared for essential organs like brain, heart and adrenal glands. Circulation changes and reduced O_2 consumption result in oxygen tension in the blood being maintained within physiological limits for some time. This O_2-conserving adaptation, however, means anaerobic glycolysis in all tissues with minimum blood supply. Anaerobic glycolysis has a great disadvantage because energy production is reduced, carbohydrate reserves are rapidly exhausted and metabolic acidosis may develop by accumulation of acid metabolites. At birth all foetuses therefore suffer from a respiratory as well as metabolic acidosis. It is the degree of acidosis that finally determines whether the foetus lives or dies. Vital cell functions cannot take place in severe acidosis, at a blood pH value of 6.7 foetal life ends. Before that, the organism's regulatory system of chemical buffering (bicarbonate) in the blood comes into operation to keep the offspring alive. Duration of survivable asphyxia always depends on the glycogen reserves in the heart muscle but it is not more than 4 to 6 minutes.

The morphology of jejunal and ileal Peyer's Patches and concentrations of serum IgG in hand-reared neonatal calves supplemented with fatty acids
Kentler, C.A., Babatunde, B.B. and Frankel, T.L., La Trobe University, Bundoora, VIC 3086, Australia; cabramley@students.latrobe.edu.au

Neonatal calves are particularly susceptible to antigen penetration of the gut, most commonly resulting in diarrhoea, dehydration and death. Jejunal and ileal Peyer's Patches (PPs) found on the mucosal surface of the small intestine are vital for innate immune defence, whilst circulating IgG levels can give an indication of passive transfer and acquired immune development. This study aimed to investigate the effect of fatty acid supplementation on the morphology of jejunal and ileal PPs and the development of acquired immune response. Twenty-four Friesian bull calves were hand-reared from 2 days of age until 2 weeks old. Calves were randomly assigned to three treatment groups: Control (CO); fed calf milk replacer (CMR) at 10% of body weight, Sunflower (SU); supplemented with sunflower oil at 5% of CMR dry weight and Palm Fat (PF); supplemented with palm shortening at 5% of CMR dry weight. Serum samples for IgG analysis were collected at less than 5, 10 and 14 days of age. Calves were euthanized at 14 days of age and intestinal samples collected. Morphological analysis was conducted using image capture and analysis technology following H&E staining and IgG concentration determined with an ELISA assay. Data was analysed statistically using either Oneway ANOVA, or Non-parametric Kruskal-Wallis tests. No significant treatment effects were detected; although calves supplemented with fatty acids demonstrated some trends in morphological features of both jejunal and ileal PPs. Treatment effects were also not evident in concentrations of serum IgG at 2 weeks of age. Dietary supplementation of neonatal calves with fatty acids does not appear to affect PP morphology or serum IgG concentration up to 2 weeks of age. This may be due to age, or genetic or management interactions. Further study is necessary to determine optimal feeding practices for calf health during this critical period of growth and development.

Improving neonatal lamb survival

Fragkou, I.A., Mavrogianni, V.S. and Fthenakis, G.C., Veterinary Faculty, University of Thessaly, P.O. Box 199, 43100 Karditsa, Greece; gcf@vet.uth.gr

Neonatal mortality of lambs includes deaths during the first week ('hebdomadal') and deaths subsequently to that and until the 28th day of life ('post-hebdomadal'). First week of life is crucial for survival of a lamb. Within this, death of a lamb is defined as 'immediate' (initial 24 hours of life), 'delayed' (24 to 72 hours of life) or 'late' (72 hours to 7 days of life). Based on their aetiology, disorders of newborn lambs can be classified as disorders of non-infectious aetiology (congenital defects, injuries during birth, hypothermia) and pathological conditions of microbial or parasitic aetiology (systemic infections: e.g. *Bibersteinia* infection, tetanus, microbial or parasitic infections of the digestive tract: e.g. clostridial infections, enteritis caused by *Escherichia coli*, viral infections, watery mouth disease, *Salmonella* infection, *Cryptosporidium* infection, respiratory infections: e.g. *Mannheimia* infection, *Bibersteinia* infection, *Pasteurella* infection, other localised infections: e.g. liver abscess syndrome, streptococcal polyarthritis. First step to improving neonatal lamb survival is correct management of pregnant ewes: proper feeding prevents pregnancy toxaemia and leads to birth of lambs with appropriate body weight, as well as to production of colostrum and milk in suitable quantity and quality; pregnancy diagnosis helps allocation of ewes in respective groups; vaccination of pregnant ewes supports prevention of diseases affecting newborns; athelmintic treatment of ewes, whether before or immediately after lambing, minimises chances for infection of lambs and leads to improved milk production of dams. Subsequently to lambing, establishment of dam-offspring bond contributes to reduced neonatal mortality and leads to improved body weight gains. Hygiene measures (e.g. disinfection of remaining umbilical cord) contribute significantly to increasing lamb survival. Finally, if mortality cases occur, early and accurate diagnosis is crucial to implement specific measures and help preventing further deaths.

Organic farrowing conditions as an example for future conventional pig husbandry?

Vermeer, H.M., Wageningen UR Livestock Research, P.O. Box 65, 8200 AB Lelystad, Netherlands; herman.vermeer@wur.nl

Sows kept under organic conditions on average farrow larger litters then sows kept under conventional conditions. Larger litters mean lower birth weights and higher mortality risks. This makes an organic system an ideal setting to study neonatal mortality with implications for future conventional systems. In the last decade experiments focussed on housing, climate, nutrition and management to improve neonatal survival. The summarized results will be presented. In short they resulted in the following conclusions: (1) higher creep use gives lower mortality, (2) a separate dunging area results in a cleaner solid floor, (3) a short period of additional heating around farrowing improves vitality, (4) no effect of extra straw, (5) flaps in creep opening reduce mortality, (6) sow water intake does not affect mortality. In present experiments more attention is paid to genetics affecting piglet vitality, rearing conditions of the sow effecting maternal behaviour and management measures to reduce piglet mortality. With the ban on individual housing of pregnant sows on Jan 2013 the pressure towards loose housing of conventional farrowing sows increases. Conventional pens now have 5 m^2 where organic farrowing pens have 7.5 m^2. Space is necessary to separate lying and dunging behaviour and to promote maternal behaviour, but has of course also financial implications. Developments in organic pig husbandry can so have an impact on future developments in conventional pig husbandry. Cooperation and exchange between organic and conventional research projects results in a cross fertilization for the pig husbandry as a whole.

Maternal investment, sibling competition, and piglet survival – the effects of increased litter size
Andersen, I.L.[1], Nævdal, E.[2] and Bøe, K.E.[1], [1]Norwegian University of Life Sciences, Animal and Aquacultural Sciences, P.O.Box 5003, 1432 Ås, Norway, [2]Frisch Centre, Gaustadalléen 21, N-0349 Oslo, Norway; inger-lise.andersen@umb.no

The aim of this study was to examine the effects of litter size and parity on sibling competition, piglet survival and weight gain. It was predicted that competition for teats would increase with increasing litter size, resulting in a higher mortality due to maternal infanticide (i.e. crushing) and starvation, thus keeping the number of surviving piglets constant. We predicted negative effects on weight gain with increasing litter size. Based on maternal investment theory we also predicted that piglet mortality would be higher for litters born late in a sow's life and thus that the number of surviving piglets would be higher in early litters. Altogether, 40 healthy, Landrace x Yorkshire sows of different parities from a commercial farm were used in the present study. For analysis of piglet mortality/survival and the proportion of live born piglets that died from different causes, a generalized model, Poisson regression, was conducted by using the GENMOD procedure in SAS. As predicted, piglet mortality increased with increasing litter size both due to an increased proportion of crushed piglets, where most of them failed in the teat competition, and due to starvation caused by increased sibling competition, resulting in a constant number of survivors. Piglet weight at day 1 and growth until weaning also declined with increasing litter size. Sows in parity four had higher piglet mortality due to starvation, but the number of surviving piglets was not affected by parity. In conclusion, piglet mortality caused by maternal crushing of piglets, many of which had no teat success, and starvation caused by sibling competition, increased with increasing litter size for most sow parities. The constant number of surviving piglets at the time of weaning suggests that 10 to 11 piglets could be close to the upper limit that the domestic sow is capable of taking care of.

Neonatal mortality in piglets: genetics to improve behavioural vitality
Dauberlieu, A.[1,2], Billon, Y.[3], Bailly, J.[3], Launay, I.[3], Lagant, H.[4], Liaubet, L.[1], Riquet, J.[1], Larzul, C.[4] and Canario, L.[1], [1]INRA UMR 444 ENVT, Animal Genetics, Laboratoire de génétique cellulaire - Chemin de Borde Rouge, 31326 Castanet-Tolosan, France, [2]Université Paris 13, 93430, Villetaneuse, France, [3]INRA, UE967 GEPA, 17700, Surgères, France, [4]INRA, UR1313 Génétique Animale et Biologie Intégrative, 78352, Jouy-en-Josas, France; laurianne.canario@toulouse.inra.fr

Maternal abilities and piglet vitality were analyzed on 24 Meishan (MS) and 24 Large White (LW) gilts. Females were inseminated with a mixture of semen from both breeds. Three MS and 3 LW boars were used to constitute 3 duos formed by mixing of MS and LW semen in equal proportions. Farrowing events were studied over 5 successive batches. The proportion of purebred and crossbred piglets within the litter varied according to the duo used and the dam breed (P<0.01). The average within-litter percentage of purebred piglets in LW and MS sows was respectively 43% and 50% with use of duo 1, 64% and 23% with duo 2 and 69% and 81% with duo 3. Gestation was shorter in MS than in LW sows (111.6 vs 114.0 days; P<0.05) and litter size tended to be larger in LW than in MS sows (14.6 vs 12.8 total born piglets; P=0.08). Over the three first days of lactation, piglet probability of survival was similar between purebred and crossbred piglets born from LW sows (94.5% vs 95.0%) and higher in purebred than crossbred piglets born from MS sows (96.6% vs 98.7%, P<0.05). In LW sows, crossbred piglets were heavier at birth and more reactive in a novel environment than purebred piglets (1.29 vs 1.21 kg, P<0.10; reactivity score: 1.38 vs 1.03 respectively). In MS sows, purebred piglets had a lower birth weight than crossbred piglets but showed similar vitality (0.86 vs 1.08 kg, P<0.001; reactivity score: 1.00 vs 1.03). Birth process and piglet behavior in early lactation will be analyzed to estimate the interaction between dam breed and piglet genetic type (purebred vs crossbred) on the expression of maternal behavior and piglet vitality (udder activity and survival).

Effect of essential fatty acid supplementation of sow diets on piglet survival in two farrowing systems

Adeleye, O.O., Brett, M., Blomfield, D., Guy, J.H. and Edwards, S.A., Newcastle University, School of Agriculture, Food and Rural Development, Agriculture Building, Newcastle University, Newcastle upon Tyne, NE1 7RU, United Kingdom; oluwagbemiga.adeleye@ncl.ac.uk

Current developments in the pig industry pose increased challenges for piglet survival as a result of selection for increased prolificacy and welfare pressures to abolish the use of farrowing crates. The effect of supplementation of the maternal diet with the essential fatty acid, docosahexaenoic acid (DHA), on the performance of sows and their piglets farrowing in two different housing conditions was studied using a 3x2 factorial experiment. A control diet was compared to 2 levels of DHA supplementation from algal biomass (0.03 and 0.3% DHA, delivered by 1.5 g/kg and 15 g/kg algal biomass) during the last 4 weeks of pregnancy and lactation, using 60 sows (mean parity 4.7 sem 0.32) in two different farrowing systems (farrowing crate and PigSAFE farrowing pen). Two-way analyses of variance showed no statistically significant interactions between dietary treatment and housing system. Piglet survival and growth did not differ between the crate and pen systems. Litter size (13.1 sem 0.42) and piglet birthweight (1.45 sem 0.047 kg) did not differ between dietary treatments, but the number of stillborn piglets per litter was reduced with increasing DHA supplementation (1.13, 0.67, 0.25, sem 0.205, P=0.014, with litter size covariate). Mortality of liveborn piglets in the first 3 days, and number weaned per litter (after fostering) were unaffected by treatment, as were sow weight and backfat loss in lactation. However, piglet weaning weight was reduced by DHA supplementation. The inclusion of DHA in the sow diet for the last month of gestation improved piglet survival during farrowing. The reasons for this are being further investigated by video recordings of farrowing duration and metabolic state on the neonate.

Effect of encapsulated calcium butyrate (GREEN-CAB-70 Coated®) on growth parameters in Iranian Holstein female calves

Karkoodi, K., Nazari, M. and Alizadeh, A.R., Department of Animal Science, Saveh Branch, Islamic Azad University, Saveh, Iran; kkarkoodi@yahoo.com

This study was conducted to evaluate the effects of encapsulated calcium butyrate (GREEN-CAB-70 Coated®, Nutri Concept Co.) on growth parameters of Holstein female calves. Sixteen female calves with mean age 5 days in a completely randomized design were divided into two equal groups (n=8) namely, experimental (starter diet and milk replacer with 3 g calcium butyrate per day) and control (starter diet and milk replacer). Body weight was measured at birth and on days 12, 24, 36 and 48. Also physical traits (rump height, withers height and hip width) were measured at the beginning, middle, and end of the experiment. Feed intake and fecal score were measured daily and respiratory rate and rectal temperature were measured weekly. Results showed that daily feed intake, body weight, average daily gain and feed conversion ratio improved significantly in the experimental group (P<0.05). Calcium butyrate also improved physical traits in the experimental group (P<0.05), but no significant effects were observed on rectal temperature, respiratory rate and fecal score (P>0.05). The study suggests that using encapsulated calcium butyrate in Holstein female calves fed before weaning, shortens rumen development period facilitating early weaning.

Accuracies of estimated breeding values from ordinary genetic evaluations do not reflect the potential for genetic gain
Bijma, P., Animal Breeding and Genomics Centre, Wageningen University, Marijkeweg 40, 6709PG Wageningen, Netherlands; piter.bijma@wur.nl

The rate of genetic improvement is proportional to the accuracy of EBVs. Accuracy is defined as the correlation between true and estimated breeding values. In ordinary genetic evaluations, accuracies are obtained from the inverse of the LHS of the MME, or are approximated when inversion is computationally prohibitive. Because such accuracies are often routinely available, they are convenient proxy's for potential genetic gain. It is well-known that the ordinary calculations of accuracy ignore selection, and are therefore approximate. It is less well-know, however, that the difference between approximate and true accuracies can be very large, and strongly depends on the type of phenotypic information. This work illustrates that approximate accuracies do not reflect the potential for genetic gain. Results show that accuracy of the pedigree index (PI) is severely overpredicted. The accuracy of a PI in dairy cattle, for example, is ~0.12; much lower than breeders are used to. As the accuracy of the PI is overpredicted, phenotypic or genomic information of the current generation is undervalued and breeding schemes are ranked incorrectly. Examples of traditional and genomic selection schemes in dairy cattle, pigs and laying hens show that the pattern in approximate accuracies of alternative schemes may not reflect the pattern in true accuracies. Besides its role in response to selection, the accuracy is also used to represent the credibility or SE of an individual EBV, and reflects the risk that the true BV may differ considerably. The results show that, in selected populations, there is no direct relationship between the SE and the correlation between true and estimated BV. This occurs because a SE is a property of a single estimate, whereas a correlation is a property of a population. It is impossible, therefore, to define a single measure of accuracy reflecting both the correlation between true and estimated BV and the credibility of individual EBVs.

Management of long-term genetic improvement and diversity using mate selection
Kremer, V.D.[1,2] and Newman, S.[1], [1]Genus/PIC, Genetic Development, 100 Bluegrass Commons Blvd., Hendersonville, TN 37075, USA, [2]The Roslin Institute, Genetics & Genomics, Roslin, Midlothian, EH25 9PS, United Kingdom; valentin.kremer@pic.com

This study addresses the management of genetic improvement and diversity in a set of relatively small simulated pig populations over long-term selection. 30 years of selection were simulated for populations of various sizes between 25 and 400 target breeding females. The number of males used every year and their mating allocation were decided dynamically, based on optimized mate selection (MS). Different intensities were applied to ongoing restriction on inbreeding. The effect of mid-term changing of the breeding objective and risk policy has been studied. MS proves to be an extremely efficient tool for managing inbreeding rate while maximizing genetic gain. MS provides the necessary flexibility needed when changes in the breeding objective or risk policy occur. It allows for a quick re-optimization of the population resources existent at the time of the change, minimizing the impact of the expected 'bottle-neck' effects. MS is an ideal tool for animal genetic resource management and conservation programs of critical and endangered breeds as it could be very effectively used in minimizing the rate of inbreeding.

Restricting inbreeding and maximizing genetic gain by using optimum contribution selection with a genomic relationship matrix

Körte, J., Hinrichs, D. and Thaller, G., Institute of Animal Breeding and Husbandry, Christian-Albrechts-University, Hermann-Rodewald-Straße 6, 24118 Kiel, Germany; dhinrichs@tierzucht.uni-kiel.de

The aim of this study was to extend the optimum contribution selection (OC) to the genomic relationship matrix and compare this genomic OC selection with the pedigree OC. The impact of this breeding schemes on pedigree and genomic rates of inbreeding, responses to selection, changes in QTL gene frequencies, loss of favorable QTL alleles and number of selected animals were evaluated for a trait (h^2=0.1) by using stochastic computer simulation. Breeding schemes were closed nuclei with 1000 animals from 100 full sib families. A Genome consists of 10 Chromosomes with a total length of 1000 cM and 500 allocating QTL were simulated. BLUP breeding values and genomic breeding values were maintained. Inbreeding followed the expectations on the scale it was constrained, either pedigree or genomic. In case of constraining pedigree inbreeding and selecting for GEBV the genomic inbreeding was around 5 times higher in Generation 10 than the constrained. This indicated that the true inbreeding was underestimated when using additive genetic relationship matrix. After 10 generations of selection the cumulative response of the GEBV breeding scheme was 44% higher than in the BLUP breeding scheme, when constraining genomic Inbreeding and 33% higher when constraining pedigree inbreeding. For all selection strategies, accuracy of the estimated breeding values decreases, whereas the decrease in pedigree OC strategies is greater. Trends in mean frequencies of favorable QTL alleles followed the response to selection whereby the pedigree OC resulted in higher mean frequencies than genomic OC. These results indicated that it is useful to restrict genomic inbreeding in genomic improvement programs. Genomic OC can maintain higher genomic variability in terms of lower frequencies of favorable QTL and lower proportion of QTL that were lost and a smaller decrease of genetic variance which led to a higher response of selection on the long term.

Non-random mating is less effective in reducing rates of inbreeding in genomic selection than in BLUP schemes

Nirea, K.G.[1], Sonesson, A.K.[2], Woolliams, J.A.[1,3] and Meuwissen, T.H.E.[1], [1]Norwegian University of Life science, Department of Animal and Aquacultural Sciences, P.O. Box 5003, 1432 Ås, Norway, [2]Nofima Marin, P.O. Box 5010, 1432 Ås, Norway, [3] University of Edinburgh, The Roslin Institute and R (D) SVS, Roslin, Midlothian, EH25 9PS, United Kingdom; kahsay.nirea@umb.no

Previous studies on non-random mating designs have relied on best linear unbiased prediction (BLUP) evaluation. Recently, genomic selection (GS) has been introduced in animal breeding in which the total genetic value of the selection candidates is predicted based on dense genotyping and estimation of SNP effects, which come from a genotyped and phenotyped training set. This study has compared the effects of non-random mating designs on rates of inbreeding (ΔF) relative to random mating (RAND) for schemes with BLUP and GS evaluations. We compared the effect of minimum coancestry (MC) and minimizing covariance of ancestral contributions (MCAC) mating for schemes, where the trait information comes from the selection candidates (CAND) or sibs of the candidates (SIB); with BLUP or with GS selection; with pair-wise family structure (PAIR) or with factorial family structure (FAC). The ΔF was substantially reduced in PAIR schemes (42-57% by MC) but less in FAC schemes (13-27% by MC). In BLUP schemes, ΔF was substantially reduced in SIB (29-57% by MC) but less in CAND (13-48% by MC). MC was more effective to reduce ΔF than MCAC. The effect on ΔF was much smaller in GS schemes (14-43% by MC) than in BLUP schemes (13-57%) regardless of the information sourceand family structure, this is because the GS evaluation accurately predicted the Mendelian sampling term so that mating proportions of animals are closer to their long-term contributions, which reduces the need for non-random mating to reduce the sum of long-term contributions, i.e. the rate of inbreeding.

Penalized relationship in multi stage selection schemes

Hinrichs, D.[1] and Meuwissen, T.H.E.[2], [1]Institute of Animal Breeding and Husbandry, Christian-Albrechts-University, Hermann-Rodewald-Straße 6, 24098 Kiel, Germany, [2]Dept. of Animal and Aquacultural Sciences, University of Life Sciences, P.O. Box 5003, 1432 Aas, Norway; dhinrichs@tierzucht.uni-kiel.de

Optimum contribution (OC) selection was extended to breeding schemes with multi-stage selection. An extreme example was considered here: the pre-selection of dairy bulls that enter a progeny test. First the penalty on average relationship in selection step 1 is assumed the same as in step 2. Thereafter, situations with different penalties in the two selection steps were analyzed. The simulation started with the generation of prior EBVs, from a truncated normal distribution. Candidates for progeny testing were selected and progeny test EBVs were simulated, where the progeny test based on 100 daughters per young bull. In situations with high accuracy of prior EBVs, high heritability and prior EBVs were available for 2000 bulls, the results were similar for both approaches, independent of the family size. However, in a situation with low accuracy of prior EBVs and low heritability it could be observed that with increasing penalty on average relationship, correction for relationship in stage 1 yielded in a higher corrected genetic level (CGL), i.e. genetic level – (Penalty*Average relationship), compared to selecting only for high prior EBV. If the number of bulls with prior EBVs increased up to 4000 an increasing penalty gave improved CGL. A further improvement, with respect to CGL and average relationship could be observed by increasing the penalty in selection step 1 above that in selection step 2. All in all this study showed that it is beneficial to use a penalty already for the selection of bulls that enter the progeny test. In case OC was applied with a constraint on the average relationship in stage 2, this may be translated into a penalty on average relationship and the current results suggested that the optimal penalty in selection stage 1 should be twice that of stage 2.

Impact of genomic selection on functional traits in a dual purpose cattle breeding program

Ytournel, F.[1], Willam, A.[2] and Simianer, H.[1], [1]Georg-August University Goettingen, Department of Animal Sciences, Albrecht-Thaer-Weg 3, 37075 Goettingen, Germany, [2]University of Natural Resources and Applied Life Sciences, Institute of Animal Sciences, Gregor-Mendel-Straße 33, 1180 Vienna, Austria; fytourn@gwdg.de

The importance of the functional traits in the breeding goals has considerably increased in the last decades. The use of genomic data to estimate breeding values should be particularly beneficial for those traits, which have a low heritability. The efficiency of genomic selection or marker assisted selection to improve the genetic gain on functional traits has been mainly shown in studies considering the traits individually. However, selection relies on an index combining many traits. Generally, the production traits, which have moderate to high heritabilities, are genetically negatively correlated with functional traits. The Austrian breeding program for the Simmental breed was modeled in the software ZPLAN+ to evaluate the improvement that could be achieved for these traits under various weightings of the functional traits relative to the production traits. Some of the functional traits benefit from the genomic selection even when keeping the economic weights to the present values, especially the functional longevity (genetic gain per year from +9 days in the conventional scheme to +29 days in the genomic breeding scheme). Under the current parameters, genomic selection would also have a positive impact on the genetic trend of the maternal fertility compared to the present breeding scheme (+90%). However, the expected changes due to genomic selection were highly diverse for other functional traits (from -67% for the paternal fertility to + 100% for the paternal stillbirth).

Two decades of selection for fat content in milk
Petersson, K.-J.[1], Franzén, G.[1], Lundén, A.[1], Bertilsson, J.[2], Martinsson, K.[3] and Philipsson, J.[1], [1]Swedish Univ. of Agric. Sci., Dept. of Animal Breeding and Genetics, Box 7023, 750 07 Uppsala, Sweden, [2]Swedish Univ. of Agric. Sci., Dept. of Animal Nutrition and Management, Kungsängens forskningscentrum, 753 23 Uppsala, Sweden, [3]Swedish Univ. of Agric. Sci., Dept. of Agricultural Research for Northern Sweden, Skogsmarksgränd, 901 83 UMEÅ, Sweden; karl-johan.petersson@slu.se

A selection experiment on consequences of breeding for high or low milk fat content in the milk of high producing Swedish Red cows was started at three experimental farms of the Swedish University of Agricultural Sciences in the end of the 1980s. The experiment went on for 15 to 20 years with different duration in the three farms, with up to six generations of selection. The selected AI bulls all had high breeding values for fat corrected milk (FCM), but with maximum divergence in genetic merit for fat content. Data from the official milk recording together with more extensive experimental data were collected. In total, 1,701 lactations were completed from the high fat line and 1,644 lactations from the low fat line. Mean FCM per lactation was 8,645 kg over generations and did not differ significantly between the two selection lines, but differed significantly for fat content with 0.59%. A correlated response in protein content was observed with a difference of 0.19% between the two lines. The high fat line showed a continuous increase in fat content until a peak of 4.95% was reached in generation three. The low fat line had a steady decrease in fat content throughout the experiment ending at 4.14%. The declining fat content in the high fat line after generation three paralleled a decreasing genetic trend in fat content for the Swedish Red breed during the same period. Differences between the selection lines in various functional traits such as fertility, udder health and other diseases were also examined. No association between fertility traits and selection for fat content was found while cows of the low fat line had a significantly higher risk of mastitis and parturient paresis.

Higher milk yield genetically correlate to more frequent milking, faster flow but less cell count in automatically milked cows
Løvendahl, P.[1], Lassen, J.[1] and Chagunda, M.G.G.[2], [1]Aarhus University, Genetics and Biotechnology, Research Centre Foulum, DK 8830 Tjele, Denmark, [2]SAC Research, Sustainable Livestock Systems Group, King's Buildings, West Mains Road, Edinburgh, EH9 3JG, Scotland, United Kingdom; Peter.Lovendahl@agrsci.dk

First lactation cows milked automatically with free access to milking units were intensively recorded for milk yield, milk composition and milkability traits in order to obtain estimates of genetic co-variation among traits and estimate correlated response to selection for higher ECM yield. Seven years of data from the Danish Cattle Research Center experimental herds included records from 556 first parity cows of three breeds (141 Red Dane, 268 Holsteins, and 140 Jersey), and herd mates in later lactations (although these were omitted from this study). Data from 280,510 voluntary milkings were used, excluding any incomplete or faulty milkings, and milkings proceeded by incomplete milkings. Of these milkings 144,751 had milk composition data. Yield per milking was extrapolated to 24 h basis to calculate ECM kg/d. Cell counts were used as an indicator for mastitis. Milkability traits included average flow rate; milking time per milking and total time in milking box per day. Data were analysed using a mixed model with repeated records within lactation; additive genetic covariance components assumed constant over the lactation. A wide range of heritability estimates were obtained: (Milking frequency h^2=0.17; Yield/milking 0.26; ECM/24 h 0.35; Cells 0.26; Flow rate 0.62; Milk time 0.46; Box time/d 0.30). Genetic correlations to ECM/24 h showed that higher yield was associated with more frequent milking and more milk per milking and also higher flow rate. However, the correlation to milking time per milking was close to zero and the daily box occupation time increased with ECM yield. Surprisingly, somatic cell count was negatively correlated to daily ECM yield. It is concluded that higher yield is associated with more milkings and more time for occupying the milking box if cows are allowed free access to the AMS.

Selection for resilience and resistance to parasitism in Creole goat

Gunia, M.[1], Bijma, P.[2], Phocas, F.[3] and Mandonnet, N.[1], [1]INRA Antilles-Guyane, UR 143, Domaine Duclos, 97170 Petit-Bourg, Guadeloupe, [2]Wageningen University, Animal Breeding and Genomics Centre, P.O. Box 338, Wageningen, Netherlands, [3]INRA, UMR 1313, Domaine de Vilvert, 78352 Jouy en Josas, France; melanie.jaquot@antilles.inra.fr

A breeding programme aiming at improving production, reproduction and adaptation traits will be implemented for Creole goats in Guadeloupe. Selection response of alternative breeding objective and index were studied using SelAction software for a closed nucleus composed of 300 does and 20 bucks. Body weight at 11 months (BW), fertility (FER) and LS (Litter Size) were the traits in the breeding objective and index constituting the 'Base' scenario. Then, the Packed Cell Volume (PCV) of does, a measurement of resilience to parasitism, was integrated in the selection objective for the 'Resilience' scenario. Different economic values for PCV were tested, as well as accounting for faecal egg count (FEC), a measurement of resistance to gastrointestinal nematodes, in the selection index. Finally, a 'Resistance' scenario combining the three traits of the 'Base' breeding scheme and FEC was considered. Selecting for resistance or resilience in Creole goat will decrease the selection response on the other traits. However, adaptation of animals to their environment is the key-point of sustainability of breeding systems in the tropics and should be integrated in the breeding goal. Selecting on resilience with a low economic value on PCV and without recording this trait directly gave very low and slow genetic progress on this trait. Increasing arbitrarily its economic value did not bring a much higher selection response for this trait, but decreased the accuracy of selection and the genetic progress of the production and reproduction traits. In the Creole goat population, selecting on FEC gives a more efficient selection response than selecting on PCV. The breeding scheme for selection on resistance improves the production and reproduction traits of Creole goats while maintaining their resistance trait.

Study of using genetic markers on a Nellore breeding program for post-weaning gain by model comparison

Rezende, F.M.[1], Ibáñez-Escriche, N.[2], Ferraz, J.B.S.[1], Eler, J.P.[1], Silva, R.C.G.[1,3] and Almeida, H.B.[3], [1]University of Sao Paulo, Av. Duque de Caxias Norte, 225, 13635-900 Pirassununga, Brazil, [2]IRTA, Animal Breeding and Genetics, Av. Alcalde Rovira Roure, 191, 25198 Lleida, Spain, [3]Merial Igenity, Av. Carlos Grimaldi, 1701, 13091-908 Campinas, Brazil; frezende@usp.br

Data of a commercial Nellore beef cattle selection program were used to evaluate MAS implementation on its breeding program by model comparison. The data file consisted of 1,088 records for post-weaning gain (PWG) adjusted for systematic, maternal and dam permanent environmental effects. A set of 106 SNP markers was analyzed to estimate the genomic breeding value. A total of three models were used to evaluate the MAS implementation on this study. Model 1 included only polygenic effects, model 2 included only markers effects and model 3 included both polygenic and markers effects. Bayesian inference via Markov chain Monte Carlo methods, performed by TM program, was used to analyze the data. Two criteria were adopted for model comparison: deviance information criteria (DIC) and k-fold cross-validation. Two k-fold cross-validation strategies were applied: 4-fold and 1-fold cross-validation. The posterior means of heritability estimated for post-weaning gain from models 1 and 3 were in agreement with the estimates reported in the literature. The model comparison results were not consistent across models. The estimates of the deviance information criterion (DIC) indicated the highest credibility of model 2, since it presented the lowest DIC value. However, model 2 was the worst evaluated model in terms of prediction ability. Considering the ability to predict the next generation performance, model 1 was the best model for the MSE criterion and model 3 for PC criterion. Therefore, it was not clear the advantage of including marker effects on the Nellore genetic evaluation for PWG. Nevertheless, further studies including the application of penalized methods (i.e. GS-BLUP, Bayes A, B, C and Bayesian Lasso) are needed in order to confirm these outcomes.

Phenotypic characterization: a tool for breed differentiation of the camel in Northeast Balochistan, Pakistan

Abdul, R., SAVES, Kakar house, st 7, Faisal Town, Brewary road Quetta, 83000, Pakistan; skydoms@hotmail.com

Camel is a device for livelihood earning and food in hostile environment of northeastern Balochistan. The camel keepers of the region keep 3 distinct camel breeds known as Kohi, Pahwali and Raigi. Among indigenous livestock keepers, a breed is a cultural notion with the other criterion as phenotype, habitat/ adaptability, breeding goals and production potential for specific economic traits. In this study, a specific methodology, i.e. surveying, discussion with the camel keepers and body measurements (biometry) of the live animals was used to differentiate between the three important camel breeds. The perspectives of the camel keepers were asked through a pre-tested questionnaire and were conducted by tape meter in cm early in the morning with empty bellies. It was knew that camel herders differentiate breeds with their own perspectives and the demarcation they had had made was proved by the characterization of other traits like production, live body weight and biometric measurements used in this study. The most important special trait for each breed was mentioned as disease resistance of Kohi camel, long walking ability of Pahwali camel and thick milk production of Raigi camel. This study portrayed a picture of the camel in the area and proved that the camel keeper's perspectives along with the other criteria mentioned above can be well use for the differentiation of the breeds on one hand and selection of important breed for conservation on the other uses.

Screening for early pregnant cows in compact calving systems

Cutullic, E.[1,2,3], Delaby, L.[2,3], Gallard, Y.[4] and Disenhaus, C.[2,3], [1]Swiss College of Agriculture, Länggasse 85, 3052 Zollikofen, Switzerland, [2]Agrocampus-Ouest, UMR1080 Dairy production, 35000 Rennes, France, [3]INRA, UMR1080 Dairy production, 35590 St-Gilles, France, [4]INRA, UE326 Domaine Expérimental du Pin-au-Haras, 61310 Exmes, France; erwan.cutullic@bfh.ch

In compact calving systems, whatever the feeding strategy, some cows are able to get pregnant early, within 3 or 6 weeks of the breeding period. The objective of the present study was to analyze milk yield and body condition score (BCS) parameters distinguishing early pregnant cows from other cows, in a breed x feeding system experiment. Milk yield and BCS change were uncoupled on the overall dataset, since high-fed cows had high milk yields but moderate BCS change over lactation, and low-fed cows had low milk yield but deep BCS change. Respectively 19 and 33% of Holstein cows were pregnant within 3 and 6 weeks (n=98 lactations), respectively 27 and 53% of Normande cows (n=105). Both variance analysis approach, including a pregnant vs. non-pregnant factor, and multivariate logistic regression approach, predicting pregnancy occurrence, were used. In both breeds, 3-week pregnant cows had lower milk yield over lactation. This effect was not observed anymore after 6 weeks of breeding. In Normande cows, BCS levels and dynamics did not appear as a problem. In Holstein cows, both low milk yield and moderate BCS loss were required to get pregnant within 3 weeks, an overall higher BCS over lactation was required to get pregnant within 6 weeks. This suggests that prolonged low BCS can penalize reproductive performance for a long time, conversely to more reversible effects of either high milk yield or large BCS loss. Abnormal *post partum* ovarian activity was associated with overall decreased pregnancy rates in Holstein cows. In conclusion, high milk yield appears as a major risk factor for the 3-week pregnancy rate in compact calving systems, although it seems less influent later. In Holstein cows, reproductive performance improvement also clearly requires to select cows with regular ovarian activity.

Genetic evaluations for birth weight: comparison of continuous and discrete definitions of birth weight under varying accuracies of recording

Waurich, B.[1], Wensch-Dorendorf, M.[1], Cole, J.B.[2] and Swalve, H.H.[1], [1]Martin-Luther-University Halle-Wittenberg, Institute of Agricultural and Nutritional Sciences, Theodor-Lieser-Str. 11, 06120 Halle, Germany, [2]Agricultural Research Service USDA, Animal Improvement Programs Laboratory, 10300 Baltimore Avenue, 20705 Beltsville, MD, USA; benno.waurich@landw.uni-halle.de

When recorded, birth weights can be used as auxiliary variables in genetic evaluations for calving ease. Subjective scores for birth weights would cost less labour but may lack the precision needed. The aim of the study was to assess this precision using a data set with recorded birth weights from dairy cattle contract herds in North-eastern Germany. Data included 67,715 calving with birth weights, of which 25,462 were from heifer calving. First, the original data were used to form categorical scores for three different schemes: 1) five classes, each 20%; 2) three classes, each 33%; and 3) three classes with 25/50/25%. Since subjective classification is prone to errors, a random error term was added to the original birth weights to simulate classification errors, and the manipulated birth weights used to form classes. Variance components and breeding values were estimated using linear univariate sire-maternal-grandsire models. For the original birth weight data, estimates of direct and maternal heritabilities were 0.30 and 0.08, respectively, for all lactations. Estimates of direct heritabilities increased for heifer calving while calving from later parity cows showed increased maternal heritabilities. Estimates were slightly lower when based on the categorical scores, and were again reduced when based on scores generated from manipulated data. The lowest estimates of direct and maternal heritabilities were 0.20 and 0.05, respectively. Rank correlations between EBVs for sires indicated only slight shifts in ranking between different trait definitions. These results suggest that subjective scoring systems may be a valuable alternative to costly weighing of calves.

Genetic parameters for somatic cell score in early lactation

Schafberg, R., Rosner, F. and Swalve, H.H., Martin-Luther-University Halle-Wittenberg, Institute of Agricultural and Nutritional Sciences, Theodor-Lieser-Str. 11, 06120 Halle, Germany; renate.schafberg@landw.uni-halle.de

In German dairy herds, intramammary infections, claw disorders and metabolic diseases are main reasons for early culling. These diseases have a demand on the immune system of the cow that additionally suffers from metabolic stress. Somatic cell score (SCS) is widely accepted as an indicator for the immune response and for its usefulness in the improvement of udder health in dairy cows. Measuring SCS in early lactation (SCSel) is suggested as a new management tool to control the health status of transition cows. The objective of this study was to examine SCSel from records specifically taken in the very early onset of lactation while accounting for between and within-cow variation. Data consisted of 2836 Holstein cows in first to third lactation from seven large herds that calved in the year 2010. Aiming at recording SCSel on day 5 of the lactation, the average for days in milk (DIM) of one single recording per cow was 5.2 and varied between 0 to 20 DIM. The collection of data within a project on neo-partus health is ongoing. In general, the lactation curve for SCS typically shows high values at the beginning of the lactation. In our data, SCSel was 4.1 to 4.6 (DIM 5) and official milk recording results for the same cows were 2.5 to 2.9 at first test day (mean DIM = 23) and 2.2 to 2.3 at second test day (mean DIM = 57). It should be noted that the high value for SCSel not only can be attributed to the colostrums phase of the lactation but also arises from the fact that cows that are culled before they reach their first test day are included. Estimates of the heritability of SCSel and SCS of the first tow test days were in the range of 0.10 to 0.12 while genetic correlations between all types of SCS measurements were on a low level.

Across breed and country validation of mid-infrared calibration equations to predict milk fat composition

Maurice - Van Eijndhoven, M.H.T.[1,2], Soyeurt, H.[3], Dehareng, F.[4] and Calus, M.P.L.[1], [1]Animal Breeding and Genomics Centre, Wageningen UR Livestock Research, P.O. Box 65, 8200 AB Lelystad, Netherlands, [2]Animal Breeding and Genomics Centre, Wageningen University, P.O. Box 338, 6700 AH Wageningen, Netherlands, [3]Animal Science Unit, Gembloux Agro-Bio Tech, University of Liège, B-5030, Gembloux, Belgium, [4]Walloon Agricultural Research Centre, Valorisation of Agricultural Products, B-5030, Gembloux, Belgium; myrthe.maurice-vaneijndhoven@wur.nl

The aim of this study was to investigate the accuracy to predict detailed fatty acid (FA) composition of milk, from a population with different characteristics like country and breed, using calibration equations. Calibration equations for predicting FA composition using mid-infrared spectrometry were developed in the European project RobustMilk and based on 1236 milk samples from multiple breeds from Ireland, Scotland, and the Walloon Region of Belgium. The validation data set contained 190 milk samples from cows in the Netherlands across 4 breeds: Dutch Friesian, Meuse-Rhine-Yssel (MRY), Groningen White Headed, and Jersey (JER). Gas-liquid chromatography (GC) was used as golden standard to measure FA composition. The capillary column of the GC was different as the one used to develop the calibration equations, therefore some groups of FA were not considered due to differences in definition. Over all breeds calibration equations gave highly accurate predictions ($R^2 > 0.80$) for 8 individual FA and 7 groups of FA. Calibration equations for 3 individual FA were moderately accurate (R^2 0.60-0.80) and for 4 individual and 2 groups of FA predictions were less accurate ($R^2 < 0.60$). Comparing the different Dutch breeds FA composition in milk from MRY was predicted most accurately and from JER with the lowest accuracy. Generally, FA with lower concentrations in milk were predicted with lower accuracy. In conclusion, the RobustMilk calibration equations are robust and can be used to predict most FA in milk from the 4 breeds in the Netherlands with only a minor loss of accuracy.

Mid-infrared predictions of cheese yield from bovine milk

Vanlierde, A.[1,2], Soyeurt, H.[1,3], Anceau, C.[2], Dehareng, F.[4], Dardenne, P.[4], Gengler, N.[1,3], Sindic, M.[2] and Colinet, F.G.[1], [1]University of Liege, Gembloux Agro-Bio Tech, Animal Science Unit, Passage des Deportes 2, 5030 Gembloux, Belgium, [2]University of Liege, Gembloux Agro-Bio Tech, Analysis, Quality and Risk Unit, Passage des Deportes 2, 5030 Gembloux, Belgium, [3]National Fund for Scientific Research, Rue Egmont 5, 1000 Brussels, Belgium, [4]Walloon Agricultural Research Centre, Valorisation of Agricultural Products Department, Ch Namur 24, 5030 Gembloux, Belgium; Frederic.Colinet@ulg.ac.be

Economically, cheese yield (CY) is very important, and empirical or theoretical formulae allow estimating the theoretical CY from milk fat and casein or protein content. It would be interesting to predict CY during milk recording directly without the need to estimate milk components. Through the BlueSel project, 157 milk samples were collected in Wallonia from individual cows and analyzed using a mid-infrared (MIR) MilkoScanFT6000 spectrometer. Individual laboratory cheese yields (ILCY) were determined for each sample and expressed as g of dry coagulum/100 g of milk dry matter. An equation to predict ILCY from MIR was developed using partial least squared regression (Winisi III). A first derivative pre-treatment of spectra was used to correct the baseline drift. To improve the repeatability of the spectral data, a file containing the spectra of samples analyzed on 5 spectrometers was used during the calibration. During calibration, 23 outliers were detected and removed from the calibration set. The ILCY mean of the final calibration set was 63.9% with a SD of 11.2%. The calibration (C) coefficient of determination (R^2) was equal to 0.76 with a standard error (SE) of calibration of 5.5%. A full cross-validation (CV) was preformed to assess the robustness. R^2_{cv} was 0.72 with a SECV of 6.0%. The similarity between R^2_c and R^2_{cv} as well as between SEC and SECV permits to consider robustness of the developed equation as good. Even if it is planned to improve the equation with addition.

Nutritional flexibility of Charolais cows

De La Torre, A.[1], Blanc, F.[2], Egal, D.[3] and Agabriel, J.[1], [1]INRA, UR1213, Theix, 63122 St Genès Champanelle, France, [2]VetAgro Sup, Campus Agronomique, 63370 Lempdes, France, [3]INRA, UE1296 Monts Auvergne, 63820 Laqueuille, France; anne.delatorre@clermont.inra.fr

In an uncertain context, beef cattle systems should seek feed autonomy. To maintain performances, cows must be able to cope with periods of restrictive feed availability and to have the ability to mobilize and quickly recover body reserves. Adaptive abilities of Primiparous (PP) and Multiparous (MP) Charolais cows facing to nutritional post-partum challenges were studied. In a 2x2 factorial experiment, 2 groups of 7 PP (768±45 kg) and 2 groups of 7 MP (821±24 kg) were fed High (H) or Low (L) post-partum energy levels from calving to turn out (117 d). Average difference in winter energy intake between H/L levels was 31.3 (PP) and 27.0 (MP) MJ/d. From turning out, cows grazed similar permanent pastures (63 d). Body reserves were estimated using adipose cell diameter and live weight (LW). Data were analyzed using Proc mixed procedure. Milk production and calf gain did not differ among groups. Dynamic winter weight changes differed between parity and feed level: L resulted in a curvilinear (PP) vs. a linear (MP) LW decrease, whereas H induced a linear weight gain in PP and no changes in MP. All cows had a linear LW recover at pasture. PPL and MPL mobilized respectively 30 vs. 28 kg of lipids. Only MPL cows recovered part of their body reserves (+7 kg lipids) at pasture. At the end of the experiment, a LW gap remained between PPL and PPH cows (46 kg), but the amount of protein accretion was similar (9.5 and 10.8 kg). Estimated energy available for maintenance (Eam) was reduced by 30% in PP and 25% in MP cows after feed restriction (0.28 vs. 0.40 MJ/d/BW$^{0.75}$ in L and H). At pasture, Eam converged to a common value close to 0.43 MJ/d/BW$^{0.75}$ which is 20% higher than the theoretical value of 0.35 MJ/d/BW$^{0.75}$. These results highlight that underfed PP and MP cows have the same ability to adapt their maintenance energy expenditure. PP cows were also able to maintain their protein accretion under moderate energy restriction.

Influence of *post partum* nutritional level on estrus behavior in primiparous Charolais cows

Recoules, E.[1,2], De La Torre, A.[1], Agabriel, J.[1], Egal, D.[3] and Blanc, F.[2,4], [1]INRA, Theix, 63122 Saint-Genès-Champanelle, France, [2]INRA, BP 10448, 63370 Lempdes, France, [3]INRA, Les Razats, 63820 Laqueuille, France, [4]Clermont Université, VetAgro Sup, BP 10448, 63000 Clermont-Ferrand, France; emilie.recoules@clermont.inra.fr

Beef cattle feeding systems mainly rely on forage resources. As feedstocks may be inadequate due to environmental constraints, cows may be underfed during winter periods. Under-nutrition could affect their reproductive performances (mostly in primiparous cows) and so economic viability of suckling farms. Oestrus expression is a key component of reproductive performance, especially when artificial insemination is used. Two experiments (trial 1: n=14 and trial 2: n=16) were successively carried out using primiparous Charolais cows. Body condition scores (scale 0-5) at calving were 2.4±0.13 in trial 1 and 2.0±0.20 in trial 2. In each trial, two energy level diets (High, H vs. Low, L) were applied from calving to turn out. Cows were reared in groups (n=7 or 8) in a loose-housing system from November to May. Weight and body condition were regularly measured during the experiment. Reproductive efficiency was assessed through physiological and behavioural variables. Cyclicity was studied from plasmatic progesterone profiles. Oestrus expression was analyzed from 24 h/24 h video camera records that were studied using The Observer® software. Sexual behaviours were significantly more expressed during oestrus than in luteal periods: 54±14% of total interactions vs. 6±9%. Among all behaviours expressed during oestrus, standing to be mounted represented only 3.7±3.5%. Oestrus duration (time between first and last standing to be mounted) was 7±6 h. Oestrus intensity (sum of all sexual behaviours occurring during oestrus) was 242±157. Duration of oestrus was longer (9±5 vs. 5±6 h, P<0.05) and intensity of oestrus expression was higher (303±182 vs. 184±100, P<0.05) in L than in H cows. This result might be explained by less time spent interacting with fellows in H than in L groups, as cows spent more time eating.

Milk and concentrate intakes in Salers calves modify body composition at weaning and feeding efficiency in young bulls' production

Garcia-Launay, F.[1], Sepchat, B.[1], Cirié, C.[2], Egal, D.[3] and Agabriel, J.[1], [1]INRA, UR1213 Herbivores, Theix, F-63122 Saint-Genès Champanelle, France, [2]INRA, UE1296 Monts Auvergne, Marcenat, F-15190 Marcenat, France, [3]INRA, UE1296 Monts Auvergne, Laqueuille, F-63820 Laqueuille, France; florence.garcia@clermont.inra.fr

Improving the gross margin of Salers young bulls' production may be achieved by increasing feeding efficiency and/or by improving carcass conformation. The pre-weaning growth trend may be a key determinant of feeding efficiency and of body composition after fattening. We aimed at determining whether the composition of the diet of suckled calves in milk, forage and concentrate modifies the changes in body mass composition and feeding efficiency. Three groups (Control, C; Concentrate, CC; Milk, M) of Salers young males (n=48) were reared from 3 months to weaning at 9 months. All the animals were suckled twice a day by their dam and had free access to permanent grassland hay. CC received an additional supply of concentrate. The M calves were suckled once a day by a dairy cow. Four calves per group were slaughtered at weaning and the 18 animals remaining were slaughtered after fattening with a common hay and concentrate diet. Repeated data were analysed with linear mixed models. Average daily live weight gain (ADG) was higher in M and CC than in C before weaning. During fattening, ADG was not different between groups. These growth trend resulted in different body compositions at weaning. The non-carcass adipose tissue and the liver proportions in body mass were higher in CC than in M and C. The development of the fore stomach was lower in M than in C and CC. The development of muscles was higher in M than in CC at weaning. After fattening, no difference in hot carcass weight and body composition was observed. Feeding efficiency was higher in M and C than in CC before weaning, resulting in lower feedstuff quantities needed to produce similar carcasses. These results will be included in dynamic growth models of young bulls from birth to slaughter.

Differentiation of carcasses of cattle fed mainly forages or concentrates on the basis of conformation, fattening and colour variables

Alvarez, R.[1], Chessa, M.[2], Margarit, C.[1], Nudda, A.[2], Cannas, A.[2] and Alcalde, M.J.[1], [1]Univ Seville, Ctra Utrera Km.1, 41013 Seville, Spain, [2]Univ Sassari, Via de Nicola 9, 07100 Sassari, Italy; ralvarez1@us.es

It could be interesting to establish differences based on the feeding system in bovine carcasses by means of variables that are easy to measure, as carcass or colour variables, to ensure the traceability of meat. 125 beef carcasses belonging to class A according to European regulation (young males under two years not castrated) were studied. The finishing diets of 73 animals were based on forages and those of 52 animals were based on concentrate. The carcass variables measured were conformation and fattening (Reg. CE 1249/2008). Colour of Trapezium muscle and subcutaneous fat was measured with a spectrophotometer according to CIE. Because of initial differences in carcass weight, this parameter was used as covariate. In meat samples only carcass parameters varied significantly (P<0.001), whereas in fat samples both carcass and colour variables varied significantly (P<0.001) due to the feeding system. In a discriminant analysis made with carcass and colour parameters on meat samples, only carcass variables and carcass weight entered to form part of discrimination, with 96.8% of the animals being accurately classified according to their feeding system. For fat samples, besides the carcass variables, the colour variables b* (yellow index) and H* also influenced discrimination, leading to an accurate classification of 97.6% of the animals. A further discriminant analysis was made considering only colour variables. The only variable considered for meat samples was L*, with 72.8% of the carcasses accurately classified, whereas for fat samples the variables H* and b* were considered, with 72.4% of the carcasses accurately classified. In conclusion, although discrimination including only colour variables is satisfactory, the use of both colour and carcass variables leads to a more accurate discrimination of meat due to feeding systems.

Genetic parameters for pathogen specific clinical mastitis in Norwegian Red cows

Haugaard, K.[1], Heringstad, B.[1,2] and Whist, A.C.[3], [1]The Norwegian University of Life Sciences, Department of Animal and Aquacultural Sciences, P.O. Box 5003, 1432 Ås, Norway, [2]Geno Breeding and A.I. Association, P.O. Box 5003, 1432 Ås, Norway, [3]Norwegian School of Veterinary Science, P.O. Box 8146 Dep, 0033 Oslo, Norway; katrine.haugaard@umb.no

The objective of this study was to estimate heritabilities for and genetic correlations among pathogen specific clinical mastitis (CM) in Norwegian Red cows. In Norway breeding values for mastitis are predicted based on records of CM. Bacteriological milk sample results from the mastitis laboratories has been recorded routinely into the Norwegian Dairy Herd Recording System since 2000, but has so far not been used in genetic analyses. This additional source of data may provide valuable information on pathogen CM. Health recordings from 234,088 first lactation Norwegian Red cows, daughters of 1,656 sires, were used for genetic analyses of four binary mastitis traits: unspecific CM and three pathogen specific CM-traits. The pathogens were *Staphylococcus aureus*, *Streptococcus dysgalactiae* and *Enterococcus* spp/*Escherichia coli* (the three most frequent pathogens causing CM). A Bayesian approach using Gibbs sampling was applied. A multivariat threshold liability model was used for the analysis. The posterior mean (s.d.) of the heritabilites were 0.061 (0.006) for liability to unspecific CM, and 0.036 (0.008), 0.021 (0.007) and 0.032 (0.010) for *S. aures*-, *Str. dysgalactiae*- and *E.coli*-mastitis respectively. The posterior mean (s.d.) of the genetic correlations were all high, ranging from 0.75 (0.14) to 0.87 (0.07). The highest correlation was found between unspecific CM and *Str. dysgalactiae*, while the lowest was found for *E.coli* and *S. aureus*. Genetic correlations lower than 1 indicates that mastitis caused by different pathogens can be considered as partly different traits. In spite of high rank correlations (0.95-0.98), some re-ranking of sires were observed. Only 3 of the top 10 sires for unspecific CM were ranked among top 10 for all the three pathogen specific CM traits.

Replacing soybean meal with protein of European origin in dairy cow feed

Froidmont, E.[1], Focant, M.[2], Decruyenaere, V.[3], Turlot, A.[3] and Vanvolsem, T.[4], [1]Walloon Agricultural Research Centre, Productions and Sectors Department, Animal nutrition and Sustainability Unit, Rue de Liroux 8, 5030 Gembloux, Belgium, [2]High School Charlemagne, Rue Saint Victor, 3, 4500 Huy, Belgium, [3]Walloon Agricultural Research Centre, Productions and Sectors Department, Animal breeding, Quality production and Welfare Unit, Rue de Liroux 8, 5030 Gembloux, Belgium, [4]Dumoulin SA, Parc Industriel 18, 5300 Seilles, Belgium; froidmont@cra.wallonie.be

The EU imports more than 30 million tons of soybean meal each year. Although it is a benchmark for high nutritional value, importing it on such a large scale is not without consequences: nitrogen enrichment of the environment, presence of GMO, economic dependence on external markets, clearing of primary forests. The growth of the biofuel industries opens up new alternatives to food manufacturers such as rapeseed meal or brewer's grains as a soybean meal substitute. The aim of this trial was to compare a Conventional Protein (CP) concentrate (57% soybean meal, 38% rapeseed meal) with a concentrate containing only European Protein (EP) sources (55% rapeseed meal, 18% sunflower cake, 18% brewer's grains and 4% maize germ). CP and EP concentrates were supplied at 15.1 and 19.1% of dry matter intake in order to balance the nutritional value of the diets. The basal diet was composed by 58% maize silage, 39% grass silage and 3% straw. Milk production (25.7 L/animal/day) and the milk butterfat (4.12%) and protein (3.59%) contents were found to be equivalent in both systems. The concentrate based on the EP sources slightly increased the content of unsaturated fatty acids in the milk (P<0.01), probably due to its high content in rapeseed meal. As regards cost, the feed price per litre of milk was similar in both cases (12 €/100 L). The results therefore confirm the production and economic feasibility of replacing soybean meal with EP sources, assuming soybean meal to cost 300 €/ton.

Supplementing live yeast to Jersey cows grazing kikuyu/ryegrass pastures

Erasmus, L.J.[1], Coetzee, C.[1], Meeske, R.[2] and Chevaux, E.[3], [1]University of Pretoria, Dept Animal and Wildlife Sciences, 0001 Pretoria, South Africa, [2]Outeniqua Research Farm, Dept Agriculture Western Cape, 6529 George, South Africa, [3]Lallemand Specialities Inc., 6120 W Douglas Ave., Milwaukee,WI, 53218, USA; lourens.erasmus@up.ac.za

Cows grazing high quality pastures are often supplemented with high levels of concentrate after milking, resulting in reduced performance due to lower rumen pH, reduced fibre digestibility and depressed intake. Supplementation with yeast products have improved performance of TMR fed cows but limited data are available for grazing cows. Our objective was to evaluate the effect of live yeast supplementation on the performance of Jersey cows grazing ryegrass/kikuyu pasture. Thirty multiparous Jersey cows were allocated to one of two concentrate treatments (Control/Control plus 1 g of Levucell SC 10 ME) in a randomized complete block design for a period of 70 days. Pasture was divided into 41 strips and cows were allocated a new pasture strip twice a day after milking. Three kg of the two concentrate treatments were fed twice daily after milking. The control concentrate contained 82% maize, 10% soybean meal, 4% molasses, 0.5% salt and a mineral mix. Milk yield (20.1 and 19.7 kg/d), 4% FCM (20.1 and 20.3 kg/d), milk protein (3.51 and 3.58%), and BW change (+37.8 and +36.4 kg) did not differ (P>0.05) for control and live yeast supplemented cows respectively. Milk fat % was increased by live yeast supplementation (3.99 and 4.24%, respectively; P<0.05). In addition an in sacco study was conducted using 5 rumen cannulated cows in a crossover design. NDF disappearance of ryegrass was 46.6 and 52.2% after 12 hours and 65.1 and 69.2% after 24 h incubation for control and live yeast supplemented cows. Yeast supplementation increased NDF disappearance at 12 h by 11.9% (P<0.05). Results suggest that live yeast supplementation to grazing cows did not affect production, live weight or BCS of cows, but increased milk fat % and in sacco NDF disappearance of ryegrass.

Effect of dietary starch source and alfalfa hay particle size on chewing time and ruminal pH in mid-lactation Holstein dairy cows

Nasrollahi, S.M., Khorvash, M. and Ghorbani, G.R., Isfahan University of Technology, Department of Animal Sciences, Isfahan, Iran, 84156-83111, Iran; ghorbani@cc.iut.ac.ir

The trial investigated the effects of, and interactions between, dietary grain source and alfalfa hay particle size on chewing time and ruminal pH of Holstein dairy cows. Eight cows (175 DIM) were allocated in a replicated 4×4 Latin square design with four 21-d periods. The experiment was a 2×2 factorial arrangement with 2 theoretical particle size of alfalfa hay (fine = 15 mm, or long = 30 mm) each combined with 2 different sources of cereal grains (barley grain alone, GB, or barley plus corn grain in a 50:50 ratio, GBC). On day 19 of each period, eating and rumination time were monitored visually for each cow over a 24-h period at 5-min intervals, assuming to be persisting for the entire 5-min interval. Samples of rumen fluid were taken on day 20 and 21 of each experimental period at 3 and 6 hours after feeding respectively, using a stomach probe and measured for pH. Data were analyzed using the mixed procedure of SAS; period, source of grain, particle size of alfalfa hay, and the interaction of grain and alfalfa hay were the fixed effects and square and cow within square were the random effects in the model. Increasing hay particle size increased eating and tended to increase rumination time which resulted in greater total chewing time (589.8 vs. 566.3 min/d; P=0.05). Rumen pH at 3 h post feeding increased as hay particle size increased (6.26 vs. 6.46; P<0.01). Ruminal pH at 6 h and eating time tended to be higher (227.3 vs. 212.4 min/d; P=0.03) for GBC than for GB. The study shows that both grain source and forage particle length have the potential to improve chewing activity and rumen pH of mid-lactation dairy cows.

Immediate and residual effects on milk yield and composition of decreasing levels of udder emptying during milking in dairy cows

Guinard-Flament, J.[1], Albaaj, A.[2], Hurtaud, C.[2], Lemosquet, S.[2] and Marnet, P.G.[1], [1]Agrocampus Ouest, UMR1080 Production du Lait, 65 Rue de Saint-Brieuc, 35042 Rennes Cedex, France, [2]INRA, UMR1080 Production du Lait, Domaine de la prise, 35590 Saint-Gilles, France; sophie.lemosquet@rennes.inra.fr

The quantity of milk accumulated in the udder, under extended milking intervals, could reduce milk yield, with residual effects on following milking intervals. Aim of the trial was to modify the quantity of milk stored in the udder over one milking interval, by decreasing udder emptying (100, 70, 40, and 0%), in order to describe the short-term effects on milk yield and composition. 16 dairy cows averaging 41 kg/d of milk were assigned to treatments 100, 70, 40, and 0% according to a Latin square design with four 7-d periods. Cows were milked twice daily at 0700 and 1730. Treatments were applied at the morning milking called M0. Changes in milk yield and composition were assessed on the 7 following milking (M1 to M7), using the mixed procedure of SAS with milking as repeated measures. The quantity of milk collected at M1 milking linearly increased as udder emptying decreased at M0. Nevertheless, because of milk accumulation in the udder, M0+M1 milk yield was quadratically depressed by -1.5, -5.3, -12.9 kg with 70, 40 and 0% treatments, respectively. Residual effects on milk yield were only observed for 40 and 0% treatments on M2 and M3 milking and did not differ between 40 and 0% treatments. Milk fat content increased only for treatment 0% at M1 and for treatments 40 and 0% at M2. Milk lactose content was lower for treatments 40 and 0% at M1 and only for treatment 0% till M4. Lower milk protein contents were observed from M2 till M5 milking for either treatment 40 or 0% or both. Till M5 milking, somatic cell score linearly increased as udder emptying decreased at M0. In conclusion, a 70%-emptying milking at one milking did not alter milk yield and composition in dairy cows. Residual effects on milk yield and composition were only observed for higher amounts of milk accumulated in the udder.

Effects of feeding whole or physically broken flaxseed on milk fatty acid profiles of Holstein lactating dairy cows

Danesh Mesgaran, M.[1], Jafari Jafarpoor, R.[1], Danesh Mesgaran, S.[1], Ghohari, S.[1] and Ghaemi, M.R.[2], [1]Ferdowsi University of Mashhad, Dept. Animal Science, P.O. Box 91775-1163, Mashhad, 91775, Iran, [2]kompania Turkaz llc, Qostanay, 1003, Kazakhstan; ghaemi.mohammadreza@gmail.com

Nine primiparous Holstein cows 495±34.5 kg of average body weight (BW) and 70±5 days in milk were assigned to a 3×3 Latin square design to determine the effects of whole (WFS) or physically broken flaxseed (GFS), used instead of extruded soybeans (ESS), on milk fatty acid composition. Three iso-nitrogenous and iso-energetic diets (crude protein: 180 g/kg DM and metabolizable energy: 12.6 MJ/kg DM) containing ESS, WFS or GFS used as 110, 90 and 90 g per kg DM, respectively, were provided. The diets were offered *ad libitum* as total mixed rations (forage to concentrate as 45:55) twice a day. Each experimental period consisted of 21 d of adaptation to the diets and 7 d for daily milk collection to determine milk fatty acid profiles. Data were statistically analyzed using GLM procedure of SAS. Milk fat concentration did not alter among the animals evaluated in this study (GFS = 32, WFS = 33 and ESS = 32 g/kg milk). There was a significant difference ($P<0.05$) in both mono- and poly-unsaturated fatty acid concentrations (GFS = 11.9 and 2.8, WFS = 10.9 and 2.2, ESS = 9.9 and 2.1 g/kg milk, respectively) among the cows fed the experimental diets. Inclusion of physically broken flaxseed in the diet caused a significant ($P<0.05$) increase in milk C18:2trans and decrease in C18:2cis concentrations (GFS = 1.3 and 1.1 vs. WFS = 0.9 and 1.0 and ESS = 0.5 and 1.3 g/kg milk, respectively). One of the finding of the study was a significant increase ($P<0.05$) of C18:3 fatty acid proportion in the milk fat of cows fed GFS compared with those fed WFS or ESS (C18:3trans: 0.13 vs. 0.11 and 0.09, C18:3cis: 0.63 vs. 0.37 and 0.27 g/kg milk, respectively). It was concluded that increasing the availability of healthy fatty acids by feeding broken flaxseed in the lactating dairy cow diets would result in milk with higher nutritive value.

Animal wise prediction of milk, protein and fat yield using solutions from the Nordic test day model
Pitkänen, T.[1], Mäntysaari, E.A.[1], Rinne, M.[2] and Lidauer, M.[1], [1]MTT Agrifood Research Finland, Biotechnology and Food Research, Myllytie 1, 31600 Jokioinen, Finland, [2]MTT Agrifood Research Finland, Animal Production Research, Tietotie 1, 31600 Jokioinen, Finland; timo.pitkanen@mtt.fi

An accurate forecast of future production of individual cows would be useful for dairy herd management purposes. A prediction model (PM) was developed, which utilizes the solutions from the Nordic test day genetic evaluation model (NTM) to predict milk, protein and fat yield for the Finnish Red Dairy cattle. The predictions can be made for one year ahead. The predictions were calculated for two-week intervals within a year. The PM will be implemented into an interactive web application where the herd owner can obtain yield predictions for a cow or for a group of cows. The predictions are made for heifers and cows alive. The NTM gives herd test day (HTD) solutions for observed HTDs. Those solutions are modeled by linear mixed model which is already used to predict the solutions for the future test days. Other NTM solutions utilized are lactation curves by calving month, calving age, within herd persistency, days carried calf, individually estimated genetic and non-genetic lactation curves. The NTM yield evaluations are updated four times a year and updated solutions will be transferred to a web application. The web application will be available for herd owners during the year 2011.

Using self-feeders for all-concentrate diets offered to weaned beef calves
Simeone, A.[1], Beretta, V.[1], Elizalde, J.[2], Cortazzo, D.[1], Viera, G.[1] and Ferrés, A.[3], [1]University of the Republic, Agronomy Faculty, Paysandu, 60000, Uruguay, [2]Consultant, Rosario, 2000, Argentina, [3]Consultant, Montevideo, 11500, Uruguay; asimeone@adinet.com.uy

A trial was conducted to evaluate the effect of feed delivery system with all-concentrate diets fed to weaned calves. Forty-eight Hereford calves (148.3±25.8 kg) were randomly allocated to 8 pens outdoors to receive 1 of 2 treatments (n=4 pens/ treatment), which consisted in an all-concentrate diet (6.3% rice hulls, 60% sorghum grain, 15% wheat bran, 14% sunflower meal, 0.9% molasses, 1.03% urea, 2.8% mineral and vitamin premix) offered *ad libitum* daily in 3 meals (DF) or using a self-feeder trough placed in each pen (SF). Animals were gradually introduced to diets and then fed for 57 days. Live weight (LW) was recorded every 14 days. Subcutaneous back fat (SCF) and Longisimus dorsi area (LDA) were determined by ultrasonography on day 57. Dry matter intake (DMI) was determined on 3 consecutive days per week. Feed to gain ratio (F:G) was calculated based on adjusted mean live weight gain (LWG) and mean daily DMI. Records of DMI and LW were analyzed as repeated measures using the Mixed Procedure of SAS, while F:G and carcass traits were analyzed with GLM procedure. Statistical model included treatment effect and the initial LW. Live weight increased linearly with time (P<0.01) and it was affected by initial LW (P<0.01) and treatment (P<0.01). DF calves showed higher LWG (1.516 vs. 1.362 kg/d; SE 0.061, P<0.01) and DMI expressed in kg/d (6.8 vs. 5.7 kg/day; SE 0.143, P<0.01) and as percentage of LW (3.72 vs. 3.15%; SE 0.073, P<0.01) compared to SF. However, SF system tended to show a lower F:G ratio (4.5:1 vs. 4.2:1; SE 0.061, P=0.08). By the end of the feeding period DF calves were heavier (P<0.05) but no differences (P>0.1) were observed in SCF (3.64 vs. 3.21 mm; SE 0.429) or LDA (40.9 vs. 39.3 cm^2; SE 1.36). Results indicate that although DF improves animal performance, SF is a viable system for all-concentrate diets given a better F:G ratio and observed high absolute LWG.

Optimum proportion of corn cob silage in total mixed rations for intensively-fed rosé veal calves

Vestergaard, M.[1], Mikkelsen, A.[2] and Jørgensen, K.F.[3], [1]Aarhus University, Faculty of Agricultural Sciences, Foulum, DK-8830 Tjele, Denmark, [2]Danish Cattle Research Centre, Foulum, DK-8830 Tjele, Denmark, [3]Danish Cattle Federation, Skejby, DK-8200 Aarhus, Denmark; mogens.vestergaard@agrsci.dk

The increases in cereal prices have encouraged rosé veal calf producers to increased utilization of home-grown high-energy forage crops. To be able to use a large proportion in total mixed rations (TMR), corn-cob silage (CCS) has proven better than whole-crop maize silage. However, the optimal use of CCS in rosé veal calf production has not been thoroughly documented: in this trial the effects of 20, 40 or 60% of NE in TMR coming from CCS was studied. The CCS had 54.1% DM and 603 g starch/kg DM. Barley was used in the substitution with CCS, and TMRs also included canola cake, soybean meal, sugar beet pulp, and mineral-vitamin mixture. TMR with 20 and 40% CCS were water-adjusted, so all TMRs had 60% DM. TMRs had 17% protein, 21-22% NDF and 36-39% starch. A total of 64 HF bull calves in two batches were blocked according to herd of origin and randomly allocated to one of 3 treatment groups. Calves were housed in 2 x 3 straw-bedded pens (10-11 per pen), and fed the experimental diets from 150 kg BW until slaughter, at 385 to 400 kg BW or before 10 month of age. DMI (5.7 kg/d) and ADG (1.54 kg/d) were not affected but starch intake was 7% lower (P<0.05) for 60% than for 20 and 40% CCS. FCE was 10% lower (P<0.05) for 40 and 60% compared with 20% CCS. With the slaughter strategy used, carcass weight (196 kg) and EUROP conformation (3.7) and fatness (2.2) were not affected by CCS proportion. However, fewer carcasses received premium in the 60% CCS group leading to a 4% lower slaughterhouse payment. The calculated feed cost was 6% lower for 60% compared with 20% CSS in TMR. It is concluded that despite tendencies to reduced performance for 60% CCS this might still be profitable due to comparative lower feed costs.

Effects of turn-out date and post-grazing sward height on dairy cross steer performance at pasture

O' Riordan, E.G., Keane, M.G. and Mc Gee, M., Teagasc, Grassland Science, Animal and Grassland Research and Innovation Centre, Dunsany, Co. Meath, Ireland; edward.oriordan@teagasc.ie

As winter feed is a major cost in Irish beef production systems, attention has focused on evaluating the effects of early turnout to pasture and on grazing practices, which are likely to increase pasture utilisation and improve animal performance. This study examined the effects of early spring turnout to pasture and of post-grazing sward height on animal performance. Sixty yearling steers (equal mix of Belgian Blue x Holstein-Friesian and Aberdeen Angus x Holstein-Friesians, mean initial LW of 280 kg) were assigned to either early (March 23rd) or late (April 12th) turnout to pasture treatments. Animals grazed to a stubble height of either 3.5 or 5 cm in a rotational grazing system. From late May onwards, the 60 steers grazed as followers in a leader (calves)-follower rotational grazing system. The grazing unit was socked at approximately 2.7 LU/ha. Animals were weighed at approximately 3 week intervals. The statistical model (GLM in SAS) included terms for breed, turnout date, grazing height slaughter date, and their interactions where appropriate. There were no significant interactions between breed, turnout date and grazing height. Compared with late turnout, early turnout steers were significantly heavier in early May, but differences were not significant later in the season. Early turnout steers had a greater total LW gain at pasture, but when expressed as daily gain at pasture there was no significant difference between the turnout dates. Steers grazing to a post grazing stubble height of 3.5 cm were significantly lighter than those grazing to 5 cm. LW differences, which increased as the grazing season progressed, reached 30 kg in late October. The performance of AA and BB were similar over the grazing season. The effect on LW of grazing to a sward height of 3.5 cm are such that it would be difficult to generally recommend such a practice for wide scale use in production systems such as these described here.

Effect of sex and frame size on carcass composition of Nellore cattle

Silva, S.L., Gomes, R.C., Leme, T.M.C., Bonin, M.N., Rosa, A.F., Souza, J.L.F., Zoppa, L.M. and Leme, P.R., Universidde de São Paulo, Animal Science, Av. Duque de Caxias Norte, 225, 13635900, Brazil; sauloluz@usp.br

The use of non-castrated males for beef meat production has been an increased practice in Brazil because they grow fast, utilize feed more efficiently and show high-yielding and leaner carcasses. Frame size (FS) also has been related to differences in carcass muscle and fat distribution. The effects of sex and FS on carcass tissues distribution and retail cuts yield in Nellore young bulls and steers finished in feedlot was evaluated. Throughout 2009 and 2010, Nellore bulls (n=76, 500±4.5 BW, 23 month old) and steers (n=79, 474±6.1 BW, 23 month old) from small (n=51), medium (n=53) and high (n=50) FS, according to BIF, were finished in feedlots receiving high-grain diets for 50 d to 140 d. Following slaughter carcasses were chilled for 24 h and then the left sides was first cut into wholesale pistola hindquarter (PHQ), forequarter (FQ) and combined plate, flank and short ribs (PA). The hump weight was recorded and added to FQ weight. Subsequently, wholesale cuts were individually weighed and broken into retail cuts (RC), bones and trimmings. Retail cuts were defined as all cuts trimmed of excess fat to a thickness of approximately 5 mm. Animals of high FS had higher percentage of PHQ and RC in PHQ than medium and small FS ($P<0.05$). Intact males showed greater percentages of PHQ, FQ and PA ($P<0.0001$). Intact males had higher percentages of RC in FQ and PA ($P<0.001$) but with no difference in RC in PHQ when compared to castrated males. Castrated males showed higher percentages of trimmings in PHQ, FQ and PA in comparison to intact males ($P<0.0001$). There was no effect of frame size in total RC but intact males showed greater RC than castrated males ($P<0.0001$). These results indicate that frame size did not have great effect on carcass traits but non-castrated males have higher percentages of RC and smaller percentages of trimmings, which can reduce costs of production with a beneficial impact in the production systems.

Effect of roughage level and particle size on the performance of cattle fed high grain feedlot diets

Beretta, V.[1], Simeone, A.[1], Elizalde, J.[2], Cortazzo, D.[1] and Viera, G.[1], [1]University of the Republic, Agronomy Faculty, Paysandu, 60000, Uruguay, [2]Consultant, Rosario, 2000, Argentina; beretta@fagro.edu.uy

The objective of this study was to evaluate the effect of diet roughage level and particle length on cattle feedlot performance. Sixty two Hereford steers (277±32 kg) were randomly allocated to 16 pens outdoors to receive 1 of 4 treatments (n=4 pens/ treatment) in a factorial arrangement 2x2: 2 roughage levels (RL, 10% or 30% sorghum silage) and 2 particle cut length of silage (PL, short or long). A total mix ration with a barley based concentrate was offered *ad libitum* in 3 meals. Animals were gradually introduced to diets and then fed for 45 days. Liveweight (LW) was recorded every 14 days and dry matter intake (DMI) determined on 3 consecutive days per week. Feed to gain ratio (F:G) was calculated based on adjusted LW gain (LWG) and mean DMI. Steers were slaughtered after feeding period; hot carcass weight was determined and subcutaneous fat thickness measured at 12th rib (SCF). Records of DMI and LW were analyzed as repeated measures using the Mixed Procedure while F:G, and carcass traits were analysed through the GLM procedure (SAS Inst. Inc., Cary, NC). Statistical model included main effects, the 2-way interaction (RLxPL) and initial LW. LW increased linearly with time ($P<0.01$) and it was affected by initial LW ($P<0.01$). Neither RL, PL nor the 2-way interaction affected LWG (1.80, 2.00, 1.97 and 1.97 kg/day for 10%-short, 10%-long, 30%-short, 30%-long, respectively; $P>0.1$). DMI and F:G were only affected by RL. Independent of PL ($P>0.1$), cattle receiving 10% silage showed lower DMI compared to 30% silage (13.2 vs. 16.0 kg/day, $P<0.01$; or 2.4 vs. 2.73 kg/100 kg LW, $P<0.01$), and a better F:G ratio (6.9:1 vs. 8.1:1, $P<0.01$). No differences between treatments were observed in carcass weight or SCF ($P>0.1$). Varying silage particle cut length does not affect finishing cattle feedlot performance when forage level varies between 10% and 30%, while F:G is improved by reducing silage level.

Relationship between grading scores and image analysis traits for marbling in Angus, Hereford and Japanese Black x Angus in Australia

Maeda, S.[1], Polkinghorne, R.[2], Kato, K.[1] and Kuchida, K.[1], [1]Obihiro University of A&VM, Obihiro Hokkaido, 080-8555, Japan, [2]Marrinya Agricultural Enterprises, Wuk Wuk Vic., 3875, Australia; s22330@st.obihiro.ac.jp

Beef carcasses are assigned based on Australian Meat Industry Classification System (Aus-meat) in Australia. Some grading components such as Marbling (MB) and Ossification (OSS) are assigned by the visual judgment. In addition, an eating quality program called Meat Standards Australia (MSA) has been developed from consumer testing to suit consumer needs. MSA grades each carcass cut and utilizes additional grading components including pH, *Bos indicus* %, etc. This study seeks to compare the relationship between grading scores and analysis traits for marbling. Digital images of the 12-13th rib for 30 heads of each breed: Angus (A), Hereford (H), Japanese Black x Angus (JBA) taken using by the mirror type camera. Each shooting date was Session1 (S1)=30th June 2010, S2=15th May, S3=16th Oct. Image analysis software (BeefAnalyzerII) analyzed image analysis traits (marbling %, coarseness index, etc.). Least square means of marbling % were A: 4.8%, H: 2.1%, JBA: 5.8% (S1); A: 4.3%, H: 1.8%, JBA: 3.2% (S2); A: 14.2%, H: 8.7%, JBA: 12.1% (S3). Significantly differences ($P>0.01$) between A and H were detected in all sessions. Similarly, A differs significantly from JBA in S2. Although Japanese Blacks commonly having high marbling %, there was no significantly difference between A and JBA in S2 and S3. It's possible that JBA had poor marbling development because the ratio of carcasses that were OSS 150 (Approx age is 20 months) or less was 90% in this survey. The fineness index of marbling showed significantly differences between H (0.31) and A (0.69), JBA (0.59). The correlation coefficients of MB (MSA) with marbling % were 0.83 (S1), 0.80 (S2), 0.86 (S3). Compared to the Aus-meat's values which were 0.82 (S1), 0.73 (S2), 0.85 (S3), MSA marbling correlations were higher than Aus-meat. An explanation of why the two standards were different relates to the grade scales.

Effects of slaughter age on the degree of marbling and fatty acid composition of marbling in Holstein steers

Ogata, M. and Kuchida, K., Obihiro University of Agriculture and Veterinary Medicine, Animal and Food Hygiene, Inada-cho Obihiro-shi, 080-8555, Japan; s18081@st.obihiro.ac.jp

In Japan, 1.2 million beef cattle are slaughtered per year. Wagyu (Japanese Black), which is famous for high levels of marbling, and Holstein (HOL) account for 50 and 25% of total slaughter numbers, respectively. However, few studies have focused on the effects of slaughter age on the resultant meat quality, i.e. marbling features, in HOL that has an average slaughter age of 20 months in Japan. 280 HOL at 12~23 months of age were used. Eleven carcass traits, eight image analysis traits (e.g.: MP: marbling percentage, CIM: coarseness index of marbling, FIM: fineness index of marbling), and two fatty acid compositions (e.g.: MUFA: mono-unsaturated fatty acid) were included in the analysis. Greater values of CIM and FIM indicate the existence of coarser marbling particles and of fine marbling particles in the rib eye (*L. dorsi*). To investigate the change of meat quality with age, the averages of each month of age for all traits were calculated. Carcass weight (CWT), rib eye area (REA), rib-thickness (RT) and subcutaneous fat thickness (SFT) increased with age. The averages of these traits by month of age were in the range of 322.9~461.3 kg, 34.6~45.8 cm^2, 3.7~6.1 cm and 0.8~2.2 cm, respectively. The correlation coefficient between the aforementioned averages and month of slaughter age were 0.85 (CWT, $P<0.01$), 0.91 (REA, $P<0.01$), 0.79 (RT, $P<0.05$) and 0.82 (SFT, $P<0.01$). Subjectively assigned BMS (marbling degree) also has the same tendency (r=0.78). MP, CIM and FIM from image analysis showed strong correlation with month of age ($R^2=0.95$, 0.92 and 0.92, $P<0.01$). The averages of C18:1 and MUFA by month of age were in the range of 43.5~49.3%, 47.2~53.1%, and correlation coefficient of C18:1 and MUFA with the month of age were 0.79 and 0.88. From the linear regression equation, CWT, MP and MUFA increased 15.9 kg, 1.2% and 0.47%, per month, respectively.

Milk production and N excretion of dairy cows fed on different forage types

Migliorati, L.[1], Boselli, L.[1], Masoero, F.[2], Abeni, F.[1], Giordano, D.[1], Cerciello, M.[1] and Pirlo, G.[1], [1]CRA-FLC, via Porcellasco,7, 26100 Cremona, Italy, [2]ISAN - UCSC, via Emilia Parmense,84, 29122 Piacenza, Italy; luciano.migliorati@entecra.it

Italian dairy cows are typically fed on a forage system based upon corn silage and alfalfa hay. The aim of this study was to assess the effect of different forage types on milk production and N excretion. Twenty-four multiparous Italian Friesian cows were divided into 3 groups according to parity, DIM (165±93) and milk production, and were given diets with 17% CP (on DM basis). Components of the three experimental diets were on % DM basis: corn silage 41, alfalfa hay 16, concentrate 43 (group CS); barley silage 24, corn silage 21, alfalfa hay 8, concentrate 47 (group BS); alfalfa hay 32, corn silage 24, concentrate 44 (group HA). Forages and concentrates were fed as TMR. Experimental design was a Latin square (3x3) with 3 treatments and 3 periods of four weeks. The first two weeks were for adaptation, whereas the last two were sampling periods. Milk production and composition were recorded three times a week for every cow, DMI and N excretion were determined daily per group. Nitrogen excretion was estimated on the difference between N intake and milk N, utilizing DMI, CP dietary, milk production and milk protein content. DMI was measured daily while BCS at ration shift. Effects of treatment and group were determined with ANOVA using the MIXED procedure of SAS. No difference was found in milk production, protein and lactose contents, milk protein and fat yield, milk chemical-parameters (NCN, NPN, pH, acidity) among treatments. Milk fat percentage was higher in BS than in HA (P<0.05) and urea content was higher in CS and BS than in HA (P<0.001). No difference was found for average DMI and BCS. Average DMI of groups were 182±14 kg/d while average BCS were 2.94±0.34. Estimate N excretion was reduced respectively by 10 and 5% in BS compared to CS and HA. Partial substitution of CS with AH or BS did not influence milk production, but introduction of barley silage contributes to reduce N excretion.

Assessing enteric methane emissions from ruminants using a calibrated tracer: effects of non-linear sulphur hexafluoride release and permeation tube rumen residence time

Deighton, M.H.[1,2], Boland, T.M.[2], O'loughlin, B.M.[1], Moate, P.J.[3] and Buckley, F.[1], [1]Teagasc, Moorepark, Fermoy, Ireland, [2]University College Dublin, Belfield, Dublin, Ireland, [3]Department of Primary Industries, Ellinbank, Victoria, Australia; matthew.deighton@teagasc.ie

This study assesses the effect of post-manufacture tube age and pre-experimental rumen residence duration upon performance of sulphur hexafluoride (SF_6) permeation tubes. SF_6 tubes are used in determination of enteric methane (CH_4) emissions via the calibrated tracer (ERUCT) technique. The SF_6 permeable surface consisted of a 0.2 mm polytetraethylene membrane (PM) and 1.6 mm sintered stainless steel frit. Fistulated cows were used for pre-dosing rumen residence. Three treatments (n=12) were applied; (1) age of 55 d and rumen duration of 7 d (Control), (2) age of 202 d and rumen duration of 159 d (A), and (3) tubes with an additional 0.1 mm external PM, age of 202 d and rumen duration of 159 d (B). Tube performance was assessed *in vivo*; duplicate tubes were monitored at 39 °C *in vitro*. Thirty six cows housed in a ventilated barn were offered ryegrass silage. Cows were dosed with a single tube 7 d before sampling. Methane emissions were determined for 6 d via ERUCT, individual feed intake was automatically recorded. Data were analysed using the MIXED procedure of SAS. Mean CH_4 emission per kg of dry matter intake determined using linear SF_6 release overestimated emissions from A and B tubes relative to Control tubes by 19% and 20% respectively (P<0.01). This overestimation is attributed to the effect of age as the 160 d *in vitro* SF_6 release rate of A and B tubes decreased from their initial 50 d calibration by 19% and 20% respectively. A curvilinear model was fitted to describe SF_6 release and enable correction of *in vivo* emission values. As *in vivo* emissions calculated using A and B tubes did not differ we conclude that rumen exposure of the frit does not inhibit SF_6 release. Tube rumen residence duration per se appears unlikely to cause error in the ERUCT technique.

Influence of yeasts supplementation on microbiome composition in liquid fraction of rumen content of lactating cows

Sandri, M.[1], Manfrin, C.[2], Gaspardo, B.[1], Pallavicini, A.[2] and Stefanon, B.[1], [1]University of Udine, Dept. Agricultural and Environmental Science, via delle Scienze, 208, 33100 Udine, Italy, [2]University of Trieste, Dept. of Biology, Via L. Giorgieri, 34127 Trieste, Italy; bruno.stefanon@uniud.it

A greater understanding of rumen microbiome genetic traits and its metabolic potential allow the improvement of feed efficiency reducing methane emissions. A gene centric metagenomic approach and next generation sequencing (NGS) technology was used to investigate the rumen microbiome composition of 12 lactating dairy cows without or with supplementation of 2 strains of dried Saccharomyces cerevisiae in a potato protein support. The cows were housed in the same barn and were feed a standard diet with corn silage (48%), forage (23%) and concentrates (21%). Three groups of 4 cows each were assigned to the experimental treatments: control (CTR, supplemented with 50 g/head/d of potato protein), SV (supplemented with strain1 50 g/head/d) and LV (supplemented with strain2 50 g/head/d). Experimental period lasted a fortnight and rumen content was collected with a rumen probe the day before the beginning and at the end of the experimental period. After separation the liquid phase was washed with buffer solution and the dried samples used to extract genomic DNA with a protocol from fecal DNA extractions. Amplification of DNA of 16S V1-V3 regions was made for eubacteria with universal primers E8F AGAGTTTGATCCTGGCTCAG and E534R ATTACCGCGGCTGCTGGC and for archaea with universal primers A2F TTCCGGTTGATCCYGCCGGA and AE532R TACCGCGGCKGCTGRCAC. Bioinformatic analysis (clustering, classification and library comparison) revealed that yeast administration produced a variation of microbial population with a shift to bacteria involved in cellulolytic activity. A great variation of microbial population was also observed between cows, independently from yeasts supplementation. Statistical differences on experimental samples are on validation.

Rumen wall morphology in low- and high-RFI beef cattle

Leme, P.R.[1], Oliveira, L.S.[1], Mazon, M.R.[1], Souza, J.L.F.[1], Silva Neto, P.Z.[1], Sarti, L.M.N.[2], Barducci, R.S.[2], Martins, C.L.[2], Gomes, R.C.[1] and Silva, S.L.[1], [1]University of Sao Paulo, Department of Animal Science, Duque de Caxias Norte 225, 13635-900, Pirassununga, SP, Brazil, [2]Sao Paulo State University, Department of Animal Breeding and Nutrition, Distrito Rubião Júnior s/n, 18618-970, Botucatu, SP, Brazil; prleme@usp.br

The aim of the trial, part of a broader project investigating the biological mechanisms affecting RFI in Nellore (Bos indicus) cattle, was to evaluate the rumen wall morphology of beef cattle with low and high residual feed intake (RFI), to test if differences in the papillary profile exist across feed efficiency classes. Nellore steers (n=30) and bulls (n=30) with 20 months and 355±26 kg BW, were fed in individual pens a diet containing 2.6 Mcal ME/kg DM and 14.6% CP for 62 days. Daily dry matter intake (DMI) and average daily gain (ADG) were recorded individually and RFI was calculated within each sex group. Cattle with RFI >0.5SD above and RFI <0.5SD below the mean were classified into high (H-RFI, less efficient) and low (L-RFI, most efficient) RFI cattle, respectively. Five L-RFI and five H-RFI cattle within each sex group was randomly chosen and slaughtered. Upon evisceration, tissue samples were obtained from the ventral sac of the rumen for assessment of papillary profile. Data was analyzed by one-way analysis of variance for a randomized block design (block=sex). L-RFI steers had similar final BW (503 vs. 508 kg, P=0.60) and ADG (1.61 vs. 1.69 kg/d, P=0.48), lower DMI (9.6 vs. 11.7 kg/d, P<0.0001), RFI (-0.79 vs. 1.18 kg/d, P<0.0001), and feed:gain ratio (6.0 vs. 7.1, P=0.02) than H-RFI cattle. There were no differences between L-RFI and H-RFI cattle on absorption surface (26.4 vs. 22.5 cm^2/cm^2 of rumen wall; P=0.53), papilla number (54 vs. 47 papillae/cm^2; P=0.39) and papillary area (96.2 vs. 95.4% of rumen wall surface; P=0.33). Although great differences exist in feed consumption, rumen wall morphology and absorption surface are similar across feed efficiency classes in Nellore cattle.

Effects of astaxanthin on cultured bovine luteal cells
Kamada, H., Hayashi, M., Akagi, S. and Watanabe, S., National Institute of Livestock and Grassland Science, Molecular Nutrition, Ikenodai-2, Tsukuba, Ibaraki, Japan, 305-0901, Japan; kama8@affrc.go.jp

The strong antioxidant effect of astaxanthin (AST) is noticed in various fields including human. The present study examined the possibility that AST directly affects the corpus luteum (CL) using primary culture of bovine CL. The minced CL tissue was digested with collagenase solution in spinner flask. The cell suspension was purified by centrifugation using Percoll discontinuous gradient to remove cell debris. 2×10^4 viable cells were seeded pcr well in Medium 199 containing 10% foetal calf serum. Cultures were incubated at 37 °C in air containing 5% CO_2. After monolayer formation, the culture medium was replaced with fresh medium containing AST (0.1~1000 nM) and subsequently changed every 2 days using same medium. Each treatment was performed in quadruplicate. The progesterone (P4) concentrations of culture medium were measured by EIA method. Addition of low level (0.1~10 nM) of AST increased the P4 production of cultured luteal cells. Supplementation of AST to the cow may improve its reproduction performance.

Bovine oocyte viability under modified maturation conditions and vitrification
Nainiene, R. and Siukscius, A., Institute of Animal Science of Lithuanian University of Health Science, R. Zebenkos 12, LT-82317 Baisogala, Radviliskis distr., Lithuania; rasa@lgi.lt

In mammals, oocyte cryopreservation success is important for establishment of genetic banks, but oocytes are particularly difficult objects to cryopreserve. Vitrification is a quick and relatively easy cryopreservation technique, but requires high cryoprotectant concentrations and extremely high cooling rates. Both immature and mature oocytes form a functional unit with cumulus cells layers. These layers reduce the speed of cryoprotectant penetration either at equilibration or at dilution. A study was conducted to determine the effects of cumulus cells layers on bovine oocyte maturation and on the post thaw viability of oocytes frozen by method of vitrification. Bovine cumulus oocytes complexes were randomly divided into 4 groups: cultivation: 1) with and 2) without cumulus cells; vitrification: 3) with and 4) without cumulus cells. For vitrification, the oocytes were exposed to the equilibration media for 10 min (TCM-199 with 20% FCS and 1.4 M glycerol) and transfer to the 4 drops of vitrification medium (30% glycerol, 20% FCS, 50% saccharose in bidistilled water) for 1-2 min in each. The study indicated that there was a positive effect of cumulus cells on oocyte maturation and fertilization. The numbers of mature and fertilized oocytes were, respectively, 3.4 and 25.1% higher for the cumulus cultured oocytes. Cumulus cells also positively affected the post thaw viability of the oocytes after vitrification: 35 oocytes out of 68 frozen with cumulus matured successfully (51.5%) and 20 oocytes out of 35 fecundated (29.4%). Removal of cumulus cells has resulted in 22.9% lower maturation and 12.3% lower fertilization of the oocytes.

Use of dinoprost tromethamine (PGF2α) at two different times in lactating Nellore cows synchronized with a CIDR-based protocol with FTAI

Biehl, M.V.[1], Pires, A.V.[1], Susin, I.[1], Nepomuceno, D.D.[1], Lima, L.G.[1], Cruppe, L.H.[2], Da Rocha, F.M.[1] and Day, M.L.[2], [1]University of São Paulo, Piracicaba-SP, 13418-900, Brazil, [2]Ohio State University, Columbus-OH, 43210, USA; alvpires@esalq.usp.br

The effect of a split-dose of prostaglandin (Lutalyse®) on reproductive performance of lactating Nellore cows (n=191) synchronized using 7d CIDR+EB program was determined. Cows were blocked according to BCS (2.53±0.32, scale 1 to 5), in a completely randomized design arrangement. The treatments were: 1xPGF-5d (e.g. 7d CIDR treatment and one injection of 25 mg PGF on 5th day); 2xPGF-7d (e.g. 7d CIDR treatment and two injections of 12.5 mg PGF at time of CIDR insertion and withdrawal). Blood samples for progesterone analysis were collected 10 d before and at CIDR insertion to classify cows as cyclic. At CIDR insertion, all cows received 2 mg of EB and cows from 2xPGF-7d received their first PGF injection. At CIDR removal, all cows received 300 IU of eCG (Novormon®) and their appropriate PGF treatments. Oestrus was detected for 60 h and FTAI was performed 50 h after CIDR removal. Beginning 16 d after CIDR withdrawal, oestrus detection was performed for 6 d and cows were AI according to the AM/PM protocol. Pregnancy diagnosis was performed by US 60 d after the first AI and at the end of the breeding season. Data were analyzed using GLIMMIX procedures of SAS. At the beginning of the protocol, 86% (164/191) of the cows were in anoestrous. Oestrus was detected in 54.5% (104/191) of cows, and onset and distribution of oestrus (mean 36.4±8.7 h after CIDR removal) did not differ between treatments. Timed AI pregnancy rates were not different (1xPGF-5d, 59.5; and 2xPGF-7d, 59.7%). Final pregnancy rates did not differ at the end of the breeding season (1xPGF-5d, 82.9; and 2xPGF-7d, 86.6%). In conclusion, the use of 12.5 mg of PGF at the initiation of the synchronization program neither improved oestrus response nor timed AI pregnancy rates and resulted in similar pregnancy rates at the end of the breeding season.

No adverse effects of vaccination against BTV-8 in Swedish dairy herds

Emanuelson, U.[1] and Gustafsson, H.[1,2], [1]Swedish University of Agricultural Sciences, Dep't Clinical Sciences, POB 7054, SE-75007 Uppsala, Sweden, [2]Swedish Dairy Association, POB 210, SE-10124 Stockholm, Sweden; ulf.emanuelson@slu.se

The bluetongue virus serotype 8 (BTV-8) appeared in Central Europe in 2006 and reappeared in 2007, spreading rapidly through many countries. The European Commission encouraged mass vaccination in member states with cases of BTV, to prevent the spread of the disease. The first case in Sweden appeared in September 2008 and a vaccination campaign was started immediately in certain areas. There were concerns among farmers that the vaccination process led to adverse effects on health and reproduction of dairy cows, and the purpose of this study was to evaluate such possible associations. Herd and individual cow data from a geographical area of Sweden with vaccinated herds bordering on an area where herds were not vaccinated were retrieved from the Swedish Dairy Association. Almost 1,600 herds and 170,000 cows were included in the analysis, with information on exact vaccination dates, calving and artificial inseminations (AI), as well as milk yield and somatic cell count (SCC) from test-day records covering a period from 20 months prior to until 10 months after the first vaccination. The association between vaccination and pregnancy chance at first AI, pregnant at 30 days after the herd voluntary waiting period (PV30), calving interval, stillbirth risk, daily milk yield and log(SCC) was analysed with generalized estimation equation models with appropriate link functions. The models also included possible confounding factors such as parity, season, breed, and stage of lactation, and accounted for clustering of observations within herds. The study did not identify any significant associations between vaccination and any of the studied outcome variables and the conclusion was thus that vaccination against BTV-8 did not lead to any adverse effects on cow health and fertility.

Goals, perspectives and key challenges of the Horse Commission Nutrition Working Group

Saastamoinen, M.[1] and Miraglia, N.[2], [1]MTT Agrifood Research, Equine Research, Opistontie 10 a 1, FI-32100 Ypäjä, Finland, [2]Molise University, Department of Animals, Vegetables, Environmental Sciences, Via De Sanctis, 86100 Campobasso, Italy; markku.saastamoinen@mtt.fi

In the past the activities of the EAAP Horse Commission (HC) were mainly focused on breeding, selection and management; sessions on nutrition were routinely organized since the eighties. In this context an informal group of European scientists involved in equine nutrition has raised and proposed increasingly new sessions: more specialised sessions or/and satellites meetings have been organized by the HC. The evolution determined an integrative approach focused to deepen animal nutrition using fundamental disciplinary approach and applied section for equine experts acting as multipliers of knowledge. Consequently, the conditions were ready to introduce a Working Group Nutrition under the umbrella of the HC to promote European competitiveness in Equine Science and industry. The group has discussed about co-work in equine nutrition research and knowledge dissemination at European level. Since 2002 the Working Group has supported an organization of European Workshop on Equine Nutrition (EWEN), a scientific congress held every second year and organized by various universities or scientific organizations of European countries. EWEN has now been held five times and can be considered one of the most important 'satellite events' of the Horse Commission. The proceedings of each congress have been published by EAAP in a special series of Wageningen Academic Publishers. The main objectives of the Working Group Nutrition are: to promote scientific research in equine nutrition in Europe; to define the type of scientific platform (scientific conference) and structure including an applied section for transfer of knowledge (Workshop); promoting young people doing research; publishing 'essentials' in equine nutrition as core of obligatory guidelines for safe nutrition.

Interstallion

Janssens, S.[1], Gomez, M.D.[2], Ducro, B.[3], Ricard, A.[4], Christiansen, K.[5] and Philipsson, J.[6], [1]KULeuven, Livestock Genetics, Dep. Biosystems, Kasteelpark Arenberg 30 - 2456, 3001 Heverlee, Belgium, [2]University of Cordoba. Cordoba, Spain, Department of Genetics, C.U. Rabanales. Ed. Gregor Mendel Pl Baja., 14071 Cordoba, Spain, [3]Wageningen University, Animal Breeding and Genomics Centre, Marijkeweg 40 Zodiac (building nr. 531), 6709 PGWageningen, Netherlands, [4]INRA, SAGA, Auzeville B.P. 52627, 31326 Castanet Tolosan Cedex, France, [5]WBFSH, secretariat, Vilhelmsborg Alle 1, 8320 Mårslet, Denmark, [6]Swedish University of Agricultural Sciences, Department of Animal Breeding and Genetics, P.O. Box 7023, S-750 0, S-750 07 Uppsala, Sweden; Steven.janssens@biw.kuleuven.be

INTERSTALLION is a working group of the EAAP horse commission. The overall objective of INTERSTALLION is to develop activities to 'Improve accessibility, understanding and comparability of foreign breeding information across countries'. These aims are underpinned by the international trade of stallions and increasing use of transported semen across countries and continents. Breeders and licensing committees need to have access to official breeding information, that allows correct interpretation of the information published. In order to achieve these objectives, INTERSTALLION initiates or supports research that improves the transparency of breeding information and/or would make comparison of breeding vaules internationally possible. Two research projects indicated favorable genetic connectedness between countries and high genetic correlations between similar traits. In 2010 a Nordic project has started (Norwegian, Swedish, Danish and Finnish Warmblood) with the aim to evaluate possibilities of a joint genetic evaluation of sport horses. By time this project may be extended to include more countries. Furthermore, INTERSTALLION organizes seminars to improve the exchange of research results related to breeding of sporthorses. The last seminar was held in Uppsala in september 2010 and covered the progress of the Nordic project, recent advances in several countries and the issue of osteochondrosis in the breeding of sport horses.

Horse production and socio-economic aspects: goals, perspectives and key challenges of the Horse Commission Working Groups in Socio-Economy
Vial-Pion, C., Institut Français du Cheval et de l'Équitation, INRA, UMR MOISA, 2 place Pierre Viala, 34060 Montpellier, France; vialc@supagro.inra.fr

In the European Union, little is known about the horse industry despite its recent growth and evolutions. However, the number of studies devoted to this sector is recently multiplying all over the world. That's why a working group in socio-economy has been created during the last EAAP conference. Today, it includes 19 members, mainly coming from European countries. The goal of this group is to share ideas and experiences but also to think about topics of interest for research and development and to build common projects. Sociologic studies are emerging, analyzing actor profiles and behaviors. The economic aspect is also an important topic of research. Each segment has its specificity. For instance, racing and betting generate high incomes, whereas breeding faces some difficulties and needs support in several countries. Another example is the Equine tourism, which is growing spontaneously across Europe but which has been studied very little and promoted even less. Important key challenges are the counting and localization of Equines and also 'the unwanted horse issue', which has been prevalent in the United States since slaughter became illegal or in Ireland. Significant challenges also arise from conflicts over land use with other users in peri-urban areas. Some countries have started to underline the role that Equines are currently playing in rural development: their multi-purpose uses represent a strong advantage in land occupation and management, they contribute to agriculture diversification thanks to horse husbandry and agritourism, they help maintaining relationships between urban citizen and cultural rural life… To conclude, we can say that horses are cultural heritages, social links, economic stakes and rural development actors. For all of these reasons, the horse industry deserves to be studied and that we join our efforts to promote its sustainable development.

The development of equestrian leisure in French rural areas: between sectorial influences and periurbanization
Vial-Pion, C., Aubert, M. and Perrier-Cornet, P., Institut Français du Cheval et de l'Équitation et Institut National de la Recherche Agronomique, UMR MOISA, 2 place Pierre Viala, 34060 Montpellier, France; vialc@supagro.inra.fr

In France, equestrian leisure activities have been growing since the 1990s. This development takes place in rural and suburban areas, while these spaces are changing. These modifications are creating an ongoing space need for city expansion and leisure activities whereas agricultural functions are retreating. In this context, leisure Equines, which use more and more space, appear to be linked to these territorial changes. Their growth is in part due to individual amateurs, who buy their own animals and either use horse livery services or take care of their horses themselves. This work examines the factors that may influence the more or less high presence of these both types of amateur's Equines in the countryside. Two suburban and two rural areas were studied in France. A logistic regression based on existing databases was carried out to characterize the presence of these Equines. The results highlight reciprocal influences between their density and the residential development of these territories. In fact the development of these amateur Equine owners and the periurbanization phenomenon seem to nourish each other. The effects of some sectors are also tested: the influence agriculture is complex, governed by both competitions and complementarities for land occupation. There is moreover a clear local ripple effect of the equestrian professional dynamic on the presence of amateur's Equines. On the contrary the local touristic attractiveness doesn't seem to have an influence. Finally this work focuses on the links between the development of equestrian leisure and the current countryside changes and point out amateur's Equines as good revealers and markers of the new rural dynamics. As a result, this research is relevant to enable all stakeholders of the horse world and of land management to understand the impact of equestrian activities growth on territorial development.

The relationship between the government and the horse industry: a comparison of England, Sweden and the Netherlands

Crossman, G.K. and Walsh, R.E., Royal Agricultural College, Stroud Road, Cirencester, Gloucestershire, GL7 6JS, United Kingdom; gkcrossman@hotmail.com

In England, Sweden and the Netherlands, the relationship between the government and the horse industry has evolved over time. At the beginning of the twentieth century the horse was considered a beast of burden, with its role closely linked to agriculture, defence and transport. Just over one hundred years later and its function has been transformed with horses now predominantly involved in recreational and sporting activities. The government's view of and connection with the industry in each country the research focuses upon reflects this change. Policy network theory was used to analyse the development of the relationship between the government and interest groups in the horse sectors examined, with evidence gathered through interviews, documentary research and participant observation providing the basis for the study. A specific organisation which acted as the mouthpiece for the industry to the government in each country was identified. However, the structure of these bodies and their mode of operation differed considerably between countries. The level of government intervention and financial support afforded to each horse industry also varied. Links between existing policy areas and the horse industry were found to be significant in the success of the industry. In addition, in Sweden and the Netherlands the connection between the equine and agricultural sectors in the development of the relationship between the government and the horse industry was also important.

Overview of Equine Assisted Activities and Therapies in the United States

Hallberg, L., Author, Licensed Professional Clinical Counselor, 6690 Sourdough Canyon Road, Bozeman, Montana 59715, USA; lhallberg1@gmail.com

Since 1969, the field of Equine Assisted Activities and Therapies has grown exponentially in the United States. Starting with the creation of the North American Riding for the Handicapped Association (NARHA) in 1969, the field has evolved to support not only riding for the physically disabled, but for those struggling with mental health issues and learning disabilities. The field boasts over 3,500 NARHA certified instructors and 800 member centers around the globe, 3,527 Equine Assisted Growth and Learning certified providers, and countless professionals who provide services but are not affiliated with either organization. Institutions of higher learning now offer both undergraduate and graduate degrees in the field. Mental health professionals are billing insurance companies for their work with horses and clients, and corporate training experts partner people and horses to provide team building and professional coaching for their clientele. Most importantly, people from all walks of life are being served through this work and thus far the reported successes seem unparalleled across all socioeconomic and environmental barriers. However, as with any new field, standards, ethics, and procedures for practice in the United States are inconsistent. Individuals and organizations operate without unified oversight or research-based efficacy models. This presentation will provide participants with information about how the Unites States is dealing with these challenges. Participants will be invited to use this information to consider how Europe may choose to address similar issues from a proactive stance.

An exploration of the increasingly diverse roles that horses are playing in society in enhancing the physical, psychological and mental well-being of people

Carey, J.V., Miraglia, N., Grandgeorge, M., Hausberger, M., Hauser, G., Cerino, S., Perkins, A. and Seripa, S., Horse Commission Working Group, EAAP, Rome, Italy; jillcarey@festinalente.ie

This session will present the preliminary findings of the Horse Commission Working Group which was set up at the EAAP Conference in Crete in 2010. In an acknowledgment of the emerging and diverse roles of equines within society – both rural and urban areas–there is a need to develop a better understanding of the range and nature of the different equine related programmes being provided across Europe. The Working Group is comprised of researchers and/or practioners from Italy, France, Austria and Ireland. The initial focus of the working group has been to define and measure the range and growth of equine assisted activities which includes therapeutic riding, riding for people with disabilities, hippotherapy, vaulting, carriage driving and equine assisted learning/ coaching/ equine assisted psychotherapy. The presentation will also include an overview of the skills set of personnel involved in the programmes together with any evaluation and research that has been conducted. Data collected from over 36 countries will be presented together with an overview of the next stages of work to be undertaken.

Influence of training strategy on horse learning and human-horse relationship

Sankey, C., Henry, S., Richard-Yris, M.-A. and Hausberger, M., Université Rennes 1 – UMR CNRS 6552, Animal & human ethology, Station biologique, 35380, France; carol.sankey@univ-rennes1.fr

Aims: Interventions on young horses may be particularly risky for professionals such as breeders or veterinarians, due to the young animals' bad/non-existent education, often leading them to use constraint on the animals, thus increasing even more the risks. Here, we investigated the effects of using positive reinforcement (food reward) to train young horses to stand still and cooperate during handling, and we evaluated the effects of such training on the horse-human relationship. Methods: 23 yearlings were trained to remain immobile on a vocal order and accept various handling procedures (brushing, picking feet, surcingle, rectal thermometer …), giving half of them a food reward (positive reinforcement group, $N_{PR}=11$) whenever they responded correctly to the command (i.e. remained immobile throughout the handling procedure), while the other half (control group, $N_C=12$) was never given any reward. Results: Results showed that using positive reinforcement promotes faster learning (P<0.001) and better memorization of the immobility task (P<0.05). Horses that received the food reward also behaved better during training (less biting, kicking…, P<0.05) than controls, and not only did they easily accept the tasks included in training, but also was it easier and safer to perform new tasks such as oral deworming or radiography, and at a later stage to perform saddle breaking (P<0.05). Moreover, rewarded animals sought and accepted more contact, both with the familiar trainer (P<0.001 and P<0.01, respectively) and with a non familiar person (P<0.01 and P<0.01, respectively), even several months after completion of training (at least 6 months later). Conclusions: This study reveals that the human-animal interactions that occur during horse training may be crucial to obtain a well-educated horse, as well as to establish a positive human-animal relationship.

Therapeutic Riding and early Schizophrenia: the Italian Equestrian Federation Pindar Project
Miraglia, N.[1], Cerino, S.[2] and Seripa, S.[3], [1]University of Molise, Animal, Vegetables and Environmental Sciences, Via De Sanctis, 86100 Campobasso, Italy, [2]F.I.S.E., Therapeutic Riding Dept., Viale Tiziano 74, 00197 Rome, Italy, [3]ASL Roma F, Mental Health Service, Via Lazio 50, 00055 Ladispoli, Italy; s.cerino@alice.it

Nowadays it is very well known, in Psychiatric field, how an early intervention for schizophrenia can achieve good results, especially in metacognitive area. Attention, memory, metacognition, social cognition, parallel thought are the basis of the rehabilitative actions. Therapeutic Riding can play an important role in treating schizophrenic patients in the first years of the disease, because it acts in lowering disability, decreasing hospitalization days and on illness remission. By the Pindar Project, Italian Equestrian Federation has established an integrated methodology approach (Therapeutic Riding Centers and Mental Health Departments) for 35 patients with a very 'hard' clinical history. Efficiency evaluation has been made by BPRS, PANSS with 8 Items, TAS, MQ, TCQ, SF-36 submitted at the beginning, middle and end of the study. The final results are presented and discussed in this paper.

Economic, social and environmental sustainability assessment in sheep and goat production systems
Gabiña, D., Mediterranean Agronomic Institute of Zaragoza - CIHEAM, Av Montañana 1005, 50009 Zaragoza, Spain; gabina@iamz.ciheam.org

This presentation is a summary of the conclusions of a Seminar of the FAO-CIHEAM Network on Sheep and Goats held in Zaragoza on 10-12 November 2010 (www.iamz.ciheam.org/sg2010). In terms of methodology, new holistic approaches for evaluation of sustainability, such as MESMIS or SAFE, were proposed. For environmental impacts, Life Cycle Assessment (LCA) seemed to be the best option, as it considers the entire life cycle of a product. For economic sustainability, productivity and optimal use of on-farm resources (self-sufficiency) were key to farm profitability, especially in a context of increasing input prices. However, there are still methodological concerns about the selection and definition of indicators or how to deal with different spatial and temporal scales. In South and East Mediterranean countries, the low productivity of production systems is limiting their economic sustainability, therefore the reduction of mortality, the selection of local breeds and a sound feeding management using local resources, seem necessary to improve this situation. Processing and marketing quality products, such as PDO or PGI cheese, at the farm are other positive factors for economic sustainability. Other important limitations for sustainability are labour conditions and social recognition, which together with the perception of low profitability, lead to uncertain generational turnover. For environmental sustainability, a high productivity based on higher use of inputs dilutes the GHG emissions per kilo of product. Figures from 27 to 56.7 CO_2 eq/kg lamb-meat were reported. However, other ecosystems services provided by low-input systems have to be taken into account, such as the conservation of biodiversity and cultural landscapes and the prevention of forest fires.

Genetic parameters of some production and reproduction traits in Raienian cashmere goat

Asgari Jafarabadi, G.[1], Nazemi, E.[1] and Aminafshar, M.[2], [1]Islamic Azad University, Varamin-Pishva Branch, Dept. of Animal Science, Agriculture Faculty, 3381774897 Varamin, Iran, [2]Islamic Azad University, Science and Research Branch, Dept. of Animal Science, Agriculture Faculty, 1477893855, Iran; gh.asgari@iauvaramin.ac.ir

Raienian cashmere goat is one of the most famous native breeds in Iran which its importance primarily comes from its ability to produce fine cashmere fiber. In order to determine the heritability of some production and reproduction traits and genetic correlations among them, total number of 4,817, 1,913, 1,515, 6,249, 1,871, 1,532, 1,208 records of birth weight, both weight of 3 and 12 month age, cashmere weight, total number of lambs born, total number of lambs weaned and lambing interval in Raienian cashmere goat were used. These records were collected in Baft station from years 1992 to 2010. Data was edited using FoxPro software and genetic evaluation was carried out by DF-REML method. Single and multiple trait animal models were used for prediction of breeding values and correlations. Heritability of birth weight, weight of 3 and 12 month age, cashmere weight, total number of lambs born, total number of lambs weaned and lambing interval was estimated 0.24, 0.07, 0.20, 0.17, 0.06, 0.08, and 0.03 respectively on the basis of the best model with highest logarithm of likelihood for each trait. Genetic correlation between birth weight and 12 month weight, birth weight and cashmere weight, 3 month weight and total number of lambs weaned, 3 month weight and total number of lambs born, 12 month weight and cashmere weight, 12 month weight and total number of lambs born, 12 month weight and total number of lams weaned, 12 month weight and lambing interval, lambing interval and total number of lambs born, lambing interval and total number of lambs weaned was estimated to be 0.27, 0.072, 0.02, -0.16, 0.28, 0.10, -0.27, -0.12, 0.30 and 0.42 respectively, using multiple trait analysis.

Genetic polymorphism of transferrin in Arabi sheep breed

Jaayid, T.A., Yousief, M.Y., Zaqeer, B.F. and Owaid, J.M., College of Agriculture, Basrah University, Animal Production Department, Animal Production Department, College of Agriculture, Basrah University, Basrah, Iraq; ymuntaha@yahoo.com

This study was carried out at the Animal Farm, Hartha Research Station, College of Agriculture, Basrah University and several farms in Basrah Provence. Polymorphism of transferrine biochemical system was examined in the blood of Arabi sheep breed using vertical polyacrylamide gel electrophoresis (PAGE) in discontinuous buffer system and staining. This is the first study on transferrin on sheep breed in Iraq. Totally 100 blood samples were analysed. Transferrin loci were found to be polymorphic among all analysed samples. Seven transferrin variants have been found in Arabi sheep breed sera which are designated AA, BB, MM, DD, AB, AM, and BM in order of decreasing mobility caused by four alleles of that locus (A, B, M and D). Sheep populations were studied which differed in gene frequencies. Differences between expected and observed number of transferrin genotypes were significant. The aim of this study was to use the PAGE as a fast, efficient and low cost method to detect the genetic variants of transferrin gene in Arabi sheep breed. The polymorphism of sheep transferrin can be used for the identification of offspring.

Pedigree analysis of French goat breeds: a panorama from dairy to fiber selection
*Danchin-Burge, C.[1,2], Allain, D.[3], Clement, V.[2], Martin, P.[4], Piacere, A.[2] and Palhiere, I.[3], [1]INRA/
AgroParisTech, UMR1313 GABI, 16 rue Claude Bernard, 75005 Paris, France, [2]Institut de l Elevage,
GIPSIE, 149 rue de Bercy, 75012 Paris, France, [3]INRA, UR631 SAGA, Auzeville, 31326 Castanet-Tolosan,
France, [4]CAPGENES, Agropole, 86550 Mignaloux-Beauvoir, France; coralie.danchin@inst-elevage.asso.fr*

Pedigree analysis have been widely used to assess the genetic variability of livestock species, however few examples can be found for the goat species. France is the only country in the world where goat selection is well implemented with successful breeding programs. Our presentation will briefly describe the French dairy goat and fiber goat selection schemes. It will present the results of a pedigree analysis, by using the PEDIG software, of two dairy goat breeds, the Alpine and Saanen, and a fiber goat breed, the Angora, in order to assess their genetic variability. The populations under study are large for the Saanen and Alpine (125,797 and 183,611 goats respectively) and small (1,229) for the Angora. The pedigree depth can be considered to be good (from 5.7 equivalent generations for the Angora to 7.8 for the Alpine). The effective number of ancestors is equal to 34, 46 and 51 for the Angora, Alpine and Saanen breeds respectively, while the effective population varies between, 66, 120 and 143 for the Angora, Saanen and Alpine breeds respectively. In light of these results, for the 2 dairy breeds there are several reasons to believe that if the genetic variability has narrowed as an effect of 30 years of efficient selection, the specific management program implemented 10 years ago to maintain genetic diversity was successful. Meanwhile for the Angora breed the narrow genetic basis of the breed appears to be caused by its thin demographic basis as well as the small number of founders used at the start of the selection scheme. These results give us a good insight on the genetic variability of intensively selected populations as well as a direction on the programs that are efficient to keep enough genetic diversity.

Genome wide association study of somatic cell counts in a grand-daughter design of French Lacaune AI rams
*Rupp, R.[1], Sallé, G.[1,2], Barillet, F.[1], Larroque, H.[1] and Moreno-Romieux, C.[1], [1]INRA, UR631, F-31326
Castanet-Tolosan, France, [2]2INRA-ENVT, UMR1225, F-31076 Toulouse, France;
rachel.rupp@toulouse.inra.fr*

A genome wide association study of mastitis resistance was carried in a grand-daughter population of 825 Lacaune AI rams from 28 families with 29 (±7) sons per ram on average. The rams and their sires were genotyped using the Illumina Ovine SNP50 BeadChip assay. The studied trait was daughter yield deviation for somatic cell score calculated from the national genetic evaluation procedure. After marker quality control, 38,007 SNP were used. Data were analyzed using linkage analyses implemented with the QTLmap software. At the 5% genome-wide threshold, three regions of OAR11, OAR16 and OAR23, respectively, showed significant association with SCC. Additional SNP reached the 5% chromosome-wide threshold on chromosomes OAR3, OAR5, OAR7, OAR8, OAR14, OAR19 and OAR26. Additional genotyping data on about 200 animals will be available in the near future to confirm these QTLs. Preliminary results highlight several regions of the genome that underlies mastitis resistance in dairy sheep.

Inbreeding coefficient and pairwise relationship in Finnsheep using genomic and pedigree data

Li, M.-H.[1], Strandén, I.[1], Tiirikka, T.[1], Sevón-Aimonen, M.-L.[1] and Kantanen, J.[1,2], [1]MTT Agrifood Research Finland, Biotechnology and Food Research, Myllytie 1, FI-31600, Jokioinen, Finland, [2]Nordic Genetic Resource Center, Farm Animals, P.O. Box 115, NO-1431 Aas, Norway; juha.kantanen@nordgen.org

Genome-wide SNP data provide a powerful tool to estimate pairwise relationship among individuals and individual inbreeding coefficients. The aim of this study was to compare methods for estimating the two parameters in a Finnsheep population based on the genome-wide SNPs and genealogies, respectively. Ninety-nine Finnsheep in Finland, which are in different coat colors (white, black, brown and grey) and from a pedigree data recorded for over 25 years, were included in the study. All the individuals were genotyped with the Illumina Ovine SNP50K BeadChip by the International Sheep Genomics Consortium. We identified three genetic subpopulations which corresponded roughly to the coat colors (grey, white, and black and brown) of the sheep. We found a significant subdivision among the color types (F_{ST}= 5.4%, P<0.05). We applied robust algorithms for the genomic estimation of individual inbreeding and pairwise relationship as implemented in the programs KING and PLINK, respectively. Estimates of the two parameters from pedigrees were computed using the program RelaX2. Results of the two parameters estimated from genomic and genealogical data were well consistent between each other, in particular for the animals with high levels of inbreeding (e.g. inbreeding coefficient F >0.0625) or the pairs of animals with high levels of relationships (e.g. the half-sibs). Nevertheless, we also detected differences in the two parameters between the approaches, mostly involving in the grey Finnsheep. This could be due to the smaller sample size and lower pedigree completeness. We conclude the genome-wide genomic data will provide useful information on a per sample or pairwise samples basis in the cases of complex genealogies or absence of genealogical data.

Comparison of different mathematical models for estimating and describing lactation curve of Saanen goat

Tahtali, Y., Ulutas, Z., Sirin, E. and Aksoy, Y., Gaziosmanpasa University, Agriculure faculty, tasliciftlik/ TOKAT, 60150, Turkey; ytahtali@gop.edu.tr

In the study, a suitable mathematical model to represent the lactation curve of Saanen goat was examined. Data of milk traits were collected from Gaziosmanpasa University Agricultural Research Farm. The Wood (WD), Cobby and Le Du (CD), Dhanoa (DH) and Wilmink (WL) functions were compared. The results of four models were discussed with advantages and disadvantages, and the best fitting model was chosen according to adjusted coefficient of determination coefficient (R2d), residual sum of squares (RSS) and coefficient of Durbin-Watson (DW). With standard errors of parameters of lactation curve were estimates as follows; for lactation length 0.000±0.056 and 0.051±0.038, for Ymax 0.055±0.087 and 0.035±0.032, for Tmax 0.017±0.066 and 0.116±0.057, respectively.

Survival analysis of longevity in Moghani and Baluchi sheep breeds of Iran
Esfandyari, H., Tahmorespour, M. and Aslaminejad, A.A., Ferdowsi University of Mashhad, Azadi Square, Mashhad, 9177948974, Iran; esfandyari.hadi@gmail.com

In sheep production, lifetime performance of female sheep is one of the most important economic traits. In this study, a Weibull model for survival analysis was applied to data of 4524 female sheep of two different breeds in Iran. This model included fixed time-independent effects of breed,age at first lambing, type of birth and time-dependent effect of number of lambings on the length of productive life (LPL). The observation period in which the sheep were born or were removed from the flocks ranged from March 2004 to February 2010. About 24% of the records were right-censored. All variables had a significant effect on LPL at a level of $P<0.001$ except type of birth. The Moghani breed showed lower risk ratio with 0.84 (SE=0.24), while the Baluchi had higher hazard rate (1.00). The relative culling risk initially decreased from the first (4.26, SE=0.11) to the seventh lambing (0.41, SE=0.16), and then increased until lambing number eight. The highest relative culling risk was calculated at an age of 400 to less than 460 days at first lambing (1.00). Animals younger than 400 days at first lambing showed the lowest risk ratio (0. 61, SE=0.08). An appropriate parameter to quantify the influence of fixed and random effects on LPL is the relative risk of being culled. This study showed the importance of a low age at first lambing. Due to the assessed differences between the breeds, management should be adapted to the demands of the single breeds. For further analysis and optimization, improved data acquisition regarding the causes of culling, death and leaving the flock is required.

Selection against null-alleles in CNS1S1 will efficiently reduce the level of free fatty acids in Norwegian goat milk
Blichfeldt, T.[1], Boman, I.A.[1] and Kvamsås, H.[2], [1]The Norwegian Association of Sheep and Goat Breeders, Box 104, 1431 Ås, Norway, [2]TINE, Førde, 6800 Førde, Norway; tb@nsg.no

Free fatty acids (FFA) in Norwegian goat milk should be reduced to improve the quality for cheese production. Two single point deletions in the CNS1S1 locus of chromosome 6 are present in Norwegian dairy goats, both associated with no synthesis of alpha-s1 casein (null-alleles). One is unique for Norway found in a very high frequency (0.7-0.8). Studies by Ådnøy *et al.* and Grindaker *et al.* show that this null-allele is associated with increased FFA levels in milk. The present study should quantify the effect of the null-alleles on dry matter production and FFA. In 8 flocks female replacements were genotyped; 103 were homozygous for the null-alleles, 145 were heterozygous and 63 did not carry the deletions. Milk production per day, fat, protein, lactose and FFA content were recorded three times per lactation. The effect of the three functional genotypes 'homozygous null', 'heterozygous null' and 'homozygous not-null' on production traits was investigated using an analysis of variance including flock (n=8), lactation number (1, 2), calendar month (n=11), month of lactation (n=9) and breed of sire (Norwegian Dairy Goats, French Alpine) as fixed effects. The effect of genotype on total dry matter production per day was not significant (least square means 304, 297 and 306 grams per day, respectively). In the 'homozygous null' group FFA was 1.39 mmol per litre, significantly different from the two other groups with FFA levels of 0.84 and 0.86, respectively. Since 2008, goat milk producers have been recommended to genotype their bucks and select against the null-alleles. In 2010, 95% of all officially approved half year old breeding bucks were genotyped. The frequency of the null-alleles in these bucks was 0.33, a major decrease from 0.78 reported on bucks born before 2007. Reducing the frequency of the null-alleles in the population will most likely lead to a profound decrease in FFA levels in milk.

Fattening performance and carcass characteristics of Awassi and their crossbred ram lambs with Charollais and Romanov in an intensive-feeding system
Momani Shaker, M.[1], Kridli, R.T.[2], Abdullah, A.Y.[2] and Sonogo, S.[1], [1]Czech University of Life Sciences Prague, Animal Science and Food Processing in Tropics and Subtropics, Suchdol, 165 21 Prague 6, Czech Republic, [2]2Jordan University of Science and Technology, Faculty of Agriculture, Irbid, Jordan, 6040, Jordan; momani@its.czu.cz

The objective of this study was to evaluate the effect of ram lambs genotype on fattening performance and carcass characteristics in an intensive-feeding system. Thirty ram lambs and two months old ram lambs of three genotypes Awassi (A) = 10, F_1 crossbreds Charollais x Awassi (ChA) = 10 and F_1 crossbreds Romanov x Awassi (RA) =10) were placed in individual closed pen. All lambs were individually fed for seventy days on same concentrate mixture and water *ad libitum*. Daily weight gain (DWG), total weight gain (WG), feed conversion ratio (FCR) and cost on 1 kg weight gain were the best in crossbreds F_1 ChA ram lambs and F_1 RA ram lambs, compared to the Awassi ram lambs ($P \leq 0.001$) The height at withers and height at back, were in favour of crossbreds F_1 ChA ram lambs ($P \leq 0.001$). The best ($P \leq 0.001$) carcass meatiness was in F_1 ChA group, while the lower carcass fatness was in F_1 RA group ($P \leq 0.001$). In dressing percentage based on the warm carcass weight, the F1 crossbreds ChA had a greater dressing percentage than pure Awassi rams and F1 crossbreds RA, while the A and and F_1 crossbreds RA were similar. The highest percentage proportions of leg and loin, which represent the prime quality of the carcass, 48.64% were in F_1 crossbreds ChA, 45.97% in F_1 crossbreds RA and 45.91% in Awassi ram lambs ($P \leq 0.001$). In conclusion, the results of this study document that F_1 ram lambs crossbreds ChA and F_1 ram lambs crossbreds RA were superior than Awassi ram lambs in daily weight gain and total weight gain, feed conversion, lower cost of 1 kg meat gain and in most of the carcass characteristics. Sheep; crossing; Awassi; Charollais; Romanov; fattening performance; carcass value.

Research on the growth and carcass characteristics of the lambs from the local breed Teleorman Black Head and from the hybrids with Suffolk breed
Ghita, E., Pelmus, R., Lazar, C., Voicu, I. and Ropota, M., National Research Development Institute for Animal Biology and Nutrition, Animal Biology, Calea Bucuresti nr.1, 077015 Balotesti, Ilfov, Romania, 077015, Romania; elena.ghita@ibna.ro

The purpose of the experiment was to study the growth performance and the carcass characteristics in the F1 lambs from the cross of the local (Romanian) Teleorman Black Head (TBH) ewes with Suffolk rams, compared with the lambs form the local breed. 48 lambs assigned to two groups (24 lambs from the local breed and 24 F1, TBH with Suffolk, lambs) were weaned when 2 months, when 8 lambs from each group were slaughtered. We determined the slaughter/commercial output, the proportion of the carcass parts/ butcher parts, the meat to bone ratio for each part and for the entire carcass, specific measurements of the carcass, the chemical composition of the meat including the fatty acids and cholesterol level. The average body weight of the hybrid lambs at lambing was 5.13 kg, compared to 4.94 kg in the local breed, reaching 23.8 kg in the hybrid lambs and 19.09 kg ($P<0.01$) in the local breed at 2 months. During the nursing period, the group of hybrid lambs had an average daily weight gain of 306 g compared to 233 g ($P<0.01$) in TBH lambs. The commercial yield was 54.72% in the hybrids and 51.75% in TBH. Carcass dimensions differed: the length was larger in the local lambs, but the breast width, carcass width at rump, thorax depth and perimeter, leg perimeter were higher in the hybrids than TBH. The meat to bone ratio was better in the rump and ribs of the hybrids than in the corresponding parts of the local breed, while the overall carcass was also better in the hybrids than in the local breed. The experimental results show that the Suffolk rams improved the meat traits in the hybrid lambs, transmitting to their offsprings a higher rate of growth, the specific conformation of the meat breeds, a better dressing with muscles, particularly at the rump and rack, a better slaughter yield and a better meat to bone ratio compared to the local TBH lambs.

Influence of breed, slaughter weight and feeding on sensory meat quality of suckling kids
Sañudo, C.[1], Campo, M.M.[1], Horcada, A.[2], Alcalde, M.J.[2] and Panea, B.[3], [1]Universidad de Zaragoza, Miguel Servet, 177, 50013, Zaragoza, Spain, [2]Universidad de Sevilla, Ctra. Utrera km 1, 41013, Sevilla, Spain, [3]CITA, Avda. Montañana, 930, 50059, Zaragoza, Spain; csanudo@unizar.es

Goat production in Europe is mainly based in milk with high specialized breeds. Also, many local populations, with low dairy performances, are been considered as meat breeds and the kids as a delicatessen by South European consumers. In this scenario, the study of kid's meat quality and how it is affected by weight or rearing system is a relevant issue for the Goat Sector. We have studied the effect of breed (B) on kid's meat organoleptic characteristics using 201 animals from 7 Spanish goat breeds. The effect of slaughter weight (SW) was assessed in 5 meat purpose breeds slaughtered with 4 or 7 kg of carcass weight, and the effect of feeding (F) was assessed in 2 dairy purpose breeds reared under natural or artificial milk conditions. Eight trained panellist evaluated 11 attributes on Longissimus dorsi muscle grilled until 70 °C of internal temperature. A GLM procedure was used to evaluate (B) in light and natural feeding animals, (SW) and (B) effects in meat purpose animals, and (B) and (F) in milk purpose animals, considering interactions between effects. No significant effect was found in goat and milk odour intensities, juiciness or fat, milk and sour flavour intensities. Tenderness, fibrousness, goat flavor, metallic flavour, acid flavor and overall acceptability were significantly affected by (B) independently of their aptitude. Tenderness was higher and fibrousness lower in light and in natural-milk fed animals. Goat flavour was higher in natural-milk fed animals, but not a clear effect of (SW) was observed. Metallic or acid flavours and acceptability were not clearly affected by (SW) or (F), but many significant interactions between effects were observed. In conclusion, kid sensory quality is significantly affected by breed and, inside breed, the increase of weight and the rearing status affects differently depending upon the considered breed.

Growth rate, carcass and meat quality of ewes finished with different linseed levels
Muela, E.[1], Sañudo, C.[1], Campo, M.M.[1], Monge, P.[1], Schneider, J.[2], Macedo, R.M.G.[3] and Macedo, F.A.F.[3], [1]Universidad de Zaragoza, Miguel Servet, 177, 50.013, Zaragoza, Spain, [2]Universidade Federal de Pelotas, Gomes Carneiro, 96010-610, Pelotas, Brazil, [3]Universidade Estadual de Maringa, Avenida Colombo, 5790, 87040-900, Maringa, Brazil; muela@unizar.es

Lamb is a highly appreciated product by consumers. However, ewe's meat is hardly consumed in Europe due to its strong odour and flavour and because their productive indexes make no rentable to fatten them. Diet supplementation would enhance carcass and meat quality, giving an added value to this product and a profit for the breeder, especially when it would be done with low cost sources, such as linseed supplementation. The aim of this study was to assess the effect of 3 linseed supplementation levels -LS- (5, 10, or 15%) during two periods -P- (30 or 51 d) on 8 animals for each LS and P, and 8 animals as control batch. With this purpose, we measured weight and body condition score at the beginning and the end of each P, estimating average daily consumption (ADC) and gain (ADG), to calculate the conversion index (CI). Hot carcass weight (HCW) and condition and fatness scores were assessed, carcass yield was calculated, and tissue composition was performed on the left shoulder. Fat thickness, loin area (13th thoracic vertebrae) and pH and meat colour (5th thoracic vertebrae) were measured. Treatment effect [(3LS x 2P)+control] was assessed using the GLM procedure of the SPSS statistical package. Production indexes, carcass and meat quality were significantly affected by the studied treatments. ADG, CI and growth costs were lower at 30 than at 51 d. Final live and carcass weights, carcass yield, fat percentage and loin thickness, were higher on supplemented animals, although only 15LS/51P was significantly different from the control batch. Lightness, yellowness, and hue were higher on the control batch. In conclusion, P has a higher influence than SL on the studied variables. With these preliminary results, it could be recommended a 10LS/30d due to its lowest costs, although more studies are required.

Sunshine meat originating from lambs meat submitted to the organic and conventional production models

Zeola, N.M.B.L., Silva Sobrinho, A.G., Borba, H.S., Manzi, G.M., Lima, N.L.L. and Endo, V., São Paulo State University, UNESP, Department of Animal Science, Via de Acesso Prof. Paulo Donato Castellane, s/n, 14884-900 Jaboticabal, SP, Brazil; nivea.brancacci@ig.com.br

Dissection of 48 lambs Ile de France legs submitted to the organic and conventional production models were used to salting process, with inclusion of salt in the proportion of 15 and 20% of the meat weight. Organic and conventional production models influenced cooking loss (22.84% for organic and 31.23% for conventional) and 2-thiobarbituric acid reactive substances (3.97 and 6.49 mg of malonaldehyde/1000 g of meat, for organic and conventional, respectively), however did not affect pH, color, water holding capacity and shear force meat during the elaboration of sunshine meat. Salt tenors (0, 15 and 20%) influenced pH (5.60 for 0% and 7.44 for 15 and 20%), color (L* 40.43, a* 13.47 and b* 1.87 for 0% and L* 35.24, a* 5.76 and b* 1.29 for 15 and 20%), cooking loss (41.07% for 0% of salt and 20.01% for 15 and 20% of salt) and 2-thiobarbituric acid reactive substances(1.38, 7.72 e 6.59 mg of malonaldehyde/1000 g of meat, to 0, 15 and 20% tenors, respectively), however it did not affect water holding capacity and shear force. In the organic model, flavor, tenderness and global acceptance were influenced by the treatments, however to color, was not observed difference by the panelists. In the conventional model, color, tenderness and global acceptance were influenced by the treatments, however, for the flavor difference nothing was observed. Similar results were observed for the studied parameters in meat that received 15 and 20% of salt, and the production model did not present significant influence on evaluated parameters. The obtained values were considered appropriate during the processing of sunshine meat, with exception of pH.

Extension on ultrasound probe for *in vivo* measurements of carcass traits in sheep

Milerski, M., Research Institute of Animal Science, Přátelství 815, 104 00 Prague 10 - Uhříněves, Czech Republic; m.milerski@seznam.cz

Real-time ultrasound technology is used for the objective *in vivo* estimation of carcass traits in sheep in many countries. In the Czech Republic this method have been used for eye-muscle depth and fat+skin thickness measurements in lambs of terminal sire breeds at 100±20 days of their age (together with their weighing) since 1999. Totally 7,891 lambs were measured in 2010. Ultrasonic measurements of lambs meatiness and fattiness are however more time consuming compared to weighing them only, because this operation demand several steps: wool combing, conductive medium (ultrasound gel) application, scanning and measurements. Newly developed extension on ultrasound probe enables to use only one tool for wool combing, conductive medium application and ultrasound scanning, what reduces measurement time hence improves the labor efficiency and the animal welfare.

Hamburger originating from lambs meat submitted to the organic and conventional production models

Zeola, N.M.B.L., Da Silva Sobrinho, A.G., Borba, H.S., Manzi, G.M., Nonato, A. and Alves De Almeida, F., São Paulo State University, UNESP, Department of Animal Science, Via de Acesso Prof. Paulo Donato Castellane, s/n, 14884-900 Jaboticabal, SP, Brazil; nivea.brancacci@ig.com.br

Dissection of 48 lambs Ile de France palette submitted to the organic and conventional production models were used to hamburger elaboration, with 20 and 30% of fat. The production models (5.93 to organic and 5.84 to conventional) and the fat tenors (5.86 to 20% and 5,90 to 30% of fat) influenced the pH of the hamburgers, the production models just affected the lightness and the redness, however did not influence water holding capacity, cooking loss and 2-thiobarbituric acid reactive substances of the hamburgers, with 62.81%, 24.64% and 1.81 mg of malonaldehyde/1000 g of meat, respectively. In organic hamburger smaller shear force (0.68 kgf/cm^2) was observed in relation to the conventional (0.97 kgf/cm^2). The fat tenors affected water holding capacity, however did not influence cooking loss, shear force and 2-thiobarbituric acid reactive substances. The qualitative parameters (color, cooking loss and shear force) of the hamburgers did not suffer alterations with the inclusion of fat (20 and 30%), being higher pH and lower water holding capacity for 30% of fat, what suggests the adoption of inclusion of 20% of fat, with views to a healthier human diet.

Effect of the combination of vacuum and modified atmosphere packaging durations on lamb quality

Muela, E.[1], Sañudo, C.[1], Medel, I.[2] and Beltrán, J.A.[1], [1]Universidad de Zaragoza, Miguel Servet 177, 50013 Zaragoza, Spain, [2]Pastores Grupo Cooperativo, Ctra. Cogullada 65, 50014, Zaragoza, Spain; muela@unizar.es

Vacuum packaging (VP) is used as a long term preservation method for refrigerated primal cuts. However, dark red colour (desoximyoglobine state) is not attractive for consumers, thus high O$_2$ modified atmosphere packaging (MAP) is used for red meats consumer's retail, where bright red colour (oximyoglobine state) is determinant on purchase intention. However, this MAP composition enhances oxidation, reducing the retail display. The combination of both methods would be used to extend self-life, getting VP hygienic safety and MAP colour advantages. The study assessed the effect of 3 VP durations (7, 9 or 14 d) and 3 subsequent MAP (O$_2$:CO$_2$:Ar) durations (0, 7, or 8 d) on lamb (Rasa Aragonesa breed) loin quality, considering on these MAP durations: instrumental colour measurements (CIE L*a*b* system), visual acceptability (7 assessors scored colour acceptance with an hedonic scale from 0 -bad- to 2 -good-, and purchase intention -yes/no-), and microbiological analysis (total viable and enterobacteriaceae counts). The effects of VP and MAP durations, and their interaction on meat quality traits were assessed using the GLM procedure of the SPSS statistical package (15.0). The mean and standard error of the difference were calculated for each variable. When the main effect was significant, we used the Duncan test, with the level for statistical significance set at $P \leq 0.05$. A frequencies analysis was performed for purchase intention. Neither VP nor MAP durations had a significant effect on instrumental colour, although the higher deviation at longer durations seems to show a hue and chroma increase from 9 to 14 d VP and a decrease from 7 to 8 d MAP. However, these differences did not affect colour acceptability, since colour was acceptable (score>1.5) at 7 d MAP after VP, reflected in a purchase intention higher than 65%. Microbiology counts showed that lamb could be preserved 8 d MAP after no more than 9 d VP and less than 7 d MAP after 14 d VP.

Influence of the rich PUFA concentrate feeds on fattening lambs performance
Ropota, M., Voicu, D., Ghita, E. and Voicu, I., National Research-Development Institute for Animal Biology and Nutrition, Laboratory of Chemistry, Calea Bucuresti No.1 Balotesti, ROMANIA, 077015, Romania; elena.ghita@ibna.ro

The improvement of meat quality and of animal production is an objective necessity, given the global demand for food, as well as a priority for animal nutrition research. The purpose of the present paper is to investigate the influence of sunflower meal replacement (20.4%) in the control group C, from the compound feeds for fattening lambs, by rapeseeds (15.4%) and camelina seeds (5%) in group E. Our experiments used two groups (control, C and experimental, E) of Karabash lambs which were fattened in an intensive system, for 97 days, from the average weight of 17.4 kg to 40 kg, with an average daily weight gain of 233.6 g. Fatty acids concentration was determined by transmethylation followed by separation in a capillary column by GC- Perkin Elmer. The determination of meat composition revealed higher concentrations of linolenic fatty acid C18:3, 1.33 g% in group E compared to 0.52 g% in group C; PUFA/MUFA ratio was favourable to group E: 0.163 compared to 0.266 for group C. The formation of a new fatty acid was observed docosahexaenoic acid, C22:6n3 in low amounts 0.24 g but, important, nevertheless, due to its 6 degrees of unsaturation. Te ratio of omega 6 to omega 3 fatty acids was 9.5:1, much closer to the ideal ratio (4:1) with favourable influences on human health, which allows us to support the use of PUFA rich dietary sources for the fattening lambs.

Effect of feeding linseed on lipogenic enzyme gene expression of Rasa Navarra lamb subcutaneous and intramuscular adipose tissues
Corazzin, M.[1], Urrutia, O.[2], Soret, B.[2], Mendizabal, J.A.[2], Insausti, K.[2] and Arana, A.[2], [1]Università di Udine., Via Sondrio 2, 33100, Udine, Italy, [2]Universidad Pública Navarra, C. Arrosadía, 31006. Pamplona, Spain; aarana@unavarra.es

9 desaturase enzyme. However, in cattle it was shown that despite increases in the VA, muscle tissue concentration of 9c11tCLA did not increase, likely due to down-regulation of the StearoilCoA desaturase (SCoA) gene attributed to PUFA. The aim of this work was to examine the effects of n-3 PUFA diet supplementation on gene expression of three lipogenic enzymes: (AcetilCoA carboxylase ACC, Lipoprotein lipase, LPL, and SCoA) in sheep. Thirty six male Navarra Breed lambs were distributed into three groups: Control (C) fed on barley and soya concentrate, L-5 and L-10 groups receiving the same feed but including 5% or 10% of linseed respectively. Animals were studied from 15 to 26 kg Live Weight. Lambs had similar growth, carcass and fattening parameters. SC and IM fat from lambs fed on linseed (L-5 and L-10) presented higher percentages of VA, linolenic acid and n-3 fatty acids than the C group but he CLA content was similar in the three groups of lambs. SC gene expression of the three enzymes was higher than IM depot (P∆Long chain n-3 polyunsaturated fatty acids (PUFA) and Conjugated Linoleic Acid (CLA) have potencial health benefits, then it would be of interest to increase its content in lamb meat. Supplementation of diets with n-3 PUFA results in an accumulation of these fatty acids and in an increase of concentrations of vaccenic acid (VA), which is catalyzed to 9c11tCLA in the adipose tissue by the <0,001), and it could be related to the higher lipid composition of the SC depot. Linseed inclusion caused a decrease in the ACC expression in both depots (P<0.001) and an increase in LPL expression on SC depot (P<0.01), reflecting that the higher content on fatty acids leads to its incorporation within adipocytes. SCoA desaturase expression on lambs fed on linseed decreased in both depots (P<0.001), confirming the n-3 PUFA inhibitor effect previously observed in cattle.

The effect of birth types on growth curve parameters of Karayaka lamb

Ulutas, Z.[1], Sezer, M.[2], Aksoy, Y.[1], Sirin, E.[1], Sen, U.[3], Kuran, M.[3] and Akbas, Y.[4], [1]Gaziosmanpasa University, Faculty of Agriculture, Animal Science, Tasliciftlik-Tokat, 60240, Turkey, [2]Karamanoglu Mehmetbey, Vocational Technical School, Department of Pets and Experimental Animal, Karaman, 70100, Turkey, [3]Ondokuzmayis University, Faculty of Agriculture, Animal Science, Samsun, 55139, Turkey, [4]Ege University, Faculty of Agriculture, Animal Science, Bornova-İzmir, 35100, Turkey; zulutas@gop.edu.tr

This study focused on the comparison of the growth characteristics of single and twin birth lambs in Karayaka sheep, which is an indigenous breed of the northern part of Turkey. Gompertz growth function was fitted to body weight–age data of 81 lambs (39 males and 42 females) from birth to 10 months of age. Single birth lamb of both sexes showed lower asymptotic weight than the twin birth ones. There were a noticeable difference in the absolute growth rate between birth types before inflection point, but decline after the inflection point was slower for twins than that for singles. Similarly, the decrease in relative growth rate was higher for singles than that for twins. The Gompertz model parameters showed similar trends for birth types in both sexes. Our results indicated that the type of the birth should be taken into account besides the sex of the individuals while working on biological modelling of sheep growth and subsequent genetic evaluations of the related traits.

Alfalfa grazing increases vitamin E content and improves fatty acid profile in *L. dorsi* from light lambs

Joy, M.[1], Molino, F.[1], Gil, C.[1], Estopañan, G.[1], Alvarez-Rodriguez, J.[2] and Blanco, M.[1], [1]CITA, Avda. Montañana, 930, 50059, Spain, [2]UdL-LLeida, Avda. alcalde -rovira Roure, 191, 21198, Spain; mjoy@aragon.es

Feeding strategy affects intramuscular fat quality and meat shelf life. Forage-based diets increase naturally the polyunsaturated fatty acids (PUFA) n-3 and α-tocopherol contents in lamb meat. The aim of this study was to assess the effects of forage inclusion (alfalfa grazing vs. concentrate-fed indoors) in the diet and lactation length (weaning at 13 kg vs. suckling until slaughter at 23 kg) on the fatty acid (FA) profile and vitamin E content in L. dorsi of Rasa Aragonesa lambs. Thirty-two single lambs were assigned to one of four treatments in a 2 x 2 factorial design. ANOVA test was performed. The effect of forage inclusion was significant on FA profile and on α-tocopherol and γ-tocopherol contents while the effect of lactation length was less clear. Alfalfa grazing lambs had greater content of α-tocopherol and lower γ-tocopherol than concentrate-fed lambs (P<0.05). Some concentrate feedstuffs (as soybean and colza) increase the γ-tocopherol content, whereas forage has a negligible content. Alfalfa grazing increased the MUFA and CLA content and decreased the PUFA n-6/n-3 ratio (P<0.05). Lactation length had a less noticeable effect on vitamin E content and on FA profile. Weaned lambs had slightly greater α-tocopherol (P=0.06) because alfalfa grazing lambs had greater content than lactating lambs whereas weaning did not affect the content in concentrate-fed lambs (P<0.001). Weaning did not affect γ-tocopherol content (P>0.05). Weaned lambs presented less SFA, CLA and PUFA n-3 and more PUFA and PUFA n-6/n-3 than the lactating lambs (P<0.05). It can be concluded that alfalfa grazing improved the FA profile and increased the α-tocopherol in light lamb meat, which could contribute to human health. Lactation length had a less clear effect on vitamin E but suckling until slaughter increased CLA content and the PUFA n-6/n-3 ratio.

Effect of milk yield on growth of lambs of Merino and Mountain sheep
Ringdorfer, F. and Veit, M., ARAC Raumberg-Gumpenstein, Sheep and Goats, Raumberg 38, 8952 Irdning, Austria; ferdinand.ringdorfer@raumberg-gumpenstein.at

The effect of milk performance of sheep (30 Merino, 30 light and 30 heavy Mountain sheep (MS)) on weight gain of their lambs was studied. Milk yield was estimated by the oxytocin-method and body weight was recorded twice weekly. Lambs had free access to concentrates after day 14. All lactations were subdivided into 6 sections (I, II, III, IV, V and VI is day of lactation 6-10, 11-15, 16-20, 21-25, 26-30 and 3 -35, respectively). Milk yield was affected by breed (P=0.001), number of suckling lambs (P=0.007), and feeding conditions (P=0.044). The effect of breed became more important as lactation progressed while the number of suckling lambs lost its influence on milk yield. The highest milk yield was estimated for light MS (3.08 kg/d) in section VI. Merino ewes had significantly lower milk yields in all sections and peaked at 2.37 kg/d. Ewes nursing twins produced 0.5 kg more milk per day than ewes with singles. This effect was greatest for light MS with 23% more milk. Regarding daily weight gain of lambs, milk yield lost significance as lactation progressed (section II and VI, P=0.014 and P=0.226, respectively), while concentrate intake gained influence until it reached significance in section V (P=0.011). Further factors affecting daily lamb weight gain were the number of suckling lambs, breed of the ram, and birth weight with P=0.001, P=0.011 and P<0.001, respectively. An increase of milk yield showed significant effects on lamb weight gain. In the interval from day 6 to day 30 of lactation, an increase of one liter of milk raised daily weight gain of twins and lambs on pasture by 54 g and 45 g, respectively, while single and barn-fed lambs reached only an increase of 10 g and 18 g, respectively. In this interval only milk yield (P=0.039), birth weight (P<0.001) and number of suckling lambs (P=0.005) had a significant effect on lamb weight gain. Knowing birth weight, number of suckling lambs and daily weight gain until day 30, an inference on ewes' milk yield can be made.

Milk yield and composition in native ojalada soriana sheep under different feeding systems
Asenjo, B., Calvo, J.L., Ciria, J. and Miguel, J.A., E.U. Ingenieiras Agrarias de Soria, Produccion Animal, Campus Duques de Soria S/N, 42001 Soria, Spain; jangel@agro.uva.es

Sixty adult ewes of Ojalada Soriana local breed were randomly allocated in three groups (n=20): group 1: an integral diet based on ground straw of cereal (small fiber); group 2: based on an integral diet in form of big bales of hay (long fiber) and group 3: the traditional grazing. Ewes of groups 1 and 2 remained indoors (1,5 m²/animal) having a court of exercise outdoors (50 m²). The main complement was straw cereal the same concentrate in both groups, which composition as well as the proportion of straw was changing depending on the productive phase: (1) Maintenance and the first two thirds of the gestation: 80% straw and 20% concentrated. (2) End of gestation and lactation: 50% straw and 50% concentrated. During the 7 weeks of lactation, a weekly evaluation was carried out as to both the quantity and the chemical composition (fat, protein, lactose and solids not fat) of the milk. The milk samplings were done in ten single-birth ewes from each group. The ewes were injected with five IU of oxytocin and were milked mechanically, followed by a second more complete milking by hand. Four hours after the first injection of oxytocin, the operation was repeated, administering the same dose of oxytocin, and the amount of milk obtained from this second milking was measured. No differences were found based on feeding, with the peak lactation being situated in the third week for all three groups. The low persistence registered throughout the lactation period stands out, having obtained a production of around 1,200 ml/day in the seventh week of lactation. No differences were found in any of the chemical components in terms of the type of handling and feeding employed during lactation. Ewes in the group 1 presented a greater content in protein in the fifth week of lactation and those in the group 3 showed the least content in lactose in the seventh week of lactation.

Genetic parameters of udder cistern size in ewes diagnosed by ultrasonography
Margetín, M.[1,2], Apolen, D.[2], Milerski, M.[3] and Debreceni, O.[1], [1]Slovak University of Agriculture Nitra, Tr. Andreja Hlinku 2, 949 76 Nitra, Slovakia (Slovak Republic), [2]Animal Production Research Centre, Hlohovecká 2, 951 41 Lužianky, Slovakia (Slovak Republic), [3]Institute of Animal Science, Přátelství 815, 104 00 Praha Uhříněves, Czech Republic; margetin@cvzv.sk

Udder cistern size was diagnosed in ewes using the ultrasonograph and 3.5 MHZ linear probe in two ways: method from side and from below. Ultrasound image was made from each scanning and subsequently the size of left and right cistern was measured using the digital technique. Both cisterns were scanned approximately 12 hours after the last milking. The size of left (SLC1; 1,933.4 mm^2) and right udder cistern (SRC1; 1,973.1 mm^2), detected by the method from side, was diagnosed repeatedly in 378 ewes (within the lactation as well as between lactations); totally were performed 1,198 measurements. The size of left (SLC2; 2137.7 mm^2) and right udder cistern (SRC2; 2,171.1 mm^2), detected by the method from below, was also diagnosed repeatedly, namely in 265 ewes; totally were performed 753 measurements. Total size of both cisterns detected by the method from side (TSBC1) was 3,904.8 mm^2, and by the method from below (TSBC2) was 4,308.8 mm^2. Multiple-trait animal models (REMLF90 and VCE programs) were used to estimate the genetic parameters. In addition to random additive genetic effect of animal and permanent effect of ewe, the models involved fixed effects of control year (7 or 5 levels), lactation stage (4 levels), breed group (9 levels) and parity (3 levels). We found higher values h^2 for the parameters diagnosed by the method from below. Heritability coefficient for SLC1 and SLC2 was 0.069 and 0.182, resp.; for SRC1 and SRC2 0.168 and 0.115, resp.; for TSBC1 and TSBC2 0.123 and 0.173, respectively. Genetic correlations between SLC1 and SRC1 or SLC2 and SRC2 were high (r_g=0.726 or 0.910). Similarly, the correlations between the size of left and/or right cistern and total size of both cisterns were high with both ways of scanning (r_g=0.896 to 0.982).

Association between transferrin polymorphism and some blood parameters in Makoei fat-tailed sheep
Moradi Shahrbabak, M.[1], Moradi Shahrbabak, H.[1], Khaltabadi Farahani, A.H.[2] and Mohammadi, H.[1], [1]University College of Agriculture and Natural Resources, University of Tehran, Department of Animal Science, Faculty of Agricultural Sciences and Engineering, P.O. Box 4111, Karaj, 31587-11167, Iran, [2]University of Arak, Department of Animal Science, Faculty of Agriculture, Arak, 31587-11167, Iran; moradim@ut.ac.ir

Transferrin (TF) is a serum glycoprotein that binds free iron ions. The objective of the present research was to determine transferrin polymorphism and to find association between transferrin polymorphism and some blood parameters of blood in the population of Makoei sheep lambs. Blood samples were collected from a total 576 sheep of both sexes in Iranian fat-tailed breed Makoei sheep from the jugular vein in tubes containing EDTA and centrifuged at 4 °C. Separate aliquots of plasma and erythrocytes were stored at -20 °C until they were analyzed. Transferrin (TF) typing was performed using the polyacrylamide gel electrophoresis (PAGE), as described by Tucker and Clarke. The levels of triglyceride blood, total protein blood, glucose blood and cholesterol blood have been measured. Significant differences in Transferrin genotypes group for three parameters were considered, those results as below: level of triglyceride (P<0.001) and total protein (P<0.0001) were statistically significant; while level for cholesterol blood (P<0.066) not significant but it approximately significanct. The AA genotype resulted in a significant increase in triglyceride (29.91 mg/dl), total protein (9.961 mg/dl) and AQ genotype significant decrease in triglyceride (18.45 mg/dl), total protein (7.075 mg/dl). No significant difference was observed between the genotypes in and glucose blood (P<0.633). These results indicate that new marker associated with blood parameters can be used in marker-assisted selection in fat-tailed sheep.

Prostaglandin synchronization allows increasing silent estrous in ewes reared at high altitude

Parraguez, V.H.[1], Díaz, F.[1], Urquieta, B.[1], Raggi, L.[1], Astiz, S.[2] and González-Bulnes, A.[2], [1]University of Chile, Faculty of Veterinary Sciences; INCAS, Sta. Rosa 11735, 8820808, Santiago, Chile, [2]INIA, Dpt. of Animal Reproduction, Av. Puerta de Hierro s/n, 28040, Madrid, Spain; vparragu@uchile.cl

Sheep breeding is an important farm activity at the Andean highlands. However, the sheep productivity is low due to the hard environmental conditions and to absence of animal improvement. Estrous synchronization is essential in reproductive programs allowing to animal improvement. Cloprostenol (PGF2α analogous) has shown to be efficient for estrous synchronization in cycling sheep at low altitude (LA). The aim of the present work was to compare the synchronizing effect of cloprostenol in both native and naïve ewes at high altitude (HA). Twenty mature cycling ewes were used: HA natives (group 'HH', n=10) and LA natives recently brought to HA (group 'LH', n=10). The ewes were maintained at an altitude of 3,600 m and treated with two doses of cloprostenol 125 µg, 10 days apart. Four chest painted vasectomized rams were joined with the ewes for estrous detection. Ovaries were daily observed by ultrasound for follicular or corpus luteum dynamics, using a 7.5 MHz probe. The studied variables were: rate of ewes in estrous after treatment, time course between treatment and start of the estrous and length of the cycle. Data were analyzed by chi-square and T test when corresponded, comparing the effect of animal origin (HA v/s LA). Few animals evidenced estrous behavior (HH=5/10; LH=2/10, P=0.16), although all of the HH and 8/10 LH ewes showed follicular growth and ovulation. The time course between treatment and start of the estrous were 72.0±16.9 h and 24.0±0 h for HH and LH ewes, respectively (P=0.013). No difference was observed for the length of the estrous cycle (HH=16.0±2.0 days, LH=16.6±0.5 days; P=0.53). It is concluded that cloprostenol is not recommended for estrous synchronization in ewes at HA, due to interference with estrous behavior expression, then increasing silent estrous. Support: Grants FONDECYT 1100189 and AECI A/023494/09.

Polymorphism of calpastatin gene in Iranian Zel sheep using PCR- RFLP

Khaltabadi Farahani, A.H.[1], Moradi Shahrbabak, H.[2], Moradi Shahrbabak, M.[2] and Sadeghi, M.[2], [1]University of Arak, Department of Animal Science, Department of Animal Science, Faculty of Agriculture, University of Arak, Arak, 3158711167, Iran, [2]University College of Agriculture & Natural Resources – University of Tehran, Department of Animal Science,Faculty of Agricultural Sciences and Engineering, University of Tehran, Karaj, 31587-11167, Iran; amfarahanikh@gmail.com

Almost all Iranian sheep breeds have large fat tails. Fat-tail and other adipose depots, from economic point of view are not important in Iran. Zel breed isn't fat-tailed breed and the point is important.Calpastatin has been known as candidate gene in muscle growth efficiency and meat quality. Zel sheep is gene pool reservation and suitable for meat and wool production that until now has not been studied using molecular markers, especially with the view of calpastatin gene. Therefore, the present study was conducted to determine the genetic diversity of calpastatin gene in Zel sheep. This gene has been located to chromosome 5 of sheep. In order to evaluate the calpastatin gene polymorphism, random blood sample were collected from 115 Zel sheep from Iran. The DNA extraction was based on DNA extraction Kit (2008 Cina Gene Kit). Exon and entron I from L domain of the ovine calpastatin gene was amplified to produce a 622 bp fragment. The PCR products were electrophoresed on 1% agarose gel and stained by etidium bromide. Then, they were digested with restriction enzyme MspI and then electrophoresed on 1.5% agarose gel with ethidium bromide and revealed two alleles, allele M and allele N. Data were analysed using PopGene32 package. In this population, MM, MN, NN genotype have been identified with the 12, 88, 0% frequencies. M and N allele's frequencies were 0.56, 0.44, respectively.

Characterization of transferrin polymorphism in the Makoei sheep

Moradi Shahrbabak, H.[1], Khaltabadi Farahani, A.H.[2] and Moradi Shahrbabak, M.[1], [1]University College of Agriculture and Natural Resources, University of Tehran, Department of Animal Science,Faculty of Agricultural Sciences and Engineering, Karaj, 31587-11167, Iran, [2]University of Arak, Department of Animal Science, Faculty of Agriculture, Arak, 31587-11167, Iran; hmoradis@ut.ac.ir

The aim of the study was to determine of the polymorphism of serum transferrin in Makoei Sheep. Blood samples were collected from a total 576 sheep of both sexes in Iranian fat-tailed breed Makoei sheep from the jugular vein in tubcs containing EDTA, kept refrigerated during the transport and centrifuged at 4 °C. Separate aliquots of plasma and erythrocytes were stored at -20 °C until they were analyzed. Transferrin (TF) polymorphism was performed using the polyacrylamide gel electrophoresis (PAGE) as described by Tucker and Clarke. It was found that the transferrin type was controlled by ten alleles TfA, TfB, TfC, TfD, TfE, TfG, TfK, TfL, TfM and TfQ in Makoei sheep. Twenty-four genotypes including six of these were homozygous, and the remaining eighteen were heterozygous. It was found that the TfBC genotype (0.1892) was predominant while TfDE genotype (0.007) was least common in the analyzed flock. From the result it was found that in whole population combined, the heterozygote genotypic frequency was more than that of homozygote genotypic frequency. Considerable variations were recognized in the frequencies of transferrin alleles. The frequency of transferrin alleles were found to be TfA=0.1597, TfB=0.2474, TfC=0.2943, TfD=0.2127, TfE=0.0295, TfG=0.0286, TfL=0.0174, TfK=0.0104, TfM=0.0043 and Tfq=0.0043. Transferrin system was not in genetic equilibrium in the studied herd. In conclusion, there were polymorphism in Transferrin locus and the presence of differences among the frequencies of the six alleles could be use as a source of genetic variation in Makoei sheep.

Semen traits of lambs fed with co-products of cotton seed (*Gossypium* ssp.)

Paim, T.[1], Viana, P.[2], Brandão, E.[2], Amador, S.[2], Barbosa, T.[2], Cardoso, C.[2], Abdalla, A.[1], Mcmanus, C.[3] and Louvandini, H.[1], [1]CENA/USP, Av. Centenário 303, 13400-970 Piracicaba, SP, Brazil, [2]FAV/UnB, Campus Darcy Ribeiro, 70910-900 Brasília, DF, Brazil, [3]UFRGS, Av. Bento Gonçalves 7712, 91.540-000 Porto Alegre, RS, Brazil; pradopaim@hotmail.com

This study aimed to evaluate the influence of supplementation with cotton co-products in reproductive system of lambs during puberty. Twenty four Santa Inês lambs with six months of age and mean live weight of 21±2.72 kg were used. These animals were divided into four treatments (T). The lambs received diets containing 50% of dry matter intake (DMI) of coast cross hay (Cynodon dactylon) and concentrate mixtures comprising 20% of DMI cottonseed (T2), 20% of DMI cottonseed meal (T3), 20% of DMI cottonseed meal with high oil (T4) and a control (T1) without use of cotton. Every 15 days for ninety days scrotal circumference (SC) was measured and semen collection was performed using electroejaculation. Semen was analyzed determining: volume (V), aspect, mass movement (MM), progressive motility (M), vigor, sperm concentration (Con) and morphology. Statistical analysis was performed with SAS®: analysis of variance to verify the influence of treatment on semen parameters and SC and Wilcoxon test for nonparametric data (MM and vigor). There were no significant effects of T on the SC. T1 showed the greater value of MM. The results of M in T1 were significantly higher than T3. T1 showed a higher Con than other treatments. T1 also had higher amounts of normal sperm than T3 and T4. Head (alone and slender at the base), irregularities in the midpiece, strongly wrapped midpiece and wrapped tail were the sperm pathologies influenced by T (P<0.05). In general, treatments containing co-products of cotton had more sperm pathologies. Therefore, diets containing co-products of cotton influenced negatively seminal parameters (qualitative and quantitative).

Population structure and genetic diversity of four Swiss sheep breeds

Burren, A.[1], Flury, C.[1], Aeschlimann, C.[2], Hagger, C.[1] and Rieder, S.[1], [1]Bern University of Applied Sciences, Swiss College of Agriculture SHL, Länggasse 85, 3052 Zollikofen, Switzerland, [2]Caprovis Data AG, Belpstrasse 16, 3000 Bern, Switzerland; alexander.burren@bfh.ch

The pedigree data of the four largest Swiss sheep breeds – Swiss White Mountain sheep (SW), Brown-headed Meat sheep (BM), Swiss Black-Brown Mountain sheep (SB) and Valais Blacknose sheep (VB) – was evaluated in the context of their population structure and genetic diversity within population. The full pedigree of all herdbook individuals born in the years 1996-2008 was analysed using known open source software tools. Pedigree completeness of 90% and above, considering 1 to 6 generations, was found for BM and SB. For VB completeness varied between 80% to above 90%, whereas completeness for SW varied between 70% to 90%. Thus, the latter being clearly inferior compared to the results given for the other three breeds. This finding is not surprising as crossbreeding with Ile-de-France rams and ewes is still common today. For all four breeds average inbreeding coefficients showed a varying increasing trend, and were found at 9.3% (VB), 4.3% (BM), 3.8% (SB) and 2.5% (SW) for the year 2008. This resulted in a decreasing trend in effective population size for all four breeds. As expected, the same trend was found for the effective number of founders, the number of founder- and ancestors genomes. The decrease in genetic diversity is further supported by the increasing marginal gene contribution of the most important ancestors. For the VB population 19.8% of the present gene pool is explained by only one animal. Based on the results specific strategies to ensure long-term genetic diversity are only proposed for VB. VB is a traditional, local sheep breed, kept in a relatively small, alpine region of Switzerland. A simple index weighting the breeding values of the most important breeding sires with the average relatedness to the active breeding dams is presented for this population. For the other three breeds a regular monitoring of genetic diversity parameters is proposed.

Genetic evaluation and genetic trends of Santa Ines hair sheep in Brazil

Ferraz, J.B.S.[1], Barreto Neto, A.D.[2], Dantas, I.C.[2], Oliveira, E.C.M.[1], Rezende, F.M.[1] and Eler, J.P.[1], [1]University of Sao Paulo/FZEA, Basic Sciences, Cx. Postal 23, 13635-900 Pirassununga, SP, Brazil, [2]ASCCO, Parque João Cleophas, 49075-030, Aracaju, SE, Brazil; jbferraz@usp.br

Santa Ines is a Brazilian hair sheep composite breed formed from Morada Nova, Bergamacia and Somalis and the composite was recognized as a breed by the Brazilian Agriculture Ministry in 1977. The animals are well adapted to tropical conditions and the breed is responsible for the growth of the Brazilian sheep industry, aiming meat production. A program of genetic evaluation was established in 2006. In the 2011 genetic evaluation records of growth traits, maternal efficiency and visual scores of 11,596 animals, from 8 states of Brazil, were used for genetic analysis, with 26,936 animals in the additive relationship matrix. Traits analyzed were weight at birth (WB), 60 (W60, weaning), 180 (W180) and 270 days of age (W270), mature weight (MTW), weight gain in the intervals, total maternal ability (MAT), maternal efficiency (EFICMAT) and visual scores for hair (HAIR) and muscle (MUS). Total maternal ability was measured by adding 0.5 the EBV for direct genetic effects on W60 to the maternal genetic effect on the same traits. EFICMAT was calculated for each parity expressing kg of lambs weaned/kg of adult weight of ewe (adjusted for body score and age of ewe). Genetic analysis were performed using single trait full mixed animal models, considering age of ewe, contemporary group and age fixed effects, and the random direct additive animal and maternal effects (only for WB and W60) and permanent environmental effect of dam. EFICMAT analysis was performed using a repeatability model. The software used was MTDFREML. Heritability estimates were 0.36 (WB), 0.28 (W60), 0.23 (W180), 0.24 (W270), 0.24 (MTW), 0.20 (EFICMAT), 0.26 (HAIR) and 0.19 (MUSC). Near zero trend was observed in WB but for all other traits the trends was positive. When expressed in % of the phenotypic mean, genetic trends were low: -0.08% (WB), 0.46% (W60), 0.29% (W180), 0.20% (W270), 0.29% (HAIR) and 0.18% (MUSC).

Clostridial infections of the small ruminants in Macedonia
Ivanovska-Zdravkovska, A., Food and veterinary agency, Animal health and welfare departement, III Makedonska brigada bb, 1000, Macedonia; aivanovska@veterina.gov.mk

Clostridia are anaerobic, spore-forming, Gram-positive bacteria, size is 3-10 micrometers. In the case of prolonged feeding with feed consisting high level of proteins including poor live conditions and care, they can multiply, create toxins and cause inflammations. Enterotoxaemia and braxy caused by *Clostridium perfringens* and *Clostridium septicum* are fare the most frequent in small ruminants and cause great damages in sheep and goat population, they are manifested in three disease forms: lamb dysentery, struck in adult sheep and pulpy kidney. Lamb dysentery, the infection occurs through skin wounds and through alimentary route. Pulpy kidney disease is caused by *Cl. perfringens* type D. The infection occurs because of pour life conditions and highly concentrated feed. Struck in adult sheep and goats is caused by *Cl. perfringens* type C. The infection occurs through fecal contamination and infected pasture. BRAXY (Malignant edema, Oedema malignum) caused by *Cl. septicum* and characterized by inflammation of the abomasal wall, toxemia and high mortality. The infection appears most frequently during winter affecting young's and weak sheep. Adults have already acquired immunity. The toxins are not stable and they disappear promptly. The best material is small intestine, kidney and muscles. The samples for this study were taken and analyzed in the Veterinary Institute–Skopje, R. of Macedonia in the period from the beginning of 1997 until June 2000. During 1997, 17 samples were received, in 1998, 93 samples, 1999, 73 samples, and until June 2000, 36 were received for bacteriological analysis. Clostridia infectiona is causing great economical looses in sheep and goat production and is most frequent during the winter period from November to March. Thru rapid development of the disease and impossibility for treatment, the regular vaccination with polyvalent vaccine is the most effective measure in control of the clostridial infections.

Effect of time road transport on some blood indicators of welfare in suckling kids
Alcalde, M.J.[1], Alvarez, R.[1], Pérez-Almero, J.L.[2], Cruz, V.[3] and Rodero, E.[3], [1]Univ Seville, Ctra Utrera Km.1, 41013 Seville, Spain, [2]IFAPA, Las Torres, 41200 Alcala del Rio (Sevilla), Spain, [3]Univ Cordoba, C. Rabanales, 14014 Cordoba, Spain; ralvarez1@us.es

Loading, pre-slaughter transportation, feed deprivation, and unloading are potent stressors than can alter the animal's physiology and accordingly welfare indexes, such as hormonal, hematological or biochemical parameters. We have analyzed blood parameters in 30 Negra Serrana suckling kids in three moments: at farm (basal levels for all animals) and at slaughterhouse after short transport (2 hours) made with 15 animals and after long transport (6 hours) for the other 15 animals. The animals were slaughtered without lairage immediately after unloading. For basal levels the average glucose (ml/dl) was 87.50±4.26, RBC (red blood cells, $10^6/mm^3$) 3.70±0.38, MCV (Mean Corpuscular Volume, fl) 26.79±0.55, HbCM (Mean Cell Hemoglobin, pg) 8.41±0.26, Leucocytes ($10^3/mm^3$) 16.05±0.74, CK (Creatine Kinase, UI/l) 258.46±29.24, NEFA (non esterifies fatty acids, mmol/l) 0.36±0.03 and cortisol (ng/ml) 23.98±3.56. Using an ANOVA test we found significant differences for the whole variables. Glucose increased with short transport but decreased with long transport till basal levels because they became adapted to transport. RBC increased with the two length of time for transport. MCV, HbCM and Leucocytes decreased with respect to basal levels but none differences were found with the time of transport. CK increased with transport with respect basal levels. And with NEFA and Cortisol parameters, the differences appeared for long transport (higher levels) respect to short transport and basal levels. So the results suggested that some parameters could differentiate basal levels from transport while other ones differentiated basal levels and short transport from long transport.

Effect of housing conditions on Iberian piglet's diarrhoea

González, F., Robledo, J., Andrada, J.A., Vargas, J.D. and Aparicio, M.A., School of Vetterinay Science, University of Extremadura, Animal Production, University Avenue, 10071, Cáceres, Spain; fgonzalej@alumnos.unex.es

At weaning, the intestinal microbial ecosystem has to adapt to new feeds and to a new environment. Iberian piglets were weaned in a wide variety of production systems. And the environmental conditions pre-dispose the development of the digestive tract. Therefore, Iberian piglets growth and post-weaning diarrhoea may be influenced by weaning system. The aim of these experiments was to assess the effect of weaning systems on susceptibility piglets' diarrhoea. A total of 360 Iberian piglet were weighed weekly from 21 day of age to 63. Piglets were weaned on three different systems: Intensive, Traditional and Out-doors (I; T; O). Same number of weaned piglets was studied in each system (n=120). Two trials were conducted for each system. All animals were weighed weekly from 21 to 63 days old, and food intake was controlled with the same frequency. Piglets' diarrhoea was recorded weekly. Data were analysed using Spss® and statistical significance were accepted at P<0, 05. Comparing the three weaning system, from weaning to 63 day of age, there were significant differences in prevalence (I: 38.8%; T: 39.7%; O: 13.9%) and incidence (I: 23.3%; T: 39.7%; O: 10.4%) of piglets' diarrhoea. However there are not differences in case of mortality (I: 2.3%; t: 2.6%; O: 0.9) and lethality (I:0.06%; T: 0.13; O: 0.06). For the whole study period there were significant differences in food consumption between weaning system (out-doors (c) intensive (a) traditional (b)), and the benefit derived from it (P<0.000). In conclusion, the weaning systems influence on Iberian piglets.

Reproductive traits, feed intake and condition of group and single housed lactating sows

Bohnenkamp, A.-L.[1], Traulsen, I.[1], Meyer, C.[2], Müller, K.[2] and Krieter, J.[1], [1]Institute of Animal Breeding and Husbandry, Christian-Albrechts-University, Ohlshausenstraße 40, D-24098 Kiel, Germany, [2]Chamber of Agriculture Schleswig-Holstein, Gutshof 1, D-24327 Blekendorf, Germany; abohnenkamp@tierzucht.uni-kiel.de

The aim of the present study was to evaluate reproductive traits, daily feed intake and body condition of group (GH) and single housed (SH) lactating sows. Data from 111 cross-bred sows with 124 litters were collected on the agriculture research farm Futterkamp of the Chamber of Agriculture of Schleswig-Holstein from March '09 to August '10 in 11 batches. GH-sows had individual pens (4.7 m^2) with farrowing crate, electronically controlled gate and a shared running area (13 m^2) between the pens. GH-sows (62 litters) could move freely except 3 d ante partum until 1 d *post partum* (p.p.). GH-piglets could leave their pens 5 d p.p. SH-sows (62 litters) were kept in conventional single pens (5.2 m^2). In contrast to GH-piglets, SH-piglets grew up separately from unacquainted litters. Piglets were weaned 26 d p.p. Number of piglets born alive (NBA), stillborn (NSB), weaned (NWP), piglet losses (NPL), piglet weight at birth (IBW) and weaning (IWW) as well as daily feed intake (DFI), loss of body condition (BC) and back fat (BF) were analysed using linear models. Reproductive traits in GH and SH did not differ significantly (P>0.05; NBA: 14.1 vs. 14.6 piglets/ litter, NSB: 1.4 vs. 1.3 piglets/ litter, NWP: 11.4 vs. 11.4 piglets/ litter, NPL: 2.1 vs. 2.2 piglets/ litter, IBW: 1.4 kg vs. 1.3 kg.) with exception of weaning weights (P<0.05; 7.1 kg vs. 7.6 kg). DFI of GH-sows (5.9 kg) was significantly higher compared to SH-sows (5.6 kg). The housing system had an effect on the loss of BC (P<0.05) but not on the loss of BF (P>0.05; 4.1 mm vs. 4.2 mm). In conclusion the fertility traits were not influenced by the housing system while decreased weaning weights, body condition and an enhanced daily feed intake were determined in the group housing.

Genetic correlations between litter weight, sow body condition at weaning and reproductive performance in next litter

Lundgren, H.[1], Grandinson, K.[1], Lundeheim, N.[1], Vangen, O.[2], Olsen, D.[3] and Rydhmer, L.[1], [1]Swedish University of Agricultural Sciences, Department of Animal Breeding and genetics, Box 7023, 75007 Uppsala, Sweden, [2]Norwegian University of Life Sciences, Department of Animal and Aquacultural Sciences, P.O. Box 5003, 1432 Aas, Norway, [3]Norsvin, P.O. Box 504, 2304 Hamar, Norway; Helena.Lundgren@slu.se

Our aim was to investigate the associations between the five traits litter weight at 3 weeks (LW), sow body condition at weaning (BC), weaning-to-service interval (1-7 d, WSI7; 1-50 d transformed, WSI50) and total number born in 2nd parity (NBT2). Data on 4,964 Norwegian Landrace sows and their piglets recorded from January 2008 to April 2010 were used. Genetic parameters were estimated with a multivariate animal model. Heritability estimates for LW (n=4,534), BC (n=4,964), WSI7 (n=2,340), WSI50 (n=2,598) and NBT2 (n=2,091) were 0.17, 0.19, 0.16, 0.07 and 0.14 respectively. Estimated genetic correlations for the trait combinations were; LW-BC r_g = -0.55; BC-WSI7 r_g = -0.13; LW-WSI7 r_g = 0.29; BC-WSI50 r_g = -0.25; LW-WSI50 r_g = 0.10; BC-NBT2 r_g = -0.38; LW-NBT2 r_g = 0.28. Ability to raise heavy litters is genetically correlated to lower body condition at weaning, longer weaning-to-service interval and increased number of piglets born in the following litter. Poor body condition at weaning seems to be genetically correlated to longer WSI and more piglets born in the following litter.

Sow history features affecting growth and feed intake in finishers

Sell-Kubiak, E.[1], Knol, E.F.[2] and Bijma, P.[1], [1]Wageningen University, Animal Breeding and Genomics Centre, P.O. Box 338, 6700 AH Wageningen, Netherlands, [2]Institute for Pig Genetics B.V., P.O. Box 43, 6640 AA Beuningen, Netherlands; ewa.sell-kubiak@wur.nl

The sow provides an environment to her offspring during gestation and lactation. Certain features in the sows' life (sow history features) may affect her ability to deliver and feed a healthy litter. In genetic analyzes of finishing traits this effects are estimated as common litter or, wider, permanent sow effect. The objective of this research was to identify sow history features that affect the growth rate (GR) and feed intake (FI) of her offspring during finishing. Data on 17,256 finishers, coming from 603 sires and 680 crossbred sows, were recorded at the experimental farm of the Institute for Pig Genetics (Beilen, The Netherlands). The finishing was divided into two phases (25-70 and 70-115 kg). The sow history features were: birth litter size, birth year and season, birth farm, weaning age, and age at 1st insemination. The sow features were added to the basic model one at a time, to study their effect on finishers' traits. Subsequently, significant sow features were fitted simultaneously in an animal model. With every extra piglet in the sow's birth litter, GR of her offspring decreased by 1 g/day and FI decreased by 4 g/day. Every extra day to the 1st insemination increased GR of finishers by 0.1 g/day. The heritability estimates for GR and FI, only in the 2nd phase of finishing, decreased after adding the sow features to the model. No differences were found in estimates of the common litter effects between the basic model and the model with all significant sow features. The estimates of the permanent sow effect changed only for FI, from 0.03 (basic model) to 0.00 (model with sow features). In conclusion, selected sow features do affect finishers' traits, but their estimates are small and explain only a small proportion of the differences in GR and FI of finishers. The sow features partially explained the permanent sow effect and did not explain the common litter effect.

Impact of initial weight heterogeneity and pen density on evolution of weight heterogeneity within groups of growing pigs
Brossard, L., Dourmad, J.Y. and Meunier-Salaün, M.C., INRA - Agrocampus Ouest, UMR1079 SENAH, 35590 Saint-Gilles, France; ludovic.brossard@rennes.inra.fr

The evolution, during the fattening period, of body weight (BW) coefficient of variation (CV), rank and hierarchy within groups of growing pigs, was studied according to initial BW CV in the pen and the number of pigs per pen. One hundred and twenty pigs (half castrates and females) were studied from 30 to 105 kg BW in two replicates. In each replicate, the effect of the interaction between pen density (10 vs 20 pigs per pen) and initial BW CV (7% vs 21%) was evaluated, with one pen by modality of the interaction. Groups were fed *ad libitum* a standard grower diet, with 1 feeder for 10 pigs. They were penned on a concrete floor with straw and with similar total space area (32 m²). Feed intake was determined daily per pen. Pigs were weighed individually once per week. Behavioural activity was recorded from 2 weeks after the entrance in the fattening building and thereafter every 2 weeks. Social rank of each animal within each group was evaluated after 2, 5 and 10 weeks of experiment, using a feeding competition test. Results were analyzed by ANOVA with group size, initial BW CV level and their interaction as fixed effects, and age for repeated measures. Average daily gain, mean feed intake and behavioural activity were influenced by neither pen density nor initial BW heterogeneity. The BW CV decreased significantly during the experiment in pens with a high initial BW CV, while it did not significantly vary in pens with a low initial BW CV. The Spearman correlations for BW decreased during the experiment and tended towards a lower level when the initial CV was low, indicating a gradual reorganization of the ranking of BW during the experiment, especially in groups with low initial CV. Social relationships, evaluated by a hierarchy index, were independent of pen size or initial BW CV, and did not influence BW changes. The present study was conducted within the EU-supported project Q-Porkchains.

Direct and associative effects for androstenone in entire boars
Duijvesteijn, N.[1,2], Knol, E.F.[1] and Bijma, P.[2], [1]Institute for Pig Genetics B.V., P.O. Box 43, 6640 AA Beuningen, Netherlands, [2]Wageningen University, Animal Breeding and Genomics Centre, P.O. Box 338, 6700 AH Wageningen, Netherlands; naomi.duijvesteijn@wur.nl

In the pig industry, male piglets are castrated early in life to prevent boar taint, as it gives a strong perspiration-like odour to the meat when heated or cooked. Boar taint is caused by androstenone (AND) and skatole. and is a pheromone that can be released from the salivary glands when the boar is sexually aroused or aggressive. The expression of and within groups composed of boars is not well understood, but social interactions between the boars exist. The objective of this study is to investigate whether and levels are affected by (non-)heritable social interactions and estimate its genetic correlation with growth (GR) and backfat (BF). Accurate pen registration is required to be able to disentangle the effect of an individual and the effect this individual has on its pen mates. In total, the data contained 6,245 boars, of which 4,455 had an and observation (68%). The average pen size was 7 and the boars were housed in 898 contemporary pens and 344 contemporary compartments of the barn. Including random pen and compartment effects (non-heritable social effects) already significantly improved the model (P<0.001), and including heritable social effects even more (P<0.009). The associative effects explained 27% of the total genetic variance. BF was analysed with an animal model including non-heritable social effects and GR was analysed using a heritable social model. The genetic correlation between BF and the total genetic variance for and was close to zero. BF and the direct and associative effects for and had genetic correlations of 0.14±0.08 and -0.25±0.18, respectively. The genetic correlation between the total genetic variances for GR and and was 0.33±0.18. The genetic correlation between the direct effects was 0.11±0.09 and between the associative effects was 0.42±0.31. Results suggest that including associative effects for and could be beneficial for selection against boar taint.

Vaccination against boar taint – control regimes at the slaughter house
Fredriksen, B.[1], Hexeberg, C.[1], Dahl, E.[2] and Nafstad, O.[1], [1]Animalia, Box 396, Økern, 0513 Oslo, Norway, [2]Norwegian School of Veterinary Science, Box 8146 Dep, 0033 Oslo, Norway; bente.fredriksen@animalia.no

A two step practical study was performed in September-December 2010 to generate a common procedure for Norwegian slaughter houses on control of pigs vaccinated against boar taint. Since good effect of the vaccine has been thoroughly documented, the main focus was on how to identify animals that were not vaccinated according to the recommendations. In the initial investigation, 160 male pigs from one herd were vaccinated. Recordings of behaviour during lairage, testis size on live animals, testis size on the carcass, testis weight and length after dissection, size of bulbourethral glands, colour of testis tissue as well as analyses for skatole and androstenone were performed on all carcasses. Based on the results, testis weight was used as an indirect measure of possible boar taint in the main study where 2,415 vaccinated pigs from 19 herds were slaughtered at 4 different slaughter houses. Fat samples were collected for androstenone and skatole analyses from all animals with testis weight >300 g. If testis weight was >500 g, the carcass was withdrawn from the fresh meat market. A total of 365 animals (15.2%) with testis size >300 g were reported, and 69 of these had testis weight >500 g. Per herd, the share of animals with testis size >300 g varied from 3-67%. Androstenone levels >=1ppm were detected in 11 pigs, while skatole levels>=0.2 ppm were detected in 2 pigs. The association between testis size and levels of androstenone and skatole was not sufficiently strong to recommend testis size as criterion for boar taint. For 8.6% of the animals, the second vaccination was performed outside the recommended interval (4-10 weeks) before slaughter. Four of these animals had high levels of androstenone. The study did not identify any parameters that could, without a confirming test, be recommended as indicators of boar taint. However, the necessity of controlling the vaccination certificate at the slaughter house was emphasized.

Cluster analysis as a tool to assess litter size in conjunction with the amount of embryonic and fetal losses in pigs
Fischer, K.[1], Brüssow, K.-P.[2], Schlegel, H.[1] and Wähner, M.[1], [1]Anhalt University of Applied Sciences, Strenzfelder Allee 28, 06406 Bernburg, Germany, [2]Leibniz Insitute for Farm Animal Biology, Wilhelm Stahl Allee 2, 18196 Dummerstorf, Germany; k.fischer@loel.hs-anhalt.de

High numbers of live born and vital piglets are required for an effective piglet production. It has been reported that embryonic losses in swine can be as high as 20 to 70%. Generally, up to 30% of embryonic losses are considered within the normal biological range. The aim of our study was first to determine embryonic and fetal losses in sows of the German Landrace breed, and second to identify sows, which are able to realize high numbers of intact embryos and fetuses in combination with low rates of embryonic and fetal losses. This study was conducted on commercial farrow-finish operation and involved 64 gilts. Gilts were synchronized for ovulation and inseminated artificially (AI) twice at fixed times. Pregnant gilts were slaughtered on Day 30 (n=34) and Day 80 (n=30) after second AI. Corpora lutea (CL) and embryos/fetuses (E/F) were counted. The length of uterine horns was measured. Based on the difference between the number of CL and the number of E/F, embryonic and fetal losses were 36.9 and 37.9%, respectively. The results obtained regarding the number of CL, the number E/F (intact and total), the rate of embryonic or fetal losses and uterine space per E/F underwent a cluster analysis. The results show that several gilts are able to realize high numbers of intact E/F in combination with low rates of losses and limited uterine space per E/F. Numbers of 15 to 20 intact fetuses on day 80 of pregnancy are considered as a minimum for an adequate litter size at the end of pregnancy. In this study only 35% of gilts at Day 30 and 37% of gilts at Day 80 demonstrate this potential. Cluster analysis is a useful mathematical tool which helps to assess results of embryonic and fetal losses and to group gilts according to their performance. As a consequence cluster analysis provides clues which gilts or family structures should be analyzed more intensively.

Genetic progress in piglet survival can be increased without decreasing genetic progress in live pigs at day 5

Berg, P.[1] and Henryon, M.[2], [1]Aarhus University, Genetics and Biotechnology, Research Centre Foulum, P.O.Box 50, 8830 Tjele, Denmark, [2]Danish Agriculture & Food Council., Pig Research Centre, Axeltorv 3, 1609 Copenhagen V, Denmark; Peer.Berg@agrsci.dk

Piglet mortality is 20-25% in most countries. This represents both a potential loss in production and an ethical concern for piglet production. In Denmark, live pigs at day 5 was included in the breeding objective in 2004, resulting in a correlated decrease in piglet mortality. Despite this, the high levels of piglet mortality still persist and has prompted calls for more effective selection. We hypothesise that piglet survival will increase by including measures of piglet survival in the breeding objective without decreasing genetic progress in live pigs at day 5. Live pigs at day 5 (LP5) is a function of total number of piglets born (TNB) and piglet survival (PS). Using stochastic simulation of a full scale Danish pig breeding programme, we compared a selection criterion that included TNB and LP5 in addition to the traits currently in the Danish breeding objective. TNB and LP5 were included in the criterion with a constant weight on LP5 or with a shift in weight from LP5 to PS. These models considered piglet mortality to be a maternal trait. We also simulated PS as a binary trait of the piglet influenced by maternal and direct genetic effects. Our results showed that genetic progress in PS can be increased at the same rate of genetic progress in LP5. More specifically, genetic progress in PS was increased when LP5 was also in the breeding objective. Considering PS as a maternal trait and using linear models, revealed that long-term predictions are unrealistic. Predictions should be based on models considering PS as a binary trait of the piglet. Increasing weight on PS had little influence on other traits in the breeding objective.

Formation and growth of secondary myofibers is impaired due to Intrauterine crowding

Pardo, C.E., Koller-Bähler, A. and Bee, G., Agroscope Liebefeld Posieux, Pig nutrition and pork quality, la tioleyre 4, 1725 Posieux, Switzerland; giuseppe.bee@alp.admin.ch

The aim of this study was to determine the impact of differences in uterine space using unilateral hysterectomized-ovariectomized (UHO) and unilateral oviduct ligated (OL) sows on reproduction performance as well as organ and muscle development of selected progeny at birth. In the study 8 UHO and 10 OL Swiss Large White sows in third parity were used. At farrowing litter size and litter birth weight (BtW) were determined. Subsequently, from each litter the 2 male and 2 female piglets with the lowest (L) and highest (H) BtW were sacrificed. Inner organs and LM were collected and weighed. Number and diameter of primary (Prim) and secondary (Sec) myofibers and Sec:Prim ratio were determined by histological analyses of the LM using mATPase staining after pre-incubation at pH 4.3 and 10.2. The litter size was similar (8.0 vs. 7.6; P>0.75) for the 2 sow groups. However, as expected progeny born from UHO sows were lighter (1.43 vs. 1.85 kg; P≤0.01) than those from OL sows. When expressed per 100 g BtW, heart, liver, kidney and spleen of UHO and OL progeny were of similar (P≥0.36) weight whereas the brain was heavier (2.52 vs. 1.92%; P<0.01) and the brain:liver ratio was greater (1.03 vs. 0.75; P<0.01) in UHO than OL piglets. Due to a numerically higher (P=0.13) number of Prim but not (P=0.92) Sec myofibers, the Sec:Prim myofiber ratio was lower (27.1 vs. 29.6; P=0.05) in the LM of UHO than OL progeny. The Sec but not the Prim myofibers were smaller (8.1 vs. 9.1 μm; P<0.01) in diameter in the LM of UHO than OL piglets. Regardless of the sow treatment, relative brain weight (2.0 vs. 2.4%; P<0.01) and brain:liver ratio (0.79 vs. 0.99; P<0.01) was lower in H than L piglets. Only Prim myofiber number, which tended (P=0.06) to be greater in L than H piglets, was affected by BtW. The current data suggest that regardless of BtW and gender, in individuals born from a crowded environment, formation as well as growth of Sec myofibers in the LM might be impaired.

The biopsy of the boar testis using ultrasonography control in pigsty conditions
Liepa, L., Auzāns, A., Brūveris, Z., Dūrītis, I., Antāne, V. and Mangale, M., Faculty of Veterinary Medicine, LUA, Clinical institute, Helmana-8, LV-3004, Latvia; laima.liepa@llu.lv

The biopsy of live animal testis is an important clinical manipulation for obtaining tissue samples to control pathologies. Our aim was to develop a method of boar testis biopsy using ultrasonography (USG) and to establish the influence of this procedure on the boar testis parenchyma and breeding ability. The biopsy was carried out in six 8-months-old boars in the vivarium of the Faculty of Veterinary Medicine, Latvian University of Agriculture. 20 days prior to 21 days after the biopsy, the quality of ejaculate was examined (volume, pH, temperature, concentration, activity and pathologies of spermatozoa) with a 10-days intervals. The general anaesthesia of boars was done with 7 ml of 10% ketamine infusion intravenously. An examination was performed with USG equipment Philips HD11 using a linear probe L12-3MHz in the middle region of testis (margo liber) to detect the site of biopsy (without large blood vessels). Prior to the biopsy, the testis region was prepared for surgery with 70% ethanol solution and 1 cm skin incision was made. USG images of the testis parenchyma were recorded three times: directly before the biopsy, 15 minutes after procedure and 21 days after biopsy. Biopsies of the boar testis were performed in the depth of 1.2-1.6 cm of parenchyma. A biopsy gun was used with a 12 cm needle 1.8 mm in diameter ('Vitesse'). The tissue samples were fixed in 12% formol saline, embedded into paraffin, cut in 4 μm thick sections and stained with haematoxylin-eosin and Masson's trichrome. 15 minutes after biopsy, in the USG image of all the boar testis macroscopic injures of the parenchyma were not detected. 21 days after the biopsy, the hyperechogenic linea diameter 0.1-0.2 cm was seen in the testis parenchyma in the depth of 1.2-1.6 cm. The biopsy of boar testis did not influence the breeding ability of boars. A perfect biopsy of boar testis is easy to perform using USG in the pigsty conditions.

Efficiency of immunocastration and use of Pietrain genetics as alternatives to ractopamine administration, in terms of carcass and meat quality in pigs
Rocha, L.M.[1,2], Bridi, A.M.[2], Bertoloni, W.[3], Vanelli, A.W.[1,4] and Faucitano, L.[1], [1]Agriculture & Agri-Food canada, 2000 College Street, J1M1Z3 Sherbrooke, Canada, [2]State University of Londrina, P.O. Box 6001, 86051990, Londrina, Brazil, [3]Federal University of Mato Grosso, Av. F. Corrêa, 78060900, Cuiabá, Brazil, [4]Laval University, 2425 rue de L'Agriculture, G1V0A6, Quebec City, Canada; luigi.faucitano@agr.gc.ca

To evaluate the effects of the interaction between ractopamine administration, genetics and immunocastration on carcass and meat quality traits in pigs, a total of 756 pigs were distributed into two main groups (376 and 380 pigs each), receiving 7.5 ppm of ractopamine (RAC) or not (NRAC) in their diet during the last 28 days of the finishing period. Within each group pigs, 377 castrates (CAS) and 379 immuno-castrates (IC), and two genotypes (379 controls, CONT, and 377 PietrainNN, PI) were represented according to a 2 x 2 x 2 factorial design. Castration took place at one day of age, while immunocastration was performed through two subcutaneous injections of Improvac® (2 ml) at 10 and 4 weeks before slaughter. Meat quality was assessed in the Longissimus dorsi (LD) muscle. Data were analyzed using the Proc Mixed procedure in SAS. Carcass yield was higher in RAC pigs (P<0.01), CAS pigs (P<0.0001) and PI pigs (P<0.0001) compared to NRAC, IC and CONT pigs, respectively. Carcasses from RAC and IC pigs were leaner compared to NRAC and CAS pigs, respectively (P<0.0001 and P<0.001, respectively). As for meat quality, feeding CAS pigs with RAC decreased drip loss in pork meat (P=0.05). Shear force (texture) values were slightly higher in pork from RAC fed PI pigs compared to PI pigs not fed with RAC (P<0.01) and in pork from IC compared to CAS pigs (P>0.05). However, these differences were so small (0.1-0.3 kg) that would not be detected by consumers. The results of this study showed that immunocastration more than Pietrain genetics can be a valid alternative to ractopamine as it increased carcass leanness without any negative effect on meat quality.

Comparison of fattening performances of boars castrated or immunized against GnRF and evaluation of the vaccination efficiency

Wavreille, J.[1], Boudry, C.[2], Romnée, J.M.[1], Froidmont, E.[1] and Bartiaux-Thill, N.[1], [1]Walloon Agricultural Research Centre, Rue de Liroux 9, 5030 Gembloux, Belgium, [2]ULg, Gembloux Agro-Bio Tech, Passage des Déportés 2, 5030 Gembloux, Belgium; Froidmont@cra.wallonie.be

The vaccination against boar taint with detection of unvaccinated pigs was studied at CRA-W in an animal husbandry-based approach. This research was conducted on 160 males, issued from a Piétrain x Belgian Landrace cross, fed *ad libitum* between 25 and 122 kg body weight and group-housed under the same conditions during the fattening period (6 pigs/pen). One group (82 pigs) was castrated (chirurgical castration) before 7 days of age, whereas the other group (78 pigs) was vaccinated against boar taint (Improvac®, Pfizer Animal Health). The growth performance and the behavior of the pigs were observed during the fattening period. At slaughter, the testis size and the boar taint were evaluated. No effect was observed on growth performance between groups but feed conversion was slightly improved in the vaccinated pigs (P<0.06) resulting in a feed saving of about 15 kg per pig over the fattening period. The second vaccination, performed at 4-6 weeks before slaughter, was a turning point for growth performance: vaccinated pigs grew faster than castrated ones. After the second vaccination, pigs spent more time lying (74% vs 67% of time) and when they were standing, they spent more time at the feeder (12.6% vs 8.5% of time). In our trial, the vaccination was totally effective in preventing boar taint. Reduction in testis size was clearly visible from the second week after the recall vaccination. Average testis weight at slaughter was 330 g, whereas other observations showed an average weight over 800 g in uncastrated males. Moreover, upon 78 vaccinated males, only one had testis weight exceeding the 600 g threshold. Sensory assessment of fat (by human nose) did not reveal any boar taint. Further work will be undertaken to update results and extend comparison of production performance, animal behavior and boar taint.

The relationship between litter size, individual piglet birth weight and piglet survival

Brandt, H., Yaroshko, M. and Engel, P., Department of Animal Breeding and Genetics, Ludwigstr. 21 B, 35390 Giessen, Germany; horst.r.brandt@agrar.uni-giessen.de

The average litter size in pigs has increased in some crossbred lines up to 14 piglets total born per litter and more. Along with this development the piglet mortality has increased dramatically, therefore the relationship between litter size, individual piglet birth weight and piglet mortality comes again into focus when discussing future breeding objectives in pigs. Detailed information on piglet losses (born dead and weaning losses, n=11,779 piglets from 966 litters) from 4 multiplier farms over the last 18 month were available as well as individual birth weights from one farm over the last year (n=1,062 piglets from 74 litters) and from a second farm over the last 4 years (n=15,407 piglets from 966 litters) to analyse the relationship between litter size, individual piglet birth weight and piglet losses. The results show a nearly linear reduction of average piglet birth weight with increasing litter size. With the reduction of individual birth weight an exponential decrease in piglet survival is observed. Nearly half of the piglets with birth weights of less than 1.2 kg have no chance of survival. As a future breeding objective in dam lines the litter quality including litter size, individual birth weight, variation in birth weight and piglet survival should be considered. From a practical point of view it would be too much effort to individually weigh all piglets at birth, so scoring systems to assess individual birth weights and the variation in birth weigh are proposed.

Modifications of biotechnical methods help to reduce seasonal variations in reproduction performances in sows

Waehner, M. and Richter, M., Anhalt university of Applied Science, Agriculture, Strenzfelder Allee 28, 06406 Bernburg, Germany; m.waehner@loel.hs-anhalt.de

In summer high temperatures are stress for sows mostly. This situation often affects the reproduction physiology in animals strongly. As a result of that the reproduction performance of sows can be reduced. High environmental temperatures and especially heat accumulation in body of the animals affect the metabolism in high performance sows. That is important in very sensitive phases in reproduction cycle of sows like heat, pregnancy and lactation. Young sows (gilts and primiparous sows) are more sensitive than older sows. This situation demands zoo- and biotechnical activities expecting assistance for reproduction endocrinology in sows. Regarding this aim, special biotechnical methods have been adapted and modified to special and individual situations in farms. Following biotechnical methods are included: (1) Biotechnical synchronization of estrus in gilts and weaned sows. Especially the time distance between last application of REGUMATE and PMSG-injection should be longer then 24 hour. Optimum is 42 hours. Generally a biotechnical stimulation by PMSG affects the reproduction performances in summer positively. (2) PMSG-dosage for cycle-stimulation in primiparous sow should be 1000 IU, in older sows only 800 IU. (3) GnRH-preparation has a higher effect than hCG in stimulation of ovulation in gilts and sows. In comparison to spontaneous estrus in sows the time of ovulation is affected by different hormonal stimulation. (4) During summer oxytocin supplement in semen stabilize pregnancy rate in sows.

Post partum dysgalactia syndrome in sows: estimation of variance components and heritability

Preissler, R.[1,2], Hinrichs, D.[1], Reiners, K.[3], Looft, H.[3] and Kemper, N.[2], [1]Institute of Animal Breeding and Husbandry, Christian-Albrechts-University Kiel, Olshausenstraße 40, 24098 Kiel, Germany, [2]Institute of Agricultural and Nutritional Science, Martin-Luther-University Halle-Wittenberg, Theodor-Lieser-Straße 11, 06120 Halle (Saale), Germany, [3]PIC Germany, Ratsteich 31, 24837 Schleswig, Germany; rpreissler@tierzucht.uni-kiel.de

The *Post partum* Dysgalactia Syndrome (PDS) represents one of the most important diseases after parturition in sows. A genetic background of this disease has been discussed and heritability from 2% up to 20% for PDS was estimated, but current studies are lacking. Therefore, a dataset of 1,680 sampled sows and their 2,001 clinically examined litters was used for variance components estimation with a threshold liability model. Sows were defined as PDS-affected through clinical examination 12 to 48 hours after parturition with emphasis on rectal temperature, mammary glands, and behavior changes of sows and piglets as well. All animals were housed on six farms with similar management conditions, animal health and hygiene standards to provide maximal comparable environmental factors. The posterior distributions of the individual variance and the permanent environmental variance for the susceptibility to PDS were determined with a single trait repeatability model. Data control and further statistical analysis was performed using the statistical software SAS (SAS System). Posterior mean of additive genetic variance was 0.10, and estimated heritability for PDS averaged 0.09 with a standard deviation of 0.03. The results are in agreement with those of other studies and emphasize the importance of optimizing hygiene and management conditions as well as considering the genetic predisposition for susceptibility to PDS.

Genetic analysis of fertility and growth traits in a Duroc x Pietrain Resource population
Bergfelder, S., Große-Brinkhaus, C., Looft, C., Cinar, M.U., Jüngst, H., Schellander, K. and Tholen, E., Institute of Animal Science, University of Bonn, Animal Breeding and Husbandry, Endenicher Allee 15, 53115 Bonn, Germany; sber@itw.uni-bonn.de

Individual birth weight of piglets has an economic impact because of its influence on piglet survival and postnatal growth. Selection for increased litter size leads to a decrease of birth weight and thereby to an increase of piglet mortality. Because of the low heritability of piglet survival selection for individual birth weight seems to be more efficient in most populations. The aims of this study were a) to estimate genetic parameters for individual birth weight (IBW), weaning weight (WW), average daily gain from birth to weaning (ADG) and number of teats (Teats) and b) to detect the QTLs affecting these traits. 8,278 animals of a Duroc x Pietrain resource population were used for the study. Piglets showed an IBW of 1.4±0.4 kg, WW of 8.0±2.3 kg and ADG of 215.2±65.3 g/d. Estimated heritabilities for IBW, ADG, WW, and teats were 0.15, 0.4, 0.3 and 0.4, respectively. High significant negative correlations between litter size and IBW, WW and ADG and a significantly high positive correlation between ADG and WW were found. On the phenotypic basis, with increasing litter size IBW decreased and piglet mortality increased. The QTL analysis was performed using 914 F_2 animals were genotyped with 113 genetic markers equally spaced across the whole genome. The statistical model comprised breed, month of birth, sex and parity as fixed class effects and litter size, age of sow and birth weight as linear fixed covariates. Under consideration of additive and dominant effects, QTLs affecting IBW were detected on SSC2 and SSC16, for WW and ADG on SSC2 and SSC14 and for number of teats on SSC3, SSC7 and SSC9. When imprinting effects were taken into account, an additional QTL for IBW on SSC4 was observed. Carrying out a 2-QTL-analyis for number of teats, two QTLs for number of teats were detected on SSC7. Further analyses try to find QTLs for piglet survival.

Breeders' visions of the role of a local pig breed in an extensive farming system: the Nustrale pig case
Lauvie, A., Maestrini, O., Santucci, M.A. and Casabianca, F., INRA LRDE, Quartier Grossetti, 20250 Corte, France; lauvie@corte.inra.fr

We present the results of a survey concerning the Nustrale pig breed in links with the finishing practices using chestnut trees and oak trees in the extensive pig farming system in Corsica. Our aims were to understand (1) how breeders took into account the uncertainty on the availability of natural resources in this extensive system and (2) what is their point of view on the role of the local breed in such a situation. We conducted 28 interviews of breeders in all Corsica Island. We chose breeders of the local breed but also farmers breeding other breeds or crossbred animals. The results describe the territories available for the different breeders for the finishing process. They present the way the breeders anticipate the local resources availability and the way they deal with years of insufficient resources. When the quantity of chestnuts or acorns is insufficient, breeders chose either to act on the stocking rate, giving priority to animals to be slaughtered soon, or to feed them with specific prepared food or barley. A few breeders choose a trade-off strategy: they finish part of their animals feeding them, reserving chestnuts and acorns for the other part. Finally we present the point of view they have on the role of the breed in this management. Most of the Nustrale breeders insist on the fact that for them it is the only breed able to use the territory and take their feeding resources on it. But some breeders consider that it is the learning process of the young animals more than the breed that is important and some breeders underline the interest of crossbred animals (for instance with Duroc) as well (growing process etc.). To finish we discuss the fact that this uncertainty on resource availability and its management, throw the use of complement food, have to be formalized in the specification for the PDO application for Corsican cured pork meat and should at the consequence be better characterized.

Connectedness between five Piétrain herdbook populations

Gonzalez Lopez, V.[1], Hinrichs, D.[1], Bielfeldt, J.C.[2], Borchers, N.[3], Götz, K.-U.[4] and Thaller, G.[1], [1]Christian-Albrechts University Kiel, Institute of Animal Breeding and Husbrandy, Olshausenstraße 40, D-24098 Kiel, Germany, [2]Pig herdbook organization Schleswig-Holstein e.V., Rendsburger Str. 178, D-24537, Germany, [3]Chamber of Agriculture Schleswig-Holstein, Futterkamp, D-24327 Blekendorf, Germany, [4]Bavarian Institute of Agriculture, Institute of Animal Breeding, Prof.-Dürrwaechter-Platz 1, 85586 Poing-Grub, Germany; dhinrichs@tierzucht.uni-kiel.de

The aim of this study was to evaluate the genetic connectedness between the five most important Piétrain herdbook populations in Germany. The herdbook populations were located in Bavaria (n=8,211), Baden-Württemberg (n=6,694), Schleswig-Holstein (n=3,389), North Rhine-Westphalia (n=2,909) and Lower-Saxony (n=2,657). Animals born between 2006 and 2008 were selected as the reference population. To calculate the connectedness between the five herdbook populations all pedigrees were combined (n=23,201), which led to an improvement of pedigree quality in comparison with each herdbook population alone. To estimate genetic connectedness, different methods were used. Common boars between herdbook populations varied from 67 to 248. The genetic similarities among the five breeding populations ranged from 5.71 to 20.36% with a mean value of 11.17%. In all populations, at least one of the five ancestors with the largest marginal contributions to the reference population originated from another herdbook population. One of the most important ancestors was common for three populations. Additionally, the average genetic relationship coefficients between the five breeding organizations confirmed that the populations are genetically linked and ranged between 2.57% and 4.52%. Over the past 30 years, the contribution of foreign founders to the populations has increased as a consequence of exchange of genetic material. The various measures of connectedness indicated that there are sufficient genetic links between the five herdbook populations to enable an overall breeding value estimation system.

Comparison of four terminal hybrids in pig meat production

Sevón-Aimonen, M.-L., Voutila, L., Niemi, J. and Partanen, K., MTT Agrifood Research Finland, Biotechnology and Food Research, Animal Production Research, Economic Research, FI-31600 Jokioinen, Finland; marja-liisa.sevon-aimonen@mtt.fi

Performance, carcass and meat quality traits of four terminal hybrids were compared. Finnish Landrace x Large White crossbred sows were mated with Finnish Landrace (FL), Norwegian Landrace (NL), Norwegian Duroc X Norwegian Landrace (NDL) or Swedish Hampshire (SH) boars. The pigs were raised at MTT Agrifood Research Finland in Hyvinkää. Altogether 20 litters per breed combination were produced. Four piglets from each litter were group-reared from 26 kg to 116 kg live weight and examined in detail. Data were analysed with SAS program using the MIXED procedure. The applied model included hybrid, sex and hybrid * sex interaction as fixed effects and boar, sow, litter, and residual as random effects. Daily gain for FL, NL, NDL, and SH were 973, 1002, 997 and 1005 g/day, feed conversion ratio 2.56, 2.53, 2.53, 2.54 FU/ kg live weight gain (FU = 9.3 MJ NE), Hennessy meat percentage 59.8, 60.3, 59.0, 59.7% and dressing percentage 72.9, 72.7, 73.3, 73.0%, respectively. The effect of hybrid on daily gain, feed conversion ratio and dressing percentage were not statistically significant but in Hennessy meat percentage P-value was 0.055. Meat colour lightness (L*) in loins (*M. longissimus dorsi*) of FL, NL, NDL, and SH hybrids were 47.3, 46.1, 47.2, 46.2 and pH 5.51, 5.57, 5.55, 5.47, respectively. The L* values in topsides (*M. semimembranosus*) of FL, NL, NDL, and SH hybrids were 54.7, 56.2, 55.6, 54.3 and pH 5.53, 5.54, 5.54, 5.48, respectively (P>0.05). Differences in production cost per kg pig meat were not significant. Despite a few differences between terminal hybrids in single traits, none of the hybrids was transcendent as a whole. When deciding the use of different hybrids in practice, it should be taken into account that the observed differences between hybrids are due both additive differences and different stage of heterosis.

Estimation of muscle tissue percentage in the live pig using bioelectrical impedance analysis
Taylor, A.E.[1], Jagger, S.[2], Toplis, P.[3], Wellock, I.J.[3] and Miller, H.M.[1], [1]University of Leeds, Faculty of Biological Sciences, Leeds, LS2 9JT, United Kingdom, [2]ABN, Peterborough, PE2 6FL, United Kingdom, [3]Primary Diets, Melmerby, N.Yorkshire, HG4 5HP, United Kingdom; bsaet@leeds.ac.uk

Bioelectrical impedance analysis (BIA) is a non-invasive, relatively inexpensive and portable method that may be used to measure body composition. The aim of this study was to determine whether BIA, using surface electrodes, could be used as a method to assess muscle percentage in the live pig by comparing BIA predictions with those obtained from computed tomography (CT), a method increasingly used to validate new techniques in the live pig. A preliminary experiment was carried out using 16 pigs (Hampshire X (Large White X Landrace) at 7 and 15 weeks of age. A four terminal BioScan 920-2 (Maltron International Ltd) with surface electrodes was used. Twelve hours post BIA data collection, pigs were scanned using a CT scanner for determination of body composition (muscle %). Statistical analysis was carried out on the BIA data using multiple model fitting with simplification to create a minimal adequate model for predicting muscle tissue percentage R v.2.12.1. Data from 7 and 15 weeks of age were analysed together. A paired t-test was carried out to compare the difference between CT muscle tissue percentage and predicted BIA muscle tissue percentage. A Bland-Altman plot was used to assess agreement between the two methods and to detect any between-method bias. The minimal adequate model produced an R^2 of 0.814. A Paired t-test showed good agreement between the two methods (P>0.98). The Bland-Altman plot failed to detect any between-method bias ($R^2=0.0461$; P>0.5). This experiment indicates that there is potential to use BIA with surface electrodes as a method to measure muscle percentage in the live pig.

Immunogenic property of the Landrace pig selected for low MPS lesion and the Large White selected for high peripheral-blood immunity, and their crossbred
Suzuki, K., Goshima, M., Satoh, T., Hosaya, S. and Kitazawa, H., Graduate School of Agricultural Science, Tohoku University, 1-1 Tsutsumidori Amemiyamachi, Taihakuku, Sendai, Japan, 981-8555, Japan; k1suzuki@bios.tohoku.ac.jp

Selection for low mycoplasma pneumonia (MPS) lesion in Landrace pig during five generations was conducted at the Miyagi Prefecture Livestock Experiment Station. Nippon Meat Packers Inc. selected for peripheral-blood immunity of Large White pigs during five generations. This study investigates the immunogenic properties of both purebreds and crossbred pigs to assess the feasibility of breeding pigs for disease resistance. We compared phagocyte activity, saliva IgA, peripheral blood T cell subset, and intestine cytokine mRNA expression after vaccination of Landrace and Large White pigs for SRBC. The same treatment was done for crossbred pigs produced by mating a Large White sire with Landrace gilts. Landrace pigs have higher mRNA expression of INF-γ and IL-10 of the ileum Peyes' patch than a Large White pig has. However, a Large White has a higher peripheral blood CD4+ T cell and CE25/CD4+ T cell percentage, and IL-4 mRNA expression of ileum Eyer's patch. These observations suggest that Landrace pigs have higher cellular immunity and lower antibody-mediated immunity than Large White pigs have. Crossbred animals have lower saliva IgA, SRBC specific IgG and IgM and CD4+ T cell percentage than control LW crossbred animals. However, the CD8+ T cell percentage and peripheral-blood mRNA of IFN-γ of these crossbred animals are higher than those of control crossbred pigs. Results show that antibody-mediated immunity was suppressed and that cellular immunity became activated in these crossbred animals. Results also suggest that their immunogenic properties are influenced strongly by the immunogenic properties of Landrace pigs selected for MPS lesion.

Analyses of daily weight and feed intake curves for growing pigs by random regression
Wetten, M.[1], Ødegård, J.[2,3], Vangen, O.[2] and Meuwissen, T.H.E.[2], [1]Norsvin, P.O. Box 504, 2304 Hamar, Norway, [2]Norwegian University of Life Sciences, Department of Animal and Aquacultural Sciences, P.O. Box 5003, 1432 Ås, Norway, [3]Nofima, P.O. Box 5010, 1432 Ås, Norway; marte.wetten@norsvin.no

Norsvin has since 1991 used FIRE stations to register data on individual feed intake on boars on-test. In 2004 the FIRE stations were upgraded to also register individual data on daily growth. Annually, Norsvin now run 3,600 purebred boars through the FIRE system, followed by off-testing using CT. In this study, data on 1,476 animals of Norwegian Landrace and 1,300 animals of Norwegian Duroc were used to estimate covariance functions between feed intake and body weight on the boars from 54 to180 days of age. Random regressions on Legendre polynomials of age were used to describe genetic and permanent environmental curves in body weight (up to second order) and feed intake (up to first order) for both breeds. Heritabilities of body weight increased over time for Landrace (0.18-0.24), but were approximately constant over time for Duroc (0.33-0.35). Average heritabilities for feed intake were about the same in both breeds (0.09-0.11), and the estimates decreased over time, most pronounced in Duroc. A decrease in the genetic correlations between test days with increasing distance in time for Landrace indicated that different genes affect feed intake through the test period. The results for Duroc however indicate that daily feed intake is seemingly controlled by the same genetic factors throughout the test period. For body weight the genetic correlations between test days were in general high, and did not go below 0.8 for any of the two breeds in the study. Plots of genetic growth and feed intake curves revealed possibilities to change the shape of these curves by selection.

The congruence of observed and expectable live weights of gilts at the end of field-test in Croatia and Germany
Sviben, M.[1] and Gnjidić, P.[2], [1]Freelance consultant, Siget 22B, HR-10020 Zagreb, Croatia, [2]Biotim KG, Aleja Matice Hrvatske 37, HR-32270 Županja, Croatia; marijan.sviben@zg.t-com.hr

It was during the 60th Annual Meeting of EAAP in Barcelona 2009, when Peter Knapp estimated that after next five years none pig would be tested at the particular testing station in Europe but during the production process in the pig farms. It had been observed, however, that gilts differed with regard to their live weights and the age at the start of field-test. The durations of the test differ and consequently the ages and the live weights of gilts at the end of field-test are different. Croatian Agricultural Agency reported that in 2009 Large White gilts at the mean age of 216 days weighed 108 kg and Swedish Landrace gilts at 206 days of age averaged 101 kg. Expectable live weights were 110.1 kg of Large White gilts and 102.9 kg of Swedish Landrace gilts according to suitable trend of intensity of growth. Umbrella Association of German Pig Production published that in 2005 the average live weight of German Landrace gilts at the age of 198 days was 108 kg. Expectable value was 109.5 kg derived from the suitable equation. According to the trend of the intensity of growth expectable mean live weight of German Landrace gilts tested in Bavaria during 2009 at the age of 194 days was 106.3 kg. It was observed 104 kg. The average live weights of Large White and German Pietrain gilts at the end of field-test in Germany during 2005 were congruent to the values derived from suitable equation: 111 kg vs. 111.5 kg at 189 days; 129 vs. 126.5 kg at 206 days. The congruence of observed and expectable live weights of gilts at the end of field-test was registered too with the data on German Landrace B and Leicoma gilts tested during 2005 in Germany and suitable equation (German Landrace B 180 days 119 kg vs. 114 kg; Leicoma 187 days 117 kg vs. 121.9 days). Trends of the intensity of growth of pigs do the breeders able to evaluate any animal at any age.

Effect of weaning age on Iberian piglet's diarrhoea

González, F., Robledo, J., Andrada, J.A., Vargas, J.D. and Aparicio, M.A., School of Veterinay Science, University of Extremadura, Animal Production, University Avenue, 10071, Cáceres, Spain; fgonzalej@alumnos.unex.es

Weaning involves changes in piglet gut morphology. After weaning, piglets have a susceptibility to enteric pathogens that may cause diseases. The physical maturity of piglets varies according to their age, and determines the way in which they adapt to weaning changes. Therefore, the choice between late or early weaning becomes a defining factor in ensuring that animals' physiological balance and productivity levels are restored as soon as possible. In order to assess the influence of weaning age on susceptibility piglets' diarrhoea, a total of 360 Iberian piglets were studied. Piglets were chosen randomly during six different breeding seasons. Same number of weaned piglets was studied in each season (n=60). Animals were weaned at two different ages: at 28 (W28) and 42 (W42) days old. Three trials were conducted for each weaning age. All animals were weighed weekly from 21 to 63 days old, and food intake was controlled with the same frequency. Piglets' diarrhoea was recorded weekly. Data were analysed using Spss® and statistical significance were accepted at $P<0,05$. There were similar prevalence (Pw28 = 27,2%; Pw42 = 31,9%) and incidence (Iw28 = 22,5%; Iw42 = 19,8%) in both weaning age groups for the whole study period. Was also the case of mortality (Mw28 = 2,9%; Mw42 = 1,1%). Although Post-weaning diarrhoea is a multifactorial condition, in our case was characterised by proliferation of enterotoxigenic *E. coli.* with a low lethality (Lw28 = 0,11%; Lw42 = 0,18%). There were no differences in weight between two weaning groups at the end of the trial (W28 = 19.82±0.35 kg; W42 = 20.18±0.30 kg). However, piglets that were early weaned ate more food, but derived less benefit from it ($P<0.000$). Therefore, weaning age has similiar influence on piglets'diarrhoea, but has a greater influence on weaning production rates.

Influence of mixing strategy on post weaning performances and agonistic behaviour of piglets

Royer, E.[1], Ernandorena, V.[2], Le Floc'h, N.[3] and Courboulay, V.[4], [1]Ifip-Institut du porc, 34 bd de la Gare, 31500 Toulouse, France, [2]Ifip-Institut du porc, Les Cabrières, 12200 Villefranche de Rouergue, France, [3]INRA, UMR 1079, 35590 St-Gilles, France, [4]Ifip-Institut du porc, BP 35104, 35651 Le Rheu, France; eric.royer@ifip.asso.fr

Mixing litters at weaning is a common practice in order to reduce weight heterogeneity within the pens. A 40 days long experiment has been undertaken to investigate the effects of grouping strategy after weaning on growth performance and social behaviour of piglets. 360 male and female piglets were weaned at 28 days of age, and placed in 24 pens according to their weight and origin. Twelve pens contained 15 piglets coming from 12.7 origin litters per pen (1 or 2 piglets per litter; MIX) whereas 12 other pens had 15 piglets from 4.0 litters (2 to 5 piglets per litter; FAM). The average body weight was 7.9±1.4 kg and 7.9±1.3 kg for MIX and FAM treatments, respectively. The initial within-pen standard deviation of body weight was of 0.5±0.2 kg in MIX group and 1.2±0.1 kg in FAM group. Piglets were housed in fully slatted pens (4.4 m²) and received *ad libitum* a phase 1 diet up to 12 kg of body weight, then a phase 2 diet. FAM piglets had a higher feed intake (+5%, $P<0.05$) from day 1 to 20. Daily gain was thus higher during this period (+7%, $P<0.05$) and from day 21 to 40 (+6%, $P<0.05$). Accordingly, FAM piglets had a higher final body weight than MIX piglets (28.5 vs. 27.3 kg, $P<0.001$). In spite of the initial difference, within-pen weight heterogeneities were similar among treatments on days 20 and 40. Direct observations of agonistic behaviour were made on days 1, 2 and 8. The results showed a higher frequency of long fights for MIX piglets ($P<0.05$). Blood samples of 16 piglets per group were taken on days 1 and 8 for the determination of plasma concentrations of haptoglobin. This criterion was not influenced by the way of mixing the piglets. It can be concluded that pens should be made of piglets from a limited number of litters in order to reduce aggressions at weaning and improve post weaning performance.

The Hampshire RN-gene: association with production traits
Serenius, T.[1], Eriksson, M.[1], Lennartsson, H.J.[1] and Lundeheim, N.[2], [1]Nordic Genetics, Råby 2003, 24292 Hörby, Sweden, [2]Swedish Univ. of Agr. Sci., Box 7023, 75007 Uppsala, Sweden; timo.serenius@qgenetics.com

Within the Swedish pig breeding organization Nordic Genetics, the fattening pig is a 3-breed cross, offspring of an L*Y sow and a Hampshire boar. In the Nordic Genetics' Hampshire population, all AI boars are genotyped regarding the dominant 'RN-gene' (genotype RN-RN-, RN-rn+ or rn+rn+). Data from on-farm test and test station was analyzed to study the associations between the RN genotype of the sire and production traits of the offspring. The study is based on data on 12127 purebred Hampshire pigs, born between 2008 and 2010. The statistical analyses were performed using the DMU package. The statistical model included the fixed effects of sex, RN-genotype of the boar, birth parity number (1, 2, 3+), and herd-year-month combination. The random effects included in the statistical model were herd-year-month-testing pen and additive genetic animal effect. The results indicate that the RN- gene does not have any significant effect on production efficiency. The clearest effect was found on feed efficiency, where the progenies of RN-RN- boars consumed 5.27 kg less feed between 35 and 100 kg of live weight, compared with progenies of rn+rn+ boars. The results indicated also that killing out percentage was slightly lower (0.33%) for progenies after rn+rn+ boars than progenies of RN-RN- boars. As conclusion, the favorable effects of the RN-gene seems stronger than the unfavorable effects of the gene. More research is needed to investigate the meat quality aspects of RN- gene.

Influence of dietary betaine supplementation on chemical composition, meat quality and oxidative status of pork
Bahelka, I., Peškovičová, D. and Polák, P., Animal Research Production Centre, Department of Animal Breeding and Products Quality, Hlohovecká 2, 951 41 Lužianky, Slovakia (Slovak Republic); peskovic@cvzv.sk

Sixty commercially produced pigs Lx(HAxPN) were involved in the trial. Experimental group (n=30) was fed standard diet supplemented by 1.25 g/kg betaine for 30 days prior to slaughter. Control pigs (n=30) received diet without betaine supplementation. Animals were slaughtered at 110±5 kg live weight. Drip loss, pH, meat colour and shear force in musculus longissimus dorsi forty-five minutes and twenty-four hours after slaughter were evaluated. TBARS values expressing incubation of muscle homogenate with Fe^{2+}/ascorbate for oxidative status of pork were determined. The results did not show any significant effect of betaine supplementation on chemical composition, quality traits and oxidative stability of pork. The tendency of decreasing drip loss and shear force was just suggested. It seems to be necessary to increase the dosage of betaine or to enlarge time of supplementation to achieve of possible positive influence of betaine.

Comparison of meat and lipid quality in different muscles of hybrid pigs

Razmaitė, V. and Šveistienė, R., Lithuanian University of Health Sciences/Lithuanian Endangered Farm Animal Breeders Association, Institute of Animal Science, R. Žebenkos 12, 82317 Baisogala, Radviliškis district, Lithuania; razmusv8@gmail.com

The aim of the study was to examine the effect of genotype (terminal breed), muscle and gender on the meat quality of hybrids from Lithuanian indigenous wattle pig and wild boar backcross to lean pig breeds. The animals used were females and castrated males from domestic Lithuanian indigenous wattle and wild boar hybrid backcross to domestic lean breeds (Norwegian Landrace and Yorkshire). The hybrids were reared indoors and slaughtered when they reached approximately 90 kg weight. The chemical composition of meat and technological qualities were evaluated in M. longissimus dorsi and M. semimembranosus. The FAMEs were analysed using a gas liquid chromatograph GC – 2010 SHIMADZU. The relative proportion of each fatty acid was expressed as the relative percentage of the sum of the total fatty acids. Statistical analysis was performed with the general linear model (GLM) procedure in Minitab. Terminal breed influenced growth rate and carcass fatness. The hybrids from terminal Norwegian Landrace breed grew faster and had carcasses with lower fat than the hybrids from terminal Yorkshire breed. However, Norwegian Landrace as a terminal breed showed a tendency towards lower meat quality traits, such as pH value in the longissimus and semimembranosus muscles and an increased thawing loss in the semimembranosus muscle. Gender showed higher effects on the chemical composition of the muscles. The main differences between the muscle type were related to differences in water holding capacity, colour and pH (higher in the semimembranosus muscle) and cooking loss (higher in the longissimus muscle). The effect of terminal on fatty acid composition was lower than that of gender. The semimembranosus muscle showed higher PUFA content and more desirable PUFA/SFA and n-6 PUFA/n-3 PUFA ratios compared to longissimus muscle.

Role of lipogenic enzyme expression in breed-specific fatty acid composition in pigs

Marriott, D.T.[1], Chevillon, P.[2] and Doran, O.[1], [1]University of the West of England, Coldharbour Lane, Bristol, BS16 1QS, United Kingdom, [2]IFIP Insitute for Pig Production, Cedex, 75595, France; duncan.marriott@uwe.ac.uk

There is growing evidence that mechanisms regulating fat partitioning in pigs are breed-specific. We have previously demonstrated that the lipogenic enzyme, stearoyl-CoA desaturase (SCD) plays a key role in intramuscular fat formation in commercial breeds and in subcutaneous fat (SF) deposition in traditional breeds. The present study investigated further the role of SCD as well as two other key lipogenic enzymes, Δ6-desaturase (D6d) and fatty acid synthase (FAS) in breed-specific SF deposition. The study was conducted on three genetically diverse commercial breeds: Pietrain, Duroc X Pietrain and Large White X Pietrain. SCD, D6d and FAS protein expression was analysed in isolated microsomes and cytosol by Western blotting. Fatty acid composition was determined by gas chromatography. It was demonstrated that FAS expression was significantly higher in the Large White X Pietrain breed when compared with Duroc X Pietrain and was positively related to the product of the FAS-catalysed reaction, saturated fatty acids (SFA), in this breed. D6D protein expression was significantly higher in Large White X Pietrain animals when compared to the other two breeds and no breed-differences were observed in SCD expression. D6D and SCD expression did not correlate with changes in polyunsaturated fatty acids (PUFA) and monounsaturated fatty acids (MUFA) respectively. It was concluded that: (1) increased level of SFA in the Large White X Pietrain breed (but not in the other two breeds) is related to enhanced expression of FAS protein and (2) between-breed differences in MUFA and PUFA content cannot be explained by variations in SCD and D6D protein expression. It has been now investigated whether the breed-specific differences in MUFA and PUFA content are related to breed specific polymorphisms in the genes encoding for the enzymes catalysing biosynthesis of these fatty acids.

High fibre concentrates for Jersey cows grazing kikuyu/ryegrass pasture

Meeske, R., Van Der Merwe, G.D. and Cronje, P., Outeniqua Research Farm, Western Cape Department of Agriculture, P.O. Box 249, George, 6530, South Africa; robinm@elsenburg.com

The aim of the study was to determine the effect of replacing maize and soybean oilcake with hemi-cellulose rich by-products like hominy chop, gluten 20 and wheat bran in the concentrate fed to Jersey cows grazing high quality ryegrass pasture from September to October. Three concentrates were formulated to contain a high (80.4%), medium (40.7%) and low (20.7%) maize grain content. Maize grain was replaced by hominychop at 25 and 35%, wheat bran at 11 and 18% and gluten 20 at 11 and 18% in the medium and low maize grain treatments. As by-products replaced maize in the concentrate the starch content decreased from 57% to 36% and the hemicellulose content increased from 6% to 18%. Forty five Jersey cows were randomly allocated to treatments resulting in 15 cows/treatment. Cows were fed 6 kg as is, of dairy concentrate per day (3 kg at each of two milkings). Milk production was recorded daily and milk composition every 14 days. Cows grazed as one group on ryegrass (cv Energa) over-sown into kikuyu during March 2008) with a 28 day grazing cycle from September to October. Cows were weighed and condition scored (1-5 scale) on two consecutive days at the start and end of the experimental period. Results showed that 4% fat corrected milk production was higher (21.3 kg/day) on the low maize concentrate compared to the high maize concentrate treatment (19.9 kg/day). The milk fat content was 3.66[b], 4.03[ab] and 4.41[a]% for the high, medium and low maize concentrate treatment. Live weight and condition score change of cows did not differ between treatments. It is concluded that lowering the starch content and increasing the hemicellulose content of a dairy concentrate by replacing 75% of maize grain with hominy chop, wheat bran and gluten 20 increased 4% fat corrected milk production and milk fat content.

Comparison among soybean seeds and rapeseed cake included in diets with two different protein levels using gas production technique

Guadagnin, M.[1], Cattani, M.[1], Bondesan, V.[2], Bailoni, L.[1] and Tagliapietra, F.[1], [1]University of Padova, Animal Science Department, Viale dell universita 16 Legnaro Padova, 35020, Italy, [2]Veneto Agricoltura, Viale dell universita 14 Legnaro Padova, 35020, Italy; matteo.guadagnin@unipd.it

Aim of the present trial was to evaluate the effect of inclusion of rapeseed cake, produced by 'on farm' extraction (20% lipids, 29% CP of DM), on *in vitro* fermentation using gas production (GP) technique. Three experimental diets, with a low CP level (L; 11%), were formulated using three different feed sources (soybean meal, L_SBM, 3.4% lipids; soybean seeds, L_SBS, 4.6% lipids; rapeseed cake, L_RSC, 4.5% lipids). Two diets, with a high CP level (H, 15%), were also formulated using SBS (H_SBS, 5.8% lipids), and RSC (H_RSC, 5.7% lipids). Five experimental diets in 4 replications and 4 blanks were incubated in 2 set of bottles (48 bottles). A set of bottles was incubated for 72 h to evaluate GP kinetics using a fully automatic system (recording frequency 1 min). A second set of bottles was incubated to estimate NDF (NDFd), true DM (TDMd) degradabilities, and ammonia content at 16 h of incubation (the time at which half of total GP was produced). Each bottle was filled with 0.5 g of diet, 10 ml of rumen fluid, and 65 ml of buffer. Data were submitted to ANOVA using the diet as single source of variation and orthogonal contrasts were estimated. Both NDFd and TDMd values were greater ($P<0.05$) for RSC diets compared to SBS ones and were lower ($P=0.04$) for L compared to H diets. Compared to SBS, RSC inclusion in the diets increased the rate of GP ($P<0.05$; ml/h/g DM) but did not affect the total amount of GP. The reduction of CP level in the diets from 15 to 11% decreased the rate of GP without effects on total GP. Ammonia content increased ($P<0.01$) with the level of dietary CP, mainly in H_RSC diet. In conclusion, when diets with low CP levels are used, the inclusion of rapeseed cake in replacement to soybean seeds could improve the rate of degradation during the first hours of fermentation.

Effect of a multiple enzyme composition on the performance and nutrient digestibility in broilers
Martinsen, T.S., Nuyens, F. and Soares, N., Kemin Agrifoods EMEA, Toekomstlaan 42, 2200 Herentals, Belgium; filip.nuyens@kemin.com

The aim of this trial was to evaluate the effect of an enzyme composition containing xylanase, beta-glucanase, cellulase, protease and alpha-amylase (KEMZYME® Plus Dry) on the performance parameters of broiler chickens. The trial was performed at the Poznan University of Life Sciences in Poland. Male Ross 308 broilers were raised for 42 days in floor pens with 10 pens per treatment and 8 birds per pen. A negative control group was fed a standard wheat-maize diet without supplementation of enzymes (T1). The experimental groups included a group fed the standard diet supplemented with KEMZYME Plus Dry at an inclusion rate of 250 ppm (T2) or 500 ppm (T3). A second control group with reduced AME_n (-150 kcal/kg), lysine and methionine (-3%) was used without enzyme addition (T4) or with supplementation of 500 ppm KEMZYME Plus Dry (T5). The body weight gain of chickens fed the diet supplemented with KEMZYME Plus Dry at 500 ppm (T3) was increased significantly compared to the control group (T1) (P<0.05) over the entire trial period. No significant differences in feed intake could be observed between the dietary treatments. When the reduced energy diet was supplemented with 500 ppm KEMZYME Plus Dry (T5), the feed conversion ratio was significantly improved compared to the control treatment (T4). Nutrient digestibilities (dry matter, crude protein, fat, neutral detergent fiber, AME_n) were assessed during the trial by use of an indigestible TiO_2 tracer in the feed. Use of 500 ppm KEMZYME Plus Dry (T1, T5) improved NDF digestibility (P<0.05) compared to the standard diet (T1) and the reduced energy diet (T4), respectively. Fat digestibilities were significantly improved in all diets supplemented with enzymes. The AME_n values of the standard diet were significantly improved with 99 kcal/kg and 137 kcal/kg for the application of KEMZYME Plus Dry at 250 ppm and 500 ppm, respectively, and with 105 kcal/kg for the reduced energy diet that included 500 ppm of the enzyme composition.

The effects of stage of lactation on milk composition in Thoroughbred mares
Younge, B., Arkins, S. and Saranée Fietz, J., University of Limerick, Life Sciences, College of Science and Engineering, University of Limerick, Limerick, Ireland; Bridget.Younge@ul.ie

This study examined the stage of lactation effects on key nutrient components in mare's milk over a 90 day lactation period. Milk samples were taken from 10 thoroughbred mares on days 1, 3, 7, 21, 42, 56, 70 and 90 of lactation. Age and parity of the mares were recorded. Fat, protein and lactose were measured using a milkoscan and amino acids and fatty acids were measured using HPLC. Data was analysed using nonparametric analysis in SPSS. The % protein recorded on day 1 (13.0±1.17%) was significantly higher compared to all other days of lactation (P<0.05). The % protein on day 3 (3.2±0.18%) was significantly higher compared to day 7 (2.9±0.12%), day 21 (2.36±0.05%) and day 90 (1.83±0.06%) (P<0.01). The total essential amino acid (EAA) concentration was higher than the nonessential amino acid (NEAA) concentration on days 1, 42 and 90. The EAA content decreased from 48.8% of the total protein on day 1 to 37.4% on day 90. The NEAA content decreased from 54.7% on day 1 to 43.7% on day 90. Methionine expressed as a proportion of the total protein concentration was significantly higher on day 3 (2.84±0.37 g/100 g protein) and day 90 (2.33±0.07 g/100 g protein) compared to day 1 (1.66±0.04 g/100 g protein), (P<0.01). Threonine expressed as a proportion of the total protein concentration was significantly higher on day 1 (6.71±0.28 /100 g protein) compared to any of the other days of lactation. The content of free amino acids was also measured. The maximum amount of fat was observed on day 3 (1.97±0.28%) and decreased linearly until day 56 (1.25±0.12%), (P<0.05). The % lactose was significantly lower on day 1 (3.59±0.26%) compared to all other days of lactation. The means % lactose on day 3 (5.87±0.09%) was significantly lower compared to day 90 (6.64±0.03%) but significantly higher compared to day 1 (3.59±0.26%), (P<0.01). This study provides baseline data on the compositional changes of mare's milk from early to late lactation.

The effect of bioconversion on the fecal amino acid digestibility of corn DDGS in roosters

Tossenberger, J.[1], Szabó, C.[1], Bata, Á.[2], Kutasi, J.[2], Halas, V.[1], Tóthi, R.[1] and Babinszky, L.[3], [1]Kaposvár University, Guba S. út 40., H7400 Kaposvár, Hungary, [2]Dr Bata Corp., Pesti úti major, H2364 Ócsa, Hungary, [3]University of Debrecen, Böszörményi út 138, H4030 Debrecen, Hungary; szabo.csaba@ke.hu

Research results demonstrated that corn dried distillers grain with solubles (DDGS) can be incorporated into poultry diets up to 15%. However variable composition and poor protein quality may ultimately limit its use in poultry diets. Therefore, the aims of our study were to investigate that how affects corn DDGS bioconversion (Solid-state fermentation, where the media included agroindustrial lignocellulose material and DDGS, and was inoculated with Thermomyces. lanuginosus) the nutrient composition and the fecal digestibility of amino acids in poultry. The untreated and bioconverted DDGS contained the following amount (g/kg DM) of crude protein, Lys, Met, Cys, Thr, Arg and Ile: 266, 4.7, 5.2, 4.5, 10.0, 9.7, 10.0 and 243, 5.6, 4.9, 4.1, 14.3, 9.0, 9.1, respectively. The digestibility trial was carried out with 64 colon fistulated adult rooster (eight per treatment). The corn-soybean meal based basal diet were supplemented with 0, 5, 10 and 15% untreated or bioconverted DDGS. The collected fecal samples were lyophilized and analyzed for amino acid content. The Lys, Met, Cys, Thr, Arg and Ile digestibility in untreated DDGS were 63.7, 85.1, 75.8, 78.1, 86.7 and 81.5%, while the same values for bioconverted DDGS were 68.1, 86.7, 77.7, 79.8, 88.1 and 81.9%, respectively. The bioconversion improved Thr and Lys content and the digestibility of lysine. In conclusion, bioconversion can improve the protein quality of corn DDGS. The research project was supported by the Hungarian National Innovation Office (TECH-BIOKONV9).

Ochratoxin in poultry feed samples

Martins, H.[1], Almeida, I.[2], Guerra, M.M.[3,4], Costa, J.[2] and Bernardo, F.[3], [1]INRB, I.P.- Laboratório Nacional de Investigação Veterinária, Serviço de Micologia, Estrada de Benfica 701, 1549-011 Lisboa, Portugal, [2]Direcção Geral de Veterinária, Direcção de Serviços de Medicamentos e Produtos de Uso Veterinário, L. Ac. Nacional Belas Artes, 2, 1249-105 Lisboa, Portugal, [3]CIISA- Faculdade de Medicina Veterinária, Pólo Universitário da Ajuda – Rua Professor Cid dos Santos, 1300-147 Lisboa, Portugal, [4]Escola Superior de Hotelaria e Turismo do Estoril, Avenida Condes de Barcelona, 2769-510, Estoril, Portugal; manuela.guerra@eshte.pt

Ochratoxin A (OTA) is a naturally occurring mycotoxin produced by Aspergillus ocraceus, mainly in tropical regions, and by Penicillum verrucosum mainly in temperate areas. Poultry are the most exposed animals to the toxicity of OTA, which is considered to be the most important of the ochratoxins. Ochratoxicosis affects poultry industry by decreasing the growth rate and worsening both feed conversion and reproductive efficiency resulting in evidenced decrease in farm productive performance. In hens, decrease egg production is observed, with poor egg shell quality, higher incidence of eggs with blood spots and reduced embryo viability and decreased hatchability. Hence, the presence of OTA in poultry feed is associated to significant economic losses for the producers. To prevent these losses, the E.U. introduced regulations about ochratoxin's limits. Therefore to ensure that mycotoxins don't enter in the feed and afterwards, in the food chain, feed must be monitorized and quality control procedures must be introduced. The aim of this study was to determine the occurrence of OTA in 111 samples of laying hens feed samples randomly collected during 2008-2009 from feed factories in Portugal. The mycotoxin was determined using high liquid chromatography (HPLC). Detection limit of the method was 2.0 µg/kg. The results of this study revealed a low frequency (5.4%), and low levels of OTA, which varied from 2.0 µg/kg to 10.0 µg/kg, values that are bellow the E.U. recommended guidance values.

Effect incremental levels of zado on cow milk production and and rumen parameters
Gado, H.[1] and Borhami, B.E.[2], [1]Faculty of Agriculture, Ain Shams University, Animal Production, Shoubra Al-Kheima, Cairo, 11241, Egypt, [2]Faculty of Agriculture (El-Shatby), Alexandria University, Animal Production, Alexandria, 11303, Egypt; gado@link.net

An experiment was carried out to determine milk production response of Holstein dairy cows to incremental levels of ZADO® (it is a patent product contains cellulases, xylanases, protease and alpha amylase) in the diet. The experiment was conducted in June and July of 2010. A complete randomize block design was set up and a repeated measurement in time model used to analyze the data. Forty eight dairy cows were blocked by parity, days in milk, milk production and body weight randomly allocated to one of the following treatments: T0 = control diet, T1 = 40 g of zado/cow/day and T2 = 60 gm of zado/cow/day. Control diet was based on alfalfa plus 5 kg DM of a commercial concentrate (CP: 17.8%, ADF: 17.8%, NDF: 25% and ASH: 7.1%) offered half at each milking. ZADO (cellulose, xylanase, protease, & alpha amylase) was mixed with the concentrate just before each milking session. Cows body weight and milk production at the beginning of the experiment were 574±42, 565±51, 560±40 and 24.2±2.9, 24.6±3.3, 25±2.3, for T0, T1 and T2, respectively. Cows fed ZADO (T1 and T2) produced 1.5 and 2.3 L extra milk, respectively than control cows (T0=24.6 L; $P<0.05$). Milk production between ZADO fed cows were not significantly different (26.1 vs. 26.9 for T1 and T2, respectively. Supplementation of enzymes also increased ($P<0.05$) rumen ammonia N in comparison to control group. An increase in acetate and propionate at 4 h post-feeding for T1 –T2 (62.0-65.0 and 17.9.3-19.8.8 mol/100 mol respectively; $P<0.05$)in comparison to control. In conclusion the 40 g ZADO presented the proper inclusion rate of ZADO.

Influence of excess lysine and methionine on cholesterol, fat and performance of broiler chicks
Bouyeh, M.[1] and Gevorgian, O.X.[2], [1]Islamic Azad University, Animal science, Islamic Azad University Rasht branch, Rasht, 4193963115, Iran, [2]State Agrarian University of Armenia, Nutrition, Yerevan,Armenia, 3468574 Yerevan, Armenia; mbouyeh@gmail.com

This study was carried out to determine the effects of excess dietary Lysine (Lys) and Methionine (Met) on some important blood parameters, cholesterol and fat content of carcass and the performance of broiler chicks. In a completely randomized design three hundred male Ross 308 broilers were allotted to five group, each of which included four replicates (15 birds per replicate). The groups received the same basal diet supplemented with Lys and Met in 0, 10, 20, 30 or 40% more than NRC recommendation. The collected data were analyzed by SPSS software and Duncan's test was used to compare the means on a value of $P<0.05$. The results indicated that the two highest levels of Lys and Met treatments (30 and 40% higher than NRC recommendation) led to significant increase in carcass efficiency, breast muscle yield, heart and liver weight and also plasma cholesterol ($P<0.05$), whereas feed conversion ratio (FCR), crude fat contents of breast and thigh muscles and plasma triglyceride were lowest in these two treatment groups ($P<0.05$). Addition Lys 40% higher than NRC tended decrease in body weight gain but there was no significant effect of treatment on cholesterol content of the breast and thigh muscle and also thigh and leg yield. The finding of this experiment showed that increasing Lys and Met to diets of today's broiler in excess of NRC and even Ross Broiler Nutrition recommendations can improve FCR, abdominal fat deposition, breast meat yield, fat content of the breast and thigh muscle and carcass efficiency.

Partially replacement of wheat factory sewage for barley grain on performance of Holstein steers
Kamaleyan, E.[1], Khorvash, M.[2], Ghoorchi, T.[1], Kargar, S.[2], Ghorbani, G.R.[2], Zerehdaran, S.[1] and Boroumand-Jazi, M.[3], [1]Department of Animal Sciences, Gorgan University of Agricultural Science and Natural Resources, Gorgan, 49138 – 15739, Iran, [2]Department of Animal Sciences, Isfahan University of Technology, Isfahan, 84156 – 83111, Iran, [3]Jahad-Agriculture Institute of Scientific-Applied Higher Education, Isfahan, 81739 – 73161, Iran; boroumand1345@yahoo.com

The objective of this study was to evaluate the effects of partial replacement of barley grain with wheat factory sewage (WFS; a by-product of starch and wheat gluten meal producing factories) in the diet of steers on DMI, body weight changes, gain, and feed efficiency. Fifteen steers (416±35 kg initial weight) were used in a replicated 3×3 Latin square designed experiment with three 22-d periods. Each period had 16 d of adaptation and 6 d of sampling. The basal diet (WFS0) was formulated with 40% forage, 0% WFS and 42.5% (DM basis) barley grain, which was partially replaced with 10% WFS (WFS23.5) or 20% WFS (WFS47). Data were analyzed using the MIXED model procedure of SAS. The effect of increasing levels of WFS in the diet was examined through linear and quadratic orthogonal contrasts using the CONTRAST statement of SAS. DMI linearly decreased with increasing the WFS in the diets (10.01, 9.23, 9.17 kg; P<0.007). Body weight changes was quadratically changed (P<0.002) to be higher for WFS47 (30.9 kg/period) than for other two diets which was lower in WFS23.5 compared with WFS0 (30.2 vs. 30.6 kg/period). Average daily gain showed a quadratic response to adding WFS (1.39, 1.36, 1.38 kg; P<0.01). Furthermore, feed efficiency linearly decreased with increasing amount of WFS in diets (7.12, 7.09, 6.50 kg; P<0.01). Results indicated that partially replacing 23.5 and 47% of barley grain with WFS in steers diet decreased DM content and reduced DMI, but improved feed efficiency. It suggests that WFS can be used as alternative to grain to feed steers.

Effect of feeding wheat factory sewage for barley grain on feed intake, feeding and chewing behavior of Holstein steers
Kamaleyan, E.[1], Ghoorchi, T.[1], Khorvash, M.[2], Kargar, S.[2], Ghorbani, G.R.[2], Zerehdaran, S.[1] and Boroumand-Jazi, M.[3], [1]Department of Animal Sciences, Gorgan University of Agricultural Science and Natural Resources, Gorgan, 49138 – 15739, Iran, [2]Department of Animal Sciences, Isfahan University of Technology, Isfahan, 84156 – 83111, Iran, [3]Jahad-Agriculture Institute of Scientific-Applied Higher Education, Isfahan, 81739 – 73161, Iran; boroumand1345@yahoo.com

Feed intake and feeding behavior of steers fed diets that varied in wheat factory sewage (WFS; a by-product of starch and wheat gluten meal producing factories) content were investigated. Fifteen steers (416 kg) were used in a replicated 3×3 Latin square designed experiment with three 22-d periods. Each period had 16 d of adaptation and 6 d of sampling. The basal diet (WFS0) was formulated with 40% forage, 0% WFS and 42.5% (DM basis) barley grain, which was partially replaced with 10% WFS (WFS23.5) or 20% WFS (WFS47). Data were analyzed using the MIXED model procedure of SAS. The effect of increasing levels of WFS in the diet was examined through linear and quadratic orthogonal contrasts using the CONTRAST statement of SAS. DMI showed a linear decrease to adding WFS (10.01, 9.23, 9.17; P<0.007). Meal patterns, including number of bouts per day, meal interval, and eating rate were not changed among the treatments, except for the meal length (12.7, 11.3, 10.9 min/meal; P<0.04) and meal size (0.65, 0.58, 0.54 kg of DM; P<0.03) that showed a linear decrease to feeding WFS. Time spent eating and ruminating and thereby total chewing time, expressed as minutes per unit of DMI was not affected by feeding WFS, however time spent for eating and ruminating, expressed as minutes per day tended to show a quadratic (P<0.09) response and linearly decreased (P<0.008) by adding WFS, respectively. Results indicate that WFS supplementation for barley grain lowered meal length and meal size and had minimal effects on rumination pattern and chewing time under current feeding situations.

Effects of dietary DL-methionine substitution with piridoxine on performance of broilers

Karimi, K.[1], Javaheri Barforoushi, A.[2], Shokrolahi, B.[3] and Irani, M.[2], [1]Animal Science Department, School of Agriculture, Islamic Azad University, Varamin-Pishva Branch, Varamin, Tehran, Iran, [2]Animal Science Department, School of Agriculture, Islamic Azad University, Ghaemshahr Branch, Ghaemshahr, Iran, [3]Animal Science Department, School of Agriculture, Islamic Azad University, Sanandaj Branch, Sanandaj, Kordestan, Iran; Dr.karimi_kazem@iauvaramin.ac.ir

Transamination activities of DL-methionine are similar to pyridoxine. To investigate the possibility of dietary DL-methionine substitution with pyridoxine an experiment was carried out with three hundred day-old broiler chickens which allocated into 5 groups (M_c, $M_{0.95c}$, $M_{0.90c}$, $M_{0.85c}$ and $M_{0.80c}$) with 4 replicate pens of 15 chicks each one. Control group (M_c) was fed a complete feed mixture with sufficient dietary methionine and without pyridoxine supplement. Other groups were fed the same diet except that their methionine levels were 0. 95, 0.90, 0.85 and 0.80% of M_c group respectively. The diets methionine deficiencies substituted by addition of $20_{mg/kg/\,day}$ pyridoxine to them. The treatments were carried out for 42 days. Feed intake (FI), weight gain (WG), feed conversion ratio (FCR) were determined at 0-2, 2-4, 4-6 and 0-6 weeks of experimental periods. Results indicated that only WG at 2-4w, 4-6w and 0-6w affected by the treatments (P<0.05). $M_{0.85c}$ and $M_{0.80c}$ treatments decreased BW compared to other treatments at 2-4w, 4-6w and 2-6w. Treatments haven't any significant effects on FI and FCR at any experimental periods (P>0.05). Thus we concluded that pyridoxine such as DL-methionine can act as a methyl donor and at most 10% of dietary DL-methionine can be substituted with 20 mg/chick/day pyridoxine without any harmful effects on the broiler performance.

Effects of combination of citric acid and microbial phytase on the serum concentration and digestibility of some minerals in broiler chicks

Ebrahim Nezhad, Y., Maheri Sisi, N., Aghajanzadeh Golshani, A. and Darvishi, A., Islamic Azad University-Shabestar Branch, Animal Science, Faculty Agriculture, Shabestar, East Azerbaijan, 5381637181, Iran; ebrahimnezhad@gmail.com

This experiment was conducted to evaluate the combined effects of citric acid (CA) and microbial phytase (MP) on the serum concentration and digestibility of some minerals in broiler chicks. This experiment was conducted using 360 Ross-308 male broiler chicks in a completely randomized design with a 3×2 factorial arrangement (0, 2.5 and 5% CA and 0 and 500 FTU MP). Four replicate of 15 chicks per each were fed dietary treatments including (1) P-deficient basal diet [0.2% available phosphorus (aP)] (NC); (2) NC + 500 FTU MP per kilogram of diet; (3) NC + 2.5% CA per kilogram of diet; (4) NC + 2.5% CA + 500 FTU MP per kilogram of diet; (5) NC + 5% CA per kilogram; and (6) NC + 5% CA + 500 FTU MP per kilogram of diet. The concentration of zinc, copper and manganese of serum and their digestibility and also digestibility of apparent metabolizable energy (AMEn) was evaluated. The results showed that interaction effect of CA and MP on concentration of copper, zinc and manganese in serum of broilers fed with low available phosphorus diets was significant (P<0.05). Adding 2.5% CA into low available phosphorus diets increased digestibility of zinc in comparison with diets of without and 5% CA in broiler. Adding CA into low available phosphorus diets increased manganese digestibility on based corn-soybean meal diets (P<0.01).

The determination of metabolizable protein of grap pomace and raising vitis leaves

Moghaddam, M.[1], Taghizadeh, A.[2] and Ahmadi, A.[3], [1]Department of Animal Science, Maragheh, 5971654547, Iran, [2]Department of Animal Science, Tbriz, 5971654547, Iran, [3]Department of Animal Science, Maragheh, 5561644968, Iran; ataghius@yahoo.com

The present study was carried out to determine the metabolizable protein (MP) of grape pomace (GP) and raisin vitis leaves (RVL), using nylon bags technique. Three fistulaed whether with average BW 45±2 kg were used. The data was analyzed using completely randomized design. The incubation times were 0, 2, 4, 6, 8, 12, 16, 24, 36, 48, 72 and 96 h. The degradability parameters of crud protein (CP) for soluble fractions (a)were 5.34 and 6.48% and fermentable fractions (b) were 32.87 and 15.90% for RVL and GP, respectively. The MP of RVL and GP were obtained 126.62 and 150.00 gkg^{-1}DM, showing a significant difference between two treatments (P<0.05). The GP had high MP compared to RVL. The amount of total tannin (TT) and total phenol (TP) in RVL (3.54 and 4.59%) that was more than GP (1.98 and 2.71%) resulted low MP in RVL compared to GP. In general, the results showed GP and RVL can be used in ruminant nutrition.

The determination of metabolizable energy of grap pomace and raising vitis leaves

Moghaddam, M.[1], Taghizadeh, A.[2] and Ahmadi, A.[3], [1]Department of Animal Science, Maragheh, 5971654547, Iran, [2]Department of Animal Science, Tabriz, 5971654547, Iran, [3]Department of Animal Science, maragheh, 5561644968, Iran; ataghius@yahoo.com

The present study was carried out to determine fermentation characteristics, metabolizable energy (ME),digestible organic matter (DOM) and short chain fatty acids (SCFA) of grape pomace (GP) and raisin vitis leaves (RVL) using gas production technique.Three fistulaed wether with average BW 45±2 kg were used.The data was analyzed using completely randomized design. The incubation times were 0, 2, 4, 6, 8, 12, 16, 24, 36, 48, 72 and 96 h. The gas production of soluble and insoluble fractions (a+b) were 289.49 and 0.015 ml g^{-1}DMand the rate of gas production prices (c) were 249.93 and 0.024 (%/h) for of RVL and GP, respectively. The ME, DOM and SCFA of RVL of these test feeds were obtained 15.74 mjkg^{-1}DM, 96.97% and 1.87 mlmol, 13.63 mjkg^{-1}DM, 87.04% and 1.65 mlmol respectively, that showed significant differences (P<0.05). The amount of gas production in RVL (126.87 ml g^{-1}DM) that was more than GP (112.10 mlLitgr^{-1}) resulted low ME in GP compared to RVL. In general, the results showed GP and RVL can be used as an energy source in ruminant nutrition.

The determination of metabolizable protein of canola meal, canola meal treated with 0.5% urea and canola meal treated with macro wave

Tahmazi, Y.[1], Moghaddam, M.[2] and Taghizadeh, A.[3], [1]Animal Science of azad university, Maragheh, 5971654547, Iran, [2]Animal Science, Maragheh, 5971654547, Iran, [3]Animal Science, Tabriz, 5971654547, Iran; you.tahmazi@gmail.com

The present study was carried out to determine the metabolizable protein (MP) of treated and untreated canola meal, using nylon bags technique. Three fistulated whether with average BW 45±2 kg were used. The data was analyzed using completely randomized design. The experimental treatments were treatments A:canola meal, B: canola meal treated with 0.5% urea and C: canola meal treated with microwave. The incubation times were 0, 2, 4, 6, 8, 12, 16, 24, 36, 48, 72 and 96 h. The degradability parameters of crud protein (CP) for soluble fractions (a) were 4.74, 15.81 and 15% and fermentable fractions (b) were 31.05, 39.62 and 65.55% for treatments of A, B and C, respectively. The MP of treatments A, B and C were obtained 283.11, 329.33 and 284.39 gkg⁻¹DM, showing a significant difference between three treatments (P<0.05). The canola meal treated urea had high MP compared to others. It is concluded that the processing of canola meal with urea caused high MP, whereas processing by microwave had no effect in MP of canola meal.

Performance of broiler chickens fed starter diet containing *Aspergillus* meal (Fermacto)

Pahlevan Afshar, K.[1,2], Sadeghipanah, H.[3], Ebrahim Nezhad, Y.[4], Asadzadeh, N.[3] and Nasrollahi, M.[5], [1]Islamic Azad University - Abhar Branch, Department of Animal Science, Abhar 3146618361, Iran, [2]Islamic Azad University - Science and Research Branch, Department of Animal Science, Tehran 3146618361, Iran, [3]Animal Science Research Institute of Iran, Departemen of Biotechnology, Beheshti street - Karaj 3146618361, Iran, [4]Islamic Azad University - Shabestar Branch, Department of Animal Science, Tehran 3146618361, Iran, [5]Islamic Azad University - Tehran Jonoub Branch, Department of Statistics, Tehran 3146618361, Iran; kp_afshar@yahoo.com

In order to investigate the effect of Aspergillus sp. meal (Fermacto) as a prebiotic in starter diet on broiler performance, this experiment was carried out in a completely randomized design. 480 one-day chickens, Ross 308 strain, (average weight, 46 gr) were randomly divided into two groups: 1- control group fed diet without Fermacto, and 2- Experimental group fed diet with 0.2% of Fermacto. Each group was allocated to four pens (repetitions) and 60 chicks assigned to each pen. Feed intake, weight gain and feed conversion ratio (FCR) were weekly measured. Addition of 0.2% of Fermacto to starter diet did not have a significant effect on the feed intake, but, average daily weight gain tended (P<0.10) to increase in chicks fed Fermacto in comparison with control group during all of different 7, 14, and 21-day periods. In the 7 day-period, there was not a significant difference in FCR between two groups, but in 14 and 21 day-periods FCR in broilers fed Fermacto (respectively, 1.060 and 1.340) was significantly (P<0.05) better than control group (respectively, 1.076 and 1.434). Totally, these results indicated that the addition of Fermacto to starter diet can improve the performance of broiler chicks.

The effects of phenylpropanoid essential oil compounds on the fermentation characteristics of a dairy cow ration *in vitro*

Mufungwe, J. and Chikunya, S., Writtle College, Chelmsford, CM1 3RR, United Kingdom; sife.chikunya@writtle.ac.uk

The potential rumen modulating effects of essential oil compounds (EOCs) are now widely recognised. This study investigated three EOCs, namely, anethole (ANE), cinnamaldehyde (CIN) and eugenol (EUG) on the fermentation of a dairy cow total mixed ration (TMR). The EOCs were added to the TMR at either 250 or 500 mg/l and incubated using the *in vitro* gas production procedure. The treatments were: control (CON), ANE250, ANE500, CIN250, CIN500, EUG250 and EUG500. Measurements were taken at 3, 6, 12, 24, 48 and 72 hrs. The type and concentration of EOC significantly affected both the rate and total cumulative gas production at all incubation times ($P<0.05$). All EOCs inhibited gas production relative to the control ($P<0.05$). The inhibition was more pronounced at the higher concentrations of EOCs. The 96 hr gas production values were: 265, 242, 261, 239, 204, 230, and 239 (ml/g OM; sed=9.9) for CON, ANE250, CIN250, EUG250, ANE500, CIN500 and EUG500 respectively. Compared to the control, all three EOCs reduced ($P<001$) ammonia nitrogen (NH3-N) concentration between 48 and 96 hours of incubation, with significant interaction between EOC type and concentration of the EOC. The concentrations of NH3-N at 96 hrs were: 483, 398, 430, 437, 429, 387, and 401 (mg/l; sed=8.6) for CON, ANE250, CIN250, EUG250, ANE500, CIN500 and EUG500 respectively. There were generally no effects of either essential oil type or dose on trichloroacetic acid precipitable-nitrogen levels. The three essential oils reduced ($P<0.001$) neutral detergent fibre digestibility (NDFD) compared to the control. After 96 hrs, the mean NDFD values were: 0.69, 0.42, 0.58, 0.31, 0.32, 0.45, and 0.10 (sed=4.2) for CON, ANE250, CIN250, EUG250, ANE500, CIN500 and EUG500 respectively. The results indicate these phenylpropanoid EOCs used in this study can potentially to reduce ruminal N losses to the environment, but might also reduce fibre digestion. These effects vary with type of essential oil and the dosage used.

The 'breed': a management category with strong social dimension

Lauvie, A.[1] and Labatut, J.[2], [1]INRA LRDE, Quartier Grossetti, 20250 Corte, France, [2]INRA UMR AGIR, BP52627, 31326 Castanet Tolosan, France; lauvie@corte.inra.fr

Domestic animal populations are managed at a collective level for breeding purposes. The 'breed' is the main category chosen to collectively manage breeding activities thanks to genetics tools. We will show how this breed category is socially constructed, and how this social and organizational process has to be taken into account to understand the way livestock populations are managed and evolve from a genetic point of view. We give illustrations to show how the social dimension of the breed can have influences on their management from a genetics point of view. We illustrate this idea with two examples where the definition of the breed and its boundaries is discussed among the stakeholders. We show the possible consequences of these processes on breeding management The first example concerns the definition of the standard (phenotypic appearance) of the Manech Tete Noire sheep breed. Defining such a standard for the collective management of the breed is of the highest importance, despite the low economic value, as illustrated in this situation where several collective of stakeholders have difficulties to build a shared view of the standard. Consequently the official breeding scheme is weakened by other forms of collective management. The second example concerns the acceptance of crossbreed animals in the Rouge Flamande cattle breed. Animals from this breed have been crossbred to Danish Red animals in the past and there are still Danish Red genes in the breed. This has led in the past to a controversy with one of the conservation stakeholders who considered that conservation could only be of animal without Danish Red genes. This controversy has been closed by this stakeholder leaving the program. To conclude, we underline the importance of considering this social construction of the 'breed' category and we discuss the fact that the new genomics tools could have consequences on the boundaries of the breed as a management category for breeding programs, which might be put under question.

Collective system analysis: analysing and breaking through obstacles to reach an integral sustainable veal calve sector

Bremmer, B., Wageningen University & Research Centre, Livestock Research, Postbus 65, 8200 AB Lelystad, Netherlands; Bart.Bremmer@WUR.nl

It is expected by Dutch society and the government that the veal calve sector in The Netherlands can make a leap in sustainability. To investigate the possibilities for realising an integral sustainable veal calve production system a system analysis has been performed. However, this system analysis showed that several important stakeholders are stuck: six arguments kept coming back as a justification why the sector cannot change its behaviour on several sustainability aspects. A collective system analysis (CSA) was performed to trigger the participants to come loose from the commonly used arguments, in the first place to investigate to what extent this is possible and subsequently to analyse what changes this would bring. Three of these arguments were discussed: one about the farming system, one about the initial quality of the calve and one about the feeding strategy. After the discussions, the participants of the CSA indicated that on all three of these aspects it turned out to be relevant to come loose from the arguments. It helped them to investigate the advantages, disadvantages and solutions for potential problems. On the basis of the results it is concluded that there is space for change and performing a CSA creates a basis for creating more space.

How agronomists and farmers describe diversity: the example of mountain grasslands in France

Ingrand, S.[1], Baumont, R.[2], Rapey, H.[1], Souriat, M.[1], Farruggia, A.[2], Carrere, P.[3] and Guix, N.[4], [1]INRA, SAD, Umr 1273 Metafort, F-63122 St Genes Champanelle, France, [2]INRA, PHASE, URH Rapa, F-63122 St Genes Champanelle, France, [3]INRA, EFPA, UREP, F-63122 St Genes Champanelle, France, [4]VETAGROSUP, 89 Avenue Europe, 63370 Lempdes, France; stephane.ingrand@clermont.inra.fr

Criteria used to describe and analyse floristic diversity should be different according to the point of view of the different actors. The aim of this study, set up in upland area in France, is to connect the farmers point of view with the description of floristic and soil characteristics in fields by agronomists. From floristic measurements made on 120 permanent grassland plots in 19 farms, we analysed the intra and inter plot diversity with a vegetation functional approach taking into account the type of production system (dairy cattle only, beef cattle only, mixed dairy and beef cattle, mixed dairy cattle and sheep), the management and the altitude of the plot. An enquiry was made in the same farms, to assess the perception of the diversity of grasslands by the farmer himself: which criteria are mentioned to describe the diversity, how diversity is perceived both for the farm management and to decrease the sensitivity of the system face to risk (climatic events and rodents). The third category of data concerned the characteristics of soils, with samples made in different areas of the farms. Grasslands were classified in 3 groups of diversity according to their functional types. The diversity is higher above 1000 m, whereas at low altitude the plots are mainly in the low-diversity group. Compared to specialized dairy cattle systems, the mixed systems present a higher inter-plots diversity of grasslands, which can be valorised by the complementary of the animal needs of the different herds and animal categories. Farmers use 21 different criteria to describe their own grasslands, mainly according to structural constraints (distance, altitude, soil, etc.). We observed a good fit between the description of soils by farmers and the results obtained from our sample analyses.

An individual reproduction model sensitive to milk yield and body condition score in Holstein dairy cows

Brun-Lafleur, L.[1,2,3], Cutullic, E.[1,3,4], Faverdin, P.[1,3], Delaby, L.[1,3] and Disenhaus, C.[1,3], [1]INRA UMR1080 Production du Lait, Domaine de la Prise, 35590 Saint-Gilles, France, [2]Institut de l Elevage, Monvoisin, 35652 Le Rheu, France, [3]Agrocampus-Ouest UMR1080 Production du Lait, rue de Saint-Brieuc, 35000 Rennes, France, [4]Swiss College of Agriculture, Länggasse 85, 3052 Zollikofen, Switzerland; laure.brun-lafleur@inst-elevage.asso.fr

To simulate the consequences of management in dairy herds, the use of individual-based herd models is very useful and has become common. Reproduction is a key-driver of milk and meat production and herd dynamics, whose influence has been magnified by the decrease in reproductive performance over the last decades. Moreover, feeding management influence milk yield (MY) and body reserves, which in turn influence reproductive performance. Our objective was therefore to build an up-to-date biological reproduction model sensitive to both MY and body condition score (BCS). A dynamic and partly stochastic individual reproduction model was mainly built from data of a single recent long term experiment. This model covers the whole reproductive process and is composed of a succession of discrete stochastic events, mainly calving, ovulations, conception and possible embryonic loss. Each reproduction stage is sensitive to MY or BCS levels or changes. The model takes into account recent evolutions of reproductive performance, particularly concerning calving to first ovulation interval, cyclicity (cycle length, prolonged luteal phase), oestrous expression and fertility (conception, late embryonic loss). A sensitivity analysis of the model to MY, BCS at calving and fertility parameters was performed and the simulated performance was compared to observed data to validate the model. Globally, the model made possible to well represent reproductive performance of French Holstein cows, both in terms of reproduction delays (calving interval) and success rate (recalving rate). This model could be used in herd models to test the effects of management strategies on herd reproductive performance.

The impact of feeding strategies on greenhouse gas emissions from dairy production in Sweden

Henriksson, M.[1] and Swensson, C.[2], [1]Swedish University of Agricultural Sciences, Box 86, 230 53 Alnarp, Sweden, [2]Swedish Dairy Association, Ideon Science Park, 223 70 Lund, Sweden; maria.henriksson@slu.se

Feed intake and milk yield have large impact on milk carbon footprint (CF). Around 43% of estimated greenhouse gas (GHG) emissions/kg milk was due to feed cultivation (incl. production of N fertiliser and diesel), according to a previous study. Further studies of the farmers' choice of feeding strategies and their impact on milk CF are important in order to lower GHG emissions from milk production. This study aims to evaluate the impact of feeding plan and the use of different feed components on milk CF. One hypothesis is that milk from a dairy herd with a large share of regional grown feed has a lower output of GHG emissions than milk produced on imported protein feed. The newly developed feed evaluation tool called NorFor (used by the Nordic countries), were used by feed advisors to create a number of possible feeding plans for 5 different regions in Sweden (conditions for feed cultivation differs as dairy farms are located between the latitude of 55° and 68°). In each region, normal as well as improved qualities of grass or grass/clover silage were used to study the importance of high quality feed. Feeding plans were adjusted to 3 levels of milk yields and cover all replacement animals as well as conventional and organic production. Methane from enteric fermentation was calculated for each feeding strategy using different models. Estimates of the different feed components CF were based on a Swedish LCA-database. Amount of N fertiliser used in feed cultivation was adjusted to N in excreta (calculated by NorFor). Emission estimates of nitrous oxide from applied manure and excreta were also made. Results will be presented as CO_2-equivalents/kg milk from enteric methane, production of feed and manure management respectively. Correlations between enteric methane and the use of different feed components will be included as well as individual feed compounds impact on the total GHG emission estimates.

Prospects for bedded pack barns for dairy cattle

Galama, P.J. and Van Dooren, H.J., Wageningen UR, Livestock Research, Edelhertweg 15, 8219 PH Lelystad, Netherlands; paul.galama@wur.nl

Bedded pack barns is a housing system for dairy cattle without cubicles. Deep bedded pack systems like Compost Dairy Barns in the United States, loose housing (bedded pack with dried manure) in Israël and the experiences in the Netherlands with different bedding materials show that the sustainability of a dairy farm can increase. Bedded pack systems offer better cowcomfort and reduce the incidence of hock lesions than free stalls. Important for animal welfare and health is a dry bedding. Therefore a model study is done to study the drying potential of bedding materials (soft composting and non composting materials) under Dutch, Israeli and Minnnesotan climate conditions and at 18 versus 9 m^2 of bedding area per animal. An economic study shows the differences between the costs of free stall and bedded pack systems. The emission of ammonia and methane is measured on three bedding materials with a dynamic box on experimental farms of Wageningen UR Livestoch Research; sand, compost (wooden chips and sawdust) and clay soil with reed. Animal health and lying time is also measured. Results of emission (ammonia, methane) smell and fine dust will be shown of these experimental farms and also of four dairy farms in practice. The layout of these farms differ. Different ways of management of the bedding, feeding systems and roof will be shown. One farmer has an aerating system to stimulate the composting process of the bedding with wooden chips. A good composting process will keep the bedding dry and will influence the emissions. Small experiments are done to keep the bedding warm. Another farmer has no feed alley. He feeds the cows with mobile feeding troughs. The building is a greenhouse. Three farmers use dry compost from a compost factory to keep the bedding dry. The different bedded pack barns may be an economically feasible alternative to existing free stall housing with positive effects on cow comfort, health and welfare and the potential for environmental advantages and better manure.

The working time in milking farms in Wallonia

Turlot, A.[1], Cardoso, C.[1], Wavreille, J.[1], Bauraind, C.[2], Bouquiaux, J.M.[3], Ledur, A.[4], Mayeres, P.[5] and Froidmont, E.[1], [1]Walloon Agricultural Research Centre, Production and Sectors Department, rue de Liroux 8, 5030 Gembloux, Belgium, [2]Walloon Agricultural Research Centre, Filière Lait et Produits Laitiers Wallonne, rue de Liroux 8, 5030 Gembloux, Belgium, [3]Direction Générale Opérationnelle Agriculture, Ressources naturelles et Environnement, Direction de l'Analyse Economique Agricole, Avenue Prince de Liège, 15, 5000 Namur, Belgium, [4]FWA, Chée de Namur, 47, 5030 Gembloux, Belgium, [5]Walloon Breeding Association, rue des Champs Elysées 4, 5590 Ciney, Belgium; a.turlot@cra.wallonie.be

Today, the farm size continuously increasing in Wallonia (Belgium). The number of cattle or hectare per farm has increased respectively by 40% and 60% in 20 years. However, the number of worker in the farm does not change (1.5 per work unit) and now, more and more spouses have another employment. How to manage these structures? The project DuraLait performs an audit on the working time in different types of farms '100% milk' (95% of dairy cows and no other crops than forages). The ''Labour assessment' method (French method) was used to do a global approach of labour organization in fifty farms in Wallonia. The method investigated the time devoted to herd and land. Three types of works are distinguished: routine work, seasonal work (on herds and lands) and return work. Different parameters can influence the working time. A high level of automation (robot milking, automatic feeder, mixer operator...) can save 21 min/100 liters of milk or 1,400 hours of work per year for a farm with a milk quota of 400,000 liters. On the other hand, a high level of feed autonomy increases the working time (57 min/100 l vs 38 min/100 l). The size of the milk quota also has an impact on working time. A farmer, with a milk quota of 200,000 liters, spends 69 min/100 l while another farmer, with a milk quota of 700,000 liters, devotes only 40 min/100 l.

Maintenance and social behaviours profile during suckling period in donkey foals
D'alessandro, A.G.[1], Casamassima, D.[2], Palazzo, M.[2] and Martemucci, G.[1], [1]University of Bari, Scienze Agro-Ambientali e Territoriali, Via G. Amendola, 165/A, 70126 Bari, Italy, [2]University of Molise, S.A.V.A., Via De Santis, 86100, Italy; martem@agr.uniba.it

Behavioural studies on equids has arousing great interest to a meliorate welfare, training and husbandry system. A little focus has been addressed on behaviour of donkey foals. The study aimed to evaluate behavioural profiles of maintenance and social behaviours in Martina Franca breed donkey foals during the suckling period (1 week to 6 months after birth). The animals were reared in Southern Italy (40° 37' N latitude) and were managed in husbandry conditions, with free access to a natural scrub area and sown pasture all year round. A total of 10 foals with their mothers which had a similar foaling date were used. The behavioural observations in the foals were performed in presence of their mothers, for 120 daylight hours. A total of 4 main activities were recorded using focal animal sampling (5 min intervals), including maintenance behaviours such as suckling, walking, standing, resting, and a whole range of less conspicuous activities relative to eliminative and social behaviour such as olfactory communication, grooming, play and vocalisations. The results indicated that maintenance behaviours differed significantly as a function of the time since birth ($0.05 > P < 0.01$). The effect of a.m. or p.m. hours of day were also significant ($0.05 > P < 0.01$). All the activities included in maintenance behaviours were markedly influenced by the individual ($P < 0.01$). This study offers some information regarding quantitative behavioral profiles of donkey foals reared under domestic conditions during the suckling period.

Growth performance of rabbit broilers HYLA
Vostrý, L., Majzlik, I., Mach, K., Hofmanová, B., Janda, K. and Andrejsová, L., Czech University of Life Sciences, FAPPZ, Kamýcká 129, Prague 6, 165 21, Czech Republic; vostry@af.czu.cz

This work was aimed to evaluate the growth performance of two hybrid genotypes HYLA of rabbit broilers (genotype 1: ♂8,000 x ♀3,070, genotype 2: ♂3,160 x ♀3,070). Rabbits in both trials were fed *ad libitum* with commercial granulated feed mixture supplied by probioticum PROBIOSTAN (Czech product – 2 kg in 1000 kg mixture) and herbal coccidiostatic EMANOX (France product – 500 g in 1000 kg mixture). Generally: the testing period lasted 42-72 days of animals' life to reach the final live weight 2,600 g. The average daily gains were 44.64 g and 44 g resp.with a daily feed consumption 150.08 g and 144.8 g, which gives a feed conversion rate 3.43 and 3.36 resp. The final carcass weight reached 1554.8 g and 1561.2 g with dressing percentage 58.72% and 58.25% resp. The mortality rate in the first group was 2.85% and 10% of rabbits did not reached the weight 2,600 g in 84 days of their life, whereas in second group mortality was 6.15% and 4.6% of animals did not reached the 2,600 g in 84 days which was the end of the trial. The trials showed positive effects of both aditiva on mortality and health status of rabbits. This project was supported by grant MSM 604 607 09 01.

Technical efficiency in dairy cattle farms in La Pampa, Argentina
Angon, E.[1], Larrea, A.[2], Garcia, A.[1], Perea, J.[1], Acero, R.[1] and Pacheco, H.[1], [1]University of Cordoba, Animal Production, Edificio Produccion Animal, Campus Rabanales, Cordoba, 14071, Spain, [2]Veterinary School, Animal Production, General Pico, La Pampa, 12431, Argentina; pa2pemuj@uco.es

The aim of this study was classified the dairy cattle farms of northeast region of La Pampa (Argentine) according to technical efficiency level. Surveys were carried out in 57 farmers to establish the main technical, economical and productive aspects of the dairy farms. Information was obtained from direct surveys of the producers. Two models were used to determine the efficiency index: the linear model derived from the Cobb-Douglas function, and the non-linear or multiplicative model. Modeling allows statistical inferences to be drawn from the results obtained, while its main drawback lies in the fact that the functional form selected and used is a pre-established hypothesis. Firstly, was modeled using a Cobb-Douglas function, obtaining the following function: annual milk production=$e^{7,1796}$*consumption of concentrate per milking cow per day^{0,108071}*cows^{1,259}. Once the model was determined, it was established the deterministic frontier of Greene, with the maximum positive residual and the respective technical efficiency index of Timmer. This index was 51%, witch permits to classify farms in three levels: under 37% (low efficiency), between 37% and 65% (medium efficiency), and more than 65% (high efficiency).

Profit function in organic dairy sheep farms in Castilla la Mancha, Spain
Toro, P.[1], Angon, E.[2], Perea, J.[2], Garcia, A.[2], Acero, R.[2] and Aguilar, C.[1], [1]Universidad Catolica de Chile, Animal Science, Vicuna Mackenna, 4860, Santiago de Chile, 6904411, Chile, [2]Universitiy of Cordoba, Animal Production, Edificio Produccion Animal, Campus Rabanales, Cordoba, 14071, Spain; pa2pemuj@uco.es

The aim of this study was to evaluate the economic efficiency of organic sheep production systems by profit function. The study area corresponded to Central Spain. Thirty-one organic dairy sheep farms were sampled during 2007. Of all variables obtained from the survey, were selected 14 variables than representing the technical and economic aspects. The analysis was based on the determination of two profit functions (PF) in terms of annual work unit (AWU) and livestock unit. (LU) The models were determined through linear regression analysis. In the model accounted for BN/LU, 5 significant variables ($P<0.01$) were able to represent 83% of the variability of the data, in the case of PF/AWU 6 significant variables ($P<0.01$) accounted for 72% of the variance. With the equations were estimated PF holdings globally and typological classified into 3 groups: Family subsistence, Commercial intensive and commercial family. The group of Commercial family was presented the highest PF (being the only group with positive profit). Statistically significant differences were observed ($P>0.01$) between this group and other groups, for both estimates (PF/SU and PF/AWU). A sensitivity analysis revealed that optimistic scenarios likely (10% increase in productivity and cost reduction of 20%) do not allow positive profits in Family subsistence and Commercial intensive.

Study, development and testing of a low cost GPS-GSM collar to control cattle rustling

Tangorra, F.M., Marchesi, G., Nava, S. and Lazzari, M., Università degli Studi di Milano, Via Celoria,10, 20133, Milan, Italy; massimo.lazzari@unimi.it

Rustling is an age-old practice, and already in Roman law was severely punished because of the damage caused to farmers for the typical connotation of the herd or flock as capital goods of their activity. The cattle rustling is an international problem that is growing in the last years. Usually in Italy cattle rustling phenomena were limited trough the territorial surveillance even if it is not always feasible, especially in grazing areas that are characterized by difficulties of access and far from the farm centre. Global Positioning System (GPS) technology is increasingly applied in livestock science to monitor pasture use and track animal paths, but the use of this system to combat cattle rustling is not reported into literature. Aim of the research was to develop and test a low cost anti-theft cattle neck collar, combining the GPS and the Global System for Mobile Communications (GSM) technologies and implementing a specific software to track animals movements within a grazing area and get alert from animal trespassing of virtual fences. The collar is also equipped by an anti-detachment sensor able to emit a GSM alarm when a potential thief is trying to remove it from the animal. Field tests carried out on a grazing herd of Romagnola cows in the Northern Apennines (Italy) showed a great potential of the GPS-GSM collar for limiting cattle rustling in extensive grazing areas. The limited battery life, due to the high frequency GPS sampling required by anti-theft monitoring, was the main problem encountered during field test. Further actions will be oriented in: limiting the energy requirements; increasing battery life; integrating the devise with third generation photovoltaic cells.

Post-grazing height in early lactation: immediate and subsequent effects on dairy cow performance and swards characteristics

Ganche, E.[1,2], O'Donovan, M.[1], Delaby, L.[3], Boland, T.[2] and Kennedy, E.[1], [1]Animal & Grassland Research and Innovation Centre, Teagasc, Moorepark, Fermoy, Co. Cork, Ireland, [2]School of Agriculture, Food Science & Veterinary Medicine, Univeristy College, Dublin, Ireland, [3]INRA, UMR Production du Lait, St Gilles, France; elodie.ganche@teagasc.ie

A grazing experiment was undertaken to investigate the effect of post-grazing sward height (PGSH) on early lactation dairy cow performance and carry-over effects on animal and sward characteristics during the subsequent 6 weeks. Seventy-two Holstein-Friesian dairy cows (mean calving date February 12±14.8 days) were randomly assigned across two PGSH treatments: 3 cm (severe, S) or 4 cm (moderate, M) from February 10 to April 18 (period 1; P1). Following P1, animals were re-randomised within P1 treatment and assigned to PGSH treatments of 3.5 cm or 4.5 cm (period 2; P2). The difference in PGSH was governed by daily herbage allowance (DHA). Pre and post-grazing sward heights were measured daily. Milk yield was recorded daily; milk composition, bodyweight (BW) and body condition score (BCS) were recorded weekly. Animal variables were analysed by covariate analysis using PROC MIXED in SAS. A higher PGSH in P1 increased (P<0.001) milk (+2.3 kg/day), fat (+145 g/day), protein (+97 g/day), lactose (+113 g/day) yields and protein concentration (+1.16 g/kg). Grazing to a low PGSH in P1 reduced milk yield (P<0.05) during the 3 first weeks of P2 (- 1.1 kg milk/cow/day). There were no carry-over effects in the second 3 weeks of P2. Mean BW and BCS were not significantly affected by P1 or P2 treatments. Grazing to a low PGSH in early spring improved pasture utilisation (+29%; P<0.01) and increased milk output per unit area by 21%. The study found similar sward characteristics between treatments. It was concluded that grazing to a low PGSH in early lactation results in milk production losses. Following a 10-week experimental period where the reduction in milk production equated to 0.11 the carryover effects only lasted for 3 weeks when DHA was increased.

Profile of leptin and bone markers in peripartal sows in relation to body condition and peripartal feeding strategy

Cools, A.[1], Maes, D.[1], Van Kempen, T.[2], Buyse, J.[3], Liesegang, A.[4] and Janssens, G.P.J.[1], [1]Veterinary Medicine, Ghent University, Heidestraat 19, 9820 Merelbeke, Belgium, [2]Provimi Research, Veilingweg 23, 5334 LD Velddriel, Netherlands, [3]Livestock Physiology, Catholic University Leuven, Kasteelpark Arenberg 30, 3001 Leuven, Belgium, [4]Vetsuissse Faculty, University of Zürich, Winterthurenstrasse 260, 8057 Zürich, Switzerland; an.cools@ugent.be

Parturition induces many metabolic changes in sows. Hence, the effect of feeding scheme and body condition on peripartal energy and bone metabolism was studied. At day 106 of gestation, back fat (BF) of 67 sows was measured to classify sows as skinny (BF<16 mm, n=11), normal (16<BF<22 mm, n=38) or fat (BF>22 mm, n=18). All sows were fed the same feed either *ad libitum* (n=33) or restricted (decreasing towards farrowing, increasing afterwards, n=34) during the peripartal period. Sows were bled on day 107, 109 and 112 of gestation and on day 1, 3 and 5 of lactation. The samples were analysed for leptin, osteocalcin (OC, bone formation) and CrossLaps (CL, bone degradation). Data were subjected to repeated measures analysis of variance using SPSS 17.0 with time as within and feeding scheme and body condition as between-subject factors. Significance over time ($P<0.05$) was identified using an LSD test with Bonferonni correction. The results show that peripartal leptin profile increases significantly over time ($P<0.001$) with a peak on day 3 *post partum*. The profile was not affected by feeding scheme, but fat sows tended ($P=0.078$) to have higher leptin levels than normal or skinny sows. Profiles of CL and OC were not affected by feeding scheme or body condition. Still, both markers significantly changed over time ($P<0.001$) with a drop of OC on day 1 of lactation and a drop in CL on day 109 of gestation, both rising again afterwards. In conclusion, there is no link between leptin and bone marker profiles in peripartal sows. Secretion of these parameters is independent of feeding scheme or body condition except for leptin concentrations being higher in fat sows.

Seasonal variation in semen characteristics and libido of male ostriches

Bonato, M.[1], Rybnik, P.K.[2], Malecki, I.A.[2], Cornwallis, C.K.[3] and Cloete, S.W.P.[1,4], [1]University of Stellenbosch, Department of Animal Sciences, P.O. Box X1, 7600 Matieland, South Africa, [2]University of Western Australia, Faculty of Natural and Agricultural Sciences, School of Animal Biology, WA 6009 Crawley, Australia, [3]University of Oxford, Department of Zoology, Edward Grey Institute, OX1 3PS Oxford, United Kingdom, [4]Institute for Animal Production, Elsenburg, Private Bag X1, 7607 Elsenburg, South Africa; mbonato@sun.ac.za

Reproductive performance of male ostriches can exert a considerable influence on the success of breeding program. Season can potentially influence the quality and quantity of semen collected for artificial insemination and therefore fertility. We thus examined semen output and male libido of seven ostriches (aged 2 to 4 years) over a period of 12 months. We collected ejaculates using a dummy female and measured semen characteristics (volume, concentration, number of spermatozoa per ejaculate and motility) and male libido (willingness to mount the dummy). The volume of semen (mean ± SEM) varied between 1.03±0.12 ml and 1.85±0.07 ml, the concentration between 3.21±0.12 x 10^9/ml and 4.16±0.74 x 10^9/ml and the number of spermatozoa between 3.42±0.28x10^9 and 7.66±0.47x10^9. The largest values for ejaculate volume were found during autumn and winter (April-September) while higher numbers of spermatozoa per ejaculate were found during the spring and summer months (October-April), with a peak in spring (October-November). Both motility and libido score peaked in spring and showed lower values in the middle of summer (January-February). Furthermore, we observed high individual variation between males for all variables tested, except for motility. These results indicate that collections conducted in spring will yield higher numbers of spermatozoa, when libido of males is at a maximum. Seasonal variation should thus be considered during the development of artificial insemination programs in this species to optimize fertility.

Semen production of Black Racka rams: effect of season and age

Egerszegi, I.[1], Sarlós, P.[1], Molnár, A.[1], Cseh, S.[2] and Rátky, J.[1], [1]Research Institute for Animal Breeding and Nutrition, Gesztenyés str 1., 2053 Herceghalom, Hungary, [2]Szent Istvan University, Faculty of Veterinary Science, István str 2., 1078 Budapest, Hungary; jozsef.ratky@atk.hu

Nowadays there is an increasing interest for preservation and conservation of old and traditional farm animal breeds. In situ maintenance of Hungarian farm animals seems to be successful; however contribution of *ex situ in vitro* methods should be strengthened. Hortobagy Racka is a native sheep breed strictly with seasonal reproductive period (August to December). Semen collection during the year could be an option for use the rams in cryo-conservation program, however more profound knowledge on reproductive characteristics of the rams is needed. Aim of this study was to determine the season and age effects on semen production of Racka rams. Nine mature Black Racka rams were included into the trial. Semen samples (n=790) were collected weekly in two consecutive years. Ejaculate volume was registered and motility was assessed under phase contrast microscope and classified 1 to 5. Concentration of the semen was determined by spectrophotometer and total sperm number was calculated. Morphology of the spermatozoa was evaluated after Cerovsky-staining. The data were statistically analyzed. The semen production of the Black Racka rams can be characterized by a mean ejaculate volume of 0.695 ± 0.0119 ml, $5.346\pm0.0513\times10^9$ cells/ml, $3.791\pm0.0814\times10^9$ total sperm number, an average 4.807 ± 0.0212 motility grade and $10.263\pm0.273\%$ defected cells. Significant differences were detected in semen parameters between seasons (P<0.05). The lowest ejaculate volumes were measured in March to May, whilst the highest in summer (June to August) and autumn (September to November) respectively. The mean concentrations of the semen samples were relatively high compared to other breeds and it became constant with the age. Percentage of morphological defects remained low during the trial. In conclusion semen parameters of Black Racka rams could be responding for semen processing.

Physiological and hormonal responses of Egyptian buffalo to different climatic conditions

Hafez, Y.M.[1], Taki, M.O.[2], Baiomy, A.A.[3], Medany, M.A.[2] and Abou-Bakr, S.[1], [1]Faculty of Agriculture, Cairo University, Animal Production, gammaa Street, 12613, Giza, Egypt, Egypt, [2]Central Laboratory for Agricultural Climate, Agriculture Research Center, MOA, Dokki, Giza, 12613, Egypt, [3]Faculty of Agriculture, South Valley University, Animal Production Department, Qena, Egypt, 12613, Egypt; yasseinhafez@yahoo.com

To investigate the effect of two different climatic conditions of Egypt on some hormonal and physiological responses of Egyptian buffaloes, A total of sixteen multiparous lactating buffalos were assigned, six from Giza and ten from Qena. The physiological measurements were taken fortnightly at 0900 h and 1600 h. Blood samples were collected from buffaloes fortnightly at 0900 h before the morning feeding from July to September which is summer time in Egypt and analyzed for plasma total thyroxine, total triiodothyronine and cortisol. Weather data were collected to calculate Temperature Humidity Index (THI) for the two climatic regions. Morning rectal temperature values were the highest (P<0.05) in buffaloes from Qena. All the measures of Skin temperature were the highest (P<0.05) in buffaloes from Qena farm. The ear temperature at the evening gave different trend than the rest of skin temperature values. The evening respiration rate was the highest in the buffaloes from Giza farm, but the morning respiration rate was the highest in buffaloes from Qena farm. The concentration of plasma total triidothyronine was the highest (P>0.05) in buffaloes from Qena farm, but there were no significant differences in the concentration of total thyroxin, T4/T3 ratio and cortisol. The THI was positively correlated with rectal temperature, skin temperature and respiration rate. In conclusion, the THI was a good indicator to the microclimate affecting the Egyptian buffaloes. Also, relevant hormonal and physiological responses of Egyptian buffalo were indicators to the homeostatic reactions due to climatic conditions.

Effect of encapsulated calcium butyrate (GREEN-CAB-70 Coated®) on some blood parameters in Iranian Holstein female calves

Karkoodi, K., Alizadeh, A.R. and Nazari, M., Department of Animal Science, Saveh Branch, Islamic Azad University, Saveh, Iran; kkarkoodi@yahoo.com

This study was conducted to evaluate the effects of encapsulated calcium butyrate (GREEN-CAB-70 Coated®, Nutri Concept Co.) on some blood parameters of Holstein female calves. Sixteen female calves with mean age 5 days in a completely randomized design were divided into two equal groups (n=8) namely, experimental (starter diet and milk replacer with 3 g calcium butyrate per day) and control (starter diet and milk replacer). Blood serum was collected for analysis of some parameters on days 1, 12, 24, 36 and 48. Serum glucose, insulin, and B-hydroxy butyric acid concentration showed significant increase in the experimental group compared with control group on days 1, 12, 24, 36 and 48 (P<0.05). Serum cortisol concentration in the experimental group decreased, although this was not statistically significant, on days 1, 12, 24, 36 and 48 (P>0.05). The study suggests that using encapsulated calcium butyrate in Holstein female calves fed before weaning increases liver gluconeogenesis, shortening pancreas and rumen development period which in turn facilitates early weaning.

Effect of breeding season and epidermal growth factor on the competence of camel (*Camelus dromedarius*) oocytes to mature *in vitro*

El-Sayed, A.[1], El-Hassanein, E.E.[2], Sayed, H.[2] and Barkawi, A.H.[3], [1]Faculty of Agriculture Research Park (FARP), Faculty of Agriculture, Cairo University, 12613 Giza, Egypt, [2]Department of Animal and poultry Production, Desert Research Center, 11753 Cairo, Egypt, [3]Department of Animal Production, Faculty of Agariculture, Cairo University, 12613 Giza, Egypt; ashrafah99@yahoo.com

The present study was conducted on 657 oocytes of camels to investigate the effect of season and culture media supplied with epidermal growth factor (EGF) on the competence of oocytes to mature *in vitro*. The oocytes were collected during breeding (n=381) and non-breeding (n=276) season from ovaries (n=115) delivered in normal saline solution at 37 °C from a local slaughterhouse. Cumulus oocyte complexes (COCs) were recovered from ovaries by slicing. Good-quality oocytes (with >1 layer of cumulus cells and homogeneous, dark cytoplasm) were selected for IVM. The COCs were matured for 30 h in TCM-199 medium. The effect of season and EGF (10 ng/ml) on cumulus expansion and maturation of oocytes was assessed. All cultured were done at 38.5 °C, under 5% CO_2 and 95% humidity. The oocytes cultured in maturation medium containing EGF (n=313) exhibited a higher (P<0.05) percentage of expansion and maturation rates (75.8%) than those without EGF (62%, n=344). There was no significant difference between breeding and non-breeding season on oocyte maturation rate either with EGF or without. In conclusion, the use of EGF as an additive in maturation medium has a positive effect on oocyte expansion and maturation rate in camels.

Cysts and ovarian activity in Holstein dairy cows in early lactation related to milk and plasma fatty acids and to plasma metabolites and hormones

Knapp, E., Dotreppe, O., Hornick, J.L., Istasse, L. and Dufrasne, I., Nutrition Unit, Veterinary Faculty, University of Liege, Bd Colonster 20, 4000 Liege, Belgium; eknapp@ulg.ac.be

Nowadays, high milk yields are achieved by dairy cows. During early lactation, there are large metabolic changes associated with negative energy balances with as results mobilization of fat reserves. There are also disturbances on reproduction such as delayed resumption of cyclicity after calving or high frequency of ovarian cysts. A survey was carried out on 32 dairy cows in early lactation from 5 private farms. Blood and milk samples were taken on 4 occasions on the day of the milk record. There was a gynecological examination. The ovarian activity was expressed by presence of a corpus luteum. The ovarian cysts were also recorded. The proportion of C4-C14 and of branched chain fatty acids in milk were higher in cows with ovarian activity (21.3 vs 19.2%, $P<0.01$ and 3.6 vs 3.2%, $P<0.01$) while that of C18:0 was lower (12.2 vs 13.7%, $P<0.01$). When expressed in concentration, there was only a difference for the C18:0 (5.3 vs 6.0 g/kg milk, $P<0.05$). The concentrations of b-hydroxybutyrate and C16:0, C18:0 and C18:1 in the plasma non esterified fatty acids were significantly lower in the cows with ovarian activity (0.7 vs 1.5, 1.1 vs 1.9, 1.0 vs 2.3 mg/dl, $P<0.01$) while the IGF1 concentration was higher (62.7 vs 38.6 mg/ml, $P<0.001$). The increases in short chain fatty acids in milk and in IGF1 along with reduction of long chain fatty acids in milk and in plasma were indications of less negative energy balances favorable for ovarian activity. As far as the cyst occurrence was concerned – 41% of the cows – there were less clear cut relationships with fatty acids, b-hydroxybutyrate or IGF1. Nevertheless cows with cysts produced milk with less poly-unsaturated fatty acids ($P<0.05$) and were characterized by higher glycemia ($P<0.05$) and lower cholesterolemia ($P<0.05$). One can thus conclude that information of interest for reproduction purposes can be obtained from individual fatty acids.

Detection of embryonic mortality by means of ovine pregnancy-associated glycoproteins after *in utero* stem cells transplantation in sheep

Terzano, G.M.[1], Barbato, O.[2], Mazzi, M.[1], Tintoni, M.[3], Riccardi, M.[3] and Noia, G.[3], [1]Animal Production Research Centre (PCM), Agricolture Research Council (CRA), Via Salaria, 31, 00015 Monterotondo (RM), Italy, [2]University of Perugia, Faculty of Veterinary Medicine, Biopatological Veterinary Science, Via S. Costanzo, 4, 06126 Perugia, Italy, [3]Catholic University of the Sacred Heart, Obstetrics and Gynecology, L.go F. Vito, 1, 00100 Roma, Italy; giuseppinamaria.terzano@entecra.it

Determination of ovine pregnancy-associated glycoproteins (Pag's) concentration can be used for pregnancy diagnosis before 45 days of conception and it could be used for detection of embryonic mortality (EM). This study aims to assay Pag's concentration before and after in utero stem cells transplantation via the intracoelomic route in the sheep. The transplantations were performed on 10 sheep foetuses at 36 days of development and plasma samples were collected at the day of conception (T0), 30 days after conception (T30), at the moment of stem cells transplantation (T36) and at 37(T37), 39(T39), 43(T43), 46(T46), 50(T50) and 57(T57) days. The Pag's measurements were performed by mean of heterologous RIA. The average of Pag's concentrations of the 10 sheep at T30 and T36 were 21.1 and 20.0 ng/ml, respectively, indicating that they were pregnant. Of the 10 treated sheep, 4 aborted after the stem cells transplantation and at T37 and T39 the Pag's concentrations were lower in aborted sheep than in pregnant ones, without significant differences. The average Pag's concentrations (ng/ml) were 18.6±8.3 vs 8.1±1.2 at T43 ($P<0.05$), 19.4±9.6 vs 7.2±1.1 at T46 ($P<0.05$), 23.6±5.9 vs 5.1±0.7 at T50 ($P<0.001$), 24.5±5.5 vs 4.0±0.5 at T53 ($P<0.0001$), 26.7±4.1 vs 2.9±0.4 at T57 ($P<0.0001$), in pregnant and in aborted sheep, respectively. Determination of Pag's concentrations by mean of heterologous RIA is a useful tool for the detection of early EM after in utero stem cells transplantation and it could be a more reliable test than the transvaginal ultrasound.

The effect of castration at 10 months of age on growth physiology of Serrana de Teruel cattle breed
Sanz, A.[1], Albertí, P.[1], Ripoll, G.[1], Blasco, I.[1] and Álvarez-Rodríguez, J.[2], [1]CITA de Aragon, Avda. Montañana 930, 50059 Zaragoza, Spain, [2]Universitat de Lleida, Avda. Rovira Roure 191, 25198 Lleida, Spain; asanz@aragon.es

Serrana de Teruel (ST) is a dark or tabby-breed raised traditionally on mountain areas of Southern Aragon (north-eastern Spain). In the framework of ST endangered breed conservation programme, beef quality differentiation through steer and bull production has been promoted. In this study, we compared productive performance and peripheral IGF-I concentration of steers and bulls from 10 (surgical castration) to 21 months old (slaughter). Fourteen male calves were managed under a feeding programme divided in 3 phases: I) 10-13 months (ad lib concentrate plus straw), II) 14-18 months (ad lib barley silage plus 3 kg concentrate) and III) 19-21 months (ad lib concentrate plus straw). Feed intake was registered daily on a group-basis. Individual live-weights (LW) were recorded weekly, and blood samples were collected monthly to analyze IGF-I concentration by a commercial kit (IMMULITE® 2000, DPC). Dry matter intake did not differ between steers and bulls, but it was greater in phase I than in phase III (79 vs. 64±4 g/kg $LW^{0.75}$/day). Steers and bulls LW were different at the start of phase III (552 vs. 623±17 kg) and at slaughter (648 vs. 743±20 kg). Average daily gain (ADG) from 10 to 21 months old was lower in steers than in bulls (0.99 vs. 1.23±0.05 kg/day), mainly due to the differences in the month following castration and the last fattening month. Concentrate supply in phase III did not counterbalance the lower ADG in steers. Average IGF-I was lower in steers than in bulls (139 vs. 192±17 ng/ml), mainly due to differences during the phase III. This response might be related to attainment of puberty, and suggests that IGF-I played a role in mediating gonadal rather than nutritional status. In conclusion, ST steers grew slowly and had lower plasma IGF-I concentration than their bull counterparts, these differences being mainly highlighted from 19 to 21 months of age.

Effect of different semen extenders on motility of liquid preserved Moo Lat boar semen – a pilot study
Keonouchanh, S.[1], Egerszegi, I.[2], Dengkhounxay, T.[1], Sarlos, P.[2] and Ratky, J.[2], [1]National Agriculture and Forestry Research Institute, Livestock Research Center, P.O. Box 7170, Vientiane, Lao Peoples Dem. Rep., [2]Research Institute for Animal Breeding and Nutrition, Gesztenyes str 1., 2053 Herceghalom, Hungary; jozsef.ratky@atk.hu

Pig production continues to be an important livelihood activity in the Lao People's Democratic Republic which is based on traditional methods. More than 80 percents of pig herds are native breeds and belong to smallholders with combined keeping systems. Moo Lat pig is one of the four indigenous swine breeds in the country. Nowadays liquid semen preservation and artificial insemination are only used for exotic breeds (e.g. Large White, Duroc) however it could be beneficial to use this method in propagation and preservation of native swine population too. The aim of this study was to determine how different extenders influence semen motility of liquid preserved Moo Lat boar semen. Semen was collected by gloved hand method from three matured Moo Lat boars. Motility and morphological defects were determined immediately after collection. Only ejaculate with more than 80% motile and less than 15% abnormal cells were included in the trial. Mixed semen samples were diluted 1:5 with three different extenders (BTS, MRA and Acromax) and preserved on 17 °C for 5 days long. Motility was assessed every day during the preservation period. Live-dead cell rate was determined after Giemsa-staining. There was no significant difference in motility and dead cell rate between extenders during preservation; however, after day 2, MRA tended to be better than the other extenders. On day 3, only MRA extended semen contained more than 60% motile cells, which declined to 45% on day 5 but remained the best motility rate. It could be concluded that all investigated extenders could be used for insemination till two days after collection and MRA 3-4 days after it. Nevertheless further experiment is needed to improve liquid preservation of Moo Lat boar semen. The study was funded by Hungarian-Laotian TéT Bilateral cooperation.

Shading effects on physical and biochemical parameters in Tunisian local goat kids during hot season

Hammadi, M.[1], Fehem, A.[1], Khorchani, T.[1], El-Faza, S.[2], Salama, A.A.K.[3], Casals, R.[3], Such, X.[3] and Caja, G.[3], [1]IRA, R Jorf, 4119 Medenine, Tunisia, [2]FST, Manar, 1060, Tunisia, [3]UAB, Bellaterra, 08193, Spain; Ahmed.Salama@uab.cat

This study aimed to evaluate the effect of shading on some physical and biochemical parameters in Tunisian local goat kids during the hot season. Fourteen female kids were allotted to 2 groups. Group 1 (6.5 ± 0.9 mo; 10.8 ± 1.1 kg) was exposed to daytime solar radiation and group 2 (6.6 ± 0.8 mo; 10.8 ± 2.0 kg) was maintained under a shade regimen throughout August. Respiratory rate per minute (RR) and rectal temperature (RT) were assessed (10:00; 13:00; 16:00 h) three times a week. Consumed water (CW) and dry matter intake (DMI) were determined and blood samples were weekly taken to analyze protein, glucose and urea contents. Data were statistically analyzed by PROC mixed of SAS. For RR and RT, model included the shading effect and day hour as main effects. For the other parameters, model included shading as main effect. Pearson coefficients were calculated between ambient temperature (AT), RR and RT. Results are presented in means $\pm$ SE. The RR was affected (55.8 ± 1.9 and 45.5 ± 1.1 in group 1 and group 2, respectively; $P<0.01$) by shade, especially at 13:00 h. The RT did not differ between the two groups. The RR, which depended on AT ($r=0.36$; $P<0.01$), correlated ($r=0.53$; $P<0.01$) with RT only in group 2. Daily DMI was similar in the two groups (34.6 ± 2.1 g/kg BW). The CW was greater ($P<0.01$) in group 1 (5.5 ± 0.9 l/kg DMI) than in group 2 (2.1 ± 0.2 l/kg). Blood protein and glucose contents did not differ, but blood urea was greater ($P<0.05$) in group 1 (0.36 ± 0.03 g/l) compared to group 2 (0.27 ± 0.02 g/l). Daily body gain tended ($P<0.10$) to be greater in group 1 than in group 2 (55 ± 8 and 37 ± 3 g, respectively). Tunisian local goats adapt to solar radiation by increasing respiratory rate, consumed water, and urea concentration in blood.

Detection of lameness and mastitis in dairy cows using wavelet analysis

Miekley, B., Traulsen, I. and Krieter, J., Institute of Animal Breeding and Husbandry, Christian-Albrechts-University, Olshausenstraße 40, D-24098 Kiel, Germany; bmiekley@tierzucht.uni-kiel.de

This investigation analyses the efficiency of wavelets, a method of digital signal analysis, for an early detection of lameness and mastitis. Data were recorded on the dairy research farm Karkendamm between January 2009 and October 2010. In total, data of 238 cows were analysed (44,837 observations for the first 200 days in milk). Lameness treatments were used to determine definitions of lameness. Mastitis was specified according to three definitions: (1) udder treatment, (2) udder treatment or somatic cell count over 400,000 ml[-1] and (3) udder treatment or somatic cell count over 100,000 ml[-1]. For detection of lameness the activity per day and animal was used, for mastitis the electrical conductivity of the milk per day was utilized. The values of both parameters were filtered by wavelets. The resulting values and residuals (deviation between the observed and filtered value) were applied to a classical and a self-starting cusum chart to identify blocks of lameness and mastitis (days of disease). First results show that block-sensitivity for lameness detection ranged between 30.2% and 49.3% for the classical chart and between 54.0% and 57.7% for the self-starting chart. The error rate varied between 94% and 97%. Concerning mastitis detection the classical chart showed better results. The block-sensitivity ranged between 67.2% and 76.3% (self-starting chart between 50.8% and 74.6%) while the obtained error rate were between 68.9% and 99.3% (self-starting chart between 73.2% and 99.6%). For both charts block-sensitivity and error rate were improved from definition (1) to (3). In conclusion, wavelet analysis seems to be appropriate for disease detection in dairy cows. Results could probably be enhanced if more traits are included in the analysis.

Owner-recorded data as source of information for genetic analyses of health traits of dairy cattle

Stock, K.F.[1], Agena, D.[1], Spittel, S.[2], Hoedemaker, M.[2] and Reinhardt, F.[1], [1]Vereinigte Informationssysteme Tierhaltung w.V., Heideweg 1, 27283 Verden / Aller, Germany, [2]University of Veterinary Medicine Hannover, Clinic for Cattle, Bischofsholer Damm 15, 30173 Hannover, Germany; friederike.katharina.stock@vit.de

Dairy performance has been improved very successfully over the last decades. However, higher long-term efficiency of milk production may require genetic improvement of functional traits. In the German project GKuh, dairy farmers are requested to record health events of any kind and independent from veterinary treatment for all female calves, heifers and cows on their farms. Numeral codes that link to a comprehensive key of diagnoses are used for documentation to standardize recording and minimize time and efforts for data collection on farm. Almost 10,000 diagnoses from 49 farms were transmitted to the health database in 2010. In this year, the total number of females on these farms was 11,327 including 6,791 animals >24 month of age. To investigate the suitability of the owner-recorded data for genetic analyses and to study the effects of trait definition (binary vs. quasi-continuous), modeling and inclusion criteria for controls, four health traits were chosen: early mastitis (MAST), i.e. mastitis recognized 10 days before until 50 days after calving, retained placenta (RET), purulent claw diseases (PCL), and non-purulent claw diseases (NPCL). Depending on trait and stringency of inclusion criteria for controls, the number of animals considered for the genetic analyses ranged between 2,293 and 4,333. Variance components were estimated in univariate linear animal models using REML. With binary coding of presence or absence of disease in a given parity, heritability estimates were 0.03-0.09 on the original scale and 0.06-0.27 using threshold model transformation. Influences of trait definition, modeling and inclusion criteria for controls on the parameter estimates were mostly small. Results indicate that owner-recorded data provide a suitable basis for genetic evaluation for health traits in dairy cattle.

Capturing variation in infectivity from binary disease data

Lipschutz-Powell, D.[1], Woolliams, J.A.[1], Bijma, P.[2] and Doeschl-Wilson, A.B.[1], [1]The Roslin Institute, Easter Bush, EH25 9RG, United Kingdom, [2]Wageningen University, Wageningen, 6701, Netherlands; debby.powell@roslin.ed.ac.uk

Reducing disease prevalence through selection for host resistance offers a desirable alternative to chemical treatment. Selection for host resistance has proven difficult, however, due to low heritability estimates. These low estimates may be caused by a failure to capture all the relevant genetic variance in disease resistance, as genetic analysis currently ignores genetic variation in infectivity. Host infectivity is the propensity of transmitting infection upon contact with a susceptible individual, and can be regarded as an associative effect to disease status. The theory of associative effects, therefore, provides a promising framework to include infectivity into genetic analyses and to investigate the design of alternative breeding strategies. Though genetic variation in infectivity is difficult to measure directly, associative effect models allows for estimates of this variance from more readily available binary disease data (infected/non-infected). We therefore generated binary disease data from simulated populations with known amounts of variation in susceptibility and infectivity to test the adequacy of traditional and associative models. Our results show that a traditional sire model does not capture the variation in infectivity inherent in the data. The current associative effects model, on the other hand, does capture some of the variation in infectivity. Moreover, selection using an index of estimated direct and associative breeding values has a greater genetic selection differential and thus reduced future disease risk than selecting by sire breeding values. However, the current associative effects model does not take individual disease status into account. This causes a biased covariance estimate and an underestimation of the associative effect variance. We therefore demonstrate how an associative effect model which takes disease status into account can be fitted.

A mobile milking robot for dairy cows at grass: effects of grazing management.
Dufrasne, I., Knapp, E., Robaye, V., Istasse, L. and Hornick, J.L., Nutrition Unit, Veterinary Faculty, University of Liege, Bd Colonster 20, 4000 Liege, Belgium; isabelle.dufrasne@ulg.ac.be

Grazing with dairy cows is associated with positive aspects such as improvements in animal health, reduced labor and reduced feeding costs. Grazing is also positively perceived by the consumers and is often used as a marketing strategy. A prototype of a mobile milking robot was developed. The milking robot and equipments were located on a trailer that can be easily moved to different places by a tractor and can be lowered to ground level. The milk tank is located on another trailer. From 20/04 till 22/06/2010, the cows were milked indoors. Both trailers were then moved to a pasture equipped with an electric point, water and facilities to collect the washing water. The trailer with the milk thank was easily accessible for milk collection by lorry. In October, on the end of the grazing season, both trailers were moved back to the barn. The number of cows varied between 47 and 50. During the 60 days of the indoor period, the cows on an average of 173 days in milk produced 29.6 kg milk per day with a mean number of milkings of 3.09 and a 1.06 milking refusal per day. On pasture, the cows were grazed on a rotational system with a total of 11 paddocks. The robot was located in a paddock of 1.33 ha. The cows were gathered twice a day at 06:00 and 18:00 h in one of the paddocks used as a waiting room located at the robot entry. During the period on grass, the cows, in their average 215 days on milk, produced 21.1 kg milk per day. The distance from the grazed paddock to the robot (100 to 400 m) tended to decrease milk yield and the number of milkings (P<0.06) and did not affect the milking refusals. The grazing cycle number – first or second paddock rotation – significantly (P<0.001) decreased milking number, milking refusals and yields. Although the differences were not significant, the day on the paddock – comparison d1-2-3-4 – tended to decrease milk yield.

Effect of milking frequency and plane of nutrition on dairy cow immune function
O'driscoll, K.[1], Llamas-Moya, S.[1], Olmos-Antillon, G.[1], Earley, B.[2], Gleeson, D.[1], O'brien, B.[1], Mee, J.[1] and Boyle, L.[1], [1]Teagasc, Animal and Grassland Research and Innovation Centre, Fermoy, Co. Cork, Ireland, [2]Teagasc, Animal and Grassland Research and Innovation Centre, Grange, Co. Meath, Ireland; keelin.odriscoll@teagasc.ie

Twice a day milking is the most common milking regime employed in Ireland. A reduction to once daily milking could have benefits for dairy cows by reducing physiological stress at the beginning of lactation associated with milk production. This study investigated how milking frequency and plane of nutrition affect dairy cow immune status. Cows (n=48) were milked either once a day (OAD) or twice a day (TAD) at one of two planes of nutrition; High (HNL), or Low (LNL), in a 2 × 2 factorial design. As well as *ad libitum* grass silage, HNL cows were provided with 7 kg/conc/cow/day until March 22, 4 kg/cow/day until 17 April, then allocated 31.3 kg DM grass/cow/day. LNL cows were provided with 4 kg/conc/cow/day, 1 kg/cow/day, and allocated 19 kg DM grass/cow/day during the same respective periods. Blood samples were collected prepartum (-7 to -1 d) and at 1-7 d, 14-21 d and 42-49 d *post partum*. Differential leukocyte %, interferon gamma (IFN-γ) production in response to Concanavalin A and Phytohaemagluttinin, and cortisol, haptoglobin (Hp) and serum amyloid A (SAA) production were evaluated. Data were analysed using SAS V9.1 (PROC MIXED). There was no effect of milking frequency or diet on total leukocyte counts. Cows milked OAD had a higher lymphocyte% (P<0.05), lower monocyte% (P<0.05), and tended to have a lower neutrophil% (P=0.09) than cows milked TAD. HNL cows had a lower eosinophil% than cows on the LNL (P=0.05). There was no effect (P>0.05) of milking frequency or diet on IFN-γ, Hp, SAA or cortisol production. The differences in leukocyte% could reflect a slight impairment in the immune function of TAD relative to OAD cows. However, the absence of differences in other indicators suggests that the stress associated with TAD milking was not enough to compromise dairy cow immune status to the extent that health might be affected.

Influence of season, physiological state and production system on dairy cows welfare assessment
Marie, M.[1,2], [1]INRA ASTER, Av. Louis Buffet, 88500 Mirecourt, France, [2]ENSAIA, INPL-Nancy, Nancy Université, B.P. 172, 54505 Vandœuvre lès Nancy, France; michel.marie@mirecourt.inra.fr

In order to evaluate the influence of season, physiological status and production system on the evaluation of animal welfare, two dairy cows herds (a 40 cows, winter-calving, fed strictly on herb one, and a 60 cows, autumn-calving, fed on herb supplemented with concentrate one) conducted in the same experimental domain have been scored (from 0 to 100) using the Welfare Quality assessment protocol during one year at 3-months intervals (March, June, September, December). Data were analyzed by the REML procedure (Genstat software) with a LSD significance level of 5%. Good feeding scores, based on percentage of lean cows, varied from 12 to 34.4, with an effect of system and time of the year, as a function of time in lactation. Comfort around resting scores varied from 58.4 to 89.2. Integument alterations (scores 58 to 100) improved in outdoors conditions, when lameness (scores 13.4 to 73.7) was more frequent in autumn, due to the cumulative walking distances. Illness scores, estimated from records of 10 pathologies, varied from 33.4 to 64.7. Aggressive behaviour (scores 66.5 to 98.6) where more intense during indoors periods, but avoidance distance (scores 41.2 to 73.2), was shorter in winter. Emotions scores (31.7 to 97.3) were globally better in outdoors situations, but with great variations due to particular situations, such as elevated temperatures. The physiological status (pregnancy, lactation) appeared more influential on body condition score and time to lie, whereas access to pasture affects (positively or negatively) cleanliness, injuries (integument and lameness), and behaviour. On the other hand, the resource- or practices-based criteria (availability of water points, animal housing, dehorning, days on pasture) gave constant scores around the year. This study shows that the welfare status of the herd is changing within a year, and that a representative evaluation should be based on more than one single control.

Effects of dietary supplementation with protection factors at a high incorporation rate on the lactoferrin content in milk and on some milk quality parameters
Robaye, V., Knapp, E., Dotreppe, O., Hornick, J.L., Istasse, L. and Dufrasne, I., Nutrition Unit, Veterinary Faculty, University of Liege, Bd Colonster 20, 4000 Liege, Belgium; isabelle.dufrasne@ulg.ac.be

Lactoferrin, a glycoprotein involved in biological processes related to defense mechanisms, is produced in many secretions. In the mammary gland it is produced by the epithelial cells but also by the polymorphonuclear neutrophils. The aim of the work was to assess effects of a supplementation with dietary protection factors on the lactoferrin content and on milk quality. A herd of 62 dairy cows were offered a diet made of maize silage (52% DM basis), sugar beet pulp silage (10%) and grass silage (13%). A compound feedstuff was used to balance the diet for a milk yield of 24 kg. The cows were divided in 2 groups. The supplemented group received daily 160,000 IU vit A, 2,100 mg vit E, 210 mg β-carotene, 1,100 mg Zn and 2.1 mg Se. The experiment lasted during the whole winter period. The milk yield was not affected by the protection factors, the average production being 27.9, 25.2 and 18.5 kg/d during the period d1-d100, d101-200 and over d201 of the lactation period. Over each of the 3 periods, the somatic cell count was higher in the control group than in the supplemented animals; the opposite was observed for the lactoferrin content. On average, the cell count was 246,800 and 196,000 cells/ml (P=0.714) and the lactoferrin content 545.0 and 373.8 mg/l (P=0.254) in the control and the supplemented animals. The proportion of mastitis was higher in the control animals than in the supplemented (27 vs 16%). Furthermore, there were no repeated cases of mastitis with the supplement. According to linear models, the lactoferrin content in milk was significantly influenced (P<0.001) by the days in milk, the milk yield, the cells count and the protein content. It could be concluded that, although an improvement in the lactoferrin content was expected but not observed when protection factors were offered, the major effects were improvements of the mammary health.

A financial cost-benefit analysis of the 'Healthier goats' program to Norwegian dairy goat farmers

Nagel-Alne, G.E.[1,2], Valle, P.S.[1,3], Asheim, L.J.[4], Hardaker, J.B.[4,5] and Sølverød, L.[2], [1]Norwegian School of Veterinary Science, P.O. 8146 Dep, 0454 Oslo, Norway, [2]TINE SA, TINE Rådgiving, Box 58, 1430 Ås, Norway, [3]Molde University College, P.O. Box 2110, 6402 Molde, Norway, [4]Norwegian Agricultural Economics Reseach Institute, P.O. 8024 Dep., 0030 Oslo, Norway, [5]School of Business, Economics and Public Policy, University of New England, Armidale, NSW 2351, Australia; gunvor.elise.nagel-alne@tine.no

In 2001 the Goat Health Services in Norway initiated a program to sanitize for caprine arthritis encephalitis, paratuberculosis (Johne's disease) and caseous lymphadenitis in dairy goat herds ('Healthier goats'). The program is largely financed by the government. A financial cost-benefit analysis (CBA) of the project has been conducted from the perspective of the dairy goat farmers by calculating the net present value (NPV) of their costs and returns over 10 to 20 years. The calculations are based on reported results showing i. a. average milk yield increased by 21 per cent (from 627 to 756 kg) per goat in sanitized herds. The farm gate milk price also increased due to lower cell counts following sanitation. Also, sanitized goats last longer resulting in lower costs of replacement. Information about farmers' costs was gathered in a questionnaire. The costs comprised farmers' work on sanitation, feed costs, farm building and other investments, and temporary production problems related to the sanitation. This information was supported with herd-level data from the Goat Herd Control Database and Goat Efficiency Control Database to arrive at total sanitation investment costs. Benefits due to increased milk yield is restricted due to the milk quota as well as affected by governmental animal premium payments. A substantial part of farmers' costs were reimbursed by the program. The importance of farmers' investments and government participation, in preventive veterinary medicine and animal disease control in the future Norwegian goat milk production, is discussed in relation to the results of the financial CBA.

Precision, repeatability and representative ability of faecal egg counts in a model host-parasite system

Das, G.[1], Savas, T.[2], Kaufmann, F.[1], Idris, A.[1], Abel, H.[1] and Gauly, M.[1], [1]University of Goettingen, Dep. of Animal Sciences, Albrecht-Thaer-Weg 3, 37075, Goettingen, Germany, [2]Canakkale Onsekiz Mart University, Dep. of Animal Science, Campus of Terzioglu, 17020, Canakkale, Turkey; gdas@gwdg.de

Estimations of gastrointestinal nematode infection intensity in the living hosts widely rely on quantification of egg concentration in faeces. However, randomly taken single faecal samples are less likely to provide a reliable quantification of the nematode infection intensity. This study investigated whether a precise, repeatable and representative quantification of nematode egg excretion can be achieved based on collection of the daily total amount of faeces from Heterakis gallinarum infected chickens as a model host-parasite system. Daily individual total faeces (N=2240) were collected from H. gallinarum infected chicks for 5 wk to determine the numbers of eggs per gram of faeces (EPG) and total number of eggs excreted within 24 h (EPD). Precision of EPG counts was not influenced by consistency (P=0.870) and total amount of faeces (P=0.088), but by the concentration of eggs in faeces (P<0.001). A segmented regression analysis indicated that precision of EPG counts was heightened up to a breakpoint (EPG ≤617) as the response to changing egg concentration of the faecal samples. Moderate repeatabilities (R=0.49) for EPG and EPD were estimated in the first week of egg excretion, whereas the estimates were higher (R= 0.67-0.84) in the following weeks. Correlations between number of female worms with daily measured EPG and EPD increased to r ≥0.70 (P<0.05) in a few days after the nematode first excreted eggs and predominantly remained so for the rest of the sampling period. It is concluded that precision of the EPG counts increases until a breakpoint as the egg concentration in the faecal sample increases. Egg excretion of H. gallinarum, quantified as EPG or EPD, is highly repeatable and closely correlated with the actual worm burden of birds starting as early as 4 wk after infection.

Spectrum, prevalence and intensity of helminth infections in laying hens kept in organic free range systems in Germany
Kaufmann, F., Das, G., Sohnrey, B. and Gauly, M., University of Goettingen, Dep. of Animal Sciences, Albrecht-Thaer-Weg 3, 37075, Goettingen, Germany; gdas@gwdg.de

A total of 740 laying hens from 18 organic free range farms in Germany were collected between 2007 and 2010. The hens were sacrificed and the gastrointestinal tracts were examined for the presence and intensity of helminth infections with standard methods. Three nematode (Ascaridia galli, Heterakis gallinarum, Capillaria spp.) and four cestode (Raillietina cesticillus, Hymenolepis cantaniana, Hymenolepis carioca, Choanotaenia infundibulum) species were found. Almost all hens (99.6%, N=737) harboured at least one helminth species. Average worm burden per hen was 218 worms. The most prevalent species were the nematodes H. gallinarum (98%) followed by A. galli (88%) and Capillaria spp. (75.3%), whereas the overall prevalence of the cestodes was 24.9%. Total worm burden was significantly higher in the summer season than in the winter season (254 vs. 191 P<0.0001). The most dominant helminth species was H. gallinarum averaging 190 worms per hen in the summer and 129 in winter season, respectively (P<0.0001). Average A.galli burden was 25 in summer and 26 in winter season, respectively (P=0.1160). Risk of being infected with any of the nematodes was 1.5 times higher in summer than in winter (Ψ=1.49, P<0.0319). Probability of infection with any of the tapeworm species was 4.5 times higher in the summer than in winter (P<0.0001). It can be concluded that the vast majority of the hens are infected with various helminth species. The prevalence as well as intensity of the helminth infections, particularly with tapeworms, considerably increases in summer. The present results indicate that it is essential to adopt alternative control strategies in order to lower infection risk and to limit the potential effects of helminth infections on production and welfare of hens kept in organic systems.

The effect of pasture feeding system on udder health of cows
Frelich, J. and Šlachta, M., University of South Bohemia, Faculty of Agriculture, Studentská 13, 37005 České Budějovice, Czech Republic; slachta@zf.jcu.cz

In the mountain areas of the Czech Republic, the breeding of two most common dairy breeds, the Holstein and the Czech Fleckvieh, relies on two feeding strategies: the seasonal pasture in May-October period followed by the silage feeding indoor in the rest of a year, and the all-year-through indoor silage feeding without any access to pasture. The dietary and the ethological factors related to pasture are suspected to act as the stressors and to increase the risk of subclinical mastitis in grazing cows. In this study, the somatic cell count (SCC) in individual cow milk was used as an indicator of the udder health of cows. The monthly test-day records on SCC in milk of total 12,788 cows from 26 herds were scrutinized in order to elucidate the impact of the seasonal grazing on occurrence of milk secretion disorders. The General Linear Models analysis was conducted in order to evaluate the seasonal, the management (farm, feeding system) and the genetic (breed) effects of SCC in individual milk samples. The interaction between the season (summer: May-October; winter: November-April) and the feeding system (seasonal grazing; permanently-indoor feeding) effects was used as a measure of the effect of pasture on SCC. In both the feeding systems, the SCC was lower in the summer season than in the winter season (P<0.01). The effect of season did not interact with the effect of feeding system (P>0.05). The results indicated a higher importance of the management and the genetic factors for the occurrence of milk secretion disorders in comparison to the seasonal factors. No negative effect of the seasonal grazing on udder health was identified.

Risk of stress associated to rumenocentesis: benefits of local anaesthesia?
Mialon, M.M., Deiss, V., Andanson, S., Anglard, F., Doreau, M. and Veissier, I., INRA, URH1213 Theix, 63122 St Genes, France; michel.doreau@clermont.inra.fr

Rumenocentesis is carried out on ruminants to collect ruminal juice to detect ruminal acidosis. We questioned whether rumenocentesis induces stress and whether local anaesthesia can limit such negative effects on animals. We compared rumenocentesis practised with (AR) vs. without local anaesthesia (R), and local anaesthesia only (A). Twenty-four dairy cows were assigned to the three treatments AR, R, and A. The cows were led to a restraining cage during three consecutive days and the anaesthesia and/or rumenocentesis was performed the second day. The same measures were performed identically on the three days. Blood samples were taken for cortisol determination and heart rate was recorded from 0.25 hour before treatment and during the subsequent four hours. Behaviour was recorded when the cows were in the restraining cage by recording head and ear positions and the number of moos. Feed intake and milk production were measured on three consecutive days the week before treatment, on the day of treatment and the day after. Statistical analyses were performed using the PROC MIXED procedure of SAS for repeated measurements. Whatever the treatment, we observed a rise in cortisol level and an increase in heart rate when the cows were in the cage (P<0.0001). The R and AR treatments did not provoke cortisol, cardiac or behavioural responses significantly more important than the A treatment (P>0.05). Cortisol level and heart rate were not different between days (P>0.05). Feed intake and milk production were not hampered the day of treatment and the day after (P>0.05). In addition, we were expecting more marked responses when the cows were led again to the restraining if they had remembered a negative experience in the cage. However such a phenomenon was not observed (P>0.05). We conclude that rumenocentesis does not seem more stressful than local anaesthesia or even more handling. Our results show that the benefits of a local anaesthesia for rumenocentesis are not confirmed.

Effect of oral treatment with bovine lactoferrin on plasma concentrations of metabolites and hormones in preruminant calves
Kushibiki, S., Shingu, H. and Moriya, N., National Institute of Livestock and Grassland Science, Tsukuba, Ibaraki, 305-0901, Japan; mendoza@affrc.go.jp

Neonate calves are highly susceptible to a variety of diseases that have an effect on performance and mortality before weaning. Lactoferrin (LF) plays a role in iron absorption and is believed to be an important component of host defense. The objectives of this study were to clarify the effects of long-term oral administration of LF on the hormone and cytokine responses in blood in the preruminant calves. Three-day-old male Holstein calves were used for 21 days. All calves were separated from their dams and were given colostrums within 3 h. They were housed individually in pens and received their dams' milk for 2 days. The calves had *ad libitum* access to fresh water and were fed warm Holstein whole milk daily at 0900 h and 1600 h. At 3 days of age (Day 0), each calf was assigned to one of three treatment groups (n=10) based on its body weight. The treatments consisting of LF 1 g/d (LF-L), LF 3 g/d (LF-H), or 10 ml of saline/d (control) were orally administered for 21 days (Day 0 – Day 20). LF and saline was added to whole milk daily for the LF and the control group, respectively. The average daily weight gain (ADG) values of the calves during the experimental period were 0.53 kg for the control, 0.65 kg for LF-L, and 0.72 kg for LF-H. The ADG in the LF-fed calves were higher (P<0.05) than in the control calves. In the calves treated with LF 3 g/d, the plasma concentration of the growth hormone increased as compared to the calves treated with the control calves, but insulin level was lower than the control calves. The LF feeding induced a strong increase in interleukin (IL)-2 and IL-18 mRNA expression in the peripheral blood mononuclear cells. Plasma interferon-g, IL-1beta, and IL-4 concentrations in the calves fed LF were higher than the control calves. These results suggest that oral LF accelerates the growth and improves the immune function of preruminant calves.

Effects of temperature humidity index (THI) on the behaviour of dairy cows

Sanker, C., Kroos, A.,-T. and Gauly, M., Livestock Production Systems, Animal Science, Albrecht-Thaer-Weg 3, 37075 Göttingen, Germany; mgauly@gwdg.de

The aim of this study was to analyze the effect of temperature (T) and humidity (H) on the preference of dairy cows in a cubicle housing system to stay in an insulated or a non-insulated area. The behaviour of 54 lactating cows was indirectly observed (video analysis) over 24 hours for a total of 16 days form March to August 2011. Inside the cubicle house the animals had the choice to move to an insulated or non-insulated area. Apart from the insulation the facilities and the environment were similar in both areas (drinkers, feed and cubicle). Videos were analysed with the continuous and the point sampling method (30 min. intervals). The following parameters were observed: no. of cows in the areas, percentage of lying animals, no. of changes between the two areas. The temperature (°C) and the relative humidity (%) was recorded every 15 min. using 4 data logger (Tinitag Plus 2, Gemini). THI was calculated using the following formula: THI = $(1.8 \times T + 32) - ((0.55 - 0.0055 \times H) \times (1,8 \times T - 26))$ (NRC, 1971). The measured values were divided into 5 classes. At THI values <55 the lowest no. of cows stayed in the non-insulated area ($53.6 \pm 19.2\%$; $P \leq 0.05$). At THI values >55 the no. of cows in the non-insulated area increased up to $70.5 \pm 18.0\%$. Independent of THI classes over 30% of the cows were lying in the non-insulated area. At THI values between 60 and 65 the highest no. of cows was lying ($44.1 \pm 16.7\%$; $P < 0.05$). Compared to the other parameters the changes between the two areas was lowest at these values ($10.7 \pm 6.4\%$; $P < 0.05$). When given the choice between insulated and non-insulated areas, cows preferred to stay in the non-insulated area independent of the THI.

Serum protein profiles as biomarker for infectious disease status in pigs

Stockhofe-Zurwieden, N.[1], Mulder, H.A.[2], Koene, M.G.J.[1], Kruijt, L.[2] and Smits, M.A.[2], [1]Central Veterinary Institute of Wageningen UR, P.O. Box 65, 8200 AB Lelystad, Netherlands, [2]Wageningen UR Livestock Research, P.O.Box 65, 8200 AB Lelystad, Netherlands; mari.smits@wur.nl

In veterinary medicine, there is a need for early warning tools for disease. In this study the potential of serum protein profiles based on SELDI-TOF MS as biomarker for infectious disease status is explored. Serum samples were analysed from 26 pigs in an experimental model for Porcine Circovirus type 2 (PCV2) infection in combination with either Porcine Reproductive and Respiratory Syndrome virus (PRRSV) (n=9) and Porcine Parvovirus (PPV) (n=8) and from non-infected control animals (n=9). Seldi-TOF MS protein profiles were generated from serum samples taken at different time points, before infection (day 0), before clinical signs became apparent (day 5) and at day 19, when animals showed clear clinical signs as a result of the infections. Statistical analysis on Seldi-TOF proteomics data of 586 proteins was performed using ridge partial least square regression and Anova. Based on protein profiles, classification accuracy of infected versus non-infected animals was very good. At day 5 and 19 post infection, 88% and 93% respectively of infected animals were identified as such. Moreover, protein profiles could distinguish between separate infection models although results for PCV2/PRRSV infected animals were slightly better compared to PCV2/PPV infected animals. Limiting the number of proteins in the profile generally had minor effects on the classification performance. Results from this survey, using data from standardized experimental settings, show that protein profiles based on MS proteomics technology can detect viral infection in pigs in early phase of the disease. Application of proteomics data may have potential for development of biomarkers for disease, but need further exploration in field settings.

Detection of the Aleutian mink disease virus infection by antibody and PCR
Farid, A.H., Hussain, I., Rupasinghe, P.P., Stephen, J., Arju, I. and Gunn, J.T., Nova Scotia Agricultural College, Plant and Animal Sciences, 58 River Road, Haley Institute, Truro, Nova Scotia B2N 5e3, Canada; hfarid@nsac.ca

Eight 10-fold serial dilutions of a 10% (W/V) passage 2 spleen homogenate from mink infected with a local strain of the Aleutian mink disease virus (AMDV) were used to inoculate 64 mink. Eight female black mink were anesthetized and inoculated intranasally with 0.5 ml of each of the dilution series. Blood samples were collected by toe-nail clipping after sedation on days 0 (prior to exposure), 20, 35, 56, 84, 140 and 196 post-inoculation (pi). Antibody was detected by counterimmunoelectrophoresis (CIEP) and the virus was tested by PCR in triplicate. The number of PCR positive mink was 27 and 35 on d 20 and d 35 pi, respectively. Of these, 14 remained PCR positive until d 196 pi, and 5, 4 and 8 became PCR negative by d 84, 140 and 196 pi, respectively. The number of CIEP positive mink was 20 and 31 on d 20 and d 35 pi, respectively, and none turned negative by d 196 pi. Four PCR and CIEP positive mink died after d 35 pi. Five PCR positive mink remained CIEP negative throughout the experiment. The time between inoculation and production of detectible levels of antibody, which was longer than 20 days in 35.5% of the infected mink, and the presence of five CIEP negative infected mink, may contribute to persistent infection of ranches that practice the test-and-kill strategy. The virus was not detected by PCR in one mink until d 140, while this animal remained CIEP-positive throughout this period. The virus might have been sequestered and triggered antibody production, which became active again by d 196 pi. Persistent antibody production and the transient viral replication and short-lived viremia have considerable ramifications for the use of PCR and CIEP as diagnostic tools for AMDV. The results also showed large variations among individual mink for the duration of viral replication and viremia. Most of the mink that received low doses of the virus remained PCR and antibody negative until d 56 pi.

Association of herd characteristics with reproductive performance in dairy herds
Brouwer, H.[1], Bartels, C.J.M.[1], Muskens, J.[2] and Van Schaik, G.[1], [1]Animal Health Service Ltd, Diagnostics, Research & Epidemiology, Arnsbergstraat 7, 7400 AA, Netherlands, [2]Animal Health Service Ltd, Department of Ruminant Health, Arnsbergstraat 7, 7400 AA, Netherlands; h.brouwer@gddeventer.com

In recent decades, herd size and milk yield per cow have increased in many parts of the world. These developments should lead to management adjustments. At the same time reproductive performance of dairy herds has declined. The objective of this study was to quantify the association of herd characteristics with reproductive performance in Dutch dairy herds. Reproductive data of 84% (n=16,603) of the Dutch dairy herds using artificial insemination (AI) were available from July 2006 to June 2009. Reproductive performance was determined as the average predicted calving interval (CI), interval calving to first AI and number of AI per animal for each herd per quarter of the year. Each of the reproductive performance indicators was taken as dependent variable in a linear regression model. Different herd characteristics such (increase of) herd size, purchase of cattle, milk yield and presence of automatic milking system (AMS) were forced into the models. In addition, season and region were included as possible confounders. The results showed that larger herd size was associated with shorter predicted CI, calving to first AI intervals and a lower number of AI per animal (P≤0.01). In addition, faster growth in herd size was associated with shorter predicted CI (P≤0.01). Moreover, herds that purchased cattle had longer predicted CI than herds that did not purchase cattle (P≤0.01). Higher milk yield was associated with a higher number of AI per animal but shorter predicted CI and calving to first AI intervals (P≤0.01). Herds with AMS had higher predicted CI than herds without AMS (P≤0.01). This study showed that several herd characteristics have associations with reproductive performance indicators. These herd characteristics are a starting point for analyzing reproductive performance in dairy herds.

Current state of the healthier goats project in Norway

Lindheim, D.[1], Leine, N.[1], Sølverød, L.[1] and Hardeng, F.[2], [1]TINE Norwegian Dairies SA, Project Healthier Goats, Department of Goat Health Services, Tine Rådgiving, Postboks 58, N-1431 Ås, Norway, [2]Norwegian School of Veterinary Science, Pb. 8146 Dep., N-0033 Oslo, Norway; dag.lindheim@tine.no

The aim of the project is to eradicate the chronic contagious diseases CAE, CLA and Johne's disease from the Norwegian goat population. In 2004 88% of the milk producing herds had CAE antibodies in bulk milk, and a questionnaire in 2003 showed >70% flock prevalence for CLA. Also, Johne's disease is enzootic in parts of southern Norway, and vaccination is compulsory in these areas. The last 20 years increasing disease problems and decreased milk yield, has been recorded. Therefore the project 'Healthier goats' was started in 2001. Kids are taken from the mother ('Snatching') and the 'infected' barn immediately after birth. They are housed in a clean barn, given cow colostrum and raised on milk replacer, water, concentrate and hay. The goats are slaughtered at the end of the lactating period. Thereafter, the barn and the near surroundings are cleaned and disinfected. Healthy goats are then moved back to the barn to start lactation. The sanitized flocks are monitored by clinical observation and antibody detection in serum and bulk tank milk. By march this year 383 (66,7%) farmers have applied to join the voluntary project and 315 flocks are sanitized. It have been observed 3 CAE reinfections, 1 Johne's disease reinfection and 8 CLA reinfections. Re-infected flocks are carefully monitored and test positive animals are slaughtered. We conclude that Healthier Goats have successfully sanitized participating flocks and these show a significant increase in milk yield. The snatching also removes other contagious agents from the flocks, potentially contributing to the observed improved milk yield. Animal welfare is improved by removal of chronic clinical diseases. We further believe that the aim of the project can be reached within the year 2018.

Identification of carriers with mutations causing citrullinaemia and factor XI deficiency in Khuzestan cattle population of Iran

Eydivandi, S.[1], Seyedabadi, H.R.[2] and Amirinia, C.[2], [1]Department of Animal Science, Behbahan Branch, Islamic Azad University, 31585 Behbahan, Khuzestan, Iran, [2]Animal Science Research Institute of Iran (ASRI), Department of Biotechnology, First Dehghan Villa, Shahid Beheshti St., 31585 Karaj, Iran; sirouseidivandi@yahoo.com

In cattle, the autosomal recessive genetic disorders are breed-specific. At least 40 such disorders have been characterized, in which the causative mutation has been identified at the DNA level. With the widespread use of artificial insemination and international trade of semen, the disorders may spread. The aim of this study is to estimate the incidence of two disorders, namely citrullinemia and factor XI deficiency (FXI), in Khuzestan native cows and Iranian Holstein cattle. Citrullinaemia is a rare metabolic disorder and characterized by the lack of argininosuccinate synthetase that is vital for urea cycle. Newborn calves with citrullinaemia appear normal. FXI-deficient calves may be asymptomatic or display several symptoms, e.g. prolonged bleeding and anaemia. Blood samples were collected from five different farms in Holstein populations (N:100) and five regions for indigenous cattle Khuzestan province in Iran (N:230). Genomic DNA was isolated from the blood and the PCR-RFLP method was performed to identify carriers of the disorders. Additionally, mutant citrullinaemia and FXI alleles were confirmed by Genetic Analyzer 3130 ABI. Results showed that none of the screened cows were carriers or mutants of citrullinaemia and FXI disorders. This is the 1st report on the screening Iranian cattle population for citrullinaemia and FXI disorders. Although some studies have reported that carrier animals with citrullinaemia and FXI in Holstein populations exist in different countries, we did not find any carrier individual. In order to prevent from these disorders, the bulls used for artificial insemination should be screened and wider screening programs are needed to prove countrywide disorder free status.

Effects of vaccum level and pulsation rate on milk flow traits in Tunisian Maghrebi camels (*Camelus dromedarius*)

Atigui, M.[1], Hammadi, M.[1], Barmat, A.[1], Khorchani, T.[1] and Marnet, P.G.[2], [1]Institut des régions arides, Laboratoire Elevage et faune sauvage, Fjé, 4119 Médenine, Tunisia, [2]INRA-AGROCAMPUS, Département sciences animales, 65 rue de St Brieuc, 35042 Rennes Cedex, France; marnet@agrocampus-ouest.fr

Machine milking of the she-camel is very recent and many difficulties were associated to morphological and anatomical characteristics of the udder. Six multiparous (12.7 ± 1.2 year of age; 502 ± 8 kg body weight) Maghrebi dairy camels in late lactation were used to study the effect of vacuum level and pulsation rate on milk flow parameters. Camels were reared in intensive system where machine milked once a day. The milking routine included teat washing and drying, machine milking, machine stripping, and teat dipping. Experimental design consisted of a 6×6 Incomplete Latin square to test the effects of 2 vacuum levels: 38 kPa and 48 kPa and 3 pulsation rates: 60, 90 and 120 cycles/min. Milking machine was equipped by pneumatic pulsator. Data were statistically analyzed by GLM and results are presented in means $\pm$ SE. Statistical analyses showed effects on milking rate measurements and milk production for vacuum level. Milk yield, average and peak milk flow rate and yield during the 3 first min of milking were higher ($P<0.05$) in 48 kPa than 38 kPa vacuum level. They were 2.88 ± 0.29 vs. 1.45 ± 0.22 kg, 1.32 ± 0.15 vs. 0.66 ± 0.09 kg/min, 2.52 ± 0.26 vs. 1.27 ± 0.15 kg/min and 2.74 ± 0.31 vs. 1.30 ± 0.23 kg, respectively. Milking duration was not affected by studied treatments. The interaction of vacuum level and pulsation rate was significant for milk yield, average and peak milk flow rate and yield during the 3 first min of milking. The highest records were registered for 48/60 and 48/90 treatments. Pulsation rate did not affect any milking rate measurements or milk yield. In conclusion, high vacuum level is recommended to ensure efficient machine milking for Tunisian Maghrebi camels.

Meat yield in double muscled Piemontese young bulls: official and commercial retail cuts quality production

Lazzaroni, C. and Biagini, D., University of Torino, Department of Animal Science, Via L. da Vinci 44, 10095 Grugliasco, Italy; carla.lazzaroni@unito.it

To valorize the Piemontese cattle breed and its meat production, reputed to be one of the higher among cattle, a commercial dissection trial was carried on 21 carcasses obtained from young bulls, evaluating the official and commercial quality of different retail cuts. Animals were reared in different farms in the same area, slaughtered at 16.3 ± 1.2 months of age and 592.5 ± 36.7 kg of live weight, at an average fattening degree. Carcasses (407.7 ± 30.8 kg) were commercial dissected following local market rules and retail cuts weights (14 from forequarter and 8 from pistola hindquarter) were collected, so as bones, fat and trimmings weights. Dressing percentage was high for all animals ($68.8\pm1.7\%$), so as edible meat in the carcass ($75.4\pm1.7\%$), with a low amount of bones ($15.4\pm0.6\%$), but especially of fat and trimmings ($9.2\pm1.7\%$). Considering the total meat production under an anatomical point of view, $53.0\pm2.4\%$ of edible meat was obtained from forequarters (162.9 ± 16.2 kg), while $47.0\pm2.4\%$ from pistola hindquarters (144.3 ± 9.8 kg). Under a traditional commercial point of view, carcass meat yield was mainly of 1st quality (12 cuts suitable for steak; 132.7 ± 8.8 kg and $43.3\pm2.0\%$), then of 2nd quality (22 roast cuts; 103.3 ± 11.2 kg and $33.6\pm1.7\%$) and of 3rd quality (10 stew cuts; 71.1 ± 5.8 kg and $23.2\pm1.0\%$). However, for the peculiar characteristic of Piemontese meat, local butchers are able to obtain from carcass more meat suitable to be broiled or pan-fried (18 cuts, 165.6 ± 11.0 kg and $54.0\pm1.6\%$, respectively $+32.9$ kg and $+10.7\%$), reducing the amount of meat to be roasted (18 cuts, 73.4 ± 8.9 kg and $23.8\pm1.6\%$) and stewed (8 cuts, 68.2 ± 5.5 kg and $22.2\pm1.0\%$). As meat price is related to retail cuts utilization (as average 7-12 €/kg for 3rd quality, 12-18 €/kg for 2nd quality, and 18-25 €/kg for 1st quality cuts), the local butchers' ability to utilize for steak also several 2nd quality cuts of Piemontese cattle increases the profitability of such production.

Selection for reduced PFAT decreases Isocitrate Dehydrogenase activity

Kelman, K.R.[1,2], Pannier, L.[1,2], Pethick, D.W.[1,2] and Gardner, G.E.[1,2], [1]Murdoch University, Division of Veterinary Biology and Biomedical Sciences, Murdoch, Western Australia, 6150, Australia, [2]Australian Cooperative Research Centre for Sheep Industry Innovation, University of New England, Arimdale, New South Wales, 2351, Australia; k.kelman@murdoch.edu.au

The Australian lamb industry uses Australian Sheep Breeding Values to select for reduced subcutaneous fat depth (PFAT) and increased lean meat yield. Selection for reduced PFAT increases oin muscle weight, with muscularity associated with reduced muscle aerobicity. As isocitrate dehydrogenase activity (ICDH) is a good indicator of oxidative metabolism, we hypothesised that selection for reduced PFAT would decrease ICDH. ICDH was measured within the loin muscle of 1397 lambs and data was analysed using a linear mixed effects model (SAS) with fixed effects for site, kill group within site, sex, birth type-rear type, age of dam, sire type and dam breed within sire type, and random terms for sire and dam. Within this model, covariates such as PFAT, intramuscular fat percentage (IMF) and weight of short loin fat and muscle were included individually to assess their phenotypic association with ICDH. Aligning with our hypothesis, selection for reduced PFAT decreased ICDH by 0.52 µmol/min/g tissue over the 4 unit PFAT range. However, neither short loin muscle nor fat weight demonstrated strong associations with ICDH. This contrasts with the premise of our initial hypothesis that selection for negative PFAT would decrease ICDH via its impact on whole body muscularity and the associated effect on muscle aerobicity. Alternatively, ICDH was strongly associated with IMF, with a 4% decrease in IMF aligning with a 0.84 µmol/min/g tissue reduction in ICDH. Selection for negative PFAT strongly reduces IMF, and when both covariates were used concurrently within the ICDH model, PFAT was not significant. This may imply that the impact of PFAT on ICDH is delivered via its negative impact on IMF, and appears to be independent of whole body adiposity or muscularity.

The effects dietary plants secondary compounds on small ruminants' products quality

Vasta, V., Priolo, A. and Luciano, G., University of Catania, DISPA, via Valdisavoia, 5, 95123 Catania, Italy; vvasta@unict.it

Ruminants' meat and milk quality can be manipulated by the dietary administration of plants secondary compounds (PSC) such as tannins, saponins and essential oils. These PSC can be administered to animals either as pure extract or with their native plant source. Meat and milk fatty acid profile is affected by dietary tannins as these PSC modify ruminal biohydrogenation of the ingested polyunsaturated fatty acids through changes in ruminal ecology and microbial activity. Conversely, administering to sheep or goats essential oils seems not to modify meat and milk fatty acid composition. Dietary tannins improve meat and milk flavour by reducing the ruminal biosynthesis of skatole (3-methyl indole) and indole and their accumulation in animal products. The addition of garlic or juniper essential oils in lamb's diets reduced the off-flavours perception while thyme or rosemary essential oils lowered the rancid-odour perception of meat under display. Moreover, due to their well known anti-microbial properties, dietary essential oils have been shown to reduce bacterial proliferation on meat thus extending its shelf life. It is proven both on lambs and kids that dietary phenolic compounds ameliorate meat oxidative stability and prevent meat form discoloration. It would be of great interest to understand whether PSC affect meat oxidative stability through direct or indirect mechanisms of action. The use of plants rich in secondary compounds or the supplementation of purified PSCs in small ruminants diets seem to be a promising tool for ameliorating products quality.

Meat and health: how to improve the nutritional quality of broiler meat

Haug, A., Nyquist, N.F., Rødbotten, R., Thomassen, M., Svihus, B. and Christophersen, O.A., The Norwegian University of Life Sciences, Department of Animal and Aquacultural Sciences, P.O. Box 5003, 1432 Ås, Norway; anna.haug@umb.no

Meat is a foodstuff that is highly rich in nutrients; it contains proteins and amino acids with the best quality, bioactive components such as glutathione, carnosine and taurine. It is rich in vitamin B12, thiamine, niacin, riboflavin and iron, zinc, potassium, phosphate and magnesium. The bioavailability of the minerals is better in meat than in plants and cereals. Meat is rich in oleic acid and also fatty acids of the n-3 and n-6 family; EPA, DPA and arachidonic acid. Meat also gives satiety. But meat has a potential for improvement and we think that the domestic animal products should be as healthy as possible. By doing simple corrections of the composition of feed; by removing the soybean oil supplement to feed and instead add rapeseed oil and linseed oil, the concentration of arachidonic acid can be halved, and EPA and DPA doubled. This results in a much better n-6/n-3 ratio. By adding organic selenium to the feed, the concentration of selenium in the meat can be equal to fish.

Extending the shelf life of fresh meat: what is technically and legally feasible?

Clinquart, A., Imazaki, P., Sanchez-Mainar, M., Delhalle, L., Adolphe, Y., Dure, R. and Daube, G., University of Liège, Sart Tilman B43b, B-4000 Liège, Belgium; antoine.clinquart@ulg.ac.be

Shelf life of fresh meat is limited mainly by intrinsic and extrinsic factors favoring undesirable microorganism's growth and oxidative reactions. It can be extended by using individual or combined 'hurdles' against deteriorative processes (removal of heat, modification of atmosphere, antimicrobial or antioxidant substances, biopreservation, ...) and by using decontamination techniques (irradiation, high hydrostatic pressure treatment, ...) reducing the initial microbial contamination. According to the EU legislation, fresh meat means meat that has not undergone any preserving process other than chilling, freezing or quick-freezing, including meat that is vacuum-wrapped or wrapped in a controlled atmosphere. No additive can be used (except some organic acids in pre-packed preparations of fresh minced meat). In spite of these constraints, a shelf life of several months can be obtained with chilled meat cuts by combining a temperature of -1 °C and vacuum atmosphere. This exceptionally long preservation can be explained by selection of specific flora producing a competitive effect against spoilage or pathogen microorganisms. More research is needed to better understand the microbial flora dynamics in relation to 'hurdle' conditions and physicochemical or sensory characteristics of the product. At the retail level, high oxygen atmosphere is often preferred in order to produce the attractive red color. In these conditions, the shelf life is reduced to 1-2 weeks due to oxidative unfavorable effects of oxygen on meat lipids and proteins. These last negative effects can be avoided by replacing oxygen with 0.3% carbon monoxide but this technique is not allowed in EU because it has been considered that, 'should products be stored under inappropriate conditions, the presence of carbon monoxide may mask visual evidence of spoilage'. Alternatively, the antioxidant status of the meat can be improved by using feed additives or plant extracts during the primary production.

Cows' feeding and milk and dairy product sensory properties: a review

Martin, B.[1], Verdier-Metz, I.[2], Ferlay, A.[1], Graulet, B.[1], Cornu, A.[1], Chilliard, Y.[1] and Coulon, J.B.[3], [1]INRA, UR1213 Herbivores, Theix, 63122 Saint Genès Champanelle, France, Metropolitan, [2]INRA, UR 545 Fromagères, 20 cote de Reynes, 15000 Aurillac, France, Metropolitan, [3]INRA, Phase, Theix, 63122 Saint Genès Champanelle, France, Metropolitan; bruno.martin@clermont.inra.fr

This review summarises the recent knowledge established on the relationships between the diet of animals and the sensory quality of cattle milk and dairy products. Feeding dairy cattle with pastured grass in comparison with diets based on concentrate or maize silage leads to butter and cheese, more yellow dairy products because of an increase in β-carotene and with a softer texture because of the increase in unsaturated FA to the detriment of the saturated FA with 10 to 18 carbons. The raw milk cheeses issuing from pasture are also generally characterised by their stronger flavour but this effect seems to be cancelled when the milk is previously pasteurised. Within the grass based diets, major differences in sensory characteristics of milk and derived products are also observed according to the preservation of the grass (pastured vs conserved) but this effect is higher in the case of pressed cheeses in comparison to soft cheeses. Conversely, the influence of the grass preservation mode concerns mainly the dairy product yellow colour and carotene content (higher when grass is preserved as silage, by comparison to hay) and also the cheese flavour in the case of large size cheese models. Several recent experiments showed a significant effect of grass botanical composition mainly on milk FA composition and on cheese texture, flavour and appearance. In addition, dietary supplements of plant oil or oilseeds have almost similar effects to pastured grass on FA composition and dairy products texture and in some cases may be responsible for off-flavours in milk.

Modeling and genetics of milk technological properties

Bittante, G., University of Padova, Department of Animal Science, Viale della Università - Agripolis, 35020 Legnaro (PD), Italy; giovanni.bittante@unipd.it

Milk coagulation properties (MCP) are usually measured by computerized renneting meters which are mechanical (e.g. Formagraph) or optical (e.g. Optigraph) devices able to record curd firmness over time (CF_t). Traditionally, MCP are single measurements of: rennet coagulation time (RCT, min), defined as the time interval between the addition of clotting enzyme and the beginning of coagulation; a_{30} (mm), defined as the CF recorded 30 min after the addition of rennet; and k_{20} (min), defined as the time interval from RCT to a CF width of 20 mm. Milk from Holstein Friesian and some Scandinavian cattle breeds exhibits higher incidence of non-coagulating (NC) samples, longer RCT, lower a_{30} and, consequently, a higher frequency of not estimable k_{20} than Brown Swiss, Simmental, and other Alpine local breeds. Exploitable additive genetic variation exists for MCP and has been assessed in different breeds and countries, and using different models (including or not NC). RCT and a_{30} are highly correlated, both phenotypically (r_P) and genetically (r_G), so that a_{30} does not add valuable information to RCT. These limitations are overcome by a new modeling of CF_t, which uses all information provided by computerized renneting meters and accounts for estimation of RCT, potential CF (CF_P), and curd firming rate (k_{CF}). Direct measures of MCP obtained from mechanical and optical devices show similar h^2, and exhibit high r_P and r_G. Moreover, mid-infrared (MIRS) spectra have some potential to predict MCP (r_P between measured and predicted MCP ranges from 50 to 80%). H^2 estimates of RCT and a_{30} predicted by MIRS are higher than those measured, and their r_G are strong (0.93 and 0.87, respectively). Therefore, MIRS provides a reliable and cheap opportunity to improve MCP at population level, as routine acquisition of spectra is already performed for cows in milk recording schemes. Further research is needed on new modeling and on the relationships between MCP and cheese yield and quality at individual/genetic level.

β-lactoglobulin genotype prediction based on milk Fourier transform infrared spectra
Rutten, M.J.M.[1], Bovenhuis, H.[1], Heck, J.M.L.[2] and Van Arendonk, J.A.M.[1], [1]Wageningen University, Animal Breeding and Genomics Centre, P.O. Box 338, 6700 AH Wageningen, Netherlands, [2]FrieslandCampina Research, P.O. Box 87, 7400 AB Deventer, Netherlands; henk.bovenhuis@wur.nl

β-lactoglobulin (β-LG) genotypes of cows are highly relevant to the dairy industry because they are associated with differences in bovine milk protein composition. First, the β-LG B variant is e.g. associated with higher cheese yields because of its association with higher casein concentrations in milk. Second, the β-LG B variant is associated with a lower concentration of β-LG in milk. The latter, in combination with a higher denaturation temperature of the β-LG B variant, results in lower fouling rates of heaters during processing of milk. Third, the β-LG B variant is associated with a change in the composition and properties of the whey: more α-Lactalbumin and less β-LG is produced. This is beneficial for the production of infant nutrition since human milk does not contain β-LG. In this study, we predicted β-lactoglobulin genotypes based on routinely recorded milk Fourier transform infrared (FTIR) spectra using 500 calibration samples. The validation results show that 60% of the cows carrying a β-LG AA genotype, 78% of the cows carrying a β-LG AB genotype and 80% of the cows carrying a β-LG BB genotype were predicted correctly. Furthermore, the prediction of β-LG genotypes based on Fourier transform infrared spectra showed a repeatability of 0.85 which means that only a limited number of repeated records per cow are required. It is discussed how the combined use of predicted β-LG genotypes, pedigree information and β-LG genotypes derived using other methods, can lead to a further improvement of the percentage correctly predicted β-LG genotypes. The presented methodology is easy and cheap and could ultimately provide β-LG genotypes at the individual cow level.

Breed, housing and feeding systems affect milk coagulation traits and cheese yield and quality
Mantovani, R., De Marchi, M., Cologna, N., Guzzo, N. and Bittante, G., University of Padua, Department of Animal Science, Viale Universita, 16, 35020, Legnaro (PD), Italy; roberto.mantovani@unipd.it

This study aimed at analyzing different source of variation on coagulation traits, cheese yield and cheese quality after ripening. Milk was obtained from 30 dairy farms (5 groups of 6 farms each) located in a typical area for cheese production in the North-east of Italy. Dairy farms were selected to represent two breeds reared (i.e. the local Rendena, RE, and the Italian Holstein, IH), and different housing and feeding systems within breed: the traditional farming with tied cows fed hay+concentrate (RE-T-HAY and IH-T-HAY); the farms with tied cows receiving maize silage (RE-T-SIL); and farms using total mixed ration with tied barns (IH-T-TMR), or free stalls (IH-F-TMR). In each farm, bulk milk samples (about 10 kg) were collected twice (i.e. 2 periods: late spring and autumn), transported to the cheese laboratory for the analysis of milk quality, coagulation properties and for cheese making. Cheese was made following a standard protocol and processing milk from 4 different farms/d in 4 11 l vats. Within period all farm-samples were processed within 7 d. After ripening for 90 d, cheeses were analyzed for quality traits. A mixed model analysis for repeated measures (i.e. the 2 periods) was carried out, and contrasts between breeds and combinations of housing-feeding systems were performed. Milk from RE farms showed better coagulation properties as respect to IH (RCT 13.58 vs. 18.65 min; a_{30} 34.16 vs. 22.67 mm), but a lower cheese yield after ripening (8.23 vs. 8.67%), due to the lower protein and fat contents. Milk from RE farms produced cheese with greater n-3 fatty acids (0.59 vs. 0.42) and CLA contents (0.79 vs. 0.59) than milk from IH farms. Moreover, the T-HAY system compared with the counterparts within breed showed a greater carry-over effects (i.e. from milk to cheese) for the n-3 fatty acids. In conclusion, both breed and feeding could be related to the production of cheese more valuable for the human health.

Genome-wide association of fatty acids from summer milk of Dutch dairy cattle

Bouwman, A.C., Visker, M.H.P.W., Van Arendonk, J.A.M. and Bovenhuis, H., Wageningen University, P.O. Box 338, 6700 AH Wageningen, Netherlands; aniek.bouwman@wur.nl

Bovine milk fat composition influences the nutritional and technological properties of milk and dairy products. Milk fat composition varies between cows and seasons. This variation can be due to several factors, e.g. factors related to nutrition and genetics or possibly an interaction between them. Previously we identified genomic regions responsible for variation in fatty acids from winter milk samples. The aim of present study is to perform a genome-wide association of fatty acids from summer milk samples. Two milk samples from 2,000 first lactation dairy cows were collected. The first milk sample was taken in winter, when Dutch cows are mainly kept indoors. The second milk sample, from the same set of cows, was taken in summer, when Dutch cows are often grazing outdoors for a part of the day. Milk fat composition was measured by gas chromatography. The most abundant fatty acids, i.e. C4:0-C18:0, C10:1-C18:1 and CLA, were analyzed. Summer milk contained more unsaturated fat and less saturated fat than winter milk, e.g. fat contained on average 29.2% C16:0 and 20.6% C18:1 in summer, and 32.6% C16:0 and 18.2% C18:1 in winter. Phenotypic correlations between winter and summer samples were low: 0.40-0.57 for short and medium chain fatty acids, and 0.36-0.40 for long chain fatty acids. The cows were genotyped for 50,000 markers. For each individual marker we determined whether there was a significant association with the milk fatty acids. Results showed that major regions detected on chromosome (BTA) 14, 19, and 26 in winter samples could be confirmed in summer samples. On BTA 14, DGAT1 showed association with several fatty acids. On BTA 26, SCD1 showed association with unsaturated fatty acids. Some additional regions could also be confirmed in the summer sample, such as the associations on BTA 13 and 17. However, not all associations could be confirmed in the summer samples, such as the associations on BTA 2 and 27. Novel regions were also detected in the summer sample.

Detection of QTL controlling cheese processing properties in a Holstein x Normande crossbred population

Xue, Y.[1], Larroque, H.[1], Barbey, S.[2], Lefebvre, R.[1], Gallard, Y.[2], Ogier, J.C.[3], Colleau, J.J.[1], Delacroix-Buchet, A.[3] and Boichard, D.[1], [1]INRA, UMR1313, GABI, 78350 Jouy en Josas, France, [2]INRA, UE326, Domaine du Pin, 61310 Exmes, France, [3]INRA, UMR1319, MICALIS, 78350 Jouy en Josas, France; rachel.lefebvre@jouy.inra.fr

A QTL detection experiment has been carried out in 'Le Pin' INRA experimental farm by crossing Holstein and Normande dairy cattle. To study cheese processing ability, 'Camembert' type cheese was produced from 25 litres milk from 649 individual F2 primiparous cows. We present results on gelification time (RS) and time for firming (K20S) at formagraph test, gel firmness 30 min after adding rennet (A30S), time between between rennet addition and milk curdling in the cheese vat(RCT), fat/dry matter ratio (FDR) and cheese yield (CY). Milk was standardized for fat to protein ratio before processing. These stringent conditions were defined to avoid confounding effects with fat and protein contents and, as a consequence, direct effects of fat or protein content on cheese making properties should not be detected. Accordingly, a much smaller number of QTL are expected. Animals were genotyped with the Illumina 54k Beadchip, as well as for caseins, beta-lactoglobulin and DGAT1 genes. Combined linkage and linkage disequilibrium analysis was used, including the effects of the known genes. The most unexpected result was the strong effect of DGAT1 (BTA14) on nearly all traits, including cheese yield, with a favourable effect of the K allele, showing the likely influence of the fat globules size on cheese processing and water retention. Not surprisingly, formagraph traits and RCT were mostly determined by the casein loci, particularly the kappa- and beta-caseins, but no effect of these loci was found on cheese yield. Additional strong QTLs were found on chromosomes 3 (42-45 cM), 23 (43 cM), 28 ((3 cM), and 29 (35 cM) for firmness and times for firming but no QTL other than DGAT1 was found on cheese yield. No clear QTL was found for FDR, in agreement with the standardization procedure.

Characteristics of herds that produce milkfat with a higher than average concentration of unsaturated fatty acids

Silva-Villacorta, D.[1], Lopez-Villalobos, N.[1], Blair, H.T.[1], Hickson, R.E.[1] and Macgibbon, A.[2], [1]Massey University, Institute of Veterinary, Animal and Biomedical Sciences, Private Bag 11-222, Palmerston North 4442, New Zealand, [2]Fonterra Co-operative Group Limited, Dairy Farm Rd, Palmerston North, New Zealand; d.silvavillacorta@massey.ac.nz

In recent years, there has been a growing interest in the manipulation of milkfat composition at the farm level, prompted by the growing trend in consumer's demands towards healthy and functional dairy products. The nutritive value and manufacturing characteristics of milk can be improved by increasing the concentration of unsaturated fatty acids (UFA) in milkfat. This research aimed to compare the characteristics of farms that produce milkfat with a higher than average concentration of UFA under a seasonal pastoral system. A stochastic simulation model based on a variance-covariance matrix of milk traits and liveweight was used to generate 2 groups of farms. The Average UFA group (30 farms) produced milkfat with a UFA concentration around the population average, the High UFA group (30 farms) produced milkfat with a higher than average UFA concentration. All simulated farms were 130 hectares, had 361 cows and calved in spring. Compared to the Average UFA farms, farms in the High UFA group produced significantly lower fat (223.1 vs 205.5 kg/cow, SEM=0.5) and protein (164.0 vs 157.8 kg/cow, SEM=0.3) yields per lactation, had lower fat (4.65 vs 4.30%, SEM=0.005), and protein (3.42 vs 3.30%, SEM=0.002), percentages, had lower feed demand per cow (5626 vs 5539 kgDM/year), but had higher liveweight per cow (533.8 vs 560.7 kg, SEM=0.7) and percentages of UFA (30.06 vs 34.02%, SEM=0.02) and CLA (0.936 vs 1.003%, SEM=0.001) in milkfat. There were not significant differences in milk yield between the High and Low UFA groups (4619 vs 4625 L/lactation, SEM=7.0). These results highlight the need to develop new payment systems to reward concentration of UFA in milk so that farm profitability is not affected due to losses in production of fat and protein.

The effect of genotypes and housing conditions on colour of chicken meat

Konrád, S. and Kovácsné Gaál, K., University of West Hungary, Faculty of Agricultural and Food Sciences, Institute of Animal Sciences, Vár 4, H-9200 Mosonmagyaróvár, Hungary; konradsz@mtk.nyme.hu

The aim of this study was to compare the colour of valuable meat parts (breast, thigh) for Yellow Hungarian, Yellow Hungarian cross with different meat type cocks (S77, Foxy Chick, Redbro, Hubbard Flex, Shaver Farm) (all bred under free-range conditions for 84 days) and Ross 308 (fattened in industrial conditions for 42 days). The instrumental colour measurement of the breast and thigh meat was carried out using a Minolta CR-300. During measurement the lightness (L*), redness (a*), yellowness (b*) values of the meat were measured for the breast samples on 6 different locations each, on the thigh samples on 12 locations each, and on the basis of the latter two parameters the croma value (C*) indicating the brightness and the density of the colour was calculated. Moreover the colour difference values were calculated at the breast and thigh meat for the genotypes examined. Based on the colour difference values it could be established to what extent the variations experienced in the measured parameters (lightness, redness, yellowness) between the different genotypes could be percepted by human vision. The results showed that the genotype and the keeping technology influence the colour of breast and thigh meat: the L*, a* and b* values were significantly lower for Ross 308 broilers than chickens fattened free-range (51,93 vs. 58,67; 1,99 vs. 3,10; 3,72 vs. 5,17). For the different genotypes as opposed to the breast meat smaller variance was measured for the parameters qualifying the colour of the thigh meat (lightness: 54,00 vs. 53,26; redness: 10,34 vs. 11,03; yellowness: 7,60 vs. 7,26). The colour difference values showed that compared to the other genotypes the breast meat of Ross 308 broilers and purebred Yellow Hungarian in mixed sexes differed well visibly and to a large extent, moreover between the crossed genotypes there was a well visible variation. The colour difference values of the thigh meat samples were smaller compared to that of the breast meat.

Possibilities to improve the lipid structure of the pig meat during the finishing period

Lefter, N.[1,2], Habeanu, M.[1], Hebean, V.[1] and Dragotoiu, D.[2], [1]National Research-Development Institute for Animal Biology and Nutrition, Laboratory of Animal Nutrition, Calea Bucuresti, nr 1, 077015, Balotesti, Romania, [2]University of Agronomic Science and Veterinary Medicine, Animal Nutrition, Bulevardul Marasti, nr. 59, 011464, Romania; ciuca_nicoleta@yahoo.com

The purpose of the experiment was to improve pig meat quality by increasing its content of polyunsaturated fatty acids, mostly omega-3. The experiment carry on 30 finishing Large White pigs (30 days) assigned to 3 groups (C, E1 and E2). The group C (n=10) received conventional feedstuffs, E1 (n=10) received ecological feedstuffs and 3% camelina oil (12.84% acid α-linolenic) as energy source; group E2 (n=10) received the same diet as group E1, supplemented with an antioxidant produced from plants. The camelina oil diet increased significantly (P=0.0045) the fatty acids C18:3n-3 in longissimus dorsi (LD) in groups E1 (2.40 times) and E2 (2.90 times), compared to group C. The content of α-linolenic acid increased highly significantly (P<0.0001) in the semitendinosus (ST) muscle both in group E1 (3.83 times) and in group E2 (3.90 times), compared to group C. The content of α-linolenic acid of the subcutaneous fat increased distinctly (P<0.0001) in both experimental groups (3.22 and 3.76 times, respectively) compared to group C. The n-6/n-3 ratio in LD muscle decreased significantly (P=0.0002) in groups E1 (2.29 times) and E2 (3.04 times), compared to group C. The n-6/n-3 ratio in LD muscle decreased highly significantly (P<0.0001) in both experimental groups (2.97 and 3.68 times, respectively) compared to group C. The n-6/n-3 ratio in the subcutaneous fat decreased significantly in both experimental groups (3.41 and 4.21 times, respectively) compared to group C. These results suggest that the administration of 3% camelina oil in the diets for finishing pigs changed significantly the omega-3 fatty acids (FA) content, as well as the n-6/n-3 ratio.

The content of benzo(a)pyrene in smoked meat products

Miculis, J., Valdovska, A. and Zutis, J., Research Institute of Biotechnology and Veterinary Medicine, Instituta street 1, LV 2150, Sigulda, Latvia; sigra@lis.lv

Polycyclic aromatic hydrocarbons (PAH`s) can significantly influence smoked meat quality and safety. Toxicological studies on individual PAHs in animals, mainly on the PAH benzo(a)pyrene, have shown various toxicological effects, such as haematological effects, reproductive and developmental toxicity and immunotoxicity. It has been concluded that benzo(a)pyrene (BaP) is a probable human carcinogen. One significant source of BaP in the human food chain is smoking of meat. Smoke not only gives special taste, colour and aroma to food, but also enhances preservation due to the dehydrating, bactericidal and antioxidant properties of smoke. Today it is supposed that this technology is applied in many forms to treat 40-60% of the total amount of meat products. Therefore the aim of our investigation was to determine the conetents of BaP in industrially smoked different meat products. Results were summarized and compared with maximum acceptable levels set by European Commision regulation (EC) No 1881/2006. Twenty two different smoked meat products samples were taken and packing according to the sampling procedure. After the samples treatment analyses were carried out using method HPLC. All data were presented as mean with standard deviation, significance was set at P<0.05. Analyzed samples contained BAP in concentrations below the EU permitted maximum limit. The highest content of BaP was detected in a breakfast ham (4.05 μg kg^{-1}) but the lowest BaP content was detected in smoked pork chop (0.11 μg kg^{-1}). This study clearly demonstrantes that the production of smoked meat products with BaP levels less than 1.00 μg kg^{-1} is possible in non-intensively smoked products.

Limiting enzymes of docosahexaenoic acid synthesis in bovine tissues
Cherfaoui, M., Durand, D., Cassar-Malek, I., Bonnet, M., Bauchart, D. and Gruffat, D., INRA, UR Herbivores, 63122 St-Genès-Champanelle, France; maya.cherfaoui@clermont.inra.fr

N-3 highly unsaturated fatty acids (n-3HUFA), notably docosahexaenoic acid (DHA; 22:6n-3), are essential for human health. However, consumption of n-3 HUFA in western countries is sub-optimal. These last decades, the focus was on the enrichment of human feed, such as meats and more precisely beef, in n-3 HUFA. Several nutritional studies investigated the effects of supplying the bovine diet with linseed rich in linolenic acid (ALA), the primary precursor of DHA. Although this diet led to a significant increase of ALA in beef, DHA content in beef stayed low, suggesting a limited conversion of ALA into DHA in ruminant. This study aimed at a better understanding of molecular mechanisms of DHA synthesis in bovine tissues. Tissue FA composition was achieved by GLC and mRNA abundances of all enzymes and nuclear factors involved in the conversion of ALA into DHA were assessed by qPCR in liver, intermuscular adipose tissue (IM-AT) and Longissimus thoracis muscle (LT) from Limousin bulls fed a concentrate-based diet. In the liver, mRNA coding for proteins involved in DHA synthesis was abundant in agreement with the significant content of DHA in this organ. In contrast, in IM-AT, the very-long chain acyl CoA synthetase (VLCS) mRNA was undetectable suggesting an interruption of synthesis from 24:6n-3. In LT, elongases 2 and 5 and VLCS mRNAs were undetectable, leading probably to an abortion of synthesis from 18:4n-3. However, these results disagree with the significant amount of DPA (a precursor of DHA) present in muscles. It can be hypothesized that 1) mRNA levels are not representative of the quantity/activity of proteins, 2) muscle DPA comes from the uptake of DPA synthesized and secreted by the liver (hypothesis that can be extended to DHA). Further studies are needed to investigate 1) the physiological and nutritional factors that can stimulate limiting steps of DHA synthesis, 2) levels of mRNA coding for proteins involved in tissue DHA uptake.

A gene expression approach to the genetics of dry-cured ham attributes and texture defects
Pena, R.N.[1,2], Guàrdia, M.D.[3], Reixach, J.[4], Amills, M.[5] and Quintanilla, R.[1], [1]IRTA, Genètica i Millora Animal, 191 Rovira Roure Av, 25198 Lleida, Spain, [2]UdL, Producció Animal, 191 Rovira Roure Av, 25198 Lleida, Spain, [3]IRTA, Tecnologia dels Aliments, Finca Camps i Armet, 17121 Monells, Spain, [4]Selección Batallé, S.A., Av Segadors, 17421 Riudarenes, Spain, [5]UAB, Ciència Animal i dels Aliments, Campus UAB, 08193 Bellaterra, Spain; romi.pena@prodan.udl.cat

The production of dry-cured meat products is notably affected by the quality of the raw material: poorly performing pork can lead to the development of texture defects (such as adhesiveness and pastiness) and cause significant economic losses to the sector. In order to study if dry-cured ham texture attributes and defects are affected by mRNA levels in fresh pork, we studied the global expression profiles of 105 fresh gluteus medius samples of commercial Duroc pigs used in the production of fine quality cured products. Dry-cured hams from these animals were sensory analysed by trained panellist using Quantitative Descriptive Analysis (including appearance, odour, taste/flavour and texture attributes). Comparison of expression data of animals with extreme values for pastiness led to the identification of differentially expressed genes related to the oxidation of phospholipids, energy balance, cell union and transmembrane calcium transport. Four of these genes (ETS2, PLCXD2, PPARGC-1 and MYF6) were scanned for polymorphism in the coding and regulatory regions, which led to the identification of 29 mutations. After genotyping 350 Duroc pigs, an association study was carried out with dry-cured ham organoleptic traits and fresh meat fat composition attributes. The most consistent results were obtained for SNPs in the MYF6 promoter, which associated with the percentage of intramuscular fat and saturated fatty acids content in raw material and with values for pastiness of dry-cured hams. Polymorphisms in the 3'UTR of the ETS2 gene were also associated with the percentage of saturated fatty acids content. We are now extending this study to the remaining validated genes.

Fatty acids profile and cholesterol content in muscles of young bulls of Polish Black and White breed and commercial crossbreds after Limousine sires
Grodzicki, T., Litwinczuk, A., Barlowska, J. and Florek, M., University of Life Sciences in Lublin, Department of Commodity Science and Raw Animal Materials Processing, Akademicka 13, 20-950 Lublin, Poland; anna.litwinczuk@up.lublin.pl

The aim of the present study was to analyze the fatty acid composition and concentration of total cholesterol concentration in meat from young bulls of Polish Holstein-Friesian breed of Black and White variety and commercial crossbreeds between Black and White cows and Limousine sires. The research material consisted of the samples collected from musculus longissimus lumborum (MLL) and musculus semitendinosus (MST) obtained from 46 carcasses of young bulls at the age of approximately 18 months. Animals represented two genotypes: the Polish Holstein-Friesian breed of Black and White variety, and crossbreeds from Black and White cows and the Limousine sires. Animals were maintained in the semi-intensive system. The basic animal feed during the summer period was the grass forage and maize silage, while in the winter period animals were fed with maize silage. Dietary dose was supplemented with pasture hay and crushed cereal meal. Statistical analysis was performed with two-way ANOVA with interaction using STATISTICA software to identify the impact of genotype and muscle. Significance of differences between mean values for the evaluated groups was determined by the LSD Fisher's test. The genotype significantly influenced (P<0.01) the higher percentage of PUFA, including CLA (all isomers), and PUFA/SFA ratio in muscles of Limousine crossbreeds (5.20, 0.32, and 0.10 for MLL, and 3.69, 0.19 and 0.07 for MST, respectively) compared with muscles of HF young bulls (2.62, 0.09 and 0.05 for MLL and 3.23, 0.05 and 0.06 for MST, respectively). Type of muscle significantly (P<0.01) affected the total cholesterol content. The higher content of cholesterol was determined in MLL (57.93 mg/100 g of HF young bulls and 51.05 mg/100 g of Limousine crossbreeds) in comparison with MST (respectively 44.30 mg/100 g and 43.33 mg/100 g).

When MIR spectrometry helps to promote a local and vulnerable bovine breed
Colinet, F.G.[1], Dehareng, F.[2], Dardenne, P.[2], Soyeurt, H.[1,3] and Gengler, N.[1,3], [1]University of Liege, Gembloux Agro-Bio Tech, Animal Science Unit, Passage des Deportes 2, 5030 Gembloux, Belgium, [2]Walloon Agricultural Research Centre, Valorisation of Agricultural Products Department, Ch Namur 24, 5030 Gembloux, Belgium, [3]National Fund for Scientific Research, Rue Egmont 5, 1000 Brussels, Belgium; Frederic.Colinet@ulg.ac.be

The dual purpose Belgian Blue breed (DP-BB) is a vulnerable breed rooted in the tradition of the Walloon Region of Belgium. DP-BB has interesting features (e.g. robustness, good longevity and ease of calving). Due to its dual purpose type, income generated by both milk and meat is more stable and more flexible in responding to market fluctuations. Registered DP-BB cows are milk recorded (one of the conditions for them to be registered as DP-BB and therefore to get AEM subsidies). Since near 4 years, during routine milk recording, nearly all mid-infrared (MIR) spectra generated at the milk labs and the information of test-day records were collected in a database. Calibration equations using the MIR spectrometry were developed permitting the prediction of several bovine milk components (e.g. fatty acids (FA)). Their application on the MIR spectral database would allow comparing milk composition from 920 DP-BB and 52,497 Holstein cows (selected cows had a proportion of Holstein or DP-BB in their breed composition of at least 90%). On average, each cows had 6 test-day records with MIR spectra in the database. MIR predictions were analyzed using GLM procedure with 5 fixed effects (breed, herd, lactation number, month of test-day recording and lactation stage); values presented are lsmeans (± s.e.) of the breed. Although milk and fat yields were lower for DP-BB, their FA proportions in fat were different from Holstein. Indeed, there are 66.0% (0.2) and 67.6% (0.0) of saturated FA in fat of milk for DP-BB and Holstein, respectively. Furthermore, the DP-BB milk fat was richer in omega-9 (20.7% (0.2) vs. 19.6% (0.0) for DP-BB and Holstein, respectively). Use of MIR predictions may help stakeholders to promote milk and future dairy products from DP-BB.

Ratio of n-6:n-3 in the diets of beef cattle: effect on growth, fatty acid composition, and taste of beef

Mcniven, M.A.[1], Duynisveld, J.L.[2], Turner, T.[1] and Mitchell, A.W.[1], [1]Atlantic Veterinary College, University of Prince Edward Island, 550 University Ave., Charlottetown, PE C1A 4P3, Canada, [2]Agriculture & Agri-Food Canada, Nappan Research Station, Nappan, NS B0L 1C0, Canada; mcniven@upei.ca

Effects of feeding heat-treated canola (C), soybean (S) and flax (F) or 1:1 mixtures on growth and slaughter characteristics, taste and fatty acid (FA) composition of beef were investigated to determine the potential of improving the nutritional quality of beef for human consumption. For Trial 1 (48 steers), dietary treatments were C, S, or F (roasted or extruded) to give equivalent lipid and protein contents. For Trial 2 (80 steers), the dietary treatments were: S:F (1:1), S:C (1:1), C:F (1:1) and S:F:C (1:1:1) (roasted or extruded). Soybean meal and soybean oil were used to give equivalent lipid and protein contents to each experimental diet. The basal diet consisted of grass silage, barley grain, vitamins and minerals. Steers were fed for a minimum of 120 d then slaughtered at a uniform degree of finish. Growth and slaughter characteristics of the steers were slightly affected by dietary treatment in that the soybean-fed steers consumed more feed and had a higher average daily gain than the canola or flax-fed animals in Trial 1. There was no difference in taste panel parameters for any of the treatments. Inclusion of flax in the diet increased the total n-3 content of meat. Similar results were found for canola and C18:1n-9 although this was not the case for soybean and the n-6 FA. For the n-6 FA in the phospholipid and neutral lipid fractions of the meat samples, levels were correlated with high dietary levels of n-6 or n-9 with low levels of n-3 while for the n-3 FA, levels were correlated with high dietary n-3 levels and low n-6 levels. Oilseed processing method did not have an effect on any fatty acid levels in the meat. It is possible to modify the FA composition of beef meat toward a healthier profile by including heat-treated oilseeds in the diet.

A novel quality denomination for beef: strategies, limits and differences between stakeholders

Olaizola, A.[1], Bernués, A.[2], Blasco, I.[2] and Sanz, A.[2], [1]Universidad de Zaragoza, Miguel Servet 177, 50013 Zaragoza, Spain, [2]CITA de Aragón, Avda. Montañana 930, 50059 Zaragoza, Spain; asanz@aragon.es

Beef production is undergoing a difficult situation in Spain due to the rising prices of feeds and energy, and the decreasing demand of red meat, among other factors. One of the strategies pursued by the industry to face this situation is the differentiation of meat by its quality. In this study we perform a prospective analysis of a novel beef quality label produced by the endangered breed 'Serrana de Teruel'. A Delphi method was used to identify the main ideas and opinions of different stakeholders (producers, processors and distributors, consumers and government) in terms of (1) most relevant production factors, (2) quality attributes of the new meat, (3) opinions on the perception of quality by consumers and (4) best strategies of marketing. The most important factors for each of these areas were: the utilization of pastures and conservation of landscape, the ageing of the meat, the figure of the butcher as a guarantee of quality for consumers, and the establishment of a commercial brand, respectively. However, the concept of quality is often understood differently by the different actors along the meat chain and by the consumers, leading to failure to transfer information effectively between them. In this study, a Kruskal-Wallis test allowed to identify significant differences among the opinions of the different stakeholders', which referred to the need to reduce costs through extensification of production, the role of pastures and landscape as extrinsic attributes of the meat, the access to grazing infrastructures, the importance of technical management (duration of fattening, carcass conformation and castration), the importance of quality certification and GM-free feeding systems for consumers, and the relevance of a new commercial brand. These aspects should be considered to improve the communication between stakeholders along the chain.

Meat lamb characteristics affected by different amounts of *Camelina sativa* feeding integration.

Nicastro, F.[1], Facciolongo, A.[2], Depalo, F.[1], Demarzo, D.[1] and Toteda, F.[1], [1]Università degli Studi di Bari 'Aldo Moro', Dipartimento di Produzione Animale, Via Amendola 165/A, 70126 Bari, Italy, [2]C.N.R., Istituto di Genetica Vegetale, Via Amendola 165/A, 70126 Bari, Italy; nicastro@agr.uniba.it

This paper deals with the development of feeding strategies for sheep and goats aimed at obtaining 'functional meat'. After weaning, part of fat feed in lambs slaughtered at 70 days was replaced by three levels of *Camelina sativa* feeding. Eighteen lambs 'Gentile di Puglia' were weaned at 40 days and divided into three homogeneous lots: (a) control; (b) level 5%; (c) level 10%. Performance *in vivo*, carcass quantity-quality characteristics, chemical and physical properties of meat and fatty acid profile of lipids of longissimus lumborum muscle were estimated. Data were processed by GLM procedure of SAS software package 2000. The lean leg proportion of lambs in the first lot is higher than that in b (70.8 vs. 69.0%, $P<0.05$) while the fat proportion increases only in lot c (6.7 vs. 4.5, $P<0.05$). The loin of the lambs receiving the highest level of seeds showed a higher proportion of fat compared to a and b lots (66.6 vs. 61.9-62.5%, $P<0.05$). The shrinkage of lamb carcasses due the fridge is major in lot b), of statistical significance only relevant to lot c (2.8 vs. 2.1, $P<0.01$). pH24 was strongly ($P<0.01$) influenced by the diet (5.8, 5.6, 5.5 respectively in lots c, a and b). The colorimetric characteristics show that the lambs meat in lot c is less bright than the other two (L*: 38.9 vs. 45.1, 45.0, $P<0.01$), while the red color of lambs meat in lot b appears brighter (a*: 10.9 vs. 8:41, $P<0.05$ and 10.9 vs 7.5, $P<0.01$ respectively for level 10 and 0) and the same is for yellow (b*: 12.2 vs. 9.4-9.5, $P<0.01$). The meat of lambs fed with a larger amount of lipidic feed resulted markedly harder than that of lambs fed with a lower amount (1.65 vs. 2.30 kg/cm², $P<0.05$). The meat of lambs fed with *Camelina sativa* has lipids richer in unsaturated fatty acids and better healthy properties.

Effects of dietary supplementation with mineral/vitamin mix on beef meat quality

Vincenti, F., Mormile, M. and Iacurto, M., Council for Research in Agriculture, Research centre for Meat Production and genetic Improvement, Via Salaria, 31, 00015, Italy; federico.vincenti@entecra.it

The object of this study was to evaluate the effects of a mineral/vitamin mix supplementation into the diets of finishing cattle on quality characteristics of meat beef. Thirty Charolais young bulls were fed with the same diet: without any supplementation, control group (C); with mineral/vitamin mix supplementation before slaughter, for either 30 days (Mvit30) or 60 days (Mvit60). Physical meat quality parameters (drip and cooking losses, pH, WBS on cooked (WBSc) and raw (WBSr) meat, colour parameters (L, a*, b*, Chrome, Hue) were investigated on Longissimus thoracis (LT) muscle after 7 days (7d) and 14 days (14d) of ageing. Statistical analysis was performed with SAS 9.1.3, using the General Linear model procedure nested with diet effect. Data showed that different mineral/vitamin mix times supplementation had a significant influence on meat colour stability. Significant differences were found, both after 7d and 14d, on L value and Hue angle; after 14d, on a* value, as reported by Vincenti *et al.* PH data confirmed meat colour differences, in fact, after 14d of ageing time, LT muscle, obtained from animals belong to the Mvit60 group, showed a lower pH value (5.54) than the others. Mineral/vitamin mix supplementation affected water losses, in fact, after 7d, LT muscle of Mvit60 animals showed a lowest drip loss (0.96%) and after 14 d, showed a lowest cooking loss (25.42%, P=0.0057). As reported by Mitsumoto *et al.* and Arnold *et al.*, no differences on both WBSc and WBSr were found. These results point out that supplementation with mineral/vitamin mix before slaughter increases meat stability during retail display, and would be an effective method for improving water holding capacity in fresh meat.

Effects of oil sources on feedlot performance and fatty acid profiles of meat lambs
Karami, M., Agriculture and Natural Resourses Research Cenet of shahrekord, Animal Science, Shahrekord-Edsfahan main Street, 415, Iran; karami_morteza@yahoo.com

A study was conducted to investigate the effects of different oil sources (palm and canola oil) on feedlot performance and mutton fatty acid profiles of native male lambs. Locally available breeds of livestock are important economic resources and needs to develop meat quality and change meat poly unsaturated fatty acids to saturated fatty acids ratios. Oil palm is one of the important oil that extracted of palm seeds. Twenty-four male lamb were assigned to a completely randomized design were two kinds of oils (palm and canola oil). They were used for the first trial which lasted 16 weeks as fattening period and two weeks of adjustment period. Rations were iso-metabolizable energy (2.5 M Cal/kg DM intake) and iso-nitrogenous (14% crud protein on dry matter basis. Diet was consisted 65% concentrate and 35% oil palm frond. The rations were mixed and fed *ad libitum*. The lambs bought of local folks and divided randomly in two treatments (12 lambs) of oil (palm and canola oils) and kept lambs individually in each box. At the end of feedlot period (16 weeks) all of the lambs were slaughtered. The longissimus muscle has been taken as sample for evaluated fatty acids profiles. Total fatty acids from meat samples were extracted using a chloroform-methanol solvent extraction system as described by Folch *et al.* Data were subjected to variance analysis by comparing the least square means by GLM procedure at a significant level of $P<0.05$. The mean of initial, final weight, daily feed intake, feed conversion ratio and daily weight gain were not significant. In general, incorporation of Palm and Canola oils into the animal diet had significant effects on the muscle fatty acid profile composition of the important commercial muscle cut, such as longissimus muscle. The polyunsaturated fatty acid (PUFA) n-6/PUFA n-3 ratio was significantly increased due to the dietary supplement of Canola oil.

Cis and trans isomers of 18:1 from *Longissimus thoracis* muscle are modified differentially in Angus, Limousin and Blonde d'Aquitaine bulls fed linseed-enriched diets
Bauchart, D., Thomas, A., Lyan, B., Gruffat, D., Micol, D. and Durand, D., INRA, INRA, UR1213, Herbivore Research Unit, Research Centre of Clermont-Ferrand/Theix, 63122 Saint-Genès-Champanelle, France; bauchart@clermont.inra.fr

Cis and trans 18:1 are generated by rumen microflora and then deposited in tissues. Cis 18:1 exists as 7 positional isomers (c9-c15) principally composed of oleic acid (c9-18:1) beneficial for human health. Trans 18:1 exists as 11 isomers (t6-t16) principally composed of t9 and t10 isomers, detrimental to human health (pro-atherogenic), and of t11 isomer beneficial (anti-atherogenic). This study aimed to compare the effects of the addition of linseed rich in 18:3 n-3 in the diet on cis and trans isomers of 18:1 in the Longissimus thoracis muscle of bulls from 3 breeds varying in lipogenesis capacity (Angus, A, > Limousine, L, > Blonde d'Aquitaine, BA). Bulls were fed, for a 100 d finishing period, either a concentrate/straw based diet (80/20) alone or the same diet enriched with linseed ± antioxidants (vitamin E and polyphenols). Cis and trans 18:1 were purified by preparative HPLC and their isomer distributions determined by GC-MS. In the control diet, c9 isomer represented 85.0-92.5% of total cis 18:1 and c11 isomer only 5-12%. Addition of linseed decreased c9 isomer with a corresponding increase in c11-c12 isomers only in BA bulls. Total cis 18:1 level was 4-5 times higher in A than in BA and L bulls ($P<0.01$), but linseed diets decreased cis 18:1 only in A bulls by 35% ($P<0.05$). Trans 18:1 was 3 to 4 times higher in A than in L and BA bulls. Linseed ± antioxidants improved their nutritional value by decreasing t9 isomer by 28-33% ($P<0.01$) with a corresponding increase in t13-t15 18:1. In conclusion, beef cis and trans 18:1 differed quantitatively between the 3 breeds (A>> L> BA) and addition of linseed ± antioxidants to the diet favored, in all breeds, trans 18:1 isomers beneficial to human health (study supported by the European ProSafeBeef program, FOOD-CT-2006-036241).

Protein markers of beef tenderness in young bulls from different breeds
Picard, B., Cassar-Malek, I., Kammoun, M., Jurie, C., Micol, D. and Hocquette, J.F., INRA, URH, 63122, Theix, France; jfhocquette@clermont.inra.fr

This study is part of the EU ProSafeBeef project on producing safe beef and beef products with enhanced quality characteristics. The aim of this work was to validate potential markers of beef tenderness identified in previous research. Two muscles Longissimus thoracis (LT) and Semitendinosus (ST) from 74 young bulls of three breeds: Angus (AN), Limousine (LI), Blonde d'Aquitaine (BA) were collected immediately after slaughter of animals in the INRA experimental abattoir, frozen in liquid nitrogen and stored at -80 °C. The abundance of 13 proteins was determined by dot-blot analysis using specific antibodies. These proteins are involved in muscle structure, contractile and metabolic properties, proteolysis, oxidative stress and apoptosis. Protein amounts were correlated with tenderness data obtained by sensory analysis and mechanical measurement (Warner-Bratzler). The main results confirmed that 9 proteins were putative biological markers of tenderness. They confirmed the importance of proteins involved in apoptosis such as heat shock proteins (Hsp: 40, 70-1B, 70-8 and αb-crystallin). They revealed an important role of proteins involved in oxidative stress such as Super-oxide dismutase: SOD1 and Peroxiredoxin: PRDX6. They showed that markers for mechanical or sensorial tenderness are different. We found more markers for mechanical tenderness. For sensorial analysis, we detected a higher number of markers in ST than in LT muscle. Lastly, differences were observed between breeds with very few correlations between the studied proteins and tenderness in Angus.

Improved nutritional quality of pork by feeding fish oil containing diets
Hallenstvedt, E.[1], Øverland, M.[2], Kjos, N.P.[2], Rehnberg, A.C.[3] and Thomassen, M.S.[2], [1]Felleskjøpet Fôrutvikling, Bromstadvn 57, N-7005 Trondheim, Norway, [2]Norwegian University of Life Sciences, Department of Animal and Aquacultural Sciences, P.O. Box 5003, N-1432 Ås, Norway, [3]Animalia, P.O. Box 396 Økern, N-0513 Oslo, Norway; elin.hallenstvedt@fkf.no

The aim of this study was to improve the nutritional quality of pork by increasing the content of very long chain (VLC) n-3 fatty acids without compromising the sensory quality. A total of 72 crossbred (LYDD) male and female pigs were restricted and individually fed with six experimental diets; two low fat diets with or without 0.5% EPA and DHA rich fish oil added, and four medium fat diets added palm kernel oil and fish oil blends from 4.1:0 to 3.4:0.7%. Fatty acid composition of Longissimus Dorsi was analysed and sensory evaluation according to ISO 6564 was performed on short-term and long-term frozen stored belly from females. Also a reheating procedure was conducted with a subsequent sensory evaluation of the long-term frozen stored belly. Increased levels of dietary fish oil significantly raised the VLC n-3 level in the pork ($P<0.001$), with 0.7% giving 1.05% VLC n-3 fatty acids in the meat. Less EPA was deposited when feeding 0.5% fish oil in low fat versus medium fat diet. Males deposited significantly more n-3 than females ($P<0.05$). No significant differences in sensory attributes were found among dietary groups of short- and long-term stored belly. After reheating only the highest fish oil group had higher scores of fish oil flavour and odour. In conclusion, increasing dietary levels of fish oil up to 0.5% gave higher deposition of the VLC n-3 fatty acids and improved the nutritional quality but had no negative effect on sensory evaluation even after long-term frozen storage and reheating procedure.

Prediction of Warner-Bratzler tenderness in cattle and genetic parameters of beef quality traits-preliminary results

Aass, L.[1], Hollung, K.[2] and Sehested, E.[3], [1]University of Life Sciences, Dept. of Animal and Aquacultural Sciences, P.O.Box 5003, N-1432 Aas, Norway, [2]Nofima AS-Norwegian Institute of Food, Fisheries and Aquacultural Research, Osloveien 1, N-1430 Aas, Norway, [3]Geno Breeding and A.I. Association, Holsetgata 22, N-2317 Hamar, Norway; laila.aass@umb.no

This study (2008-2012) will finally include data from approx. 600 NRF (Norwegian Red; a dual purpose breed) bulls, sired by 25 NRF A.I bull sires. Data have so far been recorded on 465 bulls, progeny of 20 sires. The bulls originate from six commercial beef producers and are slaughtered in batches at a commercial abattoir. Muscle tissue samples for calpain enzyme analyses are collected together with a hot-boned sample of the m.l.dorsi from the left carcass side. The muscle is conditioned and aged for 7 d prior to WB shear force (WBSF) and VIS-NIR (350-1025 nm; QualitySpec®BT) measurements, L*a*b colour and analyses of intramuscular fat (IMF%), hydroxyproline (%), iron (Fe%) and calcium (Ca%) content. The right carcass sides are subjected to a commercial cooling procedure p.m. Approx. 24 h p.m., the QualitySpec®BT is applied to the corresponding m.l.dorsi surface at the 11th thoracic vertebrae in addition to L*a*b and pH measurements. A USMARC developed VIS-NIR prediction equation for slice shear force (NIR-SSF) within the QualitySpec®BT has been used as an initial indicator for NIR prediction of tenderness. Preliminary h^2 for WBSF and NIR-SSF were of similar magnitude (0.25 and 0.20, respectively), while the heritabilities for IMF%, Fe% and Ca% were 0.20, 0.50 and 0.35, respectively. The latter results indicate that compounds associated with quality traits such as muscle colour/fiber type and enzyme activity are highly heritable. Further research will focus on development of WBSF prediction equations based on raw NIR-spectra 24 h and 7 d p.m, in addition to estimation of genetic parameters for all meat quality traits studied.

Slaughter value of Hungarian buffalo heifers

Barna, B. and Hollo, G., Kaposvár University, Guba S. street 40, 7400 Kaposvár, Hungary; hollo.gabriella@sic.hu

Besides cattle breeds during the history of Hungary the buffalo took important role several times. In Hungary the buffalo was mainly a draught animal. There were more than 150,000 buffaloes in Hungary before the World Wars. Some sources report that these animals were brought in by the Avars on the pouring area of Danube in 4th century. Nowadays the buffalo stock is about 1000 heads with 5-600 buffalo cows. The aim of this paper is to present the performance of Hungarian buffalo heifers, with special regards to meat quality traits in terms of fatty acid profile of longissimus muscle. The results were corresponded to the literature data of Hungarian Grey cattle. All cases, animals were kept under extensive condition on pasture. The buffaloes were slaughtered at a average live weight of 550 kg according to the Hungarian standards. The dressing percentage (54.7%) the lean meat (68%) content and the proportion of tendon (1%) are lower, but the content of bone (21%) and fat (10%) higher in buffalo carcass compared to cattle. The normal ultimate pH of buffalo meat varies from 5.3 to 5.5. Contrary to cattle the longissimus muscle of buffalo has more dry matter (28.3%) and protein (24.4%) content and lower fat proportion (1.2%). Buffalo intramuscular fat contained a higher saturated fatty acid percentage than bovine, owing to a greater concentration of stearic acid. In our examination buffalo intramuscular fat contained 51.28% SFA and 10.52% PUFA. At the same time the n-6/n-3 ratio (2.7) of buffalo meat is beneficially low, due to pasture feeding.

The effect of grassland management and production system on bovine milk quality in Mid-Norway
Adler, S.[1], Steinshamn, H.[1], Jensen, S.K.[2], Hansen-Møller, J.[2] and Govasmark, E.[3], [1]Bioforsk, Organic Food and Farming Division, Gunnars veg 6, 6630 Tingvoll, Norway, [2]Aarhus University, Department of Animal Health and Bioscience, Blichers Allé, P.O. Box 50, 8830 Tjele, Denmark, [3]Norwegian University of Life Science, Norway, Department of Plant and Environmental Sciences, P.O. Box 5003, 1432 Ås, Norway; SorenKrogh.Jensen@agrsci.dk

A two-year field study was carried out to assess the effect of grassland management (short-term (S; <5 years) or long-term grassland (L; >7 years)) and production system (organic (O) or conventional (C)) on bovine milk quality in Mid-Norway. Seven SO-farms were paired with seven SC-farms and seven LO-farms were paired with seven LC-farms, matching location and calving pattern. Milk was sampled every second month and feed twice each winter. Milk was analysed for fatty acid (FA) composition and concentrations of fat soluble vitamins, phytoestrogens and selenium (Se). Sensory qualities of milk and roughage botanical composition were assessed. Roughage was analysed for chemical composition and both roughage and concentrates were analysed for FA composition. Milk FA composition was more affected by production system and season than grassland management. O-farms had higher proportion of saturated FAs, trans-vaccenic acid, α-linolenic acid and long-chain omega-3 FAs in milk fat compared to C-farms. Summer milk had higher proportions of trans-vaccenic acid, conjugated linoleic acid and α-linolenic acid, and lower proportion of saturated FAs than winter milk. The vitamin concentrations were more affected by season than other factors. Milk from O-farms had higher concentrations of isoflavones, lignans and Se than C-farms. Milk sensory quality was not affected. Differences in milk quality can be explained by roughage:concentrate ratio, content of fat in concentrates, composition of FAs in the concentrates, high proportion of red clover in SO-grassland and high proportion of dicotyledones in LO-grassland. Organic milk may have a composition beneficial for human health except for the increased proportion of saturated FA.

***In vitro* algaecide effect of borate on Prototheca strains isolated from bovine mastitic milk**
Marques, S.[1,2], Silva, E.[1,2], Carvalheira, J.[1,2] and Thompson, G.[1,2], [1]ICBAS, Univ. Porto, Largo Prof. Abel Salazar, 2, 4099-003 Porto, Portugal, [2]ICETA/CIBIO, Univ. Porto, Rua Padre Armando Quintas, 4485-661 Vairão, Portugal; gat1@mail.icav.up.pt

Algae of the genus *Prototheca* are one of the few plant-like organisms that cause infections in humans and animals. This environmental bovine mastitis agent nowadays is recognized to be endemic worldwide and considered a public health issue. *Prototheca* was found to be resistant to most antimicrobial agents tested. Borax compound has several usages, e.g. present in detergents and used on buffer solutions preparation. In this study, we determined the *in vitro* effect of borate and phosphate buffers both at pH 9, on *Prototheca* isolates retrieved from bovine mastitic milk of cows from different dairy herds from the Northwest of Portugal. Four *P. zopfii* and 5 *P. blaschkeae* isolates were used. For the susceptibility tests phosphate (PB) and borate (BB) buffers at pH 9 were prepared according to European and US Pharmacopeias. The CLSI M27-A2 guidelines were followed and each test was incubated for 5 min, 24 and 48 h, and 1 week at 37 °C in a humid chamber. Student T test on CFU was used to evaluate the effects of both buffers on growth inhibition on the above *Prototheca* strains. Survival was generally inhibited with time. However, *P. zopfii* presented higher significant multiplication rate in PB when compared to *P. blaschkeae*. But, when BB was used at pH 9, a decrease on *P. zopfii* growth could be observed. Also, *P. blaschkeae* growth was equally but slightly inhibited by both buffers, with only significant differences to *P. zopfii* after 1 week of incubation. Our results demonstrated that BB at pH 9 had an algaecide effect on *P. zopfii* at 24 and 48 h, with a total inhibition of growth after 1 week of incubation. However, the susceptibility of *P. blaschkeae* to both buffers did not show any significant differences. This preliminary study indicates that BB compared to PB, have a greater inhibition effect on the growth of *Prototheca* strains, especially on *P. zopfii*.

The fatty acid composition and amino acid profile of organic buffalo milk
*Barna, B., Csapó, J. and Hollo, G., Kaposvár University, Guba S. street 40., 7400 Kaposvár, Hungary;
hollo.gabriella@sic.hu*

The aim of this paper was to investigate the fatty acid composition and amino acid composition of Hungarian
buffalo milk. The buffalo population in Hungary is very small (appr. 1000 head) and there is only one
dairy buffalo herd (n=40). The diet of cows consist of mainly roughage produced organically on farm and
buffalo cows spend grazing period outdoors. Milk samples from evening milking were collected. As soon
as samples were collected, it was cooled and stored at -20 0C until chemical analysis was carried out.
Data processing was made by SPSS 10.0. The examined buffalo milk samples has higher content of dry
matter (17.4%) and fat (7.33%), than commercial cow milk, besides this the buffalo milk is rich in protein
(4.19%). The proportion of casein fraction of buffalo milk (79.4) is high inside the whole protein content.
Saturated fatty acids (62.4%) predominated in buffalo milk fat; monounsaturated and polyunsaturated fatty
acids were 33.9% and 3.8%, respectively. Buffalo milk fat contained 1,5% n-6 fatty acids and 1.2% n-3
fatty acids, thus the ratio of n-6/n-3 fatty acids was beneficial high (1.3). The average content of conjugated
linoleic acid (1.04±0.21) was higher than the maximal values reported for dairy cow. The total amount of
essential amino acids of buffalo milk was 44.1, the main essential amino acids were Leucine (8.5), whereas
of non-essential amino acids Glutamic acid (22.3) was detected in the highest proportion in buffalo milk.
Preliminary results indicated that organic buffalo milk a valuable nutrient with high content of milk proteins
and rich in n-3 fatty acids and conjugated linoleic acid.

**Urea level in milk from various breeds of cows and its relationship with chemical composition and
individual traits of technological usability**
*Barlowska, J., Krol, J., Wolanciuk, A. and Szwajkowska, M., University of Life Sciences in Lublin,
Department of Commodity Science and Raw Animal Materials Processing, Akademicka 13, 20-950 Lublin,
Poland; joanna.barlowska@up.lublin.pl*

Incorrect balance in cow's diet regarding the energy and protein contents eventually leads to the decrease
of milk quality. The study included 2230 milk samples collected in summer and winter season from seven
breeds of cows: Polish Holstein-Friesian of Black-White and Red-White variety, Simmental, Jersey and local
breeds: Whiteback, Polish Red and Black and White. Cows of highly productive breeds were fed with the
TMR system. The local breeds of cows were grazing on pasture in summer season and fed with silages and
root vegetables in winter season. Total protein, casein, fat, lactose, urea, acidity, heat stability and rennet
coagulation time (RCT) were determined in each milk sample. Obtained results of analyses were divided into
three groups depending on the urea content in milk, i.e. 150, 151-300 and above 300 mg/l. The results were
analyzed statistically by Fisher's LSD test using STATISTICA software. It has been shown that local cattle
breeds produced milk with significantly (P≤0.01) lower concentrations of urea in milk. Milk samples with
urea content in range of 151-300 mg/l were characterized by significantly higher (P≤0.01) concentration of
fat (4.43%), protein (3.57%), including casein (2.62%), and dry matter (13.42%). The lowest concentration
of these components was found in milk samples with urea content extending 300 mg/l, respectively: 4.11%,
3.49%, 2.58% i 13.02%. RCT was significantly (P≤0.01) longer in milk samples with urea level over 300
mg/l, while the shortest RCT (6:29 min) was determined in milk with urea content in range of 151-300 mg/l.
Nevertheless, urea level did not affect heat stability of milk. Therefore, it can be concluded that milk with
urea content ranging from 151 to 300 mg/l (range commonly perceived as optimal) has the most favorable
chemical composition and traits of technological usability.

The effect of nonstructural carbohydrate and addition of full fat roasted canola seed on milk fatty acid composition in lactating cows

Sari, M.[1], Naserian, A.A.[2], Valizadeh, R.[2] and Salari, S.[1], [1]Ramin Agricultural and Natural Resources University, Animal Science, Mollasani, Ahwaz, 0098, Iran, [2]Ferdowsi university, Mashhad, 0098, Iran; mohsensare@yahoo.com

The objective of this study was to examine the effects of modifying the dietary profile of neutral detergent-soluble carbohydrate (NDSC), addition of full fat roasted canola seed (RCS) as a source of polyunsaturated fatty acid, and possible interactions on milk fatty acid composition in lactating cows. Twelve lactating Holstein cows (BW = 596±29 kg, DIM = 85±14) were used in a 4×4 Latin squares design involving dried citrus pulp as a source of neutral detergent-soluble fiber (NDSF), or barley as sources of starch (s)). Treatments were in a 2×2 factorial arrangement, and periods were 21 d. The first 15 d were used for diet adaptation. Milk samples were collected at each milking over the last three d of each treatment period. Pooled milk samples were analyzed for milk fatty acid composition. Data were analyzed using the GLM procedure of SAS for replicated latin square. The inclusion of RCS reduced proportion of short- and medium chain fatty acids in milk fat (C10:0-C17:0, (P<0.01)). Partial replacement of barley with citrus pulp significantly increased C10:0 and decreased C14:1 cis-9, C15:0, and C17:0 (P<0.05). Interaction (P<0.05) between NDSC profile and RCS were detected for trans-11 C18:1, with increase in trans-11 C18:1 for cows consumed NDSF+RCS (1.01, 2.13, 0.94 and 2.23 g/100 g of FA, for S, S+RCS, NDSF and NDSF+RCS, respectively). The CLA content increased by 30.1% in milk fat of cows fed the S+RCS diet and by 33.3% in milk fat of cows fed the NDSF+RCS diet (0.51, 0.73, 0.54, and 0.8%, respectively). Milk fat contents of cis-9, cis-12 C18:2, C18:3 n3, and C20:0 were increased by RCS addition (P<0.01), but were not affected by NDSC profile. Results of this study showed that adding RCS to dairy cow diets can improve the nutritive value of milk fat by enhancing the concentrations of health promoting FA such as CLA.

Residue concentrations of Nitroxynil in milk and product following administration to cows

Power, C.[1,2], O'brien, B.[3], Danaher, M.[4], Furey, A.[2], Bloemhoff, Y.[3], Sayers, R.[3] and Jordan, K.[1], [1]Teagasc, Food Research Centre, Moorepark, Fermoy, Co. Cork, Ireland, [2]Cork Institute of Technology, Bishopstown, Cork, Ireland, [3]Teagasc, Animal and Grassland Research and Innovation Centre, Moorepark, Fermoy, Co. Cork, Ireland, [4]Teagasc, Food Research Centre, Ashtown, Dublin, Ireland; Bernadette.obrien@teagasc.ie

Nitroxynil is an active ingredient in the veterinary product Trodax used to treat liverfluke in cattle. While this product is currently not recommended for use in lactating cows due to residues in milk, little information is available about the transfer of active ingredients such as Nitroxynil to milk products. The objective of this study was to establish the stability and persistence of Nitroxynil in the manufacturing process and product of milk powder. Six cows had 1.5 ml Trodax/50 kg body weight subcutaneously administered, during the lactation period. Two milk samples were taken daily from each cow up to Day 28 after administration and frozen at -20 °C until analysis. Samples were subsequently thawed and pooled into 6 independent aliquots, each containing milk from three cows between Days 1-9 (Period 1), 10-15 (Period 2) and 16-28 (Period 3). Each aliquot was separated into skim milk and cream fractions, and the skim milk was processed to milk powder using a bench-top laboratory-scale spray dryer. Nitroxynil was measured in the two separated fractions and the skim milk powder from each of the 6 aliquots using GC-MS. Average concentrations of Nitroxynil in the skim milk and cream fractions were 555.1 μg/1.2 litre and 9.0 μg/37 ml and 15.9 μg/1.1 litre and 0.44 μg/44 ml in Periods 1 and 3, respectively. Thus, ~98% and ~2% of Nitroxynil in milk migrated with the skim and cream fractions, respectively. Almost 100% of Nitroxynil in skim milk was transferred to the powder product, for example, 15.3 μg of Nitroxynil was recovered in the skim milk powder in Period 3. Thus, it is critical that similar research be conducted for other active ingredients in animal treatment products to ensure avoidance of risk to public health.

Chemical composition and technological usability of caprine milk from various groups of breeds

Barlowska, J., Szwajkowska, M., Grodzicki, T. and Wolanciuk, A., University of Life Sciences in Lublin, Department of Commodity Science and Raw Animal Materials Processing, Akademicka 13, 20-950 Lublin, Poland; joanna.barlowska@up.lublin.pl

The interest in production and processing of caprine milk in Poland and Europe continues to increase. Even though goat milk has composition similar to bovine milk, it exerts slightly different technological properties. It is less resistant to heat treatment and due to the longer coagulation time, the obtained curd is weaker and more susceptible to tearing. The research material consisted of 270 milk samples collected in summer season, between 60th and 150th day of lactation from goats of 3 breed groups: Saanen, Polish White Improved and White Non-Improved. Animals were grazing on the pasture and additionally fed with crushed cereal. Contents of protein, casein, fat, lactose, urea and acidity (pH), density, milk freezing point (MFP), heat stability (HS) and rennet coagulation time (RCT) were determined. The obtained results were analyzed statistically by Fisher's LSD test in STATISTICA software. It was found that Saanen goats produced significantly more ($P \leq 0.01$) milk (by 1.04 kg/day) than White Non-Improved goats. Nonetheless, milk from White Non-Improved goats was characterized by considerably better composition and higher HS. It contained more fat (by 0.31%), protein (by 0.76%), including casein (by 0.28%), lactose (by 0.54%) and dry matter (by 1.28%), additionally lower acidity (by 0.36 pH) and higher HS (by 0:48 min) were stated. Milk from Non-Improved goats had also the lowest content of urea (288 mg/l). Daily milk yield of White Improved goats and milk parameters had average range, except RCT, which was significantly ($P \leq 0.01$) the shortest. No significant differences in density and MFP were found between analyzed groups. The obtained results indicate that even though White Non-Improved goats were characterized by lower milk yield, they produced material more suitable for processing. This work was conducted as part of Ministry of Science and Higher Education project no. N N311 633838.

Effectiveness of new milking machine cleaning products in maintaining total bacterial counts on equipment surfaces

Gleeson, D.[1], O'brien, B.[1], Flynn, J.[1] and Jordan, K.[2], [1]Teagasc, Animal and Grassland Research and Innovation Centre, Moorepark, Fermoy, Co. Cork, Ireland, [2]Teagasc, Food Research Centre, Moorepark, Fermoy, Co. Cork, Ireland; Bernadette.obrien@teagasc.ie

New detergent cleaning products have been manufactured for the daily cleaning of milking equipment, however, they may result in increased total bacterial counts (TBC) on equipment and subsequently in milk. Six products were evaluated on each of three research farms for a 3-week period. The cleaning systems investigated were T1 = P3-mipCIP used cold; T2=P3-mipCIP used hot (chlorine added once weekly); T3 = P3-mipCIP used cold with a hot acid wash replacing the detergent in the afternoon; T4 =Hypral SP used hot; T5 = Liquid Gold used hot (control, containing chlorine); T6 =Multisan CF used cold; T7 = Parlosan NC used cold. Hydrogen peroxide was added once weekly to T6 and T7. A hot acid descale wash was carried out weekly for T1, T2, T4, T5, T6 and T7. Bacterial swabs were taken after a 'control' wash cycle at the start of week 1 and again at the end of week 3 from inside the plastic claw-piece bowl (n=36), inside the milk liner barrel (n=18) and internal stainless surfaces (n=18) for all treatments on each farm. Swabs were plated using IDF methods to measure TBC. Data were analysed using paired Student t-tests to identify differences in bacterial numbers. TBC increased on claw-pieces over the 3-week period (range 25,000 to 55,000cfu). A significant increase in TBC was observed with T1, T2, T6 and T7 (P<0.05). A high standard of equipment hygiene was maintained when a daily hot acid wash (T3), hot water twice daily (T4) or a detergent containing chlorine (T5) was used. TBC did not increase significantly on milk liners (range 100 to 140cfu) and stainless steel (range 100 to 260cfu) during the trial. Thus, it is more difficult to maintain low surface TBC on plastic than on rubber or stainless steel materials. Also, twice daily use of hot water, presence of chlorine and daily acid cleaning all individually reduce TBC on internal plastic surfaces of milking equipment.

Variability of ewe's milk fatty acid profile in relation to the morphometric characteristics of milk fat globules

Salari, F., Altomonte, I. and Martini, M., University of Pisa, Scienze Fisiologiche, Viale delle Piagge, 2, 56100, Pisa, Italy; iolealt@yahoo.it

Studies carried out on bovine milk have reported that the amount of fatty acids (FAs) tends to change based on the dimensions of the milk fat globules (MFG). This has a significant impact on the nutritional quality of milk and also leads to different technological and organoleptic characteristics of milk. The aim of this study was to verify how the FAs profile of ewe's milk is affected by changes in the morphometric characteristics of MFG. The study involved the collection and analysis of bulk milk from grazing Massese ewes. The samples were taken from a dairy every week during the months of March and April and were analyzed in duplicate for morphometric characteristics of MFG and milk FAs composition. To verify the relationship among the parameters considered, Pearson's correlations were applied. Correlations highlighted an increase in short chain FAs and in polyunsaturated FAs (PUFAs) with a higher percentage of MFG larger than 5 μm whereas an increase in the percentage of MFG smaller than 2 μm was linked to a decrease in PUFAs. An increase in the diameter of MFG was correlated to a greater content of FAs with a number of carbons from C6 to C12, of C18:1 trans11 and of C18:2 c9,12. From these, FAs from C6 to C10, C18:1 trans11 and C18:2 c9,12 have a beneficial effect on human health, whereas C12:0 increase blood total cholesterol. Some FAs, such as C17:0, C18:0, C20:0, C24:0 and C20:4, tend to decrease in milk with a higher percentage of MFG greater than 2 μm. Among these FAs, C18:0 and C20:4 are essential components of cellular membranes, indeed, although C20:4 has been described as an adipogenetic and pro-inflammatory factor the intake of C20:4 is necessary for membrane integrity. In conclusion, morphometric characteristics of MFG seem to influence nutritional characteristics of milk, further studies are needed to increase the current knowledge and to develop dairy products that have better health effects.

Effects of the α-lactalbumin +15 polymorphism on milk protein composition

Visker, M.H.P.W.[1], Schennink, A.[2], Van Arendonk, J.A.M.[1] and Bovenhuis, H.[1], [1]Wageningen University, Animal Breeding and Genomics Centre, P.O. Box 338, 6700 AH Wageningen, Netherlands, [2]University of California Davis, Department of Animal Science, 1 Shields Ave, Davis CA 95616, USA; marleen.visker@wur.nl

α-Lactalbumin has an essential physiological role in catalyzing the synthesis of lactose. Lactose is the major osmotic constituent in milk and, consequently, important for milk yield. A polymorphism in the promoter region of the α-lactalbumin gene, known as +15 relative to the transcription start site, has been reported to be associated with milk production traits. Because α-lactalbumin is one of the six most abundant proteins in milk, we have established whether this polymorphism is also associated with milk protein composition. The g.15G>A polymorphism was genotyped in 1,857 Dutch Holstein Friesian cows. Morning milk samples of these cows were used to determine detailed milk protein composition by means of Capillary Zone Electrophoresis. Association was analyzed using an animal model, comprising systematic environmental effects and genetic relationships among animals. Associations between the g.15G>A polymorphism and milk production traits were not significant in this study. However, g.15G>A was significantly associated with α-lactalbumin protein concentration and with relative amount of casein: animals with the A-allele had higher α-lactalbumin protein concentration and lower relative amount of casein. Given the current frequency of g.15G>A genotypes in the Dutch Holstein Friesian population, selection on this polymorphism could result in α-lactalbumin protein concentrations that are 0.06% lower or 0.16% higher than the current mean of 2.44% (w/w). Similarly, selection on this polymorphism could result in relative amounts of casein that are 0.36 lower or 0.07 higher than the current mean of 87.44. Higher concentration of α-lactalbumin protein could be of interest because of the high content of essential amino acids of this protein, while higher relative amount of casein could be of interest because of the positive effect on cheese yield.

The lipid components of ewe's colostrum
Altomonte, I., Salari, F. and Martini, M., University of Pisa, Physiological Science Department, Viale delle Piagge 2, 56124 Pisa, Italy; iolealt@yahoo.it

Colostrum is the mammary gland secretion until 3-5 days after delivery. To improve knowledge of the nutritional quality of ewe's colostrum and milk during the first 15 days *post partum*, we took individual colostrum and milk samples from seven Massese ewes. All ewes were reared on the same farm, were homogeneous in terms of parity and feed, and kept indoors at 10 days before partum. Five colostrum and milk samples were taken from each individual from the first 10 hours *post partum*. The 35 samples were analyzed in duplicate for dry matter (DM), fat, proteins, solids–not-fat (SNF), lactose, somatic cell count (SCC),the morphometric characteristics of milk fat globules, and fatty acid composition. The results were elaborated following a linear model for repeated measurements. First colostrum was characterized by high contents of DM, fat, protein, SNF, large globules, high percentages of monounsaturated (MUFAs) and essential fatty acids, whereas lactose and saturated fatty acids (SFAs) were found in lesser amounts than milk. The occurrence of LGs and the higher essential fatty acid content in the first colostrum have been suggested as being an adaptation of the secretion to the offspring needs. After the first day *post partum*, DM, fat, proteins, SNF percentages, somatic cells and the average diameter of milk fat globules decreased ($P \leq 0.01$), whereas the lactose percentage and the number of milk fat globules per ml^{-1} increased ($P \leq 0.01$). SFAs increased after the first day ($P \leq 0.05$) mostly linked to the increase in the short chains C4:0, C6:0, C8:0 and C10:0. Medium chain fatty acids, particularly C16:0, and MUFAs decreased ($P \leq 0.01$). Omega 3 decreased throughout the period of study, especially C22.6 (DHA). The nutritional characteristics of colostrum seem to be represented by a physiological adaptation to the requirements of the new born. These findings highlight the need for further research to better define the temporal modification of colostrum into mature milk and also to analyze factors affecting its composition.

Effects of storage time on technological ovine milk traits
Pistoia, A., Casarosa, L., Poli, P., Balestri, G., Bondi, G., Mani, D. and Ferruzzi, G., University of Pisa, DAGA, Via S. Michele degli Scalzi, 2, 56124 PISA, Italy; gbondi@hotmail.it

In Italy ovine milk is almost entirely used to produce cheese. Milk quality is associated with many factors as the time and storage conditions, infact after milking, ovine milk stops different period of time before cheese-making, in relation to sheep production and in relation to processing costs.The aim of this study was to evaluate how long milk can be stored without compromising cheese quality. Milk samples (A, B, C) from 3 Massese sheep herds homogeneous for number of subject and farming system, were collected and stored at constant temperature +4 °C. Subsamples (A,B,C) were taken at regular intervals (0-24-48-72-96 h) and analyzed for physical-chemical composition, cheese-making ability and hygienic-sanitary milk parameters. Data were tested statistically by ANOVA and the differences between means assessed with t-Student test. Chemical characteristics of milk samples at milking (A0, B0, C0), change in relation to stage of lactation in according to main standard values. Average values of milk, analyzed at different period of time (A, B, C: 24-48-72-96 h), show statistical differences on titratable acidity and free acidity that increases over time (72 and 96 hours), while pH decreases. At the same time of sampling (24-48-72-96 h), there are no significant differences in chemical composition. Total bacterial count (TBC) shows a significant increase from 72 hours sampling. Somatic cell count (SCC) tends to decline over time due to cell lysis process.Clotting time and curd firming time (r and k20) show a reduction in relation to storage time, while curd firmness (a30) shows no significant differences. In conclusion, milk stored at constant temperature preserves its technological and cheese-making characteristics up to 48 hours. After 48 hours the microbial and enzymatic activity affects milk quality and therefore cheese-making.

Opportunity to increase milk quality with biologically active substances

Miculis, J.[1], Jemeljanovs, A.[1], Sterna, V.[1], Konosonoka, I.H.[1], Antone, U.[1] and Zagorska, J.[2], [1]Research Institute of Biotechnology and Veterinary Medicine, Instituta street No 1, LV 2150, Sigulda, Latvia, [2]Latvia University of Agriculture, Faculty of Food Technology, Liela street 2, LV 3001, Jelgava, Latvia; sigra@lis.lv

To maintain optimum health status, cows need to consume a certain quantity of biologically available minerals and vitamins. For cows below optimum status, supplementation of a biologically available form of a nutrient should produce a positive response. Since animals cannot synthesize carotenoids and animal feed is generally poor in carotenoids, about 30-120 ppm of total carotenoids, are added to animal feed to improve animal health, and increase vitamin A levels in milk. It has been demonstrated that carotenoids and retinol are able to reduce mastitis in dairy cows. The aim of investigation was to study the impact of carotenoids as feed additives on the udder health and milk quality. The basic feed for the 3 groups was identical, e.g. silage and rapeseed bran, while they were given different additives: the control group: rapeseeds oil (100 g/cow per day); trial group 2: rapeseeds oil (100 g/cow per day), and carrots (7 kg/cow per day); group 3: red palm oil (100 g/per/cow per day). The protein content, fat content (Milkoscan), somatic cell count (Somacount), imunoglobulins and lyzocime (turbodimetric method), total plate count, content of beta-carotene, retinol, alfa-tocopherol (HPLC) were tested in milk samples. Results of the tested samples showed that the sum of imunoglobulins and content of lyzocime were higher in milk where cows received supplements rich in carotenoids. After 5 days storage at the temperature 4-6 °C total bacterial count increased more slowly in the trial groups milk compared with control groups milk. In conclusion, supplementation of feed with carotenoids is a opportunity to increase the antioxidant content and the quality of milk fat as well as to extend the shelf life.

Changes of chosen protein fractions content in bovine milk from three cows' breeds subject to somatic cell count

Litwinczuk, Z.[1], Krol, J.[2], Brodziak, A.[1] and Barlowska, J.[2], [1]University of Life Sciences in Lublin, Department of Breeding and Genetic Resources Conservation of Cattle, Akademicka 13, 20-950 Lublin, Poland, [2]University of Life Sciences in Lublin, Department of Commodity Science and Animal Raw Materials Processing, Akademicka 13, 20-950 Lublin, Poland; jolanta.krol@up.lublin.pl

The objective of the present work was to determine the changes of chosen protein fractions content in milk from cows of different breeds in relation to the bovine udder health status, indicated by somatic cell count (SCC). The studies were conducted on milk samples collected from the cows of 3 breeds, i.e. Polish Holstein-Friesian Black-White variety (PHF-HO), Simmental (SM) and Jersey (JE). The cows were fed with TMR system. A total of 1822 milk samples were examined (winter – 946; summer – 876). SCC, contents of crude protein, casein, α-lactalbumin (α-LA), β-lactoglobulin (β-LG), bovine serum albumin (BSA), lactoferrin (Lf) and lysozyme (Lz) were evaluated. The research material within each breed was assigned into four groups based on SCC, i.e. I – up to 100ths cells/ml, II – 101–400ths cells/ml, III – 401–500ths cells/ml, IV – 501–100ths cells/ml. The obtained results were analyzed statistically by multi-way analysis of variance, correlation coefficients and linear regression were also calculated. A marked decline trend was noted for casein content and SCC (r=-0.591). Elevation of SCC produced a slightly decrease of α-LA and β-LG. However, SCC increase induced a significant rise of Lf, Lz and BSA. The obtained interactions between a breed and SCC for the BSA content have indicated varying susceptibility of the analyzed cow breeds to SCC increase. This fact is confirmed by different values of correlation coefficients for these relationships, i.e. 0.711 (PHF-HO), 0.577 (SM) and 0.472 (JE). Higher sensitivity of PHF-HO cows to mammary gland infections was translated into greater declines of daily milk yields, that in turn was reflected in higher negative value of the correlation coefficients between SCC and milk efficiency (-0.245).

Effect of different amounts of extruded flaxseed in diets for dairy cows on chemical and fatty acid composition of milk and cheese
Cattani, M., De Marchi, M., Cologna, N., Bittante, G. and Bailoni, L., University of Padova, Department of Animal Science, Viale Università 16, 35020 Legnaro (PD), Italy; mirko.cattani@unipd.it

The study evaluated the effect of extruded flaxseed (EF) inclusion in diets for dairy cows on chemical and fatty acid (FA) composition of milk and cheese made by a laboratory cheese-making. Eighteen Holstein-Friesian cows, divided in 3 experimental groups (6 cows/group), received 3 isonitrogenous and isoenergetic diets containing 0 (EF0), 500 (EF500), or 1000 g/day (EF1000) of EF in 3 different periods of 14 days, following a 3x3 Latin square design. Individual milk samples were collected on the 7th and 13th day of each period for determining chemical and FA composition. Two cheese-making trials (11 l cheese vats) were carried out in each of 3 periods, using a representative milk sample obtained from the 6 cows of each experimental group. After 90 days of ripening period, cheese yield, pH, chemical and fatty acid composition were determined using official methodologies. Data were analyzed accounting for the effect of diets (n=3) and cheese making sessions (n=6). Orthogonal contrasts for diet were carried out. Cheese yield, fat and protein percentages, and pH values were not significantly affected by the diet. The concentrations of omega 3 FA in cheese resulted lower (P<0.001) for EF0 (0.31 g/100 g of FA), whereas no differences (P=0.13) were found comparing EF500 and EF1000 (0.53 and 0.64 g/100 g of FA, respectively). Omega 3 FA concentrations in milk and cheese were greatly correlated (r=0.86). The carry-over of omega 3 FA from milk to cheese decreased linearly (P<0.05) at increasing levels of EF (89.6, 80.2, and 73.5% for EF0, EF500 and EF1000, respectively). In conclusion, EF improved the nutritional properties of milk, and in particular its omega 3 FA concentration, but not proportionally to the amount included in the diet. The recovery of omega 3 FA in cheese was lowered after ripening, especially for EF1000 and, to a lesser extent, for EF500 and EF0.

Technological usability of milk from 3 breeds of cows
Barlowska, J., Wolanciuk, A., Krol, J. and Litwinczuk, A., University of Life Sciences in Lublin, Department of Commodity Science and Raw Animal Materials Processing, Akademicka 13, 20-950 Lublin, Poland; anna.litwinczuk@up.lublin.pl

The assessment of milk as a raw material for processing considers i.a. parameters such as heat stability and the rennet coagulation time of milk. Both are related mainly with the concentrations of individual milk components and their proportions. The study involved 176 milk samples obtained during the winter season, between 12th and 200th day of lactation, from cows of 3 breeds fed Total Mixed Ratio (TMR) system: Polish Holstein-Friesian Red and White variety (PHF-RW), Montbéliard (MO) and Jersey (JE). Contents of protein, casein, fat, lactose; acidity (pH), heat stability, rennet coagulation time (RCT) and calcium content were investigated. The obtained results were analyzed statistically by Fisher's LSD test and Pearson's correlation coefficients test using STATISTICA software. It was found that milk from JE breed was characterized by the least favorable ratio of protein to fat (0.76), while the most favorable ratio was found in MO milk (0.88) and slightly lower in PHF-RW (0.83). Milk of PHF-RW was the most heat-stable (2:55 min), which coincided with the longest RCT (5:22 min), despite the highest acidity (pH 6.62). Milk from JE cows was the least heat-stable (1:27 min) and had the shortest RCT (3:45 min), which may be associated with a significantly (P≤0.01) higher content of calcium (by 123.07 mg/l) compared to PHF-RW and MO milk (by 243.99 mg/l). Negative correlations were found between protein content and RCT (r=-0.21**) and between acidity (pH) and heat stability (r=-0.30***) and positive correlation was found between heat stability and RCT (r=0.32***). In conclusion, milk from PHF-RW is more suitable for the production of consumption milk and milk concentrates, while JE milk has better parameters for cheese production. The calculated correlation coefficients indicated significant relations between technological traits of analyzed milk.

Bioactive protein content in milk from local breeds of cows included in the genetic resources conservation programme
Krol, J.[1], Litwinczuk, Z.[2], Brodziak, A.[2] and Chabuz, W.[2], [1]University of Life Sciences in Lublin, Department of Commodity Science and Raw Animal Materials Processing, Akademicka 13, 20-950 Lublin, Poland, [2]University of Life Sciences in Lublin, Department of Breeding and Genetic Resources Conservation of Cattle, Akademicka 13, 20-950 Lublin, Poland; jolanta.krol@up.lublin.pl

In Poland the majority of milk is obtained in intensive system from Holstein-Friesian cows. In some regions (mountain, submountain, boggy) the local breeds, i.e. Polish Red (RP), Whiteback (BG), Polish Black and White (ZB) and Polish Red and White (ZR) are maintained. Owing to small number and genetic distinction, the animals of these breeds are included into the genetic resources conservation programme. The research included 640 milk samples collected from cows of 4 local breeds, i.e. RP: summer(S)–114, winter(W)–90; BG:S–101, W–97; ZB: S–86, W–70 and ZR: S–44, W–38. In summer season the cows' feeding was based on pasture forage, however in winter the hay silage was the basis. The control group were milk samples taken from Polish Holstein-Friesian cows (PHF), which were fed with Total Mixed Ratio (TMR) system – 136-S and 108-W. Contents of crude protein, casein, α-lactalbumin (α-LA), β-lactoglobulin (β-LG), bovine serum albumin (BSA), lactoferrin (Lf) were evaluated. The obtained results were analyzed by two-way ANOVA. The highest content of crude protein and casein was stated in milk of RP (3.59; 2.64%). Milk of RP breed was also characterized by the highest content of α-LA (1.15 g/l), β-LG (3.60 g/l) and Lf (129 mg/l). Imperceptibly smaller quantity of these proteins contained milk obtained from BG, i.e. 1.09 g/l; 3.57 g/l and 115 mg/l, respectively. Significantly less these proteins was researched in PHF milk (α-LA–0.98 g/l; β-LG–2.93 g/l and Lf–91 mg/l). It should be marked that higher content of these proteins characterized the milk obtained in summer, when cows grazed on pasture. Milk from local cows' breeds compared with PHF contained significantly more α-LA, β-LG, Lf; yet no significant differences were noted in BSA content.

Iodine concentration in Norwegian milk has declined the last decade
Haug, A., Harstad, O.M., Salbu, B. and Taugbøl, O., The Norwegian University of Life Sciences, Department of Animal and Agricultural Sciences, P.O. Box 5003, 1432 Ås, Norway; anna.haug@umb.no

In Norway milk and milk products are the main source of dietary iodine intake. A study was conducted to determine the iodine concentration in 104 dairy milk samples taken from 19 milk tours (routes) in different areas of Norway, at six times throughout the year in 2008. The milk was produced in 436 farms, and three of the milk tours were from farms with organic feeding systems. The iodine concentration in milk from the summer season was 90 µg iodine/kg milk. This concentration is similar to the concentration reported in a study in year 2000, including 85 samples taken from 19 different locations in Norway. The iodine concentration in milk from the indoor feeding season was 119 ug/kg milk. This is much lower than reported in year 2000; it was reported that the iodine concentration was 232 ug iodine/liter in the winter season in 2000. Thus the iodine concentration in milk from the summer season is at the same level as a decade ago, but iodine in milk from the winter season has been reduced to about half during a decade. This is a dramatic reduction in milk iodine concentration. Possible explanations for the reduction in iodine in milk produced in the winter season may be multi-factorial; one explanation may be related to increased concentration of rapeseed meal in the feed compared to 2000. Rapeseed contains glucosinolates that are degraded to thiocyanate that may inhibit iodine reabsorption in the kidney tubules of the cow.

The effect of fluorescent light exposure on meat colour stability from commercial Gascon calves and cull cows, produced in Catalunya and Midi-Pyrenees

Khliji, S.[1], Panella, N.[1], Chamorro, F.H.[2], Gil, M.[1], Blanch, M.[1] and Oliver, M.A.[1], [1]IRTA, Monells, 17121, Spain, [2]Univ. Autonoma, Chihuahua, 31415, Mexico; marta.gil@irta.es

Meat discoloration is a major factor affecting the consumer's purchase decision. Numerous intrinsic and extrinsic factors contribute to meat discoloration. To control the case-life, an understanding of the interactions among some factors is essential. In this context, 30 loins (*Longissimus dorsi*) from 2 commercial Gascon categories (15 cull cows -CC- produced in Midi-Pyrenees and 15 calves -Ca- from Catalunya; age at slaughter 10-11 months vs. 5-8 years) were used to study the effect of lighting conditions (darkness vs. fluorescent light) on meat colour stability during 15 days of storage at 4-5 °C. Meat colour (CIE L* and a*) was measured with Minolta CM-2002 spectrophotometer, and the wavelength ratio (630nm/580nm) was used to estimate colour changes due to the oxymyoglobin oxidation. Data were analysed using MIXED procedure with repeated statement of SAS®. Differences were considered significant at $P<0.05$. The 3 ways interaction (commercial categories*lighting conditions*display time) was significant ($P<0.05$) for a* and for the wavelength ratio, however only two double interactions were significant for L* (commercial categories*storage time and lighting conditions*storage time). L* values were more stable in darkness. CC had lower L* values, although these values were less stable than those from Ca. Higher a* values and higher wavelength ratio were observed over storage time. Under light, a* values followed the same pattern over time for both CC and Ca, however, in the darkness, a different pattern was observed between the 2 commercial categories. For the wavelength ratio, similar behavior of CC and Ca was observed in both, darkness and light conditions. Meat colour from CC was more stable than the meat from Ca. In both commercial categories, discoloration was faster and more intense with light exposure. Differences between CC and Ca were more emphasized when meat was stored in the darkness.

Design of low-cost aquaculture breeding programs: challenges and opportunities

Komen, J. and Blonk, R., Wageningen Univeristy, Animal breeding and genomics centre, Marijkweg 40, 6709PG, Netherlands; hans.komen@wur.nl

This paper reviews the challenges and opportunities that arise when designing and implementing low cost breeding programs. Such breeding programs are typically demanded by small to medium aquaculture enterprises that seek an opportunity to increase productivity and efficiency of their farm(s) through selective breeding. Design of breeding programs is given by farm dimensions and production targets. Breeding goals typically focus on growth and yield traits. Selection should take place within production facilities and non-selected fish should be marketed. In many cases, natural mating is the only way to produce offspring which puts restrictions on mating designs. Finally, the education level of the technical staff involved often requires that the breeding program is simple and easy to run. Given these parameters, we review different designs that can be implemented. Results show that a major cost factor of a breeding program is related to broodstock size and -management. Separating broodstock in production groups and employing rotational mating provide ways to maximize the fit between production and selection objectives. Selection on own performance under production circumstances with frequent grading is fairly robust to inbreeding. Fish typically show large variation in growth while estimated direct heritability's are comparable to those in livestock. However, responses to selection and realized heritability's in first generations are often higher, suggesting that part of the utilized variation results from social interactions in (feeding) behavior. In the final part of this review we discuss consequences for the producer of running a successful breeding program. Higher growth entails increased feeding with associated effects on oxygen demands and water quality. Producers are often not prepared for these effects, especially in RAS and low-input farms. We recommend that breeding programs are subjected to LCA to assess the impact of selective breeding.

Utilising genomic information in salmon breeding and disease resistance

Bishop, S.C.[1], Taggart, J.B.[2], Bron, J.E.[2] and Houston, R.D.[1], [1]University of Edinburgh, The Roslin Institute & R(D)SVS, Midlothian, EH25 9RG, United Kingdom, [2]University of Stirling, Institute of Aquculture, Stirling, FK9 4LA, United Kingdom; Stephen.Bishop@Roslin.ed.ac.uk

The salmon breeding industry has pioneered in the implementation of well-structured breeding programmes, incorporating genomic tools. Aided by the large full-sib families and natural discrete four-year generation interval of salmon, this has resulted in significant genetic progress in often difficult-to-measure traits. Parentage assignment using microsatellite, and latterly SNP, genotyping has aided trait recording (e.g. by only needing to recover tissue from mortalities in mixed-family sentinel disease challenges) and led naturally to the implementation of marker-assisted selection (MAS) in commercial breeding programmes. The foremost example is resistance to infectious pancreatic necrosis (IPN), where a single QTL on linkage group 21 has been shown to account for nearly all the variation in IPN mortality, in both Scottish and Norwegian populations. Recently we have used Restriction-site-Associated DNA (RAD) sequencing to identify SNPs in population-wide linkage disequilibrium with the causative mutation, greatly simplifying MAS, and we have used transcriptomic profiling to clarify the mechanism of resistance. Sustainable improvement of salmon performance and health will require new target traits and new tools. Salmon fatty acid composition on semi-vegetarian diets is an obvious target trait affecting both fish health and product quality, and is the focus of much current research. Further, several diseases are currently targets for the elucidation of the genomic regulation of host resistance, foremost being sea louse infections and pancreas disease. However, genomic resources are currently lacking, and to this end we are currently utilising RAD sequencing for association mapping, and we are developing a dense SNP chip based on an Affymetrix platform. This will enable us to dissect and improve a range of traditional and novel performance and health traits.

Non-random testing designs with three families per group is more efficient for genetic evaluation of social effects in aquaculture species

Ødegård, J. and Olesen, I., Nofima, P.O. Box 5010, 1432 Ås, Norway; ingrid.olesen@nofima.no

Heritable social interactions among individuals are commonly observed in both plants and animals. The social effect of an individual does not affect the phenotype of the individual itself, but rather the phenotypes of all other individuals sharing the environment. As an alternative to traditional breeding schemes solely focusing on direct breeding value affecting the phenotype of the individual itself, selection may be focused on total breeding values, including both direct and social effects. The latter may give a substantial increase in the rate of genetic gain of the population, given social genetic effects on the trait(s) considered. However, social effects can only be estimated using test data of several groups of animals. A design including many small groups consisting of unrelated individuals, e.g. by randomly allocating individuals to each group (RAN) has been recommended. However, we hypothesize that a design with few (three) families per group (3FAM), where each family is tested repeatedly in three different groups would be more advantageous. Using stochastic simulations, assuming a full-sib data structure typical for aquaculture breeding schemes, we compared the two designs with respect to accuracy of breeding values and their ability to produce accurate estimates of (co)variance components of direct and social effects. The results showed that the 3FAM design was clearly superior with respect to precision of estimated genetic (co)variance components of social effects and had substantially higher accuracy of both social and total breeding values (r_{TBV}), compared with the RAN design (16 to 105% increase). The advantage was particularly big (41-105% higher r_{TBV}) for scenarios with zero or negative genetic correlation between direct and social genetic effects. This implies that for traits with heritable social effects, a testing design with non-random allocation of families in (e.g. 3) repeating groups should be applied in breeding programs.

Households' willingness-to-pay for improved fish welfare in breeding programs for farmed Atlantic salmon

Grimsrud, K.M.[1], Nielsen, H.M.[2], Navrud, S.[3], Monsen, B.B.[2,3] and Olesen, I.[2,3], [1]University of New Mexico, NM 87131, Albuquerque, USA, [2]Nofima, P.O. Box 5010, 1432 Ås, Norway, [3]Norwegian University of Life Sciences, P.O. Box 5003, 1432 Ås, Norway; hanne.nielsen@nofima.no

A growing public concern about animal welfare has increased the demand for such concerns to be included in animal breeding goals. The weight given to different traits of animal welfare in the breeding goal can be determined by eliciting households' preferences in terms of their willingness-to-pay (WTP) for breeding programs including such traits. In an internet survey a random, representative sample of Norwegian households (n=760) were asked to choose among breeding programs for farmed Atlantic Salmon that differed with regards to costs and the following four traits related to fish welfare; frequency of deformities, frequency of injuries, resistance to salmon lice, and resistance to infection diseases. The survey participants were given six different choice sets in which they had to choose one of three salmon breeding programs. The programs differed with regards to whether each of the four traits were bred for ('yes' or 'no'), and the costs in terms of increased annual household expenditures for farmed salmon (amounts between 100 and 1,800 NOK). One of the three breeding programs was always status quo; i.e. at zero cost no animal welfare was bred for). Preliminary results from a conditional logit model show that households were willing to pay 415 and 64 NOK per household per year extra for their annual consumption of salmon filet to support salmon breeding for fish with lower frequency of deformities and injuries, respectively. For increased resistance to salmon lice and diseases WTP per household per year was 953 and 528 NOK, respectively. This study shows that households are willing to pay for improved fish welfare and that choice experiments can be used to derive economic values for traits in the breeding goal.

Stimulating sustainable aquaculture: access to and protection of aquaculture genetic resources

Olesen, I.[1], Rosendal, K.G.[2] and Pathak, A.R.[3], [1]Nofima, 1432 Ås, Norway, [2]Fridtjof Nansen Institute, 1326 Lysaker, Norway, [3]Pune, India; ingrid.olesen@nofima.no

The research question for this presentation is how to balance access to aquatic genetic resources and legal protection of improved breeding material to encourage innovation and sustainable aquaculture. The study aims to provide knowledge on how corporate strategies, technological developments, and regulatory regimes affect access to genetic resources and sustainable innovations in aquaculture. This is based on findings from Norwegian case studies on Atlantic salmon and cod with similar studies on carp and shrimp in India and tilapia in Asia and Africa. Aquaculture is experiencing pressure towards higher production efficiency and short term profits. Hence, actors face emerging difficulties pertaining to adequate funding for sustainable breeding programs and affordable access to improved genetic material. For example, India's policies with regard to the shrimp sector have undergone changes that illustrate the complexities involved in importing improved breeding material, and ensuring access to genetic resources. Historically, aquaculture in India has mainly been based on public investments to increase production, develop and widely disseminate material, rather than creating proprietary products. Greater involvement of private sector leads to biological or other IPR protection that keeps knowledge out of the public domain, which may have negative implications for aquaculture. Public-private partnerships will also find difficulties in resolving their differing priorities. Public sector will be tied up with regulation and monitoring, and will have less resources and capacity for supporting independent and freely accessible research and development. Public ownership or support seems to be important measures for ensuring sustainable development. This is particularly the case during early phases of breeding programs in which profitability may be viable only in the long term. Also, cooperative/farmers ownership of breeding programs is often worth considering.

Differences in susceptibility to the salmon louse between a high and a low lice susceptible group of Atlantic salmon
Gjerde, B. and Ødegård, J., Nofima Marin, P.O. Box 5010, 1432 Ås, Norway; bjarne.gjerde@nofima.no

An estimate of the genetic variation in susceptibility of Atlantic salmon to the salmon louse Lepeophtheirus salmonis was obtained through a controlled lice infestation test in July 2008 of 2,206 individually tagged post-smolts from 154 full-sib families; i.e. the offspring of 78 sires and 154 dams. The heritability of susceptibility was 0.26±0.05, measured as LD (lice density) = Sessile Lice Count per fish/Body weight2/3. A different subsample of individuals from the 154 families was reared in a net cage in seawater from June to October 28th 2008 at which 10-12 individuals were randomly sampled from the 10 highest (110 HLD fish) and the 10 lowest (111 LLD fish) ranking families with respect to LD in the infestation test. At sampling the sessile and motile lice count (natural infection) and body weight of each fish sampled was recorded. The sampled fish were put into two three meter diameter tanks (12 m^3) with an equal number of fish per tank. After being deloused the fish in each tank were infected with lice on December 1st (29 copepodids per fish and tank) and the number of sessile and motile lice per fish recorded on December 29th and January 15th, respectively. The average body weight was 994, 1,205 and 1,280 gram on October 28th, December 1st, and January 15th, respectively. The average lice count per fish was 4.2 sessile and 2.1 motile lice on October 28th, 10.8 sessile lice on December 1st and 12.8 motile lice on January 15th. On October 28th the LLD group had 20.8% lower sessile LD than the HLD group (0.038 vs. 0.048; P<0.01), while for motile lice the difference in LD between the two groups was not significantly different from zero (0.023 vs. 0.021; P>0.05). On December 29th the LLD group had 18.3% lower sessile LD than the HLD group (0.089 vs. 0.109; P<0.01). On January 15th the LLD group had 10.9% lower sessile LD than the HLD group (0.106 vs. 0.119; P=0.07). These results confirm a significant additive genetic variation in the susceptibility to the salmon louse in Atlantic salmon.

Simultaneous analysis of two challenge tests and body weight in fish: an example of multivariate generalized linear mixed model
Labouriau, R.[1], Wetten, M.[2] and Madsen, P.[1], [1]Aarhus University, Department of Genetics and Biotechnology, P.O. Box 50, DK-8830 Tjele, Denmark, [2]Norsvin, Post Box 504, 2304 Hamar, Norway; rodrigo.labouriau@agrsci.dk

We consider the simultaneous analysis of two challenge tests (Salmon Rickettsial Syndrome (SRS) and Infectious Pancreatic Necrosis (IPN)) in Atlantic salmon (Salmo salar) together with the weight performed at Aqua Gen. The data included 121,618 fish: 85,406 in the IPN challenge (83% mortality), 7,852 in the SRS challenge (89% mortality) and 28,060 with weight recordings. Here we simultaneously characterize the genetic effects of these 3 traits. To do so we used a 3 dimensional generalized linear mixed model with the structure of a sire model. The 2 first dimensions represented the 2 responses of the challenge tests as being marginally distributed according to a Bernoulli distribution and following a probit link; while the third dimension represented the weight modelled as normally distributed and following an identity link (i.e. a linear response). After correcting for fixed effects within each trait and the random effect of full sib group we obtained estimates for the genetic variances for SRS, IPN and weight of 0.3529 (with asymptotic standard error, a.s.e. 0.0748), 0.1918 (a.s.e. 0.0640) and 0.0025 (a.s.e. 0.0003) respectively. The genetic correlations between the 3 traits were estimated as: correlation between SRS and IPN, 0.1288 (a.s.e 0.1562, not statistically significantly different than zero); correlation between SRS and weight, 0.4984 (a.s.e. 0.0907, statistically significantly different than zero) and correlation between IPN and weight, -0.2665 (a.s.e. 0.1195, weakly statistically significantly different than zero). These results suggest that the genetic mechanisms involved in the susceptibility to the two diseases are uncoupled and a favourable positive genetic association between SRS and weight in the sense that animals with genetic loads for having larger weight tend to present also genetic loads associated with reduced mortality by SRS.

Development of 6K Illumina iSelect SNP arrays for mapping quantitative trait loci in rohu (*Labeo rohita*) and black tiger shrimp (*Penaeus monodon*)

Baranski, M.[1], Sahoo, P.K.[2], Gopikrishna, G.[3], Robinson, N.[1], Das Mahapatra, K.[2], Vinaya Kumar, K.[3], Saha, J.N.[2], Shekhar, M.S.[3], Das, S.[2], Gopal, C.[3], Mishra, Y.[2], Karthik, J.S.[3], Das, P.[2], Jothivel, S.[3], Barman, H.K.[2], Ravichandran, P.[3], Ponniah, A.G.[3] and Eknath, A.E.[2], [1]Nofima Marin, P.O. Box 5010, 1432 Ås, Norway, [2]Central Institute of Freshwater Aquaculture, Kausalyaganga, Bhubaneswar 751002, India, [3]Central Institute of Brackishwater Aquaculture, Raja Annamalai Puram, Chennai 600028, India; matthew.baranski@nofima.no

Using next-generation sequencing technology, DNA markers can now be rapidly generated for species lacking genomic resources, removing one of the 'roadblocks' to the implementation of molecular tools in breeding. Among the species cultured in India, rohu (*Labeo rohita*) and black tiger shrimp (*Penaeus monodon*) suffer substantial production losses due to bacterial and viral diseases like *Aeromonas hydrophila* and white spot syndrome virus. In this study, Illumina transcriptome sequencing was used to develop large SNP marker resources for efficient, high-throughput genotyping and mapping of QTL in these two species. For *L. rohita*, fish were sampled from a lines selected for *A. hydrophila* resistance and susceptibility, in order to maximise the chance of finding SNPs segregating for this trait. For *P. monodon*, samples were taken from a broad geographic distribution around the Indian coast. De novo transcriptome assembly was performed using CLC Assembly Cell software. Assembly of *L. rohita* ESTs resulted in 137,629 contigs and 33,000 putative SNPs, and assembly of P. monodon ESTs resulted in 136,223 contigs and 123,000 putative SNPs. From these SNP sets, 6,000 loci were selected for inclusion in Illumina iSelect genotyping assays, and genotyping was performed in challenge tested families of *L. rohita* and *P. monodon*. The results of this genome scan will be discussed in the context of marker distribution over the genome (linkage maps) and putative QTL identified.

Detection of growth-related QTLs in turbot (*Scophthalmus maximus*)

Sánchez-Molano, E.[1], Cerna, A.[2], Toro, M.A.[2], Fernández, J.[1], Hermida, M.[3], Bouza, C.[3], Pardo, B.G.[3] and Martínez, P.[3], [1]INIA, Departamento de Mejora Genética, Ctra. Coruña Km. 7.5, 28040 Madrid, Spain, [2]ETS Ingenieros Agrónomos. UPM, Departamento de Producción Animal, Ciudad Universitaria, 28040 Madrid, Spain, [3]Facultade de Veterinaria. Universidade de Santiago de Compostela, Departamento de Xenética, Avda. Carballo Calero s/n, 27002 Lugo, Spain; sanchez.enrique@inia.es

Turbot (*Scophthalmus maximus*) is one of the most important species in European aquaculture due to its high commercial value as food fish. Its production has increased from 3,000 tonnes in 1996 to 15,000 tonnes in 2011. Recently, a consensus linkage map has been developed by Bouza *et al.*, allowing the detection of QTLs related with different traits. In the present study, eight full sib families from commercial strains were genotyped for an average of 100 homogeneously distributed microsatellites per family and measured for three growth-related traits (weight, length and Fulton's condition factor). QTLs for these traits were detected using two different approaches: maximum likelihood (QTLMap) and multiple regression (GridQTL). Three QTLs were detected for weight, three for length and two for Fulton's condition factor, and some of them were common for several traits. For each detected QTL, an association analysis was performed to ascertain the microsatellite marker which showed the highest apparent effect on the trait, in order to test the possibility of using it in marker assisted selection.

Genomic variability and genome-wide linkage disequilibrium in commercial and naturalized salmon populations

Martinez, V.[1], Dettlef, P.[1], Lopez, P.[1], Jedlicki, A.[1], Lien, S.[2] and Kent, M.[2], [1]FAVET-INBIOGEN, Universidad de Chile, Avda. Santa Rosa 11735, La Pintana, Chile, [2]CIGENE, Norwegian University of Life Sciences, N-, 1432 Ås, Norway; vmartine@u.uchile.cl

Understanding the genome structure of commercial salmon populations is important for assessing the effect of natural and/or artificial selection following their introduction for aquacultural purposes. This has implications when designing breeding schemes using genomic information, ie selecting either specific alleles or maintaining the genomic variability and the extent of linkage disequilibrium (LD) surrounding the selected QTL. An Illumina SNP (5.5K) array was used to genotype a total of 500 unrelated individuals sampled from Chilean salmon populations, differing in their known ancestry and selection status. Measures of population differentiation and extent of LD decay were used for assessing the impact of selection at the genome-wide level. There is little differentiation when considering jointly all the information from valid SNPs. However, the results show that chromosomes differ significantly in mean Fst calculated between the commercial and the naturalized population. Two chromosomes showed a significantly higher mean Fst value. Some regions within chromosomes show an outlier Fst behavior, explaining either, high or low genetic divergence. The extent of linkage disequilibrium, also differ between populations and chromosomes. Haplotypes blocks were only observed over short distances and were relatively consistent across populations, when considering information from chromosomes harboring QTL related to disease resistance. These chromosomes showed complex patterns of differentiation. Altogether, the results show that commercial salmon populations showed evidence of balancing and directional selection shaping the genomic variability of the salmon genome. These findings will be crucial when using information of the salmon genome sequence currently under progress, to discover the genes underlying these complex patterns of selection.

Genomic tools to trace and monitor the genetic impact of aquaculture escapes

Jacq, C., Nofima Marin, Postboks 5010, 1432 Ås, Norway; celeste.jacq@nofima.no

Escapes of cultured fish into the natural environment are undesirable for several reasons including the spread of disease/parasites, competition for resources with native species and potential genetic interactions with native populations of the same species. Norway is the largest producer of cultured Atlantic salmon and also constitutes a significant proportion of the global wild Atlantic salmon stock. It has been shown that escaped cultured salmon make up a large component of stock caught in the wild and there is a concern that the genetic condition of wild stocks is deteriorating due to gene flow from escaped salmon. Artificial selection for desirable traits in aquaculture breeding programs results in cultured fish that have selective fitness for the farm environment. As such, heritable traits that are poorly adapted to the natural environment may be detrimental to the wild population following interbreeding with cultured escapees. To some extent, natural selection can lead to a recovery in fitness over time where the wild population possesses superior local adaptation. However, interbreeding between cultured and wild individuals may also facilitate the disruption of co-adapted gene complexes (epistasis), thus resulting in the introgression of selectively detrimental alleles into the wild population. Little is known about the extent of local adaptation and epistatic effects in relation to interactions between cultured and wild Atlantic salmon, as this has traditionally been difficult to assess. However, the field of population genetics has recently entered the genomics era and 'population genomics' can now provide us with novel insight into many aspects regarding farmed and wild genetic interactions. This presentation will provide an overview of the application of genome-wide SNP data in order to trace aquaculture escapees, detect local adaptation, and monitor the long-term effects of genetic interactions between cultured and wild Atlantic salmon; knowledge essential in facilitating sustainable management of our wild Atlantic salmon populations.

Can we obtain monosex sea bass populations through selective breeding?

Vandeputte, M.[1,2], Dupont-Nivet, M.[1], Haffray, P.[3], Chavanne, H.[4], Vergnet, A.[2], Quillet, E.[1] and Chatain, B.[2], [1]INRA, UMR1313 GABI, Domaine de Vilvert, 78350 Jouy en Josas, France, [2]Ifremer, Chemin de Maguelone, 34250 Palavas les Flots, France, [3]Sysaaf, Campus de Beaulieu, 35000 Rennes, France, [4]ISILS, Localita la Quercia, 26027 Rivolta d Adda, Italy; marc.vandeputte@jouy.inra.fr

In the European sea bass, the sex-ratio of farmed populations is usually highly male-biased (>75%). Females are preferred as they grow faster and mature one year later. This is known to be linked to larval and post-larval rearing temperature, but also to genetic factors which are presumably polygenic, and linked with growth rate. Phenotypic sex can be modeled as a threshold trait with an underlying liability trait called 'sex tendency'. From wild G0 parents, we created a G1 population where the proportion of females was 18.2%, in which we selected sea bass males for increased growth rate (proportion selected= 5%). Then, we studied sex-ratios in a G2 population made of the offspring of wild (W), unselected (but born in captivity, thus called 'domesticated' (D) and selected for growth (S) sea bass males. The sex-ratios were respectively 41.4%, 47.1% and 58.5% in the W, D, and S offspring groups. This was shown to conform to the expectations of a polygenic model with the genetic parameters estimated in a previous work: h^2=0.62 for sex tendency, h^2=0.41 for body length, and r_A=0.48 between both. Modeling selection response with such genetic parameters, we found that an equilibrium sex-ratio should be reached when the increase in sex tendency caused by selection for growth would equal the decrease linked to frequency-dependent selection towards a 50/50 sex ratio. For a 5% selection on body length, there would then be 78% of females in the population, in 10 generations. We should reach 95% females, (thus virtually monosex, without using hormones) in offspring issued from selected broodstock reared in environmental conditions mimicking natural conditions where wild fish have a sex-ratio of 50/50.

Evaluating of genetic diversity in three population of rainbow trout (*Oncorhynchus mykiss*) using molecular RAPD markers

Afzali, M.[1], Farhadi, A.[2] and Rahimi-Mianji, G.[2], [1]Islamic Azad University of Sari, Sari, 48161-19318, Iran, [2]Laboratory for Molecular Genetics and Animal Biotechnology, Department of Animal Science, Faculty of Animal Science and Fisheries, Sari Agricultural Sciences and Natural Resources University, 578, Iran; ayyoob_farhadi@yahoo.com

In the present study we evaluated the amount and distribution of genetic variation by surveying RAPD marker variation at 12 marker loci in three brood stock groups of rainbow trout. Of the total 120 amplified bands in Iranian strain, 47 were polymorphic, with an average number of bands and average number of polymorphic bands per primer was 10 and 3.92, respectively. The total detected bands in rainbow trout strain originated from French, was 120, with an average number of 10 bands per RAPD primer. A total of 117 bands were detected in Norwegian population, with an average number of bands and average number of polymorphic bands per primer was 9.75 and 2.58, respectively. Data for observed and effective number of alleles, Nei's genetic diversity, and Shannon's information index for all the three populations and their respective values were found as 1.31, 1.20, 0.120 and 0.170. The mean coefficient of gene differentiation value and the estimate of gene flow across the populations were found as 0.299 and 0.171, respectively. The Nei measures of genetic distance and identity between pairs of rainbow trout strains indicate that the strain originated from France and Iran has the highest genetic identity, while the fish originated from Norway and France showed the greatest genetic distance. The obtained low value of genetic variation at the present study indicates that a suitable breeding strategy should be selected in order to increase genetic variation in between and within studied fish populations.

Assessment of European breeding programmes from different sustainability aspects

Rydhmer, L.[1], Ryschawy, J.[2], Fabrega, E.[3] and De Greef, K.[4], [1]Sw. University of Agricultural Sciences, Dept. Animal Breeding and Genetics, Box 7023, 75007 Uppsala, Sweden, [2]INRA, UMR1079 SENAH, 35590 Saint-Gilles, France, [3]IRTA, Finca Camps i Armet, 17121 Monells, Spain, [4]Wageningen UR, P.O. Box 65, 8200 AB Lelystad, Netherlands; Lotta.Rydhmer@slu.se

Within the EU-project Q-PorkChains, pig production systems in four European countries (one conventional and two differentiated systems per country) were evaluated. Four types of genetic material linked to the production systems were studied: conventional breeds in conventional systems, traditional local breeds in traditional small-scale systems, breeds from conventional breeding programmes in alternative systems and specific breeds from own breeding programmes in alternative large scale systems. A checklist by Woolliams *et al.* was used to assess the breeding programmes. Information from 10 breeding organisations was collected through interviews or questionnaires. Farmers from each system also answered questions on e.g. preferred selection traits. The assessment is summarized in four dimensions. The 1st dimension describes whether the market for the product is well defined (including questions on sensitivity to external factors) and whether the breeding goal reflects the production system and the farmers' demands. The 2nd dimension describes selection procedures and genetic change in important traits for the system. We assessed whether the recording was sufficient to achieve the breeding goal and how the different traits were balanced within this goal. The 3rd dimension deals with genetic variation, both within breed and within the pig species. This dimension includes questions on effective population size and different stakeholders' interests in genetic diversity. The 4th dimension describes the functioning of the breeding organisation. Communication and transparency as well as available economic, technical and human resources belong to this dimension. Strong and weak points of different production systems with regard to breeding and genetic diversity will be presented.

Status and prospects for smallholder milk production: a global perspective

Hemme, T.[1] and Otte, J.[2], [1]IFCN Dairy Research Center, Schauenburgerstrasse 116, 24118 Kiel, Germany, [2]FAO, Animal Production and Health Division, Viale delle Terme di Caracalla, 00153 Roma, Italy; torsten.hemme@ifcndairy.org

It is estimated that 12-14% of the world population, or 750-900 million people, live on dairy farms or within dairy farming households. The worlds mean dairy herd size is around 2-3 cows. Against this background, the paper set out to assess whether: small-scale milk production can contribute to reducing poverty and will be able to compete with large-scale, capital-intensive 'high-tech' dairy farming systems, such as those in the USA and other developed countries. The analysis is based on the of the IFCN methodology. The IFCN – International Farm Comparison Network is a global research network attracting and connecting dairy researchers from over 80 countries. www.ifcndairy.org. The IFCN methodology is based on three elements: a) the TIPI-CAL model, the method of typical farm types selection and c) the method of collecting and validating farm data. The various analyses indicate that: small-scale milk production not only improves the food security of milk-producing households but also helps to create numerous employment opportunities throughout the entire dairy chain. As such, dairy development may serve as a powerful tool for reducing poverty and creating wealth in rural areas; and as small-scale milk producers incur low production costs, if well organized, they should be able to compete with large-scale, capital-intensive 'high-tech' dairy farming systems in developed countries. Given the ability of smallholder milk producers to participate in the dairy market in a profitable manner depends not only on their own competitiveness, mainly determined by their production costs, but also on the efficiency of the dairy chains to which they belong. Therefore, recommendations for smallholder dairy development must perforce include strategies to develop and increase competitiveness in all segments of the dairy chain.

Potential of Maremmana cattle for organic beef production

Esposito, G.[1,2], Di Francia, A.[1], De Rosa, G.[1] and Masucci, F.[1], [1]Università di Napoli Federico II, DISSPAPA, 80055 Portici, NA, Italy, [2]University of Queensland, School of Agriculture and Food Science, 4343, Gatton, Australia; giulia.esposito@unina.it

Aim of this study was to evaluate whether growth performance and meat quality of Maremmana bulls organically farmed could be significantly improved by the dietary inclusion of chickpea during the finishing period. The study was conduced in an extensive organic farm located in Viterbo province, Central Italy. Twelve Maremmana bulls (270±8.1 days of age; 239±31.4 kg body weight (BW)) were divided into two homogeneous groups: one was fed the farmer's diets, based on barley meal, maize meal and alfalfa hay, and the other was fed diets in which barley meal was substituted by chickpea meal. Composition of diets was adjusted for growth-related changes in BW until the fixed slaughter weight of 630 kg. The dietary content of chickpea ranged from 23 to 11%, as fed. Animals were weighed at the beginning of the trial and thereafter every 3 weeks. Carcasses form each animal were weighed and scored for conformation and fat grade. Meat quality assessment was performed on 7-day aged Longissimus thoraci. Average growth curves were calculated by the regression of BW against age, the regression slopes of the two curves were compared by the F- test. Carcass traits and meat quality parameters were analyzed by one-way ANOVA. Carcass conformation and fatness score were analyzed by the Kruskal-Wallis test. BW increased linearly with age and chickpea-fed bulls showed higher BW from the age of 410 days onward. There were significant differences among dietary treatments for almost all the growth parameters evaluated. Carcasses from bulls fed chickpea were significantly better conformed than carcasses from barley fed bulls, but had higher fatness score. No differences were observed for meat quality parameters except for drip loss (higher in barley group) and cooking loss (higher in chickpea group). The analysis of feed costs indicated that the use of chickpea can determine an increase in the profit per bulls over 50%.

Aspects of cattle ownership in Northern Cacheu Province, Guinea Bissau

Almeida, A.M. and Cardoso, L.A., IICT & CIISA, Centro de Veterinária e Zootecnia, Av. Univ. Técnica, 1300-477 Lisboa, Portugal; aalmeida@fmv.utl.pt

Guinea Bissau is one of the lowest income per capita countries in the world and the vast majority of the population is dedicated to very small scale subsistence farming where animal ownership has an important role in both food supply and ceremonial events. Despite such fact little is known about cattle production and genetic resources in the country. The North of the Cacheu province comprehends two sectors, Bigene and São Domingos, considered one of the poorest regions of Guinea Bissau. We have conducted a survey and enquire in both sectors aiming to briefly characterize cattle ownership patterns among the two major ethnic groups in the region, the Balantas and the Felupes. Such characterization is intended to serve as a basis to possible agricultural and animal production development projects in this area. Cattle (N'Dama and West African shorthorn breeds) are owned by the two larger ethnic groups in the area, the Felupes and the Balantas that have however very different management practices as the first live in closed forests and the latter in open river basins termed locally as 'bolanhas'. Cattle ownership is important also in other aspects external to classical animal production as they are statutory symbols, reserve for cash in emergencies or use in mourning and other ritual ceremonies.

Reproductive management of dairy herds - a bio-social approach
Przewozny, A., Humboldt-University Berlin, Animal Breeding in the Tropics and Subtropics, Philippstr. 13, 10115 Berlin, Germany; agnes.przewozny@agrar.hu-berlin.de

This study provides a comprehensive characterization of current management methods of dairy farms, focusing on herd fertility. Relations of management factors to fertility and milk performance are analyzed following a bio-social approach. In 2007 a questionnaire survey including face-to-face interviews and direct observations was conducted in 84 East German dairy farms. Questions referred to housing, stress prevention, and management of herds, reproduction and personnel. Herd performance data stem from milk performance testing in 2007. Data analysis combined qualitative and quantitative methods. Calving interval (CI) and 305-day-milk yield (MY) were used as dependent variables. Mean values of herd size, CI and MY were, respectively, 306.3 cows ($\pm$238.3), 413.2 d ($\pm$18.73) and 8,555 kg ($\pm$1132.9). CI tended to decrease with increasing MY (r=-0.188, P=0.10). MY increased with rising herd size (r=0.29, P=0.01). Floors were wet and slippery in 71.25% of farms. In these farms CI tended to be longer (+16.6 d, P=0.055) compared to farms with dry and non-slippery floors. Lying areas were in 29.9% of the farms dry and flexible. Here MY was higher (+1,110 kg, P=0.011) than in the farms with wet and hard lying areas. A clear assignment of responsibility for heat detection showed a trend of decreasing CI (-6.3 d, P=0.129). Female herd managers with academic qualification achieved a higher MY than likewise qualified men (+752.9 kg, P=0.005), with no difference in CI. Thus, herds managed by highly qualified women showed a better ratio of MY to CI. Employee motivation by material and social incentives or by allocating responsibility to workers and pursuing good communication was related to higher MY than motivation by performance pay alone (+1000 kg, P=0.009); CI remained unaffected. Performance pay had no positive effect on targeted parameters. Results underscore the need for improved housing and recommend further study into personnel management in dairy farming.

A comparison of health status and milk quality in dairy cows reared in nearby areas of Italy and Slovenia
Gaspardo, B.[1], Lavrenčič, A.[2], Volarič, S.[3] and Stefanon, B.[1], [1]University of Udine, Department of Agricultural and Environmental Sciences, via delle Scienze 208, 33100 Udine, Italy, [2]University of Ljubljana, Department of Animal Science, Groblje 3, 1230 Domžale, Slovenia, [3]KGZS, Kmetijsko gozdarski zavod Nova Gorica, Pri hrastu 18, 5000 Nova Gorica, Slovenia; bruno.stefanon@uniud.it

One hundred and seventy-three transition cows, reared in 19 different dairy farms located in the border area between Italy and Slovenia, were selected to compare the health status of animals and the chemical and hygienic characteristics of milk in the different farming systems. In spring, for two consecutive years, rations were sampled in each farm and analysed for dry matter, crude protein, ether extract, NDF and starch. Individual milk was sampled and analysed for chemical composition and somatic cell count. A sample of blood was collected and used to determine acute phase protein (total protein, albumin, globulins, haptoglobin, and ceruloplasmin), oxidative stress condition (haemoglobin, glutathione peroxidase, malondialdehyde), energy metabolism status (free fatty acids, β-hydroxybutyrate and glucose) and liver function (glutamic-oxaloacetic transaminase and gamma-glutamyl transpeptidase). In the Slovenian farms, grass silage and hay prevailed in comparison to corn silage and concentrate feeds, whereas in Italian farms hay and concentrates were the predominant components of the diet. No significant differences were observed in milk quality (fat, protein, SCC), excluding percentage of lactose that resulted higher in milk produced in Italy. Blood indexes indicated a significant lower level of oxidative stress (higher glutathione peroxidise and lower malondialdehyde) and acute phase proteins (albumin, globulin, haptoglobin, and ceruloplasmin) and a significant improve in energy balance (lower β-hydroxybutyrate and higher glucose) in Italian lactating cows in comparison to the cohort of cows reared in Slovenia, evidencing a remarkable influence of diet and farming systems for the health status of transition cows.

Genetic diversity of Russian native cattle breeds on the genes associated with milk production
Sulimova, G.[1], Lazebnaya, I.[1], Khatami, S.[2] and Lazebny, O.[3], [1]Vavilov Institute of General Genetics RAS, Gubkin Str. 3, 119991 Moscow, Russian Federation, [2]Shahid Chamran University, Golestan Blvd, Ahwaz, Iran, [3]Koltsov Institute of Developmental Biology RAS, Vavilova Str. 26, Moscow, 119071, Russian Federation; galina_sulimova@mail.ru

Polymorphisms of the kappa-casein (CSN3), prolactin (PRL), growth hormone (GH) and transcription factor Pit1 genes were investigated in four Russian native cattle breeds – Bestuzhevskaya, Yakutskaya, Kostromskaya and Yaroslavskaya. These genes are known to associate with milk production. Russian native cattle breeds have valuable traits such as unpretentiousness, high resistance to diseases, good adaptation to the extreme natural conditions and others. But their genofonds were analyzed using DNA-markers insufficiently. In this work CSN3(HindIII), Pit1(Hinfl), bPRL(RsaI), and bGH(AluI) gene single nucleotide polymorphisms in exons 2, 6, 5, and 3, respectively, were studied using a PCR-RFLP technique. Individual and cumulative effects of the allelic polymorphisms of the genes studied on milk fat and protein content are also analyzed with the use of the Statistica v. 6.0, PopGene 1.32 and ANOVA. Significant differences in allele and genotype frequencies were observed among the breeds for all genes studied. A high content of economically important alleles of CSN3, bPRL and bGH was shown in Kostromskaya and Yaroslavskaya cattle breeds. Significant effect ($P=0.0483$) of the bPRL gene polymorphism on the milk fat content was shown for Yaroslavskaya breed (as a pilot experiment). Cumulative effect of the bGH и bPRL coupled genotypes on the milk fat ($P=0.041$) and protein ($P=0.024$) content was clearly demonstrated. Thus, cumulative effect of candidate genes for milk production traits may be revealed even if effects of separate candidate genes are no significant. This approach can be used for developing more sensitive test-systems with the purpose of an estimation of dairy efficiency in cattle.

An assessment of attitude towards selling livestock among the pastoralists in Tanzania
Laswai, G.H.[1], Haule, M.E.[2], Mwaseba, D.L.[1], Kimambo, A.E.[1], Madsen, J.[3] and Mtenga, L.A.[1], [1]sokoine University of Agriculture, Department of Animal Scince and Production, P.O.Box 3004, Morogoro, Tanzania, [2]OXFARM, TZ, P.O. Box 15305, arusha, Tanzania, [3]University of Copenhagen, Large Animal Sciences, Groennegaardsvej 2, DK 1870, Frederiksberg C, Denmark; laswaig@suanet.ac.tz

Cattle in the pastoral system accounts for 14% of the 19 million cattle population in Tanzania but their contribution to the pastoral economy is generally low. A study was done in Ngorongoro District in Tanzania to determine the attitude and reasons for selling livestock by pastoralists and identify factors that could influence such attitude. Data were collected through household questionnaire, key informant interviews, focus group discussions and direct observations. Descriptive statistics were generated and the Tobit model was employed for data analysis and interpretation. The results indicated that 42.2% and 45.6% of respondents respectively had positive attitude towards selling cattle and goats. Tobit estimates gave significant ($P<0.05$) relationships between the attitude of respondents towards selling livestock and distance to livestock markets (-0.052), livestock price (0.604), household size (0.858) and sex of respondent (0.752). It is concluded that the attitude towards livestock selling increases with short distances to markets, high livestock prices, large households and male-headed households. The mindset of pastoralists towards selling livestock is changing. Given right policy on price of cattle and reliable markets could increase off takes. Promotion of feedlots around the pastoral areas could create reliable cattle markets. Market linkages between livestock producers and consumers could also enhance commercial livestock production in the pastoral areas.

Estimation and comparison of economic values for productive characters in hybrid and local cattles
Seidavi, A.R.[1], Ghanipoor, M.[2], Mirmahdavi, S.A.[3], Hosseinpoor, R.[3] and Ghorbani, A.[3], [1]Islamic Azad University, Rasht Branch, Animal Science Department, Rasht, 4185743999, Iran, [2]The University of Adelaide, Animal Science Research Group, South Australia, 5371, J.S. Davies Animal Genetics and Epigenetics Research Group, Australia, [3]Agricultural and Natural Resources Research Center of Guilan Province, Rasht, Guilan, Iran; alirezaseidavi@iaurasht.ac.ir

Information of returns and costs from 20 crossbred herds of Guilan province in Lahijan (7 herds), Amlesh (1 herd), Rezvanshahr (3 herds) and Masal (9 herds) cities of Iran were studied. In crossbred cows, the economic weights of milk production (rial/kg/ cow/year), fat percentage (rial/cow/year), protein percentage (rial/cow/year) and herd life (rial/day/cow/year) in maximum profit interest were (739, -60,261, -30,672 and 28), and minimum cost price interest (-0.24, 19.89, 10.12 and -0.01), Respectively. Economic weights of fat percentage and protein percentage were positive in maximum profit and negative in minimum cost interest. In native cows, the economic weights of these traits in maximum profit interest were 232, -6,242, -3,177 and 454 and minimum cost price interest -0.11, 3.2, 1.6 and -0.23. Economic weights of fat percentage and protein percentage were negative in minimum cost price interest and positive in maximum profit one. In both interest, economic weights of milk production and herd life had more importance than others. The relative economic weights of the traits were equal in both interests. The relative economic weight of herd life (to milk production) was positive in most cities so that increase in economic efficiency of production system for this trait was higher than others in the same direction to milk production. Results showed that the studied systems were different regarding feeding and its costs, milk yield and its costs, and milk selling. So behaviour of system profit and economic weights against change in production factors in different cities is not similar.

Vigour and performance of Angus cattle with different myostatin genotypes
Eder, J.[1,2], Wassmuth, R.[1], Von Borell, E.[2] and Swalve, H.H.[2], [1]University of Applied Sciences Weihenstephan - Triesdorf, Steingruberstrasse 2, 91746 Weidenbach/Triesdorf, Germany, [2]University of Halle, Institue of Agricultural and Nutritional Sciences, Theodor-Lieser-Strasse 11, 06120 Halle, Germany; eder.johannes@t-online.de

The objective was to evaluate vigour and performance of Angus cattle with different myostatin genotypes. A total of 952 cows, 106 breeding bulls and 839 calves of the Angus breed from 31 German herds were included in the present study. The genotypes of cows and bulls were analysed by Eurofins Medigenomix GmbH®. Two different variations of the inactive myostatin allele were found. One was typical for the Belgian Blue and Angus (nt821) breed and the other must be originated by Limousin (F94L). The proportions of free cows (no inactive myostatin allele), heterozygous cows (one inactive myostatin allele) and double muscled cows (two copies of the inactive allele) were 78.5%, 21.4% and 0.1%, respectively. The frequencies of bull genotypes were 85.8% and 14.2%, respectively, and no homozygous bull with two copies of the inactive allele was observed. The breeding values varied between homozygous free (99.3) and heterozygous cows (102.1) as expected. Further, heterozygous cows were heavier with 647.8 kg than free cows weighing 630.5 kg. Scores from 1 (easy) to 4 (difficult) were given for calving ease and free cows were slightly easier calving (1.14) then heterozygous ones (1.28). A significant difference was found between the two calve types in birth weight but not in body length. Calves from free dams were lighter (36.6 kg) and shorter (54.2 cm) than calves with a heterozygous dam (38.4 kg and 54.9 cm, respectively). Vigour of calves was judged from 1 (high activity) to 4 and no significant difference between calves with free dams (1.70) and calves with heterozygous dams (1.64) was found. Hence, the heterozygous myostatin genotype of cows led to more calving difficulties, higher birth weights and longer calves. The proposed dominant-recessive inheritance of the myostatin allele could not be observed.

Utilization of fodder area in the cattle fattening in the farms desisting from milk production
Litwinczuk, Z., Teter, W., Stanek, P. and Zolkiewski, P., University of Life Sciences in Lublin, Department of Breeding and Genetic Resources Conservation of Cattle, Akademicka 13, 20-950 Lublin, Poland; zygmunt.litwinczuk@up.lublin.pl

The aim of this study was to determine the developmental potential of farms undertaking the beef production after the cessation of milk production. The study was conducted in 79 farms that took the beef production in 2006-2008. The surveyed farms were divided into 5 groups, i.e. group I – more than 20 cows to change of the production direction; group II – from 11 to 20 cows; group III – from 1 cow to 10 cows; group IV – running the fattening on the basis of purchased calves; group V – keeping the beef cattle (control group). The agricultural area (ha), main fodder area (MFA) and livestock unit (LSU) were taken into consideration in the analysis. The obtained data were analyzed statistically using one-way ANOVA by Statistica ver. 8.0. Farms qualified to group I and V represented the production model based on the rational utilization of lasting grasslands. The agricultural area was established on: 39.56 and 58.74 ha, respectively. The share of grasslands in the group I was 53% in the structure of MFA, while in group V exceeded 82%. Shortage of fodder from the MFA mostly sensed farms from group III and IV. That was indicated by a high stocking per 1 ha of MFA, ranged from 2.08 in group III to 4.48 in IV, at a relatively low LSU/ha of grasslands (0.55 LSU/ha). Farms from group II showed the intermediate characteristics between group I, and III and IV; having the animal stocking on the level of 2.48 LSU/ha of MFA, with a similar to group I stocking per 1 ha (1.01 head LSU). Summing up, the area of grasslands and MFA was the main obstacle in development of beef cattle herds in the farms desisting from milk production. The major limitation of the development of beef cattle breeding and beef production on the farms that sold the milk quota to the Agricultural Market Agency was the records resulting from the Polish law, restricting the cattle stocking from 1.5 LSU/ha of grasslands to 1.5 LSU/ha of MFA.

Static pressure measurements in some domestic ungulates managed under extensive conditions
Parés I Casanova, P.M., Sabaté, J. and Kucherova, I., ETSEA, Animal Production, Alcalde Rovira Roure, 191, 25198 Lleida, Spain; peremiquelp@prodan.udl.cat

A wide range of bioenergetic, production, life history and ecological traits scales with body size in vertebrates. However, the consequences of differences in community body-size structure for ecological processes have not been explored. Estimates of hoof pressures could be extremely useful for land managers So the objective of this research is to study the scaling relationships between body mass and hoof area in three domestic Catalan ungulates ranging in size from the 60 kg in the 'Blanca de Rasquera' goat breed to the 364 kg in the 'Cavall Pirinenc Català' horse breed and the 442 kg in 'Bruna dels Pirineus' bovine breed. As part of a larger study, seventy five untrimmed and free-ranging animals were sampled. Live weight and hoof area were determined. No ethical permission was considered necessary since the materials used were collected post mortem (for equines and bovines) or did not represent pain or injuries (for goats). Regression values suggest varying degrees between live weight and hoof area, probably because some of the animals were sampled prior to the expected attainment of maximum body mass. The pressures exerted on the ground per unit area by each breed are not identical, ranging from 92.8 kPa in cattle to 60.3 kPa in goats. As these species have different dinamogenic capacities, it can expected different trampled areas of ground per unit distance travelled. The impacts of large herbivores are not limited to trampling. Questions about the ecological implications of community body-size structure for such variables as foraging and food intake, dung quality and deposition rates, methane production, and daily travelling distances remain clear research priorities. The community composition imposed by ranchers will affect not only the long-term sustainability of the herbivore community itself, but also the composition of communities of other organisms and the resilience of grassland ecosystems to stochastic events such as droughts and fires.

Sustainable small scale cattle production – Croatian example – possibility lost?

Stokovic, I.[1], Susic, V.[1], Karadjole, I.[1] and Kostelic, A.[2], [1]University of Zagreb Faculty of Veterinary Medicine, Department for Animal Husbandry, Heinzelova ulica 55, 10000 Zagreb, Croatia, [2]University of Zagreb Agricultural Faculty, Svetosimunska cesta 25, 10000 Zagreb, Croatia; igor.stokovic@vef.hr

Definitions of sustainable agriculture usually include references to financial, environmental, ethical, social and product quality issues but also need to address animal welfare issues. Croatian cattle production for decades was based on small units. Nowadays more than 80% of cows are kept on farms from 1 to 10 cows per herd. For last 20 years cattle production was criticized for too many small farms and lots of efforts were put into enlargement without many thoughts about it. Those efforts resulted in almost 50% less cattle (1990: 829.000; 2009: 447.000), total amount of milk produced was raising, but now is falling seriously and number of households selling milk reduced with almost 2/3, from 58,715 in 2003 to 19,902 in 2010. Lots of efforts and money were invested, but it seems that we 'gave fish instead of learned farmers how to fish'. It looks like that we are on the start again and we have to change direction to sustainable small scale production, which we had from the beginning. People were and still are producing milk on farms with mixed animal production (cattle, pigs and poultry, in some regions sheep). In the same time beef was produced in a system of cooperatives with finishing fattening on specialized farms. Croatia was exporter of so called 'baby beef', which was appreciated on the market. So now we are facing rural depopulation, social issue of working places lost in agriculture, more dependence on import and devastating impact of huge farms on the environment, which is very easy to see in some developed EU countries. Taking into account the definition of sustainable livestock production and all of the issues mentioned above, serious efforts should be put into education and technological modernization of production, especially of high quality products with local origin.

Influence of technical parameters and milking routine of high capacity milking parlours on milk quality

Mihina, S.[1], Cicka, A.[1], Zajac, L.[2], Martiska, M.[2] and Broucek, J.[3], [1]Slovak University of Agriculture, Faculty of Engineering, A. Hlinku 2, 94976 Nitra, Slovakia (Slovak Republic), [2]Agromont spol. s r.o., Juzna 7, 94901 Nitra, Slovakia (Slovak Republic), [3]Animal Production Research Centre Nitra, Institute of Livestock Systems and Animal Welfare, Hlohovecká 2, 95141 Luzianky, Slovakia (Slovak Republic); stefan.mihina@uniag.sk

Quality of milk is very much influenced by technical parameters of milking equipment, their nominal values and real ones, measured during milking and without milk in system, immediately after installation and after some time. Way of regular diagnostic and maintenance of milking equipments has the most important role. Besides technical aspects, working routine during milking, especially before attachment of milking cluster to udder teats and way of taking off teat cups from an udder, have to be considered as factors of milk quality parameters. In the same type of high capacity milking parlours in Slovakia were permanently recorded technical parameters like vacuum, vacuum pump capacity, phases of pulsation cycles, intensity and time of milk flow, intensity and quality of cleaning and disinfection, frequency of biotechnical diagnostic procedure making, bulky SCC and BCC. Correlations among all obtained data were calculated and significant ones were recorded.

Income effects of by-product economic values for slaughtered cattle and sheep
Kale, M.C., Aral, Y., Aydin, E., Cevger, Y., Sakarya, E. and Guloglu, S.C., Ankara University, Faculty of Veterinary Medicine, Department of Animal Health Economics and Management, Diskapi-Ankara, 06110, Turkey; vetaydin36@hotmail.com

In this study it is aimed to determine the economic values of by-products obtained as a result of slaughter and the income effect of these values within the purchase of carcass for slaughtered cattle-buffalo and lamb-sheep. Cutting records for the years 2007 and 2008 of a slaughterhouse pertaining to the private sector and the weight and price data concerning the carcass and by-products obtained from the slaughtered cattle-buffalo and the lamb-sheep constitute the material of the study. In analyzing the research data the Microsoft Excel 2007 program has been used. In the slaughterhouse where the research has been conducted for the year 2007 and 2008, the share of the by-product income within the average purchase price of carcass has been determined as 10.93% and 11.29% in average for lamb-sheep and 7.79% and 8.06% for cattle-buffalo, respectively. Besides, for the research period the rate of total monetary value of destroyed by-products within the total by-product income has been calculated as average 0.96% for lamb-sheep and 0.89% for cattle-buffalo. Finally, together with the increasing density of slaughter of the Industrial Meat Enterprises in Turkey and as a result of their being operated with efficient and high-capacity utilization, for all species of slaughtered animals, also the possibility of collecting, processing and evaluating economically of all kinds of edible and inedible animal by-products shall be ensured. In the study it has been determined that a substantial part of the amount the enterprises have paid for the carcass purchase has been obtained as the by-product income.

Designing equine education for rural innovation: the big picture and some missing details
Evans, R., Hogskulen for Landbruk og Bygdenaeringar, Postveien 213, 4353 Klepp Stasjon, Norway; eqrnetwork@gmail.com

Traditional equine education reflects an industry orientation which tends to look at the equine sector in isolation from the wider rural development context within which it is situated. While the focus on good horsemanship, ethology, nutrition management, breeding, training and stable keeping is as important as ever, there is an increasing need for acknowledgement that the sector sits within a set of rapidly changing rural development practices across Europe. These practices include those derived from the range of European Second Pillar policies such as farm multifunctionality; environmental, social and economic sustainability; and local supply chains and embeddedness. The changes include the growing role of eco-tourism and nature-based recreation and the changing demographics of nature-based activities. They promote new ways of seeing old assets for new development and innovation and new ways that communities can act together to increase the sustainability and vitality of their local economies. And with these new phenomena, come new challenges and new opportunities. How can we equip the new generation of equine actors and entrepreneurs to recognize and take advantage of these new challenges and opportunities across the wide range of equine activities? The Hogskulen for Landbruk og Bygdenaeringar offers a BA program in Rural Innovation and our challenge is to design a BA level course on 'Heste I Naering' which matches the innovative teaching and learning we provide to our students on this program. Through access to a wide range of international equine education courses through members of the Equine Research Network (www.eqrn.net) we are beginning to construct such a course – one which addresses the new opportunities and challenges in the equine sector and which will help our students deliver sustainable and satisfying participation in an ever-growing sector. This paper will present a sample curriculum and discuss the issues raised in the process.

A new bachelor education program in equine science in Norway

Vangen, O., Austbø, D. and Hauge, H., Norwegian university of life sciences, Department of animal and aquacultural sciences, P.O. Box 5003, N-1432 Ås, Norway; odd.vangen@umb.no

In 2010 our university launched a new bachelor program in equine science. The maximum number of students is set to 20 /yr. The educational program includes cooperation with the Norwegian Equine Center, where the students will learn the applied aspects of horse production, including reproduction, training physiology, use of horses in tourism and therapy. The study is giving basic knowledge of horse breeding, nutrition and etiology, and how this knowledge can be applied in practice. The students will learn to calculate nutritional requirements and feeding plans, how to make breeding plans and the basis for better horse management and horse welfare. Additional courses are dealing with managerial economics, managerial accounting and budgeting, introduction to organization theory, organization and leadership psychology, business start-up, nature and life quality-use of animals, plants forest and landscape. The educational program encourage half a year study abroad (cooperating universities), and the riding teacher education (level 1) at the Norwegian Equine Center is qualifying to replace one semester of the bachelor program. In this way students can combine the applied and theoretical education and be better qualified for the different jobs in the horse industry.

Undergraduate equine science education in the United States

Splan, R.K., Virginia Tech, Department of Animal and Poultry Sciences, 5527 Sullivans Mill Road, Middleburg, Virginia 20117, USA; rsplan@vt.edu

Over the last decade, colleges and universities across the United States have witnessed tremendous growth in undergraduate equine science programs. Student demand for equine-related course content is higher than ever. This comes despite, and perhaps partly in response to, an economic downturn experienced by the nation's horse industry. Concurrent with academic equine program growth is a change in undergraduate student profile. Compared to the average student a decade ago, today's equine science undergraduate is more likely to be female, more likely to come from a suburban background, and less likely to have significant prior experience in animal agriculture. Further instructional challenges result from the wide range of career options available to students with an equine interest. This paper provides an overview of various undergraduate equine science programs in the United States which have been developed to equip students with the tools needed for post-graduate success. Additionally, the paper highlights common and novel strategies used to improve the quality and cost-effectiveness of undergraduate instruction in the equine sciences.

Online teaching and learning in equine science
Murray, J.M.D., Royal (Dick) School of Veterinary Studies, Veterinary Clinical Sciences, Easter Bush, Roslin, Midlothian EH25 9RG, United Kingdom; Jo-Anne.Murray@ed.ac.uk

Technology has progressed throughout the world and in particular the use of internet functions has led to the development of on-line education. Many educational establishments have embraced on-line education, or e-learning, with on-line courses being delivered by a great number of institutions world-wide. The benefit to learners is the flexibility offered by e-learning. Individuals can participate in studying while maintaining busy professional and personal commitments as most courses allow students to work at a time and place that is convenient to them; thus, the classroom is essentially wherever the students wants it to be. Moreover, the benefits of on-line learning are also similar for teachers. However, e-learning programmes require careful planning, structure and facilitation on the part of the providers. Online communication is key to engaging the students in learning; the instructor's abilities to teach online are critical to the quality of online education and in evoking communication between individuals in order to form a sense of community, which is particularly important in online education where the capacity for students to form social networks is potentially inhibited as a result of the lack of face to face interactions. The medium used to facilitate communication online is typically asynchronous discussion forums within the virtual learning environment. More recently, synchronous discussions have been incorporated into teaching programmes using chat tools, such as Skype, and Virtual Classrooms, such as Wimba. The use of online virtual worlds, such as Second Life (SL), in online education has increased in recent years and these environments have been used to promote presence and a sense of community in online teaching. In conclusion, through careful construction of online learning environments, quality teaching provision, and encouragement of quality interactions then real, advanced learning that is of equivalent value to that achieved in face to face on-campus learning can be promoted.

Teaching and assessing ethics in equine science education
Ellis, A.D., Nottingham Trent University, School of Animal Rural and Environmental Sciences, Brackenhurst Campus, NG25 0QF Southwell, United Kingdom; andrea.ellis@ntu.ac.uk

Lecturing on ethics in equine science education poses some challenges in terms of assessing the students' understanding and knowledge. The MSc in Equine Health and Welfare at Nottingham Trent University offers the Module 'Ethics and Consultation', to allow the opportunity of putting ethical beliefs and understanding into the context of 'real life situations'. The aims of the module are to: a) review and evaluate current ethical dilemmas in equine sports disciplines and worldwide use of the horse and to b) apply consultation techniques while demonstrating an understanding of legal implications of working as a professional advisor in the equine industry. The assessment for this module is split into 2 parts: 1) Roleplay and Consultation and 2) Evaluation Report. The Roleplay situation is presented to the student (=consultant) in form of an e-mail sent by a client to the consultant. Following some further enquiries between the two parties a meeting (consultation) is set up during which the 'consultant' advises on the issues presented, translating scientific knowledge and understanding into 'lay persons' terms while showing good communication technique and awareness of underlying issues. The 'client' is played by a member of staff and each roleplay situation involves some less obvious background issues, some of which will arise during the consultation. The evaluation report, written by the consultant (student) for 'their line-manager', then discusses possible ethical dilemmas which have arisen, legal implications of the advice given and it also contains an element of self-evaluation. The student is required to show critical awareness of current problems/insights at the forefront of equine sports disciplines and the equestrian profession, and the philosophical, welfare and ethical issues related to these.

Establishing framework for the European equine master programme

Potočnik, K.[1], Cervantes, I.[2], Koenig Von Borstel, U.[3] and Holgersson, A.L.[4], [1]University of Ljubljana, Biotechnical Faculty, Zootechnical Department, Groblje 3, 1230 Domzale, Slovenia, [2]University of Cordoba, Veterinary Faculty, Department of Genetics, Carretera Madrid-Córdoba, Cordoba, Spain, [3]Gottingen University, Institute for Animal Production and Companion Animal Genetics, Albrecht Thaer weg, D-37015 Gottingen, Germany, [4]Swedish University for Agricultural Sciences, Department of Equine Studies, Uppsala, 7507 Uppsala, Sweden; klemen.potocnik@bf.uni-lj.si

In order to develop human resources that will be capable of managing the area of interests in the European Horse Sector an initiative to create an European Equine Master programme was taken within the EAAP Commission on Horse Production. The idea is to create a structure for collaboration among universities that have an interest in such a programme of education. The prime reason is that many universities do not have resources to perform standard syllabus through official procedures. After several meetings and discussion about the possibility to establish education the idea of structure of collaboration between universities has become clearer. With the aim to estimate the availability of present equine education programmes a database named 'Bag of knowledge' was created. Information was gathered from partner countries all over Europe. Collected data will be available to all partner universities and interested students. The purpose of this database is to provide information to support decisions of possible forms of curriculum 'sharing' and usefulness for completing the syllabus for each partner when starting education. In the first step the database was used to estimate the potential of know-how to create an adequate curriculum at University Master level study with 120 ECTS. In the next stage, the database information will be used to unify or merge courses leading to a suggestion of a study plan and these will therefore form a basis for future curriculum in a European Equine Master Programme.

Student service and faculty advancement in equine science at the research university in the United States and Europe

Splan, R.K.[1] and Ellis, A.D.[2], [1]Virginia Tech, Department of Animal and Poultry Sciences, 5527 Sullivans Mill Road, Middleburg VA 20117, USA, [2]Nottingham Trent University, School of Animal Rural and Environmental Sciences, Brackenhurst Campus, Southwell, NG25 0DS, United Kingdom; rsplan@vt.edu

Equine science educators at the research university strive to create and deliver exceptional teaching programs, achieve prominence in their scholarly work, and contribute meaningfully to the academic community. Balancing these missions can be challenging for junior and senior faculty alike. This paper explores how educators in equine science can advance professional aims and achievements without sacrificing instructional quality or service to students. Further, it highlights faculty demographics and institutional challenges unique to equine science curricula and university-based equine research programs.

The use of video surveillance, a horse simulator and information technology to enhance advanced equitation teaching

Taylor, E.T., White, C., York, S.J., Hall, C.A. and Ellis, A.D., Nottingham Trent University, School of Animal, Rural and Environmental Sciences, Brackenhurst Campus, Southwell, NG25 0QF, United Kingdom; andrea.ellis@ntu.ac.uk

Students studying the Equine Science Degree at Nottingham Trent University are offered Equitation Science as an option. This module provides the theoretical and practical aspects of equitation and training of horses as well as enabling the student to apply the fundamental scientific and practical principles necessary to evaluate human and equine sports performance. The use of IT, and in particular video analysis common in other sports to highlight weaknesses in technique and subsequently enhance performance. Through the use of Dartfish analysis software and I.T resources, students are now able to receive both qualitative and quantitative feedback on their performance. For equitation lectures, a large video screen is placed in the indoor riding arena and practical equitation sessions are videoed providing the students with instant feedback on their performance. The video clips are then analysed rapidly within a lecture using video gait analysis software (Dartfish, OnTrack), which enables the students to identify positional problems in the rider and to evaluate the horses' gaits. In addition, wireless pressure mapping technology (Tekscan) linked to electronic display equipment is available to assess rider balance and this can also be tested when using the Racewood Horse Simulator, which has settings for walk, trot, fast trot, canter and fast canter and contains pressure gauges which react to hand (rein) and leg movements. In addition the simulator supports work on human exercise physiology and training as well as measuring effect of equipment under fully controlled conditions. Recent research at the university also involves the use of a spectacle mounted mobile eye tracker and video recorder which matches up eye movements with the visual field of the rider, allowing for evaluation of rider vision as they, for example, jump around a course.

Genetic regulation of health and functional traits in dogs

Distl, O., University of Veterinary Medicine Hannover, Institute for Animal Breeding and Genetics, Buenteweg 17p, 30559 Hannover, Germany; ottmar.distl@tiho-hannover.de

There are more than 400 dog breeds with characteristic variation across and within-breed for functional body and health traits. Artificial selection, random drift and crossbreeding have shaped the today's dogs. Numerous population genetic analyses have been performed for skeletal diseases, eye diseases, hearing loss, heart anomalies/diseases and inborn defects. Animal models and complex segregation analyses significantly improved accuracy of estimations and were very useful to disentangle non-genetic and genetic effects. Estimation of breeding values (BV) and use of BVs has been introduced in some dog breeds in Europe since the late 90ies. With the dog genome draft sequence and high density SNP-arrays in hands, complex traits and diseases are now exploitable for identifying quantitative trait loci (QTL) and for unravelling their genetic architecture. Across-breed studies using >60,000 or >170,000 SNPs revealed >50 genomic regions correlating with breed differentiation. For some morphological traits, a small number (<4) of QTL explain a large proportion of the phenotypic variation among breeds. Search for selective sweeps indicated an enormous reduction of genetic variation and haplotype diversity in these genomic regions. Coincidence of selective sweeps with disease and functional traits will be elucidated when more genetic variants associated with diseases become available. More than 120 disease traits, in the majority with an assumed monogenic inheritance, have been mapped and the causal genetic variants have been identified. For complex traits, many genome-wide association studies (GWAS) are underway. So far, GWAS for hip and elbow dysplasia QTL have been performed and a number of QTL have been mapped. Genomic selection for dogs seems feasible if population structure within breeds can be taken into account and mixed model approaches are employed. Small population sizes, often clustered by lines, and a number of health problems are the challenges dog breeders have to solve.

Genetics of behaviour traits in dogs

Strandberg, E. and Arvelius, P., Swedish University of Agricultural Sciences, Department of Animal Breeding and Genetics, P.O. Box 7023, 75007, Sweden; Erling.Strandberg@slu.se

In general, there are two types of behaviour tests in dogs: 1) the traditional test of functionality, usually a field trial based on a practical situation, e.g. a hunting test; and 2) standardized behaviour characterisation, based on a number of defined behaviour traits, assumed to be related to working ability or functionality. We will give examples of some of these types of tests for various dog breeds and discuss the pros and cons in relation to the opportunity for breeding for improved behaviour. In the traditional field trials, the performance is often graded from bad to good. However, a dog can be bad in different ways (e.g. too low or too high intensity), which means that the genetic background will be different. This will lead to a lower heritability. In the behaviour characterisation, one tries not to pass a value judgement but only to describe the behaviour. One example of the importance of the scale the trait is measured on, can be seen in the two versions of the assessment of herding characteristics of Border Collies. The trait 'active working distance' was in the first version described in 5 intervals measured in meters (0-1, 1-2, ..., 10-) but in the second version in words (animals do not move, dog needs to move very close; ..., dog needs very long distance), the latter clearly having a more subjective description. The heritability for the trait changed from 0.50 to 0.18! Often many behaviours are measured, too many to be practically used as separate breeding values. One solution is to use factor analysis. In the Swedish dog mentality assessment, 33 scores are given, each on a 1-5 intensity scale. Factor analysis revealed 5 personality traits that were combinations of varying number of original scores. The personality traits had heritabilities of 0.18-0.32 (German shepherds and Rottweiler), which were higher than for the contributing scores (0.11-0.24). Four traits were relatively strongly positively genetically correlated but were almost uncorrelated to Aggressiveness.

Genetic evaluation of temperament traits in the Rough Collie

Grandinson, K. and Arvelius, P., Swedish University of Agricultural Sciences, Dept of Animal Breeding and Genetics, P.O. Box 7023, 750 07 Uppsala, Sweden; Katja.Grandinson@slu.se

Fear related problems are common in the Swedish Rough Collie population. The aim of this study was to investigate possibilities to develop a genetic evaluation for temperament, based on the Swedish Dog Mentality Assessment (DMA – a field test used in large scale to characterize temperament traits in dogs). Genetic analyses were performed on the individual DMA variables, and the five broader personality traits: sociability, curiosity/fearlessness, playfulness, chase-proneness and aggressiveness. Data was available from 2,550 Rough Collies, and parameters were estimated using a linear animal model including fixed effects of sex, year and month of test, and random effects of litter, judge, test site, genetic effect of the individual and residual. Heritabilities ranged from 0.05-0.31, and generally the higher heritabilities were found for the broader behavioural traits. Validation of the DMA was done using the C-BARQ questionnaire, to which some 'Collie-specific' questions were added. Owners of Rough Collies aged from 6 mo to 10 years were targeted and the questionnaire generated information about 1,766 dogs (a reply rate of 50%). Of these, 935 had information from the DMA. There were significant correlations between the broader personality traits measured in the DMA, and the everyday behaviour of the dogs as described by the owners in the questionnaire. For example, sociability in the DMA was positively correlated with stranger-directed interest, and negatively correlated with stranger-directed fear and non-social fear. Curiosity/fearlessness in the DMA was positively correlated with human-directed play interest, and negatively correlated with stranger-directed fear and non-social fear. Playfulness in the DMA was positively correlated with human-directed play interest. We conclude that selection for temperament in Rough Collie, based on data from the DMA, is possible and could reduce the frequency of fearful dogs in the breed.

Genetic analysis and effect of inbreeding on skeletal diseases and echocardiography measurements in Italian Boxer dogs

Cecchinato, A.[1], Sturaro, E.[1], Bonfatti, V.[1], Piccinini, P.[2], Gallo, L.[1] and Carnier, P.[1], [1]University of Padova, Department of Animal Science, Viale dell' Università 16, 35020 Legnaro, Padova, Italy, [2]Centre for the Screening of Skeletal Diseases, (CeLeMasche), 44100 Ferrara, Italy; alessio.cecchinato@unipd.it

The dog is the non human species for which the largest number of genetic disorders is known. The objectives of this study were to estimate genetic parameters for two skeletal diseases (SP: spondylosis deformans; HD: hip dysplasia) and one echocardiography measurement (AA: area of the aortic annulus) associated to the risk of subaortic stenosis and to assess the effect of inbreeding on these traits in the Italian Boxer dog population. Data included screening results, obtained by the Centre for the Screening of Skeletal Diseases, for SP and HD of 6,730 and 2,987 dogs respectively (examined from 1993 to 2008 for HD and from 1997 to 2009 for SP). Data on AA included measures of 1,024 random Italian Boxer dogs which were submitted to a complete echocardiographic examination from 1999 to 2004. The outcome variables were: the degree of osteophytes development (four-grade linear system) for SP and the FCI (Fédération Cynologique Internationale) 5-class linear grades for HD. A Bayesian analysis was implemented via Gibbs sampling. Effects of sex, year-month of birth, age of the dog at screening and experience of the x-raying veterinarian were assigned flat priors; kennel and animal genetic effects were given Gaussian prior distributions. Marginal posterior medians (SD) of heritabilities for HD, SP and AA were 0.20 (0.03), 0.38 (0.04) and 0.25 (0.07) respectively. The genetic correlations between these traits were 0.18, for HD and SP, -0.24 for HD and AA and -0.05 for SP and AA. These estimates indicate that breeding programs aimed to reduce prevalence of skeletal and cardiac diseases are feasible, provided that a regular screening program of Boxer dogs is performed. Regarding the effect of inbreeding, slight detrimental effects have been detected for the aforementioned traits.

Selection strategies against categorically scored hip dysplasia in dogs: a simulation study

Malm, S.[1], Sørensen, A.C.[2], Fikse, W.F.[1] and Strandberg, E.[1], [1]Swedish University of Agricultural Sciences, Dept of Animal Breeding and Genetics, P.O. Box 7023, 750 07 Uppsala, Sweden, [2]Aarhus University, Dept of Genetics and Biotechnology, P.O. Box 50, 8830 Tjele, Denmark; sofia.malm@skk.se

Decades of selective breeding to reduce the prevalence of categorically scored hip dysplasia (HD), based on phenotypic assessment of radiographic hip status, have had limited success. The aim of this study was to evaluate two selection strategies for improved hip status: truncation selection based on phenotypic records or BLUP breeding values. Stochastic simulation was used and selection scenarios resembled those in real dog populations. In addition, optimum contribution selection (OCS) was evaluated for one scenario. The selection scheme included two traits: HD (defined as a categorical trait with five classes and a heritability of 0.45 on the liability scale) and a continuous trait (with a heritability of 0.25) intended to represent other characteristics included in the breeding goal. A population structure mimicking that in real dog populations was modeled. The categorical nature of HD caused a considerably lower genetic gain compared to simulating HD as a continuous trait. When using BLUP selection, the genetic change, overall and for HD alone, was substantially larger than for phenotypic selection in all scenarios. However, BLUP selection resulted in higher rates of inbreeding. By applying OCS, the rate of inbreeding was lowered to about the same level as for phenotypic selection but with increased genetic improvement. Breeding restrictions based on the HD grade had a marginal effect on the genetic progress in HD and a negative effect on the overall genetic progress. For efficient selection against HD, implementation of breeding schemes based on BLUP breeding values should be prioritized. However, in small populations, BLUP should be used together with OCS or similar strategy to maintain genetic variation.

Connecting breeding value estimation in sheep
Lewis, R.M., Virginia Tech, Department of Animal and Poultry Sciences, Blacksburg 24061, USA;
rmlewis@vt.edu

Selection response in breeding programs depends on accurately estimating and reliably comparing breeding values. Where individual breeding units are small, which is often the case in sheep, pooling resources across flocks may accelerate gains. Across-flock genetic evaluation using BLUP has therefore become the norm. However, such evaluations may be substantially biased. When founders (i.e. animals of unknown parentage) of separate flocks differ genetically, differences in flock genetic means may be confounded with those due to environment. The extent of bias depends on the strength of genetic relationships or connectedness amongst animals in separate flocks. As connectedness improves, error in comparing breeding values among flocks falls. Sharing rams across flocks establishes connections. In sheep, where artificial insemination is limited, genetic relationships among flocks are often tenuous. Breeding cooperatives establish strong connections among flocks, yet their implementation is limited. Prediction error variance (PEV) of differences in estimated breeding values (EBV) between animals is an appropriate measure of connectedness. Advances in computing resources and algorithms allow PEV to be obtained even for breeds with large pedigree and performance data. Thus, even outside cooperatives, genetically connected flocks can be identified and sensibly combined in a single genetic evaluation. The accuracies of EBV are thereby increased. Although important, connectedness is not itself a breeding goal. Instead, establishing 'sufficient' connectedness to reduce bias to acceptable levels should be part of a more comprehensive program. Threshold values for sufficient connectedness have been established, and can be used to balance priorities. Although in its infancy in sheep, advances in genomics provide exciting opportunities to define connectedness. With dense genomic (SNP) scans, genomic relationships could augment those based on pedigree, reducing bias in EBV. With stronger connectedness among flocks, breeding value estimation can be improved, accelerating selection response.

Genetic connectedness among Norwegian sheep flocks
Eikje, L.S.[1], Boman, I.A.[1], Blichfeldt, T.[1] and Lewis, R.M.[2], [1]The Norwegian Association of Sheep and Goat Breeders, P.O. Box 104, N-1431 Ås, Norway, [2]Department of Animal and Poultry Sciences, Virginia Tech, Blacksburg 24061, USA; se@nsg.no

In the Norwegian breeding scheme, BLUP estimated breeding values (bEBV) are compared across flocks to identify high merit rams. If flocks are not genetically connected, bEBV may be biased due to confounding of flock genetic means and husbandry. Some sheep breeders join ram circles to form larger breeding groups with progeny testing of rams across flocks establishing connections. Genetic gains in ram circles are disseminated to outside flocks by use of rams from ram circles establishing genetic connections. If sufficient, data from outside flocks could be included in genetic evaluation to improve bEBV accuracies. In this study, connectedness among flocks in the Cheviot and Fur sheep breeds was estimated to determine (1) if bEBV can be fairly compared within and among ram circle flocks, and (2) whether outside flocks were sufficiently connected to add to genetic evaluation. Data consisted of 214,391 and 198,339 pedigrees, and 131,012 and 110,955 weaning weights, on Cheviot and Fur sheep, respectively. In Cheviot, there were 49 flocks in 4 ram circles, with 77 outside flocks. In Fur sheep, there were 8 flocks in 1 ram circle, with 134 outside flocks. Connectedness was measured as the average prediction error correlation of flock genetic means. A threshold value of 0.10 was used to define sufficient connectedness, corresponding with trivial bias when comparing bEBV. Within ram circles, mean connectedness among flocks was high (Cheviot: 0.39; Fur: 0.61), with all pair wise values between flocks exceeding the threshold. In Cheviot, mean connectedness between flocks in different ram circles was 0.13 to 0.15. Among outside flocks, 22 (Cheviot) and 29% (Fur) were connected to their respective ram circles. This corresponded with a 56 (Cheviot) and 161% (Fur) increase in numbers of lambs evaluated. Connectedness in ram circles in these breeds is sound, with opportunity to include outside flocks in genetic evaluation.

Estimation of breeding values for meat sheep in France

Tiphine, L.[1], David, I.[2], Raoul, J.[1], Guerrier, J.[1], Praud, J.-P.[1], Bodin, L.[2], François, D.[2], Jullien, E.[1] and Poivey, J.-P.[2,3], [1]Institut de l Elevage, 149 rue de Bercy, 75595 Paris Cedex 12, France, [2]INRA, UR 631, SAGA, 31320 Castanet-Tolosan, France, [3]CIRAD, UMR 112, SELMET, 34398 Montpellier, France; laurence.tiphine@inst-elevage.asso.fr

Reproductive performances of ewes and growth of lambs are registered in the national farm recording system, in order to estimate breeding values of rams and ewes for several traits: prolificacy, maternal ability and growth potential. Estimation of breeding values for prolificacy is based on normal scores of litter size after natural or hormone-induced estrus, considered as two different but genetically linked traits, under BLUP animal model methodology. Maternal ability combines growth between birth and 30 days of age and survival of lambs, each dependant on maternal and direct genetic effects, under BLUP animal model methodology. The evaluation model takes into account the number of lambs suckled thanks to a multiplicative factor (1 for a lamb reared as a single, 0.7 for twin-reared lambs) for the maternal genetic effect. Last trait recorded in farm is growth potential between 30 and 70 days, which is estimated using improved sire model BLUP. Young males are chosen on genetic values of their parents and then evaluated on central stations for growth, conformation and fattening traits. On these individual test stations, genetic evaluation is only made intra test group using animal model, but with limited pedigree information. In some breeds, the best males are then evaluated on progeny test for carcass and slaughter traits. Results obtained on many slaughter traits are presented in graphical spider charts. Currently, important work is being done to evaluate the economic importance of each selected trait. The first step, recently completed, was to model the production systems. We are now going to set up synthetic index at each step of the selection schemes. Furthermore, on-going researches concern parasite resistance, social behavior, fertility and semence production for a potential short or long term inclusion in the national genetic evaluation.

National genetic evaluations in dairy sheep and goats in France

Larroque, H.[1], Astruc, J.M.[2], Barbat, A.[3], Barillet, F.[1], Boichard, D.[3], Bonaïti, B.[3], Clément, V.[2], David, I.[1], Lagriffoul, G.[2], Palhière, I.[1], Piacère, A.[2], Robert-Granié, C.[1] and Rupp, R.[1], [1]INRA, UR631-SAGA, BP 52627, 31326 Castanet-Tolosan Cedex, France, [2]Institut de l Elevage, BP 42118, 31321 Castanet-Tolosan Cedex, France, [3]INRA, UMR 1313 -GABI, 78352 Jouy-en-Josas Cedex, France; helene.larroque@toulouse.inra.fr

In France, breeding schemes in dairy sheep and goats have been oriented for a long time towards the improvement of dairy traits (milk, fat and protein yields, and fat and protein contents). The estimated breeding values (EBV) for these traits are performed using a BLUP repeatability animal model assuming heterogeneous variances and constant variance ratios. In 2010, the evaluations involved 2,128,729 animals and 6,735,408 records from the five dairy sheep breeds, and 2,760,612 animals and 7,273,465 records from the two main goat breeds. These last years, to reduce production costs and take into account also milking labour and animal welfare, EBVs have been carried out for functional traits, i.e. somatic cell score (SCS) for mastitis resistance and udder type traits. These EBVs are computed with the dairy trait model for SCS and a multiple-trait BLUP animal model for type traits. According to the data availability, all these EBVs are not yet computed for all breeds. For the most advanced breeds, a total merit index has been proposed giving more or less the same weight for production and functional traits. The short-term objective is to implement genetic evaluations for SCS in all breeds and to update the breeding objectives accordingly. On-going researches on genetic evaluations concern reproduction traits, milking speed, longevity, milk production persistency and once-daily-milking ability. The availability for sheep, and soon for goats, of a pangenomic high density SNP chip allows considering genomic selection programs. In Lacaune dairy sheep breed, a life-size test of genomic selection is in progress based on genomic estimated breeding values (GEBV) for all traits with a GBLUP method. First results are presented in other papers at this congress.

Genetic parameters for live weight, ultrasound scan traits and muscling scores in Austrian meat sheep

Maximini, L.[1], Brown, D.J.[2] and Fuerst-Waltl, B.[1], [1]University of Natural Resources and Life Sciences Vienna, Department of Sustainable Agricultural Systems, Division of Livestock Science, Gregor-Mendel-Str.33, A-1180 Vienna, Austria, [2]Animal Genetics and Breeding Unit, University of New England, Armidale, NSW 2351, Australia; lina.maximini@boku.ac.at

Heritabilities and genetic correlations were estimated for live weight (lw) and average daily gain (adg) (n=13,634), ultrasound measured eye muscle depth (emd) and back fat depth (fat) as well as muscling scores for shoulder (shoul), back (back) and hindquarters (hind) (n=6,110) in Austrian meat sheep. An across breed analysis was carried out using performance records of Merinolandschaf, Suffolk, Texel, German Blackheaded Meatsheep and Jura sheep which were routinely tested for meat performance between 2000 and 2010. Genetic parameters were estimated with multivariate mixed animal models including both direct and maternal genetic effects and permanent environmental effects of the dam (pe) as well as fixed effects. Estimated direct heritabilities were 0.07, 0.16, 0.20, 0.21, 0.03, 0.01, and 0.08 for lw, adg, emd, fat, shoul, back and hind, respectively. Maternal genetic heritabilities were very low and significant only for lw and adg, whereas pe was fitted for every trait and explained between 0.05 and 0.10 of the phenotypic variance. Lw showed highly negative genetic correlations with emd (-0.87), fat (-0.57), and hind (-0.81). The genetic correlations are more strongly antagonistic than observed from published estimates. This may be a direct result of the structure of the data used in this study where many of the records were from small herd year season groups and often confounded by sire.

Work in progress on genomic evaluation using GBLUP in French Lacaune dairy sheep breed

Baloche, G.[1], Larroque, H.[1], Astruc, J.M.[2], Babilliot, J.M.[3], Boscher, M.Y.[3], Boulenc, P.[4], Chantry-Darmon, C.[3], De Boissieu, C.[2], Frégeat, G.[5], Giral-Viala, B.[4], Guibert, P.[6], Lagriffoul, G.[2], Moreno, C.[1], Panis, P.[6], Robert-Granié, C.[1], Salle, G.[1], Legarra, A.[1] and Barillet, F.[1], [1]INRA, UR631, Castanet-Tolosan, 31320, France, [2]Institut de l Elevage, Castanet-Tolosan, 31320, France, [3]Labogena, Jouy-en-Josas, 78352, France, [4]Ovitest, Onet-le-Château, 12850, France, [5]UPRA Lacaune, Rodez, 12033, France, [6]Confédération Générale de Roquefort, Millau, 12103, France; francis.barillet@toulouse.inra.fr

French Lacaune dairy sheep selection programme is based on an open nucleus totalizing 174,000 ewes and AI-progeny testing of 420 young rams per year. Breeding objectives are milk and udder functional traits, plus resistance against classical scrapie. The storage of DNA/blood of the Lacaune AI rams has been organized since the middle of the 90's. The Illumina Ovine SNP50 BeadChip available since 2009 makes feasible genomic selection. In January 2011, the French Lacaune reference population included 2,651 AI rams, born between 1998 and 2009, and genotyped by Labogena. The aim of this study was to compare results of pedigree- and genomic-based EBV (PEBV and GEBV respectively) of a validation population of 666 young AI rams born in 2007 and 2008, using a training population either of 1,742 AI genotyped rams born between 1998 and 2006 or 3,645 AI rams when enlarging the training population to ungenotyped rams of the same cohorts of birth. Daughter yield deviations for milk yield and contents, somatic cell score, and udder morphology traits have been used for PEBV and GEBV evaluation using GBLUPF90 software from the University of Georgia, USA. The results show that GEBV would be more efficient than PEBV: an average increase of accuracy of 14% has been found across traits. Thus selecting young unproven rams based on their GEBV could be possible. Acknowledgements: for French ANR & ApisGene (SheepSNPQTL project), and for FUI, Midi-Pyrénées region, Aveyron & Tarn departements, & Rodez town (Roquefort'in project).

A comparison of various methods for the computation of genomic breeding values in french lacaune dairy sheep breed

Robert-Granie, C., Duchemin, S., Larroque, H., Baloche, G., Barillet, F., Moreno, C., Legarra, A. and Manfredi, E., INRA, UR631-SAGA, 31326 Castanet-Tolosan, France; christele.robert-granie@toulouse.inra.fr

Genomic selection refers to selection based on genomic breeding values (GEBV), where the genomic breeding values are calculated from marker effects located across the whole genome. This study evaluated several statistical methods for predicting SNP effects for genomic selection. The methods included GBLUP ; a Bayesian approach -BayesCπ- used stochastic search variable selection for all SNPs ; the Partial Least Squares (PLS) and Sparse PLS (sPLS) regression, which reduce the number of variables in the final model and select the most important variables for the sPLS. We compared the ability of these methods to accurately predict GEBV in actual dairy sheep data set containing 2,651 AI rams, born between 1998 and 2009 and genotyped for 44,131 SNPs. These approaches were applied to estimate GEBV for milk production traits, somatic cell score and udder morphology traits. Phenotypes used for this study were DYD (Daughter Yield Deviations) corresponding to the average performance of a sire's daughters, adjusted for fixed and non genetic random effects and for the additive genetic value of their dam. DYD were weighted by their variance which is a function of the sire's Effective Daughters Contribution (EDC). Validation populations containing 666 young AI rams born in 2007 and 2008 were used to asses the accuracy of the GEBV by comparing the estimated GEBV with the DYD. The weighted correlation between GEBV and observed DYD was computed using EDC as weights. The genomic approaches tested in this study produced similar accuracies of the GBLUP method (results presented in another French paper and showing a better correlation between GEBV and observed DYD compared with pedigree-based BLUP). The accuracies of the genomic method were not significantly different for most traits.

Implementation of BLUP breeding values estimations into breeding programs for sheep in Slovakia and Czechia.

Milerski, M.[1], Margetín, M.[2,3] and Oravcová, M.[3], [1]Research Institute of Animal Science, Přátelství 815, 104 00 Prague 10 - Uhříněves, Czech Republic, [2]Slovak University of Agriculture in Nitra, Tr. A. Hlinku 2, 949 76 Nitra, Slovakia (Slovak Republic), [3]Animal Production Research Centre Nitra, Hlohovecká 2, 951 41 Lužianky, Slovakia (Slovak Republic); m.milerski@seznam.cz

While in the Czech Republic (CR) sheep husbandry is focused mostly on heavy lamb production, in the Slovak Republic (SR) predominantly dairy sheep are kept. Due to different specializations of production also different breeding programs are used in SR and CR. In the SR the highest attention is paid to estimation of breeding values (BVs) for milk yield and milk components. In Improved Valachian and Tsigai populations the BVs are estimated by the multitrait test-day BLUP Animal Model methodology, while in less numerous breeds Lacaune and Eastfriesian and in population of hybrids the single-trait whole lactation approach is applied. Additionally BVs for litter size and lamb weight at weaning are estimated. In the CR breeding values have been estimated by the use of BLUP methodology since 2003. In the year 2010 totally 16 breeds were involved. BVs for lamb weight at the age of 100 days (both direct and maternal genetic effects) and for litter size are estimated for all breeds. Additionally BVs for eye-muscle depth and back-fat thickness measured by ultrasound are estimated for terminal sire breeds (Suffolk, Charollais, Texel, Oxford Down, German Blackhead) and Romney. Program of estimation of BVs for milk traits is being implemented into practice currently. In both countries the main obstacle for the BVs estimations is very low level of AI usage resulting in limited amount of relationship connectedness between flocks. Also lack of links between performance recording databases in different countries and shortage of initial information about breeding animals in a case of their import are considered by sheep breeders as a serious problems especially in dairy sheep.

The prediction of breeding value in dairy sheep population using diferent test day animal models
Grosu, H., National Research & Development Institute for Biology and Animal Nutrition, Animal Breeding, Ploiesti-Bucharest way, No 1, Balotesti, Ilfov, 077015, Romania; hgrosu1962@yahoo.com

The objective of this study was to predict the breeding values in a dairy sheep population in order to find the best individuals for the next generation. The breeding values were predicted using two methods: (a) random regression test day animal model; and (b) auto-regressive test day animal model. The data set consisted of 833 TD records from 174 ewes in the first lactation. The whole population had 362 individuals, which the following structure: 48 sires, 140 dams and 174 offsprings (ewes with own performances). Totally, 174 ewes had records. The average number of TD per lactation was about 5.3. Data was edited and TD records were deleted if ewe' ID was unknown, if lactation number was not specified and if days in lactation for the TD record was <60 or >175 days. Also, a TD class must have at least 4 observations. The two methods were compared using the accuracy of prediction of breeding values and the percentage of squared bias. For the data set accounted the best model was the random regression test day with the three order Legendre polynomials.

Estimation of genetic parameters for milking goats in the United Kingdom using a random regression model
Conington, J., Mrode, R. and Coffey, M., Scottish Agricultural College, Sustainable Livestock Systems, West Mains Rd, Edinburgh Eh9 3JG, United Kingdom; jo.conington@sac.ac.uk

Commercial-scale goat milk production in the UK is carried out by a few technically proficient farmers managing around 30,000 goats supplying the majority of the UK's fresh milk. In anticipation that a formalised genetic improvement programme will be initiated, genetic parameters for daily milk yields from lactations 1-7 from a major goat herd were estimated. A population of mixed-breed goats with a pedigree file from 13 generations of 19,545 individuals was used that included 239 sires, 7,272 dams, 66 paternal grand-sires, 104 paternal grand–dams, 176 maternal grand-sires and 4,041 maternal grand-dams. The number of test-day records for lactations (L) 1 to 7 were 127,723 (L1) 77,209 (L2), 51,552 (L3), 32,819 (L4), 19,887 (L5), 11,625 (L6), 5,644 (L7). Univariate random regression analyses were used to estimate the heritability (h^2) and permanent environment (pe) effects separately for each lactation. Bivariate random regression analyses were undertaken for L1 vs L2, and for L1 vs L2-4 (combined) to estimate genetic correlations (r_g) for 4-305d and 4-520d test days. The model for milk yield included fixed effects of year-season, fixed lactation curves nested within management group and random regressions with orthogonal polynomials of order 2 for animal and pe effects. Heritabilities for 305d (520d) yield were 0.13 (0.09), 0.31 (0.26), 0.11 (0.05), 0.16 (0.13), 0.29 (0.20), 0.18 (0.17) and 0.46 (0.28) for L1 to L7 respectively. Permanent environment effects for 305d (520d) were 0.68 (0.91), 0.68 (0.74), 0.88 (0.94), 0.88 (0.87), 0.71 (0.79), 0.81 (0.82), 0.53 (0.72) for L1 to L7 respectively. Average r_g for L1 vs L2 was 0.72 and 0.79, and average pe was 0.52 and 0.17 for 305d and 520d respectively. For L1 vs L2-4, the average r_g was 0.74 and 0.76 and average pe was 0.36 and 0.23 for 305d and 520d respectively. Results indicate that selection to improve milk yield is likely to succeed and L1 is highly correlated genetically to subsequent lactations.

Genetic parameters for milk traits using fixed regression models for Pag sheep in Croatia

Špehar, M.[1], Barać, Z.[1], Gorjanc, G.[2] and Mioč, B.[3], [1]Croatian Agricultural Agency, Ilica 101, 10000 Zagreb, Croatia, [2]University of Ljubljana, Groblje 3, 1230 Domžale, Slovenia, [3]University of Zagreb, Svetošimunska 25, 10000 Zagreb, Croatia; mspehar@hpa.hr

The objective of this study was to estimate genetic parameters for daily milk, fat and protein yields, and somatic cell score (SCS) using test-day records of the Pag sheep in Croatia. Data included 38,068 test-day yields for 4,449 ewes recorded from 2003 to 2010. Pedigree file included 5,260 animals. Test-day records were modelled using a single-trait fixed regression repeatability test-day model. Fixed class effects in the model were: parity, litter size as a number of born lambs, season of lambing, and flock. Days in milk and age at lambing were treated as covariates. For yield traits, the effect of days in milk was nested within parity and number of lambs born and fitted using the Ali-Schaeffer lactation curve. Given the seasonal production regime in this breed, linear regression nested within parity was sufficient to model the age at lambing. This effect was not included in the model for SCS, while the effect of days in milk was nested within parity. Direct additive genetic effect, flock-test-day, and permanent environment effect over lactations were included in the model as random effects. Variance components were estimated using Residual Maximum Likelihood as implemented in the VCE-6 program. Comparison was also done with Markov chain Monte Carlo and Integrated Laplace methods. The estimated standard deviations for daily milk, fat and protein yields (kg), and SCS were: 0.15, 0.01, 0.008, and 0.54 for additive genetic, 0.15, 0.008, 0.009, and 0.83 for flock-test-day, and 0.14, 0.01, 0.008, and 0.24 for permanent environment effect over lactations. Fitting permanent environment effect within lactations lead to the underestimation of additive genetic variance due to the small number of test-day records per lactation and shallow pedigree. Results indicate the possibility of using test-day records for genetic evaluation of the Pag sheep in Croatia.

Comparison of the breeding value estimation of SURPRO V0.1 software and MTDFREML

Onder, H., Ondokuz Mayis University, Animal Science, Ondokuz Mayis University, Agricultural Faculty, Dep of Animal Science, 55139, Turkey; hasanonder@gmail.com

Farmers and animal breeding persons require information on animal resources for further studies and evolving realistic strategies for improvement and rearing of livestock. Record collection is the most important tool to improve of economic traits for all animal genotypes. As in other animal species, some software are used to record collection of sheep and goat breeding. In this study, SURPRO V0.1 software was compared versus MTDFREML in terms of breeding value estimations. Results were analyzed with the method of Spearman's rank correlation to determine whether there was a difference in ranking of animals with respect to their breeding values between two software. Obtained correlation coefficient was 0.984 (P<0.01). Results showed that there was a significant positive correlation between the breeding values calculated with SURPRO V0.1 software and MTDFREML.

White-flowering faba beans (*Vicia faba* L.) in sow diets - preliminary results

Neil, M.[1] and Sigfridson, K.[2], [1]Swedish University of Agricultural Sciences, Animal Nutrition and Management, Box 7024, S-75007 Uppsala, Sweden, [2]Lantmännen Lantbruk, Division Foder, von Troils väg 1, S-20503, Sweden; maria.neil@slu.se

Presently, inclusion of faba beans in sow diets is not recommended in Sweden. The basis for this is experiments carried out with high levels of high-tannin faba beans (17-34% of the diets), where the results indicated reductions in litter size and milk yield. Documentation regarding effects of inclusion of low-tannin white-flowering faba beans in sow diets is lacking. The hypothesis for the present experiment is that a moderate inclusion of white-flowering faba beans (10%) in combination with other domestic feed ingredients like cereals and rape seed meal will not cause reproductive or other disturbances. The control diet contains 7.5% soya bean meal. The experiment comprises 2 groups of 20 sows each. The sows will be monitored during 2 reproductive cycles, from service for the 1st experimental litter until heat after weaning of the 2nd experimental litter. Preliminary results from the 1st experimental litter in sows fed soya bean meal vs sows fed faba beans: Litter sizes were 13.5 at birth in both treatments, and 9.9 vs 9.8 at weaning. Piglet weights were 1.6 kg at birth in both treatments, 11.6 vs 11.0 kg at weaning, and 27.0 vs 27.4 at 9 weeks of age. The length of gestation was 116.1 vs 116.0 days, and the interval from weaning to service for the next litter was 4.5 vs 4.6 days. In conclusion, there is so far no indication of differences in results between sows fed soya bean meal and sows fed moderate levels of white-flowering faba beans. This study is funded by The Swedish Farmers' Foundation for Agricultural Research.

Effect of feed restriction during gestation on body weight and backfat depth in European-Chinese sows over two parities

Viguera, J.[1], Medel, P.[1], Peinado, J.[1], Flamarique, F.[2] and Alfonso, L.[3], [1]Imasde Agroalimentaria, S.L., C/ Nápoles 3, 28224 Pozuelo de Alarcón, Spain, [2]Grupo AN, Campo de Tajonar s/n, 31192 Tajonar, Spain, [3]Universidad Pública de Navarra, Campus de Arrosadía, 31006 Pamplona, Spain; pmedel@e-imasde.com

A total of 88 Youli (Gene +) sows were used to evaluate the effect of feed restriction during gestation on body weight and backfat depth. There were 2 treatments: Control sows (CS) were fed a 37.2 MJ DE/d diet during gestation whereas Experimental sows (ES) used two feeding levels based on backfat thickness (BT) during gestation. ES fed a 28.0 MJ DE/d diet from 1 to 26 d of gestation, and 30.0 or 34.0 MJ DE/d to sows with less or more than 26 mm of BT from 27 to 89 d of gestation, respectively, and 34.0 or 38.0 MJ DE/d to sows with less or more than 30 mm of BT from 89 d of gestation to farrowing. All sows were fed *ad libitum* in lactation, and were controlled over two consecutive reproduction cycles. Sow weights, BT and feed intake were determined at various intervals and data were analyzed using a GLM procedure of SAS with feeding regimen as main effect, with parity included in the model as covariate. During first parity, CS had more body weight (BW) and BT gain in gestation than ES (81.2 vs 68.3 kg, and 12.5 vs 7.1 mm, respectively; $P<0.05$). No significant differences were found in lactation for decrease of BW, but ES tended to lost less BT than CS (-7.2 vs -8.6 mm; P=0.08), probably due to greater feed intake of ES (82.5 vs 75.1 kg of dry matter; P=0.10). During second parity, CS showed greater BW and BT gain than ES (79.6 vs 67.5 kg, and 13.5 vs 7.7 mm, respectively; $P<0.01$). However, during lactation ES lost less BW and BT than CS (-32.6 vs -45.5 kg, and -6.6 vs -10.0 mm, respectively; $P<0.01$) due to the greater feed intake of ES during lactation compared to CS (106.2 vs 90.5 kg of dry matter; $P<0.01$). Feed restriction based on BT stabilizes the body condition of sows during gestation and lactation and increased feed intake during lactation.

Dose and form of vitamin D for sows: Impact on bioavailability, performance and bone status markers
Lauridsen, C., Halekoh, U., Larsen, T. and Jensen, S., Aarhus University, P.O. Box 50, 8830, Denmark; Charlotte.Lauridsen@agrsci.dk

The official vitamin D recommendation for sows during gestation and lactation is not based on scientific reports, and is ranging from 200 to 1,000 IU/kg feed. We conducted a dose-response trial with vitamin D_3 and $25(OH)D_3$, in order to gain information on vitamin D requirements for sows in terms of bioavailability, performance and bone condition. A total of 160 multiparous sows were randomly assigned from first day of mating until weaning to dietary treatments containing 4 concentrations of 1 of the 2 different vitamin D (200, 800, 1,400, and 2,000 IU/kg of vitamin D from cholecalciferol or corresponding doses of 5, 20, 35, and 50 µg/kg feed from HY•D). Blood samples of the sows were obtained on d 8 before expected farrowing, and on d 2, 16 and 28 after farrowing. Concurrently, samples of the suckling pigs were obtained on day 4, 16, and 28 of age. Plasma was stored at -80 °C until analysis for vitamin D ($25(OH)D_3$ concentration) and bone status markers. The statistical model involved the effect of form and dose of vitamin D, and the effect of the lactation state of the sows (or the age of the piglets). Data were analysed using the linear mixed model. The results showed that the plasma concentration of $25(OH)D_3$ was influenced by an form and dose interaction, and was also affected by lactation state of the sows. Irrespective of the dietary form and dose of vitamin D provided to the sows, very little vitamin D was transferred to the progeny. Performance of the sows was not influenced by the dietary vitamin D treatments, except for a decreased number of stillborn piglets with larger doses of vitamin D. Lactation day rather than dietary vitamin D treatments influenced the bone status markers (i.e. concentrations of osteocalcin and Ca as well as activities of total alkaline phosphatase and bone alkaline phosphatase in plasma). In conclusion, a dietary dose of approximately 1,400 IU of vitamin D is recommended for reproducing swine, and HY•D was more bioavailable than vitamin D_3.

Influence of phytate and phytase on native and supplemented zinc bioavailability in piglets
Schlegel, P.[1] and Jondreville, C.[2], [1]Agroscope Liebefeld-Posieux, Tioleyre 4, 1725 Posieux, Switzerland, [2]Nancy Université, URAFPA, Av de la Forêt de Haye 2, 54505 Vandoeuvre les Nancy, France; patrick.schlegel@alp.admin.ch

Zinc (Zn) is an essential trace element, a heavy metal and a non-renewable resource. Phytate is identified as the major dietary factor affecting Zn bioavailability in monogastrics, as phytate-zinc complexes are insoluble. Data from 5 experiments (31 observations) on weaning piglets, published between 2002 and 2010 were used to evaluate the interactions between phytate, phytase and Zn on Zn bioavailability. A GLM procedure was conducted using bone Zn content as dependent variable and dietary native Zn (Znn), added Zn sulfate (Znadd), Znadd^2 and non-hydrolyzed phytic P (PPhytNH) as independent variables. PPhytNH is a new variable and represents the residual phytic phosphorus (P) fraction remaining intact after hydrolysis by mainly vegetal and microbial phytase. It represents the difference between dietary phytic P and the quantity of released P. Bone Zn responded linearly to PPhytNH in non supplemented Zn diets (bone Zn = 91.7 (P<0.001) – 36.1 (P=0.001) * PPhytNH, R^2=0.81, RSE=7.24). Results from the meta-analysis (R^2=0.92; RSE=5.86) indicate that the positive effect on bone Zn from Znn (P<0.001) was clearly reduced by the interaction of PPhytNH (P<0.001). However, PPhytNH did not interact with Znadd (P>0.10). Znadd increased bone Zn linearly (P<0.001) and quadratically (P<0.001). The quadratic effect is probably due to the plateau of bone Zn with increasing dietary Zn. The present meta-analysis confirms that 1) neither phytic P nor phytase influences the bioavailability of supplemented Zn; 2) the release of Zn by phytase is proportional to the release of P; 3) microbial phytase offers an important possibility in adding value to native Zn. This model allows a quantification of the phytate antagonism on Zn bioavailability and could be used to review dietary Zn recommendations in pigs.

Can Hampshires tolerate low lysine diets post weaning?

Taylor, A.E.[1], Jagger, S.[2], Toplis, P.[3], Wellock, I.J.[3] and Miller, H.M.[1], [1]University of Leeds, Leeds, LS2 9JT, United Kingdom, [2]ABN, Peterborough, PE2 6FL, United Kingdom, [3]Primary Diets, Melmerby, HG4 5HP, United Kingdom; bsaet@leeds.ac.uk

Results (University of Leeds) have shown that the Hampshire has a higher average daily intake (ADI) compared to the Large White. It may therefore be hypothesised that they can perform better on a lower lysine diet due to this higher ADI. The aim of this study was to identify whether the Hampshire could perform better when fed a low lysine diet post weaning compared to the Large White. A 2x2 factorial design was used; two lysine diets (low and high) and two genotypes (Hampshire x (Large White x Landrace) and Large White x (Large White x Landrace)). A total of 264 pigs (132 of each genotype) were weaned at 28±4 days, 8.2±0.16 kg and remained on trial for 20 days. Pigs were given ad-libitum access to either a low (0.80%) or a high (1.75%) lysine diet. Diets were iso-energetic and formulated to the same amino acid:lysine ratios. Pigs were weighed at day 0, 7, 14, and 20. ADI and feed conversion ratios (FCR) were recorded. Pigs were checked for health daily. A GLM (Minitab 14) was used to analyse differences in performance. Results showed that a reduction in dietary lysine reduced growth performance in both genotypes. Pigs fed the high lysine diet were 2.1±0.12 kg heavier than pigs fed the low lysine diet at day 20 (P<0.001) due to a higher average daily gain (ADG) and a more efficient FCR (P<0.001; P<0.001). The Hampshire grew faster than the Large White due to a higher ADI (P<0.001). Lysine level did not affect ADI for either genotype. An interaction was observed for overall ADG and day 20 weight (P<0.05). The difference in weight at day 20 was greater for the Hampshire on the two diets than the Large White (2.7±0.16 kg vs. 1.5±0.16 kg; P<0.001). There was no difference in health performance between treatments. Neither genotype could increase their feed intake to compensate for the low lysine diet. The Hampshire was less tolerant to a low lysine diet; they were faster growers and therefore may have a greater requirement for lysine.

Interest of using synthetic amino acids, including L-Valine, for formulating low crude protein pig diets based on rapeseed meal

Quiniou, N.[1], Primot, Y.[2], Peyronnet, C.[3] and Quinsac, A.[4], [1]IFIP-Institut du Porc, BP 35104, 35650 Le Rheu, France, [2]Ajinomoto Eurolysine SAS, 153 rue de Courcelles, 75817 Paris Cedex 17, France, [3]ONIDOL, 12 avenue George V, 75008 Paris, France, [4]CETIOM, Rue Monge, Parc industriel, 33600 Pessac, France; nathalie.quiniou@ifip.asso.fr

One hundred forty-four group-housed growing-finishing pigs were allocated to one of the three experimental feeding strategies. Diets S were formulated with soybean meal and their dietary crude protein (CP) content averaged 15.9 and 15.0%, respectively during the growing (before 65 kg) and the finishing periods. In diets R, CP levels were reduced at 15.0 and 14.1%, respectively, and soybean meal was replaced partially or completely by rapeseed meal and balanced with L-Lysine, DL-Methionine, L-Threonine and L-Tryptophan. In diets RV, L-Valine was also incorporated (0.3 g/kg) both in the growing and the finishing diets allowing an additional reduction of CP content (14.5 and 13.2%, respectively). All diets were formulated on the same net energy basis (9.7 MJ NE/kg) and on minimum ratios between digestible lysine and other amino acids following the ideal protein profile. Over the 27-111 kg BW range, no significant differences were observed between treatments on average daily gain (S: 801, R: 801, RV: 818 g/d), feed intake (S: 2.34, R: 2.38, RV: 2.33 kg/d), feed conversion ratio (S: 2.94, R: 2.97, RV: 2.87) or carcass fatness (S: 61.3, R: 60.9, RV: 61.8% lean). These results indicate that it is possible to replace soybean meal by rapeseed meal in association with available free amino acids for a long time without any consequence on growth performance. They also show that an additional reduction of dietary CP content can be achieved with L-Valine utilisation without any consequence on growth performance, when diets are formulated on the NE basis and in agreement with the ideal protein concept. Reduced dietary CP performed through the R and RV strategies was associated with a reduction of N output by 400 and 650 g /pig, respectively.

Effect of feed grade glycerol on the meat quality of fattening pigs
Zsédely, E., Kovács, P. and Schmidt, J., University of West Hungary, Faculty of Agricultural and Food Science, Vár 2., H-9200 Mosonmagyaróvár, Hungary; zsedelye@mtk.nyme.hu

A fattening trial was conducted with 2 x 50 (Norwegian Landrace x Duroc) pigs fed 0 or 5% feed grade (86.3%) glycerol in the diet between 30-105 kg in order to evaluate the effect of dietary glycerol on several quality traits of meat. In experimental diet maize was replaced with glycerol in such a way that energy and protein content of diets should be similar. At the end of the trial all animals were slaughtered and tissue (longissimus dorsi muscle) sample was collected from 10 animals in each group to determine several quality traits (chemical and fatty acid composition, color, defrosting loss, cooking and frying losses, WB-share force and sensory properties) of meat. Statistical analysis was carried out by using GenStat 11R software (VSN International Ltd.). A single-factor variance-analysis was used to show the impact of the glycerol on the attributes of meat. Although dietary glycerol decreased the crude protein and either extract contents compared to control but the differences were not significant. Fatty acid profile of the meat samples was changed slightly. Dose of 5% glycerol increased significantly C17:0, C18:3 n-3 fatty acids and proportion of n-3 group, but decreased C18:1 n-9 compared to control samples. Color was observed lighter of meat derived from glycerol-supplemented pigs, but without adverse effect on the sensory scores for preference. The glycerol feeding reduces the defrosting loss of the meat, stored at a temperature between -12 and -20 °C degrees. Additionally cooking loss was reduced too but frying loss was similar in both groups. Although WB-share force was measured lower in control samples (58.2 vs. 60.7 N), but the 7 assessors found the meat of glycerol group more tender during sensory analysis. This group achieved higher score in the case of flavour too but odour did not differ from each other. Our results proved that the substitution of maize with feed grade glycerol did not modify adversely the quality of pork.

Effect of adding Jerusalem artichoke in feed to entire male pigs on skatole level and microflora composition
Vhile, S.G.[1], Sørum, H.[2], Øverland, M.[1,3] and Kjos, N.P.[1], [1]Norwegian University of Life Sciences, Department of Animal and Aquacultural Sciences, P.O. Box 5003, N-1432 Ås, Norway, [2]Norwegian School of Veterinary Science, P.O. Box 8146 Dep., N-0033 Oslo, Norway, [3]Aquaculture Protein Centre, CoE, P.O. Box 5003, N-1432 Ås, Norway; nils.kjos@umb.no

The main objective of the experiment was to investigate effect of different levels of dried Jerusalem artichoke (J. A) in feed to entire male pigs on skatole level in the hindgut and in adipose tissue. Additionally, effect on microflora, pH, dry matter and short chain fatty acids (SCFA) in the hindgut was evaluated. A total of 55 entire male pigs divided into five experimental groups (n=11) were used. Seven days before slaughtering the groups were given the different dietary treatments: Negative control (basal diet), positive control (basal diet + 9% chicory-inulin), basal diet + 4.1% J. A., basal diet + 8.1% J. A. and basal diet + 12.2% J. A. Samples from colon, rectum, faeces and adipose tissue were collected. Effect of dietary treatment on growth performance and carcass traits, on skatole, indole and androstenone in adipose tissue and on pH, skatole, indole, dry matter, SCFA and microbiota in the hindgut was tested by GLM (SAS, 1990). Orthogonal polynomials tested linear responses of increased levels of J. A. Feeding increasing levels of J. A. to entire male pigs reduced skatole in colon and faeces (linear, P<0.01). Adding J. A. resulted in a slightly decreased dry matter content in the hindgut, and a decreased pH in colon (linear, P<0.01). In adipose tissue, there was a tendency towards decreased levels of skatole (linear, P=0.06). The amount of *Clostridium perfringens* was reduced in both colon and rectum (linear, P<0.05). There was an increase in total amount of SCFA, acetic acid and valerianic acid in faeces (linear, P<0.05). In conclusion, adding dried J. A. in feed to entire male pigs one week before slaughtering reduced skatole and positively influenced the microflora in the hindgut.

Effects of feeding Bt (MON810) maize on the intestinal microbiota and immune status of pigs

Walsh, M.C.[1], Buzoianu, S.G.[1,2], Gardiner, G.E.[2], Rea, M.C.[3], Ross, R.P.[3] and Lawlor, P.G.[1], [1]Teagasc, Pig Development Department, Animal and Grassland, Research and Innovation Centre, Moorepark, Fermoy, Co. Cork, Ireland, [2]Waterford Institute of Technology, Department of Chemical and Life Science, Waterford, Ireland, [3]Teagasc, Food Research Centre, Moorepark, Fermoy, Co. Cork, Ireland; maria.walsh@teagasc.ie

The aim was to examine the effect of feeding genetically modified maize (GMm) on gut microbiota and immune status of pigs. Male pigs (N=72) were weaned at ~ 28 d of age, blocked by weight and litter and randomly assigned after a 12 day acclimatisation period to 1 of 4 treatments; 1) non-GMm from d 0 to 110 (T1), 2) GMm from d 0 to 110 (T2), 3) non-GMm from d 0 to 30 followed by GMm from d 30 to 110 (T3) and 4) GMm from d 0 to 30 followed by non-GMm from d 30 to 110 (T4) where d 0 was the first day experimental diets were fed. Blood and faecal samples were collected on d 0, 30, 60 and 100 for haematological analysis, immune cell phenotyping and microbiological analysis, respectively. Ileal and cecal digesta were sampled at slaughter (d 110) for microbiological analysis. Haematology and faecal microbiology data were analyzed as repeated measures using the MIXED procedure of SAS. Ileal and cecal microbiota were analyzed as a one-way ANOVA using the GLM procedure of SAS. Feeding GMm had no effect on the faecal microbial populations enumerated or ileal and cecal counts of Lactobacillus or total anaerobes. Ileal Enterobacteriaceae counts in pigs fed T4 were lower than in pigs fed T1 (P=0.02). Lymphocyte counts tended to be higher in pigs fed T4 compared to T1 and T2 (P=0.09). On d 100, white blood cell (P=0.06) and lymphocyte (P=0.04) counts were higher in pigs fed T3 compared to T1 and T2. Pigs fed T4 had higher counts of red blood cells on d 100 than pigs fed T2 and T3 (P=0.02). Monocytes tended to be higher in pigs fed T3 and T4 compared to T1 and T2 on d 100 (P=0.10). Feeding GMm to pigs alters culturable ileal microbiota and blood counts; however, the significance of these findings remains to be determined.

The effect of chitooligosaccharide supplementation on performance, intestinal morphology, selected microbial populations, volatile fatty acid concentrations and immune function in the weanling pig

Walsh, A.M., Sweeney, T., Bahar, B., Flynn, B. and O'doherty, J.V., University College Dublin, Lyons Research Farm, Newcastle, Dublin, Ireland; ann.walsh1@ucdconnect.ie

The objective of this study was to investigate the effects of supplementing different molecular weights (MW) of chitooligosaccharide (COS) on performance, intestinal morphology, selected microbial populations, volatile fatty acid concentrations and the immune status of the weaned pig. Twenty-eight piglets (24 days of age, 9.1 (± s.d. 0.80) kg live weight) were assigned to one of four dietary treatments in a complete randomised design. The pigs were fed the dietary treatments for 8 days and then sacrificed. The treatments were (1) control diet (0 ppm COS) (2) control diet plus 5-10 Kda COS (3) control diet plus 10-50 Kda COS and (4) control diet plus 50- 100 Kda COS. The COS was included in dietary treatments at a rate of 250 mg/kg. Pigs fed the 10-50 Kda COS had a higher villous height (P<0.05) and villous height: crypt depth ratio (P<0.05) in the duodenum and the jejunum in comparison to the control group. Supplementation of different MW of COS had no significant effect on the expression of the cytokines TNF-α, IL-6, IL-8, and IL-10 in the gastro-intestinal tract of the weanling pig. Supplementation of COS at the 5-10 Kda level resulted in a lower *E. coli* (P<0.05) and lactobacilli (P<0.05) number in the colon of the pig compared to the control group. Pigs offered the 5-10 Kda MW of COS had significant lower levels of acetic and valeric acid compared to the control group (P<0.05). The current results indicate that lower MW of 5-10 Kda COS possessed strong antibacterial activity while the higher MW of 10-50 Kda was optimum for enhancing intestinal structure. COS supplementation exerted no deleterious effects on immune function or growth performance of the pigs while reducing the incidence of diarrhoea.

Inhibitory action of analytical grade zinc oxide and of a new potentiated ZnO on the *in vitro* growth of Escherichia coli strains

Durosoy, S.[1], Vahjen, W.[2] and Zentek, J.[2], [1]Animine, 335 chemin du noyer, 74330 Sillingy, France, [2]Free University of Berlin, Faculty of Veterinary Medicine, Institute of Animal Nutrition, Königin-Luise-Str. 49, 14195 Berlin, Germany; sdurosoy@animine.eu

A new and potentiated feed grade zinc oxide product (ZinPot, Animine) is compared to analytical grade zinc oxide for their growth repressing effect on two pathogenic bacterial *Escherichia coli* strains: *E. coli* PS79 (O47:K88) and *E. coli* PS7 (0138:K81). Strains were incubated in Brain-Heart Infusion Medium (BHI). Saturated Zn-containing media were produced by adding 10 g Zn-source to a total of 100 ml BHI media at pH 4.6 and pH 6.5, then incubated, autoclaved, centrifugated and finally adjusted with BHI media to identical Zn-concentration at different dilution levels. Measurements during incubation were taken every 2 minutes over a period of 8 h for media at pH 6.5 and every 4 minutes for 16 h at pH 4.6. Growth curves were obtained from data and regression analysis was employed to calculate coefficients for maximum growth in the stationary phase and lag time for each individual incubation. In comparison to MIC determination which defines complete bacterial inhibition, this kinetic assay measured the influence of ZnO source on growth of enterobacterial strains at sub-inhibitory Zn-concentrations. Below the MIC of 0.32 µg/ml at pH 6.5, lag phase, exponential growth and stationary phases could be distinguished. Compared to analytical grade ZnO, the lag time was significantly ($P \leq 0.05$) higher in ZinPot for the two *E.coli* strains at all Zn concentrations and at both pH, but only at low Zn-dosage for E.coli PS7 at pH 6.5. At pH 4.6, both *E.coli* strains showed higher maximum growth for standard ZnO compared to ZinPot at all sub-inhibitory Zn-concentrations. In order to assess the impact of ZinPot on bacterial communities in the weaned piglet intestine, further *ex vivo* studies should explore the effects of these two types of ZnO on bacterial fermentation and bacterial cell number.

Influence of dose and forms of vitamin D on early reproduction and gene expression in uterus of gilts

Lauridsen, C. and Theil, P., Aarhus University, Animal Health and Bioscience, P.O. Box 50, 8830, Denmark; Charlotte.Lauridsen@agrsci.dk

In swine nutrition, little is known about the vitamin D requirement for reproduction, and currently, the vitamin D recommendation for gilts and sows is not based on scientific evidence. Reproduction in females is markedly diminished in states of vitamin D deficiency. The objective of the present research was to study the bioefficiency of vitamin D in terms of early reproduction and regulation of specific target genes associated with implantation, e.g. HoxAgenes. A total of 160 gilts were randomly assigned from the first estrus until d 28 of gestation to dietary treatments containing 4 concentrations of 1 of the 2 different vitamin D sources [cholecalciferol or corresponding doses of HY·D]. At day 28 of gestation, the gilts were slaughtered, and the reproductive organs were harvested. The number of ovulations was assessed by counting the corpora lutea and implanted embryos, and from a selected number of animals, autopsies were collected for qPRC from two different implantation sites in the uterus. The statistical model used involved the effect of form and dose of vitamin D, and data were analysed using the MIXED procedure in SAS. The results showed that no influence of dietary form or dose of vitamin D could be seen with regard to early reproductive measurements (i.e. average number of fetuses, average number of eggs in pregnant sows or percentage of implanted fetus). A tendency ($P=0.13$) to an increased upregulation of the HOXA10 was obtained in pigs fed cholecalciferol rather than HY·D. The mRNA abundance of the nuclear hormone receptor for vitamin D_3 (VDR) was not modified by dietary treatments. In conclusion, the early reproduction in gilts was not influenced by dietary vitamin D treatments, however, it should be mentioned that when the same 8 dietary treatment were provided to 160 multiparous sows from first day of gestation until weaning in a concurrent experiment, a significant decreased number of stillborn piglets with larger doses of vitamin D was obtained.

Preference for diets with L-Thr in pigs with different Thr status
Suárez, J.A.[1], Roura, E.[2] and Torrallardona, D.[1], [1]IRTA, Ctra. Reus-El Morell km 3.8, 43120 Constantí, Spain, [2]Lucta SA, Ctra. Granollers-Masnou km 12.4, 08170 Montornés del Vallès, Spain; david.torrallardona@irta.es

Individual amino acids could serve as indicators of protein in the feed. Their sensorial identification by the pigs could affect diet palatability. A double choice test was conducted to determine the preference for diets with different free L-Thr levels in pigs under different Thr status. 108 pigs (18.7±1.4 kg BW) were divided into 3 groups and adapted for 1 wk to diets that were either deficient (D), adequate (A) or excessive (E) in Thr (6.4, 7.8 or 9.2 g Thr/kg, respectively). Next, animals were taken in pairs and offered 3 double-choices (during 2d) between diet D (without free L-Thr) as reference and 3 diets with free L-Thr in excess (E1, E2 and E3), providing 9.2, 10.5 and 11.9 g Thr/kg, respectively. Diets only differed in their free L-Thr content, which was included at the expense of maize starch in diet D. For each double choice comparison and Thr status, a total of 6 observations were obtained. Preference for each L-Thr diet was expressed as its proportional (%) contribution to total feed intake. Values were analyzed with ANOVA using the GLM procedure of SAS by considering the main effects of L-Thr level (E1, E2 or E3), Thr status (D, A or E), and their interaction. Additionally, each mean was compared with the neutral value of 50% (i.e. no difference between D and L-Thr diets) using the Student's t-test. L-Thr level did not affect preference (P=0.53) and no interaction between L-Thr level and Thr status was observed (P=0.83). On the other hand, a significant effect of Thr status (P<0.01) on feed preference was observed. Pigs on the E status had a higher preference for free L-Thr than those on A or D status (66, 43 and 52±4.4%, respectively). Preference in the pigs from the E status was significantly different from 50% (P<0.05), whereas for the D and A status there was no difference. It is concluded that the pigs fed on diets with Thr in excess may develop a preference for free L-Thr.

Preference for diets with free DL-Met in pigs with different Met status
Suárez, J.A.[1], Roura, E.[2] and Torrallardona, D.[1], [1]IRTA, Ctra Reus-El Morell km 3.8, 43120 Constantí, Spain, [2]Lucta SA, Ctra. Masnou-Granollers, km 12.4, 08170 Montornés del Vallès, Spain; david.torrallardona@irta.es

The sensorial perception of individual amino acids by pigs could serve as indicator of protein in the feed and influence the palatability of the diets. A double choice test was conducted to determine preference for diets with free DL-Met in pigs under different Met status. 108 piglets (18.5±2.4 kg BW) were divided into 3 groups and adapted for 1 wk to diets that were either deficient (D), adequate (A) or excessive (E) in Met (2.65, 4.25 or 5.85 g Met/kg, respectively). After the period of adaptation, the animals were used in pairs to perform a series of double-choice tests (during 2d) between diet D (without free DL-Met) as reference and 3 diets with free DL-Met in excess (E1, E2 and E3) to provide 5.85, 7.45 and 9.05 g Met/kg, respectively. Diets differed only in their free DL-Met content, which was included at the expense of maize starch in the basal diet. For each double choice comparison and Met status, a total of 6 observations were obtained. Preference for each tested diet was expressed as its proportional (%) contribution to total feed intake. Each preference mean was compared with the neutral value of 50% (i.e. no difference between the reference and test diets) using a Student's t-test. Additionally, preference values were analysed with ANOVA using the GLM procedure of SAS by considering the main effects of DL-Met level (E1, E2 or E3), Met status (D, A or E), and their interaction. Overall, the addition of DL-Met resulted in a preference value of 55±1.8%, which was significantly higher (P<0.05) than the neutral value of 50%. No significant effects on feed preference of pig's Met status (P=0.833), DL-Met inclusion level (P=0.989) or their interaction (P=0.933) were observed. In conclusion, the addition of DL-Met improves feed palatability in pigs, independently of its inclusion level and the nutritional status of the animals.

Effects of benzoic acid and Na benzoate on the performance and composition of the gastrointestinal microbiota of weaned piglets

Torrallardona, D.[1], Badiola, J.I.[2], Vilà, B.[1] and Broz, J.[3], [1]IRTA, Ctra. Reus-El Morell km 3.8, 43120 Constantí, Spain, [2]CReSA, Campus UAB, 08193 Bellaterra, Spain, [3]DSM Nutritional Products, P.O. Box 2676, 4002 Basel, Switzerland; david.torrallardona@irta.es

The effects of feeding benzoic acid (BA) or Na benzoate (SB) on weanling pig performance and on the composition of their microbiota were evaluated. 140 newly weaned entire male piglets (6.3±1.11 kg; 3 wk of age) were distributed among 28 pens according to a randomised block design with 4 treatments and 7 blocks of BW. Treatments consisted of a control diet (CTR), and the same diet with 0.35% BA (BA35), 0.5% BA (BA50) or 0.4% SB (SB40). Performance was measured at 14, 28 and 42d of trial. At 28d one pig per pen was killed and their ileal and caecal digesta microbiota were analysed with traditional culture methods, real time quantitative PCR (RT-PCR) and restriction fragment length polymorphism (RFLP). The pH of their urine was also measured. Data were analysed according to a randomised block design with the proc GLM of SAS. Relative to CTR, BA35 improved ($P<0.05$) weight gain (ADG) and feed efficiency (FGR) at 14 and 28d and ADG at 42d. BA50 improved ($P<0.05$) ADG and FGR at 28d and ADG at 42d. Finally SB40 improved ($P<0.05$) ADG and FGR at 14d. The urine pH for BA35 and BA50 was significantly ($P<0.01$) lower than for SB40. CTR pH was intermediate and not different from the other groups. No differences among treatments were observed in the contents of *E. coli*, Lactobacilli spp., Enteroccocci spp. and *C. perfringens* measured with traditional methods, or in the contents of *E. coli*, Lactobacilli spp., *Bacteroides* spp. and total bacteria measured with RT-PCR. However, RT-PCR showed lower ($P<0.05$) counts of ileal *E. coli* K88 for CTR than the other groups. The RFLP profiles of the ileal microbiota revealed a lower ($P<0.05$) similarity for CTR than the other groups, and a higher ($P<0.05$) similarity for BA50 than SB40. The results suggest that BA and SB improve piglet performance by different mechanisms.

The performance of pigs fed diets supplemented with glycerol

Juskiene, V., Leikus, R., Juska, R. and Norvilienė, J., Lithuanian University of Health Sciences, Institute of Animal Science, R. Zebenkos 12, Baisogala, LT-82317, Lithuania; violeta@lgi.lt

Feeding trial with fattening pigs was carried out at the LUHS Institute of Animal Science to determine the effect of glycerol on pig growth intensity, feed consumption and carcass quality. Two analogous groups of 12 animals each were made up according to their parentage, age, weight, body conditions score and gender. The pigs of both groups were fed twice daily on home made compound feed. However, the pigs of Experimental group received 15% glycerol supplementation (150 g/kg) as an additional source of energy that was included in the compound feed. The food allowance was adjusted according to the feed intake. The research data were processed using Statistics for windows (version 7; Stat. Soft Inc,Tulsa, OK, USA). The study indicated that there was a tendency towards higher weight gains of pigs after glycerol inclusion into the compound feed. The average daily gain of pigs of Experimental group was by 1.5-5.6% higher than that of Control pigs but differences were insignificant. Glycerol supplementation of pig diets improved feed consumption only at growing period (up to 60 kg weight). In this period the daily feed consumption was 2.8% higher and feed consumption per kg gain was 2.9% lover than that of Control pigs. Also, the study showed that glycerol supplementation of pig diets had no influence on pig health and did not affected the carcass quality of pigs.

Effects of a short exposure period to deoxynivalenol (DON) contaminated wheat on health parameters and post weaning and fattening performances of pigs

Royer, E., Ifip-Institut du porc, 34 bd de la gare, 31500 Toulouse, France; eric.royer@ifip.asso.fr

The contamination of feed with DON decreases feed intake and consequently daily gain of pigs but also affects their immune response. An experiment has been undertaken to evaluate the effects of an initial short exposure time to a medium level of DON on the growth performances during post-weaning and fattening periods. 336 male and female piglets, weaned at an average weight of 8.0 kg (28 days of age), were blocked and affected to 3 dietary treatments. After a standard phase 1 diet, they received from 12.9 kg BW a control (C) phase 2 diet containing 70% clean wheat, or another containing 66% naturally DON contaminated wheat for 14 days (D), or for 22 days (D+). Both diets were iso-energetic and contained the same levels of amino acids. The DON content of the D/D+ diet was 1500 µg per kg. Piglets were housed in 4 fully slatted pens of 14 piglets each, per treatment and sex. During the exposure period, D/D+ treatments tend to decrease feed intake and daily gain, but from weaning to day 40 after weaning, piglets had similar DFI (mean: 831 g/d), ADG (572 g/d) and FCR (1.45 kg/kg). No negative residual effects of post-weaning dietary treatments appeared on growth performance or carcass quality of fattening pigs. The frequency of individual veterinarian treatments and the mortality rate did not differ among treatments during neither the post weaning nor the fattening periods. It is concluded that in good rearing conditions a brief exposure to a medium level of DON does not affect the health status and the growth performance until market weight.

Fate of transgenic DNA and protein from orally administered Bt (MON810) maize in pigs

Walsh, M.C.[1], Buzoianu, S.G.[1,2], Gardiner, G.E.[2], Rea, M.C.[3], Gelencsér, E.[4], Jánosi, A.[4], Ross, R.P.[3] and Lawlor, P.G.[1], [1]Teagasc, Pig Development Department, Animal and Grassland, Research and Innovation Centre, Moorepark, Fermoy, Co. Cork, Ireland, [2]Waterford Institute of Technology, Department of Chemical and Life Science, Waterford, Ireland, [3]Teagasc, Food Research Centre, Moorepark, Fermoy, Co. Cork, Ireland, [4]Central Food Research Institute, Department of Biology, Budapest, Hungary; maria.walsh@teagasc.ie

The objective of this study was to determine the fate of ingested recombinant DNA and protein in pigs, thereby facilitating a clearer assessment of the safety of Bt (MON810) maize. Twenty entire male pigs (7.5±2.5 kg) were used in a 31 day experiment to determine the fate of transgenic DNA and protein *in vivo*. At weaning, pigs were fed a non-genetically modified (GM) starter diet during a 6 day acclimatization period after which pigs were divided into two groups (n=10) and fed a diet containing 38.9% GM (Bt MON810) or non-GM isogenic parent line maize for 31 days. Pigs were fed 300 g of feed 3 h prior to slaughter on d 31. At slaughter, samples of the heart, liver, kidney, spleen and muscle were collected for detection of recombinant and endogenous DNA and protein. Blood samples together with digesta from the stomach, ileum, caecum and colon were collected for detection of recombinant and endogenous DNA and protein. The recombinant (cry1Ab) and endogenous (shrunken 2) maize genes and proteins were not detected in the organs or blood of pigs fed either GM or non-GM maize. The cry1Ab gene was detected in stomach, ileal and caecal digesta (100, 20 and 10%, respectively) of GM maize-fed pigs but not in digesta from the colon. The cry1Ab protein was found in stomach, ileal, cecal and colon digesta (30, 80, 30 and 80%, respectively) of GM maize-fed pigs. In conclusion, maize-derived DNA, either of intrinsic or recombinant origin, was largely degraded in the gastrointestinal tract. There was no evidence of cry1Ab gene or protein translocation to the organs or blood of weanling pigs following short-term exposure to Bt (MON810) maize.

Addition of seaweed polysaccharides from brown seaweed (*Laminaria digitata*) to porcine diets: Influence on the oxidative stability of organ tissues and fresh pork quality
Moroney, N.C.[1], O'grady, M.N.[1], O'doherty, J.V.[2] and Kerry, J.P.[1], [1]Food Packaging Group, School of Food and Nutritional Sciences, College of Science, Engineering and Food Science, University College Cork, Ireland, [2]School of Agriculture, Food Science, and Veterinary Medicine, College of Life Sciences, Lyons Research Farm, University College Dublin, Newcastle, Co. Dublin, Ireland; natasha.moroney@gmail.com

Seaweeds are rich in polysaccharides such as laminarin, fucoidan and alginic acid. Health benefits of laminarin and fucoidan (L/F) include antitumor, antiviral, antibacterial and antioxidant activities. *Laminaria digitata* (a brown seaweed) extract containing L/F was manufactured in a wet (L/F-WS) and spray-dried form (L/F-SD). The study aimed to assess the effect of supplementation of pig diets with the laminarin (500 ppm) and fucoidan (420 ppm) containing seaweed extract on fresh pork (*M. longissimus dorsi*) (LD) quality. Pigs (n=24) were fed one of three diets for 3 weeks pre-slaughter: control group (C) (n=8); + L/F-WS (n=8) and + L/F-SD (n=8). Susceptibility of porcine liver, kidney, heart and lung tissue homogenates to iron-induced (1 mM $FeSO_4$) lipid oxidation was investigated. Surface colour (Minolta colorimetry) and microbiology (psychrotrophic and mesophilic counts, log CFU/g pork) were unaffected by dietary L/F in LD steaks stored in MAP (75% O_2 : 25% CO_2) for 14 days at 4 °C. In general, lipid oxidation (TBARS, mg MDA/kg muscle/tissue) in LD steaks and liver tissue homogenates followed the order: C > LF-SD > L/F-WS. A significant reduction in lipid oxidation ($P<0.05$) was observed in LD steaks from 75% of pigs (n=6) fed L/F-WS compared to controls. Plasma (total antioxidant status), kidney, heart and lung tissue homogenates (TBARS) were unaffected by dietary L/F. Results demonstrate potential for incorporation of marine derived bioactive antioxidants into porcine muscle via the animal diet. Future research will examine the effects of L/F levels, form and duration of feeding on pork quality.

Effect of protein and lysine level on fat deposition in finishing pigs
Tous, N.[1], Lizardo, R.[1], Vilà, B.[1], Gispert, M.[2], Font I Furnols, M.[2] and Esteve-Garcia, E.[1], [1]IRTA, Monogastric Nutrition, Ctra. Reus-el Morell, km.3.8, 43120, Constantí (Tarragona), Spain, [2]IRTA, Food Technology, Finca Camps i Armet, 17121, Monells (Girona), Spain; marina.gispert@irta.cat

In the last decades, animals with a lower fat deposition were selected for pork production. Additionally, a reduction of intramuscular fat (IMF), which had some negative effects in sensory meat quality parameters (such as flavor or tenderness), was also observed. It was proposed that a reduction of the protein or lysine (Lys) level in finishing pig diets could increase the percentage of IMF. Accordingly, the aim of this study was to observe the effect of diminishing the level of protein (keeping the lysine level constant), lysine (keeping the protein level constant) or both in pig fat deposition, mainly IMF. Sixty-four barrows (Landrace x Duroc) from approximately 60 to 115 kg LW were fed one of the four experimental diets (n=16/diet): control (13% CP and 2.60 g Lys/Mcal NE from 60-90 kg LW or 11% CP and 2.13 g Lys/Mcal NE from 90-115 kg LW), low lysine (13% CP and 2.20 g Lys/Mcal NE from 60-90 kg LW or 11% CP and 1.72 g Lys /Mcal NE from 90-115 kg LW), low protein (11.5% CP and 2.58 g Lys/Mcal NE from 60-90 kg LW or 9.75% CP and 2.13 g Lys/Mcal NE from 90-115 kg LW) and low lysine and protein (11.5% CP and 2.20 g Lys/Mcal NE from 60-90 kg LW or 9.75% CP and 1.72 g Lys/Mcal NE from 90-115 kg LW). Performance and carcass quality parameters were determined. A tendency ($P<0.1$) to increase the feed to gain ratio, muscle depth and a significant increase of backfat was observed in animals fed the low lysine diets. On the other hand, the reduction of the protein level in pig diets produced a tendency to increase the backfat without modifying the loin depth. The percentage of IMF was only increased when protein or lysine level of the diet was reduced, otherwise when both parameters were reduced at the same time, the lowest value was observed. These results suggest that the reduction of protein in pig diets could improve the meat quality without modifying pig performance.

Influence of different dietary fat sources on pork fatty acid composition
Alonso, V., Najes, L.M., Provincial, L., Guillén, E., Gil, M., Roncalés, P. and Beltrán, J.A., University of Zaragoza, Animal Production and Food Science, Miguel Servet 177, 50013, Zaragoza, Spain; veroalon@unizar.es

This study compared the influence of dietary fat sources on intramuscular and subcutaneous fatty acid composition in pork. The study was undertaken with carcasses of 43 entire male pigs from the crossbreeding P × (LD × LW) (carcass weight 83.8±6.3 kg). The animals were fed diets containing corn, barley and wheat grain and soybean meal 44% (CP) (control diet) supplemented with different fats: animal fat (1% (AF1); 3% (AF3)), soya oil (1% (SO1)) and palm oil (1% (PO1)). The Longissimus thoracis et lumborum (LTL) muscle and a sample of subcutaneous fat (SCF) were removed from each carcass 48 h after slaughter. The fat was extracted in chloroform-methanol from LTL muscle and SCF and quantified. The fat samples were used to determine composition in fatty acids from intramuscular fat (IMF) and SCF. All data were statistically analyzed by the GLM procedure of SPSS, version 15.0. Duncan's post hoc test was used to assess differences between mean values when P≤0.05. IMF content was significantly (P≤0.05) higher in AF3 than SO1 diet. There were no significant differences in the concentration of saturated fatty acids (SFA) among diets in IMF and SCF. The proportion of monounsaturated fatty acids (MUFA) was (P≤0.05) lower in SO1 in both fats. However, this diet obtained the highest values in the concentration of n-6, n-3 and polyunsaturated fatty acids (PUFA) and PUFA/SFA index. The worst n-6/n-3 ratio value was found in AP1 group in both types of fat. The AF3 and control diets produced IMF lipids with higher content of SFA and MUFA than SO1. Diets did not appear to affect the amount of n-6 and n-3 and the content of PUFA in IMF. By contrast, the amount of n-6, n-3 and PUFA in SCF had (P≤0.001) the highest values in SO1. No significant differences were observed among diets for the content of SFA, whereas the sum of MUFA content in PO1 was (P≤0.05) higher compared to AF1 and SO1 in SCF.

Influence of different dietary fat sources on pork quality and sensory analysis
Alonso, V., Provincial, L., Gil, M., Guillén, E., Najes, L.M., Roncalés, P. and Beltrán, J.A., University of Zaragoza, Animal Production and Food Science, Miguel Servet 177, 50013, Zaragoza, Spain; veroalon@unizar.es

The aim of this study was to compare the influence of dietary fat sources on meat quality and sensory analysis in pork. The study was undertaken with carcasses of 43 entire male pigs from the mating of Pietrain sires to Landrace × Large White dams, with an average carcass weight of 83.8±6.3 kg. The animals were fed diets containing corn, barley and wheat grain and soybean meal 44% (CP) (control diet) supplemented with different fats: animal fat (1% (AF1); 3% (AF3)), soya oil (1% (SO1)) and palm oil (1% (PO1)). The Longissimus thoracis et lumborum (LTL) muscle was removed from each carcass 48 h after slaughter. Ultimate pH (pHu), instrumental colour (CIEL*a*b*), drip loss and lipid oxidation (TBARS method) were measured in LTL muscle at 72 h post-mortem. The samples were cooked in a grill until the internal temperature reached 72 °C and used for sensory panel evaluations. The samples were served randomly to a nine-member trained sensory panel, which had been trained in sensory assessment of meat. Panellists used numerical scales to quantify sensory attributes. All data were statistically analyzed by the GLM procedure of SPSS, version 15.0. Duncan's post hoc test was used to assess differences between mean values when P≤0.05. No significant differences were observed among diets for pHu and instrumental colour. The drip loss was (P≤0.05) lower in pigs fed with control and PO1 diets than pigs fed with AF1. The chops of pigs fed with control and animal fat (1%, 3%) diets had (P≤0.01) higher TBARS values than SO1 and PO1 diets. The sensory attributes of odour (pork, fat, urine and acid), flavours (pork, fat and acid), tenderness, juiciness, fibrousness, as well as overall acceptance, were not affected by dietary fat. In conclusion, there were no large differences in meat quality and sensory analysis for pigs fed with different dietary fat sources.

Comparative aspects of lactation

Knight, C.H., University of Copenhagen, LIFE, Grønnegårdsvej 7, DK 1870 Frb C, Denmark; chkn@life.ku.dk

It is 34 years since the Zoological Society of London hosted a Symposium entitled Comparative Aspects of Lactation. Then, experts from around the world gathered to discuss basic and applied aspects of lactation biology in a range of species. The emphasis was on physiological mechanisms, and no attempt was made to 'catalogue' lactation characteristics in Aardvarks, Zebras or any other species. The freedom offered by this approach even allowed for discussion of some milk-like secretions in non-mammalian species! The preface to the Proceedings emphasised the vital, complex and yet often neglected importance of lactation in all our lives. Consider, on the one hand, the vast economic and societal impact of global dairy production and, on the other hand, the fact that the mammary gland is the only major human organ that does not have a medical specialism associated with its normal function. In 1977 the virtues of comparative approaches were seen to include extrapolation of observations made in animals to man, for at that time the study of human lactation was in its infancy. In the intervening period much has been learnt about human and animal lactation, but these processes have taken divergent paths. BOLFA (Biology of Lactation in Farm Animals) and ISRHML (International Society for Research in Human Milk and Lactation) have met regularly, but never together. This is almost certainly not optimum. Having spent vast amounts of money combating mastitis in dairy cattle, would it not be sensible to apply some of the knowledge obtained to human mastitis? (which, incidentally, has a similar prevalence rate). This session within the 2011 EAAP meeting is one part of an attempt to bring 'animal' and 'human' lactation scientists together, to focus on the health of mothers and neonates. Another part of that same ambition is the recent creation of CoLact, the NordForsk Researcher Network in Comparative Lactation Biology (www.colact.net). Please visit the website!

Metabolic health of the bovine mother

Collier, R.J., University of Arizona, Animal Sciences, P.O. Box 210038, Tucson AZ, USA; rcollier@ag.arizona.edu

Management of dairy cattle during the dry and peripartum periods is now identified as key to the eventual health of the dam during lactation. Approximately 80% of adverse health events occur during the first 21 days *post partum* in lactating dairy cows. Historical and recent data indicate that opportunities to improve health and well-being of cattle exist in improved management strategies for dairy cows during the dry and peripartum periods. These opportunities include rapidity of dry-off, length of the dry period, nutritional strategies during the dry period, immune system stimulation during the dry period, milking interval during early lactation and environmental management during the dry and peripartum periods. Maximizing the management of the dairy cow during these periods improves health and well-being of dairy animals as well as lactation performance. Additionally, health and well-being of the neonate is improved as well. The following parameters are known to be affected by management of the dam during the dry period and calving; birthweight, survival during the immediate post-partum period, growth rate and eventual lifetime milk yield. Thus, there is ample evidence to demonstrate that maximizing the well being of the dairy cow during the dry and peripartum periods results in large financial returns to the producer and healthier dams and offspring.

Metabolic health of the breastfeeding mother: calcium and energy metabolism in human lactation
Goldberg, G.R., MRC Human Nutrition Research, Elsie Widdowson Laboratory, Cambridge CB22 4SE, United Kingdom; gail.goldberg@mrc-hnr.cam.ac.uk

There is a long history of comparative lactation research in Cambridge. It is a topic that was close to the heart of Dr Elsie Widdowson, after whom the building which MRC Human Nutrition Research occupies, is named. Lactation can impose a considerable nutritional burden on the breastfeeding mother. For example the average energy costs of lactation on a daily basis are about 2 MJ; exported as macronutrients and required for the synthesis of milk. With respect to micronutrients, for example, on average approximately 200 mg of calcium per day is exported in breast milk. Unlike in many other mammalian species human lactation appears to be a particularly robust process, and one that is not dependent on daily replenishment of body stores, a constant dietary supply, or a sudden and rapid transition to a higher plane of intake. A combination of physiological adaptations and behavioural changes operate to ensure that women can lactate successfully even under very marginal nutritional circumstances, and are possible reasons why many supplementation studies have not generated the expected results. This paper will consider maternal energy and calcium 'budgets' and consequences for the short- and long-term health of the mother and the short-term health and long-term development of the baby.

Early life nutrition and gut development
Thymann, T., University of Copenhagen, LIFE, Rolighedsvej 30, DK-1958 Frb C, Denmark; ttn@life.ku.dk

Birth is the most radical change in life. The newborn must meet the acute challenges of active respiration and thermoregulation, but also quickly adapt to the dietary and microbiological changes that are associated with birth. From receiving total parenteral nutrition via the umbilical cord in utero, to receiving total enteral nutrition via the digestive system ex utero, is a radical and sudden transition that requires acute adaptation. The intestinal adaptive mechanisms can however become compromised if the neonate is born premature or receives a suboptimal enteral diet such as milk replacer. Introduction of milk replacer to a premature neonate associates with increased risk of necrotizing enterocolitis. However, the negative side effects of milk replacers are only precipitated in the presence of gut bacteria, as milk replacer nutrition under germ free conditions have shown no detrimental effects. Artificial milk replacers for premature neonates are only to some extent based on bovine dairy products. In a long series of studies with premature neonatal pigs we have shown that intact bovine colostrum provides protective effects similar to that of intact porcine colostrum and human milk. For human premature neonates bovine colostrum may become an alternative to artificial milk replacer. However there is still room for improvement of artificial milk replacers, and we have shown that manipulation of the carbohydrate and fat fraction can reduce the incidence of necrotizing enterocolitis. Future diet development for premature neonates may be more focused on the interplay between diet and gut microbiota and host immune function. We have shown that early life colonization is different depending on the state of development (mature versus premature neonates). It remains a challenge to understand how we can manipulate early life colonization through the diet, but it may be most efficient way to ensure a stable diverse pool of commensal gut bacteria.

Culture-independent analysis of the bacterial communities in healthy and mastitic milk

Mcguire, M.A.[1] and Mcguire, M.K.[2], [1]University of Idaho, Department of Animal and Veterinary Sciences, Moscow, ID 83844-2330, USA, [2]Washington State University, School of Biological Sciences, Pullman, WA 99164, USA; mmcguire@uidaho.edu

Several studies have demonstrated that differences in the composition of bacterial communities inhabiting various niches in and on the human body are related to a variety of health outcomes (e.g. obesity). We investigated the microbiome of human milk by collecting 3 samples over 4 wk from 16 healthy women. Bacterial communities were characterized via universal primer amplification of the bacterial 16S rRNA gene and pyrosequencing. The most commonly observed genera in milk included Streptococcus, Staphylococcus, Serratia and Corynebacteria. The communities present were reasonably complex with 12 genera representing an average relative abundance of $\geq 1\%$. Further, bacterial communities were generally personalized and relatively stable in composition. Richness of phylotypes was found to be comparable to the gastrointestinal tract but different in composition. The complexity and stability of these bacterial communities in human milk may influence the bacterial communities present in the gastrointestinal tract of the nursing infant and mammary health of the lactating woman. It is likely that use of culture-independent methods may improve identification of mastitic pathogens. Dairy producers and veterinarians are often frustrated with their ability to identify potential pathogenic agents due to the time necessary for results of bacterial growth and high false negative rates (25-40%) even in clinically ill cows. Others have shown that use of the polymerase chain reaction (PCR) technique for detecting mastitic pathogens is a fast and reliable option to culture-dependent methods. These culture-independent methods may help improve the detection and identification of pathogens when bacterial culture results are negative in clinical mastitis cases, and may also identify bacterial communities better than methods relying on bacterial growth.

The influence of enforced or natural dry period length on periparturient mammary development in dairy cows

Agenas, S.[1] and Knight, C.H.[2], [1]Swedish University of Agricultural Sciences, Nutrition and Management, Kungsangens Research Centre, SE-753 Uppsala, Sweden, [2]University of Copenhagen, LIFE, Gronnegardsvej 7, DK-1870 Frb C, Denmark; sigrid.agenas@slu.se

Mammary biopsies were obtained during late lactation, two weeks pre-partum and two weeks post-partum from a total of 44 cows of the Swedish Red breed. The enforced dry period group (n=31) had a mean interval of 69 (range 51 to 129) days from dry-off to calving whilst the group that dried-off naturally had a mean of 33 (range 0 to 64) days. Biopsy samples were analysed for expression of 31 genes involved in mammary development. Many of these genes demonstrated developmental regulation, but only three differed significantly according to dry period length. Two weeks pre-partum there was evidence of decreased cell proliferation and increased apoptosis in the natural group (expression of proliferation-associated cyclin D1 was lower, 1.33 sd 0.54 vs 2.88 sd 0.22, P=0.01 whilst the ratio of expression of the pro-apoptotic factor BAX to the anti-apoptotic factor BCL2 was higher, 1.46 sd 0.18 vs 0.73 sd 0.07, P<0.001). Two weeks post-partum apoptosis remained higher in the natural group (BAX/BCL2 1.69 sd 0.19 vs 0.98 sd 0.12, P<0.01), but proliferation no longer differed. Curiously, expression of alpha lactalbumin was significantly increased in the natural group, even though their milk yield was lower. There were no differences between the groups in late lactation. These differences in mammary cell proliferation and apoptosis may start to explain why most cows are incapable of lactating continuously from one lactation to the next.

Milk partitioning and accumulation in the camel udder according to time elapsed after milking

Caja, G.[1], Salama, O.A.[2], Fathy, A.[2], El-Sayed, H.[2] and Salama, A.A.K.[1,2], [1]Ruminant Research Group (G2R), Animal and Food Sciences, Universitat Autònoma de Barcelona, 08193 Bellaterra, Barcelona, Spain, [2]Animal Production Research Institut (APRI), Camel and Sheep & Goat Departments, 4 Nadi El-Said, 12311 Dokki, Giza, Egypt; ahmed.salama@uab.cat

Ten Egyptian camels (Camelus dromedarius L.) at mid lactation (281 ± 41 d in milk) from the Matrouh Station of APRI were used. Camels have 10.3 ± 0.9 yr (parity, 5.8 ± 0.6) and 484 ± 29 kg LW, were adapted to be hand-milked (8:00 and 20:00 h) without calf suckling and yielded 4.4 ± 0.4 L/d of milk. They were in loose stalls and feeding was 3.5 kg/d concentrate and 6.5 kg/d dry forages. Milk accumulation in the udder was studied by randomly applying 5 milking intervals (4, 8, 12, 16, 20 and 24 h) at each udder quarter during consecutive days. Cisternal and alveolar milk were obtained by serial milking without or with previous i.v. injection of 6 IU of oxytocin (OT), respectively. Udder cisterns were explored by ultrasonography. Milk recoil was also measured by milking 90 min after OT injection. Milk partitioning between front:rear quarters was 44:56% and cisternal milk did not vary by milking interval, on average being $7.5\pm1.3\%$. Small udder cisterns were visualized by ultrasonography which dramatically engorged after OT. Quadratic increases in alveolar milk (y, ml = -2.2 x^2 + 139.3 x + 37.3; R^2=0.98) and in total milk (y, ml = -2.4 x^2 + 151.0 x + 29.2; R^2=0.98) with milking interval time (x, h) were observed and both milk fractions did not reach a plateau during the 24-h milk accumulation period. Milk secretion hourly rate decreased linearly from 4 to 24 h milking interval (150 to 93 ml/h), showing marked milk loses after 12 h of milk stasis. A dramatic milk drop (-61.8%) was also observed when milking was delayed after milk letdown, showing a strong elastic recoil as a consequence of the small-cisterned udder of the camel. Selection to large-cisterned udders should increase camel milkability and it is recommended in practice (Acknowledgement: AECID Spain-Egypt Project PCI A/025331/09).

Calcium and Phosphorus levels in milk among hybrid (LY) sows during different stages of lactation.

Thingnes, S.L.[1,2], Gaustad, A.H.[1], Rootwelt, V.[2] and Framstad, T.[2], [1]Norsvin, P.O. Box 504, 2304 Hamar, Norway, [2]Norwegian School of Veterinary Science, Department of Production Animal Clinical Sciences, P.O. Box 8146 dep, 0033 Oslo, Norway; signe-lovise.thingnes@norsvin.no

Two important minerals provided by sow milk are Calcium (Ca) and Phosphorus (P), which are essential in bone formation. Deficiency can lead to misshapen bones, lameness and stiffness. A literature review shows that Ca and P levels in sows' milk is relatively constant regardless of the maternal dietary levels of these minerals, but low levels in the diets can lead to demineralization of the bone tissue in order to meet the requirements of milk production. This can put a severe strain on the sows' skeleton, and leg abnormalities, lameness and fractures can become prominent. As a preliminary data collection for a study of the effect of mineral content in feed on sow productivity and bone demineralization, 40 milk samples from three different stages of lactation were collected and analysed for Ca and inorganic P levels. The analysis showed that Ca (mg/dl $\pm$ SEM) levels in colostrum were 76.99 ± 3.56, at 19-21 days *post partum* 168.24 ± 5.23 and 230.90 ± 7.56 at weaning. P (mg/dl $\pm$ SEM) levels were 49.21 ± 1.98 at parturition, 83.52 ± 2.48 at 19-21 days *post partum* and 93.95 ± 2.25 at weaning. Average lactation length was 33.2 days. A factorial two-way ANOVA analysis showed no significant effect of parity number on Ca and P levels at the different stages of lactation. At weaning two milk samples from each sow were collected to determine if there is a difference in Ca and P levels in samples collected from teats at the front part of the udder compared to teats at the back part. A paired-samples t test showed no significant difference with regards to mean Ca levels 0.47 ± 1.07 or P levels -0.25 ± 0.59. The data shows that Ca and P levels in milk increase throughout lactation, and a more in-depth study are being planned in order to see if the dietary mineral content may influence sow performance and bone demineralization during lactation.

Premature removal and mortality of commercial sows
Engblom, L.[1], Stalder, K.[2] and Lundeheim, N.[1], [1]Swedish University of Agricultural Sciences, Department of Animal Breeding and Genetics, P.O. Box 7023, 75007 Uppsala, Sweden, [2]Iowa State University, Department of Animal Science, Ames, IA 50011-3150, USA; Linda.Engblom@slu.se

A high proportion of early removal of young breeding sows is an undesired part of today's swine production. Similar removal patterns are found in studies from US farms keeping sows housed in stalls and crates, and in studies from Swedish farms where sows are kept in pens and in group housing systems during gestation. Annual combined culling and mortality rates often exceed 50%. The major proportion of sow removal, about 70%, is premature and unplanned culling due to reasons like reproductive disorders and lameness. Almost a third of the commercial sows are removed before third parity, i.e. before they reach peak performance and have produced enough to 'pay for themselves'. Early removal also includes a substantial proportion, about 15%, of the removed sows that are not sent to slaughter, but are mortalities or are euthanized on the farm. In our modern production system with its scheduled farrowing batches there is little room for biological variation or deviations from the time schedule. Low profit margins have resulted in modern swine production systems targeting high annual production, with little or no attention to long term production. Selection for litter size and lean tissue deposition has produced highly productive but less robust sows which cannot fully cope with the environmental challenges in present production systems and are therefore too often premature removed. Just like with an unfavourable genetic correlation, both high annual production and longevity have to be taken into consideration. Decreasing sow removal and increasing sow longevity is not only improving animal well being but is also an economically good decision, since it is more profitable to have production systems with low removal rates. Reducing sow removal has to be a joined effort including improved management, breeding for more robust sows and developing housing systems that are capable of meeting the sows' needs.

On-farm mortality of cows in Swedish dairy herds
Alvåsen, K.[1], Jansson Mörk, M.[1,2], Hallén Sandgren, C.[2], Thomsen, P.T.[3] and Emanuelson, U.[1], [1]Swedish University of Agricultural Sciences, Clinical Sciences, P.O. Box 7054, SE-75007 Uppsala, Sweden, [2]Swedish Dairy Association, P.O. Box 210, SE-10124 Stockholm, Sweden, [3]University of Aarhus, Animal Health and Bioscience, P.O. Box 50, DK-8830 Tjele, Denmark; karin.alvasen@slu.se

An increase of on-farm mortality (euthanasia and death) in dairy herds has been reported in several countries in the last decade. This does not only imply possible problems with animal welfare, but also causes economical losses for the farmer. The mortality rate in Swedish dairy herds has not been thoroughly studied previously and there is a need to investigate this emerging problem also under Swedish conditions. The objective of this study was to evaluate time-trends in on-farm cow mortality and to identify potential risk factors at herd level. Data from all Swedish dairy herds enrolled in the milk recording scheme between 2002 and 2010 were retrieved. A total of 6,898 herds with >20 cows and yearly mortality rate <40% were included in the analysis. The outcome variable was the number of euthanized and dead cows per year and season. A negative binomial regression model, adjusted for clustering within herd, was applied to the data. The fixed predictors in the model were herd size, milk yield, breed, calving interval, region, year and season. The study demonstrated that mortality rates have gradually increased from 5.1% to 6.7% during the study period. Swedish mortality rates are thus in level with or even higher than those of other comparable countries. Higher mortality was associated with larger herd size, longer calving intervals and herds with Swedish Holstein as the predominant breed. Lower mortality was observed in herds with a higher average milk yield per cow and during the autumn (September-December). An interaction between herd size and season was found. There were also regional differences in mortality. This first assessment of on-farm mortality in Swedish dairy herds confirmed that the rate has increased in the last years and identified some herd-level risk factors.

Association of herd demographics and biosecurity with cattle longevity and udder health in dairy herds

Brouwer, H., Bartels, C.J.M. and Van Schaik, G., Animal Health Service, Diagnostics, Research and Epidemiology, Arnsbergstraat 7, 7400 AA, Netherlands; h.brouwer@gddeventer.com

With the forthcoming relaxation of milk quota and the prospect of ending intervention prices for milk produced in the European Union, herd size of Dutch dairy farms increases rapidly. Such growth should lead to management adjustments which may have cattle health consequences. The objective of this study was to quantify the association of herd size, herd growth and biosecurity with cattle longevity and udder health. Dutch census data concerning longevity and udder health were available for 2007 and 2008. Biosecurity levels were defined by means of the no. of cattle introduced in one year, i.e. few (<8) or many (≥8). Five dairy herd types were defined based on herd size, growth and biosecurity: Large herds with few cattle introductions (L), fast growing herds with few cattle introductions (FG), herds with many cattle introductions (MCI), fast growing herds with many cattle introductions (FGMCI) and average herds. Each of the longevity and udder health parameters was taken as dependent variable in a linear, poisson or logistic regression. The different herd types were forced into the models. In addition, season, milk yield and region were included as fixed factors. Compared with average herds, L and FG herds had similar longevity. MCI herds had increased cow and young stock mortality (40% and 60% resp.). The culling rate was 185% higher in MCI herds. For FGMCI herds, cow and young stock mortality increased with 45% and 60%, resp. Culling rate increased with 231% compared with average herds. Compared with average herds, FG herds had similar udder health, but in L herds, MCI and FGMCI herds, the incidence of subclinical mastitis increased with 5-10% and BMSCC with 6-8%. In bulk milk of L herds and FGMCI herds antibiotics were more often detected (67% and 85% resp.) than in average herds. The results of this study indicated that the association of longevity and udder health was stronger with biosecurity than with herd size or herd growth.

Claw and foot healt: Early diagnostics and prevention of foot lesions in dairy cattle

Tamminen, P., Häggman, J., Pastell, M., Tiusanen, J. and Juga, J., University of Helsinki, Department of Agricultural Sciences, P.O. Box 28, 00014 University of Helsinki, Finland; jarmo.juga@helsinki.fi

Foot and claw disorders are painful and costly diseases that cause lameness. It is estimated to be one of the major animal welfare issues in dairy production. Lameness decreases milk production, increases involuntary culling and affects reproductive performance. Due to these costs it is the third important health trait after mastitis and infertility. We present here an ongoing research, which aims at early prediction of lameness with technological methods and preventing new cases by selection scheme, which includes genetic evaluation for claw health. Part of the variation in lameness and claw disorders is of genetic origin, which enables the selection for better claw and foot health. Many feet and leg conformation traits have also been found to be genetically correlated to lameness and claw disorders. Lameness needs to be included in the breeding goal of dairy cattle due to its high cost and since it is a severe welfare problem. We have analysed the effect of different environmental conditions in barns, which affect claw health, genetic variation of different claw problems and genetic correlations of claw disorders with feet and leg conformation traits. A model for predicting lameness will be developed using data from multiple sources: silage and water consumption, feeding duration, lying time and overall activity and data collected from milking robot. However, the aim is to develop a model in a way that it would perform well when using only few or even one data source. We have developed a system to accurately measure cows lying time and feeding behaviour in free stall. So far the results have shown significant changes in feeding behaviour for lame cows. System measuring the lying time has shown to be capable of transferring activity data wirelessly not only in stall conditions, but also from pasture over long time periods and distances.

Effect of inbreeding and estimation of genetic parameters for heifer mortality in Austrian Brown Swiss cattle
Fuerst-Waltl, B.[1] and Fuerst, C.[2], [1]University of Natural Resources and Life Sciences Vienna, Dep. of Sust. Agric. Syst., Div. Livest. Sci., Gregor Mendel-Str.33, A-1180 Vienna, Austria, [2]ZuchtData EDV-Dienstleistungen GmbH, Dresdnerstr. 89, A-1200 Vienna, Austria; waltl@boku.ac.at

Mortality of heifers before first calving results in higher replacement costs but also in reduced possibility for selection and is thus of importance in cattle breeding. The aim of this study was to explore the effect of inbreeding on postnatal mortality in calves and replacement heifers and to estimate genetic parameters for these traits in Austrian Brown Swiss cattle. The following periods were defined for analyses: P1 = 2-30 d, P2 = 31-180 d, P3 = 181 d-age at first calving or a maximum age of 1200 d if no calving was reported, P4 = 2 d-age at first calving or a maximum age of 1200 d if no calving was reported. Records of animals which were slaughtered or exported within a defined period were set to missing for this and consecutive periods while their records were kept for preceding periods. After data editing records of 69,571 Brown Swiss heifers were utilized. Their pedigree comprised of 203,894 animals. Mortality rates were 3.16%, 2.32%, 3.24%, and 9.32% for the defined periods P1-P4. For the estimation of the effect of inbreeding and the genetic parameters, a linear animal model with fixed year*month, number of lactation (lactations >5 set to 5), calving ease, the inbreeding coefficients as linear and quadratic covariates and the random herd*year and random genetic effect of animal was applied. Significant linear inbreeding depression for mortality in calves and replacement heifers was observed for all age groups; for P1, also the quadratic term was significant. When increasing the inbreeding coefficient from 0% to 10% the mortality increased by 1.1%, 1.2%, 1.6% and 3.6% for P1-P4, respectively. The estimated heritabilities were low and ranged from 0.001 (P3) to 0.014 (P4). Total calf and heifer mortality is higher than stillbirth. Consequently, its economic impact on cattle breeding should not be neglected.

Association between herd characteristics and on-farm cow mortality in Swedish dairy herds
Jansson Mörk, M.[1,2], Alvåsen, K.[2] and Hallén Sandgren, C.[1], [1]Swedish Dairy Association, Box 210, 101 24 Stockholm, Sweden, [2]Swedish University of Agricultural Sciences, Clinical Sciences, Box 7054, 750 07 Uppsala, Sweden; karin.alvasen@slu.se

As the characteristics in Swedish dairy herds are changing there is a concern that the animal welfare will be negatively affected. It has previously been shown that register-based welfare indicators could be used to identify herds with animal-based welfare remarks. The objective of the study was to evaluate associations between register-based welfare indicators and herd characteristics. Here we present the results for the analysis of on-farm mortality (death or euthanasia). Data from 4,261 dairy herds enrolled in the Swedish milk recording scheme during September 2009 to August 2010 were used in the analysis. Herds with a herd size <20 cows or a mortality rate >40 dead or euthanised cows /100 cow-years were excluded. The association between mortality and herd characteristics was evaluated using negative binomial regression, and adjusted for within-herd clustering. The fixed effects were herd size, breed, milk yield, conventional vs. organic farming, housing system, region and season. Higher mortality was seen in herds with an average milk yield below median, in conventionally managed farms and in herds with mainly Swedish Holstein cows. There was an interaction between herd size and season. In general, herds with <200 cows had lower mortality during September-December compared with the rest of the year. Also, the mortality increased with increasing herd size, although not statistically significantly during summer. There were regional differences in mortality with highest mortality in Norrland and lowest in Västra Götaland, but no effects of housing system. In Sweden there is an increase in the number of organically managed farms, of loose housing systems, use of Holstein cows and in herd size. The association between higher herd-level mortality due to death and euthanasia and herd level characteristics found here needs to be studied further.

Can eating quality genetics be incorporated into the Meat Standards Australia lamb grading system?

Pannier, L.[1], Pethick, D.[1], Ball, A.[2], Jacob, R.[3], Mortimer, S.[4] and Pearce, K.[1], [1]Murdoch University, School of Veterinary & Biomedical Sciences, Perth, Western Australia, 6150, Australia, [2]Meat & Livestock Australia, University of New England, Armidale, New South Wales, 2351, Australia, [3]Department of Agriculture & Food, South Perth, Western Australia, 6151, Australia, [4]Industry & Investment, Agricultural Research Centre, Trangie, New South Wales, 2823, Australia; d.pethick@murdoch.edu.au

The Meat Standards Australia (MSA) grading scheme for underpinning the eating quality of lamb is currently a pathways system with guidelines for best practice feeding, handling, slaughter, product aging and retail presentation of lamb cuts. This paper describes an experiment to determine the role of genetics to underpin the continuous improvement in eating quality of lamb cuts. Eating quality data was generated from 745 lambs produced from 97 sires at 2 sites within the Information Nucleus (IN) program of the CRC for Sheep Industry Innovation. Grilled steaks were prepared from 2 cuts (m. longissimus lumborum – LD; m. semimembranosis – SM) with consumers scoring steaks for tenderness (TE), juiciness, liking of flavour and overall liking (OL) on a 0-100 scale. Consumers also assigned a quality rating to each sample: unsatisfactory, satisfactory everyday quality, better than everyday quality or premium quality. The data for TE and OL were analysed using linear mixed models in ASReml with fixed effects of IN site, kill group, sex, sire breed type and dam breed type. Sire and consumer session were random terms. There were significant effects ($P<0.05$) of cut (LD >> SM), IN site, kill group and sire breed. Sire accounted for 5.3% and 3.3% of the total variance in TE and OL of both cuts with a sire range of 8-12 consumer points, sufficient to change the final consumer rating of the steaks. This preliminary study shows that genotype effects need to be considered in the development of a new MSA lamb grading model.

Prediction of beef eating quality in France using the Meat Standards Australia (MSA) system

Legrand, I.[1], Hocquette, J.F.[2], Polkinghorne, R.J.[3] and Pethick, D.W.[4], [1]Institut Elevage, Service Qualite des viandes, 87060 Limoges Cedex 2, France, [2]INRA, UR1213, 63122 Theix, France, [3]Marrinya Agricultural Enterprises, Vic., 3875, Australia, [4]Murdoch University, WA, 6150, Australia; hocquet@clermont.inra.fr

An experiment was set up to test how (1) to compare Australian and French consumer preferences to beef and (2) to see how well the MSA grading scheme could predict the eating quality of beef in France. Six muscles from 18 Australian and 18 French cattle were tested as paired samples. In France, steaks were grilled medium or rare, while in Australia medium cooking was used. In total, 360 French consumers took part in the medium cooking test, with each eating half Australian beef and half French beef and 180 French consumers tested the rare beef. Consumers scored steaks for tenderness (TE), Juiciness (JU), Flavour liking (FL) and overall liking (OL). They also assigned a quality rating to each sample: 'unsatisfactory', 'satisfactory everyday quality' (3*), 'better than everyday quality' (4*) or 'premium quality' (5*). The accuracy of the MSA weighed eating quality score (0.3TE+0.1JU+0.3FL+0.3OL) to correctly predict the final ratings by the French consumers was over 70%, which is very good according to the Australian experience. The boundaries between 'unsatisfactory', 3*, 4* and 5* were found to be ca. 38, 61 and 80, respectively. The differences between extreme classes are therefore slightly more important in France than in Australia. On average, the MSA model predicted the meat quality score relatively accurately, even though it does not have predictive equations for bull meat, with predicted scores deviating by 5 points on a 0-100 scale except for the Australian oyster blade and the French topside, rump and outside (deviating by less than 15). Overall the data indicates that it would be possible to manage a grading system in France as there is high agreement and consistency across consumers. The rare and medium results are also very similar indicating that a common set of weightings and cut-offs can be employed.

High energy supplement post-weaning does not enhance marbling in beef cattle
Greenwood, P.L., Siddell, J.P., Mc Phee, M.J., Walmsley, B.J. and Pethick, D.W., CRC for Beef Genetic Technologies, Armidale, NSW 2350, Australia; paul.greenwood@industry.nsw.gov.au

Objectives of this study were to 1) determine whether high energy supplement during the immediate post-weaning period enhances marbling; 2) determine whether nutrition and genotype interact to affect intramuscular (IM) and subcutaneous (SC) fat; 3) obtain data and samples for detailed study of fat depot development. Weaner steers (n=165) within three genotypes were studied. Targeted genotypes were high IM and high SC fat (Angus, A), low IM and high SC fat (Hereford, H) and high IM and lower SC fat (Wagyu x Angus, WA). From weaning, steers were fed pasture, or pasture plus high energy pellets (12.3 MJME/kgDM, 110 g CP/kgDM) at 1% liveweight (LW) for 168 d. Pasture-fed (P) and supplemented (S) steers were then backgrounded until feedlot entry at 18 month of age. Steers were then short (100 d) or long (250 d) feedlot fed. LW did not differ due to nutritional treatment at any stage. Base-line steers (n=15) were slaughtered at weaning, and groups slaughtered at end of nutritional treatments (n=30), prior to feedlot entry (n=30), and after short (n=30) and long (n=60) feedlotting. Genotype, Kill (1 to 5) and Post-weaning nutritional effects and interactions on carcass traits were assessed (at P<0.05) by analyses of variance, with initial LW as a covariate due to Angus being heavier. Hereford steers had more SC fat at the P8 site and less marbling than the other genotypes. Carcass weight, SC fat depths and marbling increased with kill number. Post-weaning supplement depressed Rib fat depth compared with forage only feeding. No interactions were evident. We conclude that post-weaning supplement did not enhance marbling and had a somewhat suppressive effect on SC fat. The genotypes had predicted marbling characteristics, although SC fat did not differ overall between A and WA. The phenotypic data will be used in detailed studies of fat depot development.

Contribution of Protected Denomination of Origin (PDO) beef for sustainable agriculture and meat quality in Portugal
Costa, P., Alfaia, C.M., Pestana, J.M., Bessa, R.J.B. and Prates, J.A.M., CIISA, Faculdade de Medicina Veterinária, Avenida da Universidade Técnica, Alto da Ajuda, 1300-477 Lisboa, Portugal; paulocosta@fmv.utl.pt

In the last twenty years, the UE policies, with the aim to develop sustainable agriculture systems, created the opportunity of meat obtained from Portuguese cattle breeds to be commercialized with 'certification of origin', known as Protected Denomination of Origin (PDO). PDO meats have a well defined breed, geographical distribution and production system. Thus, the combination of specific genotype and geographic region is the major basis for definition of the Portuguese PDO beef, promoting the biodiversity. The underlying assumption is that meat obtained with a specific genetic background coupled with the traditional agricultures practices used in a certain region is unique. However, this assumption is far from being scientifically validated and little is known about the genetic factors involved in meat quality in Portuguese breeds. Our research group has recently characterized the meat (longissimus dorsi muscle) of several local cattle breeds (Alentejana-1051 carcass tons produced in 2007, Mertolenga-334 tons, Barrosã-249 tons, Mirandesa-326 tons and Arouquesa-36 tons) from different regions of Portugal and produced according to PDO rules. Alentejana and Mertolenga PDO meats are obtained from purebred young bulls (16-24 months old) produced in the South of Portugal according to a semi-extensive grazing system and finished on concentrate (3-6 months). Barrosã, Mirandesa and Arouquesa PDO meats have their origins on purebred calves (6-9 months) reared under extensive grazing systems, fed with farm products and suckling from cows grazed on natural pastures. Meat from these breeds is lean (<3% intramuscular fat, IMF) and IMF composition in PDO veal (Arouquesa Barrosã and Mirandesa) have typical pasture-fed characteristics (high levels of n-3 PUFA and t11,c13 and c9,t11 CLA isomers), while in PDO beef (Alentejana and Mertolenga) the values are close to those obtained in animals raised on concentrates.

Traditional Corsican meat and dairy products move upmarket: could local consumers be excluded?
Casabianca, F. and Linck, T., INRA, SAD LRDE, Quartier Grossetti, Corte, F20250, France;
fca@corte.inra.fr

Developing added value from traditional products relies on their particular geographical, temporal and cultural anchorage. These products are deeply rooted in the rural societies of earlier centuries, when they were everyday foods for local communities. Upgrading such products' market involves projecting them onto an urbanized society with very different cultural references. A farmhouse product that was once embedded in egalitarian social relationships acquires the connotation of a luxury product reserved for an elite. In tourist regions such as Corsica, specialist shops and upscale restaurants and hotels are taking advantage of this shift to apply prices that select customers on the basis of purchasing power. To understand this gastronomic mutation, we analyze 3 aspects: (1) Symbolic: progressive standardization leads people to seeks some way of differentiating their identity from the mass. The heritage value of a food constitutes an identity marker, enabling people to identify with a place of origin and membership of a social space. (2) Technical: while production in most spheres continues to intensify, added value can be drawn from production systems based on respect for seasonal variations and making use of local knowledge, practices and resources (particularly local breeds and varieties). This adds to the cost of production. (3) Marketing: restaurants and the tourist trade reinforce the trend, driving prices higher and making these products much less accessible to their usual consumers. We will study the situations of three traditional products embedded in local Corsican food systems: a small ruminant cheese, suckling kid meat and a processed pork meat. We evaluate the shift in their gastronomic image, describe how it is expressed in each of the above three aspects and assess to what extend it may result in the exclusion of local consumers.

Consumer gastronomic evaluation of wild and farmed brown trout (*Salmo trutta*)
Pohar, J., Biotechnical faculty, Animal science, Groblje 3, 1230 Ljubljana, Slovenia; jure.pohar@bf.uni-lj.si

In contrast to production of meat by terrestrial animals where only a minor amount of total meat consumed is presented by game, the amount of fish caught (wild fish) is still larger than the amount of fish farmed. For some species consumers has the possibility to choose between wild or farmed fish; as a rule the price of wild fish is higher. The willingness of consumers to pay premium price for wild fish is caused by many factors. One of these factors could be the expectation of superior culinary quality. There are data in literature where sensorial quality of wild and farmed fish was appraised by panel of trained experts, but there is no data about appraisal by panel of consumers. In order to appraise the sensorial quality of wild and farmed fish 34 consumers assessed different traits of wild and farmed brown trout. Each consumer was served a blind sample of wild and farmed trout equally treated from catch to serving. If consumer noticed the difference for the traits mentioned he (she) was asked to decide which of the two samples is preferred for the trait under consideration. The majority of consumers reported that they noticed the difference between two samples for different traits (32 for juiciness, 31 for general impression and mouth feeling, 28 for appearance; 26 for aroma). However preference differs from trait to trait. The preference where the first figure indicate the number of consumers which preferred wild fish and the second figure the number of consumers which preferred farmed fish were as follows: 18:10 for appearance; 18:8 for aroma; 7:25 for juiciness; 13:18 for mouth feeling; 15:16 for general impression. In order to make it possible to evaluate the sensitivity of recognition of the difference, consumers were after first evaluation asked to appraise in the same way two new samples. This time both samples were represented only by farmed trout. Comparisons of results showed that the decision about preference in first trial was based on difference which was really noticed and was not just made up.

Breed effects and heritability for concentration of fatty acids in milk fat determined by Fourier transform infrared spectroscopy

Lopez-Villalobos, N.[1], Davis, S.R.[2], Lehnert, K.[2], Mellis, J.[3], Berry, S.[2], Spelman, R.J.[3] and Snell, R.G.[4], [1]Massey University, Private Bag 11222, 4442 Palmerston North, New Zealand, [2]ViaLactia Biosciences, P.O. Box 109185, 1149 Auckland, New Zealand, [3]Livestock Improvement Corporation, Private Bag 3016, 3240 Hamilton, New Zealand, [4]The University of Auckland, Private Bag 92019, 1142 Auckland, New Zealand; N.Lopez-Villalobos@massey.ac.nz

Calibration equations to predict concentration of each of the most common fatty acids (FA) in bovine milk fat, based on partial least squares using Fourier transform infrared spectroscopy (FTIR) data, were derived. Milk samples (n=848) collected in the 2003-2004 season from 348 second-parity crossbred cows during peak, mid and late lactation were used. The concordance correlation coefficients of the calibration equations for the most important FA ranged from 0.67 to 0.94. The calibration equations were then used to predict the concentration of each of the FA in 34,141 milk samples from 3,445 Holstein-Friesian (HF), 2,935 Jersey (JE) and 3,609 crossbred HFxJE cows sampled on average 3.4 times each cow during the 2007-2008 season. A repeatability animal model was used to obtain breed effects and estimates of variance components for each predicted FA. Predicted concentrations of C6:0, C8:0, C10:0, C10:1, C12:0, C12:1, C14:0 and C16:0 were significantly lower (P<0.05) in HF cows than in JE cows but concentrations of C18:0, cis-9 C18:1 (oleic acid), cis-7 C18:1, cis-9,12 C18:2 n-6 (linoleic acid), C20:0, and unsaturated FA were significantly higher in HF (P<0.05) than in JE cows. Heritability of predicted concentration of FA in milk fat ranged from 0.21 to 0.42. Fatty acid composition in milk fat of New Zealand dairy cattle can be predicted using FTIR spectroscopy. There are significant genetic variation and breed effects that can be exploited in a breeding program to alter FA composition of milk fat.

Analysis of the red meat price changes over the last 25 years and effects of import decisions in Turkey

Aydin, E., Aral, Y., Can, M.F., Cevger, Y., Sakarya, E. and Isbilir, S., Ankara University, Faculty of Veterinary Medicine, Department of Animal Health Economics and Management, Diskapi, Ankara, 06110, Turkey; yaral@veterinary.ankara.edu.tr

In this study, it is aimed to determine and examine the change occurring in red meat prices in Turkey over time within the framework of important economic events, as well as import decisions of slaughtered animals and red meat. The figures concerning livestock and red meat import for the year 2010 in Turkey, wholesale prices of red meat, livestock, and concentrated fattening breeding feed occurring in the period between the years of 1985 and 2010, the data relating to Wholesale Price Index-Producer Price Index, and Purchasing Power Index have constituted the material of the research. It was determined that Turkey imported around 100,500 tons of red meat until December of the year 2010. While the mutton price is fluctuating in the range of 1.69 TL (1.13 US$) in real terms over the average prices between the years of 2003 and 2008, in 2009 it increased at the rate of 51.4% compared to 2008 but in 2010 it increased at the rate of 68.5% compared to 2008. On the other hand, whilst the annual average beef price in real terms is fluctuating in the range of 1.26 TL (0.84 US$) between the years of 2003 and 2009, only in 2010 it increased at the rate of 37.0% compared to 2009. Finally, not taking timely political measures for removing the economic crises and their effects undergone over years has affected the livestock sector negatively and this has resulted in decreases and withdrawals from production. During the recent periods, the instability in the prices of livestock and red meat and the increased financial risk have had a negative impact on the producers' expectations and objectives. It is considered that ensuring the stability in the red meat prices in Turkey and ceasing the red meat import may be achieved on medium and long terms only if rational livestock policy measures are taken rapidly with the participation of sector stakeholders.

Industrial response to increased sustainability demand in aquaculture

Wathne, E.[1] and El-Mowafi, A.[2], [1]EWOS Group, Tollbodalmenningen 1 B, 5803 Bergen, Norway, [2]EWOS Innovation, R&D Centre, 4335 Dirdal, Norway; einar.wathne@ewos.com

The entire value chain of aquaculture is challenged by extended sustainability demands not only from customers, NGOs or regulators, but also from other stakeholders like investors and analysts. Starting with feed ingredients, the main focus in aquaculture has been on the use of marine ingredients in fish feeds. Does growth in aquaculture increase fishing pressure on global fish stocks; are fisheries managed in line with sustainability principles; is it right to feed fish with fish; or could the forage fish be used more efficiently otherwise. These are questions to which the aquaculture industry needs to respond. For a lower reliance on marine raw materials, questions turn to sustainable replacement by non marine ingredients. Is the use of GM ingredients and animal by products a viable, sustainable option in the aquaculture industry? The art of making feeds is extended, far beyond providing cost effective nutrients to the target specie. Fish health and welfare is core in all animal production, and consumers have become more and more aware of this, driving improvements in husbandry, health management and a focus on disease prevention. The aim to reduce use of medicines and chemicals in fish husbandry changes focus from cure to prevention. This means that a more systematic and integrated approach to fish health and welfare is required. Functional feeds and vaccine development are current examples of advances and innovation in the area of disease and parasite prevention. Openness is another response from the industry in order to demonstrate and measure activities in sustainability. In order to gain trust and be transparent, the aquaculture industry is more open today than ever; related to use of medicines and chemicals, feed and feed ingredients used, and escapees. Data requested from external stakeholders are commonly found in sustainability reporting today. This move to an open industry will stimulate dialogue and finally gain trust between industry and consumers.

Resource use in Norwegian salmon production

Ytrestøyl, T., Ås, T.S., Berge, G.M., Sørensen, M., Thomassen, M. and Åsgård, T., Nofima, Sjølseng, 6600 Sunndalsøra, Norway; trine.ytrestoyl@nofima.no

Resource use in Norwegian salmon production Norway During the last decade, the production of Atlantic salmon has increased by almost 70% from 900,000 tonnes worldwide to more than 1,500,000 tonnes today. Farmed salmon is the most widely consumed sea product in the industrialised world, and the growth in the salmon industry has raised concerns about the sustainability of salmon farming. The dependence on fish meal and fish oil and the potential effects on wild fish stocks are arguments often used against sustainability. Feed is the main input factor in salmon production, so an understanding of how different feed formulations affect environmental impacts and resource utilisation is vital. To sustain a population of 9-12 billion people on earth within the next 40 years the limited resource pool must be utilised effectively. Thus, understanding how different food production systems utilize the available resources is important. In a study financed by FHF the efficiency of salmon production was compared with other food productions, and methods that are used for comparing the biological and ecological efficiency of different food production systems were evaluated. The retention of nutrients and energy in the edible part of the animal and in the whole animal was calculated for salmon, pig and poultry production (protein efficiency ratio, PUFA efficiency ratio and energy efficiency ratio). Several methods commonly used for assessing the eco-efficiency and environmental impacts of a production (carbon and ecological footprints, eco-efficiency models) were evaluated in terms of strengths, weaknesses and sensitivity. LCA was used to evaluate resource use and environmental impacts of Norwegian salmon production over time and compare it with industrial pig and poultry production today. Possible resource deficiencies that could limit further growth in the salmon industry are also discussed.

Microbes: a sustainable aquafeed resource for the future
Mydland, L.T.[1], Romarheim, O.H.[1], Landsverk, T.[2], Skrede, A.[1] and Øverland, M.[1], [1]Aquaculture Protein Centre (APC), CoE, Department of Animal and Aquacultural Sciences, Norwegian University of Life Sciences, Aas, Norway, [2]Department of Basal Sciences and Aquatic Medicine, Norwegian School of Veterinary Science, Oslo, Norway; liv.mydland@umb.no

Microbial products like bacteria, yeast and micro-algae represent an exciting nutrient source in future fish feed and may thus relieve the pressure on limited and expensive protein sources such as fish meal. Bacterial meal (BM; mainly *Methylococcus capsulatus*) grown by fermentation of natural gas, ammonia and mineral salts, contains about 70% crude protein (CP) and 10% lipids and has a favourable amino acid (AA) composition for salmonid diets. BM can partially replace high-quality fish meal in diets for Atlantic salmon and rainbow trout without impairing growth performance. Recent studies from APC have also shown that BM appears to stimulate growth and integrity of intestinal tissue and to efficiently prevent soybean-induced enteritis in the distal intestine of salmon. Yeast and microalgae also represents a valuable, sustainable nutrient source for aquafeeds. Yeasts like *Rhizopus oryzae* have been shown to be efficient in fermenting co-products from wood processing and agricultural industries into a biomass rich in proteins, lipids, and nucleic acids. Rainbow trout fed diets containing *Rhizopus oryzae* obtained similar growth rate and feed efficiency as those fed a fish meal-based diet. Other yeast species (e.g. *Pichia* spp. and *Kluyveromyces* spp.) have shown a CP content of about 50-60%, with favourable AA profile and a high digestibility. Evaluation of three algae (*Nannochloropsis oceanica*, *Phaeodactylum tricornutum*, and *Isochrysis galbana*) showed a CP content ranging from 23-50% of DM, and a lipid content ranging from 8-18%. The AA profile was favourable in all products, but there was a large difference in the apparent digestibility of CP among the algae. Thus by feeding microorganisms to salmonids, we can convert inedible, non-food substrates into valuable edible products for humans.

Optimal feeding rate for genetically improved Nile tilapia
Storebakken, T., Kumar Chowdhury, D. and Gjøen, H.M., Aquaculture Protein Centre, CoE, Department of Animal and Aquacultural Sciences, Norwegian University of Life Sciences (UMB), 1432 Ås, Norway; trond.storebakken@umb.no

Genetically improved Nile tilapia (*Oreochromis niloticus*) has obtained 10% faster growth rates per generation. It is common practice to use more than 1.5 kg of feed per kg tilapia growth, and virtually no published information exists on optimal feeding of genetically improved tilapia. The current project was carried out to determine optimal feeding rates of fish from the 14th generation of a combined family-phenotype selection. The fish were kept in tanks with 26-27 °C recycled freshwater, and fed an extruded feed. Tilapias weighing from 1 to 77 g were fed a diet based on 87% plant ingredients and 7% fish meal (34% crude protein, 7% lipid, 25% starch, supplemented with lysine and methionine). The fish were fed 22 times a day at various feeding rates ranging from restricted feeding to overfeeding. Juvenile (1-g) tilapia should have their daily feeding rates gradually reduced from 8% to 7% of body weight (BW) during the first 2 weeks, and a further reduction to 5% during the next two weeks. Feed utilization ranged from less than 0.7 kg dry matter (DM) per kg gain for the most restricted feeding to 1.4 for the highest overfeeding. Nitrogen retentions declined from 68% (restricted) to 40% (overfed), and energy retentions declined from 50 to 37%. Thus, best feed conversion can be obtained at restricted compared to feeding for maximum growth. For 77-g tilapia, highest growth (150% gain in 13 d) was seen by feeding 3% BW a day. This feeding rate also coincided with the most efficient feed utilization (0.87 kg DM per kg gain), and optimal nitrogen (41%) and energy (54%) retentions. In conclusion, genetically improved Nile tilapia can grow rapidly, and efficiently utilize their feed at feeding rates matching their daily needs for nutrients and energy. Feed demands must be updated in accordance with continuously improved growth resulting from genetic selection.

In vitro assays are useful tools in the value chain of aquaculture products

Moyano, F.J., Morales, G. and Márquez, L., Department of Applied Biology, Engineering School, University of Almería, 04120 Almería, Spain; fjmoyano@ual.es

In vitro assays are extensively used in different fields of basic and applied science, like pharmacological tests, assisted reproduction, biorreactions and human and animal nutrition. In spite of their limitations, _in vitro_ assays posses some interesting advantages over experiment performed _in vivo_, since they allow simplification of processes affected by a number of different factors, this making their evaluation more accessible, and also reducing time and costs needed to obtain results. In addition, they have a better social acceptance, since ethical conflicts associated to the experimentation with humans and animals almost disappear. In the case of aquaculture products, _in vitro_ assays can be applied at different stages along the value chain, increasing the information required to develop better control operations. At the feed level, they can be used in the assessment of the possible effects that some antinutritional compounds present in some feedstuffs may exert on the live organism, as well as in determining to what extent bioavailability of macro-nutrients may be affected by physical or chemical processes during elaboration of feeds. At the organism level, they can help to estimate bioavailability of nutrients and minerals present in the feeds and how it can be affected by the inclusion of some additives. At the environmental level, they can be used in the estimation of the expected nitrogen and phosphorus loads to water after feed digestion. Finally, at the product level, they can provide information on the effect of fish processing (storage, cooking, salting, irradiation) which can affect the physicochemical state of proteins and thus the bioavailability of amino acids for human nutrition.

Interaction between physical quality, feed intake and nutrient utilization in modern fish feeds

Sørensen, M.[1,2], Åsgård, T.[3] and Øverland, M.[2], [1]Nofima, Feed and Nutrition, P.O. Box 5010, 1430 Ås, Norway, [2]Aquaculture Protein Centre, CoE, Department of Animal and Aquacultural Sciences, Norwegian University of Life Sciences, P.O. Box 5003, 1432 Ås, Norway, [3]Nofima, Feed and Nutrition, Sjølseng, 6600 Sunndalsøra, Norway; mette.sorensen@nofima.no

High feed intake of a well balanced nutrient dense diet of good quality is an assumption for fast growth and good feed utilisation in intensive aquaculture production. The feed pellets need to be durable and have physical quality that withstands handling, transportation and use of automatic feeding devices without forming small particles and dust. High quality diets, in terms of nutritional and physical quality, are produced with use of extrusion technology. Commercial diets are made from a variety of ingredients and are balanced to meet known nutrient requirements and often supplemented with immune stimulants to promote fish health. The physical quality of feed varies with ingredient composition due to differences in the functional properties, preconditioning and extrusion conditions such as screw configuration, mechanical and thermal energy input. Resent studies have shown significant correlations between physical quality of feed, feed intake as well nutrient utilization of various diets. Some studies suggest that physical quality of the feed interferes with passage rate throughout the gastrointestinal tract and feed intake. Consequently, variable feed intake in different feeding experiments may express variation in gastrointestinal retention time. These studies suggest that physical quality of the feed should be addressed when performance of fish feed based on new ingredients are evaluated.

Sodium diformate and extrusion temperature affects nutrient digestibility and physical quality of diets with fishmeal or barley protein concentrate for rainbow trout (*Oncorhynchus mykiss*)

Morken, T.[1], Kraugerud, O.F.[1], Barrows, F.T.[2], Sørensen, M.[1], Storebakken, T.[1] and Øverland, M.[1], [1]Aquaculture Protein Centre, CoE, Norwegian University of Life Sciences, Department of Animal and Aquacultural Sciences, P.O. Box 5003, 1432 Ås, Norway, [2]U.S. Department of Agriculture, Agricultural Research Service, 3059-F National Fish Hatchery Road, 83332 Hagerman, USA; thea.morken@umb.no

High protein content (>55%) combined with low starch levels (<3%) makes barley protein concentrate (BPC) a potential alternative to fishmeal (FM) for use in salmonid diets. The objectives of this experiment were to evaluate the effects of BPC, extrusion temperature, and the acid salt sodium diformate (NaDF) on apparent nutrient digestibility and physical quality of diets for rainbow trout. The experiment had a 2^3 factorial design with two dietary ingredient sources (FM and BPC), two extruder temperatures (110 and 141 °C), and two levels of NaDF (0 and 10.6 g kg^{-1}). The diets were fed to triplicate groups of fish and nutrient digestibility was determined by fecal stripping using yttrium oxide as an inert marker. Physical quality of the extruded diets was evaluated by measuring hardness, expansion ratio, durability, and water stability. Inclusion of BPC to diets improved the digestibility of protein and several NEAAs, but reduced the digestibility of lipid, starch, and all EAAs. Diets with BPC showed improved hardness, expansion ratio, and durability, but reduced water stability. Extrusion at high temperature (141 °C) gave increased digestibility of several major nutrients and AAs, and increased pellet durability. The addition of NaDF to diets improved the digestibility of all major nutrients and individual AAs, as well as the physical feed parameters such as expansion ratio, durability, and water stability. These findings demonstrate that BPC is a promising feed ingredient, and that high extrusion temperature (141 °C) and inclusion of NaDF can improve both nutrient digestibility and physical quality of diets for rainbow trout.

Emerging trends and research needs in sustainable aquaculture

Fiore, G., Natale, F. and Hofherr, J., Joint Research Centre - European Commission, Institute for the Protection and Security of the Citizen - Maritime Affairs Unit, Via E Fermi 2749, 21027 Ispra (Varese), Italy; fabrizio.natale@jrc.ec.europa.eu

Scientific research evolves constantly in order to respond to policy and economic drivers in each underlying area of application. A way for understanding these driving forces and identifying key challenges for the future is to analyze how scientific literature is evolving, which are the 'hot topics' emerging and which disciplines and technologies are receiving most attention. This approach enables also to address the need expressed in several forums, to ensure rationalization of the use of the public budget dedicated to research, by preventing multiplication of investment in similar areas and dispersing financial resources. In this study we applied quantitative methods to map and identify emerging trends in sustainable aquaculture research. The analysis was carried out on a digital library of bibliographic information related to around 18.000 articles extracted from SCOPUS and on the abstracts of around 300 research projects related to aquaculture in the EU Framework Programs 5, 6 and 7 extracted from CORDIS. Abstracts, titles, keywords and other bibliographic information have been analyzed using latent semantic analysis and co-word occurrence, which are well established methods from information retrieval and computational linguistics. The analysis allowed defining a knowledge map of aquaculture research in recent years showing the relevant patterns and emerging trends. The results from the quantitative analysis were complemented with a qualitative review of EU policy and economic drivers in relation to sustainable aquaculture looking at dedicated knowledge transfer projects and platforms and through structured interviews with key policy makers, experts and stakeholders.

Local breed farmers providing ecosystem services and sustainable development

Soini, K., MTT, Agrifood research Finland, Economic research, Latokartanonkaari 9, 00790 Helsinki, Finland; katriina.soini@mtt.fi

It is generally agreed that local breeds provide many ecosystem services: provisioning services, regulating services, supporting services and cultural services. It has also been showed that the local breeds contribute to the sustainable development of rural areas and society at large. The production of ecosystem services and sustainable development, in turn, depends on the farmers' motives and willingness to keep the local breeds. So, the question is how the farmers perceive the local breeds and biodiversity of agriculture at large, and what kind of farmers keep local breeds. The ultimate aim of the paper is to show the linkages between the ecosystem services, sustainable development and farmers' perceptions of the breeds. The paper will synthesise the results of qualitative studies related to the farmers' perceptions local breeds and biodiversity of agricultural landscapes more broadly, as well as social and cultural sustainability of rural areas. The data originates from several research projects the author has been involved in, and geographically concerns East Siberia (Yakutia), several European countries and Finland. Based on the results of these case studies comparisons between the perceptions of experts' and farmers' of the values of the local breeds, differences between developing and more advanced societies and preconditions for socially and culturally sustainable local breed farming are discussed. Also a typology of the local breed cattle farmers will be presented. The results show that there are a considerable differences how the experts and farmers perceive biodiversity of agricultural landscapes and local breeds underlining a need for community-based and participatory approaches. On the other hand there is a high diversity of the perceptions of local breeds. This can be seen as strength for the preservation of animal genetic resources, although it also constitutes a challenge for policy making. Moreover, the results also show the differences between developing and more advanced societies.

Positive effects of animal production in France: a preliminary study based on interviews of stakeholders in two contrasted territories

Disenhaus, C.[1], Le Cozler, Y.[1] and Bonneau, M.[2], [1]Agrocampus Ouest, UMR 1080, 65 rue de Saint Brieuc, 35042 Rennes, France, [2]INRA, UMR 1079, 65 rue de Saint Brieuc, 35042 Rennes, France; catherine. disenhaus@agrocampus-ouest.fr

Within the frame of a national research project on the evaluation of services provided by livestock productions to territories, this preliminary study evaluates stakeholders' perception of these services in a mountain (Chartreuse in the Alps) and in a plain territory (Brittany: intensive animal production and dense human population). Our hypothesis is that services are the same whatever the territory, but with a different hierarchy of interest according to the territory and the stakeholders. Agrocampus-Ouest master's level students conducted 67 semi-directive interviews with local councillors (14), farmers (13), researchers (10), extension workers (9), ecologists (10), food chain (9) and tourism professionals (2). Additionally, 50 consumers were surveyed when shopping. Animal production was overwhelmingly recognised as a driving force for the social and territorial dynamism, in terms of direct or indirect employment, services, animation… Besides the production function, recognised by all protagonists in both territories, a major role of animal production in rural life maintenance was underlined. Animal production was also seen as part of heritage, mostly natural heritage in Chartreuse, mostly cultural and architectural heritages in Brittany. Its role in the maintenance of natural environments was emphasised in both regions by researchers, consumers and ecologists. The various productions were perceived differently: only ruminants (and associated pastures) are positively perceived in both territories. In Chartreuse, it was systematically noticed that maintaining ruminant production is crucial for landscape preservation. In conclusion, in both territories, most stakeholders perceive positive effects of animal production even though environmental effects were more readily mentioned in Chartreuse than in Brittany.

Farm animal genetic resources and social values

Oosting, S.[1], Partanen, U.[2] and Soini, K.[2], [1]Wageningen University, Animal Sciences, P.O. Box 338, 6700 AH Wageningen, Netherlands, [2]MTT Agrifood Research Finland, Economic Research, Latokartanonkaari 9, 00790 Helsinki, Finland; simon.oosting@wur.nl

Indigenous cattle breeds comprise a group of animals that have been formed by natural selection in their production environments following the needs of cattle producers before genetic techniques were scientifically known and practiced. The numbers of indigenous cattle have diminished rapidly on a global scale during recent decades, and a large number of cattle breeds have become endangered or even extinct. The loss of breeds is particularly alarming in developing countries as about 70% of extinct breeds are from these countries. In addition to genetic values in biodiversity conservation, farm animal breeds can exhibit ecological, economic, social, cultural, political and ethical values which should be considered in breed conservation. Therefore, results from socio-economic and socio-cultural studies can provide critical information for setting conservation priorities among farm animal breeds. The role of indigenous breeds in developed societies where agriculture is already modern is to be components of multifunctional agriculture or sustainable rural development. This is achieved through rural entrepreneurship and provision of social and cultural services, the animals representing a form of rural capital. In the western societies a new argument for the conservation is the therapeutic use of the local breeds. These so called Green Care activities and especially social farming and care farming are emerging all over Europe. These activities provide a new possibility for the conservation of farm animal genetic resources. Results from an inventory of breeds at green care farms in the Netherlands and other countries and from stakeholder interviews analyzed regarding the arguments and values connected to the breeds will be presented. The governance approach has been used in order to present a coherent picture of the different levels and different actors that affect the on-farm conservation.

The farmer as a main factor of structural change and sustainability in rural development: a Slovenian case-study

Klopčič, M.[1], Kuipers, A.[2], Bergevoet, R.[3] and Koops, W.[2,4], [1]University of Ljubljana, Biotechnical Faculty, Dept. of Animal Science, Groblje 3, 1230 Domžale, Slovenia, [2]Wageningen UR, Expertise Centre for Farm Management and Knowledge Transfer, P.O. Box 35, 6700 AA Wageningen, Netherlands, [3]Wageningen UR Livestock Research, P.O. Box 65, 8200 AB Lelystad, Netherlands, [4]Wageningen UR, Animal Production Systems Group, P.O. Box 338, 6700 AH Wageningen, Netherlands; Abele.Kuipers@wur.nl

After accession to EU, farmers in the new-member states had to adjust to the EU agricultural policies and market. In Slovenia an analysis is made taking the cattle sector as case. Farm size, strategies like specialization or diversification and consolidation or expansion and operating in flat, hilly or mountainous regions have been considered as important factors to study. As tool questionnaires were used, anonymously collected from three groups of farmers: dairy farmers (1,110 questionnaires available for processing), farmers with suckler cows (121) and local Cika breed farmers (111). The results show significant differences in typologies between these three groups of farmers. About 40% of dairy farmers choose for keeping the farm business the same and 50% intend to develop the farm further, while more than half of these developing farmers choose for specialization less than half for diversification. The interest in special local products and ecological farming was far below expectations in the group of dairy farmers and much higher for the other two groups. Indeed, suckler cow and local breed farmers are environmentally friendly oriented, while dairy farmers are more business oriented. To deepen the analysis, an additional questionnaire was distributed asking for social and economic factors affecting the development plans. 525 questionnaires were received back. There appeared to be a strong relation between the farmers goals, preferred farm type, the farmers personal characteristic and his/her perception of opportunities and threats related to society and neighbourhood and the present farming system and size.

Understanding how farmers last over the long term: a typology of trajectories of change in farming systems. A French case-study.

Ryschawy, J., Choisis, N., Choisis, J.-P. and Gibon, A., INRA, UMR 1201 DYNAFOR, BP 52627, 31326 Castanet-Tolosan, France; julie.ryschawy@toulouse.inra.fr

In the current context of market fluctuation on agricultural products prices, European agriculture is endangered. In hilly areas, the orientation of CAP policy promoting specialization added to an always increasing lack of work forces challenges the future of farms. Nevertheless, farmers found how to adapt to local context to last on the long term. In this study, we try to assess the diversity of the adaptative strategies developed by farmers to last in analyzing their trajectories of change. Our study aims to understand the variety in trajectories of farms from 1950 up to now. We applied an integrated approach to the farm population of a case-study site, in the Coteaux de Gascogne. In this hilly region of south-western France, agriculture maintained with a limited specialization of production. We made a survey of the history of every farms working land in an area of about 4,000 ha. We used a two steps-analysis including: (1) a manual assessment of the trajectory of each farm and (2) a typology of farm trajectories build on a combination of multivariate analysis on a set of data composed by 20 variables for 50 farms on 10-year steps. The interpretation of the types was based on the results of the manual assessment. The resulting 6 types of trajectories reflect different objectives and strategies. Farmers found different 'paths to last' in a same local context (environmental, political and economic). In two types of trajectories, farmers became specialized, in the other ones, farmers maintained more traditional systems, based on a crop-livestock association. This typology was validated by local farmers. Our results stress out the importance to understand the systemic functioning of farms to study local change in agricultural systems. In a next step of our study these results will be used in a participatory future process with local stakeholders, through co-constructed prospective scenarios.

Socio-cultural sustainability of pig production: farm visits with citizen panels in the Netherlands and Denmark

Oosting, S.J.[1], Boogaard, B.K.[2], Boekhorst, L.J.S.[1] and Sørensen, J.T.[3], [1]Wageningen University, Animal Sciences, P.O. Box 338, 6700 AH Wageningen, Netherlands, [2]Wageningen University, Social Sciences, P.O. Box 8130, 6700 EW Wageningen, Netherlands, [3]University of Aarhus, Department of Animal Health and Bioscience, P.O. 50, DK-8830 Tjele, Denmark; simon.oosting@wur.nl

Socio-cultural sustainability of animal farming is about social perceptions and underlying values. Values, broadly defined as 'everything that matters to people' are important fundaments of culture. A possible assessment of socio-cultural sustainability is to make an inventory of social appreciation and concerns of animal production systems. In the present study such an inventory was made of pig production in the Netherlands and Denmark. We conducted farm visits with citizen panels with two panels of nine respondents each in the Netherlands and two panels of four respondents each in Denmark. In each country the panels visited a conventional and an organic pig farm. During the farm visits, respondents noted their sensory experiences i.e. what they smelled, heard, saw and felt. In addition, each respondent made pictures of six positive and six negative aspects on the farms and they had to write a motivation about his or her judgment. The qualitative analysis resulted in seven socio-cultural themes (SCT) of pig production namely: 1) meat production, 2) farm activities, 3) farm income, 4) animals, 5) housing system, 6) environment and nature, and 7) culture and landscape. Each SCT included several socio-cultural aspects (appreciations, SCA) and socio-cultural issues (concerns, SCI). We identified 31 SCAs in the Netherlands and 33 SCAs in Denmark, of which 29 were SCIs in both countries. Although many SCIs were about animal welfare, the results also showed that social concerns of pig production extended beyond animal welfare. In general it can be stated that citizens are strongly concerned about over-exploitation of animals in contemporary pig production systems, but at the same time they appreciated dynamism in a pig farm including certain modern developments.

Livestock innovation systems and networks: Findings from the smallholder dairy farmers in North Western Ethiopia

Asres, A.[1], Sölkner, J.[1], Puskur, R.[2] and Wurzinger, M.[1], [1]BOKU-University of Natural Resources and Life Sciences, Gregor-Mendel-Str. 33, 1180 Vienna, Austria, [2]ILRI-International Livestock Research Institute, P.O.Box 5689, 0010 Addis Ababa, Ethiopia; johann.soelkner@boku.ac.at

In the last years many external funded development projects were carried out with the aim to provide farmers with new technologies in order to improve their farming practices. As a case study the Integrated Livestock Development Project (ILDP), which was designed to modernize the livestock sector in North Western Ethiopia, was selected to investigate if and how farmers have changed their management practices. Over 10 years this project has been involved in a number of activities like forage resource development, animal health service, breed improvement for milk, meat and traction, marketing & cooperative promotion, capacity building and networking. The objective of this research is to assess the project innovative working systems, seeks to determine the level of adoption of improved livestock technologies and its contribution to increased productivity. A total of 224 households have been selected from four districts. Within each district, a two-stage selection process has been followed, selecting first, 2 villages, purposively on the bases of their relative importance in having more project beneficiaries, and finally selecting households from within each of the selected villages using systematic random sampling technique proportionately. Semi-structured questionnaires were applied and specific focus was on a) socio demography of the household; b) household assets; c) access to rural services; d) work groups and cooperation membership; e) participation in ILDP project; f) how government/ private service providers are organized to extend support or to administer the program; g) what challenges have been faced and how these have been surmounted. First results indicate that the majority of farmers discontinue with the new practices as the linkage to markets is very weak and therefore there is no incentive for them for more investments.

Unravel the chaos in social aspects of livestock farming systems

Eilers, C.H.A.M. and Oosting, S.J., Wageningen University, Animal Production Systems, Animal Sciences, Marijkeweg 40, 6709 PG Wageningen, Netherlands; karen.eilers@wur.nl

Social aspects of livestock farming systems are mainly defined in the context of sustainable development. Social sustainability aspects are often developed through consultation with stakeholders of a certain farming system. These aspects are important on local level but vary according to the needs and interests of the stakeholders of the farming systems in which they are developed. The needs and interests of stakeholders are partly culturally defined. The variation in farming systems in different EU countries with different cultural background causes that among EU countries different social aspects of livestock farming systems are considered important. Hot issues vary among countries and thus different social aspects are studied by animal scientists and social scientists from different countries. Within the Livestock Farming Systems group social aspects were always considered important, even though scientists from different countries were talking about different and sometime incomparable social aspects. To unravel this Babylonian confusion of tongues we asked animal scientists and social scientists from 10 EU countries to describe the most important social sustainability aspects of the livestock farming systems they study in their own country. Results show a large variation in important social aspects among countries and among scientists with different backgrounds. An effort to combine all these social aspects in one definition makes it too broad to further the sustainable development of livestock farming systems. It is recommended to develop a series of research themes around which discussions of social sustainability of livestock farming systems can be conducted; e.g. labor issues, landscape perception or regional products are discussed in separate sessions with case-studies from different EU countries.

Increasing nutritive values of raw materials and compound feed by feed technology
Van Der Aar, P., Schothorst Feed Research B.V., Meerkoetenweg 26, 8218 NA Lelystad, Netherlands; PvdAar@schothorst.nl

Technology in the feed industry has several objectives, nutritional, hygienic and commercial. These can be conflicting. The optimal solution can only be reached if the technology choice is a joint decision of the responsible persons for the parties concerned. An example is that pellet hardness of ruminant feed is mostly preferred, whereas several experiments have shown that technological treatments reduce the rumen bypass of starch and thus reducing the net energy value of the feedstuff. Experiments in which technological treatments on animal performance have been studied have often inconsistent results. Technological treatments exert variable effects on individual feedstuffs, therefore the effect depends on the composition of the diet. When nutritionists consider nutritional value they take into account not only the effect on nutrient utilization but also its effect on digestibility, anti nutritional factors, animal health, welfare and product quality. Because of the different effect on feedstuffs, it should be considered whether the envisioned effect accounts for the complete feed or for individual feedstuffs. It determines the choice whether the feed or a specific feedstuff should be treated. In the case of reduction of ANF's treatment of one feedstuff is mostly preferred. In other situation, fe the reduction of salmonella contamination the whole feed is preferred. Another important aspect in the evaluation of technology is the age, physiological stage of the animal and the animal species. Fe in the feed for young piglets reduction of rapidly fermentable NSP's might be desired to reduce colonization of *E. Coli*'s whereas these compounds in older animals can be an energy source for the animals.

Effects of roasting of barley on utilization of nutrients and performance in dairy cows
Prestløkken, E., Mcniven, M.A. and Harstad, O.M., ; egil.prestlokken@umb.no

Barley is a major source of energy in diets for ruminants. However, including large amounts of barley starch in the diet may hamper rumen environment and thereby feed digestion and utilization. Earlier experiments indicate that roasting may reduce rumen digestion of barley starch. The objective of the experiment was to study the effect of roasting barley on digestion of nutrients and performance of dairy cows. Three barley-based concentrates were produced.1 (C1) consisted of 75% ordinary hammer milled barley. In 2 (C2), the barley was roasted at 150 °C in an electric roaster for 4 minutes. In 3 (C3), 25% barley was replaced by maize. Grass silage (182 g CP and 496 g NDF/kg DM) was offered 6 rumen and intestinal fistulated cows giving ca. 45% silage and 55% concentrate in the diet (DM basis). Roasting of barley (C2) reduced rumen and total tract digestion of NDF (P<0.05) and increased rumen digestion of starch (P<0.05) compared to diets C1 and C3. No difference in rumen digestion of NDF and starch was observed between C1 and C3, whereas total tract digestion of starch was significantly lower in the diet containing maize than in the two other diets. Duodenal flow of non-ammonia nitrogen and non-ammonia non-bacterial N was numerically increased by roasting. Roasting reduced rumen digestion of N, but also total tract digestion of N. Roasting reduced milk fat content (P<0.05), but no difference on yield of energy corrected milk was observed. No differences between ordinary barley or inclusion of maize were observed except for reduced total tract digestion of starch with maize. We conclude that roasting of barley affected rumen and total tract digestion of nutrients negatively.

Towards new feed unit systems based on absorbed nutrients and animal responses in France

Sauvant, D.[1], Peyraud, J.L.[2] and Noziere, P.[3], [1]AgroParistech-INRA UMR MoSAR, 16 rue C.Bernard, Paris 5, France, [2]INRA UMRPL Rennes, Saint Gilles, 39590, France, [3]INRA URH, Theix, 63122, France; sauvant@agroparistech.fr

A steering group is now working in France to update the feed energy, protein and fill units systems by 2013. A major step of this project is to simply model the major digestive events to predict the absorbed flows of VFA, amino-acids, glucose and fatty acids. Recent studies demonstrated the value of meta-analysis of experimental databases to predict feeding practices influences on quantitative digestion of proteins, starch and glucose, fatty acids and volatile fatty acids (VFA) and from amount and composition of fermentable organic matter. Similar approaches are now being applied to obtain new models of responses or to update previously published equations. The same approach is being applied to microbial growth efficiency. Afterwards, the major responses of digestion will be integrated into a simple mechanistic model with an optimization procedure to determine the optimal values of the major parameters of digestion (outflow rates, microbial growth...) and the influences of digestive interactions due to levels of intake, proportion and nature of concentrate and rumen protein balance. Thus the most probable flows of absorbed nutrients will be predicted for every ration. A particular effort will be done to enlarge the field of application of the new unit systems according to diversity of feed resources and contexts of production. The second step concerns the updating of animal requirements and the predictions of their multiple responses to absorbed nutrients according to their nutritional and physiological status. These corresponding values and laws will be built from data bases were absorbed nutrient can be predicted with a reasonable degree of accuracy.

The characteristics of nitrogen subfractions according to the CNCPS for extracted oil seeds

Chrenková, M.[1], Čerešňáková, Z.[1], Weisbjerg, M.R.[2], Van Vuuren, A.M.[3] and Formelová, Z.[1], [1]Animal Production Research Centre Nitra, Institute for Nutrition, Hlohovecká 2, 95141 Lužianky, Slovakia (Slovak Republic), [2]University of Aarhus, Faculty of Agricultural Sciences, P.O. Box 50, DK-8830 Tjele, Denmark, [3]Wageningen UR Livestock Research, P.O. Box 65, 8200 AB Lelystad, Netherlands; Ad.vanvuuren@wur.nl

New systems for prediction of nutritive value of feedstuffs and availability of nutrients from feeds require exact information about ruminal degradation and postruminal digestibility characteristics of feed protein. This work was focused on protein quality of three groups of extracted oil seeds, rapeseed meal (RM), sunflower meal (SFM), and soybean meal (SM), all known as good and rich sources of nitrogen in cattle feeding. In the CNCPS feed protein is partitioned into five fractions, which are related with ruminal degradability and intestinal digestibility. Between and within feed groups significant differences were observed in the concentration of crude protein (CP) and nitrogen subfractions. The fraction A (soluble true protein) was highest for SFM (38.4 g.kg^{-1}CP). For non-protein nitrogen (NPN) there was significant difference (P<0.01) between the samples of SFM (3.18) and SM (8.86 g.kg^{-1}CP). Buffer-soluble N (g.kg^{-1}CP) was highest in SFM (47.3 g.kg^{-1}CP) and lowest in SM (11.7 g.kg^{-1}CP). For unavailable protein fraction (C) insoluble in acid detergent no significant differences between feed groups were observed. Subfraction B3 (that is slowly degraded in the rumen) was significantly higher for RM than for SFM and SM (51.6, 30.1 and 28.2 g.kg^{-1}CP, resp.). Fraction B2 correlated well with in situestimates of fractional rate of CP degradation ('c', P<0.01), potential degradable fraction 'b' (P<0.001), and with effective CP degradation (P<0.05). The relationship between in situparameter 'b' and protein subfraction B2 was 'b' = 24.54 + 0.852(B2), R^2=0.8023).

Effect of semi-arid native fennel (*Foeniculum vulgare*) essential oil on *in vitro* gas production parameters of various ruminant fiber source feeds

Danesh Mesgaran, M.[1], Jani, E.[2], Vakili, A.R.[1] and Solaimani, A.[2], [1]Ferdowsi University of Mashhad, Dept. Animal Science, P.O. Box 91775-1163, 91775, Iran, [2]Azad Islamic University, Dept. Animal Science, Kashmar-branch, 77235, Iran; hooman1350@yahoo.com

The effect of semi-arid native fennel (Foeniculum vulgare) essential oil (FE) on the fermentation potential of cottonseed hulls (CH) and wheat straw (WS) was evaluated. Samples were ground through a 1-mm screen, and then dried (66 °C for 48 h). Both CH and WS were subjected to an *in vitro* gas production technique as un-supplemented or supplemented with FE as 40 or 80 µl/g DM; named CHFE40, WSFE40, CHFE80 and WSFE80, respectively. Approximately 0.2 g of each feed sample (n=4) was placed in a 100 ml glass syringe containing 40 ml of buffere:rumen fluid (2:1). Rumen fluid was obtained from two rumen canullated sheep (body weight= 45.5±2 kg) before the morning feeding and immediately strained through four layers of cheesecloth. Animals were fed 1.5 kg DM alfalfa hay and 0.4 kg DM concentrate (165 g CP/ kg DM) per head per day. Syringes were incubated at 39 °C and the volume of gas produced was recorded at 2, 4, 8, 12, 24, 36, 48, 72 and 96 h. The gas production data were fitted to an exponential equation of $P= b(1-e^{-ct})$, where b= volume of gas produced, c= fractional rate constant (/h), t= incubation time (h) and P= volume of gas produced at time t. The parameters of the supplemented samples were compared with the non-supplemented samples as controls using Dunnett,s test at P<0.05. Supplemented samples of CH and WS had a significant (P<0.05) less b fraction compared with the non-supplemented samples (CH=73, CHFE40=9, CHFE80=11, WS=86, WSFE40=7 and WSFE80=10 ml/0.2 g DM). The rate constant of gas produced (c) from CH and WS (0.02 and 0.03, respectively) was significantly (P<0.05) increased by the adding of FE at the both applied rates (0.05 and 0.05, 0.05 and 0.04, respectively). The present results indicate that FE has a potential to alter the fermentation characteristics of the both feed samples evaluated.

Feeding value of carob (*Ceratonia siliqua* L.) as an industrial processing by-products

Çürek, M.[1], Işık, M.[2] and Özen, N.[3], [1]Ministry of Agricultural, Project and Statistics Department, Sedir Mah. Vatan Bulvarı, Tarım Kampüsü, Antalya, 07040, Turkey, [2]West Mediterranean Agricultural Research Institute, Administrative and Financial Department, Demircikara Mah.Paşakavaklar Cad. no.7 P.K.: 35 -130, Antalya, 07100, Turkey, [3]Akdeniz University Agricultural Faculty, Animal Science, Akdeniz Üniversitesi, Ziraat Fakültesi Zootekni Bölümü,Kampüs, Antalya, 07059, Turkey; nozen@akdeniz.edu.tr

Nutrient composition, digestibility, feed and feeding value of carob (Ceratonia siliqua L.) pulp and pulp silage were evaluated through chemical analysis and fecal collection method. Laboratory analysis indicated that carob pulp and carob silage contained 34.43 and 37.63% dry matter (DM), 96.95 and 92.47% organic materials (OM), 7.20 and 5.5% crude protein (CP), 0.66 and 0.16% ether extracts (EE), 9.24 and 12.58% crude fiber (CF), 3.05 and 7.53% crude ash (CA) 76.80 and 74.23% nitrogen-free extracts (NFE) in dry matter basis, respectively. Digestion coefficients of DM, OM, CP, EE, CF, NFE of pulp and pulp silage calculated depending upon the data collected from digestion trials conducted on three Chios rams were 55.32, 54.65, 59.89, 63.69, 66.54, 52.27%; 53.43, 51.76, 63.04, 62.14, 65.56, 49.13%, respectively. The feed value of carob pulp and pulp silage in terms of total digestible nutrients (TDN) and starch unit (SU) were calculated as 51.32 and 47.96%, 474 and 417 g/kg (47.40 and 41.70%). Similarly, DE, ME and NE contents were found to be 2258 and 1856, 1117 and 2110, 1734 and 982 kcal/kg., respectively. In addition, pH of the carob pulp silage measured and ranked as 'good' depending on organic acid contents and as 'very good' according to both sensory evaluations and Flieg scoring method. Overall results of the research indicated that carob pulp and pulp silage could be efficient alternative roughage in ruminant nutrition.

The effect of nonstructural carbohydrate and addition of full fat roasted canola seed on milk production and composition in lactating cows

Sari, M.[1], Naserian, A.A.[2], Valizadeh, R.[2] and Salari, S.[1], [1]Ramin Agricultural and Natural Resources University, Animal Science, Mollasani, Ahwaz, 0098, Iran, [2]Ferdowsi University, Mashhad, 0098, Iran; somayehsallary@yahoo.com

The objective of this study was to examine the effects of modifying the dietary profile of neutral detergent-soluble carbohydrate (NDSC), addition of full fat roasted canola seed (RCS) as a source of fat, and possible interactions on milk production and composition in lactating cows. Twelve lactating Holstein cows (BW = 596±29 kg, DIM = 85±14) were used in a 4×4 Latin squares design involving dried citrus pulp as a source of neutral detergent-soluble fiber (NDSF), or barley as sources of starch (S). Treatments were in a 2×2 factorial arrangement, and periods were 21 d. The first 15 d were used for diet adaptation. Milk yield was recorded and samples were collected at each milking over the last three d of each treatment period. Data were analyzed using the GLM procedure of SAS for replicated latin square. Daily DMI of cows fed S diets was higher than cows fed NDSF diets (24 vs 22.9 kg/d). The DMI did not affected by addition of RCS. Cows receiving diet S produced more milk (33.8 vs 32 kg/d) and animals fed NDSF diet produced milk with greater concentration of milk fat (3.20 vs 2.95%). Consequently, NFC source did not affect 3.5% FCM yield (28.42 vs 28.02 kg/d for S and NDSF, respectively). An interaction ($P<0.05$) between NDSC profile and RCS was observed for ECM because of higher production with NDSF than S diet when RCS was added; (31.1, 31.2, 29.3, and 31.5 kg/d for treatments S, S+RCS, NDSF, and NDSF+RCS, respectively). Cows fed the S diet had an increased milk protein content than when fed the NDSF diet (3.01 vs. 2.89%, respectively), and the inclusion of RCS reduced the protein content (2.95 vs. 2.78%). Production efficiency was not affected by NDSC profile, but was higher for cows fed RCS diets. Adding RCS had positive effects on milk production and production efficiency using citrus pulp as NDSF source.

Intestinal digestibility of amino acids of lucerne in ruminants influenced by maturity stage

Homolka, P.[1], Koukolová, V.[1], Němec, Z.[2], Mudřík, Z.[2], Hučko, B.[2] and Sales, J.[1], [1]Institute of Animal Science, Department of Nutrition and Feeding of Farm Animals, Přátelství 815, 104 00 Prague Uhříněves, Czech Republic, [2]Czech University of Life Sciences, Faculty of Agrobiology, Food and Natural Resources, Department of Microbiology, Nutrition and Dietetics, Kamýcká 129, 165 21 Prague 6 - Suchdol, Czech Republic; homolka.petr@vuzv.cz

Lucerne (Medicago sativa L. var. Palava), harvested at four successive dates over a 30-day period, were evaluated (project No. MZE0002701404) for chemical composition, amino acid contents, and intestinal digestibility with dairy cows. Dry matter (r=0.78), organic matter (r=0.95), crude fibre (r=0.91), neutral detergent fibre (r=0.94), acid detergent fibre (r=0.79) and acid detergent lignin (r=0.48) presented positive linear correlation coefficients (r) with growth stage, whereas crude protein (r=-0.96), ether extract (r=-0.86) and nitrogen-free extract (r=-0.70) showed negative relationships. Total essential amino acid content decreased (r=-0.94) from 84.1 to 55.3 g/kg of dry matter with maturity, with r-values of higher than -0.90 obtained between growth stage and contents of lysine, methionine, threonine and valine. With the exception of tyrosine (r=-0.68), r-values between growth stage and individual non-essential amino acids were all higher than -0.90. Total amino acid (r=-0.98) and nitrogen (r=-0.99) contents presented comparable tendencies with successive sampling times. Whereas no definite trends were detected for the amino acid composition of rumen incubated (16 hours) lucerne samples, intestinal digestibility of total essential (r=-0.78), total non-essential (r=-0.58), and total (r=-0.69) amino acids, as well as nitrogen (r=-0.99), decreased with growth. It can be concluded that, although limited in sample size, this report presents information on the decrease in amino acid contents and intestinal amino acid digestibility as growth proceeds in lucerne (var. Palava) produced in the Czech Republic, which could be utilized in the feeding of ruminants.

Characterization of organic soybean by-product (okara) as a protein supplement in organic fattening of calves

Villalba, D., Seradj, A.R. and Cubiló, D., University of Lleida, Animal Production, Avda. Rovira Roure 191, 25198 Lleida, Spain; dvillalba@prodan.udl.es

The EU organic regulation implies the reduction of concentrate use in fattening calves diets. From an economic point of view the increase in prices of soybean force the farmers to find alternative sources of protein supply. The aim of this study is to characterize the achieved by product (okara) after extracting the milk from organic soybean to use it in diets of fattening calves. Prevalent chemical composition such as crude protein (CP), crude fibre (CF), neutral and acid detergent fibre (NDF, ADF) were measured, and *in vitro* gas production technique was applied in order to provide an information on the fermentation kinetics of this organic by product (okara). A total mixed ration (TMR) including different levels of okara, forage and concentrate was offered to 4 batches of calves in growing (17% of okara in TMR) and finishing (10% okara in TMR and *ad libitum* access to the concentrate) calves. Intake was evaluated in a batch basis and average daily gain was calculated using monthly weights. On the dry matter basis, okara samples had a 28% CP, 23% CF, 36% NDF and 23% ADF. Okara showed an intermediate gas production values comparing with concentrate and forage gas production (the differences were statistically significant). Animals showed high appetence for the TMR including okara offered *ad libitum* (22 g TMR/kg LW in the growing phase) and no negative effect on intake was detected even when okara last more than 3 weeks exposure to the air before mixing it in the TMR. Growth of animals which received okara in their diet was a little bit lower than expected values according to the INRATION® predictions at the growing phase (0.9 kd/d from 250 kg to 350 kg) but similar to the expected values at the finishing phase (1.2 kg/d from 350-450 kg). Considering okara as a problematic (waste) by-product in the soybean milk factories, and its low price, makes this by product as an interesting alternative as a protein supplement in calve diets.

Comparison amounts of gas production in fresh and oven dried silage in orange pulp silage using *in vitro* gas production

Lashkari, S. and Taghizadeh, A., University of Tabriz, Animal Science, Dep. of Animal Science, University of Tabriz, Tabriz, 51664, Iran; ataghius@yahoo.com

In this study the *in vitro* gas production was used to study the difference of gas production between fresh and dried silages. The composition of silages were: 1) 73% orange pulp + 27% straw (control), 2) 74% orange pulp + 12% straw + 14% poultry by-product meal (OSP) and 3) 63% orange pulp + 25% straw + 3% urea solution (OSU). A semi-selective lactobacilli medium (MRS) was used for the isolation and enumeration of lactic acid bacteria (LAB). Blends silage were ensiled for 90 days in triplicate. Dried samples (300 mg) and fresh samples (1000 mg) placed into 100 ml serum vial and was incubated in 6 replicates with 20 ml of rumen liquor and buffer solution (1:2). Amounts of cumulative gas production were recorded at 2, 4, 6, 8, 10, 12, 16, 24, 36, 48, 72 and 96 h of incubation. The means of gas production (ml/g of DM) and counts of LAB (log cfu/g silage) were compared using Duncan's multiple range comparison tests. The pH Values in control, OSP and OSU were 4.14, 4.29 and 8.43, respectively. The means of LAB counts (log cfu/g silage) in control, OSP and OSU were 3.45, 3.48 and 3.38, respectively. The mean of LAB population were higher (P<0.05) in control and OSP than in OSU. The amounts of gas production in control silage for all time points were significantly (P<0.05) higher in fresh than dried silage with exception of 36 and 96 h. The amounts of gas production at 2, 4 and 6 h were significantly (P<0.05) higher in fresh with compared to dried OSP, whereas in 96 h was lower (P<0.05) in fresh OSP. Values of gas production in all time points for OSU were significantly (P<0.05) lower in fresh than dried OSU exception of 36 and 48 h. It will be evident that in the silages with low pH, LAB is predominant. These results showed that LAB in fresh silages has the potential to influence on rumen fermentation and undried samples ferment differently in *in vitro* than dried samples.

Effects of bovine bile powder and enzyme treatment on performance of broilers fed high lipid diets

Karimi, K.[1], Esmaeilpour, V.[2] and Rezaeipour, V.[2], [1]Animal Science Department, College of Agriculture, Islamic Azad University, Varamin-Pishva Branch, Varamin- Tehran, Iran, [2]Animal Science Department, College of Agriculture, Islamic Azad University, Ghaemshahr-Branch, Ghaemshahr, Iran; Dr.karimi_kazem@iauvaramin.ac.ir

A 2×2 factorial design was used to investigate the effects of supplementation of dietary bovine bile powder (BBP) and COMBO enzyme (CE) on performance, protein and energy intake efficiency in broilers. Bovine bile obtained from a cow slaughter house and dried under standard conditions then powdered to approach the BBP. COMBO includes lipases, proteases and carbohydrase and pectinase enzymes. Three hundred and sixty day-old broiler chickens were allocated into 4 groups(C, B, E and BE) with 9 replicate pens of 10 chicks each one. Group C was fed a complete feed mixture without supplements. Group B was fed the same diet with 3.5 g kg^{-1} BBP. Group E was loaded with 5 g kg^{-1} CE, and group BE was loaded with 3.5 g kg^{-1} BBP+5 g kg^{-1} CE. Feed intake(FI), daily feed intake(DFI),weight gain(WG), daily weight gain(DWG), feed conversion ratio(FCR), protein intake efficiency (PIE)and energy intake efficiency(EIE) in broilers were determined periodically in starter(0-3w), finisher (3-6w)and total(0-6w) experimental periods. Results indicated that enzyme supplementation increased FI and DFI in total experimental period by1.8% on average and DFI in finisher period by 2.5% on average (P<0.05). Result showed that BBP had no effects on performance traits (P>0.05). FI and DFI in finisher period significantly increased by BE dietary treatments compared to E groups by 4.2% and 5% on average, respectively (P<0.05).There were tendencies for WG (P=0.065), DWG (P=0.073) and PIE (P=0.060) in starter period to increased by BE dietary treatments. FCR and EIE haven't affected by any dietary treatments (P>0.05). Thus we concluded that diet supplementation only with BBP has no effects on performance and for better results this must be associated with a multi enzyme in broiler diets.

Effects of a probiotic and it's carrier supplementation on performance and nutrient digestibility of broiler chicks

Han, Y.K., Sungkyunkwan University, Food Science & Biotechnology, 300 Chunchun-dong, Jangan-gu, 440-746, Suwon, Korea, South; swisshan@paran.com

This experiment was conducted to determine if the performance of broilers fed diets based on corn, wheat and soybean meal could be enhanced with probiotic and two types of carriers. Two day old, female broilers were assigned to one of four treatments in a completely randomized design and housed in groups of four with six cages per treatment. The three experimental diets consisted of the basal diet supplemented with 0.1% of probiotic + carrier-I(yellow earth), 0.05% probiotic + carrier-I or probiotic 0.05% + carrier-II(malt, yellow earth, zeolite and enzyme mixture). The probiotic was composed of Rhodopseudomonas capsulata, Bacillus subtilus, Bacillus coagulance and Clostridium butyricum. Over the 42 day experiment, feed conversion of birds fed the probiotic treatment (0.1% + carrier-I and 0.05% carrier-I) was superior (P=0.02) to the control. Feed intake and daily gain did not differ among treatments (P>0.05). Compared with the control, supplementation with 0.1% probiotic + carrier-I significantly increased the digestibility of energy and methionine (P<0.05). In contrast, the digestibility of argine and methionine was lower (P<0.01) for birds supplemented with 0.05% probiotic + carrier-II than the control. In summary, the performance of broilers was significantly enhanced by the addition of a probiotic to the diet. However, under the conditions of this experiment, supplementation with a carrier containing yellow earth, zeolite and enzyme mixture failed to improve broiler performance.

Study of characterization of eggplant (*Solanum melongea*) silage with and without molasses
Karimi, N., Shirazi, H. and Forudi, F., Islamic Azad University (I.A.U)- Varamin Pishva branch, Animal Science Department, Faculty of Agriculture, Varamin-Tehran, 33817-74895, Iran; Dr.karimi@iauvaramin.ac.ir

Large quantities of eggplant (*Solanum melongea*) is produced in Iran that are usually unutilized. This study was conducted to ensiling of eggplant with and without molasses. Samples of eggplant were chopped and ensiled experimentally in plastic buckets. Samples were ensiled in three treatments and 4 replicates (1) control (samples without any additives) (2) samples with 4% molasses based on dry matter (3) samples with 16% molasses based on dry matter. This experiment carried out in a Completely Randomized Design (CRD). Raw materials (eggplant before ensiling) were analyzed for their Dry Matter (DM), Crude Protein (CP), Neutral Detergent Fiber (NDF), Acid Detergent Fiber (ADF), Dry Matter Digestibility (DMD), Organic Matter Digestibility (OMD) and Organic Matter in Dry Matter Digestibility (DOMD) and Results were 21, 10.09, 60.2, 47, 44.88, 43.46 and 39.60 percentage respectively. Silage samples were analyzed like raw materials (and for pH) and total data were analyzed using the SAS program and followed by the Duncan's test. Results of silage samples analysis (samples without any additives – samples with 4% molasses – samples with 16% molasses) were compared together respectively for pH (4.99^a, 4.83^a and 3.82^a), DM (18.30^b, 19.13^b and 20.33^a%), NDF(57.55^a, 58.95^a and 43.87^b%), ADF(44.70^a, 45.75^a and 33.20^b%), DMD (52.13^b, 48.26^b and 60.30^a%), OMD (47.49^b, 43.35^b, 57.47^a%) and DOMD (41.11^b, 37.73^b, 50.67^a%). Results indicate that changes of Dry Matter, NDF, ADF, DMD, OMD and DOMD in samples during ensiling are significant (P<0.05) and ensiling of eggplant with 16% molasses improved product properties compared to others.

QTL detection and fine mapping of birthcoat type in sheep
Allain, D.[1], Cano, M.[2,3], Foulquié, D.[4], Autran, P.[4], Tosser-Klopp, G.[3], Moreno, C.[1], Mulsant, P.[3] and François, D.[1], [1]INRA, UR631, SAGA, BP 52627, 31326 Castanet Tolosan, France, [2]INTA, Instituto de Genetica, CICVyA, CC25, 1712 Castelar, Buenos Aires, Argentina, [3]INRA, UMR444, LGC, BP 52627, 31326 Castanet Tolosan, France, [4]INRA, UE321 Domaine de La Fage, Saint Jean et Saint Paul, 12250 Roquefort, France; daniel.allain@toulouse.inra.fr

Birthcoat type is an important component of lamb survival for sheep raised under harsh environment. At birth two types of coat were observed: a long hairy coat or a short woolly one. It was shown that hairy coat lambs are more adapted to survive around lambing time due to a better coat protection with less heat losses at coat surface and show better growth performances up to the age of 10 days than woolly coat lambs. Birthcoat type was shown to be a high heritable trait. An experimental design was initiated in 2004 comprising 8 sires families (200 halfsibs/sire from a total of 547 dams) for QTL detection of birthcoat type and other adaptative traits in a sheep Romane population. A total of 1629 lambs were born and observed at birth for coat type and coat depth. All dams had been phenotyped at birth for coat type. Within this experimental population, 850 animals (842 lambs; 8 sires) were genotyped using the Illumina OvineSNP50 bead Chip and analysed for QTL detection. We report that birthcoat type could be under a simple genetic determinism by analysing phenotype data within sire family and according to the lamb coat type of the dam. Proportion of hairy coat lambs was about 2/3 (67.8%) in 5 sire families (SG1), ½ (48.7%) in 2 sire families (SG2) and 1/3 (35.8%) in one sire family (SG3). When the dam bore an hairy coat at birth, proportion of hairy coat lambs was 77.6%, 64.1%, 48.0% in SG1, SG2 and SG3 respectively. When the dam bore a woolly coat at birth, proportion of hairy coat lambs was 44.9%, 13.2% and 7.6% in SG1, SG2 and SG3 respectively. A highly significant QTL affecting birthcoat type was found on chromosome 25 suggesting that a gene with major effect is located on this chromosome. Fine mapping is in progress.

Estimation and comparison of phenotypic value changes in silkworm pure lines during consecutive generations under selection pressure
Seidavi, A.R., Islamic Azad University, Rasht Branch, Animal Science Department, Rasht, 4185743999, Iran; alirezaseidavi@yahoo.com

This experiment was conducted in order to investigation on the phenotypic value changes of silkworm economic traits including cocoon weight, cocoon shell weight and cocoon shell percentage under individual selection based on cocoon weight. All stages of rearing and data record were performed over four rearing periods. Each pure line was contained two groups as selected and random groups. It were compared the effect of selection methods, the effect of pure line, and generation effect on the phenotypic values. Main effects of pure line, generation, sex and group on cocoon weight, cocoon shell weight and cocoon shell percentage were significant (P<0.01). Interactions of the pure line × generation, pure line × group (except for cocoon shell percentage), generation × group, pure line × generation × group, pure line × sex, generation × sex, pure line × generation × sex, group × sex (except for cocoon shell percentage), pure line × group × sex (except for cocoon weight and cocoon shell weight), generation × group × sex (except for cocoon shell percentage) and the pure line ×generation × group × sex have been significant on the cocoon traits characteristics (P<0.01). Obtained results were indicated cocoon weight and cocoon shell weight in selected group are higher than control or non-selected group. Both selected and non-selected groups had the the lowest cocoon weight, cocoon shell weight and cocoon shell percentage in the fourth generation due to unfavorable environmental conditions.

Historical relationships between sheep breeds according to their macroscopical fleece traits
Parés I Casanova, P.M.[1] and Perezgrovas Garza, R.[2], [1]ETSEA, Animal Production, Alcalde Rovira Roure, 191, 25198 Lleida, Spain, [2]Universidad Autónoma de Chiapas, Instituto de Estudios Indígenas, Centro Universitario Campus III. San Cristóbal de Las Casas, 29264 Chiapas, Mexico; peremiquelp@prodan.udl.cat

Wool data were collected from 666 animals belonging to 15 American breeds: Blanca Colombiana, Chiapas Blanca, Chiapas Café, Chiapas Negra, Criolla Boliviana, Crioula Brasileña, Latxa Chilena, Linca, Mora Colombiana, Navajo Churro, Oaxaca, Oaxaca Mixteca, Socorro, Tarahumara and Zongolica. Samples were taken randomly from the mid-lateral part of each ewe before shearing. Seven wool macroscopic characteristics were investigated for each sample: fibre length and percentage of each type of fibre (long-coarse fibres short-fine fibres and kemp), and yield after alcohol scouring. Characteristics were determined using standard methods at the Wool Quality Laboratory at the Institute for Indigenous Studies, Univ. Autónoma de Chiapas, in San Cristóbal de Las Casas, Chiapas (Mexico). With the aim of obtaining an aggregate of breeds and relationship among population accessions, a principal coordinate analysis (PCoA) was generated from the correlation matrix. Different PAST – 'Paleontological Statistics Software Package for Education and Data Analysis' programs were utilized to perform the analysis. In the PCoA it should be noted that Chiapas Blanca differed clearly from the rest of breeds. The only logical explanation is that its selection has been based on the staple length, unlike other breeds. With its more abundant long fibres, the Chiapas Blanca has been selected as a 'long fine wool' sheep with a clear double coat whereas the sympatric Chiapas Negra, for example, has been selected for fibre colour (black). Moreover, Chiapas Blanca seems to have been originated from the Churra, whereas the Chiapas Negra from the Manchega and Castellana. Other studied breeds have been selected for a meat purpose. The data reported here provide valuable assess for inter-breed comparison and moreover can help in the study of sheep historical relationships.

Genetic parameters for fiber diameter at different shearings

Pun, A.[1], Morante, R.[2], Burgos, A.[2], Cervantes, I.[1], Pérez-Cabal, M.A.[1] and Gutiérrez, J.P.[1], [1]Faculty of Veterinary, Complutense University of Madrid, Department of Animal Production, Avda. Puerta de Hierro s/n, 28040 Madrid, Spain, [2]Pacomarca S.A., Av. Parra 324, Arequipa, Peru; apun@vet.ucm.es

Alpaca is the most important fiber producer of South American camelid species, being an important income for the Andean communities. Nowadays, the fiber diameter is considered as the main selection objective in alpaca populations all over the world. However, fiber diameter increases with age of animals, and its value at consecutive shearings might be affected by different genetic and environmental conditions. The goal of this study was to estimate the genetic parameters for fiber diameter depending on the age of the animal. The data set consisted of a pedigree of 4,173 individuals of Huacaya breed, and 8,405 records of fiber diameter corresponding to 3,257 individuals. Shearing for four age groups (one year old animals, two years old, between three and four years old, and five or more years old) were defined. The model for each group includes age in days as a covariable, sex, fiber color, and month-year as fixed effects, as well as a permanent environmental effect (for the second group onwards), and additive genetic effect as random effects. REML estimates were obtained using VCE6 program under a multitrait procedure. Heritabilities ranged from 0.54 to 0.70, and were higher than that estimated when all shearings were considered as a unique trait. The highest heritability was estimated for two years old animals, indicating that shearing at two years of age is the best showing the genetic value of the animal. Environmental permanent variance was only relevant for the last group with a ratio of 0.11. Genetic correlations ranged between 0.76 and 0.98 but those involving the first shearing were lower and decreased as the time elapsed between shearings increased. First shearing seemed to be the worst to represent the fiber diameter of the animal along its life.

A deletion in exon 9 of the LIPH gene is responsible for the rex hair coat phenotype in rabbits

Diribarne, M.[1], Mata, X.[1], Chantry-Darmon, C.[1], Vaiman, A.[1], Auvinet, G.[2], Bouet, S.[1], Deretz, S.[2], Cribiu, E.P.[1], Rochambeau, H.[3], Allain, D.[3] and Guérin, G.[1], [1]INRA, UMR1313, GABI, Domaine de Vilvert, 78350 Jouy en Josas, France, [2]INRA, UE967, GEPA, Le Magneraud, 17700 Surgères, France, [3]INRA, UR631, SAGA, BP 52627, 31326 Castanet Tolosan, France; daniel.allain@toulouse.inra.fr

Hair follicles are complex structures which naturally play an essential role in the protection of mammals against climatic variations. In addition, it is of utmost importance in certain domestic species bred for the quality their coat such as in rabbit. Among these is the rex coat which confers to rabbits a soft plush-like fur of high economic value. The coat of common rabbits is made of 3 types of hair differing in length and diameter while that of rex animals is essentially made up of amazingly soft down-hair. Rex short hair coat phenotype in rabbits was shown to be controlled by a mutation segregating at an autosomal-recessive locus. A positional candidate gene approach was used to identify and primo-localized the rex gene within a 40 cM region on rabbit chromosome 14 by linkage analysis in large experimental families. Then, fine mapping refined the first defined interval region by genotyping 359 offspring for 94 microsatellites. Comparative mapping pointed the LIPH (Lipase Member H) as a candidate gene. The rabbit gene structure was established and a deletion of a single nucleotide was found in LIPH exon 9 of rex rabbits (1362delA). This mutation results in a frameshift and introduces a premature stop codon potentially shortening the protein by 19 amino acids. The association between this deletion and the rex phenotype was complete as determined by its presence among a panel of 60 rex and its absence in all 60 non-rex rabbits. This strongly suggests that this deletion, in a homozygous state, is responsible for the rex phenotype in rabbits. This result represents for breeders a good marker for introgressing this character in other rabbit strains for quality pelt improvement.

Opportunities to improve value and utilization of sustainable animal fibre production in Europe
Renieri, C.[1], Gerken, M.[2], Allain, D.[3], Antonini, M.[1], Gutierrez, J.P.[4] and Galbraith, H.[5], [1]University of Camerino, School of Environmental Sciences, Via Gentile III da Varano s/n, 62032 Camerino, Italy, [2]Gerog-August-Universitat, Nutztierwissenschaften, Albrecht Thaer Weg 3, 37075 Gottingen, Germany, [3]INRA, SAGA, BP 52627, 31320 Castanet Tolosan, France, [4]Universidad Complutense de Madrid, Produccion Animal, Avda Puerta de Hierro s/n, 28040 Madrid, Spain, [5]University of Aberdeen, School of Biological Sciences, 23 St Machar Drive, AB24 3RY Aberdeen, United Kingdom; carlo.renieri@unicam.it

The natural 'wool' product of farmed fibre-bearing animals is underutilised in Europe. It is a natural product of domesticated ruminant/pseudo-ruminants such as sheep (wool), and small numbers of goats (mohair, cashmere) and South American camelids (SAC, alpaca). The scale of production is substantial. For the major animal species, sheep and 62 m breeding ewes (Eurostat) alone, we can estimate up to 150,000 tonnes of raw wool. Production, marketing and record-keeping vary among countries. Wool prices are low and frequently unprofitable. Production of higher value goat and SAC fibres is small by comparison. Chemical outputs of production include fibre (keratin-based protein) milk, meat and greenhouse gases (GHG: carbon dioxide, methane and nitrogen oxides). The specific contribution of fibre to GHG synthesis requires evaluation. Better sustainability may be attained by more effective production, typically achieved by improvements in husbandry, knowledge of hair follicle biology and regulation, application of new genetic tools for selection and breeding and development of more desirable end-products. The role of newly developing commercial breeds and cross breeds, and husbandry such as 'easy care' systems requires attention. The production chain may also be enhanced by development of new techniques for evaluation of wool quality on-farm, improved collection and marketing systems and better integration of processing and end-user and marketing interests.

Quantitative variation of melanins in alpaca
Cecchi, T., Valbonesi, A., Passamonti, P. and Renieri, C., University of Camerino, Environmental Sciences, Via Gentile III da Varano, 62032, Italy; carlo.renieri@unicam.it

The amount of melanin pigments was investigated in 95 Peruvian alpaca, representative of six different fleece colours, by means of spectrophotometric assays: SpEM (Spectrophotometric Eumelanin), SpPM (Spectrophotometric Pheomelanin), SpASM (Spectrophotometric Alkali Soluble Melanin), and SpTM (Spectrophotometric Total Melanin). It was found that these melanin pigments were suitable for identifying three homogeneous groups, each consisting of two closely related colours. A low, an intermediate, and a high amount of SpASM, SpTM, and SpPM characterise pinkish grey and light reddish brown, brown and reddish brown, dark reddish brown and black fleeces, respectively. SpEM and SpTM provide a further split within this latter group; high concentration of these pigments distinguish black fleece from dark reddish brown. From this preliminary survey it results that the usual fleece colour classification, based on about 22 different colours, should be reviewed, and a new one may be proposed on the base of the objective parameter, amount of melanin pigments.

Pelt follicle characteristics and their relationship with pelt quality in fat-tailed gray Shirazy sheep
Safdarian, M. and Hashemi, M., Fars Research Centre for Agriculture, Shiraz, P.O.Box 716, Shiraz 71555, Iran; safdarian@farsagres.ir

The purpose of the present investigation was to measure the pelt follicle characters after birth and determine relationship among pelt characters, primary and secondary follicle density and secondary/primary (S/P) ratio in fat-tailed gray shirazy (Karakul) sheep. Skin samples were taken from 150 lambs at 1 and 120 days old and their parents by a biopsy punch (8 mm diameter) from right mid-side after the wool had been sheared off, then 8 μm sections prepared, mounted and stained on glass slides. Data were analyzed by statistical procedures of the Statistical Analysis Systems (SAS). The least square mean of primary follicle density/mm2 skin in male lambs were 14.6 and 7.1 and in female lambs were 15.4 and 7 in one and 120 days old respectively. The least square mean of secondary follicle density/mm2 skin in male and female lambs were 33.3 and 34.7 in one day old and 21.7 and 20 in 120 days old respectively. S/P ratio in male lambs was 2.9 and 3.6 and in female lambs were 2.7 and 3.4 in one and 120 days old respectively. There was no significant difference between sex of lambs for follicle density/mm2 skin and S/P ratio in one and 120 days old (P>0.05). The correlation coefficient between one day old follicle characters and final score of pelt were not significant (P>0.05), but for 120 days old were moderate and significant (P<0.05). Correlation coefficient between pelt traits in newborn lambs were significant (P<0.01). The results of the present study indicated that although the density of follicle and S/P ratio was moderate in fat-tailed pelt lamb but there was no correlation between it and final score of pelt. Further studies are suggested to find whether other follicle characters like follicle curvature, follicle depth and follicle group size would have a high correlation with final score of pelt for assessed objectively skin characters.

Results from genomic selection in Pietrain pig breeding
Bennewitz, J.[1], Wellmann, R.[1], Neugebauer, N.[1], Tholen, E.[2] and Wimmers, K.[3], [1]Institute of Animal Husbandry and Breeding, University Hohenheim, Garbenstrasse 17, D-70599 Stuttgart, Germany, [2]Institute of Animal Science, University of Bonn, Endenicher Allee 15, D-53115 Bonn, Germany, [3]Leibniz Institute for Farm Animal Biology, Wilhelm-Stahl-Allee 2, D-18196 Dummerstorf, Germany; j.bennewitz@uni-hohenheim.de

Compared to dairy cattle breeding, genomic selection is much more difficult to implement in pig breeding schemes. Problems arise from usually much smaller reference populations, from lower and heterogeneous accuracy of estimated breeding values of animals in the reference population and from the ratio of genotyping costs and economic value of animal. This study presents the first results from genomic breeding value estimation using a porcine 60K chip genotyped for currently 500 Pietrain boars. After various genotype quality checks around 48,000 SNPs remained in the data set. Growth traits are considered so far but additional traits will be included. Marker effects were estimated using a random regression BLUP approach assuming equal marker variance across the genome. Phenotypes were conventional BLUP EBVs based on progeny testing. These BLUP EBVs were de-regressed. The heterogeneous accuracy was considered in the mixed model equations for estimating the marker effects by using different residual weights. Because of the currently low sample size, model validation was done using a repeated cross validation approach ignoring the age structure of the animals. First results indicate an accuracy of genomic selection in between 0.45 and 0.5 for the growth traits. The reference population will soon be increased by additional 300 progeny tested boars. Part of the ongoing research is testing alternative models for genomic breeding value estimation, testing alternative SNP sets with reduced densities and calculation of population genetic parameters.

Genomic selection in maternal pig breeds

Lillehammer, M.[1], Meuwissen, T.H.E.[2] and Sonesson, A.K.[1], [1]Nofima Marin, P.O. Box 5010, 1432 Ås, Norway, [2]Norwegian University of Life Sciences, Department of Animal and Aquacultural Sciences, P.O. Box 5010, 1432 Ås, Norway; marie.lillehammer@nofima.no

Within maternal pig breeds, maternal traits like litter size and litter weight are economically important, but hard to improve because of low heritability and lack of information in male selection candidates. Aim of this study was to compare different implementations of genomic selection (GS) to a conventional breeding scheme (CONV) with respect to their ability to improve maternal traits. Schemes were compared on their accuracy of selection (r), genetic gain (ΔG) and rate of inbreeding (ΔF) over 10 years. The comparison was performed by stochastic simulation of a pig nucleus population of 2,775 sows. Twice a year, selection of 1,200 replacement sows (based on conventional breeding values in GS and CONV) and 25 boars (based on conventional breeding values in CONV and genomic breeding values in GS) was performed based on a trait with heritability 0.1, measured on females after first litter. The total number of genotyped animals was kept constant at 1,800 per selection, but the ratio of genotyped females (dams with records) and genotyped males (selection candidates) was varied. A GS-scheme where 1,200 females and 600 males were genotyped more than doubled r, compared to CONV, which caused ΔG to be increased by 69%, and reduced ΔF by 64%. If the number of genotyped dams was reduced in order to genotype more males, thereby increasing the number of male selection candidates, ΔG was reduced and ΔF increased due to a lower accuracy of selection and an increased co-selection of full-sib males. However, all GS-schemes outperformed CONV. Even the schemes where only males were genotyped increased ΔG by 35% and reduced ΔF by 34%, relative to CONV. In conclusion, genomic selection can greatly increase ΔG for maternal traits in boar selection. To genotype dams with records is essential to obtain high r and utilize the full potential of genomic selection and should have a high priority.

Genomic selection using Bayesian mixture models for reduction of aggressive behaviour in pigs

Kapell, D.N.R.G.[1], Ashworth, C.J.[2], Turner, S.P.[1], D'eath, R.B.[1], Lundeheim, N.[3], Rydhmer, L.[3], Lawrence, A.B.[1] and Roehe, R.[1], [1]Scottish Agricultural College, Main Road, EH9 3JG Edinburgh, United Kingdom, [2]The Roslin Institute and R(D)SVS, Midlothian, EH25 9PS Edinburgh, United Kingdom, [3]Swedish University of Agricultural Sciences, Box 7023, S-750 07 Uppsala, Sweden; rainer.roehe@sac.ac.uk

The objective of this study was to compare models for genomic selection to reduce aggressive behaviour in pigs at regrouping. Duration spent in reciprocal aggression (RA), delivery of non-reciprocal aggression (DNRA) or its receipt (RNRA) were available on 1,657 pigs of which 552 purebred Yorkshire pigs were genotyped using the porcine SNP60 panel. The average linkage disequilibrium between adjacent SNPs was 0.36. Bayesian approach was used fitting three models, which, besides fixed effects, included polygenetic effects only [1], genomic effects only [2] and both of those effects [3]. For models [2] and [3], five sub-models were fitted within 100% to 1% (mixture model) of SNPs considered to have an effect on the traits. Predictive ability of the genotype was determined by cross validation within and between families. The genomic model [2] explained, depending on the trait, 61 to 67% of the variance obtained by using the polygenic model [1]. The ratios of genomic over phenotypic variance using model [2] were 0.23±0.03, 0.21±0.02 and 0.08±0.01 for RA, DNRA and RNRA respectively. The decrease in the number of SNPs considered the have an effect, using different mixture sub-models, had hardly any influence on the predictive ability of the genotype. Inclusion of both genomic and polygenetic effects [3] captured substantially more genetic variance than model [2] but did not improve the predictive ability of the genotype. Therefore, genomic selection appears to be very robust in its predictive ability even when ignoring polygenic variance. In addition, results using Bayesian mixtures models indicated that low density SNP panels could be used to improve these behavioural traits without reducing their predictive ability.

Most of the benefits from genomic selection can be realised by genotyping a proportion of selection candidates

Henryon, M.[1], Berg, P.[2], Sørensen, A.C.[2], Ostersen, T.[1] and Nielsen, B.[1], [1]Pig Research Centre, Axeltorv 3, 1609 Copenhagen V, Denmark, [2]Aarhus University, Department of Genetics and Biotechnology, P.O. Box 50, 8830 Tjele, Denmark; maa@lf.dk

We reasoned that there is a diminishing marginal return from genomic selection when the number of genotyped selection candidates is increased and BLUP breeding values are used to choose the candidates that are genotyped. We tested this premise by stochastic simulation. We estimated the genetic gain and inbreeding realised by genomic selection when 100, 40, 20, 10, 5, and 0% of the selection candidates with highest BLUP breeding values are genotyped. Phenotypic-marker information was generated for the genotyped candidates and assumed to predict true breeding values with an accuracy of 0.71. Ten sires and 300 dams were selected each generation, each sire was randomly mated to 30 dams, and each mating produced 5 offspring. Selection was for a single trait with heritability 0.20. We found diminishing marginal returns for genomic selection as the number of genotyped selection candidates was increased. Genotyping 40% of the candidates realised 90% of the genetic gains and 92% of the reduction in inbreeding that was generated by genotyping 100% of the selection candidates. Genotyping 20, 10, and 5% of the selection candidates realised 79, 73, and 63% of the genetic gains and 84, 75, and 62% of the reduction in inbreeding. Our findings demonstrate that only a proportion of selection candidates need to be genotyped to obtain most of the benefits from genomic selection when a priori information is available. Such genotyping strategies are readily applicable to breeding schemes, such as those used in pig breeding, where selection candidates are phenotyped before major selection decisions are carried out.

Potential of genomic selection for traits with a limited number of phenotypes

Van Grevenhof, E.M., Van Arendonk, J.A.M. and Bijma, P., Wageningen University, Animal Breeding and Genomics Centre, P.O. 338, 6700 AH Wageningen, Netherlands; ilse.vangrevenhof@wur.nl

In the last 10 years, genomic selection has developed enormously. Simulations and results of real data suggest that breeding values can be predicted with high accuracy using genetic markers alone. To reach high accuracies, large reference populations are needed. In many livestock populations, these cannot be realized when traits are difficult or expensive to record, or when population size is small. The value of genomic selection (GS) becomes questionable then. In this study, we compare traditional breeding schemes based on own performance or progeny information to genomic selection schemes, when the number of phenotypic records is limiting. For this goal, deterministic simulations were performed using selection index theory. Results showed that genomic selection schemes suffer more from the Bulmer effect than traditional breeding schemes, so that comparison of traditional versus genomic breeding schemes should focus on Bulmer-equilibrium response to selection. To maximize the accuracy of genomic EBVs when the number of phenotypic records is limiting, the phenotyped individuals, rather than progeny tested individuals, should be genotyped. When the generation interval cannot be decreased when implementing GS, large reference populations are required to obtain equal response as with own performance selection or progeny testing. The accuracy of genomic EBVs, however, increases non-linearly with the size of the reference population, showing a diminishing-return relationship. As a consequence, when a GS scheme has a small decrease in generation interval, relatively small reference population sizes are needed to obtain equal response as with own performance selection or progeny testing. When the trait of interest cannot be recorded on the selection candidate, GS schemes are very attractive even when the number of phenotypic records is limited, because traditional breeding would have to rely on progeny testing schemes with long generation intervals.

Accuracy of imputation in a sparsely-genotyped pig pedigree

Cleveland, M.A.[1], Kinghorn, B.P.[2] and Hickey, J.M.[2], [1]Genus plc, 100 Bluegrass Commons Blvd. Suite 2200, Hendersonville, TN 37075, USA, [2]School of Environmental and Rural Science, University of New England, Armidale, NSW 2351, Australia; matthew.cleveland@pic.com

Imputation of high-density genotypes for individuals in a population that have not been genotyped, or have been genotyped at lower densities, is desired for cost-effective implementation of genomic selection. An imputation approach that takes advantage of known pedigree relationships has been developed and applied to a pig pedigree with varying levels of relatedness among high-density genotyped animals. The pedigree consisted of 6,473 records, of which 3,559 individuals were genotyped at 60k. SNPs with known positions on SSC01 were selected for testing (M=3,129). High-density genotypes for pigs at the bottom of the pedigree (N=300), with no progeny, were masked to simulate low-density genotyping at 1, 5, 10, 15, 25 and 50% of the total SNPs. Masked genotypes for the test animals were imputed using the AlphaImpute software and compared to their known genotypes. When both parents of the test individuals were genotyped at high density, the percent of correctly imputed genotypes (and percent error) was 93.6 (5.3), 97.9 (1.0), 98.6 (0.4), 98.8 (0.3), 98.8 (0.2) and 99.0 (0.09) for the 1, 5, 10, 15, 25 and 50% scenarios respectively. When both parents were not genotyped the imputation accuracy decreased (and error increased), but the addition of genotyped ancestors yielded accuracy nearly equivalent to having both parents genotyped. These results were compared to those obtained from Impute2 for the same data. AlphaImpute yielded a higher percentage of correct genotypes, and lower error, at all densities when both parents were genotyped. The results indicate that high-density genotypes can be imputed from low-density panels with high accuracy when multiple generations are genotyped, but also when target individuals are less related to the genotyped population. This approach makes it possible to genotype selection candidates at very low densities, with little expected information loss.

Effect of genotype imputation errors on the accuracy of genomic selection

Neugebauer, N., Wellmann, R. and Bennewitz, J., University of Hohenheim, Institute of Animal Husbandry and Breeding, Garbestr. 17, 70599 Stuttgart, Germany; N.Neugebauer@uni-hohenheim.de

In order to reduce costs of genomic selection a combination of low density SNP chips (LD-chip) and chips with a higher density (e.g. 50k chips, HD-chip) is frequently proposed, especially for species where the economic value of a single breeding individual is limited (e.g. chicken and pigs). This study investigated a combined use of LD- and HD-chips for a typical sire line pig breeding population by stochastic simulations. A mutation-drift population was simulated. Male individuals from five successive generations are HD-genotyped (50K) and progeny-tested. Marker effects were estimated by G-BLUP and BayesA. Young selection candidates are offspring from this reference population and are LD-genotyped with 0.3k, 0.5k,... or 10k. The markers of the LD-chip were equal spaced along the genome. Missing genotypes of the selection candidates were imputed using a published method that combined family information and linkage disequilibrium (i.e. LDMIP). Imputation errors were recorded. The accuracy of gEBV of the selection candidates (i.e. the correlation between gEBV and TBV) was analysed and compared across different imputation error rates. In particular it was investigated at which error rate it is beneficial to use only markers of the LD-chip, also in the reference population. The first results showed that for low to moderate imputation error rates (up to 20%, as observed for a 3k LD-chip) all effects of all markers should be used, which resulted in moderate loss of accuracies of 14% compared to a situation were all genotypes were known without error. If the imputation error rate was higher (e.g. ~55%, as observed for a 0.5k LD-chip) the loss of accuracy was around 53%. In this situation the use of only LD-marker effects might be more appropriate.

Application of generalized BayesA and BayesB in the analysis of genomic data

Strandén, I.[1], Mrode, R.[2] and Berry, D.P.[3], [1]MTT Agrifood Research Finland, Biotechnology and Food Research, 31600 Jokioinen, Finland, [2]Scottish Agricultural College, Sir Stephen Watson Building, Bush, Penicuik, EH260PH, United Kingdom, [3]Moorepark Dairy Production Research Center, Fermoy, Co. Cork, Ireland; ismo.stranden@mtt.fi

BayesA and BayesB are commonly used to analyse genomic data. Predictive ability using these models are known to be affected by the prior information provided. It can be shown that the prior density of the genetic marker effects in these models follows a Student-t density with known dispersion and degrees of freedom parameters. A natural generalisation of these models is to use Student-t density with unknown dispersion and degrees of freedom for the marker effects. In this study models that generalise BayesA (GtA) and BayesB (GtB) were investigated. In GtA and GtB, the dispersion parameter was always unknown. The degrees of freedom parameter was either fixed to 4.01 or was unknown. In addition, a simple Gaussian model with common genetic and residual variances unknown was considered, named common Gaussian (Gc). Following editing genotype data on 41,739 SNPs from the Illumina Bovine50 Beadchip from 1,096 Holstein-Friesian AI sires with daughters in Ireland were available. The 1,096 AI sires were divided into a training and validation datasets with 755 and 254 individuals, respectively. Deregressed PTAs of milk, fat, and protein yield were analysed using adjusted reliability as the weighting factor. All inference was based on Bayesian methods using MCMC. The MCMC chains were run for 400,000 cycles with the first 20,000 discarded as burn-in. The proportion of SNP variances set to zero in BayesB and GtB was 0.67. According to the validation reliabilities, BayesA and BayesB performed better than Gc only for fat. The GtA and GtB models were always as good or better than the Gc model. The best validation reliabilities were for GtB with known degrees of freedom where reliabilities from 0 to 12 percentages higher were observed compared to the Gc model.

Genomic prediction on pigs using the single-step method

Christensen, O.F.[1], Madsen, P.[1], Nielsen, B.[2], Ostersen, T.[2] and Su, G.[1], [1]Aarhus University, Genetics and Biotechnology, Blichers Alle 20, 8830 Tjele, Denmark, [2]Dansih Agriculture & Food Council, Pig Research Centre, Breeding & Genetics, Axeltorv 3, DK-1609 Copenhagen V, Denmark; OleF.Christensen@agrsci.dk

The single-step method combines pedigree, genomic and phenotypic information for genomic prediction. This study investigates accuracy of genomic prediction using the single-step method on pig data. The data consist of 2,000 Duroc boars genotyped using a 60K SNP panel. For a number of traits, comparisons are made between the single-step method, traditional pedigree-based BLUP and GBLUP with deregressed proofs as response variable. The results show that the two genomic prediction methods are superior to the pedigree-BLUP method, and that the sigle-step method performs slightly better than the GBLUP method. In addition, for pedigree-based BLUP and the single-step method, comparisons are made between univariate models and multivariate models. The results show that when improvements can be obtained using a multivariate model for pedigree-based BLUP, then improvements also exist using a multivariate model for the single-step method.

Effect of different genomic relationship matrices on accuracy and scale
Misztal, I.[1], Chen, C.Y.[1], Aguilar, I.[1], Vitezica, Z.G.[2], Legarra, A.[2] and Muir, W.M.[3], [1]University of Georgia, Athens, GA 30602, USA, [2]INRA, Castanet-Tolosan, 31326, France, [3]Purdue University, West Lafayette, IN 47907, USA; ignacy@uga.edu

Phenotypic data on body weight (BW) and breast meat area (BM) were available on up to 287,614 broilers. A total of 4,113 animals were genotyped for 57,636 SNP. The records were analyzed by a single step genomic BLUP (ssGBLUP), which accounts for all phenotypic, pedigree and genomic information. The genomic relationship matrix (G) in ssGBLUP was constructed using either equal (0.5; GE) or current (GC) allele frequencies, and with either all SNP or SNP with minor allele frequencies (MAF) below multiple thresholds (0.1, 0.2, 0.3, and 0.4) ignored. Additionally, a pedigree based relationship matrix for genotyped animals (A_{22}) was available. The matrices and their inverses were compared with regard to average diagonal (AvgD) and off-diagonal (AvgOff) elements. In A_{22}, AvgD was 1.00 and AvgOff was 0.01. In GE, both averages decreased with the increasing thresholds for MAF; AvgD decreasing from 1.37 to 1.02 and AvgOff decreasing from 0.72 to 0.03. In GC, AvgD was around 1.01 and AvgOff was 0.00 for all MAF. For inverses of relationship matrices, all AvgOff were close to 0; AvgD was 2.4 in A_{22}, varied from 11.6 to 13.0 for GE, and increased from 8.7 to 12.9 for GC as the threshold for MAF increased. Predictive abilities with all GE and GC were similar. Compared with BLUP, EBVs of genotyped animals in ssGBLUP were, on average, biased up to 1 additive SD higher with GE and down by 2 additive SD with GC. The bias was eliminated by adding constant 0.014 (equal to AvgOff in A_{22}) to GC. This constant is equivalent to twice the mean relationship between gametes in the genotyped and ungenotyped populations. Reduction of SNP with low MAF has a low effect on the realized accuracy. Unbiased evaluation in ssGBLUP may be obtained with GC scaled for compatibility with A_{22}.

Flexible prior specification for the genetic covariance matrix via the generalized inverted Wishart distribution
Munilla, S. and Cantet, R.J.C., Universidad de Buenos Aires, Facultad de Agronomía, Departamento de Producción Animal, Av. San Martín 4453, 1417 Buenos Aires, Argentina; munilla@agro.uba.ar

Consider the estimation of genetic (co)variance components from a maternal animal model (MAM) using a conjugated Bayesian approach. Usually, more uncertainty is expected a priori on the maternal additive variance than on the direct additive variance. However, it is not possible to model such differential uncertainty using the standard approach based on assuming an inverted Wishart (IW) distribution for the genetic covariance matrix. Instead, consider setting a generalized inverted Wishart (GIW) distribution. The GIW is essentially characterized by a larger set of hyperparameters, a feature that offers more flexibility while specifying prior knowledge. Taking advantage of this property, we present an elicitation method based on using previously estimated values of the (co)variance components to assess the hyperparameters on the next round. This Bayesian updating strategy stems naturally from the standard practice of genetic evaluations, where genetic parameters are frequently re-estimated as data is accrued over the years. A stochastic simulation study using the MAM as the data generator process was carried out to test the procedure. Genetic parameters were estimated through a Bayesian analysis via the Gibbs sampler. Posterior means, posterior standard deviations, and autocorrelations for the direct heritability, the maternal heritability and the direct-maternal genetic correlation were used as the criteria for comparison against more standard prior specifications. The elicitation method returned on average accurate estimates and reduced standard errors compared with non informative prior settings, while improving the convergence rates. In general, faster convergence was always observed when a stronger weight was placed on the prior distributions. However, analyses based on the IW distribution and initialized with over-dispersed starting values produced biased estimates with respect to the true simulated values.

Genome-wide association study for inguinal- and umbilical hernias using the 60K Porcine SNParray

Grindflek, E.[1], Hansen, M.H.S.[1,2], Hamland, H.[1,2] and Lien, S.[2], [1]Norsvin, P.O. Box 504, 2304 Hamar, Norway, [2]Norwegian University of Life Sciences, Department of Animal and Aquacultural Sciences and CIGENE, P.O. Box 5003, 1432 Ås, Norway; eli.grindflek@umb.no

Hernia is shown to be a serious problem in pig production, causing severe economic loss for the pig producers. The most common types are inguinal (and/or scrotal) hernia and umbilical hernia. The aim of this study is to identify and fine map QTL regions affecting inguinal- and/or umbilical hernias using a high density SNP panel, and to use this information in practical breeding to reduce incidence of hernias in Norwegian pig populations. Genotyping was carried out with the PorcineSNP60 array, using the iScan (Illumina, USA) according to manufacturer's instructions. Additionally, characterizations and association analysis of 65 biological candidate genes were performed. A total of 220 Landrace pigs with inguinal and 385 with umbilical hernias, together with healthy full sibs and parents, were collected from nucleus herds. Genome-wide association between SNPs and hernia phenotypes were investigated using the ASSOC function of the program package PLINK. For candidate genes the analysis of variance procedures (GLM procedure) were carried out using the statistical software package SAS. Significant (P<0.01) genetic effects were detected on all autosomal chromosomes for both inguinal and umbilical hernia. However, the most convincing regions were found on the chromosomes (SSC) 2, 8, 9, 13 and 15 for inguinal hernia and on SSC 5, 11 and 14 for umbilical hernia. For inguinal hernia the SSC 2 is particularly interesting and results indicate that several genes on the chromosome are involved. We sequenced a large amount of candidate genes to find SNPs within the genes. After analyzing the candidates two genes on SSC 2 were found to be significantly associated with inguinal hernia. The analysis to find the best combinations of genotypes for implementation in the breeding scheme is currently in progress.

Genome-wide association study for carcass and meat quality traits in commercial French Large White pigs

Shumbusho, F., Le Mignon, G., Bidanel, J.-P. and Larzul, C., INRA, UMR1313 GABI, F-78350 Jouy-en-Josas, France; jean-pierre.bidanel@jouy.inra.fr

A Genome-Wide Association Study for carcass and pork quality was conducted in a commercial French Large White pig population. The mapping population consisted of 428 male offspring from 98 sows inseminated by 56 boars. A total of 15 traits related to growth, carcass composition and meat quality were analysed. Piglets and sires were genotyped with the Illumina PorcineSNP60 Bead Chip. Association analyses were performed using GRAMMAR-GC procedures implemented in GenABEL and the Haploview software was used for Linkage Disequilibrium (LD) and haplotype studies. A total of 54 SNPs were found to have a significant (P-value <10-4) association with one of the traits investigated: 2 SNPs for ADG, 9 SNPs for carcass composition traits, 5 SNPs for water holding capacity, 20 SNPs for colour and 18 SNPs for ultimate pH measurements. On SSC15, 16 markers associated with quality traits were mapped within a narrow region including the RN gene. Of these, 4 markers had pleiotropic effects on colour, WHC and/ or pH. Haplotype analyzes confirmed that detected regions had significant effects on carcass and meat quality. The LD observed demonstrated dependences among neighbouring markers. Our results indicate that purebred populations retain a significant amount of genetic variation and their use in marker-assisted selection will offer opportunities for improving pork quality by within-line selection.

A genome-wide association scan for loci affecting osteochondrosis in German Warmblood horses
Tetens, J.[1], Wulf, I.R.[1], Kühn, C.[2] and Thaller, G.[1], [1]Christian-Albrechts-University, Institute of Animal Breeding and Husbandry, Hermann-Rodewald-Str. 6, 24118 Kiel, Germany, [2]Leibniz Institute for Farm Animal Biology, Wilhelm-Stahl-Allee 2, 18196 Dummerstorf, Germany; jtetens@tierzucht.uni-kiel.de

The aim of the current study was to identify quantitative trait loci affecting osteochondrosis and osteochondrosis dissecans (OC/OCD) in the German Warmblood population. We performed a genome-wide association scan using the Illumina Equine SNP50®BeadChip comprising a total of 54,602 SNP markers. The study included 944 German Warmblood stallions licensed in the years 2005-2008 and belonging to four different breeding organisations. The OC/OCD phenotypes for the hock, fetlock, interphalangeal and stifle joints were derived from the protocols of the radiological examination routinely conducted at licensing. The affection status was coded as a binary trait meaning that at least one radiological finding had to be present in affected animals. The frequncies of radiological alteration corresponded well to those reported in literature. As our population was structured due to the different breeding organisations and the presence of some large halfsib groups, we applied a principal component based approach as implemented in the R-package GenABEL to correct for this stratification. Genome-wide significance was determined applying the False Discovery Rate. As a major result, we identified a single genome-wide significant ($q<0.05$) quatitative trait locus affecting OCD in the fetlock joint and the interphalangeal joints at 28 Mb on equine chromosome 20. These and additional results for the other phenotypes and localisations will be presented in detail at the meeting. This is for the first time to our knowledge that the Illumina Equine SNP50® BeadChip has been applied to a genome-wide association scan for loci affecting OC/OCD in such a large cohort of animals probably representing the entire German Warmblood population.

Using Richards mixed effects growth model in different goose lines
Vitezica, Z.G.[1], Lavigne, F.[2], Dubois, J.P.[2] and Auvergne, A.[1], [1]Université de Toulouse, UMR 1289 TANDEM, INRA/INP-ENSAT/ENVT, F-31326 Castanet Tolosan, France, [2]La Ferme de l Oie, F-24400, Coulaures, France; zulma.vitezica@ensat.fr

Knowledge about the growth pattern of geese is necessary to optimize the goose production system (e.g. by selection). Growth models describe body weight (BW) changes over time. Richards model, a non-linear model (NLM), has been used extensively in different species to describe growth. Other possible models are the mixed effects nonlinear functions, which are able to deal with correlated errors and heterogeneous variances (heteroscedasticity) among the BW measures. These models have recently been used in ducks and in Japanese quail. In goose, there is little recent information in the literature on the modeling of growth curve variables. The best growth models for geese are yet to be found. The aim of this study was to use Richards model (fixed and mixed effects) to describe the growth of 3 lines of geese. From hatching to 22 weeks of age, geese (n=174) were fed *ad libitum* and the BW data were collected biweekly. The asymptotic weight and the inflection point age were considered random variables associated with individual in Richards mixed effects model. Fixed and mixed models were compared by using AIC. All statistical analyses were performed using R, by maximum likelihood. The within-line SD for BW increased with age. A goose that is heavier than average at one age will be even heavier at the next measure. These observations point out the problem of heteroscedasticity and correlated errors which are ignored in fixed effects models. In all lines, the Richards mixed effects model provided the best goodness of fit based on AIC values. The residual standard deviation (RSD <0.14 kg) was lower than fixed effect model for all lines. The improvement in the accuracy of estimated parameters is shown by a reduction of about 60% in RSD. Mixed effects NLM are a more flexible framework to model growth because they take into account the within-individual correlated nature of repeated measures.

Association between cholecystokinin type A receptor gene haplotypes and growth traits in Hinai-dori chicken cross

Rikimaru, K.[1,2], Komatsu, M.[1], Suzuki, K.[2] and Takahashi, H.[3], [1]Livestock Experiment Station, Akita Prefectural Agriculture, Forestry, and Fisheries Research Center, Daisen, Akita, 019-1701, Japan, [2]Graduate School of Agricultural Science, Tohoku University, Sendai, Miyagi, 981-8555, Japan, [3]National Institute of Livestock and Glassland Science, Tsukuba, Ibaraki, 305-0901, Japan; Rikimaru-Kazuhiro@akita.pref.lg.jp

The Hinai-dori is a slow-growing breed of chicken native to Akita Prefecture, Japan. We previously identified quantitative trait loci (QTL) for body weight and average daily gain in a common region between MCW0240 (chr 4: 69.9 Mb) and ABR0622 (chr 4: 86.3 Mb) on chicken chromosome 4 in an F_2 resource population produced by crossing low- and high-growth lines of the Hinai-dori breed. Cholecystokinin type A receptor (CCKAR) is a positional candidate gene affecting growth traits in the region. In this study, we analyzed DNA polymorphisms of the CCKAR gene and examined the association of CCKAR with growth traits in an F_2 intercross population of the Hinai-dori breed. PCR-amplified products were obtained for all exonal regions of the CCKAR gene of the parent individuals, nucleotide sequences were determined, and CCKAR haplotypes were identified. To distinguish resultant diplotype individuals in the F_2 population, a mismatch amplification mutation assay was performed. Five haplotypes (Haplotypes 1-5) were accordingly identified. Six genotypes produced by the combination of three haplotypes (Haplotype 1, 3, and 4) were examined in order to identify associations between CCKAR haplotypes and growth traits. The data indicate that Haplotype 1 is superior to Haplotype 3 and 4, for example, with regard to average daily gain between 10 and 14 weeks, and 0 and 14 weeks of age, both in F_2 males and females. We conclude that CCKAR is a useful marker of growth traits and is used to develop strategies for improving growth traits in the Hinai-dori breed.

Candidate gene analysis and fine mapping of quantitative trait loci for intramuscular fat content on SSC 13 in a Duroc purebred population

Nakano, H.[1], Uemoto, Y.[2], Kikuchi, T.[2], Sato, S.[2], Shibata, T.[3], Kadowaki, H.[3], Kobayashi, E.[2] and Suzuki, K.[1], [1]Tohoku University, Graduate School of Agricultural Science, 1-1 Amamiya-machi, Tsutsumidori, Aoba-ku, Sendai, 981-8555, Miyagi, Japan, [2]National Livestock Breeding Center, 1 Odakurahara, Nishigomura, Nishi-shirakawa-gun, 961-8511, Fukushima, Japan, [3]Miyagi Prefecture Animal Industry Experiment Station, 1 Iwadeyama, Minamisawa-aza, Hiwatashi, Osaki, 989-6445, Miyagi, Japan; nh@bios.tohoku.ac.jp

Intramuscular fat content (IMF) plays an important role in the eating quality of meat and heritable traits in pig populations. Quantitative trait loci (QTL) for IMF have been located on SSC13 in a Duroc purebred population, as described by Soma *et al.* In this region, peroxisome proliferator-activated receptor γ (PPARγ) and ghrelin (GHRL) genes are regarded as candidate genes for these QTL. This study was conducted to identify single nucleotide polymorphisms (SNPs) of these genes and to examine the effects of these genes on IMF in the same population. Furthermore, fine mapping surrounding this QTL region on SSC 13 was done. The Duroc purebred population used for this study was selected for IMF during seven generations: 1004 pigs were typed and 543 had the IMF data. In candidate gene analysis, a total of 2 non-synonymous SNPs were detected in exon 2 of the PPARγ gene (c.278A>G) and exon 3 of the GHRL gene (c.358C>A) in base pigs. They were then genotyped in all pigs. However, no significant association was found between these SNPs and IMF in this population. In fine mapping QTL analysis based on the variance component method, a total of 18 microsatellite markers on SSC 13 were genotyped. Then QTL analyses were performed to evaluate the effects of QTL and the possible causality of these genes. This study detected the significant QTL with a LOD score of 3.87. However, the QTL was not in the same region as the PPARγ and GHRL genes. Therefore, the QTL found in this study presents the possibility of other candidate genes for IMF.

Genome-wide mapping and SCD gene effects for fatty acid composition and melting point of fat in a Duroc purebred population

Uemoto, Y.[1], Nakano, H.[2], Kikuchi, T.[1], Soma, Y.[2], Sato, S.[1], Shibata, T.[3], Kadowaki, H.[3], Kobayashi, E.[1] and Suzuki, K.[2], [1]National Livestock Breeding Center, Odakurahara, Nishigo-mura, Nishi-shirakawa-gun, Fukushima, 961-8511, Japan, [2]Graduate School of Agricultural Science, Tohoku University, 1-1 Amamiya-machi, Tsutsumidori, Aoba-ku, Sendai, Miyagi, 981-8555, Japan, [3]Miyagi Prefecture Animal Industry Experiment Station, 1 Iwadeyama, Minamisawa, Hiwatashi, Osaki, Miyagi, 989-6445, Japan; y0uemoto@nlbc.go.jp

The fatty acid composition and melting point of fat are among the most important economic traits in pig breeding. The stearoyl-CoA desaturase (SCD) gene is located on SSC14, and is considered a candidate gene for fatty acid composition. We conducted whole-genome quantitative trait locus (QTL) analysis for fatty acid composition and melting point of inner and outer subcutaneous fat and inter- and intramuscular fat in a Duroc purebred population. We also performed positional candidate gene analysis to fine map the detected QTL on SSC14, to identify polymorphisms of SCD gene and to examine the effects of SCD gene on these traits. The pigs were evaluated for fatty acid composition and melting point of 4 different tissues, and the number of pigs with phenotypes ranged from 479 to 521. A total of 129 markers on all autosomal SSC were genotyped and used for QTL analysis. In addition, 13 new additional markers on SSC14 were genotyped for fine map. In genome-wide QTL analysis, 10 significant QTLs and 30 suggestive QTLs were detected. On SSC14, significant QTLs for C18:0 and C18:1 of outer subcutaneous fat and intramuscular fat and melting point were detected in the same region. In candidate gene analysis, a total of 2 SNPs (g.-353C>T and g.-233T>C) in the promoter region of the SCD gene were identified and were completely linked in this population. In addition, a significant association was found between SCD gene and C18:0, C18:1, and melting point of fat. These results indicated that the SCD gene have a strong effect for fatty acid composition and melting point of fat.

Estimation of genetic parameters for udder traits using a Random Regression Model on age at classification for Holsteins in Japan

Osawa, T.[1], Hagiya, K.[1] and Oka, T.[2], [1]National Livestock Breeding Center, Fukushima, 961-8511, Japan, [2]The Holstein Cattle Association of Japan, Tokyo, 164-0012, Japan; t0ohsawa@nlbc.go.jp

The objective of this study is to estimate genetic parameters for the udder traits (i.e. overall udder (UDDER), fore udder (FU), rear udder height (RUH), rear udder width (RUW), udder support (US), udder depth (UD) and front teat placement (FTP)). The data consisted of 1,066,185 records from 818,488 Holstein cows provided by The Holstein Cattle Association of Japan. The Random Regression model included herd-classifier-date at classification, parity-age group at classification and stage of lactation as fixed effect, additive genetic and permanent environmental effects with random regression on age at classification (21 to 100 mo of age) using the second-order Legendre polynomials and residual effects assuming heterogeneous variances. The GIBBS3F90 program was applied to estimate the genetic parameters. The heritability estimates for UDDER, FU, RUH and US increased dramatically with cow's age (0.15 to 0.56, 0.19 to 0.37, 0.24 to 0.39 and 0.18 to 0.36, respectively). In contrast, heritability estimates for RUW varied slightly (0.20 to 0.25). UD and FTP, on the other hand, did not show monotonic behavior; it increased until 65 mo of age (0.42 to 0.47 and 0.36 to 0.40), and later decreased down to 0.26 and 0.29. The larger genetic correlation was observed between groups with smaller difference in ages. When we plotted the genetic correlations of the 29 mo of age group against the other age groups varying from 41 to 81 mo of age groups, we observed clear downward tendency. In particular, it decreased from 0.92 to 0.74, 0.95 to 0.86, 0.90 to 0.72, 0.91 to 0.74, 0.96 to 0.87, 0.96 to 0.76 and 0.98 to 0.93 for UDDER, FU, RUH, RUW, US, UD and FTP, respectively. Effect of aging was different for each traits. Amongst all the udder traits, FTP exhibited the highest correlations and the most stable behavior. On the other end of the spectrum, RUW achieved the lowest value and showed the most erratic behavior.

Desired genetic gains for a breeding objective: a novel participatory approach

Sae-Lim, P.[1], Komen, H.[1], Kause, A.[1], Van Arendonk, J.A.M.[1], Barfoot, A.J.[2], Martin, K.E.[2] and Parsons, J.E.[2], [1]Animal Breeding and Genomics Centre, Wageningen University, 6700 AH Wageningen, Netherlands, [2]Troutlodge, Inc., 12000 McCutcheon Rd., Sumner, 98390 WA, USA; panya.sae@wur.nl

Distributing animals from a single breeding program to a global market may not satisfy all producers' product and market objectives. To solve this, producers can be asked to rank the most important breeding traits. However, most participatory methods do not provide information on how much gain in one trait is desired compared to other traits in the breeding goal. Here we present a novel participatory method to define a global breeding objective using analytical hierarchy process (AHP) and weighted goal programming (WGP). The method was applied among customers of a rainbow trout breeding company. Questionnaire A (Q-A) was sent to 178 farmers from five continents. Questionnaire B (Q-B) was sent to 53 farmers who responded Q-A. For Q-A, farmers ranked six most important traits out of 13 alternatives. These were thermal growth coefficient (TGC), survival (S), FCR, condition factor (CF), fillet percentage (F), and late maturation (LM). For Q-B, preferences for these six traits were compared using pairwise comparison. Preference intensity (score 1-9) was given for comparing how much expected genetic gain (in % of a trait mean; G%) in one trait is more preferred than that in the other trait. Individual preferences on six traits were estimated using AHP and were aggregated to society preferences using WGP. Multiplication between the society preferences and G% produces desired genetic gains. Society preferences for FCR, S, TGC, LM, F, and CF was 0.258, 0.257, 0.213, 0.114, 0.094, and 0.064, resp. Corresponding desired gains were 1.96%, 1.54%, 1.45%, 1.63%, 0.07%, and 0.31%, resp. Relative selection index weights can be derived when mean values and phenotypic and genetic (co)variances of the traits are known. In conclusion, AHP and WGP can be used to define a breeding objective for any breeding program which serves large and potentially diverse markets.

Effects of GH, FASN and SCD gene polymorphisms on fatty acid composition in Japanese Black cattle

Yokota, S.[1], Sugita, H.[1], Suda, Y.[2], Katoh, K.[1] and Suzuki, K.[1], [1]Tohoku University, Graduate School of Agricultural Science, 1-1 Amamiya-machi, Tsutsumidori, Aoba-ku, Sendai, 981-8555, Miyagi, Japan, [2]Miyagi University, School of Food, Agricultural and Environmental Sciences, 2-2-1 Hatatate, Tihaku-ku, Sendai, 982-1215, Miyagi, Japan; yokota@bios.tohoku.ac.jp

The meat quality of Japanese Black cattle is generally superior in terms of marbling. Recently, the fat quality has become an important factor of delicious beef production in Japan. Fat quality is determined mainly by the fatty acid composition. Especially, mono-unsaturated fatty acids (MUFA) such as oleic acid (C18:0) contribute positively to beef flavor and tenderness; C18:1 is a major MUFA in beef fat. Heritability of fatty acid production is generally high. Several genes, such as growth hormone (GH), fatty acid synthase (FASN) and stearoyl-CoA desaturase (SCD) reportedly affect fatty acid composition. This study evaluated the effects of polymorphisms in these genes (GH, FASN, and SCD) on the fatty acid composition of the longissimus muscle in 1260 Japanese Black cattle. Polymorphisms in the GH, FASN, and SCD genes significantly affected the fatty acid composition. However, the respective contributions of GH, FASN, and SCD gene polymorphisms to total genetic variance were 5.13%, 6.31%, and 6.78% for C18:1. These three genes collectively contributed 18.22% of the total genetic variation in C18:1, implying that only 0.08 of heritability of C18:1 (0.43) was explained by them. The present results suggest that polymorphisms in other genes related to lipid metabolism affect fatty acids composition because these polymorphisms of GH, FASN, and SCD gene contributed only approximately 18% of the total genetic variation in C18:1. To use these gene polymorphisms directly for marker-assisted selection, gene frequencies of these polymorphisms should be estimated more exactly in other populations and in larger numbers of animals.

Estimation of genetic parameters for milkability using threshold methodology in a Bayesian framework
König, S.[1], Tong, Y.[1] and Simianer, H.[2], [1]University of Kassel, Nordbahnhofstr. 1a, 37213 Witzenhausen, Germany, [2]University of Göttingen, Albrecht-Thaer-Weg 1a, 37075 Göttingen, Germany; tyin@gwdg.de

The pattern of phenotypic daughter records for milkability measured as milking speed (kg milk/minute) relatively often follows a bimodal structure instead of a Gaussian distribution. Consequently, the basic requirement for the application of linear mixed models to infer genetic parameters is strongly violated. Furthermore, due to antagonistic relationships with udder health, milking speed should be considered as a trait with an intermediate optimum. As an alternative to standard mixed model theory, threshold methodology in a Bayesian framework by setting one or two arbitrarily chosen thresholds has been evaluated for a dataset of 2,742,140 Holstein cows. Setting one threshold implies two categories of data, i.e. a small fraction of cows with low milking speed and a large pool of remaining observations; or a small fraction of cows with high milking speed and the remaining dataset. Setting two thresholds allows a differentiation in a) high milking speed, b) average milking speed, and c) low milking speed, and a more pronounced selection of cows with intermediate values. Irrespective of the applied statistical model, posterior distributions of heritabilities for milkability were relatively equal, with a posterior mean ranging from 0.26 to 0.28. Genetic correlations between milkability defined as a Gaussian trait and milkability defined as a categorical trait were throughout higher than 0.98. Also solutions for fixed effects, i.e. calving year, calving season, milking frequency, and days in milk, as well as estimated breeding values of influential sires, were unaffected by the choice of the statistical model. In conclusion, milkability can be evaluated by a broad variety of statistical models without changes in estimated genetic parameters. However, a higher impact of the statistical model is assumed for small datasets and limited observations within sire.

Association of polymorphisms in the Leptin gene with carcass and meat quality traits in Italian Friesian young bulls
Catillo, G., Grandoni, F., Vincenti, F., Antonelli, S., Iacurto, M., Moioli, B. and Napolitano, F., CRA - Animal Production Research Centre, Via Salaria, 31, 00015 Monterotondo - Roma, Italy; gennaro.catillo@entecra.it

The purpose of this study was to evaluate the effects of polymorphisms in the leptin gene (coding and regulating sequences) on beef performance traits measured at slaughter and dissection, and on some meat rheological and colorimetric characteristics. Twenty Italian Friesian young bulls, reared in the experimental farm to the age of 15 months, were slaughtered (body weight = 544±9.9 kg). DNA of each animal was extracted from blood and amplified for genotyping 31 Single Nucleotide Polymorphisms (SNP) in the leptin gene. Seventeen of these SNP (14 in the promoter region, 2 in exon 2 and 1 in exon 3) were not in linkage disequilibrium; therefore, for them, the allele substitution effect, on each trait, was estimated by regressing the number of copies of each allele, using a linear model, which included the age and the weight at the start of the trial. All the 17 SNP were in agreement with Hardy-Weinberg equilibrium, except rs29004468T>A and rs29004474C>T. Carcass weight was affected by the SNP in the promoter region: g.1540G>A (AB070368), rs29004173G>A, rs29004474C>T; and by the SNP in exon 3: rs29004508C>T ($P<0.03 \div 0.002$). Fat score was affected by g.1540G>A, g.1935_1937delGTT (AB070368) and rs29004171G>A ($P<0.05 \div 0.03$). Among meat quality parameters, cooking loss, that contributes to defining the juiciness of meat, is influenced by SNP: rs29004468T>A, rs29004171G>A, rs29004469delG, rs29004470C>T and rs29004476T>C of the promoter region ($P<0.05\div0.02$). Considering that leptin hormone regulates energy metabolism and therefore beef attitude in cattle, our results confirm the importance of the leptin gene, where some polymorphisms showed their influence on some important beef traits. The detected polymorphisms, if confirmed by further investigations on different cattle populations, could be used as markers in selection programs to improve meat quality.

Role of LEP gene on porcine productive traits related to appetite, interaction with LEPRc.1987C>T SNP

Pérez-Montarelo, D.[1], Fernández, A.[1], López, M.A.[1], Folch, J.M.[2], Pena, R.N.[3], Óvilo, C.[1], Rodríguez, C.[1], Silió, L.[1] and Fernández, A.I.[1], [1]SGIT-INIA, Mejora Genetica Animal, Ctra. Coruña Km7.5, 28040, Spain, [2]Facultad Veterinaria, UAB, Ciència Animal i dels Aliments, Bellaterra, Barcelona, 08193, Spain, [3]IRTA, Genètica i Millora Animal, Av. Alcalde Rovira Roure, 191, Lleida, 25198, Spain; avila@inia.es

Leptin (LEP) interacts with the leptin receptor (LEPR) and acts as a signal of satiety affecting feed intake, energetic balance and body composition. In this study we have analyzed the LEP gene sequence to find polymorphisms associated with productive traits in F2 and F3 intercrosses between Iberian and Landrace pigs. Moreover, previous studies in these and other pig populations confirmed the effects of a polymorphism in LEPR gene (c.1987C>T) on growth and fatness, raising the interest of studying the interaction effects between LEP and LEPR polymorphisms on these traits. The sequencing of the LEP gene in purebred Iberian and Landrace parental animals revealed 35 SNPs and 4 indels, located mostly in introns. Two SNPs (LEP g.1444 C>T and LEP g.1782A>G) were genotyped in the 486 F2 and F3 animals and association analyses were conducted. The results showed significant additive effects of LEP g.1444 C>T on growth and unexpected dominant effects on backfat thickness and weight of belly bacon. The T allele, fixed in Iberian sires, showed positive effects on live weight nearest to 100 kg and carcass weight, probably mediated by an increased appetite, in the same sense as the LEPR SNP. No interaction could be detected between the effects of both LEP g.1444 C>T and LEPRc.1987C>T on growth and fatness traits. However, significant interaction effects on the weights of hams, shoulders and ribs adjusted by carcass weight were evidenced between both LEP and LEPR SNPs. The present study highlights the relevant role of LEP gene on pig production traits, although previous analysis on SSC18 in the same material did not show significant QTL regions for these traits, probably due to the low marker density used in the scan.

Variance component estimates for growth traits in South African Brangus cattle

Neser, F.W.C.[1], Van Wyk, J.B.[1], Fair, M.D.[1], Lubout, P.C.[2] and Crook, B.[3], [1]University of the Free State, Animal Wildlife and Grassland Sciences, P.O. Box 339, 9300, Bloemfontein, South Africa, [2]SA Brangus Society, P.O. Box 12465, 9324 Brandhof, South Africa, [3]Agricultural Business Research Institute, UNE, 2351 Armidale NSW, Australia; neserfw@ufs.ac.za

A combination of multivariate and repeatability models were used to estimate genetic parameters for birth weight (BW), weaning weight (WW), yearling weight (YW), eighteen month weight (FW) and mature weight (MW – three measurements) on 21,673 records obtained from the South African Brangus Cattle Breed Society. The data covered a period of 7 generations from 1985-2010. Direct heritability estimates obtained were 0.21 ± 0.024, 0.23 ± 0.021, 0.23 ± 0.025, 0.29 ± 0.029 and 0.24 ± 0.019 for BW, WW, YW, FW and MW respectively. Maternal heritability was also estimated for birth- (0.05 ± 0.01) and weaning weight (0.11 ± 0.001). The direct genetic correlations between the different traits were all positive, ranging from moderate (0.43 ± 0.081–YW and MW) to high (0.99 ± 0.043– for WW and FW).

Introgressed background removal using genetic distances

Amador, C.[1], Toro, M.A.[2] and Fernández, J.[1], [1]INIA, Ctra. A Coruña Km 7.5, 28040 Madrid, Spain, [2]ETSIA UPM, Ciudad Universitaria, 28040 Madrid, Spain; amador.carmen@inia.es

Some populations require to be maintained pure because they have an economical or conservational interest linked to their particular genetic background (Iberian pigs, dog breeds, endangered natural populations…). When one of these groups of individuals is introgressed by another genetically differentiated, we have to look for a method to recover the original information, depending on the available data. In this study we analyzed through computer simulations the potential for the recovery of an original population background using different sets of molecular markers (microsatellite like). The strategy is to determine the contributions (i.e. number of offspring) of current available individuals to the next generation in such a way that the expected allele frequencies for the markers are as close as possible to the already known frequencies in the original non-introgressed population. The goodness of fit between present situation and ideal one was tested using different genetic distance measurements: Kullback-Leibler, Nei minimum distance… We evaluated the efficiency of the process through the percentage of population information coming from non exogenous founders (Native-founder Genomic Representation) and other parameters like inbreeding or mean coancestry related to the general objectives of a conservation program. Results showed that the success in the recovery of information was limited by two factors: the percentage of foreign individuals introgressed and the number of generations elapsed until the management started. Of course, the number of markers has an influence, but it is less important than the previous factors. It is crucial to have a very clear reference population so the frequencies are accurately known.

Quantitative trait loci study in Baluchi sheep

Dashab, G.R.[1], Aslaminejad, A.A.[2], Nassiri, M.R.[2], Esmailizadeh Koshkoih, A.[3], Sahana, G.[1] and Saghi, D.A.[2], [1]Aarhus University, Genetic and Biotechnology, Blichers Allé 20, Postboks 50, DK-8830 Tjele, Denmark, [2]Ferdowsi University of Mashhad, Animal science, Azadi square, 9177948974-Mashhad, Iran, [3]Shahid Bahonar University, Animal science, 22 Bahman Blvd, 76169-133, Kerman, Iran; golamr.dashab@agrsci.dk

We selected three sheep chromosomes (OAR1, 5 and 25) based on published literature to study quantitative trait loci (QTL) segregating in Baluchi sheep breed of Iran. A total of 14 microsatellite markers were genotyped in 505 sheep belong to 13 half-sib families. The selected markers were located in the chromosomal regions previously reported to harvour QTL for wool production and wool quality traits in sheep. The average number of progeny per sire was 40 and ranged between 24 and 58. A total of 18 traits (16 wool production and quality traits along with birth weight and litter size) were analyzed. The phenotypic data were collected from Research Centre on Baluchi breed from 2009-2010. The phenotypes were corrected for fixed effects of birth-year, sex, herd, birth-type using software SAS and DMU. In addition, wool traits were also corrected for the shearing-date. The QTL analyses were performed using GridQTL software. We identified QTL on OAR1 for litter size, birth weight, wool type traits, fiber diameter, greasy fleece weight and clean wool trait. We also identified QTL on OAR5 for greasy and clean fleece traits, and on OAR25 for wool type traits and greasy fleece weight. As the marker density was sparse, the QTL intervals were very large. More markers are needed to refine the QTL locations before QTL information can be utilized for selection of sheep for breeding purpose.

Association between somatic cell count and functional longevity in Polish Holstein-Friesian cows
Morek-Kopec, M.[1] and Zarnecki, A.[2], [1]University of Agriculture in Cracow, Mickiewicza 24/28, 30-059 Cracow, Poland, [2]National Research Institute of Animal Production, Balice, 32-083 Cracow, Poland; rzmorek@cyf-kr.edu.pl

A Weibull proportional hazard model was applied to examine the relationship of somatic cell count and functional longevity in Polish HF cattle. Data were records of 646,635 cows from 3043 herds, daughters of 4245 sires, calving for the first time between 1997 and 2010. Functional longevity was defined as length of productive life measured as the number of days from first calving to culling (uncensored records) or last test day (censored records) adjusted for production. Mean length of productive life was 956 days for 334,003 uncensored cows (48%), and 887 days for 312,632 censored cows. Test-day somatic cell count (SCC) was transformed to somatic cell score (SCS). Lactation SCS (LSCS) was calculated as an arithmetic mean of test-day SCS. Cows were grouped into ten classes according to first lactation LSCS. The statistical model included time-independent fixed effects of LSCS classes and age at first calving, time-dependent fixed effects of year-season, parity-stage of lactation, annual change in herd size, relative fat yield and protein yield, time-dependent random herd-year-season effect and sire effect. Likelihood ratio tests showed a highly significant overall impact of LSCS on risk of culling. LSCS values between 0.5 and 4.5, corresponding to low and average somatic cell count level, were not associated with increased risk of culling. For LSCS in the interval between 5.0 to 8.0 the relative risk rose gradually with the increasing score. Cows with highly elevated somatic cell counts (LSCS>8.0) were almost four times more likely to be culled than cows with low or average LSCS. Increased risk of culling was also observed for extremely low SCC (LSCS≤0.5). Some authors have attributed a high risk of culling of cows with low SCC to small number of leukocytes and consequently a weaker defense mechanism.

Preliminary studies on mitochondrial DNA control region partial sequence in Polish native Swiniarka sheep
Krawczyk, A., Słota, E. and Rychlik, T., National Research Institute of Animal Production, Krakowska Str. 1, 31-047 Balice, Poland; akrawczyk@izoo.krakow.pl

The worldwide research on Ovis aries mtDNA of Asian and European breeds allowed to classify them into five phylogenic clusters. Until now three sequenced elsewhere DNA samples of Swiniarka breed were assigned to haplogroup B. Swiniarka is a primitive breed originated from the Central and Western Europe. In Poland at the end of XIX century the breed was assumed to be nearly extinct. Restoration work succeeded in 14 flocks comprising in the 2010 year over 600 ewes. The material of research comprised 21 DNA probes isolated from blood of Swiniarka sheep. The PCR and sequence primers were designed according to the reference sequence of ovine haplogroup B (GeneBank accession no. AF010406). The post PCR products were directly sequenced. Using BioEdit tool the obtained 21 sequences were aligned with 3 previously published in GeneBank sequences for Swiniarka sheep. The number of tandem repeats was established by Tandem Repeats Finder Program whereas sequence variation sites and haplotypes were determined by DnaSP Version 5. Obtained 24 sequences were 338bp long, running from 15561 to 15898bp in the reference sequence. All 15 identified variable sites represent transitions. Two of them were singletons, whereas 13 were parsimony informative sites. There was found no insertion/deletion type polymorphism. The obtained sequences were clustered into 6 haplotypes of which degree of diversity was 0.7754. All sequences contain four copies of 75bp long tandem repeats of consensus: 5'GTACATAGTATTAATGTAATATAGACATTATATGTATAAAGTACATTAAATGATT TACCCCATGCATATAAGCAC. The results on mtDNA variability in Swiniarka breed compared with worldwide literature data suggest lower genetic diversity in the Polish breed. This might be caused by small number of ewes from which Swiniarka was restored and a low genetic diversity of females taken to breeding.

Genetic parameters for birth weight and its environmental variability in a divergent selection experiment for birth weight variability in mice
Pun, A., Cervantes, I., Nieto, B., Salgado, C., Pérez-Cabal, M.A. and Gutiérrez, J.P., Faculty of Veterinary, Complutense University of Madrid, Deparment of Animal Production, Avda. Puerta de Hierro s/n, 28040 Madrid, Spain; apun@vet.ucm.es

The control of the environmental variability by genetic selection represents a new alternative to positive selection for productive traits. Genetic parameters for birth weight trait and its variability were estimated applying both homogeneous and heterogeneous variance models using data from a divergent selection experiment for birth weight variability in mice. A number of records of 6,271 and 8,138 individuals in the pedigree were used. The model included period of birth, litter size and sex as fixed effects, and the litter and additive genetic effect as random effects besides the residual. The heterogeneous variance model was solved by using the GSEVM program while the homogeneous variance model was solved by using the TM program. The mean of the posterior distribution for the heritability of birth weight using homogeneous variance model was found to be low, 0.103, being the litter component much more important, 0.471. The mean of the posterior distribution for the genetic correlation between the trait and its variability was extremely high and negative, -0.980, suggesting an artefact of the model. A model considering null the genetic correlation was carried out providing similar values for the rest of the parameters, and it was used to obtain the breeding values for the selection process. The skewness of the trait might be influencing the results and further research is needed.

Linkage disequilibrium and haplotype blocks in chicken layer lines
Lipkin, E.[1], Fulton, J.[2], Arango, J.[2], O'sullivan, N.P.[2] and Soller, M.[1], [1]The Hebrew University of Jerusalem, Genetics, Dept. of Genetics, The Hebrew University of Jerusalem, Jerusalem, 91904, Israel, [2]Hy-Line Int, Dallas Center, IA, 50063, USA; lipkin@vms.huji.ac.il

Linkage Disequilibrium (LD) was measured as r^2 and Haplotype blocks (HTB) were defined in three chicken chromosomal regions on GGA2 (4.86 Mb), GGA5 (3.43 Mb) and GGA8 (2.51 Mb), in two chicken layer lines: Line 1, an elite, partially inbred, White Leghorn layer line, genotyped for a set of 107 SNPs; and Line 2, coming from a Brown-egg layer line (BES) selected for production traits while maximizing retention of genetic variation, genotyped by Illumina 42K chicken microarray. Quality control filter was based on missing genotypes, deviation from Hardy-Weinberg equilibrium and minimum minor allele frequency (MAF). LD between markers was inversely proportional to the bp distance and MAF difference between them. Syntenic LD was much higher in Line 1 regions (mean 0.341, 0.001-1.000) than Line 2 (mean 0.082, 0.000-1.000), while non-syntenic LD was virtually the same for both lines (mean 0.002, 0.000-0.019 or 0.017). Very clear HTB definition was obtained for Line 1, with only 2 main SNP haplotypes segregating in each block. A few small blocks of two adjacent SNPs defined by LD>0.5 were obtained for Line 2. However, for each of the three regions in this line, we observed one major haplotype, 3-4 four minor haplotypes and many rare recombinant haplotypes. Thus, for Line 1 only one or two tagging SNPs are required per block for applications, while for Line 2 mini-haplotypes of 3 or 4 markers should be able to tag the major and minor haplotypes, and capture most of the LD in the population. These results imply that 42K array is easily adequate for GWS and GWA for Line 1; while for Line 2 the 42K array is probably adequate for GWS but far from adequate for GWA based on single marker association. However, in this line, GWA based on multi-marker haplotypes may be more successful, but will not provide high resolution mapping.

Genetic profile of Sztumski and Sokolski cold blooded horses in Poland

Polak, G.M.[1], Mackowiak, M.[2], Zabek, T.[3] and Fornal, A.[3], [1]National Research Institute of Animal Production, National Focal Point, Wspolna, 30, 00-930 warsaw, Poland, [2]University of Life Science, Division of Horse Breeding, Wolynska, 33, 60-637 Poznan, Poland, [3]National Research Institute of Animal Production, Independent Laboratory of Genomics, Krakowska 1, 32-283 Balice, Poland; grazyna.polak@minrol.gov.pl

The cold blooded horses were breed in Poland since the beginning of XIX century. They have been developed from the oldest local European native breeds descending from Tarpan (e. g. Polish Konik and Zemaitukai) with strong influence of other European cold blooded breeds. The most important influence of Ardennes significantly increased the height and body weight. The number of cold blooded horses in Poland decreased dramatically in 2nd part of XX century. That time, all local types of cold blooded horses were included in a single Polish Cold Blood Stud Book. At present, it is possible to reestablish only two local types: Sztumski and Sokolski, but the selection of horses with 'pure' pedigree is very difficult. In 2007, the horse genetic recourses conservation programme for these two local types has been initiated. It permits to characterize and to conduct the monitoring and research. The aim of genetic analysis including four horse populations was, in the first step, to establish the genetic distance between Sztumski and Sokolski cold blooded horses and, in the second step, between Polish Cold Blooded commercial population and protected population. The a third step included comparison between Ardennes horses which, in fact were contributing to development of all Polish types of cold blooded horses. In the analysis, conducted in two laboratories, 12 microsatellites of 2nd class genetic markers, were used on material consisting of: 230 Sztumski, 317 Sokolski, 101 Polish Cold Blooded and 286 Ardennes horses. The results show that genetic distance between these four breeds is not very significant, what indicates considerable exchange of genetic material between breeds and require most accurate selection to ensure maintaining differentiation of these populations.

Estimation of inbreeding based on effective number for Baluchi, Arman and Iran Black breed of sheep

Hosseinpour Mashhadi, M.[1], Vakili, A.[2], Jenati, H.[3], Jafari, M.[3], Shokrolahi, B.[4] and Tabasi, N.[5], [1]Department of Animal Science, Mashhad Branch, Islamic Azad University, Iran, [2]Islamic Azad University- Kashmar Branch, Kashmar, Iran, [3]Animal Breeding Center of Abas Abad, Mashhad, Iran, [4]Islamic Azad University-Sanandaj Branch, Sanandaj, Iran, [5] Bu-Ali Research institute, mashhad, Iran; Mojtaba_h_m@yahoo.com

Most programs involving domestic animals may try to minimize accumulation of inbreeding and quantify the increase by calculating the change in inbreeding per generation (ΔF). It is important to account effects of inbreeding in populations undergoing selection to properly adjust the breeding process for the potential reduction in performance. The purpose of this study was to monitor current rates of inbreeding in the Baluchi, Arman, and Iran Black sheep. Data are from animal breeding center (Abbas Abad) in North East of Iran. Baluchi sheep consist of two flocks that named Baluchi 1 and 2. The number of male and female in mating program from 1985 until 2010 was used in this study. The number of generation was approximately 26. The population size and the ratio of males to females are two important factors that have an effect on ΔF. The effective number ($N_e = (4 N_m N_f)/(N_m + N_f)$)of each generation calculate based on male (N_m)and female (N_f) number and harmonic mean of N_e estimated. The inbreeding of each breed of sheep estimated by this equation: $\Delta F = 1/(2 N_e)$. The estimated inbreeding rate (ΔF) for Baluchi1, Baluchi2, Arman, and Iran Black bred of sheep were 1.09, 1.19, 1.09, and 1.10% respectively. In general the results involve that inbreeding at present is not a serious problem in the animal breeding center (Abbas Abad) in North East of Iran.

Genetic selection response in scenarios of maximization of genetic variance in small populations: preliminary results

Cervantes, I.[1], Gutiérrez, J.P.[1] and Meuwissen, T.H.E.[2], [1]Department of Animal Production.Faculty of Veterinary.University Complutense of Madrid, Avda. Puerta de Hierro s/n, 28040, Spain, [2]Department of Animal and Aquacultural Sciences.Norwegian University of Life Sciences, UMB, P.O. Box 5003, Ås, 1432, Norway; icervantes@vet.ucm.es

The preservation of the maximum genetic diversity in a population is one of the main objectives within a breed conservation program. This ensures more genetic variation for extreme traits, which could be useful in case the population needs to adapt to a new environment/production system. The aim of this study was to test the usefulness of the genetic variance maximization analyzing the genetic selection response. Six scenarios were created with 100 discrete generations. A total of 50 males and 50 females were born in each generation. We simulated a random scenario, a full-sib scenario, a scenario using the MVT (Maximum Variance Total) method using the annealing algorithm, a MVT+DF scenario limiting the increase in inbreeding or the individual increase in inbreeding (MVT+DF*) and a scenario of minimum coancestry selection (MCS). We used these different scenarios to study the performance of the population after 10 generations of selection using two selection intensities (20 and 50% best individuals kept for breeding). The best genetic response was for the FS and MVT scenarios. Both were very similar until fourth generation, after that FS scenario was better. Regarding the other scenarios, in the first generation of selection the genetic gain was higher in the scenario MVT+DF than in MCS and random scenario. After second generation, MCS attained higher response but not significantly different from MVT+DF. When the selection intensity was higher the MVT+DF and MCS showed more differences. The FS and MVT scenario ensured a better genetic response but, in both there will be a lot of inbreeding within the sublines. In that case we can limit the inbreeding depression using the scenarios with restrictions in inbreeding.

Frequency and effect of the bovine DGAT1 (K232A) variants on milk production traits in Iranian Holstein dairy cattle

Ahani, S.[1], Hosseinpour Mashhadi, M.[2], Nassiry, M.R.[3] and Aminafshar, M.[4], [1]Islamic Azad University-Kashmar Branch, Department of Animal Science, Kashmar, Iran, [2]Islamic Azad University-Mashhad Branch, Department of Animal Science, Mashhad, Iran, [3] Ferdowsi University of Mashhad, Department of Animal Science, Mashhad, Iran, [4]Islamic Azad University-Science and Research Branch, Department of Animal Science, Tehran, Iran; mojtaba_h_m@yahoo.com

Most traits in livestock are controlled by multiple genes and are therefore called genetically complex traits. Candidate gene approaches provide tools for identifying and mapping genes affecting quantitative traits. In this study QTL affecting milk production traits (milk yield, protein yield, fat yield, protein content and fat content) were mapped in Iranian Holstein dairy cattle. One of bovine QTL affecting milk yield and composition was identified to be located on bovine chromosome 14 near the centromeric end. A very important role of Acyl-CoA: diacylglycerol acyltransferase 1 (DGAT1) enzyme catalyzes the final step is the synthesis of triglyceride. The blood samples of 100 Dairy Cattle from the Animal Agriculture Center of Mashhad-Iran were used to genotype for DGAT1 gene. The DNA of the blood extracted by GuSCN-Silica Gel method. A 411 bp fragment from exon VIII amplified with standard PCR method. The amplified fragment was digested by restriction enzyme Cfr1 with PCR-RFLP method. Frequency of alleles K and A were 0.515 and 0.485. The frequencies of KK, KA and AA were 0.31, 0.41 and 0.28 respectively.Until now, 2 alleles, the lysine variant and the alanine variant were postulated at DGAT1K2322.The lysine encoding variant was associated with the high milk fat yield as well as fat and protein contents, whereas the alanine allele was related to the increased milk yield.The most significant result was obtained for the fat content of milk.

Relationship between the leptin LEP/Sau3AI polymorphism and fatty acid profile in muscle tissue of beef cattle

Oprzadek, J.M., Urtnowski, P.S., Dymnicki, E., Sender, G. and Pawlik, A., Institute of Genetics and Animal Breeding, Animal Genetic, Jastrzebiec, ul Postepu 1, 05-552 Wolka Kosowska, Poland; j.oprzadek@ighz.pl

Meat consumption including beef has increased significantly throughout the decades. Beef is one of a main source of protein and essential long chain fatty acids, n-3 as well as n-6, important for normal physiological functioning of humans. The aim of this work was to evaluate association between LEP/Sau3AI polymorphism and fatty acid profile in muscles tissue as well as health promoting properties of beef. The leptin is key enzyme in para- and auto- endocrine regulation of lipid metabolism. It curbs triglycerides synthesis on the one, and enhances β-oxidation of fatty acids in liver on the other hand. The investigation was conducted in experimental barn in Institute of Genetics and Animal Breeding. 40 yearling Polish Holstein-Friesian (phf) bulls were slaughtered and dissected into retail cuts. Fatty acid profile in meat was determined according to Kramer method. LEP/Sau3AI polymorphism was conducted according to Pomp's methodology using PCR-RFLP method. The 1820 nucleotides product was restricted with use of Sau3AI enzyme. Four genotypes were identified in investigated group. Among four genotypes we distinguished two homozygous and two heterozygous, AA, BB and AC, BC respectively. The frequencies of alleles were 0.706 for A allele, 0.166 for B allele and 0.128 for C allele. Statistical analysis shows genetic link between LEP/Sau3AI polymorphism and long chain fatty acids. The highest level of C22_0, C20_5, C22_4, C24_0, C22_5, C22_6 fatty acids was observed in AC animals. No differences was find in accordance to CLA content and leptin polymorphism.

Characterization of the ovine lipoprotein lipase (LPL), Scavenger Receptor BI (SCARB1) and alpha-tocopherol transfer protein (TTPA) genes: it is role in vitamin E content

González-Calvo, L.[1], Dervishi, E.[1], Serrano, M.[2], Joy, M.[1] and Calvo, J.H.[1], [1]CITA, Tecnología en Producción animal, Avda. Montñana 930, 50059-Zaragoza, Spain, [2]INIA, Mejora Genética animal, Ctra. La Coruña Km 7,5, 28040-Madrid, Spain; lgonzalezc@ext.aragon.es

Vitamin E is the collective name for a group of fat-soluble compounds with distinctive antioxidant activities. Antioxidants protect cells from the damaging effects of free radicals, which damage cells and might contribute to the development of cardiovascular disease and cancer. Furthermore, antioxidants increase shelf life of animal products improving colour and fat stability. To obtain this effect vitamin E can be added directly to meat or milk, but this may not be acceptable to consumers, or included in animals' diets. LPL, SCARB1 and TTPA in the metabolism of vitamin E. This work focuses on the characterization and evaluation of LPL, SCARB1 and TTPA as a candidate genes related to the vitamin E content. Genomic DNA from animals with extreme values for the vitamin E content (n=4, Rasa aragonesa) and four domestic sheep breeds (Churra Tensina, Assaf, Manchega, Churra) was used to search polymorphisms. Studies of putative regulatory elements within the promoter and potential target sites for miRNA within the 3' UTR regions were performed using TF Search and microinspector softwares. Promoter region of LPL was isolated finding a SNP affecting a CdxA and Tst-1 consensus sites. The total coding region sequence, UTRs and promoter regions of SCARB1 and TTPA genes were isolated. In SCARB1 14 SNPs were isolated: 5 conservative SNPs in the coding region, 5 SNPs in the UTRs, and 7 intronic polymorphisms. For TTPA, 6 polymorphisms were identified: 1 conservative SNP (exon 2), 1 Leucine for Phenilalanine substitution SNP (exon 5), and 4 intronic polymorphisms. Finally, the specific expression of these genes in two groups of animals with extreme values for vitamin E content was investigated by RT-PCR.

Cattle breed discrimination based on microsatellites markers
Vrtkova, I., Stehlik, L., Putnova, L. and Bartonova, P., Mendel University in Brno, Laboratory of Agrogenomics, Zemedelska 1, 613 00, Brno, Czech Republic; irenav@mendelu.cz

The aim of this study was utilize microsatellites for routine parentage testing and identification in breed cattle discrimination. The analysis was performed in three beef cattle (Aberdeen Angus n=93, Hereford n=125, Charolais n=101) and one dairy cattle breed (Czech Fleckvieh n=46). In total, 365 animals were genotyped with using Finnish Bovine Genotypes™ Panel 3.1 for 18 microsatellite markers recommended by both ISAG for parentage testing and FAO for genetic studies of domestic animals (TGLA227, BM2113, TGLA53, ETH10, SPS115, TGLA126, TGLA122, INRA23, ETH3, ETH225, BM1824, BM1818, and SPS113, RM067, CSRM60, MGTG4B, CSSM66, ILSTS006, respectively). The STRUCTURE (version 2.2) program based on a Bayesian clustering method was used to obtain population structure. The software detected four clusters corresponding to the four breeds without using prior information. The genetic markers used in study were suitable for good clustering ability by evidencing good discrimination power. Differentiation among individuals of each breed characterized by F_{st} values range from the lowest 0.091 in Aberdeen Angus, 0.099 in Czech Fleckvieh, 0.122 in Charolais, to highest 0.146 in Hereford cattle population. Average distances (expected heterozygosity) between individuals in same cluster were for Czech Fleckvieh 0.684, Charolais 0.702, Aberdeen Angus 0.722, and Hereford 0.683. Research data certified the possibility of using the Finnish Bovine Genotypes™ Panel 3.1 of DNA microsatellites markers for the cattle breed discrimination. Supported by project of MŠMT ČR 2B08037.

A frameshift mutation within LAMC2 is responsible for Herlitz type junctional epidermolysis bullosa in Black Headed Mutton sheep
Mömke, S., Kerkmann, A. and Distl, O., University of Veterinary Medicine Hannover, Institute for Animal Breeding and Genetics, Buenteweg 17p, 30559 Hannover, Germany; ottmar.distl@tiho-hannover.de

Junctional epidermolysis bullosa (JEB) is a hereditary mechanobullous skin disease in humans and animals. Abnormalities of macromolecules which anchor the dermis to the epidermis lead to diminished cohesion of the skin layers, blister formation, and fragility. In German Black Headed Mutton (BHM) sheep, ovine HJEB occurs in several flocks and shows a monogenic autosomal recessive mode of inheritance. In this study, a total of 717 BHM sheep from different farms were included. Of these animals, 21 were affected by HJEB. Using linkage and association analyses, LAMC2 was identified as the most likely candidate gene. The complete coding sequence of LAMC2 was obtained for two unaffected and two HJEB-affected sheep. Comparing these sequences, we detected 14 single nucleotide polymorphisms (SNPs) and a 2-bp deletion. Of these polymorphisms, six were non-synonymous and led to an alternate amino acid sequence. All non-conservative polymorphisms were analysed in all 21 HJEB-affected lambs and their relatives. Only the 2-bp deletion showed a perfect co-segregation with HJEB. Testing the 2-bp deletion in all 717 BHM sheep confirmed the perfect association with HJEB. Further 327 individuals of other breeds including Suffolk, East Friesian and Leine sheep were genotyped, but none of these animals showed the 2-bp LAMC2 deletion. In conclusion, we identified the causal mutation for ovine HJEB in BHM sheep. This deletion of two base pairs is located within LAMC2 and leads to a frame shift of the open reading frame with a premature stop codon. The genetic test can be employed for eradication of this lethal mutation in the BHM sheep breed. Furthermore, this study provides a suitable animal model for therapeutic approaches of HJEB.

Phylogenetic study of the Iranian buffalo population using ISSR markers

Aminafshar, M.[1], Amirinia, C.[2], Vaez Torshizi, R.[3] and Hosseinpour Mashhadi, M.[4], [1]Department of Animal Science, Faculty of Agriculture and Natural Resources, Science and Research Branch, Islamic Azad University, Ashrafi Esfahani Highway, Tehran, Iran, [2]Animal Science research Institute, Karaj, Tehran, Iran, [3]Tarbiat Modares University, Faculty of Agriculture, Department of Animal Science, Tehran, Iran, [4]Islamic Azad University, Mashhad Branch, Mashhad, Iran; aminafshar@srbiau.ac.ir

The aim of this study was to investigate the phylogenetic relationship between buffalo population of the North (Guilan & Mazandaran), North West (Azerbaijan) and South West (Khuzestan) of Iran. Genomic DNA was extracted from 360 blood samples and a panel of 14 ISSR was amplified by PCR reactions. Population genetic parameters and phylogenetic tree was drawn according to neighbor-joining (NJ) and UPGMA methods. The correlation of the genetic distance methods that were used, (Nei, Cavalli-Sforza and Edwards and Rynold, Weir and Cockerham) was 98.41 to 99.95. The lowest and highest genetic distance was observed between North-Azerbaijan and North-Khuzestan respectively. Result of the genetic similarity, estimated by total allelic relationship method (TAR), showed the highest genetic similarity between North and Azerbaijan (58.5) and the lowest between North and Khuzestan (53.8) buffalo population. The most and the least genotype similarity, were also observed within the animals of the Khuzestan (64.4) and the Azerbaijan (60.8) population according to the TAR. Amount of genetic variation within population (0.88 to 0.90) was significantly greater than genetic variation between populations (0.10 to 0.12) according to AMOVA. Genetic variation between geographical regions (North and western south), between population within regions (North and Azerbaijan population in Northern region and Khuzestan population in western south region) and within population was equal to 0.11, 0.03 and 0.86 respectively based on the AMOVA (P<0.05). The Northern and Azerbaijan buffalo population were located together at one cluster and Khuzestan buffalo population at another cluster based on phylogenetic tree.

Genome-wide associations for fertility traits in Holstein-Friesian cows using data from four European countries

Berry, D.P.[1], Bastiaansen, J.W.M.[2], Veerkamp, R.F.[3], Wijga, S.[2], Strandberg, E.[4], Wall, E.[5] and Calus, M.P.L.[3], [1]Teagasc, Fermoy, Co. Cork, Ireland, [2]Wageningen University, Wageningen, Wageningen, Netherlands, [3]Wageningen UR Livestock Research, Lelystad, Lelystad, Netherlands, [4]Swedish University of Agricultural Sciences, Uppsala, P.O. Box 7023, S-75007, Sweden, [5]Sustainable Livestock Systems Group, Scottish Agricultural College, (SAC), United Kingdom; donagh.berry@teagasc.ie

The objective was to utilise data on primiparous Holstein-Friesian cows from experimental farms in Ireland, the UK, the Netherlands and Sweden to identify genomic regions associated with fertility. Traditional fertility measures were days to first heat, days to first service, pregnancy rate to first service, number of services and calving interval; post-partum interval to the commencement of luteal activity (CLA) was derived using routine milk progesterone assays. Phenotypic and genotypic data on 37,590 single nucleotide polymorphisms were available on up to 1,570 animals. Genetic parameters were estimated using linear animal models and univariate and bivariate genome-wide association analysis was undertaken using Bayesian stochastic search variable selection performed using Gibbs sampling. Heritability of the traditional fertility traits varied from 0.03 to 0.16; the heritability for CLA was 0.13. The posterior QTL probabilities for the traditional fertility measures were all <0.021. Posterior QTL probability of 0.060 and 0.045 was observed for CLA on BTA2 and BTA21, respectively, in the univariate analyses; these probabilities increased when CLA was included in bivariate analyses with the traditional fertility traits. For example, in the bivariate analysis with calving interval, the posterior QTL probability of the two aforementioned SNPs were 0.66 and 0.12. The results from this study suggest that the power of genome-wide association studies in cattle may be increased by sharing of data and also possibly by using physiological measures of the trait under investigation.

Life-time genetic profiles for animal price
Mc Hugh, N.[1,2], Evans, R.D.[3] and Berry, D.P.[1], [1]Animal & Grassland Research and Innovation Centre, Teagasc, Moorepark, Fermoy, Co. Cork, Ireland, [2]School of Agriculture, Food Science & Veterinary Medicine, College of Life Sciences, University College Dublin, Belfield, Dublin 4, Ireland, [3]Irish Cattle Breeding Federation, Highfield House, Bandon, Co. Cork, Ireland; noirin.mchugh@teagasc.ie

Animal price contributes to the revenue in beef and dairy production systems. Previous studies have shown that large phenotypic and genetic variation exists for animal price across an animal's lifetime. The use of random regression models allows covariance functions to be calculated between any two ages facilitating the estimation of breeding values for every day across an animal's life. The objective of this study was to estimate (co)variance components for animal price across an age trajectory. A total of 2,967,791 animal price records from 2,506,110 beef and dairy animals sold at 71 livestock auctions in Ireland, were available. The fixed effects included in the random regression model were contemporary groups, gender, calving ease, heterosis, recombination loss, and a polynomial on age in weeks interacting with animal gender. The sire genetic and animal permanent environmental effect was modeled using random Legendre polynomials. The most parsimonious model was a second order random regression on the additive genetic component and a first order regression on the permanent environmental component. Residual variances were estimated separately for ages between two days and twelve weeks of age, 12 weeks to 6 months, six and twelve months and 12 and 36 months of age. The genetic variance for animal price varied from a minimum of 3,484 euro^2 at 1 week to 10,500 euro^2 at 156 weeks of age. The heritability for animal price varied from 0.33 (125 weeks) to 0.63 (11 weeks of age). Genetic correlations between animal prices at each week of age decreased as the interval between ages increased, but were always positive. Exploitable genetic variation clearly exists for animal price and can be modeled using random regression models to account for the longitudinal nature of the data.

Genetic parameters for age specific lamb survival, birth and weaning weights in South African Merino sheep
Matebesi-Ranthimo, P.A.M.[1], Cloete, S.W.P.[2,3], Van Wyk, J.B.[4], Olivier, J.J.[2] and Zishiri, O.T.[2], [1]National University of Lesotho, P.O. Roma 180, Roma, Lesotho, [2]Institute for Animal Production: Elsenburg, Private Bag X1, Elsenburg 7609, South Africa, [3]University of Stellenbosch, Private Bag X1, Matieland, 7602, South Africa, [4]University of the Free State, P.O. Box 339, Bloemfontein 9300, South Africa; vanwykjb@ufs.ac.za

High levels of lamb mortality have direct economic implications for the primary producers of lamb and wool. These lamb losses are also increasingly becoming an emotive issue, with a direct bearing on animal welfare. Data collected from ~16,700 Merino lambs, maintained at the Tygerhoek Experimental farm, were used to derive heritability estimates (h^2) for birth weight (BW), weaning weight (WW) and age specific lamb survival using the ASREML program. The fixed effects of selection line (clean fleece weight, fine wool, wet and dry and an unselected control), year (1969 to 2009), sex (male vs. female), age of dam (2-7+ years) and birth type (singles vs. pooled multiples) were significant. Significance for other random sources of variation was assessed by using the likelihood ratio test. Direct h^2 amounted to 0.16±0.02 for BW, 0.28±0.02 for WW, 0.23±0.02 for overall survival (TS), 0.12±0.02 for survival at birth (SB), 0.26±0.02 for survival between birth and 3 days (S3) and 0.11±0.02 for survival between three days and weaning (SW). Corresponding maternal heritability estimates were 0.19±0.02 for BW, 0.11±0.02 for WW, 0.04±0.01 for TS, and 0.03±0.01 for SW. Dam permanent environmental effects amounted to 0.12±0.01 for BW, 0.08±0.02 for WW, 0.02±0.01 for TS, 0.06±0.01 for SB and 0.03±0.01 for S3. Correlations between direct additive effects and maternal additive effects were significant for BW and WW and amounted to respectively -0.26±0.07 and -0.37±0.08. Age specific lamb survival exhibited significant genetic variation, with some estimates exceeding 0.20. Genetic progress in these survival traits may thus be achievable, as was recently demonstrated in another resource flock.

Whole genome association study for energy balance and fat protein ratio in German Holstein bull dams

Seidenspinner, T., Buttchereit, N., Tetens, J. and Thaller, G., Institute of Animal Breeding and Husbandry, Christian-Albrechts-University, Hermann-Rodewald-Str. 6, 24118 Kiel, Germany; nbuttchereit@tierzucht.uni-kiel.de

The metabolic status of cows is important to health and fertility, especially in early lactation and energy balance and fat protein ratio are considered as appropriate indicators. The aim of this study was to detect SNPs associated with energy balance and fat protein ratio. Phenotypes were breeding values estimated via random regression animal models for lactation days 11, 20, 30 and 42 for energy balance (n=431) and fat protein ratio (n=459). Negative energy balance became positive after 42 days on average, therefore, other stages of lactations were not considered. All investigated bull dams (n=459) belonged to the research herd Karkendamm and were genotyped using the Illumina BovineSNP 50K Bead chip® comprising 54001 SNPs. 43,593 SNPs and 398 (energy balance) and 419 (fat protein ratio) bull dams passed the quality control criteria. Whole genome association analyses were done using a principal component approach to adjust for genetic substructures. Permutation tests were applied to estimate genome-wise significance on a 5% level. Across all observed lactation days, significant SNPs associated with fat protein ratio could be detected on chromosomes 2 and 14, most of them in the region of DGAT1. Significant hits on chromosome 16 were identified for lactation days 11, 20 and 30. One SNP was significant on chromosome 27 for lactation days 20, 30 and 42. SNPs significantly associated with energy balance in all lactation stages were detected on chromosomes 5 and 18. Furthermore, SNPs located on chromosome 14 showed significance on lactation days 11, 20 and 30. None of them was in the DGAT1 region. In addition, SNPs on chromosomes 7 and 22 passed the significance level on day 42. Despite, both traits are considered as indicators for the cows metabolic status results for energy balance and fat protein ratio were inconsistent. This might be explained by the small data sets and the low reliabilities of the breeding values.

Effect of age of dam on the carcass performance of progeny in Japanese Black cattle

Kato, K., Nakahashi, Y. and Kuchida, K., Obihiro Univeersity of A&VM, 11, 2 Nishi, Inadacho, Obihiro, Hokkaido, 080-8555, Japan; s19091@st.obihiro.ac.jp

Production system of Japanese Black cattle (known as Wagyu) could be generally divided into breeding farmer and fattening farmer. Calves from the breeding farmers are purchased by the fattening farmers through the calf market. In the market, it is commonly believed that the performance of progeny from aged dam tend to be inferior than that from younger dam. Thus the progeny from aged dam are lowly competitive in price, so it is possible that the dam would be culled early even if the dam's ability is higher. The objective of this study was to investigate the effect of age of dam on the carcass performance of the progeny in Japanese Black cattle. Data consisted of 14,074 (steer: 9,852, heifer: 4,222) carcass records collected between 2000 and 2009 in Hokkaido, Japan. At the same time, digital images around the rib eye area at the 6th and 7th rib were taken by photographing equipment for image analysis. The age of the dam was classified into 8 groups (under 3 years old, 3-5 years, ..., 13-15 years, over 15 years) to confirm the influence by the difference of age of dam. Data were analyzed by the GLM procedure using SAS for analysis (1). Furthermore, genetic analysis was performed by the REML method using AIREML programs for analysis (2). From analysis (1), least squares means of the age class of dam ranged from 407.4 to 417.8 kg for carcass weight (CW). Class 7 and 8 had smaller values than that of class 6 significantly. For beef marbling standard (BMS) and marbling percent (MP), they ranged from 4.0 to 5.4 and from 36.4 to 41.9%, respectively. From analysis (2), BLUE of age class 1 and 8 for CW were lower than that of other class. The estimates for BMS and MP decreased with the increment of the age class. Although these results suggested the effect of the age of dam, other factors such as genetic trend or feeding control might affect the result.

A whole genome association study using the 60K porcine SNP beadchip reveals candidate susceptibility loci for scrotal hernia in pigs

Stinckens, A.[1], Janssens, S.[1], Spincemaille, G.[2] and Buys, N.[1], [1]KULeuven, Laboratory of Livestock Physiology, Immunology and Genetics, Biosystems, Kasteelpark Arenberg 30, 3001 Heverlee, Belgium, [2]Rattlerow-Seghers Holding N.V., Oeverstraat 21, 9160 Lokeren, Belgium; anneleen.stinckens@biw.kuleuven.be

Congenital genetic defects are quite common in swine and cover a range of conditions. One of the most important congenital genetic defects that occur in piglets is scrotal hernia, a condition which is phenotypically characterised by an uncontrolled prolapse of the small intestine into the scrotum. Hernia scrotalis occurs within the pig population at an average frequency from 1.7% to 6.7%, but in practice percentages of up to 10% are recorded in offspring of particular boars. Heritability of scrotal hernia in swine is estimated from 0.2 to over 0.6 and it was found that the susceptibility for scrotal hernia in pigs is probably due to a small number of genes, amongst which some major genes. In order to map the loci that are responsible for the susceptibility for scrotal rupture in swine, in this study a whole genome scan was performed, using marker data obtained by the 60K porcine SNP beadchip. In total 144 animals (80 parent-offspring trios) of 5 commercial lines were genotyped using the porcine 60K SNP chip and analysed using Genomestudio Data Analysis Sofware (Illumina) and the free whole genome association toolset PLINK (http://pngu.mgh.harvard.edu/purcell/plink/). 141 animals showed a call rate of over 98% and 51467 SNP showed a genotyping rate of over 90% and these data were included in the whole genome association study. Based on the genetic variance between the animals, the samples were found to be divided into 5 clusters coinciding with the commercial lines. Using a stratified analysis/permutation approach, two significant loci for the susceptibility for scrotal hernias in pigs could be detected.

Genetic parameters for somatic cell score in Croatian Holstein cattle

Špehar, M.[1], Ivkić, Z.[1], Gorjanc, G.[2], Bulić, V.[1], Mijić, P.[3] and Barać, Z.[1], [1]Croatian Agricultural Agency, Ilica 101, 10000 Zagreb, Croatia, [2]University of Ljubljana, Groblje 3, 1230 Domžale, Slovenia, [3]University of J.J. Strossmayer, Trg Sv. Trojstva 3, 31000 Osijek, Croatia; mspehar@hpa.hr

Somatic cell score is an important indicator of udder health and the prevalence of clinical and subclinical mastitis in dairy herds. The objective of this study was to estimate genetic parameters for somatic cell score in the Croatian Holstein cattle. Data consisted of 656,272 test-day records for 45,953 Holstein cows. Production data was recorded using the AT4 or BT4 method and taken from the central database of the Croatian Agricultural Agency. The number of animals in pedigree was 94,294. In order to obtain a normal distribution, logarithmic transformation for somatic cell score was performed. A single-trait repeatability fixed regression test-day model was used to estimate genetic parameters. Fixed effects in the model were: parity, region, and calving season. Days in milk was fitted using Ali-Schaeffer lactation curve nested within parity, while age at first calving was modelled as quadratic regression. Direct additive genetic effect, herd-year of test-day, and permanent environmental effect of cow within parity were included in the model as random effects. Variance components were estimated using Residual Maximum Likelihood method as implemented in the VCE-6 program. The estimated heritability was 0.182±0.002. Permanent environmental effect explained 20% of phenotypic variation, while herd-year of test accounted for another 7% of variability. Results provide genetic parameters for the application of genetic evaluation for somatic cell score in Croatian Holstein cattle.

Estimation of genetic parameters for SEUROP carcass traits in Czech beef cattle
Veselá, Z., Vostrý, L., Přibyl, J. and Šafus, P., Institute of Animal Science, Genetics and breeding of farm animals, Přátelství 815, 10400 Praha Uhříněves, Czech Republic; vesela.zdenka@vuzv.cz

The objective of this study was to estimate genetic parameters for SEUROP carcass traits in beef cattle in Czech Republic. Genetic parameters were calculated in a set of 4276 animals of eleven beef breeds slaughtered in 2005-2008. Fixed effect of classificator, fixed regression on age at slaughter by means of Legendre polynomial separately for the each breed and sex and fixed regression on heterosis coefficient were included in a model equation. The effect of HYS was considered as random. Genetic parameters were estimated by multitrait animal model using a linear model and a linear-threshold model in which carcass weight (CW) was considered as linear trait and carcass conformation (CC) and carcass fatness (CF) as threshold traits. Heritability coefficient for CW differed only moderately according to the method (0.295 in linear model and 0.306 in linear-threshold model). Heritability coefficient for CC was 0.187 and 0.237. Heritability coefficient for CF was 0.089 and 0.146. Genetic correlation between CW and CC was high (0.823 and 0.959), the correlation between CW and CF was intermediate (0.332 and 0.328) and it was low between CF and CC (0.071 and 0.053).In addition we used fixed regression on carcass weight in model for estimation of genetic parameters for CC and CF by using two-trait model. There was negative correlation between CC and CF (-0,430 and -0.429) and markedly lower heritability for CC (0,077 and 0.78) and CF (0.086 and 0.123) when carcass weight was used as fixed effect in model. The scores for carcass conformation are strongly predetermined by carcass weight. Taking into account carcass weight as fixed regression, genetic evaluation will be obtained only for pure carcass conformation, which however brings about a decrease of heritability and reliability of breeding value estimation.

Genetic trends for milk yield, milk components and milk coagulation properties in Italian Holstein Friesian and Estonian Holstein dairy cattle population
Pretto, D.[1], Vallas, M.[2,3], Kaart, T.[2,3], Ancilotto, L.[1], Cassandro, M.[1] and Pärna, E.[2,3], [1]Department of Animal Science, University of Padova, Viale dell'Università 16, 35020 Legnaro (PD), Italy, [2]Bio-Competence Centre of Healthy Dairy Products, Kreutzwaldi 1, 51014 Tartu, Estonia, [3]Institute of Veterinary Medicine and Animal Sciences, Estonian University of Life Sciences, Kreutzwaldi 1, 51014 Tartu, Estonia; denis.pretto@unipd.it

Aim of this study was to analyze genetic trend for milk yield (MY), milk components and milk coagulation properties (MCP) in Italian Holstein Friesian (IHF) and Estonian Holstein (EH) dairy cattle population. The dataset for IHF included a total of 1,592 individual milk samples, collected from April to November 2007 in 130 herds located in Northern Italy. EH dataset included a total of 17,577 individual repeated milk samples from 4,191 first lactation cows, collected from April 2005 to January 2009 in 73 herds located in Estonia. EBVs were estimated with two different models: for IHF – single trait animal model with the fixed effect of herd-test day, class of parity, class of days in milk (DIM) and additive genetic effect of animal (45,413 animals in pedigree); for EH – single trait repeatability animal model with the polynomial effect of DIM, linear effect of age at calving, fixed effect of sampling year-season and calving year-season, random effect of herd, permanent environment and additive genetic effect of animal (17,185 animals in pedigree). EBVs for bulls with at least 5 daughters with records (63 IHF and 126 EH) and cows with records were considered. Birth years (1991-2003 for bulls and 1993-2006 for cows) were grouped in 3 and the means of EBVs within groups where analyzed. Youngest group for bulls in IHF had higher EBV for fat and protein %, but lower for MCP, while accordant EBVs in EH were lower for MY and protein % but higher for fat % and MCP. Cow population showed the same trend as bulls for MCP but an opposite trend for the other traits in IHF, while in EH cows have deterioration in MCP traits.

Genetic diversity and similarity between the Amiata donkey breed and a donkey population native from Lazio

Matassino, D.[1], Bramante, A.[2], Cecchi, F.[2], Ciani, F.[1], Incoronato, C.[1], Occidente, M.[1], Pasquariello, R.[1] and Ciampolini, R.[2], [1]ConSDABI - FAO Italian Sub National Focal Point (Sub NFP.I-FAO) (Mediterranean biodiversity), Località Piano Cappelle, 82100 Benevento, Italy, [2]University of Pisa, Department of Animal Pathology Prophylaxis and Food Hygiene, Viale delle Piagge 2, 56124 Pisa, Italy; consdabi@consdabi.org

Since 13th century Italian domestic autochthonous donkey population has been characterized by Mediterranean grey mousy ancestral phenotype, currently typical of Amiata donkey. This phenotype persisted up to 16th century when a marked introduction of Hispanic and French big sized and dark bay or darkish coloured sires occurred. The aim of this research was to evaluate the genetic diversity and similarity between the Amiata donkey breed and an autochthonous donkey population native from Lazio, using molecular markers. A total of 108 animals (50 Amiata donkey and 58 autochthonous donkey from Lazio) were genetically characterized by using 10 STR markers. Genetic similarities between breeds were calculated by performing all possible pair-wise comparisons between the individual multilocus genotypes. F-statistics, molecular coancestry coefficients and kinship distances were obtained using MolKin v.3.0. A high genetic differentiation between populations was observed ($F_{ST} = 0.158$; $P<0.01$). The between-breed genetic similarity was low (0.233 ± 0.085) while kinship distances (D_k) and the mean coancestry values (f_{ij}) were 0.315 and 0.290 respectively. Considering that the between-population coancestry would represent the between-breed genetic relationships at the moment of separation we could suppose that the Amiata reflects more closely the genetic composition of the ancestral population of the donkey native from Lazio. In addition, the reproductive isolation of the last centuries may explain the differentiation observed by F_{ST} and genetic similarities. In the limits of the observation field, these preliminary results would confirm the timescale racial evolution.

Application of genomic evaluation for Canadian Yorkshire pigs: a preliminary study

Jafarikia, M.[1], Schenkel, F.[2], Maignel, L.[1], Fortin, F.[3], Wyss, S.[1], Sargolzaei, M.[2,4] and Sullivan, B.[1], [1]Canadian Centre for Swine Improvement, Ottawa, ON, K1A 0C6, Canada, [2]University of Guelph, Guelph, ON, N1G 2W1, Canada, [3]Centre de développement du porc du Québec, Québec City, QC, G1V 4M7, Canada, [4]LAlliance Boviteq, Saint-Hyacinthe, QC, J2T 5H1, Canada; mohsen@ccsi.ca

Simulation and empirical studies have shown the potential of genomic information for increasing the reliability of the Estimated Breeding Values (EBV). The aim of this preliminary study was to investigate the impact of using genomic information on the reliabilities of backfat thickness EBV. A total of 744 Canadian Yorkshire pigs were genotyped using the Illumina 60K Single Nucleotide Polymorphism (SNP) panel. In total, 38,275 of the 64,232 SNPs on the panel were used in the study. There were 17,382 SNPs excluded from analysis since they were either located on the sex chromosomes or not yet mapped to a specific chromosome. A further 8,575 SNPs were excluded because of minor allele frequency of less than 0.05. For assessing the increase in reliability of genomic EBV (GEBV), SNP genotypes from 619 pigs were used, all having reliabilities of backfat thickness EBV >0.60. Out of these animals, 546 were assigned to the training set and 73 animals, born after 2009, to the validation set. The gebv software was used to estimate the GEBV, following the method proposed by VanRaden. The GEBV were based on a combination of parent average EBV (PA) of young pigs prior to performance testing and the 38,275 SNP effects estimated from the training set. In the validation set, the mean theoretical reliability of GEBV was higher than average reliability of PA. However, the results of this preliminary study indicate that a larger training set might be required for a successful application of genomic evaluation for backfat thickness.

Calpain and calpastatin markers associations on growth, ultrasound measures and feed efficiency traits in *Bos indicus* cattle (Nellore)

Ferraz, J.B.S.[1], Gomes, R.C.[1], Carvalho, M.E.[1], Santana, M.H.A.[1], Silva, S.L.[1], Leme, P.R.[1], Rossi, P.[2], Rezende, F.M.[1] and Eler, J.P.[1], [1]University of Sao Paulo/FZEA, Basic Sciences, Cx. Postal 23, 13635-900 Pirassununga, SP, Brazil, [2]Federal University of Parana, Animal Science, Rua dos Funcionários, 1540. Bairro Juvevê, 80035-050, Curitiba, PR, Brazil; jbferraz@usp.br

The calpain-calpastatin system is one of the most relevant proteolytic systems in mammals. *In vivo*, those proteases can affect protein synthesis and degradation processes, thereby energy expenditure and body protein gain can be altered. The objective of this research was to verify the association of genetic markers discovered in the calpain and calpastatin genes with growth, feed efficiency and ultrasound carcass traits in Nellore beef cattle. Phenotypic and genotypic information was obtained of 290 Nellore bulls and steers. Animals were raised under grazing (primarily *Brachiaria* spp.) conditions until approximately 18 months of age and then were allotted to feedlots and fed medium energy level diets, individually, for 56 to 84 days. Thirteen traits related to growth, feed efficiency and ultrasound carcass measurements were taken for all animals, that were, also, genotyped for CAPN4751 (C/T, 6545 bp of AF248054) and UOGCAST (C/G, 282 bp of AY008267) in the calpain and calpastatin genes, respectively. Gene and genotypic frequencies were calculated and associations between SNPs and traits were carried out using a mixed model methodology, considering contemporary group, sex, age at measurement as fixed and sire as a random effect. Additive and dominance effects were, also, estimated. No effects of CAP4751 were observed for the majority of traits, but on final body weight (486 vs. 473 kg, P=0.03), Average daily gain (ADG, 1.53 vs. 1.44 kg/d, P=0.04) and ultrasound backfat thickness (7.5 vs. 6.70 mm, P=0.04). UOGCAST did affect only ADG (-0.0685±0.026 kg/d, P=0.0093). The UOGCAST and CALP4751 polymorphisms can affect growth and carcass traits in Nellore cattle but not feed efficiency.

Adult Merino ewes can be bred for live weight change to be more tolerant to climate change

Rose, G.[1], Kause, A.[1], Van Der Werf, J.H.J.[2], Thompson, A.N.[3], Ferguson, M.B.[3] and Van Arendonk, J.A.M.[1], [1]Wageningen University, Animal Breeding and Genomics, Marijkeweg 40, 6709PG, Wageningen, Netherlands, [2]University of New England, Division of Animal Science, CJ Hawkins Homestead, Armidale, NSW 2351, Australia, [3]Department of Agriculture and Food Western Australia, 3 Baron Hay Court, South Perth, WA 6151, Australia; gus.rose@wur.nl

Climate change is going to make managing sheep in Mediterranean climates more difficult due to increased variation in the supply of pasture and crop stubbles for grazing during summer and autumn. Farmers will rely more on providing supplementary feed which is expensive. Therefore liveweight loss during periods of low nutrition and subsequent liveweight gain are likely to be economically important traits. We estimated the genetic parameters for liveweight loss and liveweight gain on 2700 fully pedigreed 2 to 4 years old Merino ewes. When data for ewes from all ages was analysed together with age fitted as a fixed effect, liveweight gain had a heritability of 0.18 whilst liveweight loss had a heritability of 0.08. When liveweight change is analysed to be different at each age using a multivariate model, heritability for live weight gain increases to 0.37 for ewes ages 2 and 0.20 for ewes aged 3 and 4. Heritability for live weight loss increases to around 0.15 for all ages. These results suggest that liveweight change could be included in breeding programs to breed adult Merino ewes that are more tolerant to variations in feed supply.

Exploring trend of bias to prediction of genomic breeding values using Bayesian variable selection in the populations undergoing selection
Ehsani, A.[1,2], Janss, L.[1] and Christensen, O.F.[1], [1]Aarhus university, Blichers Allé 20, Postboks 50, 8830 Tjele, Denmark, [2]Tarbiat modares University, Faculty of Agricultural Science, Dept. of Animal Science, 14115, Tehran, Iran; alireza.ehsani@agrsci.dk

In this study we explored the size and trend of bias in a population undergoing selection using a Bayesian variable selection model that assumes heterogeneous variances for different markers and a prior distribution of scaling factors for QTLs. The study was based on 3 scenarios each with 2000 individuals in every generations and heritability of 0.25 with intensity of selection equal to 0.8. Scenario 1 started with a random reference population and operates selection from generation 1.The results showed that even under strong selection, there is no bias and the regression coefficient is close to 1 and remains constant over generations. We tested this scenario for stronger selection intensity equal to 1.755 and low heritability of 0.05 and the result was similar.Scenario 2 started with a selected reference population and continued selection each generation. It showed that bias is decreased as selection goes on and ultimately negligible after 5 generation. The regression coefficient was reduced from 1.64 in the first generation to 1.09 in generation 5 and finally to 1.04 in generation 10. In scenario 3, we investigated a new random base population descended from preselected parents to recalibrate the model. In this scenario the bias disappeared in the immediate generation after the recalibration, but introduces a small opposite bias thereafter. This last effect is due to decline in genetic variance by selection. Recalibrations of models will ultimately be necessary, but it is not yet clear how to do this the best possible way.

Isolation and characterization of microsatellite loci in Iranian River buffalo using PIMA
Shokrollahi, B.[1], Amirinia, C.[2], Dinparast Djadid, N.[3] and Hosseinpour, M.[4], [1]Islamic Azad University, Sanandaj Branch, Department of Animal Science, Pasdaran St, Se Rahe Adab, 618 Kurdistan, Iran, [2]Animal Science Research Institute, Department of Biotechnology, Heidar Abad, 12214090 Karaj, Iran, [3]Pasteur Institute of Iran, Malaria and Vectors Research Group, Pasteur Sq., 44543 Tehran, Iran, [4]Islamic Azad University, Mashhad Branch, Department of Animal Science, 1243151, Iran; Borhansh@yahoo.com

Microsatellite markers can be use in many study fields, as they are hypervariable and widely dispersed through genome. Researchers used cattle microsatellite markers for defining the genome make up in buffalo because no systematic studies have been undertaken to develop polymorphic DNA markers specific to this species. Cattle microsatellite markers have many disadvantages such as low polymorphism and loss of amplification. In this study PCR-based Isolation of microsatellite Array (PIMA) methodology was used for isolation of microsatellite loci in Iranian river buffalo. Blood samples of eighty unrelated individuals from four buffalo populations (Khuzestan, Mazandaran, Guilan and Azarbayejan) were taken and following DNA extraction, isolation of microsatellite loci initiated using enrichment with random amplified polymorphic DNA (RAPD) primers. RAPD-PCR fragments were ligated into TA cloning vectors and transformed into DH5? competent cells. Obtained colonies were screened for presence of repetitive elements by repeat-specific and M13 forward and reverse primers. After designing primer pairs for repeat containing fragments, they were tested in all buffalo populations. Two microsatellite loci (RBBSI and RBBSII) were informative and polymorphic. Number of alleles for RBBSI and RBBSII in 80 individuals was 5 and 6, respectively. Expected heterozygosity ranged from 0.65 to 0.81. Two newly isolated microsatellite loci could be useful for population genetic studies in B. bubalis.

Isolation of three SINLE-like sequences in Iranian River Buffalo
Shokrollahi, B.[1], Karimi, K.[2], Aminafshar, M.[3] and Hosseinpour, M.[4], [1]Islamic Azad University, Sanandaj Branch, Department of Animal Science, Kurdistan, 618, Iran, [2]Islamic Azad University, Varamin-Pishva Branch, Department of Animal Science, Tehran, 584, Iran, [3]Islamic Azad University, Science and Research Branch, Department of Animal Science, Tehran, 674, Iran, [4]Islamic Azad University, Mashhad Branch, Department of Animal Science, Khorasan Razavi, 784, Iran; Borhansh@yahoo.com

The aim of this research was to identify repeat sequences in Iranian river buffalo. Short Interspersed Element (SINE) is one of repeat sequences that is mainly used for evolutionary and phylogenetic studies. While some SINEs are derived from 7SL RNA or 5S rRNA, most SINEs are derived from tRNA. Hence, the tRNA-like secondary structure as well as the conserved RNA polymerase III–specific internal promoter sequences (designated A and B boxes) allows new SINE elements to be distinguished from other repetitive elements in the genome. In this study three Short Interspersed Elements (SINEs)-like sequences referred to as ART2A, Bov-tA1 and Bov-tA2 were developed. The isolation of these sequences began with RAPD-PCR enrichment in extracted DNA from four populations of Iranian water buffalo (n=40). Several RAPD primers were used to amplify fragments in separate reactions. Many obtained intense band were gel-extracted and ligated into pDrive vector and the plasmids transformed into Top10F' cells. Plasmid DNA from each cloned fragment was purified and then DNA inserts were sequenced. Sequences masked with repeat masker program, and result showed that three sequences are short interspersed elements including ART2A, Bov-tA1 and Bov-tA2. RNA polymerase III internal promoter, tRNA-related and unrelated sequences which can be find in all tRNA derived SINEs, exist in Bov-tA1 and Bov-tA2. BLAST search reveals significant homology with other previously described SINEs.

SNP association analysis of meat quality traits in the Finnish Landrace pig breed
Uimari, P. and Sevón-Aimonen, M.-L., Agrifood Research Finland, MTT, Biometrical Genetics, 31600 Jokioinen, Finland; pekka.uimari@mtt.fi

Pork meat quality is important in food processing and in home cooking. The Finnish national breeding value estimation includes three meat quality traits: lightness (L*), redness (a*) and pH measured in the muscles Longissimus dorsi (LD) and Semimembranosus (SM). The purpose of this study was to identify the SNPs associated with these traits in Finnish Landrace breed using Illumina's PorcineSNP60 BeadChip. The genotypic data included 366 AI boars with a call rate of over 0.90. Of the 62,163 SNPs in the chip, 57,868 SNPs had call a rate of over 0.9 and 7,632 SNPs were monomorphic. Deregressed EBVs were available for 340 animals. The association of each SNP with different meat quality traits was tested with a weighted linear model, using SNP as a regression coefficient and animal as a random variable. Deregressed EBV was used as the dependent variable in the linear model. Weights were based on the reliabilities of the EBVs. The analyses were performed using the AI-REML method in the DMU program package. Based on Bonferroni correction SNPs with P-value<2.0E-06 were considered as statistically significant results and SNPs with P value<4.0E-06 as suggestive findings. Two SNPs on chromosome 7 (ASGA0037025, 130470366 bp, P-value=8.6E-07 and MARC0045334, 133942365 bp, P-value=1.8E-06) were significant for a* measured in LD and one SNP on chromosome 15 (INRA0050276, 114169040 bp, P-value=7.1E-07) was significant for pH measured from SM. The estimated SNP effects were 0.35 units and 0.03 units for a* and pH, respectively. Suggestive findings were also obtained for a* measured in SM (DRGA0016597) and for a* measured in LD (ASGA0045619). As a conclusion, two putative chromosomal regions having effect on meat quality (a* and pH) were detected. After these initial findings have been confirmed in a larger population study, the specified SNPs can be used in national breeding value evaluation using either marker-assisted selection or genomic selection.

Authors index

A

Bertoni, G.	131, 203	Borba, H.S.	243, 244
Bertrand, G.	157	Borchers, N.	262
Bertrand-Michel, J.	119	Borchersen, S.	15
Berzal, B.	185	Borhami, B.E.	83, 271
Bessa, R.J.B.	38, 368	Boroumand-Jazi, M.	272, 272
Biagini, D.	298	Boscher, M.Y.	345
Bickell, S.L.	55	Boselli, L.	228
Bidanel, J.	14	Bosi, P.	158
Bidanel, J.-P.	11, 180, 190, 395	Bostad, E.	130
Biehl, M.V.	231	Boudry, C.	259
Bielfeldt, J.C.	262	Boudry, P.	9
Bienefeld, K.	12	Bouet, S.	387
Bigler, A.	91	Bouix, J.	200
Bijma, P.	20, 21, 174, 182, 211, 215, 254, 255, 289, 391	Boulenc, P.	345
Billon, Y.	180, 190, 209	Bouquet, A.	6
Birģele, E.	206	Bouquet, P.	86
Biscarini, F.	2	Bouquiaux, J.M.	279
Bishop, S.C.	168, 323	Bourdillon, Y.	200
Bister, J.L.	151	Bourjade, M.	93
Bittante, G.	174, 301, 302, 320	Bouvier, F.	57
Bjornsdottir, S.	38	Bouwman, A.C.	303
Blair, H.T.	304	Bouyeh, M.	173, 271
Blanc, F.	219, 219	Bouza, C.	326
Blanch, M.	152, 154, 155, 156, 322	Bovenhuis, H.	15, 179, 302, 303, 317
Blanco, M.	101, 246	Boyle, L.	290
Blasco, I.	287, 308	Boyle, L.A.	23
Blichfeldt, T.	240, 343	Bøe, K.E.	46, 92, 93, 94, 150, 209
Bloemhoff, Y.	315	Bracher, A.	127
Blomfield, D.	210	Bradley, A.J.	146
Blonk, R.	322	Bramante, A.	414
Blum, Y.	109, 119	Brandão, E.	250
Bocquier, B.F.	198	Brandt, H.	46, 181, 259
Bodin, L.	12, 344	Brebels, M.	143, 146
Boekhorst, L.J.S.	377	Bremmer, B.	277
Bohnenkamp, A.-L.	253	Brett, M.	17, 210
Boichard, D.	3, 15, 28, 303, 344	Bridi, A.M.	258
Boiti, C.	12	Brito, N.V.	187
Bokkers, E.A.M.	202	Britto, F.O.	79
Boland, T.	282	Brockmann, G.A.	31
Boland, T.M.	228	Brodziak, A.	319, 321
Boleckova, J.	34	Bron, J.E.	323
Bolhuis, J.E.	20, 21	Brossard, L.	105, 255
Boman, G.M.	184	Broucek, J.	335
Boman, I.A.	240, 343	Brouwer, H.	296, 365
Bonafos, L.	151	Brown, D.J.	345
Bonaïti, B.	344	Broz, J.	356
Bonato, M.	283	Brøkner, C.	37
Bondesan, V.	268	Brøndum, R.F.	26, 27
Bondi, G.	97, 318	Bruce, A.	85
Bondt, N.	192	Bruckmaier, R.M.	67
Bonfatti, V.	342	Brun, A.	104
Bonin, M.N.	226	Brun-Lafleur, L.	278
Bonneau, M.	153, 375	Brunschwig, P.	201
Bonnet, M.	180, 306	Brüssow, K.-P.	256
Boogaard, B.K.	377	Brüveris, Z.	258
		Buckley, F.	60, 228

Esmaeilkhaniani, S.	160	Folch, J.M.	121, 122, 401
Esmaeilpour, V.	384	Font I Furnols, M.	104, 152, 152, 155, 156, 358
Esmailizadeh Koshkoih, A.	190, 402	Formelová, Z.	380
Esposito, G.	330	Fornal, A.	405
Esquivelzeta, C.	8	Forni, S.	5, 178
Esteve-Garcia, E.	358	Fortin, F.	104, 414
Estopañan, G.	246	Forudi, F.	385
Evans, R.	336	Foroughi, A.R.	53
Evans, R.D.	129, 410	Foulquié, D.	385
Eydivandi, S.	297	Fourichon, C.	170, 171
Ezanno, P.	170	Fradinho, M.J.	38
Ezra, E.	125	Fragkou, I.A.	208
		Framstad, T.	363
F		Francàs, C.	104
Fàbrega, E.	22	Francois, D.	200
Facciolongo, A.	309	Françoso, R.	43
Fair, M.D.	401	Frankel, T.L.	207
Fancote, C.R.	48	Frankena, K.	145
Fantinato Neto, P.	52, 52	Franzén, G.	214
Fantova, E.	147	Fredriksen, B.	256
Farhadi, A.	183, 328	Frégeat, G.	345
Farid, A.H.	296	Freitas, A.	185
Farkas, J.	111	Frelich, J.	293
Farruggia, A.	277	Frieden, L.	155
Fassier, T.	200	Friggens, N.C.	82, 162
Fathy, A.	363	Fritz, S.	3, 7, 26, 28
Faucitano, L.	18, 258	Frkonja, A.	176
Faverdin, P.	278	Froidmont, E.	221, 259, 279
Federici, C.	130	Fthenakis, G.C.	208
Fehem, A.	288	Fuchs, K.	197
Feitsma, H.	14	Fuerst, C.	125, 143, 176, 366
Ferencakovic, M.	176	Fuerst-Waltl, B.	125, 143, 345, 366
Ferguson, M.B.	415	Fulton, J.	183, 404
Ferlay, A.	301	Funk, D.	77
Fernández, A.	401	Fureix, C.	94
Fernández, A.I.	121, 401	Furey, A.	315
Fernández, J.	326, 402	Furre, S.	134
Ferrari, A.	131		
Ferraz, J.B.S.	215, 251, 415	**G**	
Ferreira, E.M.	57	Gabiña, D.	236
Ferreira-Dias, G.	38	Gado, H.	83, 271
Ferrés, A.	224	Galama, P.J.	68, 279
Ferruzzi, G.	318	Galbraith, H.	167, 169, 388
Fikse, W.F.	108, 134, 135, 177, 342	Galindo, C.E.	65
Filangi, O.	119	Gallard, Y.	216, 303
Fiore, G.	374	Gallardo, D.	121
Fioretti, M.	35	Gallo, L.	342
Fischer, K.	256	Galmeus, D.	69, 69
Flamarique, F.	349	Galon, N.	125
Florek, M.	307	Ganche, E.	201, 282
Flury, C.	91, 136, 251	Gandini, G.	13
Flynn, B.	353	Gandorfer, M.	163
Flynn, J.	316	Gandra, J.R.	43
Flysjo, A.	59	Gao, H.	28, 32
Focant, M.	221	Garcia, A.	281, 281
Folch, J.	147, 189, 199	García, J.	159

Garcia-Cortes, L.A.	175	Gomes, R.C.	36, 226, 229, 415
Garcia-Launay, F.	220	Gomez, M.D.	132, 232
Gardemin, C.	155	Gomez-Brunet, A.	58
Gardiner, G.E.	353, 357	Gomez-Raya, L.	177
Gardner, G.E.	299	Gonzaga, I.V.F.	40
Garmo, R.T.	103	González, F.	253, 265
Gaspardo, B.	229, 331	Gonzalez, J.	62
Gault, E.	153	Gonzalez Lopez, V.	262
Gauly, M.	22, 142, 149, 149, 292, 293, 295	González-Bulnes, A.	249
Gaustad, A.H.	363	González-Calvo, L.	407
Geary, U.	99	Gonzalez-Recio, O.	5, 178
Gelencsér, E.	357	González-Rodríguez, A.	188
Gengler, N.	14, 27, 102, 140, 218, 307	Gopal, C.	326
Gentil, R.S.	57	Gopikrishna, G.	326
Georgoudis, A.	44	Gorjanc, G.	33, 45, 111, 348, 412
Geraldo, A.C.A.P.M.	52, 100	Goselink, R.M.A.	66
Gerken, M.	388	Goshima, M.	263
Gevorgian, O.X.	271	Götz, K.-U.	30, 262
Ghaemi, M.R.	223	Goues, T.	70
Ghanipoor, M.	333	Govasmark, E.	313
Ghita, E.	49, 49, 241, 245	Goyache, F.	123
Ghohari, S.	223	Grainger, C.	60
Ghoorchi, T.	272, 272	Grandgeorge, M.	235
Ghorbani, A.	333	Grandinson, K.	254, 341
Ghorbani, G.R.	222, 272, 272	Grandoni, F.	400
Ghoreishi, S.F.	81	Gras, M.	49
Giambra, I.J.	46	Graulet, B.	301
Gianola, D.	5	Gredler, B.	3, 176, 176
Gibbs, R.	117	Green, L.E.	146
Gibon, A.	377	Greenwood, P.L.	368
Gil, C.	246	Gremmen, H.G.J.	161
Gil, J.	152	Grevle, I.S.	103
Gil, J.M.	154, 155, 156	Grimsrud, K.M.	324
Gil, M.	155, 156, 322, 359, 359	Grindflek, E.	90, 179, 395
Gilbert, H.	180	Grodzicki, T.	307, 316
Gillon, A.	102	Große-Brinkhaus, C.	261
Giordano, D.	228	Grossi, P.	203
Giral-Viala, B.	345	Grosu, H.	347
Giraud, D.	201	Groves, J.	104
Girlea, M.	138, 138	Grøva, L.	150
Gispert, M.	104, 358	Gruffat, D.	306, 310
Gitterle, T.	168	Grzybek, W.	64
Givens, D.	166	Guadagnin, M.	268
Gjerde, B.	168, 325	Guàrdia, M.D.	121, 306
Gjerlaug-Enger, E.	107	Guérin, G.	387
Gjøen, H.M.	372	Guerra, M.M.	270
Glass, E.J.	168	Guerrier, J.	344
Gleeson, D.	290, 316	Guéry, E.	57
Gnjidić, P.	264	Guiatti, D.	131
Gobesso, A.A.O.	40, 43	Guibert, P.	345
Goddard, M.	114	Guillaume, F.	3, 7, 26, 28
Godet, E.	122	Guillén, E.	359, 359
Goetz, K.-U.	184	Guillou, H.	119
Goldberg, G.R.	361	Guinard-Flament, J.	65, 223
Golghasemgharehbagh, A.	81	Guix, N.	277
Golian, A.	83	Guldbrandtsen, B.	24, 26, 27, 29, 120, 175, 177

Guloglu, S.C.	336	Helberg, A.	69, 69
Gunia, M.	215	Hemme, T.	62, 124, 329
Gunn, J.T.	296	Hennessy, D.	100
Gunnarrson, L.I.	78	Henriksson, M.	278
Gustafsson, H.	231	Henry, S.	93, 235
Gutiérrez, J.P.	123, 140, 387, 404, 406	Henryon, M.	257, 391
Guy, J.H.	17, 210	Henryon, M.A.	17
Guzzo, N.	302	Heringstad, B.	29, 134, 184, 221
		Hermansen, J.E.	59
H		Hermida, M.	326
Hååard, M.	14	Herskin, M.S.	19
Haase, B.	136	Heuven, H.	12
Habeanu, M.	305	Heuven, H.C.M.	167, 170
Häberli, M.	158	Hexeberg, C.	256
Hadjipavlou, G.	96	Hickey, J.M.	392
Hafez, Y.M.	284	Hickson, R.E.	304
Haffray, P.	9, 328	Hiemstra, S.J.	13
Hagger, C.	251	Hietala, P.	86
Häggman, J.	365	Higuera, M.A.	11
Hagiya, K.	398	Hinrichs, D.	212, 213, 260, 262
Halas, V.	270	Hirayama, T.	33
Halekoh, U.	350	Hoang, C.	192
Hall, C.A.	340	Hocquette, J.F.	311, 367
Hallberg, L.	234	Hoedemaker, M.	289
Hallén Sandgren, C.	364, 366	Hofer, A.	107
Hallenstvedt, E.	311	Hofherr, J.	374
Haman, J.	34	Hofmanová, B.	41, 280
Hamland, H.	395	Höglund, J.K.	120
Hammadi, M.	288, 298	Holand, Ø.	95
Hammami, H.	102	Holgersson, A.L.	339
Hamzic, E.	176	Hollo, G.	312, 314
Han, Y.K.	384	Hollung, K.	312
Handel, I.G.	168	Holm, B.	11
Hanigan, M.	81	Holmes, C.W.	68
Hanrahan, J.P.	198	Homolka, P.	382
Hansen, M.H.S.	395	Hoofs, A.I.J.	16
Hansen-Møller, J.	313	Horbańczuk, J.O.	64
Hansson, M.	19	Horbańczuk, K.	64
Hardaker, J.B.	292	Horcada, A.	242
Hardeng, F.	297	Hornick, J.L.	65, 286, 290, 291
Harms, J.	102	Hosaya, S.	263
Harstad, O.M.	69, 69, 321, 379	Hosseinpoor, R.	333
Hashemi, M.	389	Hosseinpour, M.	416, 417
Hasler, H.	136	Hosseinpour Mashhadi, M.	405, 406, 409
Hatami, M.	183	Houston, R.D.	323
Haug, A.	43, 99, 300, 321	Houwers, H.W.J.	62
Haugaard, K.	221	Hoving, I.E.	58
Hauge, H.	337	Hozé, C.	3
Haule, M.E.	332	Høøk Presto, M.	203
Hausberger, M.	93, 94, 235, 235	Hučko, B.	382
Hauser, G.	235	Huijps, K.	197
Hayashi, M.	230	Huisman, A.E.	182
Hebean, V.	305	Humblot, P.	14
Heck, J.M.L.	302	Huntley, S.	146
Hecold, M.	142	Hurtaud, C.	223
Hedebro Velander, I.	17	Hussain, I.	296

Ravichandran, P.	326	Romero, C.	159
Razmaite, V.	35, 90	Romero, G.	56, 148
Rea, M.C.	353, 357	Romnée, J.M.	259
Recoules, E.	219	Roncalés, P.	359, 359
Récoursé, O.	201	Rönnegård, L.	108
Reents, R.	77	Rootwelt, V.	363
Rehn, T.	94	Ropota, M.	241, 245
Rehnberg, A.C.	73, 74, 311	Rosa, A.	38
Reimert, I.	21	Rosa, A.F.	226
Reiners, K.	260	Rosa, G.J.M.	5
Reinhardt, F.	115, 289	Rosati, A.	13
Reinsch, N.	116, 133	Rose, G.	415
Reist, S.	13	Rosendal, K.G.	324
Reixach, J.	121, 306	Rosner, F.	217
Renieri, C.	137, 388, 388	Ross, R.P.	353, 357
Reusken, C.B.E.M.	169	Rossi, P.	415
Revell, D.R.	55	Rossoni, A.	183
Rezaei, S.A.	81	Roura, E.	355, 355
Rezaeipour, V.	384	Rovai, M.	47
Rezende, F.M.	215, 251, 415	Royer, E.	265, 357
Ribaud, D.	172	Rødbotten, R.	300
Ricard, A.	232	Rundgren, M.	203
Riccardi, M.	286	Rupasinghe, P.P.	296
Richard-Yris, M.A.	94, 235	Rupp, R.	57, 238, 344
Richards, S.	117	Ruść, A.	119
Richter, M.	260	Rutten, M.J.M.	302
Rieder, S.	136, 251	Ruz, J.M.	147
Rikimaru, K.	397	Ruzicka, Z.	34
Rincon, G.	117	Ryan, M.T.	204
Ringdorfer, F.	45, 247	Rybnik, P.K.	283
Rinne, M.	224	Rychlik, T.	403
Ripoll, G.	287	Rydhmer, L.	19, 254, 329, 390
Ripoll-Bosch, R.	61	Rye, M.	168
Riquet, J.	180, 209	Ryschawy, J.	62, 329, 377
Risdal, M.	69		
Rius-Vilarrasa, E.	177	**S**	
Robaye, V.	290, 291	Saastamoinen, M.	232
Robert-Granié, C.	3, 7, 344, 345	Saatchi, M.	25
Robic, A.	190	Sabaté, J.	334
Robinson, N.	326	Sabatier, R.	163, 164
Robledo, J.	253, 265	Sadeghi, M.	249
Roca, A.	148	Sadeghipanah, H.	47, 275
Rocha, L.M.	258	Sae-Lim, P.	399
Rochambeau, H.	387	Safdarian, M.	389
Roche, B.	163	Šafus, P.	413
Rodero, E.	252	Saghi, D.A.	190, 402
Rodríguez, C.	401	Saha, J.N.	326
Rodriguez, P.	104	Sahana, G.	118, 120, 175, 402
Roehe, R.	87, 390	Sahoo, P.K.	326
Roep, D.	112	Sakarya, E.	336, 370
Roessler, R.	91	Salama, A.A.K.	288, 363
Roh, S.-G.	33	Salama, A.K.K.	47
Rojas-Olivares, M.A.	47	Salama, O.A.	363
Rollmann, S.	117	Salari, F.	317, 318
Román-Ponce, S.I.	34	Salari, S.	315, 382
Romarheim, O.H.	372	Salau, J.	102

Salbu, B.	321	Sebek, L.B.J.	124
Sales, J.	382	Seegers, H.	170, 171
Salgado, C.	404	Sehested, E.	126, 312
Salle, G.	345	Seidavi, A.R.	333, 386
Samorè, A.B.	34	Seidenspinner, T.	411
Sánchez, M.J.	39, 42	Sell-Kubiak, E.	254
Sanchez, M.P.	180	Seltenhammer, M.H.	120
Sanchez-Mainar, M.	300	Sen, U.	246
Sánchez-Molano, E.	326	Sender, G.	407
Sandri, M.	229	Sendra, E.	56
Sanker, C.	295	Sepahvand, M.	205
Sankey, C.	235	Sepchat, B.	220
Santana, M.H.A.	415	Seradj, A.R.	383
Santiago, J.	58	Serenius, T.	266
Santolaria, P.	50, 50	Seripa, S.	235, 236
Santucci, M.A.	261	Serrano, M.	160, 185, 407
Sañudo, C.	242, 242, 244	Sevón-Aimonen, M.-L.	239, 262, 417
Sanz, A.	287, 308	Seyedabadi, H.R.	297
Saranée Fietz, J.	269	Sezer, M.	246
Sargolzaei, M.	414	Sgorlon, S.	131
Sari, M.	315, 382	Shalloo, L.	60, 99
Sarlós, P.	284	Shekhar, M.S.	326
Sarti, L.M.N.	229	Shibata, T.	397, 398
Sartin, J.	166	Shingu, H.	294
Sartin, J.L.	165, 166	Shirali, M.	87
Sarto, P.	185	Shirazi, H.	385
Sartori, C.	112	Shokrolahi, B.	273, 405
Sato, S.	397, 398	Shokrollahi, B.	416, 417
Satoh, T.	263	Shumbusho, F.	190, 395
Sattler, T.	154	Siddell, J.P.	368
Sauvant, D.	380	Sigfridson, K.	349
Savar Sofla, S.	191, 191	Sigurdsson, Á.	135
Savas, T.	292	Sijtsma, S.R.	193
Sayed, H.	285	Silió, L.	401
Sayers, R.	315	Sillanpää, M.J.	6
Sæther, N.	73, 74, 74	Silva, E.	313
Schaefer, A.	39	Silva, R.C.G.	215
Schafberg, R.	217	Silva, S.L.	36, 226, 229, 415
Schellander, K.	189, 261	Silva Neto, P.Z.	229
Schenkel, F.	414	Silva Sobrinho, A.G.	243
Schennink, A.	317	Silva-Villacorta, D.	304
Schlather, M.	117	Simeckova, M.	34
Schlegel, H.	256	Simeone, A.	224, 226
Schlegel, P.	63, 127, 350	Simianer, H.2, 76, 88, 115, 115, 117, 136, 213, 400	
Schmidely, P.	82	Simon, A.J.	193
Schmidt, J.	352	Simongiovanni, A.	158
Schmitt, E.	40	Sindic, M.	218
Schmoll, F.	154	Sirin, E.	239, 246
Schneider, J.	242	Siukscius, A.	42, 230
Schnyder, U.	91, 176	Sklaviadis, T.	44
Scholten, M.C.T.	161	Skrede, A.	372
Schrooten, C.	25	Skuce, R.A.	168
Schulman, N.	177	Šlachta, M.	293
Schulman, N.F.	118	Słota, E.	403
Schurink, A.	145	Small, R.W.	73
Schwarzenbacher, H.	3, 125, 176	Smith, E.M.	146

White, C.	340	Zeron, Y.	125
Wicke, M.	155	Zervas, G.	156
Wickham, B.	10, 75	Zgur, S.	127
Wickramasinghe, S.	117	Zillig Neto, P.	36
Widjaja, A.	66	Zishiri, O.T.	410
Wijga, S.	32, 409	Zolkiewski, P.	334
Willam, A.	125, 213	Zom, R.L.G.	66
Williams, I.H.	48	Zoppa, L.M.	226
Williams, J.	84	Zsédely, E.	352
Wimmers, K.	389	Zutis, J.	305
Windig, J.J.	6, 89		
Winters, M.	76		
Wittenburg, D.	116		
Wolanciuk, A.	314, 316, 320		
Woolliams, J.	2		
Woolliams, J.A.	4, 109, 168, 212, 289		
Wu, X.-L.	5		
Wulf, I.R.	396		
Wurzinger, M.	378		
Wyss, S.	104, 414		

X

Xue, P.	167
Xue, Y.	303

Y

Yániz, J.L.	50, 50
Yaroshko, M.	259
Yazdi, M.H.	168
Yin, T.	29, 88
Yokota, S.	399
Yonekura, S.	33
York, S.J.	340
Yosef, E.	161
Younge, B.	269
Yousief, M.Y.	237
Ytournel, F.	213
Ytrestøyl, T.	371
Yu, X.	7
Yvon, J.-M.	40

Z

Zabek, T.	405
Zagorska, J.	319
Zajac, L.	335
Zaqeer, B.F.	237
Zare Shahneh, A.	47
Zargarian, B.	25
Zarnecki, A.	403
Zavadilova, L.	34
Zebeli, Q.	84
Zehetmeier, M.	163
Zeilmaker, F.	9, 10, 10, 12, 12, 13, 16
Zentek, J.	354
Zeola, N.M.B.L.	243, 244
Zerehdaran, S.	272, 272

Printed in the United States
by Baker & Taylor Publisher Services